This is a continuation of a very successful recent book by the same authors that aims to explain the mathematical methods using a pragmatic and simple approach. It concentrates on how to get the answers in the simplest and most transparent physical way instead of trying to stick to the mathematical rigour. This makes this book extremely valuable for a great majority of researchers in natural sciences and engineering where the barrier to understanding the subject is frequently associated with mathematics. Despite the fact that the topics considered in this volume are more complex than in the first volume, the book is amazingly simple and elegant. Using multiple examples the authors show how specific mathematical questions can be solved. But what is even more important, they show how to understand the mathematical answers from a common-sense point of view. I was also delighted to see multiple references to famous scientists who developed these methods. Another aspect of this book that helps readers go through the relatively complex material is its light, emotional and almost entertaining style. You can easily imagine this book as a friendly discussion with your mentor who patiently explains to you various mathematical tricks that help you understand the subject. I wanted to thank the authors for writing such an important book that, in my opinion, many future generations will continue to explore and enjoy!

**Anatoly B. Kolomeisky**
**Chair, Department of Chemistry**
**Professor of Chemistry**
**Professor of Chemical and Biomolecular Engineering**
**Professor of Physics and Astronomy**
**Primary Investigator, Center for Theoretical Biological**
**Physics, Rice University**

The scientific and higher-educational volume title, *How to Derive a Formula*, sounds almost too attractive. Who, young or more mature actors in the physical, chemical, and perhaps even biological sciences, would not be thrilled by such a title? Attractive analytical mathematical formulae that immediately connect different physical quantities, say rate constants with driving forces in chemical processes, do not capture all the details that large-scale computational work provides. In contrast to the latter they offer, however, immediate transparency, clear views of what happens, when parameters are changed including limiting cases of large and small parameter values, and not in the least, easy discovery of "bugs".

In two impressive volumes, Alexei A. Kornyshev and Dominic O'Lee at Imperial College London have set out to exploit this philosophy by combining their long-time scientific and teaching competence at the highest-level of pure and applied mathematical physics into not only a single, but very recently also a second, *How to Derive a Formula* volume. Such an enterprise is highly meritorious. Can young and mature scientists in the physical and chemical sciences really learn to derive their own formulae from these two volumes, or is formula deriving something you are, kind of, born with? Alexei A. Kornyshev and Dominic O'Lee have taken heroic efforts to help along the way.

Following a "lively style" introduction in Volume 1, where the metaphor of climbing Mt. Everest stage-wise is introduced, the authors set out to explain, light-heartedly but in tutorial detail, core concepts and methodology of mathematical physics. These include: mathematical functions and variables; differential and integral calculus; probability theory; vectors; and complex functional theory, presented gently and with a glint in the eye. The chapters are completed by applied notes and problems to solve. Throughout the text they ascertain that both experienced and not-so-experienced readers are with them. This volume is a delight.

Volume 2 attempts to maintain the "lively style" of Volume 1, but rises to more demanding and breath-taking as well as less relaxing and anecdotic levels. This is a natural evolution, moving from the light-hearted to the tougher in formula deriving, or from Camp 2 and onwards in the authors' Mt. Everest climbing metaphor. Chapter headlines illuminate readers' stamina needed, at the same time unravelling the prodigious strength of mathematical physics (and chemistry): multiple integrations; 3D surfaces and vectors; differential and partial differential equations; variational calculus; and, not in the least, contour integration and approximation techniques, the latter prominent in converting complex phenomena into palatable analytical representations.

In their two-volume magnum opus, the authors offer detailed overviews of the strength and beauty of analytical mathematical physics, and how to use it in a wealth of exciting physics contexts. A further message is that beautiful analytical formalism can pave the way subsequently for large-scale computational efforts. In their depth and enthusiastic spirit these two volumes will offer new and higher-level guidance than hitherto to both newcomers and those who are already somewhat accustomed to deriving their own formulae, in many forthcoming exciting efforts.

**Jens Ulstrup**
**Emeritus Professor of Chemistry**
**Technical University of Denmark**

As computers become increasingly prevalent, many students believe that analytical calculations are becoming obsolete and therefore do not attempt to perform them. However, this belief is misguided. Analytical and numerical approaches are complementary, and their combined use aids in understanding, analyzing and predicting complex phenomena. Furthermore, analytical equations allow us to overcome limitations of numerical simulations that are currently restricted to short time and length scales.

Students often not only miss the knowledge of analytical methods, but the technology to apply that knowledge, based very much on the culture of using common sense. They often lack the ability to determine if the result obtained is correct or incorrect. This can be done by checking obvious limiting cases, dimensional analysis, or simply considering the order of magnitude. If students lack these skills (not speaking of the knowledge!), they may develop a negative attitude towards mathematics, resulting in statements such as "I hate math", or more polite forms of it: "I am not good at math". To address this issue, this two-volume book has a goal to humanize, 'demystify' mathematics and to do it in an engaging, easy manner. That has certainly been achieved in the Volume 1 devoted to basic issues and skills necessary for deriving formulae. Humour, historical diversions, all the distracting-from-stress tricks have made that volume truly unique.

Obviously, it was much harder to achieve this goal in Volume 2, the task of which was to complete this project by introducing the methods that inevitably require more cumbersome derivations. Despite the increased difficulty, the methods are illustrated using examples, each solved to a closed-form end result. Some physical examples may look a bit more contrived than in Volume 1, as the authors designed them to demonstrate the corresponding methods in their full power, rather than simply applying to solve a needed problem for which not all aspects of the methods may be needed.

There are particularly beautifully written chapters that are still very close in their spirit to Volume 1, with a fresh look and introduction to contour integration and complex variable calculus, summation of series, partial differential equations and variational calculus. Introduction to path integrals, presented in a relatively easy manner, is also very useful, although it does not touch upon all available methods of calculating them, as that would have required a whole new book (and there are such books). That particular chapter as well as the even lighter "integral equations" were intended to introduce the reader to the subject, so that if needed, they could later deal with specialized books. The same refers to the last, 'entertaining' chapter on fractional mathematics, the goal of which is again to bring the reader back to the ideas on dimensionality and scaling that were first introduced in Volume 1. This chapter teaches students that in certain situations, complexity can lead to simplicity, if one thinks about the problem differently. Overall, this project's heroic efforts will undoubtedly be rewarded. While the mathematical methods covered in these two volumes do not exhaust the entire field, the authors provide readers with a list of powerful techniques and skills that they would benefit from, on the road to deriving their own beautiful formulae. These two volumes serve as an excellent base for mastering these methods.

**Michael Urbakh**
**Emeritus Professor of Chemistry**
**Tel Aviv University**

# How to
# Derive a Formula

**Volume 2:**
**Further Analytical Skills and**
**Methods for Physical Scientists**

# Essential Textbooks in Physics

Print ISSN: 2059-7630
Online ISSN: 2059-7649

The *Essential Textbooks in Physics* explores the most important topics in Physics that all Physical Sciences students need to know to pass their undergraduate exams (years 1, 2 and 3 of the BSc). Some topics are run-of-the-mill topics, others introduce students to more applied areas (e.g. Quantum Optics, Microfluidics…).

Written by senior academics as well lecturers recognised for their teaching skills, they offer in around 200 to 250 pages a theoretical overview of fundamental concepts backed by problems and worked solutions at the end of each chapter.

Their lively style, focused scope and pedagogical material make them ideal learning tools at a very affordable price.

Most authors are based at prestigious universities: Imperial College London, Oxford, UCL, Ecole Polytechnique.

*Published*

*How to Derive a Formula*
*Volume 2: Further Analytical Skills and Methods for Physical Scientists*
   by Alexei A. Kornyshev and Dominic O'Lee

*Electromagnetism — Principles and Modern Applications:*
*With Exercises and Solutions*
   by Chris D. White

*Application-Driven Quantum and Statistical Physics: A Short Course for Future*
*Scientists and Engineers*
*Volume 3: Transitions*
   by Jean-Michel Gillet

*How to Derive a Formula*
*Volume 1: Basic Analytical Skills and Methods for Physical Scientists*
   by Alexei A. Kornyshev and Dominic O'Lee

More information on this series can also be found at https://www.worldscientific.com/series/etip

*(Continued at end of book)*

Essential Textbooks in Physics

# How to
# Derive a Formula

**Volume 2:**
**Further Analytical Skills and**
**Methods for Physical Scientists**

**Alexei A. Kornyshev**
**Dominic O'Lee**

*Imperial College London, UK*

 **World Scientific**

NEW JERSEY · LONDON · SINGAPORE · BEIJING · SHANGHAI · HONG KONG · TAIPEI · CHENNAI · TOKYO

*Published by*

World Scientific Publishing Europe Ltd.

57 Shelton Street, Covent Garden, London WC2H 9HE

*Head office:* 5 Toh Tuck Link, Singapore 596224

*USA office:* 27 Warren Street, Suite 401-402, Hackensack, NJ 07601

**Library of Congress Cataloging-in-Publication Data**
Names: Kornyshev, A. A. (Alexei A.), author. | Lee, Dominic J. O'., author.
Title: How to derive a formula / by Alexei Kornyshev (Imperial College London, UK) and
    Dominic O'Lee (Imperial College London, UK).
Description: New Jersey : World Scientific, 2018–    | Series: Essential textbooks in physics |
    Contents: v. 1. Basic analytical skills and methods for physical scientists --
    v. 2. Further analytical skills and methods for physical scientists
Identifiers: LCCN 2018043878 | ISBN 9781786346346 (hc : alk. paper) |
    ISBN 9781786346445 (pbk : alk. paper)
Subjects: LCSH: Mathematics--Formulae.
Classification: LCC QA41 .K66 2018 | DDC 510.2/12--dc23
LC record available at https://lccn.loc.gov/2018043878

**Volume 2: Further Analytical Skills and Methods for Physical Scientists**
ISBN 9781800612792 (hardcover)
ISBN 9781800612976 (paperback)
ISBN 9781800612808 (ebook for institutions)
ISBN 9781800612815 (ebook for individuals)

**British Library Cataloguing-in-Publication Data**
A catalogue record for this book is available from the British Library.

For any available supplementary material, please visit
https://www.worldscientific.com/worldscibooks/10.1142/Q0377#t=suppl

Desk Editors: Nandha Kumar/Adam Binnie/Shi Ying Koe

Typeset by Stallion Press
Email: enquiries@stallionpress.com

# Preface

This is the second volume of the textbook *How to Derive a Formula*. We will not spell out again here the philosophy of this project; this has been done in detail in the Preface and Introduction to Volume 1 (to which, on various occasions, we refer the readers of this volume). We will focus here only on several new aspects, specific to this volume.

**Different atmosphere.** Using the metaphor of an ascent to Everest, Parts I and II of Volume 1 paralleled the trips between the Base Camp to Camp 1, and from Camp 1 to Camp 2, respectively. Now in this volume, Parts III and IV correspond to the climbs from Camp 2 to Camp 3, and from Camp 3 to Camp 4. With air growing thinner, these sectors of the ascent become increasingly difficult. So, as much as we tried to maintain the relaxed and enjoyable spirit of Volume 1, this was not easy to do in Parts III and IV. Some chapters turn out to be... tedious, not because they are not interesting or of not much use, but because presenting them in a clear manner requires substantial algebra. Indeed, we often had to go through a large number of equations to obtain the final formulae, when considering *physical applications* or *maths practice examples*. Still, we think that we managed to achieve what we wanted in many other chapters, making mathematical methods transparent and easy to understand, which means—*easy to love*. With that said, we still want to apologize for this compelled unevenness of the text in this volume. Whenever we felt we could relax, as the reader could have expected to enjoy the lesson, we accentuated it with a joke, metaphor, historical anecdote and/or cartoon. But this did not come about as often as in Volume 1; please, be prepared for that.

**From problems to exercises.** Whereas Parts I and II of Volume 1 were each completed by sets of problems to solve, to practise the attained knowledge (with their solution outlines appended on the web), Parts III and IV of Volume 2 are not. To keep the reader fit, instead, we have inserted many exercises actually in the text. These are either (i) intermediate derivations of the complicated equations, (ii) problems that are similar to the considered examples, (iii) variations of the theoretical expressions for changed boundary or initial conditions, or (iv) suggestions for finding asymptotic expressions of the obtained results, if they were are not presented straightaway. The text contains thousands of equations needed to set the methods and derive what we wanted to, so such an approach is understandable.

**Proceedings.** Work on this volume had a dramatic history. It was substantially delayed by the tragic accident that happened to Dominic. It happened shortly before the final proof reading of Volume 1, which made him unable to work on the completion of Volume 2. Dominic, however, has drafted many of its sections. Those passed my editing, but he could not cross-edit the material that was drafted by me. Neither could he see how I trimmed and amended his drafts (when preparing Volume 1 we passed the text and equations back and forth between each other in many rounds). Needless to say, he could not take part in the proof reading of the book, a very important stage for a volume of this size and number of equations. Equally, he could not filter or rephrase the jokes, diversions or cartoons, which he always carefully edited to avoid any tone that may be found too straightforward or patronizing. So, the responsibility for any deficiencies in the final outcome of this book are entirely on me.

**Misprints**. Several bugs noticed in Volume 1 have been put on the web at https://www.worldscientific.com/doi/suppl/10.1142/q0189/ suppl_file/q0189_errata.pdf.

Alas, the Devil never sleeps, and those misprints appeared, 'of course', in the *most beautiful places* of that volume. If our diligent readers find any slips in Volume 2 (their appearance is *statistically inevitable* in a book with so many equations), we would be grateful if those cases could be kindly communicated to me: https://www. imperial.ac.uk/people/a.kornyshev. They will be checked and put on the web as Errata to Volume 2.

**Authenticity**. It must have been obvious in Volume 1 that all the material discussed there is common knowledge, available in the public domain, but its presentation, style, the way of explanations and composition is, or so we believe, original. And of course there are a number of invented examples. The former part of this statement may at a first glance be less obvious in Volume 2. So, we must stress that it is a textbook, and *neither Dominic nor I invented anything in the mathematics presented here. We tried to share with you our understanding of these methods after we had ourselves sometimes suffered but mostly enjoyed learning how to apply them in our work.* When the text referred to mathematicians or physicists who pioneered and developed those methods, we have named them, giving their portraits and providing some historical information about them. When dealing with more modern things, usually developed very much 'collectively', we often present the material from the 'common knowledge point of view', without such quotations. For any of the chapters, interested readers can further go to specialized literature—textbooks or monographs—and those will contain all the needed references to the main contributors to the development of those methods, or specific results obtained by them.

**Acknowledgements**. The first people to acknowledge in connection with preparation of this volume are Dominic's mother Arabella Lee and his brother Ben Lee, who helped to rescue from his laptop the chapters drafted by him before his accident. I also thank my wife, Mila, for her enormous patience for 'stolen' evenings and 'lost holidays' due to my work on this volume. I thank also many of my internal and external collaborators on the running projects, as well as the students involved in them, for their understanding and tolerance while waiting for my response and contributions, delayed by the work on this book. The same refers to the teaching administration of my department.

I am indebted to my daughter Daria Kornysheva for the enlivening cartoons that she has designed and drawn for this volume.

I thank postgraduate students, Zac Goodwin, Ehud Haimov, and Cristian Zagar (at Imperial) and Pedro de Souza (at MIT) who helped with proof reading of several sections. Speaking of patience and support, I am greatly obliged to a fantastic production-team of World Scientific, feeling their shoulder at every stage of the work on this project; here very special thanks go to Stephen Cashmore, Editor, with whom I worked days and nights not only on the language and grammar: he meticulously checked every line of the text and equations, to make sure that they all make perfect sense and could not be formulated differently. If some caveats still remain, they are to be charged to me only.

**Readership.** Volume 1 should have interested not only undergraduate and postgraduate students, and some postdocs but, as we have explained, also eager high-school students, since we started there practically from scratch. To expect this for Volume 2 would be too optimistic: it is the next level of the learning curve. But both volumes, we hope, will be equally useful for those who wish to teach 'practical mathematics' to physical scientists, and most importantly to teach their students not only how to obtain their results but also to deeply understand them.

With all the dramas behind the preparation of this volume, we both think (I write 'we' because I know that Dominic would have had the same opinion, had he been able to express it) that this second volume will radically strengthen the muscles of the readers if they find the strength and persistence to go all the way to the end of it.

Of course, as will be obvious to readers, some chapters of this volume could be read separately, or in groups. But what is most essential for getting through Volume 2 or its parts, is an understanding of the spirit and practical approaches of Volume 1.

From here—ready, steady, go!

Alexei Kornyshev
London
March 2023

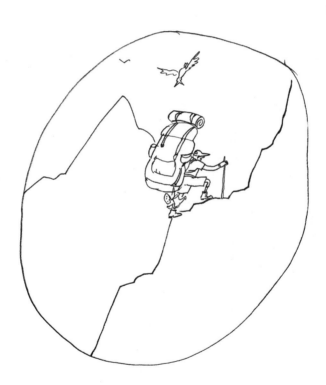

# About the Authors

 **Alexei A. Kornyshev** has been a Professor of Chemical Physics at Imperial College London for the last 20 years. There, among various subjects, he has taught mathematical methods to undergraduate and graduate students of physics and chemistry. A theoretician, with research spanning widely over the interdisciplinary fields of electrochemistry, nanoscience, photonics, energy, and biophysics, he has published more than 350 articles, many of which are highly cited.

 **Dominic O'Lee**, now retired due to health problems, was a senior member of Kornyshev's research group, and later a Teaching Fellow at Imperial College, with a passion of teaching the 'technology' of understanding the results of mathematical calculations. His research focussed on applications of field theory in condensed matter physics (superconductivity), statistical physics of biopolymers and biophysics. He has 50 publications to his name.

Please see the front matter of Volume 1 for a more detailed biography of each author.

# Contents

# Part III

# From Camp 2: In the Multidimensional and Complex World

# Chapter 1

# Multiple Integration in More Depth

In Section 1.6 of Part II (Vol. 1), we introduced the concept of multiple integration over areas and volumes. Here, we are going to study multiple integration in more detail and build on what we learnt previously. First, we extend the concept of integration over areas (area or surface integrals) and volumes (volume integrals) to shapes more complicated than rectangles and cuboids. Next, we state Green's theorem in the plane, without proof, and explain how it can be used for the evaluation of integrals. Then, we discuss how to make variable changes both in *area* and *volume* integrals and how such integrals can be written in terms of polar coordinates. These more sophisticated mathematical manipulations are invaluable for solving many physical problems. Our goal here is to demonstrate how they work, with minimal mathematical rigour. Although they may look cumbersome and complicated at a first glance, we want to show you that they are in fact quite trivial and straightforward to use.

## 1.1.  Area and Volume Integrals over Different Shapes

Here, we will consider two different types of integrals. The area integral of a function $f(x, y)$ is equal to the volume under a surface described by that function. Indeed, $z = f(x, y)$ gives the height of the surface at a point $(x, y)$, which is required to be single-valued. The volume integrals integrate over a volume, or region of space, any three variable function $f(x, y, z)$, which should be integrable (making the resulting integral not diverging). These definitions can be extended to higher dimensions, though it becomes hard to visualize then what it going on.

## *Area integrals*

Let us can consider area $S$ to be region bounded by two curves $y = y_1(x)$ and $y = y_2(x)$; and, from this, calculate the volume under a surface traced out by the function $f(x, y)$, as illustrated in Fig. III.1.1. In this case the area integral, i.e. the volume, bounded by the curves, is defined as

$$V = \int_S f(\mathbf{R}) d\mathbf{R} = \int_b^a dx \int_{y_1(x)}^{y_2(x)} dy f(x, y)$$

$$= \lim_{\substack{\Delta x \to 0 \\ \Delta y \to 0}} \sum_{j=\text{nint}(b/\Delta x)}^{\text{nint}(a/\Delta x)} \sum_{k=\text{nint}(y_1(j\Delta x)/\Delta y)}^{\text{nint}(y_2(j\Delta x)/\Delta y)} f(j\Delta x, k\Delta y) \Delta x \Delta y \quad \text{(III.1.1)}$$

The sums on the right side of this equation sum up the contribution from small squares that cover the region of integration. Here, we have introduced a new function to the reader, as the limits of those summations must be integers. This is $n = \text{nint}(x)$, the *nearest integer function*. This function takes its argument (input) $x$ and rounds it to the closest integer as the output $n$ (pictorially how this works for a finite number of squares is shown in Fig. III.1.1). In (Eq. III.1.1), we have started by summing the squares up (over $k$) along the $y$-direction, which are limited by the two functions $y = y_1(x)$ and $y = y_2(x)$ that describe two curves that bound the area of integration. The limits of the final sum over $j$ depend only on the end points $a$ and $b$, the maximum and minimum values of $x$ for the area of integration. Over each square the function $f(x, y)$ is taken to be a constant $f(j\Delta x, k\Delta y)$. Then, by making the area of each square infinitesimally small, increasing their number infinitely so that they still cover

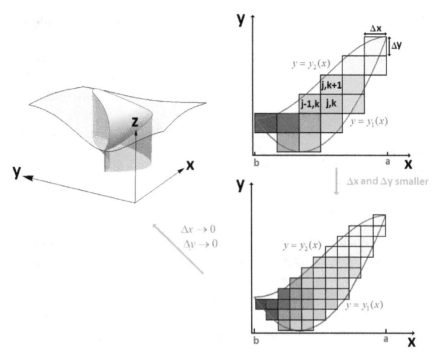

**Fig. III.1.1** Diagrams illustrating how area integration works. The procedure for finding the volume under a surface $z = f(x, y)$ bounded by two curves $y_1(x)$ and $y_2(x)$. In the top right diagram we approximate the volume by dividing up the region enclosed by curves into small squares or rectangles with dimensions $\Delta x$ and $\Delta y$. In each of the small rectangles we approximate $f(x, y)$ to have a constant value, i.e. the 'height' of the function does not change within it (colour coding here is different from usual maps: the lighter colours refer to higher heights and the darker colours to lower ones). An estimate of the volume can be found by summing up all the heights for each of the rectangles and multiplying by $\Delta x \Delta y$. A systematic way of doing this is to give each rectangle labels $j$ and $k$, the former referring to the $x$-position of the rectangle, such that $x = j\Delta x$, and the latter its $y$-position, $y = k\Delta y$. We may first sum the rectangles over $k$ between the two bounding curves, where $k$ is restricted to lie between $\text{nint}(y_1(j\Delta x)/\Delta y)$ and $\text{nint}(y_2(j\Delta x)/\Delta y)$ (the nearest integer function, $\text{int}(x)$, rounds to the nearest integer of the input $x$, see the text). Then the final sum over $j$ is taken between $\text{nint}(b/\Delta x)$ and $\text{nint}(a/\Delta x)$. If we make $\Delta x$ and $\Delta y$ smaller and increase the number of rectangles to cover the region integration, as shown in the bottom right picture we improve our estimate of the volume. In the limit where $\Delta x \to 0$ and $\Delta y \to 0$, the number of rectangles to cover the region tends to infinity, and we obtain the area integral given by Eq. (II.1.1). This gives exactly the volume under the surface illustrated by the left-hand picture.

the region, an area integral over an area of arbitrary shape can be calculated (as illustrated in Fig. III.1.1). Quite simple it is!

To evaluate any area integral we perform the $y$-integral normally, but keep $x$ as a fixed constant; we then use the result of this integration for the integrand for the $x$-integration, which is again performed normally. It is sometimes convenient with many integrations to write $dx$ and $dy$ with the integral signs before the function we want to integrate over, as is done in the middle of Eq. (III.1.1); the reader should not get confused by this.

## Volume integrals

Likewise we can integrate a function $f(x, y, z)$ over a volume bounded by two surfaces $z = z_1(x, y)$ and $z = z_2(x, y)$, between two curves $y = y_1(x)$ and $y = y_2(x)$. A diagram showing a region of integration is given in Fig. III.1.2; the shape can be divided up into small cuboids labelled by numbers $j, k$ and $l$, in the $x$, $y$ and $z$ directions, respectively. These cuboids have dimensions $\Delta x$, $\Delta y$ and $\Delta z$, and

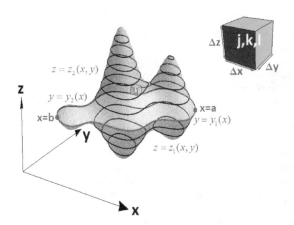

**Fig. III.1.2** An illustration of a region of integration for a volume integral. The region of integration is enclosed by two bounding surfaces $z = z_1(x, y)$ (shown in pale pink) and $z = z_2(x, y)$ (shown in pale yellow). The two surfaces meet in two bounding curves (shown as red lines) described by the functions $y = y_1(x)$ and $y = y_2(x)$ that run between the points $x = b$ and $x = a$. The region of integration can be divided up into small cuboids labelled by $j, k$ and $l$, which specifies their positions, i.e. $x = j\Delta x$, $y = k\Delta y$, and $z = l\Delta z$. One such cuboid element contributing to the region of integration is shown as a green cube. In the limit $\Delta x \to 0$, $\Delta y \to 0$ and $\Delta z \to 0$, taking the number of cuboid elements to infinity, we obtain the volume integral.

within each one the value of the function $f(x, y, z)$ is set to be constant equal to $f(j\Delta x, k\Delta y, l\Delta z)$. Then, such a volume integral is defined as

$$V = \int_V f(\mathbf{r})d\mathbf{r} = \int_b^a dx \int_{y_1(x)}^{y_2(x)} dy \int_{z_1(x,y)}^{z_2(x,y)} dz f(x, y, z)$$

$$= \lim_{\substack{\Delta x \to 0 \\ \Delta y \to 0 \\ \Delta z \to 0}} \sum_{j=\text{nint}(b/(\Delta x))}^{\text{nint}(a/(\Delta x))} \sum_{k=\text{nint}(y_1(j\Delta x)/(\Delta y))}^{\text{nint}(y_2(j\Delta x)/(\Delta y))} \sum_{l=\text{nint}(z_1(j\Delta x,k\Delta y)/(\Delta z))}^{\text{nint}(z_2(j\Delta x,k\Delta y)/(\Delta z))}$$

$$f(j\Delta x, k\Delta y, l\Delta z)(\Delta x \Delta y \Delta z) \qquad \text{(III.1.2)}$$

To perform the volume integral we perform the $z$-integral first in the normal way, treating both $x$ and $y$ as constants, then use the result of the integration as the integrand for the $y$-integral. The $y$-integral would then be performed by keeping $x$ fixed to determine the integrand for the final $x$-integration.

### *Interchange of the order of integration*

We are not restricted in our choice of the order of integration. For instance, for area integrals, when approximating our integral with the sum, one should realize that we don't have to sum over $k$ (in the $y$-direction) first. In fact, if we know the inverse functions that also define the bounding curves, as $x = y_1^{-1}(y)$ and $x = y_2^{-1}(y)$ (the superscript "−1" denotes the *inverse function*), we can sum over $j$, in the $x$-direction first. So we have the following relationship for area integrals

$$V = \int_S f(\mathbf{R})d\mathbf{R} = \int_b^a dx \int_{y_1(x)}^{y_2(x)} dy f(x, y) = \int_d^c dy \int_{y_1^{-1}(y)}^{y_2^{-1}(y)} dx f(x, y)$$

$$\text{(III.1.3)}$$

where

$$\int_d^c dy \int_{y_1^{-1}(y)}^{y_2^{-1}(y)} dx f(x, y) = \lim_{\substack{\Delta x \to 0 \\ \Delta y \to 0}} \sum_{k=\text{nint}(d/(\Delta y))}^{\text{nint}(c/(\Delta y))}$$

$$\sum_{j=\text{nint}(y_1^{-1}(j\Delta y)/(\Delta x))}^{\text{nint}(y_2^{-1}(k\Delta y)/(\Delta x))} f(j\Delta x, k\Delta y)\Delta x \Delta y$$

$$\text{(III.1.4)}$$

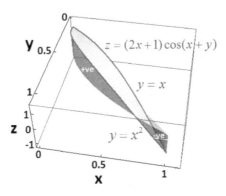

**Fig. III.1.3**    Here, we show the volume under the surface $z = f(x, y) = (2x+1)\cos(x+y)$ (with respect to the $z = 0$ plane) bounded by the curves $y = x^2$ and $y = x$. We calculate this as an area integral described by Eq. (III.1.5) of the text. In this particular case, the surface falls below the $z = 0$ plane, and the volume enclosed becomes a negative contribution (analogous to areas under curves when the curve falls below the $x$-axis).

Obviously, $c$ and $d$, the limits of integration over $y$, correspond to minimum and maximum values, respectively, that $y$ can take in the integration region.

For volume integrals we are also free to choose the order of integration, provided that we can express the surfaces enclosing the region of integration as $y = \tilde{y}_1(x, z)$ and $y = \tilde{y}_2(x, z)$ or $x = \tilde{x}_1(y, z)$ and $x = \tilde{x}_2(x, z)$, and find suitable choices for the two bounding curves in terms of the remaining variables.

**Example III.1.1 (Maths Practice ♫♪).** Let us find the volume under the surface $z = f(x, y) = (2x + 1)\cos(x + y)$ with the $x$–$y$ plane bounded between the curves $y = x^2$ and $y = x$. The curves cross at the coordinates $(0,0)$, $(1,1)$, as illustrated in Fig. III.1.3. Therefore, we have that $a = 1$ and $b = 0$, $y_1(x) = x^2$ and $y_2(x) = x$ in our general expression, Eq. (III.1.1). Thus, the area integral (the volume under the surface) is given by

$$V = \int_A f(\mathbf{R})d\mathbf{R} = \int_0^1 dx \int_{y=x^2}^{y=x} dy(2x + 1)\cos(x + y) \qquad \text{(III.1.5)}$$

We start by performing the $y$-integration, yielding

$$V = \int_A f(\mathbf{R})d\mathbf{R} = \int_0^1 dx[(2x+1)\sin(x+y)]_{y=x^2}^{y=x}$$

$$= \int_0^1 dx(2x+1)\sin(2x) - \int_0^1 dx(2x+1)\sin(x^2+x) \qquad \text{(III.1.6)}$$

The first of these integrals can be evaluated by parts

$$I_1 = \int_0^1 dx(2x+1)\sin(2x) = \left[-\left(x+\frac{1}{2}\right)\cos(2x)\right]_0^1 + \int_0^1 dx\cos(2x)$$

$$= \frac{1}{2} - \frac{3}{2}\cos(2) + \left[\frac{\sin(2x)}{2}\right]_0^1 = 0.5 - 1.5\cos(2) + 0.5\sin(2) \approx 1.58$$

$$\text{(III.1.7)}$$

The second integral appearing in Eq. (III.1.6) can be evaluated by a substitution $u = x^2 + x$ so that

$$I_2 = \int_0^1 dx(2x+1)\sin(x^2+x)$$

$$= \int_1^3 du\sin u = [-\cos u]_1^3 = \cos(1) - \cos(3) \approx 1.53 \qquad \text{(III.1.8)}$$

Thus, the total 'volume' under the surface is given by $V = 1.58+1.53 = 3.11$. Note, that in Eq. (III.1.6), $I_2$ enters with the negative sign, but obviously if we calculate the volume, we should change the sign of this contribution to a positive one, as we did. In this example, strictly speaking $x$ and $y$ are dimensionless and so must $z = f(x, y)$ be. This is because the argument of the cosine of $x+y$ must be dimensionless. If we wanted to make the $V$ a true volume, and $z$ a true height, we would introduce dimensional parameters.

**Exercise.** To make everything dimensionally consistent, if $x$, $y$ and $z = f(x,y)$ have dimensions of length, modify $y_1(x) = x^2$ and $f(x,y) = (2x+1)\cos(x+y)$, by introducing three dimensional parameters, $a, b$ and $c$ which have the dimensions of length. Then find the correct form for $V$. **Hint:** See example below.

**Example III.1.2 (Physical Application ▶▶).** Let us consider a shape that can be described by the interception of the $x-y$ plane

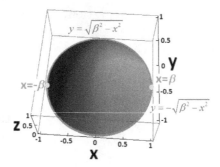

**Fig. III.1.4** We plot the integration region for the volume integral described in Example III.1.2 which delivers the mass of an object with this shape and density $\rho(x, y, z) = \alpha|x||y|z$. The two bounding surfaces are $z/\beta = 1 - (x/\beta)^2 - (y/\beta)^2$ (shown in orange-red) and $z = 0$. They intercept along the bounding curves $y = \sqrt{\beta^2 - x^2}$ and $y = -\sqrt{\beta^2 - x^2}$ (shown as blue lines), which terminate at the points $x = -\beta$ and $x = \beta$, for which $y = 0$.

with the surface $z/\beta = f(x, y) = 1 - (x/\beta)^2 - (y/\beta)^2$, where $x, y$, and $z$ have the dimensionality of length. The shape has a (mass) density approximated by the function $\rho(x, y, z) = \alpha|x||y|z$ (**Think:** What should be the units of $\alpha$ and $\beta$?).

The region of integration, the shape, is shown in Fig. III.1.4. The curves of interception between the $x$–$y$ plane and the surface are given by $0 = 1 - (x/\beta)^2 - (y/\beta)^2$, so that $y = \pm\sqrt{\beta^2 - x^2}$. Let's now try to find the mass of this shape. This is given by the volume integral

$$M = \int_V \rho(\mathbf{r})d\mathbf{r} = \alpha \int_{-\beta}^{\beta} |x|dx \int_{y=-\sqrt{\beta^2-x^2}}^{y=\sqrt{\beta^2-x^2}} |y|dy$$

$$\times \int_0^{z=(1-(x/\beta)^2-(y/\beta)^2)} z\,dz \tag{III.1.9}$$

Before doing anything, let's introduce the dimensionless variables $X = x/\beta$, $Y = y/\beta$ and $Z = z/\beta$. On changing variables, rescaling, we find (you should check this) that

$$M = \alpha\beta^6 \int_{-1}^{1} |X|dX \int_{Y=-\sqrt{1-X^2}}^{Y=\sqrt{1-X^2}} |Y|dY \int_0^{Z=1-X^2-Y^2} Z\,dZ$$

Let's evaluate each integral in turn, starting with $Z$-integration, which yields

$$\int_0^{z=1-X^2-Y^2} Z dZ = \left[\frac{Z^2}{2}\right]_0^{1-X^2-Y^2} = \frac{(1-X^2-Y^2)^2}{2} \quad \text{(III.1.10)}$$

So that we can write for the mass:

$$M = \alpha\beta^6 \int_{-1}^1 X dX \int_{Y=-\sqrt{1-X^2}}^{Y=\sqrt{1-X^2}} |Y| dY \frac{(1-X^2-Y^2)^2}{2} \quad \text{(III.1.11)}$$

Through the substitution $s = Y^2$ we evaluate the $Y$-integral as

$$\int_{Y=-\sqrt{1-X^2}}^{Y=\sqrt{1-X^2}} |Y| dY \frac{(1-X^2-Y^2)^2}{2} = \frac{1}{2}\int_0^{1-X^2} ds (1-X^2-s)^2$$

$$= \left[-\frac{(1-X^2-s)^3}{6}\right]_0^{1-X^2} = \frac{1}{6}(1-X^2)^3 \quad \text{(III.1.12)}$$

(writing the first line of Eq. (III.1.12), we have used that the function $|Y|(1-X^2-Y^2)$ is even). Substituting for $u = X^2$, we can, finally, write Eq. (III.1.11) as

$$M = \frac{\alpha\beta^6}{6}\int_{-1}^1 |X| dX (1-X^2)^3 = \frac{\alpha\beta^6}{6}\int_0^1 du(1-u)^3$$

$$= \frac{\alpha\beta^6}{24}\left[-(1-u)^4\right]_0^1 = \frac{\alpha\beta^6}{24} \quad \text{(III.1.13)}$$

**Exercise.** Repeat the calculation in Example III.1.2 using the density function $\rho(\mathbf{r}) = \alpha|x|y^2$.

**Example III.1.3 (Physical Application ▶▶).** In this example, we'll look at applying area integration to calculate the moment of inertia of a triangular sheet of constant density ($\rho$) and constant thickness ($l$). First, let us recall that in circular (rotational) motion, about the $z$-axis the kinetic energy is given by $T = I_z\omega^2/2$, where $\omega$ is the angular velocity of rotation and $I_z$ is the moment of inertia about the $z$-axis which in our case is given by

$$I_z = l\rho \int_V (x^2 + y^2) d\mathbf{r} \quad \text{(III.1.14)}$$

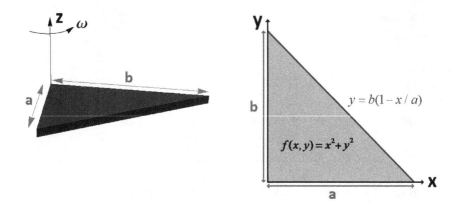

**Fig. III.1.5** In Example III.1.3, we calculate the moment of inertia of a uniformly dense right-angled triangular slab with sides $a$ and $b$ adjacent to the right angle, and the thickness $l$ of the triangular 'slab', for an axis of rotation perpendicular to right angle vertex. This situation is shown in the left figure. The moment of inertia can be calculated by taking the function $f(x,y) = x^2 + y^2$ and integrating it over the area bounded by the line $y = b(1-x/a)$ and a second 'bounding curve' made up of the $x$- and $y$-axes.

We will consider the case that the axis of rotation passes through a right-angled vertex of the triangle, which is formed by two of sides length $a$ and $b$ (shown in Fig. III.1.5). We can choose for these sides to line (length $a$) on the $x$-axis and $y$-axis (length $b$). The triangle can be described as an area bounded by the $x$-axis, $y$-axis and the line $y = b(1-x/a)$ (see Fig. III.1.5). We can write

$$I_z = l\rho \int_0^a dx \int_0^{b(1-x/a)} dy(x^2 + y^2)$$

$$= l\rho \int_0^a dx \left[ x^2 y + \frac{y^3}{3} \right]_0^{b(1-x/a)}$$

$$\Rightarrow \quad I_z = l\rho \int_0^a dx \left[ \frac{b}{a}(ax^2 - x^3) + \frac{b^3(a-x)^3}{3a^3} \right]$$

$$= l\rho \left[ \frac{b}{a}\left( \frac{ax^3}{3} - \frac{x^4}{4} \right) - \frac{b^3(a-x)^4}{12a^3} \right]_0^a$$

$$\Rightarrow \quad I_z = l\rho \left( b\left( \frac{a^3}{3} - \frac{a^3}{4} \right) + \frac{b^3 a}{12} \right) = \frac{l\rho ab}{12}\left( a^2 + b^2 \right) \qquad \text{(III.1.15)}$$

The mass of this sheet is $m = \rho abl/2$, so, in terms of the mass we can write the moment of inertia as

$$I_z = \frac{m}{6}\left(a^2 + b^2\right) \tag{III.1.16}$$

Therefore, the rotational kinetic energy of the triangle is given by

$$T = \frac{m}{12}\left(a^2 + b^2\right)\omega^2 \tag{III.1.17}$$

This result makes perfect sense, and to the accuracy of the numerical coefficient, it could be derived via dimensionality analysis (Chapter 10, Part I, Vol. 1)). Indeed, the dimensionality of kinetic energy is $[T] = [MV^2] = [ML^2t^{-2}]$. The only parameter of the dimensionality of the mass is the mass of the triangle, the length parameter is the hypotenuse, $\sqrt{a^2 + b^2}$, and the parameter of dimensionality of inverse time is the frequency of rotation. Hence, we could have written $T \sim m(a^2 + b^2)\omega^2$. But these arguments would not give us the important order-of-magnitude proportionality factor $1/12$. To get it we needed the whole calculation.

## 1.2. Green's Theorem in the Plane

### Green's theorem
An important theorem, called Green's theorem states that

$$\oint_C [P(x,y)dx + Q(x,y)dy] = \int_A \left[\frac{\partial Q(x,y)}{\partial x} - \frac{\partial P(x,y)}{\partial y}\right] dxdy \tag{III.1.18}$$

The l.h.s of Eq. (III.1.18) is the line integral around a closed curve $C$ (circulating in the anticlockwise direction), and the area integral on the r.h.s is evaluated over the area $A$ bounded by the curve (as illustrated in Fig. III.1.6). We won't prove Eq. (III.1.18), but just point out that it is only guaranteed to hold if both $P(x,y)$ and $Q(x,y)$, and their derivatives are well defined in

George Green
(1793–1841)

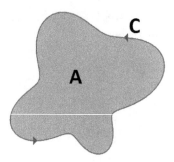

**Fig. III.1.6**   Diagram showing the area of integration, $A$ on the r.h.s of Eq. (III.1.18) with the integration path $C$ of the line integral on the l.h.s of Eq. (III.1.18) which encloses $A$.

the area $A$ (no particular singularities or discontinuities). At the end of this section, we'll show an example where Eq. (III.1.18) fails, because of a singularity within $A$.

### Green's theory consistency with state functions

Now, let's check one interesting aspect of Eq. (III.1.18). Let us suppose that we have a state function $f(x,y)$ such that

$$P(x,y) = \frac{\partial f(x,y)}{\partial x} \quad \text{and} \quad Q(x,y) = \frac{\partial f(x,y)}{\partial y} \qquad \text{(III.1.19)}$$

As long as $f(x,y)$ has no discontinuity at any point along the integration path,

$$\oint_C \left[ \frac{\partial f(x,y)}{\partial x} dx + \frac{\partial f(x,y)}{\partial y} dy \right] = 0 \qquad \text{(III.1.20)}$$

We see that Green's theorem will be consistent with Eq. (III.1.20), since in this case

$$\frac{\partial P(x,y)}{\partial y} = \frac{\partial f(x,y)}{\partial x \partial y} \quad \text{and} \quad \frac{\partial Q(x,y)}{\partial x} = \frac{\partial f(x,y)}{\partial x \partial y} \qquad \text{(III.1.21)}$$

Substituting in Eq. (III.1.21) into the integrand, we see that the r.h.s of Eq. (III.1.18) will indeed be zero.

## Another special case of Green's theorem

Now, let's consider a special case where we choose

$$P(x,y) = \frac{\partial f(x,y)}{\partial y} \quad \text{and} \quad Q(x,y) = -\frac{\partial f(x,y)}{\partial x} \qquad \text{(III.1.22)}$$

In this case Eq. (III.1.18) becomes

$$\oint_C \left[ \frac{\partial f(x,y)}{\partial y} dx - \frac{\partial f(x,y)}{\partial x} dy \right] = - \int_A \left[ \frac{\partial^2 f(x,y)}{\partial y^2} + \frac{\partial^2 f(x,y)}{\partial x^2} \right] dx dy$$

$$\text{(III.1.23)}$$

if then

$$\frac{\partial^2 f(x,y)}{\partial y^2} + \frac{\partial^2 f(x,y)}{\partial x^2} = 0 \qquad \text{(III.1.24)}$$

everywhere in $A$, then again the r.h.s of Eq. (III.1.18) is zero. Since the function $f(x,y)$ satisfies (III.1.24), it is called a *harmonic function*.

## Green's theorem holds for areas with more complicated topologies

It is also easy to show that the theorem holds for one closed curve inside another such that

$$\oint_{C_1} [P(x,y)dx + Q(x,y)dy] - \oint_{C_2} [P(x,y)dx + Q(x,y)dy]$$

$$= \int_{A'} \left[ \frac{\partial Q(x,y)}{\partial x} - \frac{\partial P(x,y)}{\partial y} \right] dx dy \qquad \text{(III.1.25)}$$

where $A'$ is the area bounded by the two curves (see Fig. III.1.7).

**Example III.1.4 (Maths Practice ♪♪).** Though we won't prove Green's, theorem, let's still check that it works for a simple case.

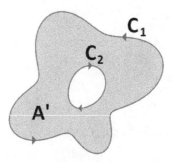

**Fig. III.1.7** We can apply Green's theorem to more complicated areas of integration, such as $A'$. The line integral which is equivalent to the area integral over $A'$ is a sum of line integrals about the paths $C_1$ and $C_2$. The integration path $C_2$ circulates in the opposite direction to $C_1$ to cut out a hole in the integration region.

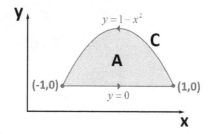

**Fig. III.1.8** We show the form of the integration region $A$, bounding curves, and end points for Example III.1.4. Note that the Cartesian equation $y = 1 - x^2$ can be found by eliminating $t$ from the bottom line of Eq. (III.1.27).

Let's consider the line integral

$$I = - \oint_C (y\,dx - x\,dy) \qquad \text{(III.1.26)}$$

where the closed curve (contour $C$) is specified as

$$y = 0 \quad x = t \quad \text{for} \quad -1 \le t < 1$$
$$\text{and} \quad y = 1 - (2 - t)^2 \quad x = 2 - t \quad \text{for} \quad 1 \le t \le 3 \qquad \text{(III.1.27)}$$

The path of integration is shown in Fig. III.1.8.

Let's first evaluate the line integral using the methods that we learnt from Chapter 1 of Part I, Vol. 1.

$$I = -\int_0^2 \left[ y(t)\frac{dx(t)}{dt} - x(t)\frac{dy(t)}{dt} \right] dt \tag{III.1.28}$$

Now, let's first evaluate the integrand of Eq. (III.1.28):

$$x(t)\frac{dy(t)}{dt} - y(t)\frac{dx(t)}{dt} = 0 \quad \text{for } -1 \le t < 1$$

$$x(t)\frac{dy(t)}{dt} - y(t)\frac{dx(t)}{dt} = 1 + (2-t)^2 \quad \text{for } 1 \le t \le 3 \tag{III.1.29}$$

Thus, we find that

$$I = \int_1^3 (1 + (2-t)^2)dt = \int_{-1}^1 (1 + t^2)dt = \left[ t + \frac{t^3}{3} \right]_{-1}^1 = \frac{8}{3} \tag{III.1.30}$$

Now, let's see if we get the same result using Green's theorem

$$I = -\oint_C (ydx - xdy) = 2\int_A dxdy \tag{III.1.31}$$

The area $A$ is bounded by the $x$-axis and the curve $y = 1 - x^2$ with $-1 \le x \le 1$. Thus, doing the area integration we again get

$$I = 2\int_A dxdy = 2\int_{-1}^1 dx \int_0^{1-x^2} dy = 2\int_{-1}^1 (1 - x^2)dx = \frac{8}{3} \tag{III.1.32}$$

So, we see that in this case Green's theorem works.

**Example III.1.5 (Maths Practice ♫♪).** Let's evaluate the following line integral using Green's theorem:

$$I = \oint_C \left[ -y\cos^2\left(\frac{x}{b}\right) dx + \left(\frac{x}{2} - \frac{1}{2}\sin\left(\frac{2x}{b}\right)\right) dy \right] \tag{III.1.33}$$

where the curve $C$ is an ellipse specified by $x(t) = D\cos t$ and $y(t) = E\sin t$, as shown in Fig. III.1.9. Using Green's theorem we get the

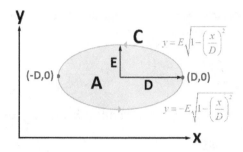

**Fig. III.1.9**  Figure showing the integration region $A$, bounding curves and endpoints for area integral in Example III.1.5, as well as the path $C$ about $A$. Note here that the parametric equations $x(t) = D\cos t$ and $y(t) = E\sin t$ defining $C$ can be written as Cartesian equations for two bounding functions, namely $y = \pm E\sqrt{1 - (x/D)^2}$.

answer immediately:

$$I = \int_A [\sin^2(x/b) + \cos^2(x/b)]dxdy = \int_A dxdy = \pi DE \quad \text{(III.1.34)}$$

where we have used the fact that the area of an ellipse is $A = \pi DE$, with the half lengths of its two axes being $D$ and $E$. Note that when $D = E = R$, we get $A$ (and so $I$) to be $\pi R^2$, as expected.

**Example III.1.6 (Maths Practice ♪♪).** Whereas in the previous example we have shown the benefits of going from line integration to area integration by using Green's theorem, the present example shows the case when it is simpler to evaluate an area integral via transforming it to a line integral. Consider

$$I = \int_A [2y\cos(x^2 + y^2) - 2y^3\sin(x^2 + y^2) + y\cos(x^2 + y^2)$$
$$- 2yx^2\sin(x^2 + y^2)]dxdy \quad \text{(III.1.35)}$$

where $A$ is the area bounded by a circle of radius 2 centred at the origin ($x = y = 0$). Let us identify the two parts of the integrand of Eq. (III.1.35) as

$$2y\cos(x^2 + y^2) - 2y^3\sin(x^2 + y^2) \equiv -\frac{\partial P(x, y)}{\partial y} \quad \text{(III.1.36)}$$

and

$$y \cos(x^2 + y^2) - 2yx^2 \sin(x^2 + y^2) \equiv \frac{\partial Q(x,y)}{\partial x} \qquad \text{(III.1.37)}$$

By inspection it can be seen that

$$P(x,y) = -y^2 \cos(x^2 + y^2) \quad \text{and} \quad Q(x,y) = yx \cos(x^2 + y^2)$$
$$\text{(III.1.38)}$$

—just differentiate these expressions for $P$ over $y$ and $Q$ over $x$ (**Exercise.** If you have patience, get this by integration of the l.h.ss of these two equations over, respectively, $y$ and, $x$). Then, using Green's theorem we can write $I$ as a line integral

$$I = - \oint \left[ y^2 \cos(x^2 + y^2)dx - xy \cos(x^2 + y^2)dy \right] \qquad \text{(III.1.39)}$$

where the path $C$ is specified by $x(t) = 2\cos t$ and $y(t) = 2\sin t$, where $0 \le t \le 2\pi$. Hence,

$$I = - \int_0^{2\pi} y(t) \cos(x(t)^2 + y(t)^2) \left[ y(t) \frac{dx(t)}{dt} - x(t) \frac{dy(t)}{dt} \right] dt$$

$$I = 8\cos(4) \int_0^{2\pi} \sin t \left[ \sin^2 t + \cos^2 t \right] dt = 8\cos(4) \int_0^{2\pi} \sin t \, dt = 0$$
$$\text{(III.1.40)}$$

So we see that such a nasty beast described by Eq. (III.1.35) is cut right down to zero using Green's theorem!

**Example III.1.7 (Maths Practice ♫♪).** This example is an object lesson on how sometimes we have to be very careful in the application of Green's theorem. Suppose that we want to evaluate the line integral around any closed path

$$I = \oint_C \frac{1}{x^2 + y^2} [y\,dx - x\,dy] \qquad \text{(III.1.41)}$$

Let's start by blindly applying Green's theorem. We first identify

$$P(x,y) = \frac{y}{x^2 + y^2} \quad \text{and} \quad Q(x,y) = -\frac{x}{x^2 + y^2} \qquad \text{(III.1.42)}$$

Thus, we find

$$\frac{\partial P(x,y)}{\partial y} = \frac{1}{x^2 + y^2} - \frac{2y^2}{(x^2 + y^2)^2} = \frac{x^2 - y^2}{(x^2 + y^2)^2} \qquad \text{(III.1.43)}$$

$$\frac{\partial Q(x,y)}{\partial x} = -\frac{1}{x^2 + y^2} + \frac{2x^2}{(x^2 + y^2)^2} = \frac{x^2 - y^2}{(x^2 + y^2)^2} = \frac{\partial P(x,y)}{\partial y}$$

$$\text{(III.1.44)}$$

Thus, the integral is expected to be zero by Green's theorem.

Now, let's specify a path $C$ around a unit circle centred about the origin so that

$$x(t) = \cos t \quad \text{and} \quad y(t) = \sin t, \quad \text{where } 0 \le t \le 2\pi \qquad \text{(III.1.45)}$$

Then we have that for $I$

$$I = \int_0^{2\pi} \frac{1}{x(t)^2 + y(t)^2} \left[ y(t) \frac{dx(t)}{dt} - x(t) \frac{dy(t)}{dt} \right] dt$$

$$= -\int_0^{2\pi} \left[ \sin^2 t + \cos^2 t \right] dt = -\int_0^{2\pi} dt = -2\pi \qquad \text{(III.1.46)}$$

So what's gone wrong? The problem is that both $P(x,y)$ and $Q(x,y)$ have a pole, or singularity, at $x = y = 0$. **Remember the disclaimer: Green's theorem may not hold when the path of integration encloses such a point, where the functions $P(x,y)$ and $Q(x,y)$ and their derivatives *are not defined*.** Note that if we had chosen a circle of any radius $b$ we would have got the same result! In fact, if we had chosen any path about the singularity we would have got the same result as Eq. (III.1.46). Also, if we had considered the difference in two line integrals for two closed paths specified by $C_1$ and $C_2$ (both circulating in the anticlockwise direction), both about the singularity, we would have got zero. Any area bounded by curves that contains the singularity will evaluate to zero. In fact Eq. (III.1.41) is a rather important example that comes up in Ampere's Law in Magnetism (which we'll discuss later), when there is current flowing through the point $x = y = 0$ and nowhere else. Such a situation will also occur when we come to consider complex contour integrals in Chapter 8 of

Part III of Vol. 2 (these are line integrals in the complex plane with a few more rules about functions of complex variables) around poles in $f(z)$.

**Exercise.** Consider line integral

$$I = \oint_C \frac{1}{x^2 + y^2}[x\,dx + y\,dy] \qquad (\text{III.1.47})$$

Would we have the same problem as in the previous example? Evaluate both the line integral and the appropriate area integral from Eq. (III.1.18) to find out.

## 1.3. Changing Variables and Coordinate Systems

We can change variables in volume and surface integrals, just as in ordinary integrals.

### *Area integrals*

Consider the area integral

$$I = \int_A f(\mathbf{R})d\mathbf{R} = \int_{x_2}^{x_1} dx \int_{y_2=y_2(x)}^{y_1=y_1(x)} dy\, f(x, y) \qquad (\text{III.1.48})$$

We can make a variable change to Eq. (III.1.48) described by two functions that interrelate $x$- and $y$-variables with $u$- and $v$-variables, such that $x = \tilde{x}(u, v)$ and $y = \tilde{y}(u, v)$. In terms of the new variables the integral reads as

$$I = \int_{A'} \tilde{f}(u, v)du\,dv = \int_{u_2}^{u_1} du \int_{v_2=v_2(u)}^{v_1=v_1(u)} dv\, f(\tilde{x}(u, v), \tilde{y}(u, v)) \mathscr{J}(u, v)$$

$$(\text{III.1.49})$$

where $A'$ is the new area of integration for the new coordinates $u$ and $v$. Here, $\mathscr{J}(u, v)$ is called the Jacobian after a Prussian mathematician Jacobi (famous in the first place as the 'father of elliptic functions'), which has the form

$$\mathscr{J}(u,v) = \begin{Vmatrix} \dfrac{\partial \tilde{x}}{\partial u} & \dfrac{\partial \tilde{x}}{\partial v} \\[2mm] \dfrac{\partial \tilde{y}}{\partial u} & \dfrac{\partial \tilde{y}}{\partial v} \end{Vmatrix} = \left| \dfrac{\partial \tilde{x}}{\partial u} \dfrac{\partial \tilde{y}}{\partial v} - \dfrac{\partial \tilde{x}}{\partial v} \dfrac{\partial \tilde{y}}{\partial u} \right|$$

$$(\text{III.1.50})$$

Carl Gustav Jacob
Jacobi(1804–1851)

Here the double bars in Eq. (III.1.50) mean that we must take the modulus of the determinant. The functions $v_1(u)$ and $v_2(u)$ can be determined through the boundary curve equations $y = y_1(x)$ and $y = y_2(x)$ which can be written in terms of $x = \tilde{x}(u,v)$ and $y = \tilde{y}(u,v)$ as

$$y_1(\tilde{x}(u,v_1(u))) = \tilde{y}(u,v_1(u)) \quad y_2(\tilde{x}(u,v_2(u))) = \tilde{y}(u,v_2(u))$$

$$(\text{III.1.51})$$

Solving Eq. (III.1.51) for $v_1(u)$ and $v_2(u)$ will give us their functional forms. The end points $u_1$ and $u_2$ can be obtained through the intercepts of $v_1(u)$ and $v_2(u)$, i.e. $\tilde{v}_1 = v_1(u_1) = v_2(u_1)$ and $\tilde{v}_2 = v_1(u_2) = v_2(u_2)$. This is because the end points $x_1$ and $x_2$ are determined by the interception of the bounding curves $y = y_1(x)$ and $y = y_2(x)$, and we have 'mapping' between $(x,y)$ and $(u,v)$. Note that this mapping should be 1 to 1, i.e. single-valued, over the whole area of integration. If this is not the case, considerable care must be taken: one would need to split up the integration areas $A$ and $A'$ into parts, so as to provide us with unambiguous mappings in between the confining curves of each of the specified sectors.

Below we show a schematic picture of the variable change and the mapping (Fig. III.1.10), and also clarify the procedures with few examples. Do not, however, expect too much—this technique is only going to be really useful, when it is easy to find $u_1$, $u_2$, $v_1(u)$, $v_2(u)$, and if it simplifies things, as, for example, in the examples shown.

The proof of Eq. (III.1.49) involves Green's theorem in the plane to convert the integral into a line integral, making a change of variables and converting back; it's given in many other standard higher mathematics books, and we won't bother you with proving it here.

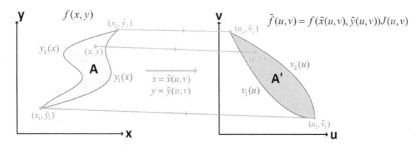

**Fig. III.1.10** Here, we illustrate the change of variables of an area integral. On the l.h.s we show the original area of integration $A$ with integrand $f(x, y)$. The original area of integration is confined by the bounding curves $y_1(x)$ and $y_2(x)$, which have the end points $(x_1, \tilde{y}_1)$ and $(x_2, \tilde{y}_2)$, where $\tilde{y}_1 = y_1(x_1) = y_2(x_1)$ and $\tilde{y}_2 = y_1(x_2) = y_2(x_2)$. The variable change through the functions $x = \tilde{x}(u, v)$ and $y = \tilde{y}(u, v)$ maps a point $(x, y)$ in the $x - y$ plane to a new point $(u, v)$ in the plane specified by the new variables $u$ and $v$ (as shown by one of long green lines with arrows). Points that lie on the curve $y_1(x)$ are mapped on to the new bounding curve $v_1(u)$, and those lying on $y_2(x)$ are mapped on to $v_2(u)$. For some particular variable change these curves are shown on the right. The end points are mapped from $(x_1, \tilde{y}_1)$ and $(x_2, \tilde{y}_2)$ to $(u_1, \tilde{v}_1)$ and $(u_2, \tilde{v}_2)$, respectively (the mappings are shown as the upper and lower thin green lines). The new end points and bounding curves define a new area of integration $A'$. So that the area-integral remains the same, the integrand is transformed to $\tilde{f}(u, v) = f(\tilde{x}(u, v), \tilde{y}(u, v)) \, \mathcal{J}(u, v)$, where $\mathcal{J}(u, v)$ is the Jacobian defined by Eq. (III.1.50).

**Example III.1.8 (Maths Practice ♫♪).** Although we are not proving them here, let us put Eqs. (III.1.49) and (III.1.50) to the test with a simple example. This will also give us the opportunity see how the changing variables procedure works for a specific case. Let's evaluate

$$I = \int_A (x - y)^2 x \, dx \, dy \qquad (\text{III}.1.52)$$

over the area $A$ bounded by the two bounding curves $y = y_1(x)$ and $y = y_2(x)$ with end points $x = -2$ and $x = 2$. We will consider the case with the following bounding curves:

$$y_1(x) = x \qquad -2 \le x < 0$$
$$y_1(x) = 2x \qquad 0 \le x \le 2 \qquad (\text{III}.1.53)$$

and

$$y_2(x) = 2x + 2 \qquad -2 \le x < 0$$
$$y_2(x) = x + 2 \qquad 0 \le x \le 2 \qquad (\text{III}.1.54)$$

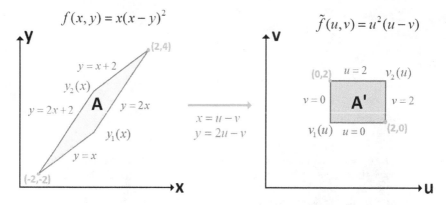

**Fig. III.1.11**  In this figure, we show the variable change for the area-integral dealt with in Example III.1.11. Here the integrand in the old variables is explicitly given by $f(x,y) = x(x-y)^2$, the bounding curves $y_1(x,y)$ and $y_2(x,y)$ are given by Eqs. (III.1.53) and (III.1.54), which are labelled on the l.h.s panel. The end points are $(-2,-2)$ and $(2,4)$; these and the bounding curves define the integration area $A$. The variable change $x = u - v$ and $y = 2u - v$ then map $x, y$ to the new variables $u, v$. The end points $(-2,-2)$ and $(2,4)$ map to $(0,2)$ and $(2,0)$, respectively. The new area of integration $A'$ is a square bounded by the lines $u = 0$, $v = 0$, $u = 2$ and $v = 2$. The integrand is then transformed to $\tilde{f}(u,v) = u^2(u - v)$.

The integration region is shown on the l.h.s of Fig. III.1.11. Let's first evaluate Eq. (III.1.52) by changing variables through the choice

$$y = x + u \quad \text{and} \quad y = 2x + v \qquad \text{(III.1.55)}$$

Before going through the procedure of changing variables, we should first ask ourselves why we think this is a good choice. Varying $u, v$ in Eq. (III.1.55) describes sets of lines that run parallel to the bounding curves, dividing up the integration region into a regular grid. Importantly, if we set $u = 0$ and $v = 0$ we recover $y_1(x)$, and when $u = 2$ and $v = 2$ we recover $y_2(x)$.

Now, let's blindly go through the changing the variables procedure on Eq. (III.1.52), applying the variable choice from Eq. (III.1.55). Rearranging Eq. (III.1.55) we have that

$$x = u - v \quad \text{and} \quad y = 2u - v \qquad \text{(III.1.56)}$$

Now, let us find the bounding curves on the plane of new variables $u$ and $v$. Let's use Eq. (III.1.51) to write, using $y_1(x)$ (given by Eq. (III.1.53))

$$y \equiv 2u - v = x \equiv u - v \quad \Rightarrow \quad u = 0$$

followed by

$$y \equiv 2u - v = 2x \equiv 2u - 2v \quad \Rightarrow \quad v = 0 \tag{III.1.57}$$

and to write using $y_2(x)$ (given by Eq. (III.1.54)

$$y \equiv 2u - v = 2x \equiv 2u - 2v + 2 \quad \Rightarrow \quad v = 2$$

followed by

$$y \equiv 2u - v = x \equiv u - v + 2 \quad \Rightarrow \quad u = 2 \tag{III.1.58}$$

Thus, the new integration region is a square (shown on the r.h.s of Fig. III.1.11). Also, the end points are trivially determined by the interception between the lines $u = 0$ and $v = 2$, as well as $u = 2$ and $v = 0$.

Therefore, we can write

$$I = \int_0^2 \int_0^2 u^2(u - v)\mathscr{J}(u, v)\,du\,dv \tag{III.1.59}$$

What remains is to evaluate the Jacobian

$$\mathscr{J}(u, v) = \left\| \begin{matrix} \frac{\partial x}{\partial u} & \frac{\partial x}{\partial v} \\ \frac{\partial y}{\partial u} & \frac{\partial y}{\partial v} \end{matrix} \right\| = \left\| \begin{matrix} 1 & -1 \\ 2 & -1 \end{matrix} \right\| = 2 - 1 = 1 \tag{III.1.60}$$

Equation (III.1.59) becomes

$$I = \int_{A'} u^2(u - v)\,du\,dv = \int_0^2 \int_0^2 u^2(u - v)\,du\,dv$$

$$\Rightarrow \quad I = \int_0^2 \left[ uv - \frac{v^2}{2} \right]_0^2 u^2\,du = \int_0^2 (2u^3 - 2u^2)\,du$$

$$\Rightarrow \quad I = 2\left[ \frac{u^4}{4} - \frac{u^3}{3} \right]_0^2 = 2\left( 4 - \frac{8}{3} \right) = \frac{8}{3} \tag{III.1.61}$$

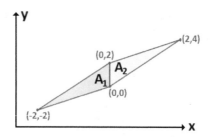

**Fig. III.1.12**    To evaluate the integral without changing variables, we divide $A$ up into $A_1$ and $A_2$ as shown in the figure.

Now, let's evaluate the same integral in the old way (as in Section 1.1)—sticking with coordinates $(x, y)$—to see if changing variables works. First, we need the intersection points of the four lines; these are $(0, 0)$, $(0, 2)$, $(-2, -2)$ and $(2, 4)$. We see that the first two interception points lie on the $y$-axis; this suggests that we should divide the area $A$ up into two areas $A_1$, for $x \leq 0$ and $A_2$ for $x \geq 0$ (as illustrated in Fig. III.1.12) so that

$$I = \int_{A_1} (x - y)^2 x \, dx \, dy + \int_{A_2} (x - y)^2 x \, dx \, dy$$

$$= \int_{-2}^{0} dx \int_{y=x}^{y=2x+2} dy (x-y)^2 x + \int_{0}^{2} dx \int_{y=2x}^{y=x+1} dy (x-y)^2 x$$

$$\text{(III.1.62)}$$

We can then evaluate the $y$-integrals

$$I = \int_{-2}^{0} dx \, x \left[ \frac{(y-x)^3}{3} \right]_{y=x}^{y=2x+2} + \int_{0}^{2} dx \, x \left[ \frac{(y-x)^3}{3} \right]_{y=2x}^{y=x+2}$$

$$= \int_{-2}^{0} dx \, x \left( \frac{(x+2)^3}{3} \right) + \int_{0}^{2} dx \, x \left( \frac{8}{3} - \frac{x^3}{3} \right) \qquad \text{(III.1.63)}$$

Thus, we are left with

$$I = \int_{-2}^{0} dx \left( \frac{8x}{3} + 4x^2 + 2x^3 + \frac{x^4}{3} \right) + \int_{0}^{2} dx \left( \frac{8x}{3} - \frac{x^4}{3} \right)$$

$$= \left[ \frac{4x^2}{3} + \frac{4x^3}{3} + \frac{x^4}{2} \right]_{-2}^{0} + \left[ \frac{4x^2}{3} \right]_{0}^{2} = \frac{8}{3} \qquad \text{(III.1.64)}$$

We get the same answer, so indeed the change of variables method works. Which way is more convenient—you are to decide!

**Example III.1.9 (Maths Practice ♪♪).** Here, we'll look at a particularly nasty monster that can actually be cut down to size by change of variables. Consider

$$\int_A f(\mathbf{R})d\mathbf{R} = \int_A \frac{xy^{3/2}}{(x^2+1)^{3/2}} \exp\left(-\frac{x^2 y}{(x^2+1)}\right) dxdy \qquad \text{(III.1.65)}$$

where $A$ is the area confined by the bounding curves $y = y_1(x)$ and $y = y_2(x)$, which are defined as

$$y_1(x) = (x^2+1)/x^2 \quad 0.5 \le x < 1$$
$$y_1(x) = x^2+1 \qquad 1 \le x \le 2 \qquad \text{(III.1.66)}$$

and

$$y_2(x) = 4(x^2+1) \qquad 0.5 \le x < 1$$
$$y_2(x) = 4(x^2+1)/x^2 \quad 1 \le x \le 2 \qquad \text{(III.1.67)}$$

The area of integration is displayed in the l.h.s panel of Fig. III.1.13.

Evaluating Eq. (III.1.65) without doing it numerically may look impossible at a first glance, but there is a way, if we smartly choose new variables! Let's go for $u$ and $v$ such that

$$y = u^2(x^2+1)/x^2$$
$$y = v^2(x^2+1) \qquad \text{(III.1.68)}$$

Here, we won't blindly determine the limits of integration using Eq. (III.1.51), but instead use some common sense. We can see (in Fig. III.1.13) that the curves defined by Eq. (III.1.68) form a distorted grid, with the bounding curves corresponding $u = 1$, $u = 2$ and $v = 1$ and $v = 2$. This immediately suggests that a transformed region of integration $A'$ will be a square for which $1 \le u \le 2$ and $1 \le v \le 2$.

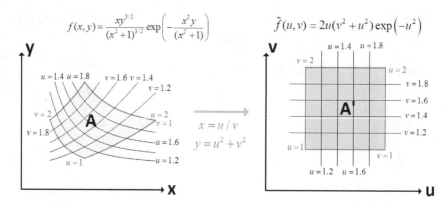

**Fig. III.1.13** This is an illustration of the transformation of variables for evaluating the integral given by Eq. (III.1.65. On the l.h.s, as well as showing the integration region, we show the families of curves $y = u^2(x^2+1)/x^2$ and $y = v^2(x^2+1)$. The bounding curves are reproduced by the values $u = 0$ and $v = 0$ (for $y_1(x)$), as well as $u = 2$ and $v = 2$ (for $y_2(x)$). On making the transformation, $x = u/v$ and $y = x^2+y^2$ the integration area $A$ transforms into the square area $A'$. The integrand $f(x,y) = xy^{3/2}/(x^2+1)^{3/2} \exp\left(-x^2 y/(x^2+1)\right)$ transforms to $\tilde{f}(u,v) = u(v^2+u^2)\exp(-u^2)$, and it only remains to calculate the Jacobian, to get the new integrands. Each of the illustrated curves, on the left, is mapped onto a regular grid on the r.h.s panel. The area of integration is bounded by the lines $u = 1$, $v = 1$, $u = 2$ and $v = 2$.

Now, rearranging, we find $\tilde{x}(u,v)$ and $\tilde{y}(u,v)$:

$$\tilde{x}(u,v) = \frac{u}{v} \quad \text{and} \quad \tilde{y}(u,v) = u^2 + v^2 \qquad \text{(III.1.69)}$$

This allows us to write

$$\int_A f(\mathbf{R})d\mathbf{R} = \int_{A'} uv^2 \exp(-u^2)\mathscr{J}(u,v)dudv$$

$$= \int_1^2 \int_1^2 uv^2 \exp(-u^2)\mathscr{J}(u,v)dudv \qquad \text{(III.1.70)}$$

Using Eqs. (III.1.50) and (III.1.69) we get the Jacobian,

$$J(u,v) = \begin{Vmatrix} \dfrac{1}{v} & -\dfrac{u}{v^2} \\ 2u & 2v \end{Vmatrix} = 2\left|1 + \dfrac{u^2}{v^2}\right| \qquad \text{(III.1.71)}$$

We can drop the modulus sign, as $1 + \frac{u^2}{v^2} > 1$. Therefore, we find

$$\int_A f(\mathbf{R})d\mathbf{R} = 2\int_1^2 \int_1^2 u(v^2+u^2)\exp(-u^2)dudv \qquad \text{(III.1.72)}$$

We can easily do the integration over $v$ first:

$$\int_A f(\mathbf{R})d\mathbf{R} = 2\int_1^2 \left[\frac{v^3}{3} + vu^2\right]_1^2 u\exp(-u^2)du$$

$$= 2\int_1^2 \left(\frac{7}{3} + u^2\right)u\exp(-u^2)du \qquad (III.1.73)$$

Making here the substitution $s = u^2$, we finally obtain

$$\int_A f(\mathbf{R})d\mathbf{R} = \int_1^4 \left(\frac{7}{3} + s\right)\exp(-s)ds$$

$$= [-s\exp(-s)]_1^4 + \frac{10}{3}\int_1^4 \exp(-s)ds$$

$$= \exp(-1) - 4\exp(-4) - \frac{10}{3}[\exp(-4) - \exp(-1)]$$

$$= \frac{13}{3}\exp(-1) - \frac{22}{3}\exp(-4) \approx 1.46 \qquad (III.1.74)$$

**Exercise.** We owe you a drink of your choice, if you find a better way to obtain this! But what you for sure could do is to check it by a numerical integration of the integral, Eq.(III.1.65).

### Area integrals in polar coordinates

The examples above were a bit of 'showing off', although they clearly demonstrate the principle of variable change. Now, let's get to business and look at one classic variable change—from 2D Cartesian coordinates to 2D polar coordinates. Suppose we want to evaluate the area-integral of a function $f(x, y)$

$$I = \int_A f(\mathbf{R})d\mathbf{R} = \int_A f(x,y)dxdy \qquad (III.1.75)$$

with the area $A$ described by a circle of radius $b$. The natural choice here to change the $x$- and $y$-variables for polar coordinates variables, $r$ and $\theta$. That is:

$$x = r\cos\theta \quad y = r\sin\theta \qquad (III.1.76)$$

where the area of a circle can be covered by changing $\theta$ from 0 to $2\pi$ and $r$ from 0 to $b$.

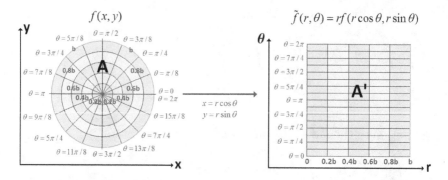

**Fig. III.1.14**   Here, we consider a circular area of integration, $A$, of radius $b$. For such an area of integration applying polar coordinates $x = r\cos\theta$ and $y = r\sin\theta$ are a good choice. On the left we divide the integration region into lines of constant $\theta$ where $0 \le \theta \le 2\pi$ and constant $r$ for which $0 \le r \le b$. Here, we think of this change of coordinates as a variable change where the new area of integration, $A'$, is a rectangle and the transformed integrand is $rf(r\cos\theta, r\sin\theta)$. Note that at the point $r = 0$ there is no '1-1 mapping' because the point on the left graph transforms to a line on the right one. This is not a problem, because the contribution of this one point to the integral is negligible, if an integrand is not singular at that point (no pole). Also, the lines $\theta = 0$ and $\theta = 2\pi$ are separated in the transformed graph, so function $\tilde{f}(r,\theta)$ is required to be periodic in $\theta$. Here, the transformed domain reflects the topology of the original circular integration region in polar coordinates with the origin ill-defined in $\theta$. In certain cases when the area of integration is on a surface, identifying the correct topology is important.

In Fig. III.1.14 we show how the integration $A$ region is covered for lines of constant $\theta$ and $r$. The l.h.s panel looks a bit like a dart board, reader, if you are familiar with darts, a typical English pub game, but this one is based a tad more on rational principles... on a real dart board only God knows why the number 19 is near the bottom, whereas the 20 is at the top! But let's not get distracted. The area of integration is effectively converted using the variable changes shown in Eq. (III.1.76) to the rectangle in the r.h.s panel (if we were to choose $r$ and $\theta$, as horizontal and vertical axes, respectively). In this case we can write

$$I = \int_0^{2\pi} d\theta \int_0^b dr\, f(x(r,\theta), y(r,\theta)) \mathcal{J}(r,\theta) \qquad \text{(III.1.77)}$$

with the Jacobian

$$\mathcal{J}(r,\theta) = \begin{Vmatrix} \dfrac{\partial x}{\partial r} & \dfrac{\partial x}{\partial \theta} \\ \dfrac{\partial y}{\partial r} & \dfrac{\partial y}{\partial \theta} \end{Vmatrix} = \begin{Vmatrix} \cos\theta & -r\sin\theta \\ \sin\theta & r\cos\theta \end{Vmatrix} = r\left|\cos^2\theta + \sin^2\theta\right| = r \qquad \text{(III.1.78)}$$

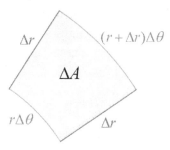

**Fig. III.1.15** Instead of thinking of a transformation between Cartesian and polar variables, we can consider an area element $\Delta A$. This is shown here in terms of small changes $\Delta r$ and $\Delta \theta$. We can divide a circular region (or any other region, for that matter) up in terms of these elements. Considering these elements makes for a far more intuitive understanding of area integration in polar coordinates as opposed to the formal variable transformation method.

Thus, in polar coordinates

$$I = \int_0^{2\pi} d\theta \int_0^b r\,dr\,\tilde{f}(r,\theta) \tag{III.1.79}$$

where $\tilde{f}(r,\theta) = f(r\cos\theta, r\sin\theta)$.

The Jacobian method is all very well for getting the form of Eq. (III.1.79), but let's see if we can justify Eq. (III.1.79) using some common sense. Let's consider a small segment of area in polar coordinates of radial width $\Delta r$ and angular width $\Delta\theta$ centred about $(r,\theta)$ (in polar coordinates), as illustrated in Fig. III.1.15. The area of this element is $\Delta A = ((r+\Delta r)^2\Delta\theta - r^2\Delta\theta)/2 \approx r\Delta\theta\Delta r$ (this is simply got by realizing that $\Delta A$ is the difference in area between two segments of circles with radius $r+\Delta r$ and $r$, and taking $\Delta r$ to be small). To cover the area of integration $A$ we can fill the region of integration with such small segments (just as we considered small rectangles of dimension $\Delta x \Delta y$), setting $r\tilde{f}(r,\theta)$ to be constant, taken at an average value within each small interval.

When we take $\Delta r \to 0$ and $\Delta\theta \to 0$, as well as the number of these elements to infinity, we do indeed recover Eq. (III.1.79) (as $r$ appears in $\Delta A$). We will talk more about such common-sense analysis when we come to consider volume integrals in polar coordinates, after introducing the Jacobian method for volume integrals.

**Example III.1.10 (Physical Application ▶▶).** Let's look at a physical example of integrating using polar coordinates. The current of a wire, $I(t)$, is related to the current density, $J(x, y, t)$, through the rather obvious relationship

$$I(t) = \int_A J(x, y, t) dx dy \qquad \text{(III.1.80)}$$

where $A$ is the cross-sectional area of the wire. For a wire with a circular cross section of radius $b$ we can write in polar coordinates

$$I(t) = \int_0^{2\pi} d\theta \int_0^b r dr \tilde{J}(r, \theta, t) \qquad \text{(III.1.81)}$$

where $\tilde{J}(r, \theta, t) = J(r \cos \phi, r \sin \phi, t)$ is the current density in polar coordinates. If the wire is of uniform resistivity, in response to alternating electromotive force (e.m.f.) of frequency $\omega$, the current density is, as suggested by the theory (see physics/electrical engineering textbooks), given by

$$\tilde{J}(r, \theta, t) = \bar{J}_s \text{Re} \left[ \exp \left\{ i\omega t - \frac{(1+i)(b-r)}{\delta(\omega)} \right\} \right] \qquad \text{(III.1.82)}$$

where we have supposed that current density near the surface of the wire is varying as $\sim \cos(\omega t)$ in time. This formula is valid when $b \gg \delta(\omega)$, and $\delta(\omega)$ is a decay length called the skin depth: see the sketch and discussion in Fig. III.1.16. The skin depth decreases with increasing frequency, but we won't quote its full dependence here. We will only state that, for low frequencies, it reduces in frequency as $1/\sqrt{\omega}$. In the case where Eq. (III.1.82) is valid, the current will be given by the integral

$$I(t) = 2\pi \bar{J}_s \text{Re} \int_0^b dr\, r \left[ \exp \left\{ i\omega t - \frac{(1+i)(b-r)}{\delta(\omega)} \right\} \right] \qquad \text{(III.1.83)}$$

We'll evaluate Eq. (III.1.83), but as always first we need to tidy things up. To do so, we rescale $b(1 - s) = r$ and introduce a dimensionless parameter $\kappa = b/\delta(\omega)$:

$$I(t) = 2\pi \bar{J}_s b^2 \text{Re} \int_0^1 (1 - s) \exp(-(1 + i)\kappa s + i\omega t) ds \qquad \text{(III.1.84)}$$

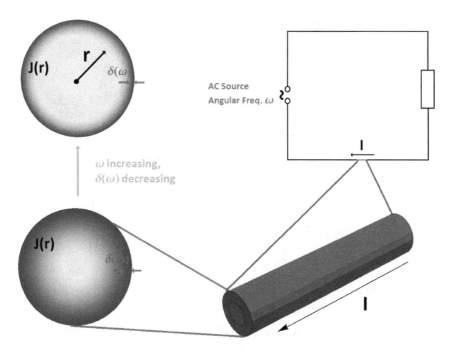

**Fig. III.1.16** A schematic picture of the skin effect for alternating currents in conductors. We consider a wire attached to an alternating current source in a simple circuit, at some instance $t$, when a current $I$ is flowing. We can look at the current density flowing through a cross-section of a uniform wire (assuming it to be perfectly cylindrical), at that time, $J(r)$ (strictly speaking $\tilde{J}(r, \theta, t)$ for all times and variables, but for a uniform wire there is no $\theta$-dependence), as function of the distance $r$ away. The dark blue area represents where there is a large current density, while white is where there is little current passing through. Most of the current is concentrated in a cylindrical layer of thickness $\delta(\omega)$ (called the skin depth) away from the surface of the wire. As the frequency of the A-C current is increased, the skin depth becomes smaller and the current more and more concentrated near the surface of the wire.

Now, since $\kappa \gg 1$, we can approximate the top limit of integration, such that

$$I(t) \approx 2\pi \bar{J}_s b^2 \mathrm{Re} \int_0^\infty (1-s)\exp(-(1+i)\kappa s + i\omega t)ds \qquad (\text{III.1.85})$$

This can be rewritten as

$$I(t) \approx 2\pi \bar{J}_s b^2 \mathrm{Re}\left[\left(1 + \frac{1}{1+i}\frac{d}{d\kappa}\right)\int_0^\infty \exp(-(1+i)\kappa s + i\omega t)ds\right]$$

$$(\text{III.1.86})$$

We can then evaluate the integral:

$$I(t) \approx 2\pi \bar{J}_s b^2 \mathrm{Re}$$

$$\left[ -\left( 1 + \frac{1}{1+i}\frac{d}{d\kappa} \right) \frac{1}{(1+i)\kappa} \left[ \exp\left( -(1+i)\kappa s + i\omega t \right) \right]_0^\infty \right]$$

$$\Rightarrow \quad I(t) \approx 2\pi \bar{J}_s b^2 \mathrm{Re}\left[ \left( 1 + \frac{1}{1+i}\frac{d}{d\kappa} \right) \frac{1}{(1+i)\kappa} \exp(i\omega t) \right]$$

$$= \frac{2\pi \bar{J}_s b^2}{\kappa} \mathrm{Re}\left[ \left( \frac{1}{(1+i)} - \frac{1}{(1+i)^2 \kappa} \right) \exp(i\omega t) \right]$$

$$\Rightarrow \quad I(t) \approx \frac{2\pi \bar{J}_s b^2}{\kappa} \mathrm{Re}\left[ \left( \frac{1-i}{2} - \frac{1}{2i\kappa} \right) (\cos\omega t + i\sin\omega t) \right]$$

$$\Rightarrow \quad I(t) \approx \frac{\pi \bar{J}_s b^2 \delta(\omega)}{b} \left[ \cos\omega t + \left( 1 - \frac{\delta(\omega)}{b} \right) \sin\omega t \right] \qquad \text{(III.1.87)}$$

**Example III.1.11 (Maths Practice ♫♪).** We can also use polar coordinates in area integrals to evaluate single integrals. For instance, we can show, as had been first done, presumably by Euler, that

Leonard Euler
(1707–1783)

$$\sqrt{2\pi} \equiv I = \int_{-\infty}^{\infty} \exp\left( -\frac{x^2}{2} \right) dx \qquad \text{(III.1.88)}$$

This is an important result in physical sciences, because it determines the normalization constant for Gaussian probability distribution. Furthermore, we find it incredibly elegant: how could one know that the area behind the Gaussian function is related to the square root of the ratio of the perimeter of a circle to its radius? Why on earth this should be so? Here is how it comes about.

First we note that $I^2$ can of course be written as

$$I^2 = \left\{ \int_{-\infty}^{\infty} dx \exp\left( -\frac{x^2}{2} \right) \right\} \cdot \left\{ \int_{-\infty}^{\infty} dy \exp\left( -\frac{y^2}{2} \right) \right\} \qquad \text{(III.1.89)}$$

But looking attentively at the structure of Eq. (III.1.89), we can see that we can rewrite its r.h.s as an area integral over all $x, y$ plane, such that

$$I^2 = \int f(\mathbf{R})d\mathbf{R} = \int_{-\infty}^{\infty} dy \int_{-\infty}^{\infty} dx f(x, y)$$

$$\text{where } f(x, y) = \exp\left(-\frac{x^2 + y^2}{2}\right) \tag{III.1.90}$$

By realizing that integration over the whole $x, y$ plane can be represented by integrating over a circle of infinite radius, we can use Eq. (III.1.79) (in the limit $b \to \infty$) to write

$$I^2 = \int f(\mathbf{R})d\mathbf{R} = \int_0^{2\pi} d\theta \int_0^{\infty} r dr \tilde{f}(r, \theta) \tag{III.1.91}$$

where

$$\tilde{f}(r, \theta) = f(r\cos\theta, r\sin\theta) = \exp\left(-\frac{r^2}{2}\right) \tag{III.1.92}$$

This then becomes

$$I^2 = \int_0^{2\pi} d\theta \int_0^{\infty} r dr \exp\left(-\frac{r^2}{2}\right)$$

$$= 2\pi \int_0^{\infty} r dr \exp\left(-\frac{r^2}{2}\right) = 2\pi \int_0^{\infty} \exp(-s)ds = 2\pi \tag{III.1.93}$$

Thus, we indeed find that $I = \sqrt{2\pi}$. Isn't it beautiful, dear reader?

### Changing variables for volume integrals

We can also look at variable changes for volume integrals. In general, given a volume integral of the form

$$I = \int_V f(\mathbf{r})d\mathbf{r} = \int_{x_2}^{x_1} dx \int_{y=y_2(x)}^{y=y_1(x)} dy \int_{z=z_2(x,y,z)}^{z=z_1(x,y,z)} dz f(x, y, z) \tag{III.1.94}$$

we can make a variable change

$$x = \tilde{x}(u, v, w) \quad y = \tilde{y}(u, v, w) \quad z = \tilde{z}(u, v, w) \tag{III.1.95}$$

Then the integral can be written

$$I = \int_{V'} \tilde{f}(\mathbf{r})d\mathbf{r} = \int_{u_2}^{u_1} du \int_{v=v_2(u)}^{v=v_1(u)} dv$$

$$\int_{w=w_2(u,v)}^{w=w_1(u,v)} dw \tilde{f}(u,v,w) \mathscr{J}_{3D}(u,v,w) \qquad \text{(III.1.96)}$$

Here,

$$\tilde{f}(u,v,w) = f(\tilde{x}(u,v,w), \tilde{y}(u,v,w), \tilde{z}(u,v,w)) \qquad \text{(III.1.97)}$$

The volume $V'$ is bounded by the surfaces $w_1(u,v)$ and $w_2(u,v)$, the curves $v_1(u)$ and $v_2(u)$, with endpoints $u_1$ and $u_2$. The bounding surfaces can be found through the functional relations

$$\tilde{z}(u,v,w_1(u,v)) = z_1(\tilde{x}(u,v,w_1(u,v)), \tilde{y}(u,v,w_1(u,v)))$$

$$\tilde{z}(u,v,w_2(u,v)) = z_2(\tilde{x}(u,v,w_2(u,v)), \tilde{y}(u,v,w_2(u,v))) \ \text{(III.1.98)}$$

and the curves $v_1(u)$ and $v_2(u)$, with endpoints $u_1$ and $u_2$, through the interception of the surfaces. As for $\mathscr{J}_{3D}(u,v,w)$, the 3D Jacobian, it reads

$$\mathscr{J}_{3D}(u,v,w) = \begin{Vmatrix} \dfrac{\partial \tilde{x}}{\partial u} & \dfrac{\partial \tilde{x}}{\partial v} & \dfrac{\partial \tilde{x}}{\partial w} \\[2mm] \dfrac{\partial \tilde{y}}{\partial u} & \dfrac{\partial \tilde{y}}{\partial v} & \dfrac{\partial \tilde{y}}{\partial w} \\[2mm] \dfrac{\partial \tilde{z}}{\partial u} & \dfrac{\partial \tilde{z}}{\partial v} & \dfrac{\partial \tilde{z}}{\partial w} \end{Vmatrix} \qquad \text{(III.1.99)}$$

In many cases, it is difficult to find $w_1(u,v)$ and $w_2(u,v)$, and the curves $v_1(u)$ and $v_2(u)$, with endpoints $u_1$ and $u_2$; however, in simple cases it is clear what they should be. Here, we have a mapping between the point $(x,y,z)$ to the point $(u,v,w)$ which should be 1 to 1 within the integration volume. If this is not the case, considerable care must be taken. Then, we would need to split up either the integration volume, or include only a part of the transformed volume $V'$ to provide us with 1 to 1 mappings.

The process of changing variables is illustrated in Fig. III.1.17.

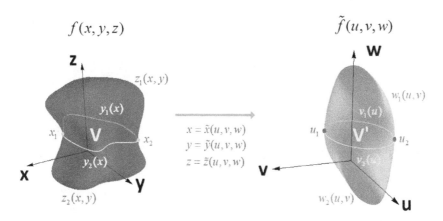

**Fig. III.1.17** An illustration of the changing variables process for volume integrals. In the left panel, we have an integration volume $V$ over which we integrate the function $f(x, y, z)$. This volume is defined by two bounding surfaces $z_1(x, y)$ and $z_2(x, y)$ that intercept along the bounding curves $y_1(x)$ and $y_2(x)$ that join together at the end points $x_1$ and $x_2$ on those curves. A 3D variable transformation $x = \tilde{x}(u, v, w)$, $y = \tilde{y}(u, v, w)$ and $z = \tilde{z}(u, v, w)$ provides a mapping between the space spanned by $x, y$ and $z$ values to a new space spanned by the $u, v$ and $w$ values. In this mapping, the bounding surfaces $z_1(x, y)$ and $z_2(x, y)$ are mapped to the surfaces $w_1(u, v)$ and $w_2(u, v)$, respectively; the bounding curves $y_1(x)$ and $y_2(x)$ are mapped to $v_1(u)$ and $v_2(u)$, which join at the end points $u_1$ and $u_2$ (mapped from $x_1$ and $x_2$, respectively). This results in a new integration region $V'$, in the new space. To preserve the value of the volume integral the integrand then transforms to $\tilde{f}(u, v, w)$, which contains the 3D Jacobian as shown in Eq. (III.1.96).

### Volume integrals in polar coordinates

In Part II of Vol. 1 we introduced cylindrical and spherical polar 3D coordinates. Recall that for cylindrical polar coordinates, we have the following relationships:

$$x = R \cos\phi \quad y = R \sin\phi \quad z = z \qquad \text{(III.1.100)}$$

As the $z$-coordinate remains unchanged in this transformation, to convert cylindrical into polar coordinates we need only to make 2D variable changes in the way we did previously for 2D polar coordinates.

Consider the case when the volume of integration is a cylinder of radius $b$:

$$I = \int_V f(\mathbf{r})d\mathbf{r} = \int_{z_2}^{z_1} dz \int_0^{2\pi} d\phi \int_0^b R dR \tilde{f}(R,\phi,z) \qquad \text{(III.1.101)}$$

The 3D Jacobian in this case

$$\mathcal{J}_{3D}(R,\phi,z) = \begin{Vmatrix} \cos\phi & -R\sin\phi & 0 \\ \sin\phi & R\cos\phi & 0 \\ 0 & 0 & 1 \end{Vmatrix}$$

$$= \begin{Vmatrix} 0 & 0 & 1 \\ \cos\phi & -R\sin\phi & 0 \\ \sin\phi & R\cos\phi & 0 \end{Vmatrix} = \mathcal{J}_{2D}(R,\phi) = R \quad \text{(III.1.102)}$$

is not different from the 2D Jacobian. Obviously, variables that don't change do not need to be included in the Jacobian.

Recall that for spherical polar coordinates we have that

$$z = r\cos\theta \quad y = r\sin\theta\sin\phi \quad x = r\sin\theta\cos\phi \qquad \text{(III.1.103)}$$

In this case we have for the 3D Jacobian:

$$\mathcal{J}_{3D}(r,\theta,\phi) = \begin{Vmatrix} \sin\theta\cos\phi & r\cos\theta\cos\phi & -r\sin\theta\sin\phi \\ \sin\theta\sin\phi & r\cos\theta\sin\phi & r\sin\theta\cos\phi \\ \cos\theta & -r\sin\theta & 0 \end{Vmatrix} \qquad \text{(III.1.104)}$$

**Exercise.** Show that Eq. (III.1.104) evaluates to $\mathcal{J}_{3D}(r,\theta,\phi) = r^2\sin\theta$.

For the case of calculating a volume integral in polar coordinates we may have

$$I = \int_V f(\mathbf{r})d\mathbf{r} = \int_0^{2\pi} d\phi \int_0^\pi \sin\theta d\theta \int_0^{r=r_1(\theta,\phi)} r^2 dr \tilde{f}(r,\theta,\phi)$$

$$\text{(III.1.105)}$$

where we suppose that we have a integration region encompassing the origin and the volume is simply connected (no holes within it). In this case we only need, in spherical polar coordinates, one completely closed bounding surface $r = r_1(\theta,r)$ to enclose it (this could be a composite one made up of different functions in regions, i.e. joining together distinctly different surfaces $z_1(x,y)$ and $z_2(x,y)$; however, the values of $\theta$ and $\phi$ that trace out the bounding curves that joins them need to be known and those integrals split appropriately).

If the integration volume $V$ is a sphere of radius $a$, then $r = a$, and Eq. (III.1.105) immediately simplifies to

$$I = \int_V f(r)dr = \int_0^{2\pi} d\phi \int_0^{\pi} \sin\theta d\theta \int_0^a r^2 dr \tilde{f}(r,\theta,\phi) \quad \text{(III.1.106)}$$

If we take $a \to \infty$ we have an integral over all space in spherical polar coordinates.

### Volume integrals in polar coordinates through common sense arguments

As well as considering the change of variables to both cylindrical and spherical polar coordinates from Cartesian ones, it's worth deducing the correct forms for the volume integrals from some common sense, to put even greater faith in Jacobians... (not to be confused with faith in Jacobins, those once upon a time bloodthirsty French Revolutionaries—reader, all we ask is, please, don't lose your head with them). Let's reconsider integrating a function $f(R,\phi,z)$ over a cylindrical volume $V = \pi b^2 l$. First, we 'divide' our cylinder up into small disks of width $\Delta z$; the volume of each of these disks will be $\pi b^2 \Delta z$. Each of these small disks can be divided into the small area segments, as we did for 2D polar coordinates, such that $\Delta A = R \Delta R \Delta \phi$. The two stages of this division are illustrated in Fig. III.1.18.

We can write for a volume element making up the cylinder:

$$\Delta V = R \Delta \phi \Delta R \Delta z \quad \text{(III.1.107)}$$

If we make these segments small enough we can approximate a volume integral as

$$\int_V f(\mathbf{r})d\mathbf{r} \approx \sum_i f(R_i,\phi_i,z_i)\Delta V_i = \sum_i f(R_i,\phi_i,z_i)R_i \Delta\theta \Delta R \Delta z$$

$$\text{(III.1.108)}$$

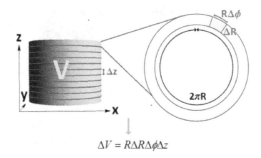

$$\Delta V = R\Delta R\Delta\phi\Delta z$$

**Fig. III.1.18**  We can divide up a cylindrical integration volume into small disks of thickness $\Delta z$ as shown on the left. Then each disk can be split up into rings of thickness $\Delta R$ and circumference $2\pi R$ as shown on the right. We can then split the ring up into segments of angular width $\Delta\phi$. One such segment is shown shaded in yellow (on the right). It is possible to show (using previous arguments) that each segment or element has a volume $\Delta V = R\Delta R\Delta\phi\Delta z$. To approximate the volume integral we can set the function $f(R,\phi,z)$ that we are integrating over to a constant value within each of these elements.

where the values $R_i$, $\phi_i$ and $z_i$ are the values of $R$, $\phi$ and $z$, at the midpoint of the $i$-th segment making up the cylinder. The limit where $\Delta z \to 0$, $\Delta\phi \to 0$, and $\Delta R \to 0$, with the number of segments taken to infinity, defines the integral. Thus we should write

$$\int_V f(\mathbf{r})d\mathbf{r} = \lim_{\substack{\Delta R \to 0 \\ \Delta\phi \to 0 \\ \Delta z \to 0}} \sum_{j=0}^{b/\Delta R} \sum_{k=0}^{2\pi/\Delta\phi} \sum_{l=0}^{l/\Delta z} j\Delta R f(j\Delta R, k\Delta\phi, l\Delta z)\Delta R\Delta\phi\Delta z$$

$$(\text{III.1.109})$$

where we have divided up the coordinates, namely $R = j\Delta R$, $\phi = k\Delta\phi$ and $z = l\Delta z$. From the fundamental definition of definite integration, Eq. (III.1.109) must become

$$\int_V f(\mathbf{r})d\mathbf{r} = \int_0^b RdR \int_0^{2\pi} d\phi \int_0^l dz\, \tilde{f}(R,\phi,z) \qquad (\text{III.1.110})$$

Indeed, this is what the Jacobian method gave us.

Now, we'll reconsider the integral of the function $\tilde{f}(r,\theta,\phi)$ over a sphere of radius $a$, as in Fig. III.1.19.

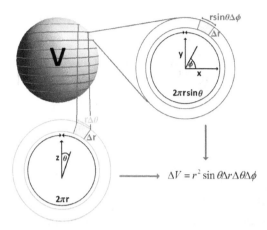

**Fig. III.1.19** For an integration volume that is a sphere we can divide the integration region up into shells of thickness $\Delta r$. Each shell can then be divided into strips as shown by blue lines on the sphere in the top left-hand picture. The bottom left-hand picture shows the vertical cross-section of a shell, a ring of circumference $2\pi r$. In this picture the yellow area represents a particular strip which has angular width $\Delta\theta$, whose cross-section area is roughly $r\Delta r\Delta\theta$. In the top right-hand figure we look along this strip that, when looking down on it, forms a ring of circumference $2\pi r \sin\theta$. This ring can then be divided into segments of angular width $\Delta\phi$. It's not hard to see that the volume contained by such a segment will be $\Delta V \approx r^2 \sin\theta \Delta r \Delta\theta \Delta\phi$.

Firstly, we can divide the sphere up into concentric shells of radius $r$ and width $\Delta r$. The volume of each shell will be

$$\Delta V_{shell} = \frac{4}{3}\pi(r + \Delta r)^3 - \frac{4}{3}\pi r^3 \approx 4\pi r^2 \Delta r \qquad \text{(III.1.111)}$$

Then we can divide each shell into strips of angular width $\Delta\theta$. We note that the volume contained by each strip will depend on its circumference, which will be $2\pi r \sin\theta$, and its radial thickness $\Delta r$, as well as its breadth $r\Delta\theta$, such that the amount of volume contained in each strip will be

$$\Delta V_{stp} \approx 2\pi r^2 \Delta r \sin\theta \, \Delta\theta \qquad \text{(III.1.112)}$$

Dividing each strip up into segments of constant angular width $\Delta\phi$ we obtain for our volume element:

$$\Delta V \approx r^2 \sin\theta \, \Delta r \Delta\theta \, \Delta\phi \qquad \text{(III.1.113)}$$

This division is illustrated in Fig. III.1.19. We then deduce that the volume integral should take the form

$$\int_V f(\mathbf{r})d\mathbf{r} = \lim_{\substack{\Delta R \to 0 \\ \Delta \phi \to 0 \\ \Delta z \to 0}} \sum_{j=0}^{b/\Delta r} \sum_{k=0}^{2\pi/\Delta \theta} \sum_{l=0}^{l/\Delta \phi} j^2 \Delta r^2 \sin(k\Delta \theta)$$

$$\tilde{f}(j\Delta r, k\Delta \theta, l\Delta \phi) \Delta r \Delta \theta \Delta \phi$$

$$= \int_0^a r^2 dr \int_0^\pi \sin\theta d\theta \int_0^{2\pi} d\phi \tilde{f}(r, \theta, \phi) \quad \text{(III.1.114)}$$

Thus, our faith in Jacobians has been tested and found justified, and we should now better understand volume integration in terms of polar coordinates.

**Example III.1.12 (Physical Application ▶▶).** We'll take an example of volume integration in finding the centre of mass of an object and then apply it to mechanics. Consider a conically shaped object, a the segment of a sphere of radius $a$, with the angle $0 \le \alpha \le \pi/2$ between the rays from the centre of the sphere to the apex of the bottom cap (the 'south pole') and the circle forming the base of the bottom cap. Let the bottom of a segment resting on a rough flat desk. This object itself has a mass $M$, the density of which has radial symmetry about the axis of the cone (see Fig. III.1.20).

Let us first prepare a general expression for the centre of mass of such an object. Because of the described symmetry of the distribution of mass of this object, its centre of mass should lie along the axis, at some position $r = r_{cm}$ away from its vertex. Thus, we can calculate it through the integrals

$$r_{cm} = \frac{\int_V r\rho(\mathbf{r})d\mathbf{r}}{\int_V \rho(\mathbf{r})d\mathbf{r}} \quad \text{(III.1.115)}$$

In this example we will consider the object of an *inhomogeneous* density. We'll suppose that it is less dense at the bottom than close the vertex; and it also denser closer to its surface, but will consider the density independent of azimuthal angle. Specifically, we'll

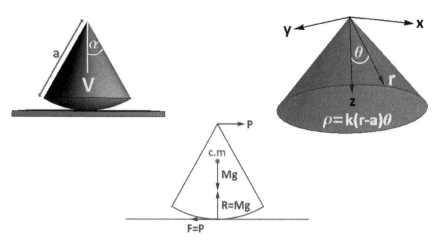

**Fig. III.1.20** On the top left is a picture of a conical object of volume $V$, which is the segment of a sphere of radius $a$. The surface of the cone makes an angle $\alpha$ with the axis of the cone. Note that an angle $\alpha = \pi/2$ would correspond to a hemisphere. The top right diagram shows the system of coordinates we use to calculate the position of the centre of mass. A point anywhere in the cone or its surface can be described in terms of the distance $r$ away from the vertex and $\theta$, the angle its position vector makes with the $z$-axis. These form part of a spherical polar coordinate system $\{r, \theta, \phi\}$. In this coordinate system we suppose the density can be described by the function $\rho(r, \theta, \phi) = k\theta \cdot (a - r)$ $[0 \leq \theta \leq \alpha]$, i.e. it is radially symmetric about the cone axis. When the object is standing upright, the picture on the bottom shows the forces acting. In this position, the object is unstable and tips when a force $P$ is applied to its tip, as this results in a torque about its centre of mass. Note that for illustrative purposes the centre of mass has not been quite drawn in the correct position; it should lie closer to the base.

consider the density to be $\rho(r, \theta, \phi) = k\theta \cdot (a - r)$, $0 \leq \theta \leq \alpha$, in terms of the spherical polar coordinate system that we show in the upper right panel of Fig. III.1.20, for a non-tilted object.

In spherical polar coordinates Eq. (III.1.115) can be written as

$$r_{cm} = \frac{k \int_0^{2\pi} d\phi \int_0^\alpha \sin\theta d\theta \int_0^a r^3 dr(a-r)\theta}{k \int_0^{2\pi} d\phi \int_0^\alpha \sin\theta d\theta \int_0^a r^2 dr(a-r)\theta} = \frac{\int_0^a r^3 dr(a-r)}{\int_0^a r^2 dr(a-r)} \tag{III.1.116}$$

To proceed further we need to consider integrals of the form

$$\Omega_n = \int_0^a r^n(a-r)dr \tag{III.1.117}$$

This evaluates to

$$\Omega_n = \left[\frac{ar^{n+1}}{n+1} - \frac{r^{n+2}}{n+2}\right]_0^a = a^{n+2}\left(\frac{1}{n+1} - \frac{1}{n+2}\right) = \frac{a^{n+2}}{(n+1)(n+2)}$$

$$(III.1.118)$$

Equation (III.1.116) then yields the following result:

$$r_{cm} = \frac{\Omega_3}{\Omega_2} = \frac{\frac{a^5}{4\times5}}{\frac{a^4}{3\times4}} = \frac{3a}{5} \qquad (III.1.119)$$

Now, we need to establish a relationship between the mass of the object and the other parameters of the system. This is given by the volume integral

$$M = k\int_0^{2\pi} d\phi \int_0^\alpha d\theta \int_0^a r^2 dr (a-r)\theta\sin\theta = 2\pi k M_2 \int_0^\alpha \theta\sin\theta d\theta$$

$$= \frac{\pi k a^4}{12}\left([-\theta\cos\theta]_0^\alpha + \int_0^\alpha \cos\theta d\theta\right) = \frac{\pi k a^4}{12}(\sin\alpha - \alpha\cos\alpha)$$

$$(III.1.120)$$

Let us consider first the configuration in its basic position, shown in Fig. III.1.20. The force of gravity through the centre of mass is given by $W = Mg$, so that the magnitude of the reaction force from the desk must be $R = Mg$. If, as in Fig. III.1.20, we apply a pushing force $P$, the friction force $F$ must resist it, such that $P = F$, otherwise the object would slide. This will happen provided that $P \le F_L = \mu Mg$, where $F_L$ is the limiting static friction force, and $\mu$ is the friction coefficient. At the value $P \ge F_L$, the object will slide when we try to pull it.

Let us next show that the object is unstable and will tilt from the initial position into an intermediate position (illustrated in Fig. III.1.21), when we apply a force to its vertex acting, for definiteness, to the right. When the force is strong enough it can topple the object over, and we'll find the minimum force to do this in terms of geometric parameters of the cone.

In any case, when applying the pushing force $P$ we can expect that we will be able to tilt the cone from its initial position. But by

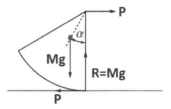

**Fig. III.1.21** Here, we show a position where the cone is about to completely topple over with the various forces acting. Again, note that for illustrative purposes the centre of mass has not been quite drawn in the correct position; it should lie closer to the base.

how much? Evaluating moments about the centre of mass, we can calculate a torque $\tau$ which is a resultant sum of them. In this case one comes from the tilting force and the other from the friction. As $P = F$, we obtain

$$\tau = r_{cm}P + (a - r_{cm})F = aP \tag{III.1.121}$$

For non-zero positive $P$ we see that we always have $\tau > 0$; and the object will tilt. (Of course, this is true, unless we give the object a flat base!)

Now, let's examine the case in which the object is in the intermediate position, and is about to topple over, as shown in Fig. III.1.21; again we have that $P = F$ and $R = W = Mg$. In this case taking moments about the centre of mass we find that the resultant torque is

$$\tau = r_{cm}R\sin\alpha - r_{cm}P\cos\alpha - (a - r_{cm}\cos\alpha)P$$

$$= r_{cm}R\sin\alpha - aP \tag{III.1.122}$$

When the object is just on the verge of toppling, the torque must be zero. Setting $\tau = 0$ in Eq. (III.1.122) we find that this will occur at a critical force $P = P_c = \frac{r_{cm}}{a}R\sin\alpha$. Noting now that $R = Mg$ and substituting Eq. (III.1.120) for $M$, we obtain

$$P_c = \frac{\pi k a^4 g}{20}\left(\sin^2\alpha - \alpha\sin\alpha\cos\alpha\right) \tag{III.1.123}$$

Let us examine $P_c(\alpha)$. The graph of the function $f(\alpha) = \sin^2\alpha - \alpha\sin\alpha\cos\alpha$ is shown in Fig. III.1.22 in the full interval of angles

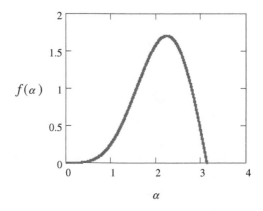

**Fig. III.1.22**   Graph of the function $f(\alpha) = \sin^2 \alpha - \alpha \sin \alpha \cos \alpha$ in the interval $0 \leq \alpha \leq \pi$.

$0 \leq \alpha \leq \pi$ ($\alpha = 0$ corresponds to a stick, $\alpha = \pi/2$ to a hemisphere, and $\alpha = \pi$ to a full sphere).

This function is a monotonic increasing function $\alpha$ in the domain $0 \leq \alpha \leq \pi/2$. When $\alpha = \pi/2$, the function is at a maximum. This case corresponds to a hemisphere standing on its edge where the pushing force is taken through its centre. When $\alpha = 0$ the object is completely unstable to toppling, as $P_c = 0$. This makes sense, as the object is now a stick! Also, one should note that if we push with $P < P_c$, and then let go, the object rocks back and forth.

**Exercise.** Think and explain: what does it physically mean, that after the maximum at $\alpha = \pi/2$, the function $f(\alpha)$ goes down and vanishes at $\alpha = \pi$?

**Exercise.** For the object to slide before it tips completely over, the critical pushing force has to be larger than maximal (limiting) friction, so that $P_c > \mu M g$. Show that for the considered object to slide we require that

$$\mu < \frac{r_{cm}}{a} \tan \alpha = \frac{3}{5} \tan \alpha \qquad (\text{III}.1.124)$$

**Exercise.** Redo Example III.1.12 considering the object with density of $\rho(r, \theta, \phi) = \tilde{k}\theta^2 \cdot (a - r)^2$.

Note that whereas $k$ had dimensionality of $[M/L^4]$ (for notations, see Chapter 10, Part I, Vol. 1), in terms of dimensionality $[\tilde{k}] = [k/L]$. Before the calculation try to guess how the results obtained in the previous example will change for this model of mass distribution in the cone.

**Example III.1.13 (Physical Application ▶▶).** This is an example of volume integration from quantum mechanics. In an atom, electrons can be characterized by quantum states that are distinguished by three numbers $n$, $l$ and $m$. An electron in the $n = 2$, $l = 1$ state forms the so called p-orbital configuration, where there are three p-orbital states corresponding to the values $m = -1, 0, 1$. It is described by the approximate wave function (which is assumed to have functional form as for hydrogen)

$$\psi_{1,1,1}(r,\theta,\phi) = -\left(\frac{Z_{eff}}{2R_B}\right)^{5/2}\left(\frac{1}{2\pi}\right)^{1/2}$$
$$\times r\exp\left(-\frac{Z_{eff}\,r}{2R_B}\right)\exp(i\phi)\sin(\theta) \quad (III.1.125)$$

where $R_B$ is the Bohr radius that we came across in Vol. 1 and $Z_{eff}$ the effective atomic number. The latter takes account of the nuclear charge and the contribution of other electrons in the atom.

The probability distribution function to find an electron at a certain position, in such a state, is given by in spherical polar coordinates by

$$P(\mathbf{r}) = P(r,\theta,\phi) = \psi_{1,1,1}^*(r,\theta,\phi)\psi_{1,1,1}(r,\theta,\phi)$$
$$= \left(\frac{Z_{eff}}{2R_B}\right)^5\frac{r^2}{2\pi}\exp\left(-\frac{Z_{eff}\,r}{R_B}\right)\sin^2\theta \quad (III.1.126)$$

The probability must be normalized to 1:

$$1 = \int P(\mathbf{r})d\mathbf{r} = \int_0^{2\pi}d\phi\int_0^{\pi}\sin\theta d\theta\int_0^{\infty}r^2 dr\,P(r,\theta,\phi) \quad (III.1.127)$$

But the form given by Eq. (III.1.126) already satisfies this. Indeed,

$$\int P(\mathbf{r})d\mathbf{r} = \int_0^{2\pi} d\phi \int_0^{\pi} \sin\theta d\theta$$

$$\int_0^{\infty} r^2 dr \left(\frac{Z_{eff}}{2R_B}\right)^5 \frac{r^2}{2\pi} \exp\left(-\frac{Z_{eff}r}{R_B}\right) \sin^2\theta$$

$$= \left\{\int_0^{\infty} dr \left(\frac{Z_{eff}}{2R_B}\right)^5 r^4 \exp\left(-\frac{Z_{eff}r}{R_B}\right)\right\}$$

$$\cdot \left\{\int_0^{\pi} \sin^3\theta d\theta\right\} \tag{III.1.128}$$

Once we do the $\theta$ integral:

$$I = \int_0^{\pi} \sin^3\theta = \int_0^{\pi} \sin\theta(1 - \cos^2\theta)$$

$$= \int_{-1}^{1} (1 - s^2)ds = 2\left[1 - \frac{s^3}{3}\right]_0^1 = \frac{4}{3} \tag{III.1.129}$$

we get:

$$\int P(\mathbf{r})d\mathbf{r} = \frac{1}{24} \int_0^{\infty} \left(\frac{Z_{eff}}{R_B}\right)^5 r^4 \exp\left(-\frac{Z_{eff}r}{R_B}\right) dr \tag{III.1.130}$$

Next, we can rescale $s = Z_{eff}r/R_B$ so we find that

$$\int P(\mathbf{r})d\mathbf{r} = \frac{1}{24} \int_0^{\infty} s^4 \exp(-s)ds = \frac{4!}{24} = 1 \tag{III.1.131}$$

Let's now find the expectation value $\langle x^2 \rangle$. The integral for this is given by

$$\langle x^2 \rangle = \int x^2 P(\mathbf{r})d\mathbf{r} = \int_0^{2\pi} \cos^2\phi d\phi \int_0^{\pi} \sin^3\theta d\theta \int_0^{\infty} r^4 dr P(r,\theta,\phi) \tag{III.1.132}$$

Note that $x = r\sin\theta\cos\phi$. Then, using Eq. (III.1.126), we have

$$\langle x^2 \rangle = \left\{\int_0^{2\pi} \cos^2\phi d\phi\right\}\left\{\int_0^{\pi} \sin^5\theta d\theta\right\}$$

$$\times \left\{\int_0^{\infty} r^4 dr \left(\frac{Z_{eff}}{2R_B}\right)^5 \frac{r^2}{2\pi} \exp\left(-\frac{Z_{eff}r}{R_B}\right)\right\} \tag{III.1.133}$$

We thus have a product of three independent integrals. For the $\phi$ integral we get

$$K = \int_0^{2\pi} \cos^2\phi\, d\phi = \frac{1}{2}\int_0^{2\pi}(\cos^2\phi + \sin^2\phi)d\phi = \frac{1}{2}\int_0^{2\pi} d\phi = \pi$$

$$\text{(III.1.134)}$$

For the $\theta$ integral we have

$$I = \int_0^\pi \sin^5\theta\, d\theta = -\int_0^\pi (1 - \cos^2\theta)^2 d(\cos\theta) = \int_{-1}^1 (1 - u^2)^2 du$$

$$= 2\int_0^1 (1 - 2u^2 + u^4)du$$

$$\Rightarrow \quad I = 2\left[u - \frac{2u^3}{3} + \frac{u^5}{5}\right]_0^1 = 2\left(1 - \frac{2}{3} + \frac{1}{5}\right) = \frac{16}{15} \qquad \text{(III.1.135)}$$

Thus, we may write

$$\langle x^2 \rangle = \frac{8}{15}\int_0^\infty r^6 dr \left(\frac{Z_{\textit{eff}}}{2R_B}\right)^5 \exp\left(-\frac{Z_{\textit{eff}}\, r}{R_B}\right)$$

$$= \frac{8}{15 \times 32}\left(\frac{R_B}{Z_{\textit{eff}}}\right)^2 \int_0^\infty s^6 \exp(-s)$$

$$= \frac{8 \times 6!}{15 \times 32}\left(\frac{R_B}{Z_{\textit{eff}}}\right)^2 = 12\left(\frac{R_B}{Z_{\textit{eff}}}\right)^2 \qquad \text{(III.1.136)}$$

The expectation value $\sqrt{\langle x^2 \rangle} = \frac{2\sqrt{3}}{Z_{\textit{eff}}}R_B$ effectively measures the extent of the wave function in the $x$-direction. We see that as $Z_{\textit{eff}}$ increases, the value of this extent reduces as the electron is brought closer the nucleus. This makes perfect sense, but neither the power of that dependence, nor the $2\sqrt{3}$ coefficient could be guessed before the calculation (write us if you can!).

**Exercise.** In a similar way evaluate expectation values $\langle y^2 \rangle$, $\langle z^2 \rangle$ and $\langle r^2 \rangle$ using Eq. (III.1.126).

## 1.4.   Key Points

- Area integration of a function of two variables $f(x, y)$ over different shaped regions can be written as a double integral, where each of the integrations can be performed normally, but in order. The limits on the (first) $y$-integration are two bounding curves described by the single-valued functions $y_2(x)$ and $y_1(x)$ that fully enclose the region of integration. For the (second) $x$-integration the limits are the maximum and minimum possible values of $x$ for the region of integration.

- Volume integration of a 3D function (scalar field) $f(x, y, z)$ over different shaped volumes can be written as a triple integral, where each of the integrations can again be performed normally, in order. The limits on the (first) $z$-integration are two bounding surfaces $z_2(x, y)$ and $z_1(x, y)$ that fully enclose the region of integration. For simplicity, we consider the $z_2(x, y)$ and $z_1(x, y)$ to be single-valued functions (for more complicated situations we have to split the integration region up!). These surfaces meet at two bounding curves $y_2(x)$ and $y_1(x)$ (single-valued functions) which forms a closed area, and form the limits of the (second) $y$-integration. Finally, the (last) $x$-integration has limits of integration from the smallest to the largest $x$ value in the integration volume.

- Clearly, an area integral can be written as the sum of area integrals with integration regions that add up, with no overlap, to make the region integration of the original integral. Likewise, we can split up the integration volume of volume integrals.

- Using Green's Theorem, Eq. (III.1.18), it is possible to convert a line integral about a closed path into an area integral over the area enclosed by this path. Alternatively, we may use it to convert area integrals into line integrals. This can be useful in the evaluation of either type of integral.

- However, particular care must be taken using Green's Theorem. If the integrand contains poles or discontinuities, Green's Theorem may not work; we showed one example of it not working.

- In calculating an area integral, we may change variables from $x$ and $y$ to $u$ and $v$. Within the chosen region(s) of integration this should be a 1 to 1 mapping (for one value of $(x, y)$ there is one value of $(u, v)$). This mapping is provided by the functions $x = \tilde{x}(u, v)$ and $y = \tilde{y}(u, v)$. The bounding curves $y_1(x)$ and $y_2(x)$ map to new bounding curves $v_1(u)$ and $v_2(u)$, which can be found, in general, through Eq. (III.1.51). These new curves should be chosen to intercept at the minimum and maximum values of $u$.

- In transforming from $x$ and $y$ to $u$ and $v$, to get the same value for the integral we must multiply $f(\tilde{x}(u,v), \tilde{y}(u,v))$ by the Jacobian $\mathcal{J}(u,v)$ to get the new transformed integrand. The formula for the Jacobian is given by Eq. (III.1.50).

- The new bounding curves $v_1(u)$ and $v_2(u)$ should be single-valued functions; otherwise the transformed integration region should be divided up, to make it single-valued in the divided regions.

- If the mapping of points $(x,y)$ gives two sets of $(u,v)$ values, the new region of integration needs to be restricted to give only one $(u,v)$ set. Equally, if two sets of $(x,y)$ points map into a single value $(u,v)$, the initial integration region needs to be divided up (to be summed later) to ensure 1 to 1 mappings from the original regions to the transformed region of integration.

- In transforming an area integral to polar coordinates, $x = r\cos\theta$ and $y = r\sin\theta$ and the integrand transforms from $f(x,y)$ to $rf(r\cos\theta, r\sin\theta)$. Take note of the factor $r$ that comes from the Jacobian.

- For a volume integral, we may change variables from $x$, $y$, $z$ to $u$, $v$ and $w$ through the functions $x = \tilde{x}(u,v,w)$, $y = \tilde{y}(u,v,w)$ and $z = \tilde{z}(u,v,w)$. The bounding surfaces $z_1(x,y)$ and $z_2(x,y)$ map to $w_1(u,v)$ and $w_2(u,v)$; the bounding curves $y_1(x)$ and $y_2(x)$ map to $v_1(u)$ and $v_2(u)$, respectively. As in surface integrals, within our chosen volumes(s) of integration this should be a 1 to 1 mapping.

- When changing variables from $x$, $y$ and $z$ to $u$, $v$ and $w$, the transformed integrand is $f(\tilde{x}(u,v,w), \tilde{y}(u,v,w), \tilde{z}(u,v,w))$ $\mathcal{J}_{3D}(u,v,w)$. The 3D Jacobian, $\mathcal{J}_{3D}(u,v,w)$ is calculated through Eq. (III.1.99).

- Similar considerations as for area integrals apply to changing the variables of volume integrals.

- In transforming a volume integral to cylindrical polar coordinates, $x = R\cos\theta$, $y = R\sin\theta$, $z = z$ and the integrand transforms from $f(x,y,z)$ to $Rf(R\cos\theta, R\sin\theta, z)$.

- In transforming a surface integral to spherical polar coordinates, $x = r \sin \theta \sin \phi$, $y = r \sin \theta \sin \phi$, $z = r \cos \theta$ and the integrand transforms from $f(x, y, z)$ to $r^2 \sin \theta \, f(r \sin \theta \cos \phi,\ r \sin \theta \sin \phi, r \cos \theta)$.
- We have justified the last two points through common-sense arguments.

# Chapter 2

# Surfaces and Vectors

We dealt with elementary vector equations and vector calculus in Chapter 7 of Part I, Vol. 1. Based on that treatment, we will now advance to more complicated concepts, some of which we will use in later chapters of this volume.

## 2.1. The Gradient Operator and Basic Notions of Curvature

### *Introducing the gradient operator*

Let's return to functions of two variables: $z = f(x, y)$. As we saw in Part II of Vol. 1, the gradient of the surface described by such a function in the $x$ and $y$ directions can be characterized in terms of the partial derivatives of $f(x, y)$ with respect to $x$ and $y$. To encapsulate this information it is useful to define a vector

$$\vec{\nabla} f(x, y) = \frac{\partial f(x, y)}{\partial x} \hat{\mathbf{i}} + \frac{\partial f(x, y)}{\partial y} \hat{\mathbf{j}} \qquad (\text{III}.2.1)$$

and an operator denoted by $\vec{\nabla} = \hat{\mathbf{i}} \frac{\partial}{\partial x} + \hat{\mathbf{j}} \frac{\partial}{\partial y}$ which, when acting on the function $f(x, y)$, gives us this vector. An immediate reason why such vector is useful is because with its help we can define the gradient of the surface along any direction specified by a vector $\hat{\mathbf{s}} = \cos \theta \hat{\mathbf{i}} + \sin \theta \hat{\mathbf{j}}$.

The value of the gradient along $\hat{s}$ is given by

$$\hat{s} \cdot \vec{\nabla} f(x, y) = \cos\theta \frac{\partial f(x, y)}{\partial x} + \sin\theta \frac{\partial f(x, y)}{\partial y} \qquad \text{(III.2.2)}$$

Let's try to understand a bit more about the gradient operator and the gradient vector of $f(x, y)$. Consider a path on our surface, parametrized by variable $s$, so that we have $z(s) = f(x(s), y(s))$. This path can be described by the vector equation $\mathbf{r}(s) = x(s)\hat{\mathbf{i}} + y(s)\hat{\mathbf{j}}$. As considered in Chapter 1, of Part II, Vol. 1, we can write

$$\frac{dz(s)}{ds} = \frac{\partial x(s)}{\partial s} \frac{\partial f(x, y)}{\partial x} + \frac{\partial y(s)}{\partial s} \frac{\partial f(x, y)}{\partial y} \qquad \text{(III.2.3)}$$

We can also rewrite Eq. (III.2.3) as

$$\frac{dz(s)}{ds} = \frac{d\mathbf{r}(s)}{ds} \cdot \vec{\nabla} f(x, y) \qquad \text{(III.2.4)}$$

Now, let us move along a path where $z(s) = z_0$, a constant, the vector equation of which we'll call $\mathbf{r}_c(s) = x_c(s)\hat{\mathbf{i}} + y_c(s)\hat{\mathbf{j}}$. We find that

$$0 = \frac{d\mathbf{r}_c(s)}{ds} \cdot \vec{\nabla} f(x, y) \qquad \text{(III.2.5)}$$

We see that the geometric interpretation of Eq. (III.2.5) is that the vector $\vec{\nabla} f(x, y)$ lies always perpendicular to the tangent vectors described by such trajectories (if $f(x, y)$ is a potential, the curves defined by $\mathbf{r}_c(s)$ are lines of equal potential—the so called *equipotential lines*—and the direction $\vec{\nabla} f(x, y)$ is the direction of the field lines).

**Example III.2.1 (Maths Practice ♫♪).** Consider the function

$$f(x, y) = \exp(-x^2 - 3y^2) \qquad \text{(III.2.6)}$$

Here, the lines along which $f(x, y)$ is constant are curves where $x^2 + 3y^2 = c^2$, which describe ellipses. Each ellipse is given through a vector equation

$$\mathbf{r}_c(t) = c\{\hat{\mathbf{i}}\cos(t) + \hat{\mathbf{j}}\sin(t)/\sqrt{3}\} \qquad \text{(III.2.7)}$$

where $0 \leq t < 2\pi$ and $\mathbf{r}_c(t)$ is the position vector at a point $t$ along the curve. Some of these curves are illustrated in a contour plot of

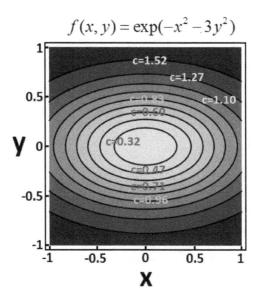

$$f(x, y) = \exp(-x^2 - 3y^2)$$

**Fig. III.2.1** The contour plot of the function $f(x, y) = \exp(-x^2 - 3y^2)$. Each line represents an elliptical curve $x^2 + 3y^2 = c^2$, labelled with its particular value of $c$. The lightest colour corresponds to output values of the function close to 1 ($c$ closest to zero), while the darkest colour corresponds to outputs closest to 0 ($c$ closest to 1). These curves may be generated through the vector equation, Eq. (III.2.7).

this function in Fig. III.2.1. The tangent vector of these curves is given as

$$\tilde{\mathbf{t}}_c(t) = \frac{d\mathbf{r}_c(t)}{dt} = c\{-\hat{\mathbf{i}}\sin t + \hat{\mathbf{j}}\cos(t)/\sqrt{3}\} \qquad \text{(III.2.8)}$$

Using $x = c\cos(t)$ and $y = \frac{c}{\sqrt{3}}\sin(t)$ we can write a vector function $\mathbf{t}(x(t), y(t)) = \tilde{\mathbf{t}}_c(t)$, which describes all the tangents of these curves, in what is called a tangent field:

$$\mathbf{t}(x, y) = -y\sqrt{3}\hat{\mathbf{i}} + \frac{x}{\sqrt{3}}\hat{\mathbf{j}} \qquad \text{(III.2.9)}$$

We plot the vectors described by Eq. (III.2.9) in the right-hand panel of Fig. III.2.2.

Now, let's calculate the gradient:

$$\mathbf{g}(x, y) = \vec{\nabla} f(x, y) = \frac{\partial}{\partial x}\exp\left(-x^2 - 3y^2\right)\hat{\mathbf{i}} + \frac{\partial}{\partial y}\exp\left(-x^2 - 3y^2\right)\mathbf{j}$$

$$= -2x\exp\left(-x^2 - 3y^2\right)\hat{\mathbf{i}} - 6y\exp\left(-x^2 - 3y^2\right)\hat{\mathbf{j}} \qquad \text{(III.2.10)}$$

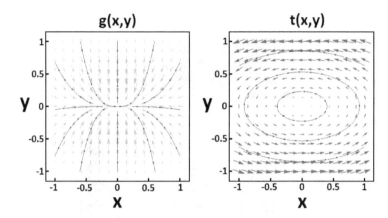

**Fig. III.2.2** We show plots of both $\mathbf{g}(x, y)$, which is the gradient (field) of the surface $f(x, y) = \exp(-x^2 - 3y^2)$ (l.h.s), and $\mathbf{t}(x, y)$, the tangent (field) of the curves $x^2 + 3y^2 = c^2$ (r.h.s ). The arrows in both plots are vectors described by $\mathbf{g}(x, y)$ and $\mathbf{t}(x, y)$ at specific points in a grid. The lines denote 'field lines', for $\mathbf{g}(x, y)$ and $\mathbf{t}(x, y)$. The 'field lines' for $\mathbf{t}(x, y)$ are curves described by $x^2 + 3y^2 = c^2$. From both plots we see visually that $\mathbf{g}(x, y)$ is perpendicular to $\mathbf{t}(x, y)$, which we show in the text.

We show a plot of this gradient field in the left-hand panel in Fig. III.2.2. We see that the vector $\mathbf{g}(x, y)$ lies perpendicular to the tangents of the curves $\mathbf{r}_c(t)$, but let's check this. Along the curve $\mathbf{r}_c(t)$ the gradient is given by

$$\tilde{\mathbf{g}}(t) = \mathbf{g}(x(t), y(t)) = -2c \cos(t) \exp(-c^2)\hat{\mathbf{i}} - 2\sqrt{3}c \sin(t) \exp(-c^2)\hat{\mathbf{j}}$$

$$(\text{III.2.11})$$

Multiplying this expression by the r.h.s of Eq. (III.2.8), and remembering that $\hat{\mathbf{i}} \cdot \hat{\mathbf{i}} = \hat{\mathbf{j}} \cdot \hat{\mathbf{j}} = 1$, $\hat{\mathbf{i}} \cdot \hat{\mathbf{j}} = 0$, one can easily show that indeed $\mathbf{t}_c(t) \cdot \tilde{\mathbf{g}}(t) = 0$.

### The 3D gradient operator

Now, let us move to the functions of three variables. Very similarly to the previous section, we can define the vectorial "gradient operator" as

$$\vec{\nabla} = \hat{\mathbf{i}}\frac{\partial}{\partial x} + \hat{\mathbf{j}}\frac{\partial}{\partial y} + \hat{\mathbf{k}}\frac{\partial}{\partial z}$$

$$(\text{III.2.12})$$

and the gradient of a function as

$$\vec{\nabla}f(x,y,z) = \frac{\partial f(x,y,z)}{\partial x}\hat{\mathbf{i}} + \frac{\partial f(x,y,z)}{\partial y}\hat{\mathbf{j}} + \frac{\partial f(x,y,z)}{\partial y}\hat{\mathbf{k}} \quad \text{(III.2.13)}$$

The gradient of a function may have important physical meaning. For instance, if we have potential energy of an object as a function of position $V(\mathbf{r}) = V(x,y,z)$, then the force on that object is given by $\mathbf{F}(\mathbf{r}) = -\vec{\nabla}V(\mathbf{r})$. This force is an example of what is called a vector field, which we'll discuss later. Similarly we can define a relationship between electrostatic potential $\phi(\mathbf{r})$ and field strength $\mathbf{E}(\mathbf{r})$.

**Example III.2.2 (Physical Application ▶▶).** Calculate the electric field strength from two spherical charges $Q_1$ and $Q_2$ positioned at the points $(x_0, 0, 0)$ and $(-x_0, 0, 0)$ as shown in Fig. III.2.3. To do this we can first work out the electrostatic potential. It is the sum of the potentials created by each charge

$$\phi(\mathbf{r}) = \frac{Q_1}{4\pi\varepsilon_0 r_1} + \frac{Q_2}{4\pi\varepsilon_0 r_2} \quad \text{(III.2.14)}$$

where the distances away from the charges are given by $r_1 = \sqrt{(x-x_0)^2 + y^2 + z^2}$ and $r_2 = \sqrt{(x+x_0)^2 + y^2 + z^2}$ so that

$$4\pi\varepsilon_0\phi(\mathbf{r}) = \frac{Q_1}{\sqrt{(x-x_0)^2 + y^2 + z^2}} + \frac{Q_2}{\sqrt{(x+x_0)^2 + y^2 + z^2}} \quad \text{(III.2.15)}$$

To find the electric field we take the minus-gradient, $\mathbf{E}(\mathbf{r}) = -\vec{\nabla}\varphi(\mathbf{r})$, which to start with gives

$$4\pi\varepsilon_0\mathbf{E}(\mathbf{r}) = -\frac{\partial}{\partial x}\left[\frac{Q_1}{\sqrt{(x-x_0)^2 + y^2 + z^2}} + \frac{Q_2}{\sqrt{(x+x_0)^2 + y^2 + z^2}}\right]\hat{\mathbf{i}}$$

$$-\frac{\partial}{\partial z}\left[\frac{Q_1}{\sqrt{(x-x_0)^2 + y^2 + z^2}} + \frac{Q_2}{\sqrt{(x+x_0)^2 + y^2 + z^2}}\right]\hat{\mathbf{j}}$$

$$-\frac{\partial}{\partial y}\left[\frac{Q_1}{\sqrt{(x-x_0)^2 + y^2 + z^2}} + \frac{Q_2}{\sqrt{(x+x_0)^2 + y^2 + z^2}}\right]\hat{\mathbf{k}}$$

$$\text{(III.2.16)}$$

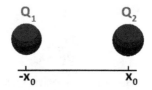

**Fig. III.2.3** A schematic picture of two spherical charges positioned at the points $(x_0, 0, 0)$ and $(-x_0, 0, 0)$ for which we calculate the field strength.

Differentiating,

$$-\frac{\partial}{\partial x}\left[\frac{Q_1}{\sqrt{(x-x_0)^2+y^2+z^2}}+\frac{Q_2}{\sqrt{(x+x_0)^2+y^2+z^2}}\right]$$

$$=\frac{Q_1(x-x_0)}{((x-x_0)^2+y^2+z^2)^{3/2}}+\frac{Q_2(x+x_0)}{((x+x_0)^2+y^2+z^2)^{3/2}} \quad \text{(III.2.17)}$$

$$-\frac{\partial}{\partial y}\left[\frac{Q_1}{\sqrt{(x-x_0)^2+y^2+z^2}}+\frac{Q_2}{\sqrt{(x+x_0)^2+y^2+z^2}}\right]$$

$$=\frac{Q_1 y}{((x-x_0)^2+y^2+z^2)^{3/2}}+\frac{Q_2 y}{((x+x_0)^2+y^2+z^2)^{3/2}} \quad \text{(III.2.18)}$$

$$-\frac{\partial}{\partial z}\left[\frac{Q_1}{\sqrt{(x-x_0)^2+y^2+z^2}}+\frac{Q_2}{\sqrt{(x+x_0)^2+y^2+z^2}}\right]$$

$$=\frac{Q_1 z}{((x-x_0)^2+y^2+z^2)^{3/2}}+\frac{Q_2 z}{((x+x_0)^2+y^2+z^2)^{3/2}} \quad \text{(III.2.19)}$$

And so we obtain for the field:

$$4\pi\varepsilon_0 \mathbf{E}(\mathbf{r}) = \left[\frac{Q_1(x-x_0)}{\sqrt{(x-x_0)^2+y^2+z^2}}+\frac{Q_2(x+x_0)}{\sqrt{(x+x_0)^2+y^2+z^2}}\right]\hat{\mathbf{i}}$$

$$+\left[\frac{yQ_1}{\sqrt{(x-x_0)^2+y^2+z^2}}+\frac{yQ_2}{\sqrt{(x+x_0)^2+y^2+z^2}}\right]\hat{\mathbf{j}}$$

$$+\left[\frac{zQ_1}{\sqrt{(x-x_0)^2+y^2+z^2}}+\frac{zQ_2}{\sqrt{(x+x_0)^2+y^2+z^2}}\right]\hat{\mathbf{k}}$$

$$\text{(III.2.20)}$$

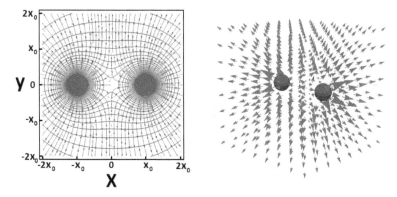

**Fig. III.2.4** On the l.h.s we plot equipotential surfaces (red contours) and the electric field vectors at particular points (green arrows) and the field lines (blue lines), along which the electric field is the tangent vector, for the two spherical charges with $Q_1 = Q_2$, both positively charged. Here the blue circles are the actual charges, inside which we do not map the field or potential. On the r.h.s we show what the electric field looks like in 3-D; at each point in a grid the arrow shows the magnitude and direction of the electric field at that point.

In Fig. III.2.4, we plot equipotential lines (lines of constant $\phi$), as well as the field $\mathbf{E}(\mathbf{r})$ in the $z = 0$ plane, for $Q_1 = Q_2 = Q > 0$, both charges positive; we see exactly what we were taught in physics at school. The maths works! It is also important to note here that the field lines are indeed perpendicular to the equipotential surfaces that they cut, as seen in the left panel of Fig. III.2.4. Also, for further illustration we show the field vectors in 3D in the right panel of Fig. III.2.4.

This was all elementary so far. Let us slowly move to more complicated issues.

### Basic notions of curvature

First, let's consider a simple notion of (explicit) curvature of a surface, further extending on it in the next section. In Chapter 4 of Part II of Vol. 1, we introduced the idea of a tensor product $\otimes$, by using which we can form an operator $\vec{\nabla} \otimes \vec{\nabla}$ which, when acting on a function $f(x, y)$, computes a matrix

Otto Hesse
(1811–1874)

(or tensor) called the *Hessian* (named after another Prussian mathematician Otto Hesse, one of the most renowned students of Jacobi), which is defined as

$$\mathbf{S}(x,y) = \vec{\nabla} \otimes \vec{\nabla} f(x,y) = \begin{pmatrix} \dfrac{\partial}{\partial x} \\[2mm] \dfrac{\partial}{\partial y} \end{pmatrix} \otimes \left( \dfrac{\partial f(x,y)}{\partial x}, \dfrac{\partial f(x,y)}{\partial y} \right)$$

$$= \begin{pmatrix} \dfrac{\partial^2 f(x,y)}{\partial x^2} & \dfrac{\partial f(x,y)}{\partial x \partial y} \\[4mm] \dfrac{\partial^2 f(x,y)}{\partial x \partial y} & \dfrac{\partial^2 f(x,y)}{\partial y^2} \end{pmatrix} \tag{III.2.21}$$

This matrix is useful as it carries information about the local curvature of the surface $z = f(x,y)$.

To understand this better, let's first consider a simple case of $f(x,y) = g(x) + h(y)$, for which the Hessian simplifies to

$$\mathbf{S}(x,y) = \begin{pmatrix} \dfrac{d^2 g(x)}{dx^2} & 0 \\[4mm] 0 & \dfrac{d^2 h(y)}{dy^2} \end{pmatrix} \tag{III.2.22}$$

If both gradients increase or decrease when we move away from the point $(x,y)$ we find that $S = \det(\mathbf{S}) > 0$, whereas if one is increasing while the other decreases we have that $S = \det(\mathbf{S}) < 0$. At a turning point where

$$\frac{dg(x)}{dx} = \frac{dh(y)}{dy} = 0 \tag{III.2.23}$$

we have a rather obvious interpretation; if $S = \det(\mathbf{S}) > 0$ we have a maxima or minima and when $S = \det(\mathbf{S}) < 0$ we have a saddle point (as shown in Fig. III.2.5).

Now, let's return to the general form of $\mathbf{S}(x,y)$. We come across it when we write a Taylor expansion of $f(x,y)$ about the point $x = x_0$

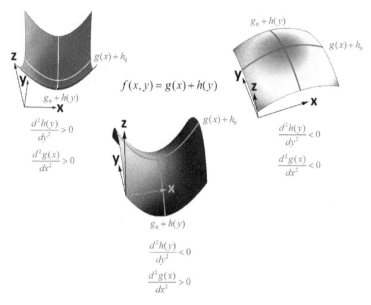

$$f(x,y) = g(x) + h(y)$$

**Fig. III.2.5** Here, we illustrate the three types of turning points we get from the function $f(x,y) = g(x) + h(y)$. The turning point is given by the coordinates $(x_0, y_0)$. Along each surface we consider the curves $z = g(x) + h_0$ and $z = g_0 + h(y)$, where $h_0 = h(y_0)$ and $g_0 = g(x_0)$. Either of these curves lies in the directions of maximum positive or negative curvature. We can classify the turning points by the second derivatives of these curves. If the second derivatives of both $g(x)$ and $h(y)$ with respect to their argument are both positive we have a minimum (shown top left); if both are negative we have a maximum (shown top right); and if one is positive and the other negative we have a saddle point (shown bottom).

and $y = y_0$ (for a more conventional form for writing such an expansion, see Chapter 1 of Part II, Vol. 1):

$$f(x,y) \approx f(x_0, y_0) + \left( \frac{\partial f(x_0, y_0)}{\partial x_0} \quad \frac{\partial f(x_0, y_0)}{\partial y_0} \right) \cdot \begin{pmatrix} \delta x \\ \delta y \end{pmatrix}$$

$$+ \left( \frac{1}{2} \right) (\delta x \ \delta y) \cdot \mathbf{S}(x_0, y_0) \cdot \begin{pmatrix} \delta x \\ \delta y \end{pmatrix} + \cdots \qquad \text{(III.2.24)}$$

where $\delta x = x - x_0$ and $\delta y = y - y_0$. Now, we can choose a set of local coordinates $\delta x' = x' - x_0'$ and $\delta y' = y' - y_0'$, where we can actually diagonalize **S**. To do so, we make a rotation of coordinates about the point $(x_0, y_0)$ so that

$$\begin{pmatrix} \delta x' \\ \delta y' \end{pmatrix} = \begin{pmatrix} \cos\alpha & -\sin\alpha \\ \sin\alpha & \cos\alpha \end{pmatrix} \begin{pmatrix} \delta x \\ \delta y \end{pmatrix} = \mathbf{R}(\alpha) \begin{pmatrix} \delta x \\ \delta y \end{pmatrix} \qquad \text{(III.2.25)}$$

Note that we encountered $\mathbf{R}(\alpha)$ in Chapter 4 of Part II, Vol. 1. Then, we may rewrite the Taylor expansion, Eq. (III.2.24) as

$$f(x, y) = \tilde{f}(x', y') \approx f(x_0, y_0)$$

$$+ \left( \frac{\partial f(x_0, y_0)}{\partial x_0} \quad \frac{\partial f(x_0, y_0)}{\partial y_0} \right) \mathbf{R}^{-1}(\alpha) \begin{pmatrix} \delta x' \\ \delta y' \end{pmatrix}$$

$$+ \left( \frac{1}{2} \right) \left( \delta x' \quad \delta y' \right) \cdot \mathbf{R}(\alpha) \cdot \mathbf{S}(x_0, y_0) \cdot \mathbf{R}^{-1}(\alpha) \begin{pmatrix} \delta x' \\ \delta y' \end{pmatrix} + \cdots$$

$$\text{(III.2.26)}$$

Now, the key is to realize is that we may also perform the Taylor expansion of

$$\tilde{f}(x', y') \approx \tilde{f}(x_0', y_0') + \left( \frac{\partial \tilde{f}'(x_0', y_0')}{\partial x_0'} \quad \frac{\partial \tilde{f}'(x_0', y_0')}{\partial y_0'} \right) \cdot \begin{pmatrix} \delta x' \\ \delta y' \end{pmatrix}$$

$$+ \left( \frac{1}{2} \right) \left( \delta x' \quad \delta y' \right) \cdot \mathbf{S}'(x_0', y_0') \cdot \begin{pmatrix} \delta x' \\ \delta y' \end{pmatrix} + \cdots \quad \text{(III.2.27)}$$

where

$$\mathbf{S}'(x_0', y_0') = \begin{pmatrix} \dfrac{\partial^2 \tilde{f}(x_0', y_0')}{\partial x_0'^2} & \dfrac{\partial^2 \tilde{f}(x_0', y_0')}{\partial x_0' \partial y_0'} \\ \dfrac{\partial^2 \tilde{f}(x_0', y_0')}{\partial x_0' \partial y_0'} & \dfrac{\partial^2 \tilde{f}(x_0', y_0')}{\partial y_0'^2} \end{pmatrix} \qquad \text{(III.2.28)}$$

By equating Eq. (III.2.26) with Eq. (III.2.27), we are able to write for each order in the expansion: $f(x_0, y_0) = \tilde{f}(x_0', y_0')$, as well as

$$\left( \frac{\partial f(x_0, y_0)}{\partial x_0} \quad \frac{\partial f(x_0, y_0)}{\partial y_0} \right) \mathbf{R}^{-1}(\alpha) = \left( \frac{\partial \tilde{f}(x_0', y_0')}{\partial x_0'} \quad \frac{\partial \tilde{f}(x_0', y_0')}{\partial y_0'} \right)$$

$$\text{(III.2.29)}$$

and

$$\mathbf{S}'(x, y) = \mathbf{R}(\alpha) \cdot \mathbf{S}(x_0, y_0) \cdot \mathbf{R}^{-1}(\alpha) \tag{III.2.30}$$

**Exercise.** For our rotation through $\alpha$ we can change variables to

$$x' = x \cos \alpha - y \sin \alpha \quad y' = y \cos \alpha + x \sin \alpha \tag{III.2.31}$$

By considering the appropriate chain rule, relate the (first and second) partial derivatives with respect to $x_0'$ and $y_0'$ with those of $x_0$ and $y_0$. Thus, show that Eqs. (III.2.29) and (III.2.30) do indeed hold.

The key point to understand is we can indeed choose $\alpha$ so that $S'_{x,y} = 0$, such that for $\delta x'$ and $\delta y'$ the Hessian is diagonalized. For such rotated coordinates close to $x_0$ and $y_0$, we then may write

$$\tilde{f}(x', y') = f(x_0, y_0) + \frac{\partial \tilde{f}}{\partial x_0'} \delta x' + \frac{\partial \tilde{f}}{\partial y_0'} \delta y' + \frac{1}{2} \frac{\partial^2 \tilde{f}}{\partial x_0'^2} \delta x'^2 + \frac{1}{2} \frac{\partial \tilde{f}}{\partial y_0'^2} \delta y'^2$$

$$\tag{III.2.32}$$

Thus, we see that $f(x, y) = \tilde{f}(\delta x', \delta y')$ can be written roughly in the form $f(\delta x', \delta y') \approx g(\delta x') + h(\delta y') + c$ (up to this order in the Taylor expansion). In Fig. III.2.6 we illustrate this process when $x_0$ and $y_0$ lie at a saddle point. We see that the $\delta x'$ and $\delta y'$ axes lie along the maximum positive and negative curvatures.

As we saw in Chapter 4 of Part II, Vol. 1, to obtain the entries of the diagonalized matrix $\mathbf{S}'$, and the specific form that $\mathbf{R}(\alpha)$ should take, we need to solve the eigenvalue equation

$$\mathbf{S} \begin{pmatrix} a \\ b \end{pmatrix} = \lambda \begin{pmatrix} a \\ b \end{pmatrix} \tag{III.2.33}$$

The entries of $\mathbf{S}'$ are the eigenvalues, and are obtained through the so-called *secular equation*

$$\begin{vmatrix} \dfrac{\partial^2 f}{\partial x_0^2} - \lambda & \dfrac{\partial^2 f}{\partial x_0 \partial y_0} \\ \dfrac{\partial^2 f}{\partial x_0 \partial y_0} & \dfrac{\partial^2 f}{\partial y_0^2} - \lambda \end{vmatrix} = \left( \frac{\partial^2 f}{\partial x_0^2} - \lambda \right) \left( \frac{\partial^2 f}{\partial y_0^2} - \lambda \right) - \left( \frac{\partial^2 f}{\partial x_0 \partial y_0} \right)^2 = 0$$

$$\tag{III.2.34}$$

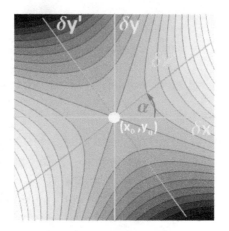

**Fig. III.2.6**   We may consider the Taylor expansion of a function $f(x, y)$ about a saddle point at $(x_0, y_0)$ in terms of the variables $\delta x = x - x_0$ and $\delta y = y - y_0$. As illustrated here, the Hessian matrix $\mathbf{S}(x_0, y_0)$ is not diagonal. We can rotate by an angle into coordinates $\delta x'$ and $\delta y'$ where the Hessian matrix is diagonal. Along these transformed axes both the negative and positive curvature are maximized.

Solving Eq. (III.2.34), we find

$$\lambda = \lambda_1 = \frac{1}{2} \left( \frac{\partial^2 f}{\partial x_0^2} + \frac{\partial^2 f}{\partial y_0^2} \right) + \frac{1}{2} \sqrt{ \left( \frac{\partial^2 f}{\partial x_0^2} - \frac{\partial^2 f}{\partial y_0^2} \right)^2 + 4 \left( \frac{\partial^2 f}{\partial x_0 \partial y_0} \right) }$$

(III.2.35)

$$\lambda = \lambda_2 = \frac{1}{2} \left( \frac{\partial^2 f}{\partial x_0^2} + \frac{\partial^2 f}{\partial y_0^2} \right) - \frac{1}{2} \sqrt{ \left( \frac{\partial^2 f}{\partial x_0^2} - \frac{\partial^2 f}{\partial y_0^2} \right)^2 + 4 \left( \frac{\partial^2 f}{\partial x_0 \partial y_0} \right) }$$

(III.2.36)

The normalized eigenvectors $((a_1, b_1)$ and $(a_2, b_2))$, where $a_1^2 + b_1^2 = a_2^2 + b_2^2 = 1$, are found through the relationships (for each of the two eigenvalues)

$$a_1 \left[ \frac{\partial^2 f}{\partial^2 x_0} - \lambda_1 \right] = -b_1 \frac{\partial^2 f}{\partial x_0 \partial y_0}$$

(III.2.37)

$$b_2 \left[ \frac{\partial^2 f}{\partial^2 y_0} - \lambda_2 \right] = -a_2 \frac{\partial^2 f}{\partial x_0 \partial y_0}$$

(III.2.38)

where we can then write

$$R(\alpha) = \begin{pmatrix} a_1 & a_2 \\ b_1 & b_2 \end{pmatrix} \qquad \text{(III.2.39)}$$

Thus setting $a_1 = b_2 = \cos\alpha$ and $b_1 = -a_2 = \sin\alpha$ we can find $\alpha$.

**Exercise.** Find an explicit expression for $\alpha$ in terms of $\frac{\partial^2 f}{\partial^2 x_0}$, $\frac{\partial^2 f}{\partial x_0 \partial y_0}$ and $\frac{\partial^2 f}{\partial^2 y_0}$, and so $a_1$, $a_2$, $b_1$, and $b_2$.

Indeed, $\mathbf{S}'$ reads as

$$\mathbf{S}' = \begin{pmatrix} \lambda_1 & 0 \\ 0 & \lambda_2 \end{pmatrix} \qquad \text{(III.2.40)}$$

Thus, the interpretation of the sign of the product of the eigenvalues, namely

$$\lambda_1 \lambda_2 = S = \det \mathbf{S} = \frac{\partial^2 f}{\partial x_0'} \frac{\partial^2 f}{\partial y_0'} - \left( \frac{\partial^2 f}{\partial x_0' \partial y_0'} \right)^2 \qquad \text{(III.2.41)}$$

is obvious. If $\lambda_1 \lambda_2 > 0$, the gradients in the $x'$ and $y'$ away from $(x_0', y_0')$ are either both increasing or decreasing, but when $\lambda_1 \lambda_2 < 0$ one gradient is increasing and the other is decreasing in size. At a turning point where

$$\frac{\partial f}{\partial x_0} = \frac{\partial f}{\partial y_0} = \frac{\partial f}{\partial x_0'} = \frac{\partial f}{\partial y_0'} = 0 \qquad \text{(III.2.42)}$$

we can have three options. When $\lambda_1 \lambda_2 < 0$, we must have a saddle point. When $\lambda_1 \lambda_2 > 0$, we have when

$$\frac{\partial^2 f}{\partial x_0^2} > 0 \quad \frac{\partial^2 f}{\partial y_0^2} > 0 \text{ that } \lambda_1 > 0 \text{ and } \lambda_2 > 0 \Rightarrow \text{ a minima} \quad \text{(III.2.43)}$$

and when

$$\frac{\partial^2 f}{\partial x_0^2} < 0 \quad \frac{\partial^2 f}{\partial y_0^2} < 0 \text{ that } \lambda_1 < 0 \text{ and } \lambda_2 < 0 \Rightarrow \text{ a maxima} \quad \text{(III.2.44)}$$

This indeed is the test that we introduced in Chapter 1 of Part II, which we have now justified. When $S = 0$ or one of the second derivatives is zero, this test is inconclusive, and higher order derivatives need to be considered (if any common sense cannot be used), but this is beyond the scope of this book, which is not a textbook on differential geometry.

**Example III.2.3 (Maths Practice ♫♪).** Here, we'll look at a case of a function where one turning point is a saddle point, although naive inspection of its second order partial derivatives with respect to purely $x$ and $y$ would suggest otherwise. For the saddle point, we'll also compute the eigenvectors of the Hessian matrix, to find along which lines the gradient changes fastest (the curvature biggest). We'll examine the function

$$f(x,y) = 2x + 2y - 6xy + x^2 + y^2 + 2y^3 \tag{III.2.45}$$

the contour plots of which are shown in Fig. III.2.7. But we first investigate the properties of this function. Let's find the turning points, which are given by the partial derivatives

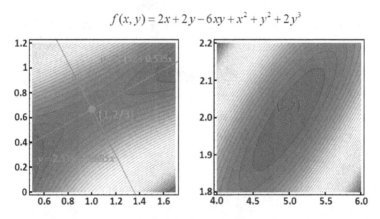

**Fig. III.2.7** We show contour plots of the function $f(x, y)$, shown above. The left panel highlights the region about the saddle point at $(1, 2/3)$, where we show the two lines and their equations that pass through the saddle point along the directions of maximum negative and positive curvature. On the right, we show the region around the other turning point, a minimum at $(5, 2)$. We leave it as an exercise to find the perpendicular lines that run through this point, corresponding to the eigenvalues of the Hessian matrix, one of these is in the direction of maximum curvature.

$$\frac{\partial f(x,y)}{\partial x} = 2 - 6y + 2x = 0 \quad \text{and}$$

$$\frac{\partial f(x,y)}{\partial y} = 2 - 6x + 2y + 6y^2 = 0 \tag{III.2.46}$$

From the first of these equations, we find that $x = 3y - 1$ and substituting that into the second equation yields

$$8 - 16y + 6y^2 = 0 \tag{III.2.47}$$

The two roots to Eq. (III.2.47), yield turning points at both $(1, 2/3)$ and $(5, 2)$. Let's now compute the second order derivatives. These are

$$\frac{\partial^2 f(x,y)}{\partial x^2} = 2 \quad \frac{\partial^2 f(x,y)}{\partial y^2} = 2 + 12y \quad \text{and} \quad \frac{\partial^2 f(x,y)}{\partial y \partial x} = -6$$

$$\tag{III.2.48}$$

Now, naively since the first two second order derivatives in Eq. (III.2.48) are positive at the turning points we might expect that we have both minimums. At the turning points, we either have

$$\frac{\partial^2 f(x,y)}{\partial y^2} = 10 \quad \text{or} \quad \frac{\partial^2 f(x,y)}{\partial y^2} = 26 \tag{III.2.49}$$

However, as the contour plot shows in Fig. III.2.7, the point $(1, 2/3)$ is not a minimum; it's a saddle point! Let's calculate $S$ for both turning points and see whether its values correspond to this.

$$\text{For } x = 1, \ y = 2/3 \text{ we have } S = 2 \times 10 - 36 = -16$$

$$\Rightarrow \text{ saddle point} \tag{III.2.50}$$

$$\text{For } x = 5, \ y = 2 \text{ we have } S = 2 \times 26 - 36 = 16$$

$$\Rightarrow \text{ minimum} \tag{III.2.51}$$

So, indeed our test gives the right result.

Now, let's compute the eigenvectors for the saddle point. For the saddle point the curvature or Hessian matrix is

$$\mathbf{S} = \begin{pmatrix} 2 & -6 \\ -6 & 10 \end{pmatrix} \tag{III.2.52}$$

The eigenvalue equation can be written as

$$\begin{pmatrix} 2 - \lambda & -6 \\ -6 & 10 - \lambda \end{pmatrix} \begin{pmatrix} a \\ b \end{pmatrix} = 0 \tag{III.2.53}$$

and so the secular equation is

$$(2 - \lambda)(10 - \lambda) - 36 = 0 \ \Rightarrow \ \lambda^2 - 12\lambda - 16 = 0 \tag{III.2.54}$$

Therefore,

$$\lambda = \lambda_1 = 6 + 2\sqrt{13} \quad \text{and} \quad \lambda = \lambda_2 = 6 - 2\sqrt{13} \tag{III.2.55}$$

Thus, we find that either

$$b = (\sqrt{13} - 2)a/3 \quad \text{or} \quad b = -(\sqrt{13} + 2)a/3 \tag{III.2.56}$$

Then the eigenvectors are

$$\begin{pmatrix} 1 \\ \dfrac{\sqrt{13} - 2}{3} \end{pmatrix} a \quad \text{and} \quad \begin{pmatrix} 1 \\ -\dfrac{\sqrt{13} + 2}{3} \end{pmatrix} a \tag{III.2.57}$$

The geometrical interpretation of the eigenvectors is indeed that they are the directions in the $(x, y)$ plane along which the curvature is greatest. The equations of the lines in the directions of these vectors take the form of

$$y \simeq c_1 + 0.535x \quad y \simeq c_2 - 1.8685x \tag{III.2.58}$$

(check this yourself). The offsets $c_1$ and $c_2$ are found by realizing that a saddle point must lie on both lines. Therefore,

$$c_1 \simeq 2/3 - 0.535 \approx 0.132 \quad \text{and} \quad c_2 \simeq 2/3 + 1.865 \simeq 2.535 \tag{III.2.59}$$

These lines are also plotted about the saddle point in Fig. III.2.6, and we do indeed see that they point in the directions of maximum negative and positive curvature.

**Exercise.** Find the rotation matrix $\mathbf{R}(\alpha)$ that corresponds to these eigenvectors, Eq. (III.2.57). **Hint:** You need to normalize the eigenvectors so that $a^2 + b^2 = 1$. Also, find the eigenvalue and vectors, lines of maximum curvature, and $\mathbf{R}(\alpha)$ of the minimum at $(5, 2)$.

## *Hessians for functions of many variables*

The analysis can be extended to functions of three variables or more, so that the Hessian matrix of the function $f = f(x_1, x_2 \ldots x_{n-1}, x_n)$ is

$$\mathbf{S} = \vec{\nabla} \otimes \vec{\nabla} f(x_1, x_2 \ldots x_{n-1}, x_n)$$

$$= \begin{pmatrix} \dfrac{\partial^2 f}{\partial x_1^2} & \dfrac{\partial^2 f}{\partial x_1 \partial x_2} & \cdots & \dfrac{\partial^2 f}{\partial x_1 \partial x_{n-1}} & \dfrac{\partial^2 f}{\partial x_1 \partial x_n} \\[2mm] \dfrac{\partial^2 f}{\partial x_1 \partial x_2} & \dfrac{\partial^2 f}{\partial x_2^2} & \cdots & \dfrac{\partial^2 f}{\partial x_2 \partial x_{n-1}} & \dfrac{\partial^2 f}{\partial x_2 \partial x_n} \\[2mm] \vdots & \vdots & \ddots & \vdots & \vdots \\[2mm] \dfrac{\partial^2 f}{\partial x_1 \partial x_{n-1}} & \dfrac{\partial^2 f}{\partial x_2 \partial x_{n-1}} & \cdots & \dfrac{\partial^2 f}{\partial x_{n-1}^2} & \dfrac{\partial^2 f}{\partial x_{n-1} \partial x_n} \\[2mm] \dfrac{\partial^2 f}{\partial x_1 \partial x_n} & \dfrac{\partial^2 f}{\partial x_2 \partial x_n} & \cdots & \dfrac{\partial^2 f}{\partial x_{n-1} \partial x_n} & \dfrac{\partial^2 f}{\partial x_n^2} \end{pmatrix}$$

$$(\text{III.2.60})$$

In this case, to find the actual eigenvalues we need to solve the eigenvalue equation

$$\mathbf{S r} = \lambda \mathbf{r} \qquad (\text{III.2.61})$$

where $\mathbf{r}$ are $n$ component eigenvectors. For a function of $n$ variables we have $n$ eigenvalues that solve the secular equation

$$|\mathbf{S} - \lambda \mathbf{I}| = 0 \qquad (\text{III.2.62})$$

Now, suppose that we have a turning point of the function $f = f(x_1, x_2 \ldots x_n)$—all the partial first derivatives vanish at this point. Then, if all the eigenvalues are positive, we have a minimum value of the function; on the other hand, if all eigenvalues are negative, we have a maximum value of the function. In all other cases we have *different types* of saddle point.

**Example III.2.4 (Physical Application ▶▶).** Let's look at a physical example, putting some of what we've learnt into practice. Suppose that we have a system described by a free energy which is a function of three variables $\alpha$, $\beta$ and $\gamma$, which close to the values $(\alpha = 0, \beta = 0, \gamma = 0))$ can be approximated as

$$F(\alpha, \beta, \gamma) = 2\alpha^2 + \beta^2 + 2\gamma^2 + 2\alpha\beta - \alpha\gamma - 2\beta\gamma \qquad \text{(III.2.63)}$$

By differentiating this expression over $\alpha$, over $\beta$ and over $\gamma$, and equating these partial derivatives to zero, we can see that these equalities will be satisfied at $\alpha = 0, \beta = 0, \gamma = 0$, so that we have a turning point here. We now want to determine what this point corresponds to: is it a minimum or a saddle point? If it is saddle point, the free energy will certainly be unstable and the state that the turning point describes will not be realized (in Thermodynamic equilibrium).

We start by computing the Hessian matrix (Eq. (III.2.60)). In this case it reads as

$$\mathbf{S} = \begin{pmatrix} 2 & 2 & -1 \\ 2 & 1 & -2 \\ -1 & -2 & 2 \end{pmatrix} \qquad \text{(III.2.64)}$$

Then, to find the eigenvalues we consider the secular equation

$$\begin{vmatrix} 2-\lambda & 2 & -1 \\ 2 & 1-\lambda & -2 \\ -1 & -2 & 2-\lambda \end{vmatrix} = 0 \qquad \text{(III.2.65)}$$

Let's evaluate the determinant:

$$(2-\lambda)\begin{vmatrix} 1-\lambda & -2 \\ -2 & 2-\lambda \end{vmatrix} - 2\begin{vmatrix} 2 & -2 \\ -1 & 2-\lambda \end{vmatrix} - \begin{vmatrix} 2 & 1-\lambda \\ -1 & -2 \end{vmatrix}$$

$$= -5 + \lambda + 5\lambda^2 - \lambda^3 = 0 \qquad (\text{III.2.66})$$

As we straightaway see that $\lambda = 1$ is a solution of Eq. (III.2.66), we can factor it, so that

$$(1+\lambda)(\lambda - 1)(\lambda - 5) = 0 \qquad (\text{III.2.67})$$

Thus, the eigenvalues are $\lambda = -1$, $\lambda = 1$ and $\lambda = 5$. As they are of different sign, we have some kind of saddle point; therefore, the free energy about the point $(\alpha = 0, \beta = 0, \gamma = 0)$ is unstable.

## 2.2. Vector Coordinate Geometry in 3D Space: Equations of Surfaces, Their Normal Vectors and Curvature

Here, we examine the vector equations of some surfaces and how a surface can be characterized by a normal vector: a unit vector that points perpendicularly to the surface. And finally, we show how to define the intrinsic curvature of a surface, a *measure of curvature that remains independent of the parametrization or coordinate system* used.

### Vector equations of surfaces

As discussed previously, in Volume 1, any surface in 3D can be defined through a vector

$$\mathbf{r}(s,t) = x(s,t)\hat{\mathbf{i}} + y(s,t)\hat{\mathbf{j}} + z(s,t)\hat{\mathbf{k}} \qquad (\text{III.2.68})$$

where the two input variables $s$ and $t$, which can be quite general, trace out the surface. We should recall that a vector equation of a line is described by one running variable (Chapter 6 of Part I of Vol. 1).

Imagine that we kept $s$ fixed: then indeed we would describe a line traced out by $t$. If we now vary $s$ but consider a set of their values, we would get a family of lines that would sweep out a surface.

**Example III.2.5 (Maths Practice ♫♪).** Let's consider the vector equation of the surface sphere of radius $a$ centred at the origin. The constraint here on $\mathbf{r}$ is that $|\mathbf{r}| = a$; the vector equation describing the surface can be obtained by rotating a vector $\mathbf{r}_0 = a\hat{\mathbf{k}}$ an angle $\theta$ in the $x$-$z$ plane followed by an angle $\phi$ in the $x$–$y$ plane (these transformations are discussed in Chapter 4 of Part II, Vol. 1). The result is that as we vary $\theta$ and $\phi$ as functions of two parameters $s$ and $t$, respectively, we trace out the surface of a sphere—provided that $\theta$ goes from 0 to $\pi$, and $\phi$ goes from 0 to $2\pi$. The vector equation for such a sphere reads

$$\mathbf{r}(s,t) = a\sin\left(\theta(s)\right)\cos\left(\phi(t)\right)\hat{\mathbf{i}} + a\sin\left(\theta(s)\right)\sin\left(\phi(t)\right)\hat{\mathbf{j}}$$
$$+ a\cos\left(\theta(s)\right)\hat{\mathbf{k}}$$
$$= x(s,t)\hat{\mathbf{i}} + y(s,t)\hat{\mathbf{j}} + z(s,t)\hat{\mathbf{k}} \tag{III.2.69}$$

Note that we can also write the equation of such a surface in the form

$$a^2 = x^2 + y^2 + z^2 \tag{III.2.70}$$

We can alternatively describe the sphere in the form of two branches of a function of two variables $z = f_+(x,y)$ and $z = f_-(x,y)$, where $f_\pm(x,y) = \pm\sqrt{a^2 - x^2 - y^2}$.

### The surface tangents and normal

For any vector describing a surface, we can define the following partial derivatives that define tangent vectors:

$$\mathbf{t}_s(s,t) = \frac{\partial \mathbf{r}(s,t)}{\partial s} = \lim_{\Delta s \to 0}\left\{\frac{\mathbf{r}(s+\Delta s,t) - \mathbf{r}(s,t)}{\Delta s}\right\}$$
$$\tag{III.2.71}$$
$$\mathbf{t}_t(s,t) = \frac{\partial \mathbf{r}(s,t)}{\partial t} = \lim_{\Delta t \to 0}\left\{\frac{\mathbf{r}(s,t+\Delta t) - \mathbf{r}(s,t)}{\Delta t}\right\}$$

Both of these vectors lie parallel to the surface at the point $(x(s,t), y(s,t), z(s,t))$. The tangent vectors point in different directions on the surface (not necessarily perpendicular to each other) and characterize the local orientation of the surface. As we shall see, it is the change in these tangent vectors that can characterize the (intrinsic) curvature of a surface. Note that this definition of curvature has wider applications than the one presented previously in this chapter. It allows us to consider surfaces that cannot be written as a single-valued function $z = f(x, y)$; furthermore, this definition does not depend on how the surface is orientated.

This new definition helps us to define the local orientation of a surface through the unit normal vector, which will be very useful when we come to consider integrals over a curved surface. The unit normal vector always points perpendicularly to the surface; locally, at the point $(s, t)$, it lies perpendicular to the plane spanned by the two tangents, $\mathbf{t}_s(s, t)$ and $\mathbf{t}_t(s, t)$. The normal is therefore given by the normalized cross product of these two tangent vectors:

$$\hat{\mathbf{n}} = \frac{\mathbf{t}_s(s,t) \times \mathbf{t}_t(s,t)}{|\mathbf{t}_s(s,t) \times \mathbf{t}_t(s,t)|} \tag{III.2.72}$$

The three vectors and their relationship to the surface are sketched in Fig. III.2.8.

**Example III.2.5 (Continued).** Let's return to the case of the sphere. Here, the tangent vectors are given by

$$\mathbf{t}_s(s,t) = a\frac{d\theta(s)}{ds}\left[\cos\theta(s)\cos\phi(t)\hat{\mathbf{i}} + \cos\theta(s)\sin\phi(t)\hat{\mathbf{j}} - \sin\theta(s)\hat{\mathbf{k}}\right] \tag{III.2.73}$$

and

$$\mathbf{t}_t(s,t) = a\frac{d\phi(t)}{dt}\sin\theta(s)\left[-\sin\phi(t)\hat{\mathbf{i}} + \cos\phi(t)\hat{\mathbf{j}}\right] \tag{III.2.74}$$

We see that for a sphere, we have $\mathbf{t}_s(s,t).\mathbf{t}_t(s,t) = 0$. Also, we can make the tangent vectors unitary:

$$\hat{\mathbf{t}}_s(s,t) = [\cos\theta(s)\cos\phi(t)\hat{\mathbf{i}} + \cos\theta(s)\sin\phi(t)\hat{\mathbf{j}} - \sin\theta(s)\hat{\mathbf{k}}]$$
$$\hat{\mathbf{t}}_t(s,t) = [-\sin\phi(t)\hat{\mathbf{i}} + \cos\phi(t)\hat{\mathbf{j}}] \tag{III.2.75}$$

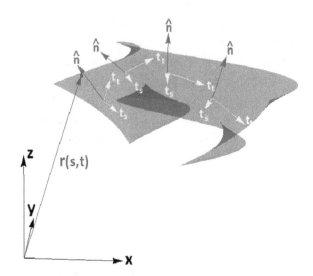

**Fig. III.2.8**  For an arbitrary surface described by the position vector $\mathbf{r}(s,t)$ we show in a few selected positions of the tangent vectors $\mathbf{t}_s(s,t)$ and $\mathbf{t}_t(s,t)$ which lie parallel to the surface at the point $(s,t)$. Mutually perpendicular to the two tangent vectors is the normal vector $\hat{\mathbf{n}}(s,t)$ which points away from the surface.

Using Eq. (III.2.75), we can compute the normal vector of a sphere (note that $|\hat{\mathbf{t}}_t(s,t) \times \hat{\mathbf{t}}_s(s,t)| = 1$ as the vectors are perpendicular to each other):

$$\hat{\mathbf{n}} = \begin{vmatrix} \hat{\mathbf{i}} & \hat{\mathbf{j}} & \hat{\mathbf{k}} \\ \cos\theta(s)\cos\phi(t) & \cos\theta(s)\sin\phi(t) & -\sin\theta(s) \\ -\sin\phi(t) & \cos\phi(t) & 0 \end{vmatrix}$$

$$\Rightarrow \hat{\mathbf{n}} = \sin\theta(s)\cos\phi\hat{\mathbf{i}} + \sin\theta(s)\sin\phi(t)\hat{\mathbf{j}} + \cos\theta(s)\hat{\mathbf{k}} \quad (III.2.76)$$

We show how the three vectors lie on the sphere in Fig. III.2.9. The vectors $\{\hat{\mathbf{t}}_s, \hat{\mathbf{t}}_r, \hat{\mathbf{n}}\}$ form an orthogonal set, and a basis set for polar coordinates; they're conventionally written as $\{\hat{\boldsymbol{\theta}}, \hat{\boldsymbol{\varphi}}, \hat{\mathbf{r}}\}$. We can also define the gradient operator in polar coordinates in terms of this basis set of unit vectors. It reads as

$$\vec{\nabla} = \hat{\mathbf{r}}\frac{\partial}{\partial r} + \frac{\hat{\boldsymbol{\theta}}}{r}\frac{\partial}{\partial\theta} + \frac{\hat{\boldsymbol{\varphi}}}{r\sin\theta}\frac{\partial}{\partial\varphi} \quad (III.2.77)$$

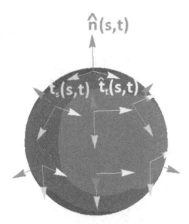

**Fig. III.2.9** Here, we show the two unit tangent vectors $\hat{\mathbf{t}}_s(s, t)$, $\hat{\mathbf{t}}_t(s, t)$ and the normal vector $\hat{\mathbf{n}}(s, t)$ of a sphere of radius $a$; parametrized as in Eqs. III.2.73–III.2.75.

**Example III.2.6 (Maths Practice ♫♪).** Let's consider the vector equation

$$\mathbf{r}(s, t) = (x_0 + a\cos(s^2 + t^2))\hat{\mathbf{i}} + s\hat{\mathbf{j}} + t\hat{\mathbf{k}} \tag{III.2.78}$$

This can also be written as the equation

$$x = x_0 + a\cos(y^2 + z^2) \tag{III.2.79}$$

A plot of Eq. (III.2.79) is shown in Fig. III.2.10.

The tangent vectors to the surface are

$$\mathbf{t}_s(s, t) = \frac{\partial \mathbf{r}(s, t)}{\partial s} = -2as\sin(s^2 + t^2)\hat{\mathbf{i}} + \hat{\mathbf{j}}$$

$$\mathbf{t}_t(s, t) = \frac{\partial \mathbf{r}(s, t)}{\partial t} = -2at\sin(s^2 + t^2)\hat{\mathbf{i}} + \hat{\mathbf{k}} \tag{III.2.80}$$

Let's compute the surface normal:

$$\mathbf{t}_s(s, t) \times \mathbf{t}_t(s, t) = \begin{vmatrix} \hat{\mathbf{i}} & \hat{\mathbf{j}} & \hat{\mathbf{k}} \\ -2as\sin(s^2 + t^2) & 1 & 0 \\ -2at\sin(s^2 + t^2) & 0 & 1 \end{vmatrix}$$

$$= \hat{\mathbf{i}} + 2as\sin(s^2 + t^2)\hat{\mathbf{j}} + 2at\sin(s^2 + t^2)\hat{\mathbf{k}} \tag{III.2.81}$$

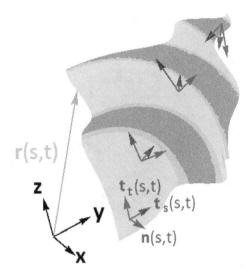

**Fig. III.2.10**   A plot of the surface described by Eq. (III.2.81). The two tangent vectors along the surface, $\mathbf{t}_s(s,t)$ and $\mathbf{t}_t(s,t)$, and the normal vector $\mathbf{n}(s,t)$, are shown here at certain points.

Thus, the unit normal is given by

$$\hat{\mathbf{n}}(s,t) = \frac{1}{\sqrt{1 + 4a^2(s^2 + t^2)\sin^2(s+t)}}$$
$$\times \left(\hat{\mathbf{i}} + 2as\sin(s^2 + t^2)\hat{\mathbf{j}} + 2at\sin(s^2 + t^2)\hat{\mathbf{k}}\right) \quad \text{(III.2.82)}$$

We also show these vectors in Fig. III.2.10.

**Example III.2.7 (Maths Practice ♫♪).** Consider the general equation of a surface of a tube which can generally be bent. To consider such a surface we start with an equation of a line that forms the axis of the tube

$$\mathbf{r}_A(s) = x_A(s)\hat{\mathbf{i}} + y_A(s)\hat{\mathbf{j}} + z_A(s)\hat{\mathbf{k}} \quad \text{(III.2.83)}$$

The way to construct the equation that describes the surface is to consider two vectors normal to each other, and both perpendicular to the tangent of this line. The unit tangent of the line described by Eq. (III.2.83) is given by

$$
\hat{\mathbf{t}}_A(s) = \left| \frac{d\mathbf{r}_A(s)}{ds} \right|^{-1} \frac{d\mathbf{r}_A(s)}{ds} = \left( \left( \frac{dx_A(s)}{ds} \right)^2 + \left( \frac{dy_A(s)}{ds} \right)^2 \right.
$$

$$
\left. + \left( \frac{dz_A(s)}{ds} \right)^2 \right)^{-1/2} \left( \frac{dx_A(s)}{ds} \hat{\mathbf{i}} + \frac{dy_A(s)}{ds} \hat{\mathbf{j}} + \frac{dz_A(s)}{ds} \hat{\mathbf{k}} \right)
$$

$$
\text{(III.2.84)}
$$

For one unit vector perpendicular to $\hat{\mathbf{t}}_A(s)$ we can choose

$$
\hat{\mathbf{u}}(s) = \left( \left( \frac{dx_A(s)}{ds} \right)^2 + \left( \frac{dy_A(s)}{ds} \right)^2 \right)^{-1/2} \left( \frac{dy_A(s)}{ds} \hat{\mathbf{i}} - \frac{dx_A(s)}{ds} \hat{\mathbf{j}} \right)
$$

$$
\text{(III.2.85)}
$$

The second vector will then be obtained from the cross product

$$
\hat{\mathbf{v}}(s) = \hat{\mathbf{u}}(s) \times \hat{\mathbf{t}}_A(s) \tag{III.2.86}
$$

Using Eqs. (III.2.84), (III.2.85) and (III.2.86) we find

$$
\hat{\mathbf{v}}(s) = \left( \left[ \left( \frac{dx_A(s)}{ds} \right)^2 + \left( \frac{dy_A(s)}{ds} \right)^2 \right]^2 + \left( \frac{dz_A(s)}{ds} \right)^2 \right.
$$

$$
\times \left[ \left( \frac{dx_A(s)}{ds} \right)^2 + \left( \frac{dy_A(s)}{ds} \right)^2 \right] \right)^{-1/2}
$$

$$
\times \left[ -\frac{dz_A(s)}{ds} \frac{dx_A(s)}{ds} \hat{\mathbf{i}} - \frac{dz_A(s)}{ds} \frac{dy_A(s)}{ds} \hat{\mathbf{j}} \right.
$$

$$
\left. + \left[ \left( \frac{dx_A(s)}{ds} \right)^2 + \left( \frac{dy_A(s)}{ds} \right)^2 \right] \hat{\mathbf{k}} \right] \tag{III.2.87}
$$

The equation of the tube is then given by

$$
\mathbf{r}(s,t) = \mathbf{r}_A(s) + (a \cos t)\, \hat{\mathbf{u}}(s) + (a \sin t)\, \hat{\mathbf{v}}(s) \tag{III.2.88}
$$

The parameter $t$ rotates out a circle spanned by the vectors $\hat{\mathbf{u}}(s)$ and $\hat{\mathbf{v}}(s)$, covering the full surface of a tube, as illustrated in Fig. III.2.11.

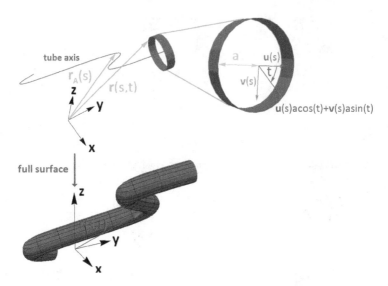

**Fig. III.2.11**  A sketch of how the equation describing the surface of a tube may be constructed. We start by considering the tube axis, as shown by the red curve in the top left. The curve can be traced out through the position of the axis-line vector $\mathbf{r}_A(s)$, where the parameter $s$ describes a position along that curve. To construct the surface we then consider two vectors, $\hat{\mathbf{u}}(s)$ and $\hat{\mathbf{v}}(s)$ perpendicular to the tangent of the tube's axis, $\mathbf{t}_A(s)$, and to each other. As shown in the top right part of the figure, the vector $\hat{\mathbf{u}}(s)a\cos t + \hat{\mathbf{v}}(s)a\sin t$ traces out a circle when $t$ is varied between 0 and $2\pi$, with the tube axis at the centre. When we add this vector to $\mathbf{r}_A(s)$, and vary both $s$ and $t$, we obtain the tube surface, where a position on it is described by $\mathbf{r}(s,t)$ (as in Eq. (III.2.88)).

**Exercise.** Calculate the tangent vectors to the surface using Eq. (III.2.71) and show that the surface normal (using Eq. (III.2.72)) is indeed $\hat{\mathbf{n}}(s,t) = (\cos t)\,\hat{\mathbf{u}}(s) + (\sin t)\,\hat{\mathbf{v}}(s)$.

**Example III.2.7 (Continued).** Let's look at a simple case where

$$\mathbf{r}_A(s) = b\cos(\omega s)\hat{\mathbf{i}} + b\sin(\omega s)\hat{\mathbf{k}} \qquad (\text{III.2.89})$$

with $b > 2a$. The trajectory of the axis is a closed circle, and the surface of the tube will be of a toroid. In this case, Eqs. (III.2.84), (III.2.85) and (III.2.87) reduce to

$$\hat{\mathbf{t}}_A(s) = -\sin(\omega t)\,\hat{\mathbf{i}} + \cos(\omega t)\,\hat{\mathbf{k}} \qquad (\text{III.2.90})$$

and

$$\hat{\mathbf{u}}(s) = -\hat{\mathbf{j}}, \quad \hat{\mathbf{v}}(s) = -\cos(\omega t)\,\hat{\mathbf{i}} - \sin(\omega t)\,\hat{\mathbf{k}} \qquad (\text{III.2.91})$$

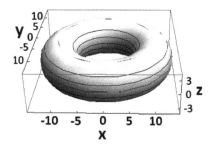

**Fig. III.2.12** Picture of a doughnut, the surface of which is described through Eq. (III.2.92), plotted with the parameter values $b = 10$ and $a = 4$ and $\omega = 1$.

Hence, the equation that describe the surface of this torus is given by

$$\mathbf{r}(s,t) = (b - a\sin t)[\cos(\omega s)\,\hat{\mathbf{i}} + \sin(\omega s)\,\hat{\mathbf{k}}] + a\cos(t)\,\hat{\mathbf{j}} \quad \text{(III.2.92)}$$

We plot this surface in Fig. III.2.12 for $b = 10$, $a = 4$ and $\omega = 1$.

We have constructed the torus diagram as an icing decked doughnut for doughnut devotees: maths can conceptualize such sugary treats! We hope that from now on, you'll consider doughnuts in a different way in your future coffee breaks, dear reader.

In general, tubes when undulating cannot be described in the simple form $z = f(X, y)$ without having to identify numerous branches of the function $f(x, y)$: already for the torus we would have to identity two: one for the top and the other for the bottom of the torus. This example illustrates the power of vectors in describing surfaces. Think about what other interesting surfaces you could map out using the power of vectors.

## The intrinsic curvature of surfaces

It is sometimes possible to define special unit arc-length parametrizations for a surface, in terms of $s_0$ and $t_0$, such that $\tilde{\mathbf{r}}(s_0, t_0) = \tilde{\mathbf{r}}(s_0(s,t), t_0(s,t)) = \mathbf{r}(s,t)$, when the topology of the surface allows for it (certain surfaces may not allow a continuous 1-1 mapping of $s$

and $t$ into $s_0$ and $t_0$ over the whole surface). In such parametrizations, $s_0$ and $t_0$ are defined such that

$$1 = |\hat{\mathbf{t}}_1(s_0, t_0)| = \left|\frac{\partial \tilde{\mathbf{r}}(s_0, t_0)}{\partial s_0}\right| \quad 1 = |\hat{\mathbf{t}}_2(s_0, t_0)| = \left|\frac{\partial \tilde{\mathbf{r}}(s_0, t_0)}{\partial t_0}\right|$$

$$\text{and } \hat{\mathbf{t}}_1(s_0, t_0) \cdot \hat{\mathbf{t}}_2(s_0, t_0) = 0 \tag{III.2.93}$$

This mysterious statement has, actually, a simple meaning: the tangents defined here form the basis vectors of a grid of two sets of curves, both of unit arc-length, which lie perpendicular to each other. Had we flattened the surface, we would have obtained a square grid as illustrated in Fig. III.2.13, where $s_0$ and $t_0$ are the same as $x$ and $y$, respectively. On a surface, flat or curved, the lines in the grid should cross each other perpendicularly, forming what is referred to as an orthonormal coordinate system (another example would be in the flattened surface $s_0$ and $t_0$ forming a polar coordinate system). The normal vector of the surface is given by $\hat{\mathbf{n}} = \hat{\mathbf{t}}_s(s_0, t_0) \times \hat{\mathbf{t}}_t(s_0, t_0)$.

Parametrization of a surface, through $s_0$ and $t_0$, is particularly convenient when discussing the curvature of a surface. This is because the tangent vectors (as well as normal) are all unitary, and so changes in them with respect to $s_0$ and $t_0$ can only be changes in their direction (curvature), and not changes in length. The derivatives of the tangent vectors with respect to $s_0$ and $t_0$ yield three vectors:

$$\boldsymbol{\kappa}_{1,1} = \frac{\partial^2 \tilde{\mathbf{r}}(s_0, t_0)}{\partial s_0^2} = \frac{\partial \hat{\mathbf{t}}_1(s_0, t_0)}{\partial s_0} \quad \boldsymbol{\kappa}_{2,2} = \frac{\partial^2 \tilde{\mathbf{r}}(s_0, t_0)}{\partial t_0^2} = \frac{\partial \hat{\mathbf{t}}_2(s_0, t_0)}{\partial t_0}$$

$$\boldsymbol{\kappa}_{2,1} = \frac{\partial^2 \tilde{\mathbf{r}}(s_0, t_0)}{\partial s_0 \partial t_0} = \frac{\partial \hat{\mathbf{t}}_1(s_0, t_0)}{\partial t_0} = \frac{\partial \hat{\mathbf{t}}_2(s_0, t_0)}{\partial s_0} = \boldsymbol{\kappa}_{1,2} \tag{III.2.94}$$

The intrinsic curvature of the surface at point $(s_0, t_0)$ is determined by the rate of change of the tangent vectors in the direction of the normal of the surface, i.e. out of the surface at that point. To see how this works look at Fig. III.2.14.

The components that determine the curvature of the surface (the so-called normal curvature parameters) are given by

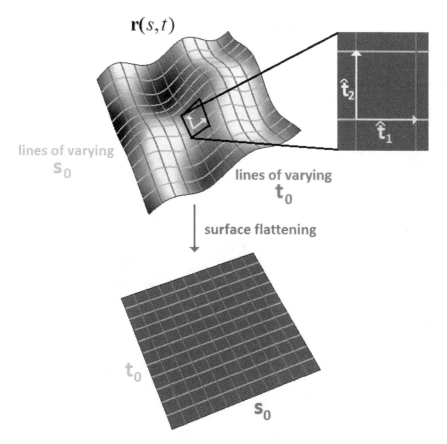

**Fig. III.2.13** Grid-line representation of a surface. In this figure we illustrate the special parameter choice $s_0$ and $t_0$ discussed in the text. In this particular parameter choice a surface can be described by a regular grid of lines of varying $s_0$ (with $t_0$ kept fixed) shown as green curves, and lines of varying $t_0$ (with $s_0$ kept fixed) shown as red curves. Tangential to these curves are the unit tangent vectors $\hat{\mathbf{t}}_1(s_0, t_0)$ and $\hat{\mathbf{t}}_2(s_0, t_0)$, as magnified on the right. When we flatten out the surface we get a regular square grid, such that $s_0$ and $t_0$ become Cartesian. This choice is not exclusive; we could make a choice where $t_0$ and $s_0$ form a polar coordinate system or any other orthonormal one (the lines of constant $s_0$ and $t_0$ being perpendicular to each other).

$$\kappa_{1,1}^{n}(s_0, t_0) = \hat{\mathbf{n}}(s_0, t_0) \cdot \boldsymbol{\kappa}_{1,1}(s_0, t_0)$$

$$\kappa_{2,2}^{n}(s_0, t_0) = \hat{\mathbf{n}}(s_0, t_0) \cdot \boldsymbol{\kappa}_{2,2}(s_0, t_0)$$

$$\kappa_{1,2}^{n}(s_0, t_0) = \hat{\mathbf{n}}(s_0, t_0) \cdot \boldsymbol{\kappa}_{1,2}(s_0, t_0)$$

$$= \hat{\mathbf{n}}(s_0, t_0) \cdot \boldsymbol{\kappa}_{2,1}(s_0, t_0) = \kappa_{2,1}^{n}(s_0, t_0) \quad \text{(III.2.95)}$$

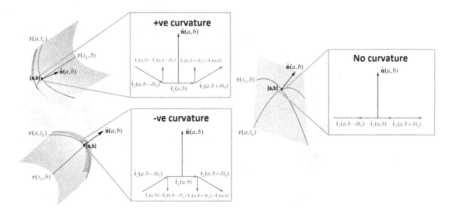

**Fig. III.2.14** Here, we illustrate how the intrinsic curvature of a surface works. We consider the curvature on the surface at a point $(s_0, t_0) = (a, b)$. Key to defining the curvature of the surface is the surface normal at that point, $\hat{n}(a, b)$, about which we consider how the tangent vectors along the curves $r(a, t_0)$ and $r(s_0, b)$ change in the vicinity of $(a, b)$. In the top left-hand picture we show both curves exhibiting positive curvature about $(a, b)$. In the green box we examine in more detail how the tangent vector $\hat{t}_2(a, t_0)$ (tangent to $r(a, t_0)$) changes about the point $(a, b)$. We see that changes in the tangent vector $\hat{t}_2(a, b + \delta t_0) - \hat{t}_2(a, b)$ and $\hat{t}_2(a, b) - \hat{t}_2(a, b - \delta t_0)$ that lie parallel to the normal vector define the curvature; and as they are in the same direction as the normal, this curvature is positive. In the bottom left-hand picture, both curves exhibit negative curvature about $(a.b)$. In the green box beside it we examine the changes $\hat{t}_2(a, b + \delta t_0) - \hat{t}_2(a, b)$ and $\hat{t}_2(a, b) - \hat{t}_2(a, b - \delta t_0)$ about that point parallel to the normal vector. In this case the changes are in the opposite direction to the normal, defining negative curvature. In the last case, the picture on the right, we show the surface flat about $(a, b)$. Here, for a particular choice of parameters, the curves that lie in the surface, $r(a, t_0)$ and $r(s_0, b)$, have tangent vectors that are changing with $s_0$ and $t_0$, but the surface is flat and has no curvature. Again, the key here is to consider how $\hat{t}_2(a, b + \delta t_0)$ and $\hat{t}_2(a, b) - \hat{t}_2(a, b - \delta t_0)$ change in the direction parallel to $\hat{n}(a, b)$. The changes are zero, so we can indeed quantify the curvature, or lack of it, through the rate of changes of the tangent vectors in the direction of $\hat{n}(a, b)$.

They define the components (elements) of an intrinsic curvature tensor (matrix):

$$\mathbf{K} = \begin{pmatrix} \kappa_{1,1}^n(s_0, t_0) & \kappa_{1,2}^n(s_0, t_0) \\ \kappa_{1,2}^n(s_0, t_0) & \kappa_{2,2}^n(s_0, t_0) \end{pmatrix} \tag{III.2.96}$$

We should point out that it is also possible to define the same curvature matrix by considering the rate of change of the surface normal $\hat{n}(s_0, t_0)$ with respect to both $s_0$ and $t_0$. As for the simple (explicit) curvature that we have discussed in the previous section for surfaces defined through $z = f(x, y)$, we can diagonalize the curvature matrix in the same way. Again, this is through a rotation matrix $\mathbf{R}(\alpha)$ rotating both $t_0$ and $s_0$ (see the previous subsection) into new values $t_0'$ and $s_0'$. Here, we choose $\alpha$ to make $\kappa_{1,2}^n(s_0', t_0') = 0$ in the new 'coordinate frame'. In this new frame, the curvature matrix is made up of its two eigenvalues $k_1$ and $k_2$; these are called the principal curvatures of the surface. Namely, we have for the diagonalized form of the curvature matrix

$$\mathbf{K}_D = \mathbf{R}(\alpha) \begin{pmatrix} \kappa_{1,1}^n(s_0, t_0) & \kappa_{1,2}^n(s_0, t_0) \\ \kappa_{1,2}^n(s_0, t_0) & \kappa_{2,2}^n(s_0, t_0) \end{pmatrix} \mathbf{R}^{-1}(\alpha) = \begin{pmatrix} k_1 & 0 \\ 0 & k_2 \end{pmatrix}$$

$$(\text{III.2.97})$$

where the principal curvatures are given by

$$k_1 = \frac{1}{2} \left\{ \begin{array}{l} \kappa_{1,1}^n(s_0, t_0) + \kappa_{2,2}^n(s_0, t_0) \\ + \sqrt{\left(\kappa_{1,1}^n(s_0, t_0) - \kappa_{2,2}^n(s_0, t_0)\right)^2 + 4\kappa_{1,2}^n(s_0, t_0)^2} \end{array} \right\}$$

$$(\text{III.2.98})$$

$$k_2 = \frac{1}{2} \left\{ \begin{array}{l} \kappa_{1,1}^n(s_0, t_0) + \kappa_{2,2}^n(s_0, t_0) \\ - \sqrt{\left(\kappa_{1,1}^n(s_0, t_0) - \kappa_{2,2}^n(s_0, t_0)\right)^2 + 4\kappa_{1,2}^n(s_0, t_0)^2} \end{array} \right\}$$

$$(\text{III.2.99})$$

It is useful to define two new quantities arising from these equations. The product of the eigenvalues, or the determinant of the curvature matrix, defines the *Gaussian Curvature* of a surface and is an intrinsic property of any surface. We can also define the *mean curvature* which is the arithmetic mean of $k_1$ and $k_2$. These two quantities are given, respectively, by

$$K = \det(\mathbf{K}) = k_1 k_2 = \kappa_{1,1}^n(s_0, t_0)\kappa_{2,2}^n(s_0, t_0) - \kappa_{1,2}^n(s_0, t_0)^2$$

$$(\text{III.2.100})$$

$$\bar{k} = \frac{1}{2} Tr\, \mathbf{K} = \frac{1}{2}(k_1 + k_2) = \frac{\kappa_{1,1}^n(s_0, t_0) + \kappa_{2,2}^n(s_0, t_0)}{2} \quad (\text{III.2.101})$$

## Calculating curvature from the first and second fundamental forms of surfaces

Now, the problem is that it is not always possible, or at all easy, to make a special parameter choice, $s_0$ and $t_0$. However, the Gaussian curvature and the curvature matrix can still be calculated for an arbitrary choice of $s$ and $t$, which parametrize a surface. We'll show now how this is done.

Firstly, we define two matrices for our surface, which are referred to as the *first* and *second fundamental forms*, $\mathbf{K}_1$ and $\mathbf{K}_2$, respectively. These are given by

$$\mathbf{K}_1 = \begin{pmatrix} \mathbf{t}_s(s,t) \cdot \mathbf{t}_s(s,t) & \mathbf{t}_s(s,t) \cdot \mathbf{t}_t(s,t) \\ \mathbf{t}_s(s,t) \cdot \mathbf{t}_t(s,t) & \mathbf{t}_t(s,t) \cdot \mathbf{t}_t(s,t) \end{pmatrix}$$

$$\mathbf{K}_2 = \begin{pmatrix} \hat{\mathbf{n}}(s,t) \cdot \mathbf{\kappa}_{s,s}(s,t) & \hat{\mathbf{n}}(s,t) \cdot \mathbf{\kappa}_{t,s}(s,t) \\ \hat{\mathbf{n}}(s,t) \cdot \mathbf{\kappa}_{t,s}(s,t) & \hat{\mathbf{n}}(s,t) \cdot \mathbf{\kappa}_{t,t}(s,t) \end{pmatrix}$$

(III.2.102)

where

$$\mathbf{\kappa}_{s,s}(s,t) = \frac{\partial \mathbf{t}_s(s,t)}{\partial s}, \quad \mathbf{\kappa}_{t,t}(s,t) = \frac{\partial \mathbf{t}_t(s,t)}{\partial t}$$

$$\mathbf{\kappa}_{t,s}(s,t) = \frac{\partial \mathbf{t}_t(s,t)}{\partial s} = \frac{\partial \mathbf{t}_s(s,t)}{\partial t}$$

(III.2.103)

and the tangents are defined by Eq. (III.2.71) [remember that the two tangent vectors point at different directions on the surface but are not necessarily perpendicular to each other] and the normal by Eq. (III.2.72). In general, the curvature matrix (or tensor) and Gaussian and mean curvature can then be calculated through these two matrices by

$$\mathbf{K} = \mathbf{K}_2 \mathbf{K}_1^{-1}, \quad K = \det(\mathbf{K}) = \frac{\det(\mathbf{K}_2)}{\det(\mathbf{K}_1)}, \quad \bar{k} = \frac{1}{2} Tr\, \mathbf{K} \quad \text{(III.2.104)}$$

We won't attempt to prove the expressions in Eq. (III.2.104). Instead, we consider below two examples, demonstrating how this notion and rules can be applied practically.

It is worth noting in conclusion that if we were to choose $s = s_0$ and $t = t_0$, the tangent vectors would be orthogonal and unitary.

In this case we would have $\mathbf{K}_1 = \mathbf{K}_1^{-1} = \mathbf{I}$, where $\mathbf{I}$ is the identity matrix, and $\mathbf{K} = \mathbf{K}_2$.

**Example III.2.8 (Maths Practice ♫♪).** Let's calculate the curvature matrix and Gaussian curvature for a sphere—the next simplest surface after a plane! First, using Eqs. (III.2.73), (III.2.74) and (III.2.102), we get $\mathbf{K}_1$, which reads as

$$
\mathbf{K}_1 = \begin{pmatrix} a^2 \left( \dfrac{d\theta(s)}{ds} \right)^2 & 0 \\ 0 & a^2 \left( \dfrac{d\phi(t)}{dt} \right)^2 \sin^2 \theta(s) \end{pmatrix} \tag{III.2.105}
$$

Next, applying Eq. (III.2.103) we get $\kappa_{s,s}$, $\kappa_{t,t}$ and $\kappa_{s,t}$:

$$
\kappa_{s,s} = a \left( \frac{d^2\theta(s)}{ds^2} \right) [\cos\theta(s)\cos\phi(t)\hat{\mathbf{i}} + \cos\theta(s)\sin\phi(t)\hat{\mathbf{j}} - \sin\theta(s)\hat{\mathbf{k}}]
$$

$$
- a \left( \frac{d\theta(s)}{ds} \right)^2 [\sin\theta(s)\cos\phi(t)\hat{\mathbf{i}} + \sin\theta(s)\sin\phi(t)\hat{\mathbf{j}} + \cos\theta(s)\hat{\mathbf{k}}]
$$

$$
\tag{III.2.106}
$$

$$
\kappa_{t,t} = a \left( \frac{d^2\phi(t)}{dt^2} \right) \sin\theta(s)[-\sin\phi(t)\hat{\mathbf{i}} + \cos\phi(t)\hat{\mathbf{j}}]
$$

$$
- a \left( \frac{d\phi(t)}{dt} \right)^2 \sin\theta(s)[\cos\phi(t)\hat{\mathbf{i}} + \sin\phi(t)\hat{\mathbf{j}}] \tag{III.2.107}
$$

$$
\kappa_{s,t} = a \left( \frac{d\phi(t)}{dt} \right) \left( \frac{d\theta(s)}{ds} \right) \cos\theta(s)[-\sin\phi(t)\hat{\mathbf{i}} + \cos\phi(t)\hat{\mathbf{j}}]
$$

$$
\tag{III.2.108}
$$

**Exercise.** Derive Eq. (III.2.105) from Eqs. (III.2.73), (III.2.74) and (III.2.102). Then get Eqs. (III.2.106)–(III.2.108) from Eq. (III.2.103). Finally, from all of this obtain Eq. (III.2.109) below.

**Example III.2.8 (Continued).** Thus using the form of the unit normal for the sphere (Eq. (III.2.76)), we find that

$$K_2 = \begin{pmatrix} -a\left(\dfrac{d\theta(s)}{ds}\right)^2 & 0 \\ 0 & -a\left(\dfrac{d\phi(t)}{dt}\right)^2 \sin^2\theta(s) \end{pmatrix} = -\dfrac{1}{a}K_1$$

$$\text{(III.2.109)}$$

Hence, subject to Eq. (III.2.104), the curvature matrix and Gaussian curvature are simply

$$\mathbf{K} = -\dfrac{1}{a}\mathbf{I}, \quad K = \dfrac{1}{a^2} \qquad \text{(III.2.110)}$$

Equation (III.2.110) makes sense, as we all understand that curvature about a sphere must be constant, it must have positive sign, and the dimensionality of $K$ tells us that it must go as $1/a^2$, as no other dimensional parameters characterize a sphere. This conclusion cannot and it does not depend on any parameter choice that we would make for $\theta(s)$ or $\phi(t)$; it's an intrinsic property of a sphere of radius $a$.

**Example III.2.9 (Maths Practice ♫♪).** Let's calculate the Gaussian and mean curvature of a surface defined by $z = f(x,y)$. The vector equation of such a surface can be defined as

$$\mathbf{r}(s,t) = s\hat{\mathbf{i}} + t\hat{\mathbf{j}} + f(s,t)\hat{\mathbf{k}} \qquad \text{(III.2.111)}$$

First we calculate the tangents:

$$\mathbf{t}_s(s,t) = \dfrac{\partial \mathbf{r}(s,t)}{\partial s} = \hat{\mathbf{i}} + \dfrac{\partial f(s,t)}{\partial s}\hat{\mathbf{k}}$$
$$\mathbf{t}_t(s,t) = \dfrac{\partial \mathbf{r}(s,t)}{\partial t} = \hat{\mathbf{j}} + \dfrac{\partial f(s,t)}{\partial t}\hat{\mathbf{k}} \qquad \text{(III.2.112)}$$

**Exercise.** Show that the first fundamental form and its determinant are then determined by

$$K_1 = \begin{pmatrix} 1 + \left(\dfrac{\partial f(s,t)}{\partial s}\right)^2 & \dfrac{\partial f(s,t)}{\partial s}\dfrac{\partial f(s,t)}{\partial t} \\ \dfrac{\partial f(s,t)}{\partial s}\dfrac{\partial f(s,t)}{\partial t} & 1 + \left(\dfrac{\partial f(s,t)}{\partial t}\right)^2 \end{pmatrix}$$

$$\text{(III.2.113)}$$

$$\det K_1 = 1 + \left(\dfrac{\partial f(s,t)}{\partial s}\right)^2 + \left(\dfrac{\partial f(s,t)}{\partial t}\right)^2$$

Also show that the inverse of the fundamental form is given by

$$\mathbf{K}_1^{-1} = \frac{1}{1 + \left(\frac{\partial f(s,t)}{\partial s}\right)^2 + \left(\frac{\partial f(s,t)}{\partial t}\right)^2}$$

$$\times \begin{pmatrix} 1 + \left(\frac{\partial f(s,t)}{\partial t}\right)^2 & -\frac{\partial f(s,t)}{\partial s}\frac{\partial f(s,t)}{\partial t} \\ -\frac{\partial f(s,t)}{\partial s}\frac{\partial f(s,t)}{\partial t} & 1 + \left(\frac{\partial f(s,t)}{\partial s}\right)^2 \end{pmatrix} \qquad \text{(III.2.114)}$$

**Example III.2.9 (Continued).** Next, we compute $\kappa_{s,s}$, $\kappa_{s,t}$, and $\kappa_{t,t}$, which are

$$\kappa_{s,s}(s,t) = \frac{\partial \mathbf{t}_s(s,t)}{\partial s} = \frac{\partial^2 f(s,t)}{\partial s^2}\hat{\mathbf{k}}$$

$$\kappa_{s,t}(s,t) = \frac{\partial \mathbf{t}_t(s,t)}{\partial s} = \frac{\partial^2 f(s,t)}{\partial s \partial t}\hat{\mathbf{k}} \qquad \text{(III.2.115)}$$

$$\kappa_{t,t}(s,t) = \frac{\partial \mathbf{t}_t(s,t)}{\partial t} = \frac{\partial^2 f(s,t)}{\partial t^2}\hat{\mathbf{k}}$$

We can then calculate the surface unit normal vector by taking the cross product of the two tangents:

$$\mathbf{t}_s(s,t) \times \mathbf{t}_t(s,t) = \begin{vmatrix} \hat{\mathbf{i}} & \hat{\mathbf{j}} & \hat{\mathbf{k}} \\ 1 & 0 & \frac{\partial f(s,t)}{\partial s} \\ 0 & 1 & \frac{\partial f(s,t)}{\partial t} \end{vmatrix} = \left(-\frac{\partial f(s,t)}{\partial s}\hat{\mathbf{i}} - \frac{\partial f(s,t)}{\partial t}\hat{\mathbf{j}} + \hat{\mathbf{k}}\right)$$

$$\text{(III.2.116)}$$

Correspondingly, we can write down the unit normal as

$$\hat{\mathbf{n}}(s,t) = \frac{1}{\sqrt{1 + \left(\frac{\partial f(s,t)}{\partial s}\right)^2 + \left(\frac{\partial f(s,t)}{\partial t}\right)^2}} \left(-\frac{\partial f(s,t)}{\partial s}\hat{\mathbf{i}} - \frac{\partial f(s,t)}{\partial t}\hat{\mathbf{j}} + \hat{\mathbf{k}}\right)$$

$$\text{(III.2.117)}$$

Then, through Eqs. (III.2.102) and (III.2.115), we obtain for the second fundamental form:

$$\mathbf{K}_2 = \frac{1}{\sqrt{1 + \left(\frac{\partial f(s,t)}{\partial s}\right)^2 + \left(\frac{\partial f(s,t)}{\partial t}\right)^2}} \begin{pmatrix} \dfrac{\partial^2 f(s,t)}{\partial s^2} & \dfrac{\partial^2 f(s,t)}{\partial s \partial t} \\[2mm] \dfrac{\partial^2 f(s,t)}{\partial s \partial t} & \dfrac{\partial^2 f(s,t)}{\partial t^2} \end{pmatrix}$$

$$\text{(III.2.118)}$$

with its determinant

$$\det \mathbf{K}_2 = \frac{1}{\left(1 + \left(\frac{\partial f(s,t)}{\partial s}\right)^2 + \left(\frac{\partial f(s,t)}{\partial t}\right)^2\right)}$$

$$\times \left(\frac{\partial^2 f(s,t)}{\partial s^2}\frac{\partial^2 f(s,t)}{\partial t^2} - \left(\frac{\partial^2 f(s,t)}{\partial s \partial t}\right)^2\right) \quad \text{(III.2.119)}$$

Combining Eqs. (III.2.113) and (III.2.119) with the middle equation of (III.2.104), we obtain the formula for the Gaussian curvature as

$$K = \frac{1}{\left(1 + \left(\frac{\partial f(s,t)}{\partial s}\right)^2 + \left(\frac{\partial f(s,t)}{\partial t}\right)^2\right)^2}$$

$$\times \left(\frac{\partial^2 f(s,t)}{\partial s^2}\frac{\partial^2 f(s,t)}{\partial t^2} - \left(\frac{\partial^2 f(s,t)}{\partial s \partial t}\right)^2\right) \quad \text{(III.2.120)}$$

The mean curvature is obtained through the combination of Eqs. (III.2.114) and (III.2.118), by taking the trace of the product of these two matrices (as in Eq. (III.2.104)):

$$\bar{k} = \frac{1}{2\left(1 + \left(\frac{\partial f(s,t)}{\partial s}\right)^2 + \left(\frac{\partial f(s,t)}{\partial t}\right)^2\right)^{3/2}}$$

$$\times \left[\frac{\partial^2 f(s,t)}{\partial s^2}\left(1 + \frac{\partial f(s,t)}{\partial t}\right)^2 + \frac{\partial^2 f(s,t)}{\partial t^2}\left(1 + \frac{\partial f(s,t)}{\partial s}\right)^2 \right.$$

$$\left. - 2\frac{\partial f(s,t)}{\partial t}\frac{\partial f(s,t)}{\partial s}\frac{\partial^2 f(s,t)}{\partial t \partial s}\right] \quad \text{(III.2.121)}$$

Alternatively, we can write Eqs. (III.2.120) and (III.2.121) in terms of the $x$ and $y$ components of Eq. (III.2.111), by substituting $x = s$ and $y = t$. This simply describes the equation $z = f(x, y)$ that we met in Chapter 1 of Part II, Vol. 1.

In this case, it's worth commenting on the relationship between the determinant of the Hessian matrix, $S = S(x, y)$, and Gaussian curvature, $K = K(x, y)$. We see that at a turning point $(x_0, y_0)$, as the first derivatives of $f(x, y)$ vanish, the definitions of curvature are identical, i.e $S(x_0, y_0) = K(x_0, y_0)$. However, at other points, they are quite different! Importantly, *for some surfaces, how we choose/orientate our coordinate system will change what particular points on the surface will be the turning points*. The definition of Gaussian curvature captures this.

**Think:** Consider a sphere in Cartesian $(x, y, z)$, where we have two branches of the function $z = f(x, y)$, each having a turning point. Where are the turning points? Can we choose our coordinates so as to obtain a different choice of turning points on the surface of the sphere?

We'll use these results for a physical example in the next section, where we'll consider the elastic energy of curved elastic sheet.

## 2.3.   Surface Integrals of Functions

### The surface integral of a scalar function

Suppose we want to integrate a function $f(\mathbf{r}) = f(x, y, z)$ that depends on positions in space (a scalar field) over a surface defined by $\mathbf{r}(s, t) = x(s, t)\hat{\mathbf{i}} + y(s, t)\hat{\mathbf{j}} + z(s, t)\hat{\mathbf{k}}$. Then the integral of $f(\mathbf{r})$ over the surface is given by

$$E = \int_S f(\mathbf{r})d\mathbf{R} = \int_{t_2}^{t_1} \int_{s=s_2(t)}^{s=s_1(t)} f(x(s, t), y(s, t), z(s, t))$$
$$\times |\mathbf{t}_s(s, t) \times \mathbf{t}_t(s, t)|dsdt \qquad \text{(III.2.122)}$$

We can also, if we need to, integrate a function $g(\mathbf{R}) = g(x, y)$ that is only defined upon the surface. In this case, the integral reads as

$$E = \int_S g(\mathbf{R}) d\mathbf{R} = \int_{t_2}^{t_1} \int_{s=s_2(t)}^{s=s_1(t)} g(x(s), y(t)) \, |\mathbf{t}_s(s, t) \times \mathbf{t}_t(s, t)| \, ds dt$$

(III.2.123)

In both Eqs. (III.2.122) and (III.2.129), $|\mathbf{t}_s(s, t) \times \mathbf{t}_t(s, t)| \, ds dt$ is an element of the surface area. Let's see why this makes sense by considering small changes in the vector that traces out the surface

$$\Delta_s \mathbf{r}(s, t) = \mathbf{r}(s + \Delta s, t) - \mathbf{r}(s, t), \quad \Delta_t \mathbf{r}(s, t) = \mathbf{r}(s, t + \Delta t) - \mathbf{r}(s, t)$$

(III.2.124)

Recollect that the area of the parallelogram swept out by these two vectors is

$$\Delta A = |\Delta_s \mathbf{r}(s, t) \times \Delta_t \mathbf{r}(s, t)| = \left| \frac{\Delta_s \mathbf{r}(s, t)}{\Delta s} \times \frac{\Delta_t \mathbf{r}(s, t)}{\Delta t} \right| \Delta s \Delta t$$

(III.2.125)

Over such a small area the function $f(x, y, z)$ can be considered constant. Thus we can write

$$E = \int_S f(\mathbf{r}) d\mathbf{R} \approx \sum_{j,k} f(x(j\Delta s, k\Delta t), y(j\Delta s, k\Delta t), z(j\Delta s, k\Delta t))$$

$$\times \left| \frac{\Delta_s \mathbf{r}(j\Delta s, k\Delta t)}{\Delta s} \times \frac{\Delta_t \mathbf{r}(j\Delta s, k\Delta t)}{\Delta t} \right| \Delta s \Delta t \qquad \text{(III.2.126)}$$

When we take the limit $\Delta s \to 0$, $\Delta t \to 0$ we do indeed recover Eq. (III.2.122). We illustrate how surface integration works in Fig. III.2.15.

### The link between surface and area integrals

Let's try to link things back with area integrals and Jacobians of the previous chapter. An area integral is equivalent to integrating over the flat surface described by $\mathbf{r}(s, t) = x(s, t)\hat{\mathbf{i}} + y(s, t)\hat{\mathbf{j}}$, a plane at $z = 0$. The tangent vectors are given by

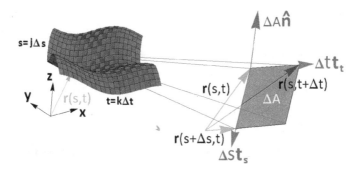

**Fig. III.2.15** To integrate a function $f(x, y, z)$ over a surface described by $\mathbf{r}(s, t)$ we may divide the surface into a grid of small regions, bounded by the lines at which $s = j\Delta s$ and $t = k\Delta t$. These regions may be approximated by parallelograms, the approximation becoming better as we increase the number of regions and decrease their size. On the l.h.s we show a surface divided up in such a way, coloured in a checkboard pattern to highlight these regions, each shown in either red or blue. Three of the vertices of each of region have position vectors $\mathbf{r}(s, t)$, $\mathbf{r}(s + \Delta s, t)$, $\mathbf{r}(s, t + \Delta t)$, as illustrated in an enlarged region on the r.h.s. Then, a region is approximated by a parallelogram of area $\Delta A$, spanned by the vectors $\mathbf{r}(s + \Delta s, t) - \mathbf{r}(s, t) \simeq \mathbf{t}_s(s, t)\Delta s$ and $\mathbf{r}(s, t + \Delta t) - \mathbf{r}(s, t) \simeq \mathbf{t}_t(s, t)\Delta t$. From what we learnt previously about the vector cross product, we are able to find $\Delta A$ through $\Delta A \hat{\mathbf{n}}(s, t) \simeq \mathbf{t}_s(s, t) \times \mathbf{t}_t(s, t)\Delta t \Delta s$. At a particular small region, we may approximate $f(x, y, z)$ by constant value $f(x(j\Delta s, k\Delta t), y(j\Delta s, k\Delta t), z(j\Delta s, k\Delta t))$ taken at the corner of the parallelogram (with position vector $\mathbf{r}(s, t) = \mathbf{r}(j\Delta s, k\Delta t)$). This allows us to approximate a surface integral by summing up all the contributions from such regions as in Eq. (III.2.128). When we take $\Delta s \to 0$ and $\Delta t \to 0$ we may describe the surface integral through the definite integrals given in Eq. (III.2.122).

$$\mathbf{t}_s(s, t) = \frac{\partial x(s, t)}{\partial s}\hat{\mathbf{i}} + \frac{\partial y(s, t)}{\partial s}\hat{\mathbf{j}} \quad , \quad \mathbf{t}_t(s, t) = \frac{\partial x(s, t)}{\partial t}\hat{\mathbf{i}} + \frac{\partial y(s, t)}{\partial t}\hat{\mathbf{j}}$$

$$(\text{III.2.127})$$

Thus, we have

$$|\mathbf{t}_s(s, t) \times \mathbf{t}_t(s, t)| = \left\| \begin{array}{ccc} \hat{\mathbf{i}} & \hat{\mathbf{j}} & \hat{\mathbf{k}} \\ \dfrac{\partial x(s, t)}{\partial s} & \dfrac{\partial y(s, t)}{\partial s} & 0 \\ \dfrac{\partial x(s, t)}{\partial t} & \dfrac{\partial y(s, t)}{\partial t} & 0 \end{array} \right\|$$

$$= \left| \frac{\partial x(s, t)}{\partial s}\frac{\partial y(s, t)}{\partial t} - \frac{\partial x(s, t)}{\partial t}\frac{\partial y(s, t)}{\partial s} \right| = \mathcal{J}(s, t)$$

$$(\text{III.2.128})$$

Again, here the double bars mean that we must take the modulus of the determinant. We get back the 2D Jacobian! Furthermore, if we parametrize our plane Cartesian-wise as $\mathbf{r}(s,t) = x\hat{\mathbf{i}} + y\hat{\mathbf{j}} = s\hat{\mathbf{i}} + t\hat{\mathbf{j}}$, we get $|\mathbf{t}_s(s,t) \times \mathbf{t}_t(s,t)| = 1$. Therefore, if we reparametrize our plane described by $\mathbf{r}(s,t)$ this way, and move to another coordinate system, we will indeed recover Eq. (III.1.49).

### *An additional formula for calculating surface integrals*

It is also possible to show an equivalent formula for calculating surface integrals:

$$E = \int_S f(\mathbf{r})d\mathbf{R} = \int_{t_2}^{t_1} \int_{s=s_2(t)}^{s=s_1(t)} f(x(s,t), y(s,t), z(s,t))$$
$$\times \sqrt{\det \mathbf{K}_1} ds dt \qquad \text{(III.2.129)}$$

**Example III.2.10 (Physical Application ▶▶).** Let's apply what we've just learnt. To do so we'll calculate the bending elasticity of an elastic sheet or a membrane (for simplicity, we consider the case when it does not have a spontaneous curvature). The free energy of such system is indeed given by a surface integral, where the function that we integrate over that surface is the elastic energy per unit area of the surface. This integral reads as

$$E = \int_S \left(2B_1\bar{k}^2 + B_2 K\right) d\mathbf{R} = \int_{t_2}^{t_1} \int_{s=s_2(t)}^{s=s_1(t)} \left(2B_1\bar{k}^2 + B_2 K\right)$$
$$\times \sqrt{\det \mathbf{K}_1} ds dt \qquad \text{(III.2.130)}$$

Suppose that we have a bent sheet that can be described in terms of a function of the form

$$z = f(x,y) = a(\cos(\omega x) + \cos(\omega y)) \qquad \text{(III.2.131)}$$

where $0 < x < 2\pi/\omega$ and $0 < y < 2\pi/\omega$ (see Fig. III.2.16). In this case $x = s$, $y = t$.

**Exercise.** Show from Eqs. (III.2.102) and (III.2.131) that the determinant of the first fundamental form is

$$f(x, y) = a(\cos(\omega x) + \cos(\omega y))$$

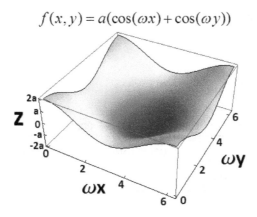

**Fig. III.2.16**   A plot of an elastic sheet described by the function $f(x, y) = a(\cos(\omega x) + \cos(\omega y))$.

$$\det \mathbf{K}_1 = 1 + a^2\omega^2 \left(\sin^2(\omega x) + \sin^2(\omega y)\right) \tag{III.2.132}$$

Thus, show, using Eqs. (III.2.120) and (III.2.121) that the Gaussian and mean curvatures take the forms—

$$K = \frac{a^2\omega^4 \cos(\omega x)\cos(\omega y)}{\left(1 + a^2\omega^2\left(\sin^2(\omega x) + \sin^2(\omega y)\right)\right)^2} \tag{III.2.133}$$

and

$$\bar{k} = -\frac{a\omega^2 \cos(\omega x)(1 - a\omega\sin(\omega y))^2 + a\omega^2\cos(\omega y)(1 - a\omega\sin(\omega x))^2}{2(1 + a^2\omega^2(\sin^2(\omega x) + \sin^2(\omega y)))^{3/2}} \tag{III.2.134}$$

**Example III.2.10 (Continued).**   Thus, our expression for the elastic energy becomes

$$E = \int_0^{2\pi/\omega} dy \int_0^{2\pi/\omega} dx$$

$$\times \left(\frac{B_1(a\omega^2\cos(\omega x)(1 - a\omega\sin(\omega y))^2 + a\omega^2\cos(\omega y)(1 - a\omega\sin(\omega x))^2)^2}{2\left(1 + a^2\omega^2\left(\sin^2(\omega x) + \sin^2(\omega y)\right)\right)^{5/2}}\right.$$

$$\left. + \frac{B_2 a^2\omega^4 \cos(\omega x)\cos(\omega y)}{\left(1 + a^2\omega^2\left(\sin^2(\omega x) + \sin^2(\omega y)\right)\right)^{3/2}}\right) \tag{III.2.135}$$

We can rescale the integrals $\bar{x} = \omega x$ and $\bar{y} = \omega y$, and rearrange so that

$$E = a^2 \omega^2 \int_0^{2\pi} d\bar{y} \int_0^{2\pi} d\bar{x}$$

$$\times \left( B_1 \frac{\cos^2(\bar{x})\left(1 - a\omega \sin(\bar{y})\right)^4 + \cos^2(\bar{y})\left(1 - a\omega \sin(\bar{x})\right)^4}{2\left(1 + a^2\omega^2\left(\sin^2(\bar{x}) + \sin^2(\bar{y})\right)\right)^{5/2}} \right.$$

$$+ B_1 \frac{\cos(\bar{x})\cos(\bar{y})(1 - a\omega \sin(\bar{x}))^2(1 - a\omega \sin(\bar{y}))^2}{\left(1 + a^2\omega^2\left(\sin^2(\bar{x}) + \sin^2(\bar{y})\right)\right)^{5/2}}$$

$$\left. + \frac{B_2 \cos(\bar{x})\cos(\bar{y})}{\left(1 + a^2\omega^2\left(\sin^2(\bar{x}) + \sin^2(\bar{y})\right)\right)^{3/2}} \right) \qquad \text{(III.2.136)}$$

Next, we use the fact that the integral is symmetric under the interchange of $\bar{x}$ and $\bar{y}$, and we can swap them around to write

$$E = a^2 \omega^2 \int_0^{2\pi} d\bar{y} \int_0^{2\pi} d\bar{x} \left( B_1 \frac{\cos^2(\bar{y})(1 - a\omega \sin(\bar{x}))^4}{2(1 + a^2\omega^2(\sin^2(\bar{x}) + \sin^2(\bar{y})))^{5/2}} \right.$$

$$+ B_1 \frac{\cos(\bar{x})\cos(\bar{y})(1 - a\omega \sin(\bar{x}))^2(1 - a\omega \sin(\bar{y}))^2}{(1 + a^2\omega^2(\sin^2(\bar{x}) + \sin^2(\bar{y})))^{5/2}}$$

$$\left. + \frac{B_2 \cos(\bar{x})\cos(\bar{y})}{(1 + a^2\omega^2(\sin^2(\bar{x}) + \sin^2(\bar{y})))^{3/2}} \right) \qquad \text{(III.2.137)}$$

The integral of the last term evaluates to zero, as clarified graphically in Fig. III.2.17, but you of course can show it immediately by changing variables, using that $\cos(x)dx = d(\sin(x))$.

The trick to simplifying the other two terms is to divide the $\bar{x}$ integration up into two regions, $0 \leq \bar{x} \leq \pi$ and $\pi < \bar{x} \leq 2\pi$. For second region we make the variable change $\bar{x} \to \bar{x} + \pi$. Thus, we can rewrite Eq. (III.2.137) as

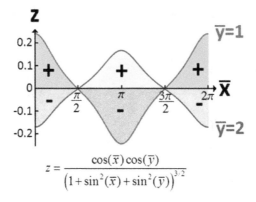

$$z = \frac{\cos(\bar{x})\cos(\bar{y})}{\left(1 + \sin^2(\bar{x}) + \sin^2(\bar{y})\right)^{3/2}}$$

**Fig. III.2.17** Here, we consider the function which determines the last term in the ingrand of the integral described by Eq. (III.2.139). We plot it, for example, for $z = f(\bar{x}, 1)$ and $z = f(\bar{x}2)$, both at, $a\omega = 1$. The integral over $\bar{x}$ of these functions is the total area "under the curve" from 0 to $2\pi$. Looking at this figure, we can see that the positive area contribution "most likely" cancels the negative one, so the integral should be zero. But this is just a visual perception. By using the properties of trigonometric functions one can prove this generally, for any value of $\bar{y}$, the integral vanishes, and the last term of Eq. (III.2.139) is zero.

$$E = a^2\omega^2 \int_0^{2\pi} d\bar{y} \int_0^{\pi} d\bar{x} \left( B_1 \frac{\cos^2(\bar{y}) \left[(1 - a\omega\sin(\bar{x}))^4 + (1 + a\omega\sin(\bar{x}))^4\right]}{\left(1 + a^2\omega^2 \left(\sin^2(\bar{x}) + \sin^2(\bar{y})\right)\right)^{5/2}} \right.$$

$$\left. + B_1 \frac{\cos(\bar{x})\cos(\bar{y})(1 - a\omega\sin(\bar{y}))^2 \left[(1 - a\omega\sin(\bar{x}))^2 + (1 + a\omega\sin(\bar{x}))^2\right]}{\left(1 + a^2\omega^2 \left(\sin^2(\bar{x}) + \sin^2(\bar{y})\right)\right)^{5/2}} \right) \tag{III.2.138}$$

This simplifies to

$$E = 2a^2\omega^2 \int_0^{2\pi} d\bar{y} \int_0^{\pi} d\bar{x}$$

$$\times \left( B_1 \frac{\cos^2(\bar{y}) \left[1 + 6a^2\omega^2 \sin^2(\bar{x}) + a^4\omega^4 \sin^4(\bar{x})\right]}{\left(1 + a^2\omega^2 \left(\sin^2(\bar{x}) + \sin^2(\bar{y})\right)\right)^{5/2}} \right.$$

$$\left. + B_1 \frac{\cos(\bar{x})\cos(\bar{y})(1 - a\omega\sin(\bar{y}))^2 \left[1 + a^2\omega^2 \sin^2(\bar{x})\right]}{\left(1 + a^2\omega^2 \left(\sin^2(\bar{x}) + \sin^2(\bar{y})\right)\right)^{5/2}} \right) \tag{III.2.139}$$

The second term of the integrand vanishes, as can be seen in Fig. III.2.18.

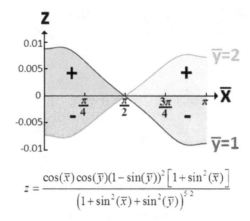

$$z = \frac{\cos(\bar{x})\cos(\bar{y})(1-\sin(\bar{y}))^2\left[1+\sin^2(\bar{x})\right]}{\left(1+\sin^2(\bar{x})+\sin^2(\bar{y})\right)^{5/2}}$$

**Fig. III.2.18**  Here, we consider the behaviour of $f(\bar{x},\bar{y}) = \cos(\bar{x})\cos(\bar{y})(1-\sin(\bar{y}))^2[1+ \sin^2(\bar{x})] \times (1+\sin^2(\bar{x})+\sin^2(\bar{y}))^{-5/2}$ which determines the last term in the integrand of Eq. (III.2.141), with $a\omega = 1$. Here, for the curves and $z = f(\bar{x},1)$ and $z = f(\bar{x},2)$ integrating over $\bar{x}$ from 0 to $\pi$ is zero, as the positive area contribution cancels the negative one. The reader should be able to show this in general, for any value of $\bar{y}$, so that the last term in Eq. (III.2.141) contributes zero.

Thus, we are left with

$$E = 4B_1 a^2 \omega^2 \int_0^\pi d\bar{y} \int_0^\pi d\bar{x}$$

$$\times \frac{\cos^2(\bar{x})\left(1 + 6a^2\omega^2 \sin^2(\bar{y}) + a^4\omega^4 \sin^4(\bar{y})\right)}{\left(1 + a^2\omega^2 \left(\sin^2(\bar{x}) + \sin^2(\bar{y})\right)\right)^{5/2}} \quad \text{(III.2.140)}$$

This formula tells us that the energy is proportional to $B_1$ and is a function of $(a\omega)^2$, the function itself defined through a double integral. We won't evaluate the integral in Eq. (III.2.140), as it is rather a tricky task (try it, if you wish!). However, we'll examine the case where $a\omega$ is small, i.e. the surface undulations are relatively smooth. In this case, we can approximate the integrand:

$$E \approx 4B_1 a^2 \omega^2 \int_0^\pi d\bar{y} \int_0^\pi d\bar{x} \cos^2(\bar{x}) \left(1 + 6a^2\omega^2 \sin^2(\bar{y})\right)$$

$$\times \left(1 - (5/2)a^2\omega^2(\sin^2(\bar{x}) + \sin^2(\bar{y}))\right)$$

$$\approx 4B_1 a^2 \omega^2 \int_0^\pi d\bar{y} \int_0^\pi d\bar{x} \cos^2(\bar{x})(1 + (7/2)a^2\omega^2 \sin^2(\bar{y})$$

$$- (5/2)a^2\omega^2 \sin^2(\bar{x}) \quad \text{(III.2.141)}$$

The integrals in Eq. (III.2.141) are easy to evaluate, namely

$$\int_0^\pi d\bar{y}\sin^2(\bar{y}) = \int_0^\pi d\bar{x}\cos^2(\bar{x}) = \pi/2$$

$$\int_0^\pi d\bar{x}\cos^2(\bar{x})\sin^2(\bar{x}) = \frac{1}{4}\int_0^\pi d\bar{x}\sin^2(2\bar{x}) = \frac{\pi}{8}$$

(III.2.142)

so that we obtain for the elastic energy a simple formula

$$E \approx 2B_1 a^2\omega^2\pi^2\left(1 + \frac{9}{8}a^2\omega^2\right)$$

(III.2.143)

**Example III.2.11 (Physical Application ▶▶).** Let's look at another physical example. Suppose that we have an electric potential due to a uniform spherical charge $\phi(x, y, z) = \alpha/(x^2 + y^2 + z^2)^{1/2}$ (where $\alpha = Q/4\pi\varepsilon_0$ in SI units, $Q$ is the charge), and a cone-like surface of uniform surface charge density $\sigma$, both embedded in the same dielectric. The surface of the cone is described by the equation $\mathbf{r}(s, t) = s\cos(t)\,\hat{\mathbf{i}} + (a + \gamma s)\,\hat{\mathbf{j}} + s\sin(t)\,\hat{\mathbf{k}}$ where $0 \le s \le l$, $0 \le t \le 2\pi$. Here, $l$ is the radius of the cone at the base, and base to tip distance is $\gamma l$; the larger $\gamma$, the sharper the cone (see Fig. III.2.19).

We want to calculate the electrostatic energy, the work required to bring the charge from $-\infty$ to a distance $a$ to the tip of the cone along its axis (the $y$-axis). If the sign of the charge $Q$ is of the same sign as $\sigma$ we get electrostatic repulsion and the energy goes up as one decreases $a$.

We plot the surface in relation to the spherical charge in Fig. III.2.19 and how it cuts the equipotential lines in the $z = 0$ plane. This is given by the integral

$$E = \sigma\int_S \phi(\mathbf{r})d\mathbf{R} = \sigma\int_0^l ds\int_0^{2\pi} dt\phi(x(s, t), y(s, t), z(s, t))$$

$$\times |\mathbf{t}_t(s, t) \times \mathbf{t}_s(s, t)|$$

(III.2.144)

The tangents of the surface are given as

$$\mathbf{t}_s(s, t) = \cos(t)\,\hat{\mathbf{i}} + \gamma\hat{\mathbf{j}} + \sin(t)\,\hat{\mathbf{k}}, \quad \mathbf{t}_t(s, t) = -s\sin(t)\,\hat{\mathbf{i}} + s\cos(t)\,\hat{\mathbf{k}}$$

(III.2.145)

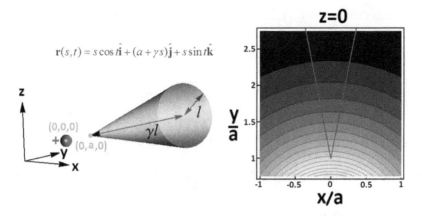

**Fig. III.2.19**  In Example III.2.10 we want to calculate the electrostatic energy due to a uniform charge density on a cone of radius $l$ and base to tip distance $\gamma l$, excluding its base which is not charged. This cone sits in the potential produced by a uniform spherical charge placed at the origin (left panel). The shape of the cone is described by the vector equation $\mathbf{r}(s,t) = s\cos(t)\,\hat{\mathbf{i}} + (a+\gamma s)\hat{\mathbf{j}} + s\sin(t)\,\hat{\mathbf{k}}$ with $s > 0$ and the tip of the cone is at the coordinates $(0, a, 0)$. In the right panel, we show a $(z = 0)$ cross-section of the cone, shown as red lines, cutting through the equipotential lines of the electrostatic potential created by the spherical charge. To calculate the energy we consider the surface integral of the potential multiplied by the surface charge density of the cone, $\sigma$, given by Eq. (III.2.146).

Then, we have

$$\mathbf{t}_t(s,t) \times \mathbf{t}_s(s,t) = \begin{vmatrix} \hat{\mathbf{i}} & \hat{\mathbf{j}} & \hat{\mathbf{k}} \\ -s\sin t & 0 & s\cos t \\ \cos t & \gamma & \sin t \end{vmatrix}$$

$$= \hat{\mathbf{i}}\left(-\gamma s\cos t\right) + s\hat{\mathbf{j}} + \hat{\mathbf{k}}\left(-\gamma s\sin t\right) \qquad \text{(III.2.146)}$$

so that

$$|\mathbf{t}_t(s,t) \times \mathbf{t}_s(s,t)| = s\left(1 + \gamma^2\right)^{1/2} \qquad \text{(III.2.147)}$$

Over the surface, the electrostatic potential is:

$$\phi(x(s,t), y(s,t), z(s,t)) = \frac{\alpha}{((\gamma s + a)^2 + s^2)^{1/2}} \qquad \text{(III.2.148)}$$

Then, the integral for the energy becomes

$$E = \alpha\sigma \left(1 + \gamma^2\right)^{1/2} \int_0^l ds \int_0^{2\pi} dt \frac{s}{((\gamma s + a)^2 + s^2)^{1/2}}$$

$$= 2\pi\alpha\sigma l \left(1 + \gamma^2\right)^{1/2} \int_0^1 du \frac{u}{((\gamma u + a/l)^2 + u^2)^{1/2}} \qquad \text{(III.2.149)}$$

The integral in Eq. (III.2.149) in dimensionless variable $u = s/l$ can be performed analytically, but it leads to a rather cumbersome expression, which we will not bother to show. Let's instead make an approximation, supposing that $\gamma \gg 1$, i.e. the cone is sharp. In this case we may neglect $u^2$ in the denominator of the integrand and get

$$E \approx 2\pi\alpha\sigma l \left(1 + \gamma^2\right)^{1/2} \int_0^1 du \frac{\gamma u}{(\gamma u + a/l)}$$

$$= 2\pi\alpha\sigma l \left(1 + \gamma^2\right)^{1/2} \left(1 - \frac{a}{l\gamma} \ln\left(\frac{\gamma + a/l}{a/l}\right)\right) \qquad \text{(III.2.150)}$$

Strictly speaking, for $\gamma \gg 1$ we should further simplify the factor $(1+\gamma^2)^{1/2} \approx \gamma$, but this is not important. In Fig. III.2.20 we compare this approximation (for $a/l = 1$) with the exact result calculated from

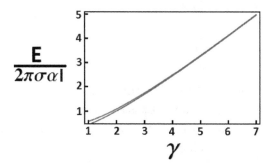

**Fig. III.2.20** We plot as functions of $\gamma$ in units of $2\pi\sigma/\alpha$ the exact electrostatic energy $E_{ext}(\gamma)$ given by Eq. (III.2.148), red curve, and the approximate result, $E_{app}(\gamma)$ for it given by Eq. (III.2.149), blue curve. In this plot we used parameter value $a/l = 1$. We see the approximation perfectly reproduces the result for $\gamma \geq 4$ and works well already for $\gamma = 3$.

Eq. (III.2.149) as function of $\gamma$. We see that agreement between the exact integral and the approximation works very well when $\gamma > 1$, and it should improve with increasing $a/l$.

To find the limiting case where $a/l \gg 1$ we expand out the logarithm in powers of $l/a$, yielding

$$E \approx 2\pi\alpha l\sigma \left(1+\gamma^2\right)^{1/2} \left(1 - \frac{a}{l\gamma}\left(\frac{l\gamma}{a} - \frac{l^2\gamma^2}{2a^2}\right)\right)$$

$$\approx \pi\alpha\sigma \left(1+\gamma^2\right)^{1/2} \frac{l^2}{a} \qquad\qquad \text{(III.2.151)}$$

This formula makes perfect sense if we realize that $\pi(1+\gamma^2)^{1/2}l^2$ is the surface area of the cone (without base), the charge of which is uniformly distributed. Therefore, the net charge on the cone is $Q_{cone} = \pi\sigma(1+\gamma^2)^{1/2}l^2$, and we have that $E \approx \alpha Q_{cone}/a = QQ_{cone}/(4\pi\varepsilon_0 a)$. We see that at very large values of $a$ we have the formula for the electrostatic energy between two point charges, when the distance between the spherical charge and cone is much larger than the dimensions of the cone.

Now, what if $a/l = 0$ in Eq. (III.2.149)? This limiting case neglects the radius of the spherical charge, supposing that it is a point charge. Equation (III.2.149) then simplifies to

$$E \approx 2\pi\alpha\sigma l \left(1+\gamma^2\right)^{1/2} \int_0^1 du \frac{u}{((\gamma^2 u^2 + u^2)^{1/2}}$$

$$= 2\pi\alpha\sigma l \int_0^1 du = 2\pi\alpha\sigma l \qquad\qquad \text{(III.2.152)}$$

This also makes sense. We realize that when $a/l = 0$, all points along the cone where $s$ is the same but $t$ is different lie at the same distance $x = s(\gamma^2 + 1)^{1/2}$ away from the 'point' charge $Q$. We can define a linear charge density (charge per unit length) of $\eta(x) = 2\pi s\sigma = 2\pi x\sigma/(1+\gamma^2)^{1/2}$ for the cone, and summing up all the electrostatic energy contributions as different values of $x$ leads to

$$E \approx \alpha\sigma \int_0^{l(1+\gamma^2)^{1/2}} dx \frac{\eta(x)}{x} = 2\pi\alpha\sigma l \qquad\qquad \text{(III.2.153)}$$

which is exactly as given by Eq. (III.2.152). Here, note that the length along the cone is $\gamma s$. Remarkably, the cone sharpness, $\gamma$, cancels out from the answer. **Think:** Why is this so?

Note that this example cannot be literally applied to various physical contexts. In reality the material of the cone would very likely have a different dielectric constant than the environment that the charge $Q$ sits in. To account for this, the problem becomes more complicated, requiring the solution of a Poisson equation; this equation we'll examine later in this book.

**Exercise.** Find approximations to Eq. (III.2.149) when $\gamma$ is small.

## 2.4.  Key Points

- Information about the gradient (rate of change) of a function $f(x, y)$ is contained in the vector $\vec{\nabla} f(x, y) = \frac{\partial f}{\partial x}\hat{\mathbf{i}} + \frac{\partial f}{\partial y}\hat{\mathbf{j}}$.
- The vector $\nabla f(x, y)$ is perpendicular to tangent vectors of the curves $f(x, y) = c$, the lines where $f(x, y)$ is constant.
- Similarly, to describe the rate of change of a 3D function $f(x, y, z)$ we may act on it with the grad-operator $\vec{\nabla} = \hat{\mathbf{i}}\frac{\partial}{\partial x} + \hat{\mathbf{j}}\frac{\partial}{\partial y} + \hat{\mathbf{k}}\frac{\partial}{\partial k}$.
- The vector $\vec{\nabla} f(\mathbf{r}) = \vec{\nabla} f(x, y, z)$ is perpendicular to tangent vectors of the surfaces $f(x, y, z) = c$, the surfaces where $f(x, y, z)$ is constant.
- The curvature of a function $f(x, y)$ may be described by a $2 \times 2$ Hessian matrix comprised of the second order partial derivatives given by Eq. (III.2.21).
- At a turning point in $f(x, y)$, which is a maximum, the eigenvalues of the Hessian matrix are both negative; at a minimum, both are positive; and at a saddle point one is negative and the other is positive.
- The Hessian matrix can be generalized to functions of $n$ variables $f(x_1, x_2...x_n)$, where it is a $n \times n$ matrix given by Eq. (III.2.60). Again, any maxima (minima) correspond to all negative (positive) eigenvalues of that matrix.

- The vector equation defining a surface is $\mathbf{r}(s,t) = x(s,t)\hat{\mathbf{i}} + y(s,t)\hat{\mathbf{j}} + z(s,t)\hat{\mathbf{k}}$, where $s$ and $t$ are independent parameters.

- The orientation of a surface at a particular point $\mathbf{r}(s,t)$, in space, may be characterized by the tangent vectors $\mathbf{t}_s = \frac{\partial \mathbf{r}}{\partial s}$, $\mathbf{t}_t = \frac{\partial \mathbf{r}}{\partial t}$ and the normal vector $\hat{\mathbf{n}} = \mathbf{t}_s \times \mathbf{t}_t / |\mathbf{t}_s \times \mathbf{t}_t|$. The normal vector points perpendicularly to the surface at a particular point.

- Surfaces have an intrinsic curvature associated with them. The intrinsic curvature at a point is defined as how the unit tangent vectors about a point change in the direction of the normal vector at the point.

- Two useful quantities defining intrinsic curvature are *Gaussian* and *mean* curvatures. They have application in the elastic theory of surfaces and membranes.

- When the Gaussian curvature is positive at a point, but the mean curvature is negative, the surface bends in all directions away from the normal at that point. When both the Gaussian curvature and mean curvature are positive, the surface bends towards the normal. When Gaussian curvature is negative, in one direction the surface there bends away, downwards from the normal, and in another direction, perpendicular to the first one, it bends up, towards the normal.

- A recipe of calculating Gaussian and mean curvature is given by Eqs. (III.2.102)–(III.2.104).

- We may integrate a function $f(x,y,z)$ over a surface $\mathbf{r}(s,t)$ using Eq. (III.2.122): a surface integral. This integration is useful; for instance, we may use it to calculate the electrostatic energy of a charged surface in an electric potential.

- Area integrals are specific cases of surface integrals where the surface is a plane. In this case, as expected, the area element $|\mathbf{t}_s \times \mathbf{t_t}|\, ds dt$ reduces to $\mathcal{J}(s,t) ds dt$, where $\mathcal{J}(s,t)$ is the 2D Jacobian. The reader should get some insight from this.

## Chapter 3

# Vector Fields and More Vector Calculus

### 3.1. What is a Vector Field? The Divergence, Curl and Gradient

We've already come across the idea of a vector field, when we considered the gradient operator. Let's define it in general in 3D. Essentially, it is a vector which has components which are functions of the coordinates $(x, y, z)$ (in the Cartesian coordinate system) such that

$$\mathbf{F}(\mathbf{r}) = \mathbf{F}(x, y, z) = F_x(x, y, z)\hat{\mathbf{i}} + F_y(x, y, z)\hat{\mathbf{j}} + F_z(x, y, z)\hat{\mathbf{k}} \quad \text{(III.3.1)}$$

At each point in space the vector field has a direction and magnitude that are decided by these three functions. We can represent the vector field as an array of arrows on a 3D grid, as shown in Fig. III.3.1. We'll now consider some important operations we can do on vector field with the gradient operator $\nabla$, which we defined in Eq. (III.2.12).

### The divergence of a vector field

The divergence of a vector field is given by

$$\vec{\nabla} \cdot \mathbf{F} = \left( \hat{\mathbf{i}} \frac{\partial}{\partial x} + \hat{\mathbf{j}} \frac{\partial}{\partial y} + \hat{\mathbf{k}} \frac{\partial}{\partial z} \right) \cdot (F_x(x, y, z)\hat{\mathbf{i}}$$
$$+ F_y(x, y, z)\hat{\mathbf{j}} + F_z(x, y, z)\hat{\mathbf{k}})$$

$$= \frac{\partial F_x(x,y,z)}{\partial x} + \frac{\partial F_y(x,y,z)}{\partial y} + \frac{\partial F_z(x,y,z)}{\partial z} \tag{III.3.2}$$

Now, let's try to understand physically what the divergence of vector field means, and derive an important physical equation: the continuity equation. To do this we'll consider a fluid which has a concentration $c(x,y,z;t)$, at a particular time $t$, that is moving with a flow described by a vector field, which is the current density. This is given by

$$\mathbf{J}(x,y,z;t) = J_x(x,y,z;t)\hat{\mathbf{i}} + J_y(x,y,z;t)\hat{\mathbf{j}} + J_z(x,y,z;t)\hat{\mathbf{k}} \tag{III.3.3}$$

Here, $J_x(x,y,z;t)$, $J_y(x,y,z;t)$ and $J_z(x,y,z;t)$ are the number of particles moving per second per unit area through an area perpendicular to the x-axis, y-axis and z-axis, respectively, at a particular time moment $t$ (the semi colon is here simply to distinguish the time as a separate variable type from the position variables).

The total number of particles contained in a small cuboid of $\Delta x$, $\Delta y$, $\Delta z$ dimensions centred at $x,y,z$, at a time $t$, is then given by

$$N(x,y,z;t) = c(x,y,z;t)\Delta x \Delta y \Delta z \tag{III.3.4}$$

Through each face of that cuboid we can define a current (the number of particles moving through that surface per second). For instance,

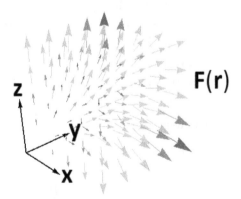

**Fig. III.3.1** How a vector field, $\mathbf{F}(\mathbf{r})$ can be represented as an array of arrows, representing the direction and magnitude of the field.

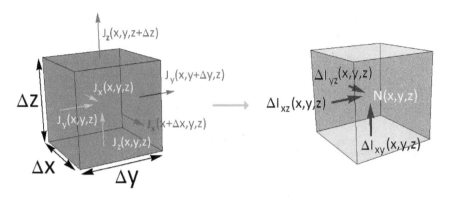

**Fig. III.3.2** We consider a small cuboid of dimensions $\Delta x$, $\Delta y$ and $\Delta z$. The number of particles flowing into the cube per second per unit area through three of the faces of cuboid are given by the current densities $J_x(x,y,z)$, $J_y(x,y,z)$ and $J_z(x,y,z)$. These are shown as yellow arrows on the left-hand diagram. Through the other three faces, particles are flowing out. Here, the current densities $J_x(x+\Delta x, y, z)$, $J_y(x, y+\Delta y, z)$ and $J_z(x, y, z+\Delta z)$ are the number of particles per second per unit area flowing out of each of the faces, shown by the red arrows in the left-hand diagram. The cube is considered small enough that the currents can be considered constant across the faces of the cube. From these current densities we may obtain the net currents $\Delta I_{xy}(x,y,z)$, $\Delta I_{xz}(x,y,z)$ and $\Delta I_{yz}(x,y,z)$(see Eq. III.3.5), which are defined as the amount of particles flowing per second into the cube in each of the three directions (shown by the purple arrows in the r.h.s diagram). The sum of these three currents gives us the rate of change in the number of particles in the cuboid $N(x,y,z)$. These considerations and the analysis in the text yields the continuity equation (Eq. III.3.9).

for $(x, y)$-face we have (see Fig. III.3.2)

$$I_{xy}(x,y,z;t) = J_z(x,y,z;t)\Delta x\Delta y \qquad \text{(III.3.5)}$$

In this case only the $z$ component of the current density matters, as this alone determines how many particles at any moment will cross this cuboid face. Next, we can consider the net number of particles flowing per unit second into three faces of the cube, and out of the three opposite faces; these are given by the three current differences of 'in' minus 'out' flows:

$$\Delta I_{xy}(x,y,z;t) = (J_z(x,y,z;t) - J_z(x,y,z+\Delta z;t))\Delta x\Delta y$$
$$\Delta I_{xz}(x,y,z;t) = (J_y(x,y,z;t) - J_y(x,y+\Delta y,z;t))\Delta x\Delta z$$
$$\Delta I_{yz}(x,y,z;t) = (J_x(x,y,z;t) - J_x(x+\Delta x,y,z;t))\Delta y\Delta z \qquad \text{(III.3.6)}$$

We can make Taylor expansions for small $\Delta x$, $\Delta y$ and $\Delta z$ in these expressions, so that we may write

$$\Delta I_{xy}(x,y,z;t) = -\frac{\partial J_z(x,y,z)}{\partial z}\Delta z \Delta x \Delta y$$

$$\Delta I_{xz}(x,y,z;t) = -\frac{\partial J_y(x,y,z)}{\partial y}\Delta x \Delta y \Delta z \qquad \text{(III.3.7)}$$

$$\Delta I_{yz}(x,y,z;t) = -\frac{\partial J_x(x,y,z)}{\partial x}\Delta x \Delta y \Delta z$$

Note that the rate of change of $N(x,y,z;t)$ must be the sum of these three currents to conserve the number of particles. Therefore, we can write

$$\frac{\partial N(x,y,z;t)}{\partial t} = \Delta I_{xy}(x,y,z;t) + \Delta I_{xz}(x,y,z;t)$$

$$+ \Delta I_{yz}(x,y,z;t) \qquad \text{(III.3.8)}$$

$$\Rightarrow \frac{\partial c(x,y,z;t)}{\partial t}\Delta x \Delta y \Delta z \approx -\left(\frac{\partial J_x(x,y,z;t)}{\partial x} + \frac{\partial J_y(x,y,z;t)}{\partial y}\right.$$

$$\left. +\frac{\partial J_z(x,y,z;t)}{\partial z}\right)\Delta x \Delta y \Delta z$$

$$\text{(III.3.9)}$$

When we take the limit $\Delta x \to 0$, $\Delta y \to 0$, $\Delta z \to 0$, this relationship becomes exact, so that we may write

$$-\frac{\partial c(x,y,z;t)}{\partial t} = \frac{\partial J_x(x,y,z;t)}{\partial x} + \frac{\partial J_y(x,y,z;t)}{\partial y} + \frac{\partial J_z(x,y,z;t)}{\partial z}$$

$$= \vec{\nabla} \cdot \mathbf{J}(x,y,z;t) \qquad \text{(III.3.10)}$$

Equation (III.3.10) is called the *continuity equation*. We see that this equation defines how $c(x,y,z;t)$ and $\mathbf{J}(x,y,z;t)$ must relate to each other if the total number of particles is conserved.

Now, let's consider what $\nabla \cdot \mathbf{F}(x,y,z) = 0$ means for any vector field. Let's go back to our cube. We can define a quantity called a flux in the same way as we defined currents (a current is simply a type of flux) in Eq. (III.3.5), for instance, the flux through one side of area $\Delta x \Delta y$ of the small cube is defined by $\Phi_{xy}(x,y,z) = F_z(x,y,z)\Delta x \Delta y$; it's the area of the cube times the component of the field perpendicular to the cube face (along the direction of the surface normal). From the previous derivation of the continuity equation, $\nabla \cdot \mathbf{F}(x, y, z) = 0$ states that the flux going into a small cube at $(x, y, z)$ is the same as the flux going out. So when $\nabla \cdot \mathbf{F}(x, y, z) \neq 0$ the flux is changing in space. This understanding is very useful when we come to consider surface integrals of vectors and understanding Gauss's divergence theorem.

**Example III.3.1 (Maths Practice ♫♪).** Here, we'll use the continuity equation to work out the form of a concentration, knowing a current density. The example—the form of the flux—is contrived, but its purpose is to show how the fluxes can determine time- and space-dependent concentration in a system, and how we can calculate it. Suppose we have a flow of particles with current density

$$\mathbf{J}(x, y, z; t) = x^2 \exp(-\kappa t)\hat{\mathbf{i}} + xy^2 \exp(-2\kappa t)\hat{\mathbf{j}} + \exp(-\alpha z)\hat{\mathbf{k}} \quad \text{(III.3.11)}$$

Let's compute the concentration with the initial condition that $c(t = 0) = c_0$. To do this we need to use Eq. (III.3.10) and find the divergence:

$$\vec{\nabla} \cdot \mathbf{J}(x, y, z; t) = \frac{\partial}{\partial x}\left(x^2 \exp(-\kappa t)\right) + \frac{\partial}{\partial y}\left(xy^2 \exp(-2\kappa t)\right)$$

$$+ \frac{\partial}{\partial z}\left[\exp(-\alpha z)\right]$$

$$= 2x \exp(-\kappa t) + 2xy \exp(-2\kappa t) - \alpha \exp(-\alpha z)$$
$$\text{(III.3.12)}$$

Hence, the rate of concentration change is

$$\frac{\partial c(x, y, z; t)}{\partial t} = -(2x \exp(-\kappa t) + 2xy \exp(-2\kappa t) - \alpha \exp(-\alpha z))$$
$$\text{(III.3.13)}$$

Then

$$c(x, y, z; t) = -\int (2x \exp(-\kappa t) + 2xy \exp(-2\kappa t) - \alpha \exp(-\alpha z))dt$$

$$+ C(x, y, z) \quad \text{(III.3.14)}$$

where the constant of integration, $C(x, y, z)$ can be a function of $x, y$ and $z$, but independent of $t$. Integration yields

$$c(x, y, z; t) = \left(\frac{2x}{\kappa} \exp(-\kappa t) + \frac{xy}{\kappa} \exp(-2\kappa t) + \alpha t \exp(-\alpha z))\right)$$
$$+ C(x, y, z) \tag{III.3.15}$$

At $t = 0$ we require

$$c_0 = \left(\frac{2x}{\kappa} + \frac{xy}{\kappa}\right) + C(x, y, z) \tag{III.3.16}$$

Therefore, the solution is

$$c(x, y, z; t) = \left(\frac{2x}{\kappa}(\exp(-\kappa t) - 1) + \frac{xy}{\kappa}(\exp(-2\kappa t) - 1)\right.$$
$$\left. + \alpha t \exp(-\alpha z))\right) + c_0 \tag{III.3.17}$$

### Introducing the Laplacian operator

We can also take the divergence of the gradient of a function $\phi(x, y, z)$. In this case we have

$$\vec{\nabla} \cdot \vec{\nabla}\phi(x, y, z) = \left(\hat{\mathbf{i}}\frac{\partial}{\partial x} + \hat{\mathbf{j}}\frac{\partial}{\partial y} + \hat{\mathbf{k}}\frac{\partial}{\partial z}\right)$$
$$\cdot \left(\frac{\partial\phi(x, y, z)}{\partial x}\hat{\mathbf{i}} + \frac{\partial\phi(x, y, z)}{\partial y}\hat{\mathbf{j}} + \frac{\partial\phi(x, y, z)}{\partial y}\hat{\mathbf{k}}\right)$$
$$\Rightarrow \quad \vec{\nabla}^2\phi(x, y, z) = \frac{\partial^2\phi(x, y, z)}{\partial x^2} + \frac{\partial^2\phi(x, y, z)}{\partial y^2} + \frac{\partial^2\pi(x, y, z)}{\partial z^2}$$
$$\tag{III.3.18}$$

The operator $\vec{\nabla}^2 = \vec{\nabla} \cdot \vec{\nabla}$ is called the Laplacian operator, and we'll talk more about it when we come to consider partial differential equations.

### The vorticity and the curl of a vector field

We can also define the curl of a vector field:

$$\vec{\nabla} \times \mathbf{F} = \left( \hat{\mathbf{i}} \frac{\partial}{\partial x} + \hat{\mathbf{j}} \frac{\partial}{\partial y} + \hat{\mathbf{k}} \frac{\partial}{\partial z} \right) \times (F_x(x,y,z)\hat{\mathbf{i}}$$

$$+ F_y(x,y,z)\hat{\mathbf{j}} + F_z(x,y,z)\hat{\mathbf{k}})$$

$$\Rightarrow \quad \vec{\nabla} \times \mathbf{F} = \begin{vmatrix} \hat{\mathbf{i}} & \hat{\mathbf{j}} & \hat{\mathbf{k}} \\ \dfrac{\partial}{\partial x} & \dfrac{\partial}{\partial y} & \dfrac{\partial}{\partial z} \\ F_x(x,y,z) & F_y(x,y,z) & F_z(x,y,z) \end{vmatrix} \quad \text{(III.3.19)}$$

$$\Rightarrow \vec{\nabla} \times \mathbf{F} = \left( \frac{\partial F_z(x,y,z)}{\partial y} - \frac{\partial F_y(x,y,z)}{\partial z} \right) \hat{\mathbf{i}}$$

$$+ \left( \frac{\partial F_x(x,y,z)}{\partial z} - \frac{\partial F_z(x,y,z)}{\partial x} \right) \hat{\mathbf{j}}$$

$$+ \left( \frac{\partial F_y(x,y,z)}{\partial x} - \frac{\partial F_x(x,y,z)}{\partial y} \right) \hat{\mathbf{k}}$$

The curl of a vector field is a measure of the vorticity (or circulation) per unit area. Essentially, each vector component of Eq. (III.3.19) is how much the field, $\mathbf{F}(r)$, vectors' direction circles in space around an axis in the direction associated with that component; for example, the first component in Eq. (III.3.19) is the amount of circulation about the x-axis.

To understand more about curl, let's consider the vorticity about the z-axis around a small circle of radius $\Delta r$ in the $x$–$y$ plane centred about the point $(x, y, z)$, as demonstrated in Fig. III.3.3. As we move around this circle the change in $\mathbf{F}$ is given by the vector

$$\Delta \mathbf{F}(x,y,z) = \mathbf{F}(x + \Delta r \cos\theta, y + \Delta r \sin\theta, z) - \mathbf{F}(x,y,z)$$

$$\approx \frac{\partial \mathbf{F}(x,y,z)}{\partial x} \Delta r \cos\theta + \frac{\partial \mathbf{F}(x,y,z)}{\partial y} \Delta r \sin\theta \quad \text{(III.3.20)}$$

Now, how $\mathbf{F}$ curves/curls or rotates around the circle, is characterized by the *vorticity*. The latter is related to the total change in $\mathbf{F}$ perpendicular to vector $\Delta \mathbf{r} = \mathbf{i}\Delta r \cos\theta + \mathbf{j}\Delta r \sin\theta$ (that lies in the plane of

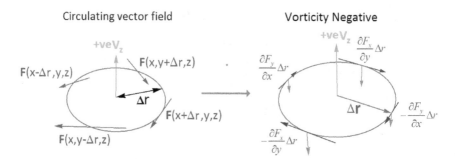

**Fig. III.3.3** This diagram describes the concept of vorticity in the z-direction about a small circle in the $x$–$y$ plane of the vector field **F**. When the circle is taken to be very small we may expand out **F** as a Taylor expansion in $\Delta r$. Specifically, the right of the two panels shows the anatomy of vorticity—the way it is mathematically defined and calculated. Then, to calculate the vorticity, we are interested in the parts of these changes that are perpendicular to the vector $\Delta \mathbf{r}$ that lies in the $x$–$y$ plane and traces out the circle. These are shown by purple arrows, next to expressions for them, shown here for the four points on the circle, obtained from the Taylor expansion (a more general expression is given by Eq. (III.3.20)). Each contribution to the vorticity is given by cross product of $\Delta \mathbf{r}$ with the vectors described by the purple arrows. These contributions are shown by the orange arrows; we sum these up from all positions on the circle as an integral, to obtain the total vorticity (see Eq. (III.3.22)). In this case, as the overall circulation of the vector field is clockwise, so that the orange arrows point downwards, the vorticity is negative. When we divide the vorticity by the area of the circle and take $\Delta r \to 0$ we obtain the z-component of $\nabla \times \mathbf{F}$. Similarly, by considering circles in the x-z plane and y-z plane, we obtain, respectively, the vorticities in the $y$ and $x$ direction and the other components of the curl.

the circle), when one circulates once around the circle. In this case, this is given by the z-component of the cross product $\Delta \mathbf{r} \times \Delta \mathbf{F}(x, y, z)$ integrated over $\theta$ from 0 to $2\pi$. The z-component of the cross product is given by

$$\hat{\mathbf{k}} \cdot (\Delta \mathbf{r} \times \Delta \mathbf{F}(x, y, z))$$

$$= \Delta r^2 \hat{\mathbf{k}} \cdot \begin{vmatrix} \hat{\mathbf{i}} & \hat{\mathbf{j}} & \hat{\mathbf{k}} \\ \cos\theta & \sin\theta & 0 \\ \dfrac{\partial F_x}{\partial x}\cos\theta + \dfrac{\partial F_x}{\partial y}\sin\theta & \dfrac{\partial F_y}{\partial x}\cos\theta + \dfrac{\partial F_y}{\partial y}\sin\theta & \dfrac{\partial F_z}{\partial x}\cos\theta + \dfrac{\partial F_z}{\partial y}\sin\theta \end{vmatrix}$$

$$\Rightarrow \hat{\mathbf{k}} \cdot (\Delta \mathbf{r} \times \Delta \mathbf{F}(x, y, z))$$

$$= \Delta r^2 \begin{vmatrix} 0 & 0 & 1 \\ \cos\theta & \sin\theta & 0 \\ \dfrac{\partial F_x}{\partial x}\cos\theta + \dfrac{\partial F_x}{\partial y}\sin\theta & \dfrac{\partial F_y}{\partial x}\cos\theta + \dfrac{\partial F_y}{\partial y}\sin\theta & \dfrac{\partial F_z}{\partial x}\cos c\theta + \dfrac{\partial F_z}{\partial y}\sin\theta \end{vmatrix}$$

$$\Rightarrow \hat{\mathbf{k}} \cdot (\Delta \mathbf{r} \times \Delta \mathbf{F}(x,y,z))$$

$$= (\Delta r)^2 \left[ \cos^2\theta \frac{\partial F_y}{\partial x} - \sin^2\theta \frac{\partial F_x}{\partial y} + \sin\theta\cos\theta \left( \frac{\partial F_y}{\partial y} - \frac{\partial F_x}{\partial x} \right) \right] \qquad \text{(III.3.21)}$$

Then, the vorticity (as a vector pointing out of the circle, like angular momentum) about the circle is given by

$$V_z \hat{\mathbf{k}} = (\Delta r)^2 \int_0^{2\pi} d\theta \left[ \cos^2\theta \frac{\partial F_y}{\partial x} - \sin^2\theta \frac{\partial F_x}{\partial y} + \sin\theta\cos\theta \left( \frac{\partial F_y}{\partial y} - \frac{\partial F_x}{\partial x} \right) \right] \hat{\mathbf{k}}$$

$$= \pi(\Delta r)^2 \left( \frac{\partial F_y}{\partial x} - \frac{\partial F_x}{\partial y} \right) \hat{\mathbf{k}} \qquad \text{(III.3.22)}$$

The vorticity per unit area is then obtained by dividing this equation by the area of our small circle $\pi\Delta r^2$, which gives

$$[V_z/\pi\Delta r^2]\hat{\mathbf{k}} = \left( \frac{\partial F_y}{\partial x} - \frac{\partial F_x}{\partial y} \right) \hat{\mathbf{k}} \qquad \text{(III.3.23)}$$

We note that this is indeed looking like the last component of Eq. (III.3.19). Now, for a vector field we are not just interested in the vorticity per unit area in the $x$–$y$ plane, we are generally interested in it in the $y$–$z$ plane and $x$–$z$ plane. By constructing similar small circles in each plane we can calculate $V_x/\pi\Delta r^2$ (for the $y$–$z$ plane) and $V_y/\pi\Delta r^2$ (for the $x$–$z$ plane). Combining all three vorticities, and taking the radii of our circles to zero per unit area we obtain the curl

$$\vec{\nabla} \times \mathbf{F} = \lim_{\Delta r \to 0} \left\{ 1/(\pi\Delta r^2)(V_x\hat{\mathbf{i}} + V_y\hat{\mathbf{j}} + V_z\hat{\mathbf{k}}) \right\} \qquad \text{(III.3.24)}$$

**Example III.3.2 (Maths Practice ♪♪).** Let's compute the curl of the vector field

$$\mathbf{F}(x,y,z) = \frac{-y\hat{\mathbf{i}}}{\sqrt{x^2+y^2}} + \frac{x}{\sqrt{x^2+y^2}}\hat{\mathbf{j}} + z\hat{\mathbf{k}} \qquad \text{(III.3.25)}$$

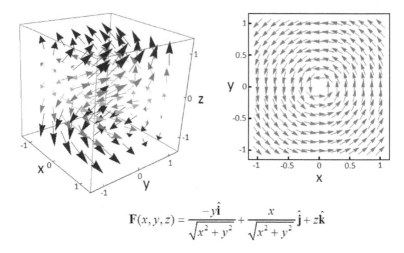

$$\mathbf{F}(x,y,z) = \frac{-y\hat{\mathbf{i}}}{\sqrt{x^2 + y^2}} + \frac{x}{\sqrt{x^2 + y^2}}\hat{\mathbf{j}} + z\hat{\mathbf{k}}$$

**Fig. III.3.4** The vector field given by Eq. (III.3.25). The left-hand panel shows it, in 3D space, as an array of arrows representing the direction and magnitude of the vector field at that point. The right-hand panel displays the projection of the vector field in the $x$–$y$ plane, $\mathbf{F}_{xy} = -y/\sqrt{x^2 + y^2}\hat{\mathbf{i}} + x/\sqrt{x^2 + y^2}\hat{\mathbf{j}}$. Here, we clearly see that we have a vortex, about which the direction field circulates.

We illustrate the vector field in Fig. III.3.4. Its curl is given by

$$\nabla \times \mathbf{F} = \left( \underbrace{\frac{\partial z}{\partial y} - \frac{\partial}{\partial z}\left[ \frac{x}{\sqrt{x^2 + y^2}} \right]}_{0} \right)\hat{\mathbf{i}} + \left( \underbrace{\frac{\partial}{\partial z}\left[ -\frac{y}{\sqrt{x^2 + y^2}} \right] - \frac{\partial z}{\partial x}}_{0} \right)\hat{\mathbf{j}}$$

$$+ \left( \frac{\partial}{\partial x}\left[ \frac{x}{\sqrt{x^2 + y^2}} \right] - \frac{\partial}{\partial y}\left[ -\frac{y}{\sqrt{x^2 + y^2}} \right] \right)\hat{\mathbf{k}} \qquad \text{(III.3.26)}$$

$$\Rightarrow \nabla \times \mathbf{F} = \left( \frac{2}{\sqrt{x^2 + y^2}} - \frac{x^2}{(x^2 + y^2)^{3/2}} - \frac{y^2}{(x^2 + y^2)^{3/2}} \right)\hat{\mathbf{k}}$$

$$= \frac{1}{\sqrt{x^2 + y^2}}\hat{\mathbf{k}} \qquad \text{(III.3.27)}$$

For this example the curl 'looks' in the $z$-direction. It becomes singular at the centre of a vortex $(x = 0, y = 0)$ formed by $\mathbf{F}(\mathbf{r})$. Fig. III.3.5 illustrates this visually.

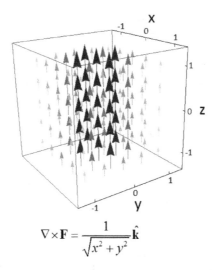

$$\nabla \times \mathbf{F} = \frac{1}{\sqrt{x^2 + y^2}}\hat{\mathbf{k}}$$

**Fig. III.3.5** The curl of the vector field given in Example III.3.3. Here, as there is only one non-zero component of the curl, in the $z$-direction, all the arrows point in this direction, see Eq. (III.3.27). The magnitude gets larger as we move towards the $z$-axis ($x = 0, y = 0$), where there curl becomes singular.

**Example III.3.3 (Physical Application ▶▶).** In magnetism it is convenient to define the magnetic flux density $B$ in terms of a vector potential such that $\mathbf{B} = \nabla \times \mathbf{A}$. Let's suppose we have vector field $\mathbf{A} = A(x, y)\hat{\mathbf{k}}$; then the magnetic flux density is

$$\mathbf{B} = \vec{\nabla} \times \mathbf{A} = \begin{vmatrix} \hat{\mathbf{i}} & \hat{\mathbf{j}} & \hat{\mathbf{k}} \\ \frac{\partial}{\partial x} & \frac{\partial}{\partial y} & \frac{\partial}{\partial z} \\ 0 & 0 & A(x,y) \end{vmatrix} = \frac{\partial A_z(x,y)}{\partial y}\hat{\mathbf{i}} - \frac{\partial A_z(x,y)}{\partial x}\hat{\mathbf{j}}$$

$$(\text{III.3.28})$$

### The gradient of a vector field

We can define the gradient of a vector as

$$\vec{\nabla} \otimes \mathbf{F} = \vec{\nabla}F_x(x,y,z)\hat{\mathbf{i}} + \vec{\nabla}F_y(x,y,z)\hat{\mathbf{j}} + \vec{\nabla}F_z(x,y,z)\hat{\mathbf{k}} \qquad (\text{III.3.29})$$

Equation (III.3.29) means that to calculate the gradient of the vector-field we need to take the gradient of each of its components. This describes what is called *a tensor*, which can be written as

$$\vec{\nabla} \otimes \mathbf{F} = \frac{\partial F_x(x,y,z)}{\partial x}\hat{\mathbf{i}} \otimes \hat{\mathbf{i}} + \frac{\partial F_x(x,y,z)}{\partial y}\hat{\mathbf{j}} \otimes \hat{\mathbf{i}} + \frac{\partial F_x(x,y,z)}{\partial z}\hat{\mathbf{k}} \otimes \hat{\mathbf{i}}$$

$$+ \frac{\partial F_y(x,y,z)}{\partial x}\hat{\mathbf{i}} \otimes \hat{\mathbf{j}} + \frac{\partial F_y(x,y,z)}{\partial y}\hat{\mathbf{j}} \otimes \hat{\mathbf{j}} + \frac{\partial F_y(x,y,z)}{\partial z}\hat{\mathbf{k}} \otimes \hat{\mathbf{j}}$$

$$+ \frac{\partial F_z(x,y,z)}{\partial x}\hat{\mathbf{i}} \otimes \hat{\mathbf{k}} + \frac{\partial F_z(x,y,z)}{\partial y}\hat{\mathbf{j}} \otimes \hat{\mathbf{k}} + \frac{\partial F_z(x,y,z)}{\partial y}\hat{\mathbf{k}} \otimes \hat{\mathbf{k}}$$

$$(\text{III.3.30})$$

This looks more compact and transparent in matrix notation,

$$\vec{\nabla} \otimes \mathbf{F} = \begin{pmatrix} \dfrac{\partial}{\partial x} \\[2mm] \dfrac{\partial}{\partial y} \\[2mm] \dfrac{\partial}{\partial z} \end{pmatrix} \otimes \begin{pmatrix} F_x & F_y & F_z \end{pmatrix} = \begin{pmatrix} \dfrac{\partial F_x}{\partial x} & \dfrac{\partial F_y}{\partial x} & \dfrac{\partial F_z}{\partial x} \\[2mm] \dfrac{\partial F_x}{\partial y} & \dfrac{\partial F_y}{\partial y} & \dfrac{\partial F_z}{\partial y} \\[2mm] \dfrac{\partial F_x}{\partial z} & \dfrac{\partial F_y}{\partial z} & \dfrac{\partial F_z}{\partial z} \end{pmatrix} \quad (\text{III.3.31})$$

although the notation of Eq. (III.3.30) is also commonly used. This operation gives us all the information about the rate of change of the vector field in the three directions in space.

We can also take the Laplace operator of a vector field, which reads as

$$\Rightarrow \vec{\nabla} \cdot (\vec{\nabla} \otimes \mathbf{F}) = \begin{pmatrix} \dfrac{\partial}{\partial x} & \dfrac{\partial}{\partial y} & \dfrac{\partial}{\partial z} \end{pmatrix} \cdot \begin{pmatrix} \dfrac{\partial F_x}{\partial x} & \dfrac{\partial F_y}{\partial x} & \dfrac{\partial F_z}{\partial x} \\[2mm] \dfrac{\partial F_x}{\partial y} & \dfrac{\partial F_y}{\partial y} & \dfrac{\partial F_z}{\partial y} \\[2mm] \dfrac{\partial F_x}{\partial z} & \dfrac{\partial F_y}{\partial z} & \dfrac{\partial F_z}{\partial z} \end{pmatrix}$$

$$\vec{\nabla} \cdot (\vec{\nabla} \otimes \mathbf{F}) = \begin{pmatrix} \dfrac{\partial^2 F_x}{\partial x^2} + \dfrac{\partial^2 F_x}{\partial y^2} + \dfrac{\partial^2 F_x}{\partial z^2} \\[2mm] \dfrac{\partial^2 F_y}{\partial x^2} + \dfrac{\partial^2 F_y}{\partial y^2} + \dfrac{\partial^2 F_y}{\partial z^2} \\[2mm] \dfrac{\partial^2 F_z}{\partial x^2} + \dfrac{\partial^2 F_z}{\partial y^2} + \dfrac{\partial^2 F_z}{\partial z^2} \end{pmatrix} \quad (\text{III.3.32})$$

which can be defined more compactly as

$$\vec{\nabla} \cdot (\vec{\nabla} \otimes \mathbf{F}) = \vec{\nabla}^2 \mathbf{F} = \vec{\nabla}^2 F_x(x,y,z)\hat{\mathbf{i}} + \vec{\nabla}^2 F_y(x,y,z)\hat{\mathbf{j}}$$
$$+ \vec{\nabla}^2 F_z(x,y,z)\hat{\mathbf{k}} \qquad \text{(III.3.33)}$$

### Vector calculus identities

To end this section we'll state some identities of vector derivative calculus. Two of the most important are

$$\vec{\nabla} \cdot (\vec{\nabla} \times \mathbf{F}) = 0 \quad \text{and} \quad \vec{\nabla} \times (\vec{\nabla}\varphi) = 0 \qquad \text{(III.3.34)}$$

These identities mean that the divergence of a curl of vector field and the curl of the gradient of a function are both zero.

**Exercise.** The identities in Eq. (III.3.34) are not hard to prove. Check them using the definitions of the gradient operator, divergence and curl.

We give a few more identities below

$$\vec{\nabla}(\phi\psi) = \phi\vec{\nabla}\psi + \psi\vec{\nabla}\phi$$
$$\vec{\nabla}(\mathbf{A} \cdot \mathbf{B}) = (\mathbf{A} \cdot \vec{\nabla})\mathbf{B} + (\mathbf{B} \cdot \vec{\nabla})\mathbf{A} + \mathbf{A} \times (\vec{\nabla} \times \mathbf{B}) + \mathbf{B} \times (\vec{\nabla} \times \mathbf{A})$$
$$\vec{\nabla} \cdot (\psi\mathbf{A}) = \psi\vec{\nabla} \cdot \mathbf{A} + \mathbf{A} \cdot \vec{\nabla}\psi$$
$$\vec{\nabla} \cdot (\mathbf{A} \times \mathbf{B}) = \mathbf{B} \cdot (\vec{\nabla} \times \mathbf{A}) - \mathbf{A} \cdot (\vec{\nabla} \times \mathbf{B})$$
$$\vec{\nabla} \times (\psi\mathbf{A}) = \psi\vec{\nabla} \times \mathbf{A} + \vec{\nabla}\psi \times \mathbf{A}$$
$$\vec{\nabla} \times (\mathbf{A} \times \mathbf{B}) = \mathbf{A}(\vec{\nabla} \cdot \mathbf{B}) - \mathbf{B}(\vec{\nabla} \cdot \mathbf{A}) + (\mathbf{B} \cdot \vec{\nabla})\mathbf{A} - (\mathbf{A} \cdot \vec{\nabla})\mathbf{B}$$
$$\psi\vec{\nabla}^2\phi - \phi\vec{\nabla}^2\psi = \vec{\nabla} \cdot (\psi\vec{\nabla}\phi - \varphi\vec{\nabla}\psi)$$
$$(\vec{\nabla} \cdot \mathbf{A}) - \vec{\nabla} \times (\vec{\nabla} \times \mathbf{A}) = \vec{\nabla}^2 \mathbf{A} \qquad \text{(III.3.35)}$$

Here, $\psi = \psi(\mathbf{r})$ and $\phi = \phi(\mathbf{r})$ are arbitrary scalar fields; $\mathbf{A} = \mathbf{A}(\mathbf{r})$ and $\mathbf{B} = \mathbf{B}(\mathbf{r})$ are arbitrary vector fields. Note here that the operator $\mathbf{A} \cdot \vec{\nabla}$ is defined as any scalar product

$$\mathbf{A} \cdot \vec{\nabla} = A_x \frac{\partial}{\partial x} + A_y \frac{\partial}{\partial y} + A_z \frac{\partial}{\partial z} \qquad \text{(III.3.36)}$$

where, $\mathbf{A} = A_x\hat{\mathbf{i}} + A_y\hat{\mathbf{j}} + A_z\hat{\mathbf{k}}$, (and similarly for $\mathbf{B} \cdot \vec{\nabla}$).

## 3.2. Line and Surface Integrals of Vectors

### *The line integral of a vector field*

For any 3D vector field $\mathbf{F}(\mathbf{r}) = \mathbf{F}(x, y, z)$ we can define the line integral

$$W = \int_C \mathbf{F}(\mathbf{r}) \cdot d\mathbf{r}$$

$$= \int_C [F_x(x, y, z)dx + F_y(x, y, z)dy + F_z(x, y, z)dz] \quad \text{(III.3.37)}$$

We specify the integration path, as we did previously for line integration, but this time in 3D. The path is now specified by a 3D vector equation of a line given by $\mathbf{r}(s) = x(s)\hat{\mathbf{i}} + y(s)\hat{\mathbf{j}} + z(s)\hat{\mathbf{k}}$, with end points $\mathbf{r}_A = \mathbf{r}(s_1)$ and $\mathbf{r}_B = \mathbf{r}(s_2)$. How such line integration works is illustrated in Fig. III.3.6.

We can suppose our curve to be made up of a chain of vectors joining points on the curve, such that $\Delta\mathbf{r}_i = \mathbf{r}_{i+1} - \mathbf{r}_i$, where $\mathbf{r}_i$ is the

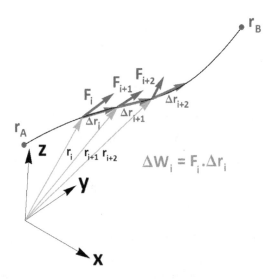

**Fig. III.3.6**  How the line integration described by Eq. (III.3.37) works. We can divide the line that we are interested in calculating the integral along into a series of many points, each labelled with the index $i$ and position vector $\mathbf{r}_i$. Two adjacent points, $i$ and $i+1$ are connected by the vector $\Delta\mathbf{r}_i = \mathbf{r}_{i+1} - \mathbf{r}_i$. We may approximate $W$, by calculating at each point $\Delta W_i = \mathbf{F}_i.\Delta\mathbf{r}_i$, where $\mathbf{F}_i$ is the value of the vector field at that point, and summing over $i$. In the limit $\Delta\mathbf{r}_i \to 0$, this leads to the *line integral*.

position vector of the $i$-th point. The line integral can be approximated as

$$W \simeq \sum_i \Delta W_i = \sum_i \mathbf{F}_i \cdot \Delta \mathbf{r}_i \qquad \text{(III.3.38)}$$

When we make the distance between points to be infinitesimally small, $\Delta \mathbf{r}_i \to 0$, and Eq. (III.3.38) becomes exact; this defines the line integral in Eq. (III.3.37).

We can simply evaluate the line integral by computing the tangent vector of the curve, so that

$$W = \int_{s_1}^{s_2} \mathbf{F}(\mathbf{r}(s)) \cdot \frac{d\mathbf{r}}{ds} ds \qquad \text{(III.3.39)}$$

We can also generalize the line integral to consider a $\mathbf{F}$ vector field that does not just depend on position, but on the tangent vector as well:

$$W = \int_{s_1}^{s_2} \mathbf{F}\left(\mathbf{r}(s), \frac{d\mathbf{r}(s)}{ds}\right) \cdot \frac{d\mathbf{r}}{ds} ds \qquad \text{(III.3.40)}$$

We will do this in one of the physical examples below, where we consider air resistance.

**Example III.3.4 (Physical Application ▶▶).** A classical example of the use of line integrals of vector field is calculating the current (the rate of flow of charge) that flows through an area $A$. Ampere's law tells us that we can compute it by considering a line integral of the magnetic flux density about a path $C$ around the perimeter of that area; namely

Andre-Marie Ampere
(1775–1836)

$$\oint_C \mathbf{B} \cdot d\mathbf{r} = \mu_0 I_A \qquad \text{(III.3.41)}$$

Here, $\mu_0$ is the permeability of the vacuum. This is illustrated in Fig. III.3.7.

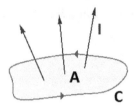

**Fig. III.3.7**   For the calculation of electric current, $I$, flowing through an area $A$ using Ampere's law we need to take a line integral of the magnetic flux density $B$ about the curve $C$, fully enclosing this area.

Imagine that someone has measured the magnetic flux density and found it to be the following form

$$\mathbf{B}/\mu_0\lambda = \left[\frac{z}{(x^2 + z^2)} - \frac{y}{(x^2 + y^2)}\right]\hat{\mathbf{i}} + \frac{x}{(x^2 + y^2)}\hat{\mathbf{j}} - \frac{x}{(x^2 + z^2)}\hat{\mathbf{k}}$$

$$(\text{III.3.42})$$

where $\lambda$ is a constant scaling the intensity of the magnetic flux. Which electrical current could be responsible for that? We'll use Ampere's law to solve this puzzle.

Let's compute the line integral in Eq. (III.3.41) over a closed path:

$$\mathbf{r}(t) = R(\sin t\hat{\mathbf{i}} + \cos\alpha\cos t\hat{\mathbf{j}} + \sin\alpha\cos t\hat{\mathbf{k}}) \quad \text{where } 0 \le t \le 2\pi \quad (\text{III.3.43})$$

The path is a circle tilted by an angle $\alpha$ away from the $x$–$y$ plane. This line integral is defined as

$$\oint_C \mathbf{B}\cdot d\mathbf{r} = \int_0^{2\pi}\left[\mathbf{B}(x(t),y(t),z(t))\cdot\frac{d\mathbf{r}(t)}{dt}\right]dt \qquad (\text{III.3.44})$$

where

$$\mathbf{B}(x(t),y(t),z(t))/\mu_0\lambda$$

$$= \frac{1}{R}\left[\frac{\sin\alpha\cos t}{(\sin^2 t + \sin^2\alpha\cos^2 t)} - \frac{\cos\alpha\cos t}{(\sin^2 t + \cos^2\alpha\cos^2 t)}\right]\hat{\mathbf{i}}$$

$$+ \frac{\sin t}{R(\sin^2 t + \cos^2\alpha\cos^2 t)}\hat{\mathbf{j}} - \frac{\sin t}{R(\sin^2 t + \sin^2\alpha\cos^2 t)}\hat{\mathbf{k}}$$

$$(\text{III.3.45})$$

and

$$\frac{d\mathbf{r}(t)}{dt} = R(\hat{\mathbf{i}}\cos t - \hat{\mathbf{j}}\cos\alpha\sin t - \hat{\mathbf{k}}\sin\alpha\sin t) \qquad \text{(III.3.46)}$$

So that

$$\mathbf{B}(x(t), y(t), z(t)) \cdot \frac{d\mathbf{r}(t)}{dt}$$

$$= \mu_0\lambda\left[\frac{\sin\alpha\cos^2 t}{(\sin^2 t + \sin^2\alpha\cos^2 t)} - \frac{\cos\alpha\cos^2 t}{(\sin^2 t + \cos^2\alpha\cos^2 t)}\right.$$

$$\left. - \frac{\cos\alpha\sin^2 t}{(\sin^2 t + \cos^2\alpha\cos^2 t)} + \frac{\sin\alpha\sin^2 t}{(\sin^2 t + \sin^2\alpha\cos^2 t)}\right]$$

$$\Rightarrow \mathbf{B}(x(t), y(t), z(t)) \cdot \frac{d\mathbf{r}(t)}{dt}$$

$$= \mu_0\lambda\left[\frac{\sin\alpha}{(\sin^2 t + \sin^2\alpha\cos^2 t)} - \frac{\cos\alpha}{(\sin^2 t + \cos^2\alpha\cos^2 t)}\right]$$

$$\text{(III.3.47)}$$

Let us substitute Eq. (III.3.47) into Eq. (III.3.44). Note that the integrand that we deal with is symmetric about $\pi/2$, $\pi$ and $3\pi/2$. This allows to write

$$I_A = 4\lambda\int_0^{\pi/2}\left[\frac{\sin\alpha}{(\sin^2 t + \sin^2\alpha\cos^2 t)} - \frac{\cos\alpha}{(\sin^2 t + \cos^2\alpha\cos^2 t)}\right]dt$$

$$\text{(III.3.48)}$$

We can further rewrite it in the following way:

$$I_A = 4\lambda\int_0^{\pi/2}\left[\frac{\sin\alpha}{\cos^2 t(\tan^2 t + \sin^2\alpha)} - \frac{\cos\alpha}{\cos^2 t(\tan^2 t + \cos^2\alpha)}\right]dt$$

$$\text{(III.3.49)}$$

This allows us to make a substitution $u = \tan t$ so that $du = \sec^2 t\,dt$, so that

$$I_A = 4\lambda\int_0^\infty\left[\frac{\sin\alpha}{(u^2 + \sin^2\alpha)} - \frac{\cos\alpha}{(u^2 + \cos^2\alpha)}\right]du \qquad \text{(III.3.50)}$$

Introducing a rescaled variable $s = u|\sin\alpha|$ (or $s = u|\cos\alpha|$), we get

$$I_A = 4\lambda \left( \frac{\sin\alpha}{|\sin\alpha|} - \frac{\cos\alpha}{|\cos\alpha|} \right) \int_0^\infty ds \frac{1}{s^2 + 1} \qquad \text{(III.3.51)}$$

The remaining integral is just a number and it evaluates to $\pi/2$. This yields the following result:

$$I_A = 2\pi\lambda \left( \frac{\sin\alpha}{|\sin\alpha|} - \frac{\cos\alpha}{|\cos\alpha|} \right) \qquad \text{(III.3.52)}$$

Let's try to make sense of it. The first thing we realize is that the radius of our curve does not matter! Amazing? Well, this suggests that the current is concentrated to run through a single point at the centre of our curve. Now, let's further investigate the behaviour of Eq. (III.3.52). When $\sin\alpha$ and $\cos\alpha$ are of the same sign, the net current is zero. Whereas, when they are of opposite signs $I_A$ is positive or negative.

What does this all mean in terms of $\alpha$? When $\pi/2 < \alpha < \pi$ we have that $I_A = 4\pi\lambda$ and when $3\pi/2 < \alpha < 2\pi$, $I_A = -4\pi\lambda$; it is zero for all other values of $\alpha$. So what's going on? This can only be the case if our bounding curve $C$ has been taken about the intersection of two thin wires each carrying a current $I_A = 2\pi\lambda$ at right angles to each other. At the values of $\alpha$ where we have $I_A = 0$, the currents are flowing in *opposite* directions through the area bounded by our curve. When $I_A \neq 0$ they are flowing in the *same* direction through the area bounded by our curve. For an illustration of this see Fig. III.3.8.

**Example III.3.5 (Physical Application ▶▶).** Let us train ourselves by computing the work done by a force **F** through line integral given in Eq. (III.3.40). We'll consider two trajectories of motion of the same object as a function of time $t$. These can be written in terms of a dimensionless variable $s = t/t_0$

$$\mathbf{r}(s) = s\hat{\mathbf{i}} \quad \text{and} \quad \mathbf{r}(s) = s\hat{\mathbf{i}} + (1 - \cos(2\pi s))\hat{\mathbf{j}} + (s - s^2)\hat{\mathbf{k}}$$

$$\text{where } 0 \leq s \leq 1 \qquad \text{(III.3.53)}$$

They start at the same point of time, $t = 0$, and end at the same point at the same time, $t = t_0$. Their respective paths are shown in Fig. III.3.9. Obviously if the second object, travelling a longer

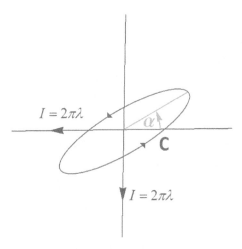

**Fig. III.3.8** Illustration of the situation encountered in Example III.3.4, on applying Ampere's law. The integration contour is a circle tilted an angle $\alpha$. The flux density along the integration contour is produced by two currents, of magnitude $I = 2\pi\lambda$, flowing along the z-axis and y-axis.

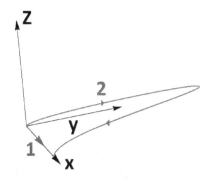

**Fig. III.3.9** Here, we illustrate the two paths given by Eq. (III.3.53) in Example III.3.5. The shorter path described by $\mathbf{r}(s) = s\hat{\mathbf{i}}$ is labelled 1, whereas the other path described by $\mathbf{r}(s) = s\hat{\mathbf{i}} + (1 - \cos(2\pi s))\hat{\mathbf{j}} + (s - s^2)\hat{\mathbf{k}}$ is labelled 2.

trajectory, gets to the same point at the same time as the first object, it moves faster! Let's consider a force acting on each of the moving objects due to air resistance, and see which of the objects has to do more work against that force. Such a force (or laminar flow) can be written as

$$\mathbf{F}_R = -\gamma \frac{d\mathbf{r}}{dt} = -\frac{\gamma}{t_0} \frac{d\mathbf{r}}{ds} \tag{III.3.54}$$

Note that in Eq. (III.3.54) we have used the chain rule $dr/dt = (ds/dt) \cdot (d\mathbf{r}/ds)$, by considering differentiation of each of the components of vector we can see that this works. Since $s = t/t_0$, $ds/dt = 1/t_0$. The force $\mathbf{F}_R$ opposes the direction of motion and its magnitude is proportional to the speed. In this case the work done against this force is given by

$$W = \frac{\gamma}{t_0} \int_0^1 \left(\frac{d\mathbf{r}}{ds}\right) d\mathbf{r} = \frac{\gamma}{t_0} \int_0^1 \left(\frac{d\mathbf{r}}{ds}\right)^2 ds \qquad \text{(III.3.55)}$$

Let's compute this work for the two paths. For the first path we simply have

$$W = \frac{\gamma}{t_0} \int_0^1 ds = \frac{\gamma}{t_0} \qquad \text{(III.3.56)}$$

What about the other path? We have that

$$\frac{d\mathbf{r}(s)}{ds} = \hat{\mathbf{i}} + 2\pi \sin(2\pi s)\hat{\mathbf{j}} + (1 - 2s)\hat{\mathbf{k}}$$

$$\left(\frac{d\mathbf{r}(s)}{ds}\right)^2 = 1 + (2\pi)^2 \sin^2(2\pi s) + (1 - 2s)^2 \qquad \text{(III.3.57)}$$

**Exercise.** Show by using Eq. (III.3.55) that one can obtain

$$W = \frac{\gamma}{t_0} \int_0^1 (1 + (2\pi)^2 \sin^2(2\pi s) + (1 - 2s)^2) ds \qquad \text{(III.3.58)}$$

and, after integration, find the result:

$$W = \frac{\gamma}{t_0} \left(\frac{4}{3} + 2\pi^2\right) \qquad \text{(III.3.59)}$$

**Example III.3.5 (Continued).** What we see is that the work done is not the same; it depends on the path! And naturally we do more work moving quicker around the longer path, as this leads to more air resistance. This is a classic example of what is called a *non-conservative* force; it does not conserve energy, rather it dissipates it. Here, we learn that forces that depend on velocity are dissipative. But there is an exception: when the force is always perpendicular to the direction of motion, as in a magnetic Lorenz force $\mathbf{F} = q\mathbf{v} \times \mathbf{B}$.

Here, this would be not true. **Think:** Why is this the case? **Hint:** Consider if there is any work done by such a force.

## *Conservative vector fields*

If a vector field can be expressed as the gradient of a function $\phi$ (a scalar field) so that $\mathbf{F} = \vec{\nabla}\phi$, Eq. (III.3.39) becomes

$$W = \int_{s_1}^{s_2} \vec{\nabla}\varphi \cdot \frac{d\mathbf{r}}{ds} ds \qquad \text{(III.3.60)}$$

Note that

$$\vec{\nabla}\phi \cdot \frac{d\mathbf{r}}{ds} = \frac{\partial\phi(x,y,z)}{\partial x}\frac{dx}{ds} + \frac{\partial\phi(x,y,z)}{\partial y}\frac{dy}{ds} + \frac{\partial\phi(x,y,z)}{\partial z}\frac{dz}{ds}$$

$$= \frac{d\phi(x(s),y(s),z(s))}{ds} \qquad \text{(III.3.61)}$$

So, in this case we simply get

$$W = \int_{s_1}^{s_2} \frac{d\phi}{ds} ds = \phi(x(s_2),y(s_2),z(s_2)) - \phi(x(s_1),y(s_1),z(s_1))$$

$$= \phi(\mathbf{r}_1) - \phi(\mathbf{r}_2) \qquad \text{(III.3.62)}$$

In this case, the end result depends only on the end points and is independent of what path we take, provided that $\phi(\mathbf{r})$ remains continuous and without singularities. We already talked about such an idea for 2D line integrals, in Chapter 1 of Part II, Vol. 1, but here we have now shown it with vectors in 3D.

**Example III.3.6 (Physical Application ▶▶).** Using line integration, let's test that gravity is a conservative force. For a conservative force the work done does not depend on the path taken, only starting and finishing positions, and any conservative force field can be expressed as the gradient of a potential energy that depends only on position. So let's check this for gravity. Consider two rockets of exactly the same mass taking off from the moon (and we neglect that their mass changes from using up any fuel). Suppose one goes straight up into space and the other goes up at an angle $\alpha$ from the surface of the moon. Both reach a distance $R$ away from the surface,

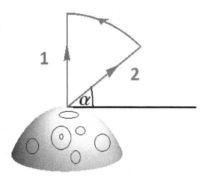

**Fig. III.3.10** Here, we consider two rockets taking off from a point on the moon. One goes straight up a height $R$, while the other one takes off at an angle $\alpha$, again moving the same distance, $r_M + R$ away from the Moon's centre before orbiting around (at fixed $r_M + R$) to meet the first rocket ($r_M$ is the moon radius).

but the second rocket orbits around the moon to meet the first. We draw these two paths in Fig. III.3.10. Let's calculate the work. The force due to gravity on the rockets is

$$\mathbf{F}(r) = G\frac{mM}{r^2}\hat{\mathbf{r}} \tag{III.3.63}$$

where $m$ is the mass of the rocket and $M$ is the mass of the moon; $G$ is the gravitational constant. The unit vector $\hat{\mathbf{r}}$ can be written as

$$\hat{\mathbf{r}} = \frac{1}{\sqrt{x^2 + y^2 + z^2}}(x\hat{\mathbf{i}} + y\hat{\mathbf{j}} + z\hat{\mathbf{k}}) \tag{III.3.64}$$

and so we can write the force as

$$\mathbf{F}(\mathbf{r}) = \frac{GmM}{(x^2 + y^2 + z^2)^{3/2}}(x\hat{\mathbf{i}} + y\hat{\mathbf{j}} + z\hat{\mathbf{k}}) \tag{III.3.65}$$

The path the first rocket takes is $\mathbf{r}_1(s) = (r_M + s)\hat{\mathbf{k}}$ where $0 \le s \le R$. Here, we have

$$\frac{d\mathbf{r}_1(s)}{ds} = \hat{\mathbf{k}} \quad \text{and} \quad \mathbf{F}(\mathbf{r}(s)) = G\frac{mM}{(s + r_M)^2}\hat{\mathbf{k}} \tag{III.3.66}$$

and we have for the work done against gravity

$$W = GMm \int_0^R ds \frac{1}{(r_M + s)^2} = GMm\left(\frac{1}{r_M} - \frac{1}{r_M + R}\right) \tag{III.3.67}$$

Now, let's consider the second rocket. Initially, the path the second rocket takes is

$$\mathbf{r}(s) = s \cos \alpha \hat{\mathbf{j}} + (r_M + s \sin \alpha)\hat{\mathbf{k}} \qquad \text{(III.3.68)}$$

where $0 \le s \le \tilde{R}$, where we need to choose $\tilde{R}$ so that $|\mathbf{r}(\tilde{R})| = R + r_M$ which means that

$$2r_M \tilde{R} \sin \alpha + \tilde{R}^2 = R^2 + 2r_M R \qquad \text{(III.3.69)}$$

The force along this path is

$$\mathbf{F}(\mathbf{r}(s)) = \frac{GmM}{(s^2 + 2r_M s \sin \alpha + r_M^2)^{3/2}} \left( s \cos(\alpha)\, \hat{\mathbf{j}} + (r_M + s \sin \alpha)\, \hat{\mathbf{k}} \right)$$
$$\text{(III.3.70)}$$

and the tangent vector is

$$\frac{d\mathbf{r}(s)}{ds} = \cos(\alpha)\, \hat{\mathbf{j}} + \sin(\alpha)\, \hat{\mathbf{k}} \qquad \text{(III.3.71)}$$

So the integral for the work is

$$W = GMm \int_0^{\tilde{R}} ds \frac{s + r_M \sin \alpha}{(s^2 + 2r_M s \sin \alpha + r_M^2)^{3/2}} \qquad \text{(III.3.72)}$$

We can make the substitution $t^2 = s^2 + 2sr_M \sin \alpha + r_M^2$, and thus obtain

$$W = GMm \int_{r_M}^R dt \frac{1}{t^2} = GMm \left( \frac{1}{r_M} - \frac{1}{r_M + R} \right) \qquad \text{(III.3.73)}$$

The second part of the second rocket's path is

$$\mathbf{r}(s) = \sqrt{r_M + R}(\sin(s_0 - s)\hat{\mathbf{j}} + \cos(s_0 - s)\hat{\mathbf{k}}) \quad 0 \le s \le s_0$$
$$\text{(III.3.74)}$$

where $\tan s_0 = \tilde{R} \cos \alpha / (r_M + \tilde{R} \sin \alpha)$. The tangent vector and the force along this path are

$$\mathbf{t}(s) = \frac{d\mathbf{r}(s)}{ds} = \sqrt{r_M + R}(\cos(s_0 - s)\hat{\mathbf{j}} - \sin(s_0 - s)\hat{\mathbf{k}}) \quad \text{(III.3.75)}$$

$$\mathbf{F}(\mathbf{r}(s)) = \frac{GmM}{(r_M + R)^2}(\sin(s - s_0)\hat{\mathbf{j}} + \cos(s - s_0)\hat{\mathbf{k}}) \qquad \text{(III.3.76)}$$

Using Eqs. (III.3.75) and (III.3.76), it is easy to show that along this path $\mathbf{F}(\mathbf{r}(s)) \cdot \mathbf{t}(s) = 0$ no work is done against gravity along this part of the rocket's trajectory. Thus, indeed, the two work values of the rockets are the same, and the work is independent of path. Therefore, from the previous discussion we should be able to write $\mathbf{F}(\mathbf{r}) = -\vec{\nabla}V(\mathbf{r})$, where $V(\mathbf{r})$ is a function that describes gravitational potential energy:

$$V(\mathbf{r}) = G\frac{mM}{(x^2 + y^2 + z^2)^{1/2}} \qquad \text{(III.3.77)}$$

**Exercise.** Show that $-\vec{\nabla}V(\mathbf{r})$, where $V(\mathbf{r})$ is given by Eq. (III.3.77) indeed yields $\mathbf{F}(\mathbf{r})$ described by Eq. (III.3.65).

### *Flux as the surface integral of a vector field*

The flux as the surface integral of a vector is defined as

$$\Phi = \int_S \mathbf{F}(\mathbf{r}) \cdot d\mathbf{S} = \int_{s_2}^{s_1} \int_{t=t_2(s)}^{t=t_1(s)} \mathbf{F}(x(s,t), y(s,t), z(s,t)) \cdot \hat{\mathbf{n}}(s,t)$$
$$|\mathbf{t}_s(s,t) \times \mathbf{t}_t(s,t)| ds dt \qquad \text{(III.3.78)}$$

Equation (III.3.78) computes $\Phi$, the total flux that goes through the surface (if $\mathbf{F}$ was a current density, $\Phi$ would be the current flowing through the surface). Here, $\hat{\mathbf{n}}(s,t)$ is the unit vector of the surface, and $\mathbf{t}_s(s,t)$ and $\mathbf{t}_t(s,t)$ are the surface tangents, as was described earlier on. We describe such surface integration in Fig. III.3.11.

Again, we can divide the surface up into small elements, and we can look at the vector from the vector field going through each element. To compute the contribution to the flux from this element we are only interested in the component of the vector field parallel to the normal of the surface, which is found by taking the dot product $\mathbf{F} \cdot \hat{\mathbf{n}}$ (if $\mathbf{F}$ is a current density this is the net flow of particles per unit area through the surface at particular point). The contribution of the element to the flux (the total current for flowing particles) is then obtained by multiplying this component by the area of the element $\Delta A$. The total flux, described by Eq. (III.3.78), is then obtained by

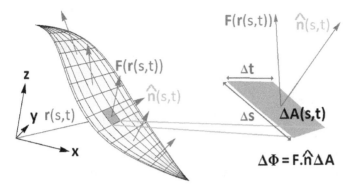

**Fig. III.3.11** This is an illustration of how the surface integration described by Eq. (III.3.78), which computes the flux, works. On the l.h.s we consider a surface described (this one, as it is drawn, might remind the reader of a banana skin) by position vector $\mathbf{r}(s,t)$. The field $\mathbf{F}(\mathbf{r})$ cuts through the surface, with vectors $\mathbf{F}(\mathbf{r}(s,t))$ on the surface, some of which are shown by red arrows. The surface can be divided up into small elements of dimensions $\Delta s$ and $\Delta t$ which form a variable grid as drawn on the surface (in black). Highlighted in green is one element, with a green arrow showing the surface normal at its centre. We magnify this element on the r.h.s. It has area $\Delta A = |\mathbf{t}_s \times \mathbf{t}_s| \Delta s \Delta t$ (roughly for finite size). The flux contribution of the element $\Delta \Phi$ is found by taking the dot product of the field, $\mathbf{F}$ at that point on the surface, with the normal vector and multiplying by the area $\Delta A$. The surface integral described by Eq. (III.3.78) is obtained by adding up all the flux contributions; as well as making each element size infinitesimal, but so increasing their number to an infinite amount to cover the surface that we want to integrate over.

summing up all these contributions, taking $\Delta A \to 0$ and taking the number of elements to infinity to fully cover the surface. The form of Eq. (III.3.78) is almost the same as when we integrated a scalar function over a surface; here the scalar function is the component of the vector field normal to the surface, close to the surface, namely $\mathbf{F}(x(s,t), y(s,t), z(s,t)) \cdot \hat{\mathbf{n}}(s,t)$.

**Example III.3.7 (Physical Application ▶▶).** We'll consider a 'physical example' of using surface integrals to calculate the charge enclosed in a surface from the electric field at its surface. We must confess that it is rather contrived electric field...excuse us, please...chosen for the sake of demonstrating how vector calculus works, but not for any specific physical relevance (it actually does not have any). We just wanted you to practise what we've learnt in applying surface integrals where the field is not necessarily parallel to surface normal.

Gauss's law states that

$$\oint_S E(\mathbf{r}) \cdot d\mathbf{S} = \frac{Q}{\varepsilon_0} \tag{III.3.79}$$

Here, the circle on the integral sign denotes that we integrate over a closed surface, $\mathbf{E}(\mathbf{r})$ is the electric field and $Q$ is the charge contained within that surface. This law is a consequence of the inverse square law for the electric field (one can derive a similar law that relates gravitational field to mass). Let's use Eq. (III.3.79) to work out a particular value of $Q$ contained within a region, given an absolutely crazy form of the electric field—even though we know that Richard Feynman hated such examples, demonstrating only teachers' perverse fantasies. Nevertheless, let's take our region to be a sphere of radius $a$ (this is at least simple!), but conceive of the nightmare field

$$E(\mathbf{r}) = \alpha \left[ x^2 \exp\left(-\kappa\sqrt{x^2 + y^2 + z^2}\right) \hat{\mathbf{i}} + xy \exp\left(-\kappa\sqrt{x^2 + y^2 + z^2}\right) \hat{\mathbf{j}} \right.$$
$$\left. + z\kappa^{-1} \exp\left(-\kappa\sqrt{x^2 + y^2 + z^2}\right) \hat{\mathbf{k}} \right] \tag{III.3.80}$$

As we already know and have used a number of times, the surface of the sphere can be described by

$$\mathbf{r}(s,t) = a \sin s \cos(t) \hat{\mathbf{i}} + a \sin s \sin(t) \hat{\mathbf{j}} + a \cos(s) \hat{\mathbf{k}} \tag{III.3.81}$$

We show how the vector field cuts the sphere in Fig. III.3.12.

It's not hard to show that $|\mathbf{t}_s(s,t) \times \mathbf{t}_t(s,t)| = a^2 \sin s$, from computing the tangent vectors from Eq. (III.3.81)—this is essentially the 3D Jacobian for spherical polar coordinates with $r = a$. In the case of the sphere, as we found out previously, the surface normal is directed along $\mathbf{r}(s,t)$ so that

$$\hat{\mathbf{n}}(s,t) = \sin s \cos(t) \hat{\mathbf{i}} + \sin s \sin(t) \hat{\mathbf{j}} + \cos(s) \hat{\mathbf{k}} \tag{III.3.82}$$

On the surface

$$\mathbf{E}(\mathbf{r}(s,t)) = \alpha a^2 \exp(-\kappa a)\left[ \sin^2 s \cos^2(t) \hat{\mathbf{i}} + \sin^2 s \cos(t) \sin(t) \hat{\mathbf{j}} \right.$$
$$\left. + (a\kappa)^{-1} \cos(s) \hat{\mathbf{k}} \right] \tag{III.3.83}$$

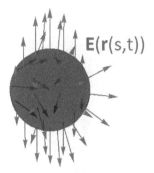

**Fig. III.3.12** We show how the electric field given by Eq. (III.3.80), orientates itself at the surface of the sphere described by Eq. (III.3.81) (with $a\kappa = 1$).

Thus, in this case, the surface integral is given by

$$\oint_S \mathbf{E}(\mathbf{r}) \cdot d\mathbf{S} = a^2 \int_0^\pi \sin s ds \int_0^{2\pi} dt \mathbf{E}(\mathbf{r}(s,t)) \cdot \hat{\mathbf{n}}(s,t)$$

$$= \alpha a^4 \exp(-\kappa a) \int_0^\pi \sin s ds \int_0^{2\pi} [dt \sin^3 s \cos^3 t$$

$$+ \sin^3 s \cos t \sin^2 t + (a\kappa)^{-1} \cos^2 s] \qquad \text{(III.3.84)}$$

$$\Rightarrow \frac{Q}{\varepsilon_0} = \alpha a^4 \exp(-\kappa a) \int_0^\pi \sin s ds \int_0^{2\pi} dt$$

$$\times [\sin^3 s \cos t + (a\kappa)^{-1} \cos^2 s]$$

$$= 2\pi\alpha a^3 \kappa^{-1} \exp(-\kappa a) \int_0^\pi \sin s \cos^2 s ds \qquad \text{(III.3.85)}$$

To evaluate the remaining integral, we make the substitution $t = \cos(s)$ so that

$$\frac{Q}{\varepsilon_0} = 2\pi\alpha a^3 \kappa^{-1} \exp(-\kappa a) \int_{-1}^1 t^2 dt = 2\pi\alpha a^3 \kappa^{-1} \exp(-\kappa a) \left[\frac{t^3}{3}\right]_{-1}^1$$

$$= \frac{4\pi}{3} \alpha a^3 \kappa^{-1} \exp(-\kappa a) \qquad \text{(III.3.86)}$$

**Example III.3.8 (Physical Application ▶▶).** More commonly we use Gauss's law to work the other way and deduce the field from this law, if the system possesses symmetries in space. For instance, for the field from a single uniformly charged surface, the field lines point

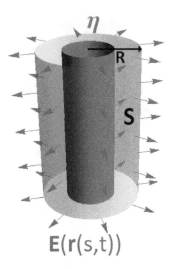

**Fig. III.3.13** Here, we construct a Gaussian surface about a cylinder (red) of uniform charge density, with charge per unit length along the cylinder, $\eta$. We choose the Gaussian surface to be a coaxial cylinder (transparent) of arbitrary radius $R$ about the charged cylinder. The advantage of constructing such a cylinder is that the field lines at the surface lie parallel to the surface normal.

in the direction of the surface normal. If we construct an appropriate surface to utilize Gauss's law, where the field lines also lie along the surface normal, we can deduce the distance behaviour of the dependence of the field. For instance, we can find this dependence about a uniformly charged cylinder (neglecting edge effects) by constructing a cylindrical surface of radius $R$ and height $h$ about that cylinder, where the field lines lie along the normal of the curved part of the cylinder so that $\mathbf{E}(\mathbf{r}) \cdot \hat{\mathbf{n}} = E(R)$. We show this situation in Fig. III.3.13.

The cylindrical surface is described by

$$\mathbf{r}(s,t) = R\cos(s)\hat{\mathbf{i}} + R\sin(s)\hat{\mathbf{j}} + t\hat{\mathbf{k}} \tag{III.3.87}$$

with tangents

$$\mathbf{t}_s(s,t) = \frac{\partial \mathbf{r}(s,t)}{\partial s} = -R\sin(s)\hat{\mathbf{i}} + R\cos(s)\hat{\mathbf{j}},$$

$$\mathbf{t}_t(s,t) = \frac{\partial \mathbf{r}(s,t)}{\partial t} = \hat{\mathbf{k}} \tag{III.3.88}$$

In this case we get $|\mathbf{t}_t(s,t) \times \mathbf{t}_s(s,t)| = R$, so that we have

$$\oint_S \mathbf{E}(\mathbf{r}) \cdot d\mathbf{S} = \int_0^h dt \int_0^{2\pi} RE(R)ds = 2\pi h RE(R) = \frac{Q}{\varepsilon_0} \quad \text{(III.3.89)}$$

where $E(R) = |\mathbf{E}(\mathbf{r})|$. Therefore, rearranging, we get

$$E(R) = \frac{Q}{2\pi h R \varepsilon_0} = \frac{\eta}{2\pi \varepsilon_0 R} \quad \text{(III.3.90)}$$

Noting that vector field $\mathbf{E}(r)$ is parallel to $\hat{\mathbf{n}} = \hat{\mathbf{i}} \cos \phi + \hat{\mathbf{j}} \sin \phi$, we obtain

$$\mathbf{E}(\mathbf{r}) = \frac{\eta}{2\pi \varepsilon_0} \frac{1}{R} (\hat{\mathbf{i}} \cos \phi + \hat{\mathbf{j}} \sin \phi) \quad \text{(III.3.91)}$$

**Exercise.** Consider a sphere of radius $a$ with uniform charge surface density $\sigma$ (charge per unit surface area). By constructing a surface $S$ described as a sphere of radius $r$, concentric with the charged sphere, use Gauss's law (Eq. (III.3.79)) to show that outside the sphere the magnitude of the electric field vector reads as

$$E(r) = \frac{\sigma}{4\pi \varepsilon_0} \left(\frac{a}{r}\right)^2 \quad \text{(III.3.92)}$$

What is the electric (vector) field $\mathbf{E}(\mathbf{r})$ in this case? **Think:** What is the electric field inside the sphere?

## 3.3. Stokes' and Gauss's Divergence Theorems in Action

### Stokes' theorem

Stokes' theorem relates a line integral around a closed path to a surface integral. It states that

$$\oint_C F(\mathbf{r}) \cdot d\mathbf{r} = \int_S \vec{\nabla} \times \mathbf{F}(\mathbf{r}) \cdot d\mathbf{S}$$

$$= \int_{s_1}^{s_2} ds \int_{t=t_1(s)}^{t=t_2(s)} dt \vec{\nabla}$$

$$\times \mathbf{F}(\mathbf{r}(s,t)) \cdot \hat{\mathbf{n}}(s,t) |\mathbf{t}_s(s,t) \times \mathbf{t}_t(s,t)| \quad \text{(III.3.93)}$$

Indeed Stokes' theorem is a 3D vector generalization of Green's theorem. We won't rigorously prove Stokes' theorem, but we can argue that it makes intuitive sense when we think back about our interpretation of $\vec{\nabla} \times \mathbf{F}(\mathbf{r})$ (see Section 3.1). Let us consider the change in $\mathbf{F}(\mathbf{r})$ about a small circle on the surface $S$. This small circle is traced out by $\Delta\mathbf{r}$, the vector from the centre of the circle to its edge, and it is perpendicular to $\hat{\mathbf{b}}$. Let us consider the line integral of $\mathbf{F}(\mathbf{r})$ about such a small circle (for simplicity we'll suppose it lies in the $x{-}y$ plane such that $\hat{\mathbf{n}} = \hat{\mathbf{k}}$):

$$\oint_{C_{\Delta r}} \mathbf{F}(\mathbf{r} + \Delta\mathbf{r}) \cdot d\Delta\mathbf{r} = \int_0^{2\pi} \mathbf{F}(\mathbf{r} + \Delta\mathbf{r}) \cdot \frac{d\Delta\mathbf{r}}{d\theta} d\theta$$

$$\Rightarrow \oint_{C_{\Delta r}} \mathbf{F} \cdot d\Delta\mathbf{r} = \Delta r \int_0^{2\pi} -F_x(x + \Delta r \cos\theta, y + s\Delta r \sin\theta, z) \sin\theta d\theta$$

$$+ \Delta r \int_0^{2\pi} F_y(x + \Delta r \cos\theta, y + \Delta r \sin\theta, z) \cos\theta d\theta \qquad (\text{III.3.94})$$

Here, we have used the fact that for our circle

$$\frac{d\Delta\mathbf{r}}{d\theta} = -\Delta r \sin\theta\hat{\mathbf{i}} + \Delta r \cos\theta\hat{\mathbf{j}} \qquad (\text{III.3.95})$$

and that about the circle we may write $F(\mathbf{r}+\Delta\mathbf{r}) = \mathbf{F}(x+\Delta r \cos\theta, y+ \Delta r \sin\theta, z)$ (also using the definition of a vector field in Eq. (III.3.1)). As the radius of the circle is taken to be very small, we can then use Taylor expansions:

$$F_x(x + \Delta r \cos\theta, y + \Delta r \sin\theta, z)$$

$$\approx F_x(x, y, z) + \frac{\partial F_x}{\partial x}\Delta r \cos\theta + \frac{\partial F_x}{\partial y}\Delta r \sin\theta$$

$$F_y(x + \Delta r \cos\theta, y + \Delta r \sin\theta, z)$$

$$\approx F_y(x, y, z) + \frac{\partial F_y}{\partial x}\Delta r \cos\theta + \frac{\partial F_y}{\partial y}\Delta r \sin\theta \qquad (\text{III.3.96})$$

Then

$$\oint_{C_{\Delta r}} \mathbf{F} \cdot d\Delta \mathbf{r} \approx -\Delta r \int_0^{2\pi} F_x(x, y, z) \sin\theta d\theta$$

$$- \Delta r^2 \int_0^{2\pi} \frac{\partial F_x(x, y, z)}{\partial x} \cos\theta \sin\theta d\theta$$

$$- \Delta r^2 \int_0^{2\pi} \frac{\partial F_x(x, y, z)}{\partial y} \sin^2\theta d\theta$$

$$+ \Delta r \int_0^{2\pi} F_y(x, y, z) \cos\theta d\theta$$

$$+ \Delta r^2 \int_0^{2\pi} \frac{\partial F_y(x, y, z)}{\partial x} \cos^2\theta d\theta$$

$$+ \Delta r^2 \int_0^{2\pi} \frac{\partial F_y(x, y, z)}{\partial y} \cos\theta \sin\theta d\theta \qquad \text{(III.3.97)}$$

On evaluation of the integrals (note that some of them vanish) we obtain

$$\oint_{C_{\Delta r}} \mathbf{F} \cdot d\Delta \mathbf{r} \approx \pi \Delta r^2 \left( \frac{\partial F_y(x, y, z)}{\partial x} - \frac{\partial F_x(x, y, z)}{\partial y} \right)$$

$$= \pi \Delta r^2 \hat{\mathbf{k}} \cdot \vec{\nabla} \times \mathbf{F} \qquad \text{(III.3.98)}$$

Then, for a small circle at any point on the surface, perpendicular to any surface normal vector $\hat{\mathbf{n}}$, this result can be generalized to $\pi \Delta r^2 \hat{\mathbf{n}} \cdot \vec{\nabla} \times \mathbf{F}$. Now, imagine that we can cover our surface with such small circles, as illustrated in Fig. III.3.14.

Note that integration along paths of adjacent circles will (roughly) cancel out, so that we can approximate

$$\oint_C F(\mathbf{r}) \cdot d\mathbf{r} \approx \sum_j \oint_{C_{\Delta r_j}} \mathbf{F}(\mathbf{r}_j + \Delta \mathbf{r}_j) \cdot d\Delta \mathbf{r}_j$$

$$\approx \sum_j \pi \Delta r^2 \hat{\mathbf{n}}_j \cdot \vec{\nabla} \times F(\mathbf{r})|_{\mathbf{r}=\mathbf{r}_j} \qquad \text{(III.3.99)}$$

where in the sum we have circles about the points $\mathbf{r}_j$ all over the surface, and $\Delta \mathbf{r}_j$ are the vectors that trace out the circles from their centres. When we take $\Delta r \to 0$, we should indeed recover Stokes'

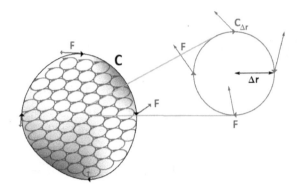

**Fig. III.3.14** An illustration of how Stokes' Theorem works. On the l.h.s we consider an arbitrary surface bounded by the closed curve $C$. The surface can be split up into small circles, shown in blue and red, on which the blue and red arrows show which direction we integrate around the red and blue circles, respectively. Each circle, as illustrated on the right, has a small radius $\Delta r$. Each integration around a small circle roughly (for finite $\Delta r$) evaluates to $\pi \Delta r^2 \hat{n} \cdot \nabla \times \mathbf{F}$ where $\hat{n}$ is the surface normal at the centre of the circle. It is important to realize that the contribution from each circle which is completely surrounded by neighbouring circles is effectively cancelled out by that of its nearest neighbours. The integration path of circles that lie on the edge of the surface then effectively sums to the integration curve $C$, as these are not cancelled by other circles. This suggests that Eq. (III.3.97) holds, and so in the limit $\Delta r \to 0$ that Stokes' Theorem (Eq. (III.3.91)) holds.

theorem, as the r.h.s of this equation will converge to the r.h.s of Eq. (III.3.93).

**Example III.3.9 (Physical Application ▶▶).** Here, we'll reformulate Ampere's law using the Stokes' theorem. Using Stokes' theorem we can write

$$\oint_C \mathbf{B}(\mathbf{r}) \cdot d\mathbf{r} = \int_S \nabla \times \mathbf{B}(\mathbf{r}) \cdot d\mathbf{S} \qquad \text{(III.3.100)}$$

We can also write the current passing through the surface $S$ in terms of the current density:

$$I = \int_S \mathbf{J}(\mathbf{r}) \cdot d\mathbf{S} \qquad \text{(III.3.101)}$$

Thus, combining Eqs. (III.3.100) and (III.3.101) we can rewrite Ampere's law, Eq. (III.3.41) as

$$\vec{\nabla} \times \mathbf{B}(\mathbf{r}) = \mu_0 \mathbf{J}(\mathbf{r}) \qquad \text{(III.3.102)}$$

Let us consider a magnetic field of the form

$$\mathbf{B}(\mathbf{r}) = -\chi(y\sqrt{x^2 + y^2}\,\hat{\mathbf{i}} - x\sqrt{x^2 + y^2}\,\hat{\mathbf{j}}) \qquad \text{(III.3.103)}$$

In this case, the current density can be computed from Eq. (III.3.102) and (III.3.103):

$$\mathbf{J}(\mathbf{r}) = \frac{1}{\mu_0}\vec{\nabla} \times \mathbf{B}(\mathbf{r}) = \frac{\chi}{\mu_0}\left(\frac{\partial}{\partial y}[y\sqrt{x^2 + y^2}] + \frac{\partial}{\partial x}[x\sqrt{x^2 + y^2}]\right)\hat{\mathbf{k}}$$

$$\Rightarrow \mathbf{J}(\mathbf{r}) = \frac{\chi}{\mu_0}\left(2\sqrt{x^2 + y^2} + \frac{x^2 + y^2}{\sqrt{x^2 + y^2}}\right)\hat{\mathbf{k}} = \frac{3\chi\sqrt{x^2 + y^2}}{\mu_0}\hat{\mathbf{k}}$$

$$\text{(III.3.104)}$$

**Exercise.** Around any closed path the electric field satisfies

$$\oint_C \mathbf{E} \cdot d\mathbf{s} = 0 \qquad \text{(III.3.105)}$$

Using $\mathbf{E} = -\vec{\nabla}\phi$, show that Stokes' law given by Eq. (III.3.93) is consistent with this. **Hint:** You may need to consider one of the vector calculus identities that we considered in Section 3.1.

### Gauss's divergence theorem

Another important law is *Gauss's law*, also known as Gauss's divergence theorem. This states that

$$\Phi = \oint_S \mathbf{F}(\mathbf{r}) \cdot d\mathbf{S} = \int_V \vec{\nabla} \cdot \mathbf{F}(\mathbf{r})dV \qquad \text{(III.3.106)}$$

This is more intuitive than Stokes' theorem. The left-hand integral of Eq. (III.3.106) defines the net flux through the surface—the total flux flowing out minus the total flux flowing in. We'll now justify the right-hand integral as being also the total flux.

We can consider the volume enclosed by the surface as divided into many small cubes of volume $\Delta x \Delta y \Delta z$, as shown in Fig. III.3.15.

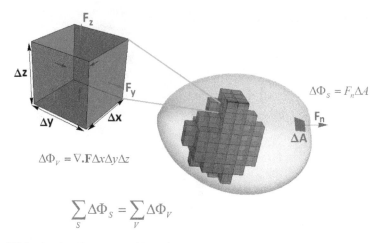

$$\sum_S \Delta\Phi_S = \sum_V \Delta\Phi_V$$

**Fig. III.3.15**   An illustration of how Gauss's theorem works. The total flux flowing out of a closed surface may be calculated by summing the flux $\Delta\Phi_S$ from surface elements of area $\Delta A$. For each element of the flux we have $\Delta\Phi_S = F_n\Delta A$, where $F_n = \hat{n}\cdot\mathbf{F}$ is the component of $\mathbf{F}$ in the direction of the surface normal $\hat{n}$ at the centre of each element. Summing all these elements in the limit $\Delta A \to 0$ yields the surface integral in Eq. (III.3.103) for computing the total flux. Alternately, we may consider the volume enclosed by the surface as being divided up into small cuboids, each with dimensions $\Delta x$, $\Delta y$ and $\Delta z$. The total net flux flowing out of one cuboid (as we saw before) is given by $\Delta\Phi_V \approx \vec{\nabla}\cdot\mathbf{F}\Delta x\Delta y\Delta z$, when the cuboid dimensions are small. Now, importantly, when we consider two adjacent cuboids, the flux flowing out of the face of one of the cuboids cancels that flowing into the face of an adjoining cuboid. Thus, if we add up the flux contributions from a volume of cuboids packed together, we get the net flux flowing out of the total surface area enclosing the volume formed by these cuboids. Thus, we may estimate the flux flowing out of the surface we are interested in by completely filling the volume it encloses with cuboids (not shown in this figure) and summing their flux contributions $\Delta\Phi_V$. As we make the cuboid dimensions smaller, this estimate becomes better, as well as the estimate of the flux contribution of each cuboid $\Delta\Phi_V \approx \vec{\nabla}\cdot\mathbf{F}\Delta x\Delta y\Delta z$. In the limit $\Delta x \to 0$, $\Delta y \to 0$ and $\Delta z \to 0$, (along with $\Delta A \to 0$), and we calculate exactly the flux, $\Phi = \sum_V \Delta\Phi_V = \sum_S \Delta\Phi_S$, which can be written as Eq. (III.3.106).

The net flux flowing through each cube is (as we saw previously in Section 3.1) given by

$$\Delta\Phi_V = \vec{\nabla}\cdot\mathbf{F}(\mathbf{r})\Delta x\Delta y\Delta z \qquad\text{(III.3.107)}$$

The flux flowing through the volume which is the sum of the cubes is given by

$$\Phi = \sum_j \vec{\nabla}\cdot\mathbf{F}(\mathbf{r}_j)\Delta x\Delta y\Delta z \qquad\text{(III.3.108)}$$

In the limit where $\Delta x \to 0$, $\Delta y \to 0$ and $\Delta z \to 0$, this gives the r.h.s of Eq. (III.3.106). This is as it should be, because it is the same as the total flux flowing through the surface, and is represented by the surface integral in Eq. (III.3.106).

**Example III.3.10 (Physical Application ▶▶).** Gauss's law for a magnetic field is given by

$$\oint_S \mathbf{B} \cdot d\mathbf{S} = 0 \tag{III.3.109}$$

(provided there are no Dirac monopoles—no-one has ever found one!)

This states that, for any magnetic field, the amount of magnetic flux entering any closed surface is the same as the amount of magnetic flux leaving it—for every south-pole there is a north-pole that cannot be separated from it. The Gauss divergence theorem (Eq. (III.3.106)) states that

$$\oint_S \mathbf{B} \cdot d\mathbf{S} = \oint_V \vec{\nabla} \cdot \mathbf{B} dV = 0 \tag{III.3.110}$$

This implies that always $\vec{\nabla} \cdot \mathbf{B} = 0$. To satisfy this this condition, it is possible to rewrite the latter in terms of another vector field $\mathbf{A}$ such that

$$\mathbf{B} = \vec{\nabla} \times \mathbf{A} \text{ since always } \vec{\nabla} \cdot (\vec{\nabla} \times \mathbf{A}) = 0 \tag{III.3.111}$$

The field $\mathbf{A}$ is called the vector potential and it is very useful in describing magnetism. For instance, we can write (as we saw from Example III.3.9)

$$\vec{\nabla} \times \mathbf{B} = \vec{\nabla} \times \vec{\nabla} \times \mathbf{A} = \mu_0 \mathbf{J} \tag{III.3.112}$$

We can re-express Eq. (III.3.112), using one of the vector calculus identities given by Eq. (III.3.35):

$$\vec{\nabla}^2 \mathbf{A} - \vec{\nabla}(\vec{\nabla} \cdot \mathbf{A}) = \mu_0 \mathbf{J} \tag{III.3.113}$$

The choice of $\mathbf{A}$ is not unique. Indeed, $\mathbf{A}'(\mathbf{r}) = \mathbf{A}(\mathbf{r}) + \vec{\nabla}\chi(\mathbf{r})$ will give the same $\mathbf{B}$-field, since we have that $\vec{\nabla} \times \vec{\nabla}\chi(\mathbf{r}) = 0$. The particular choice of the function $\chi(\mathbf{r})$ is commonly referred to as a *gauge*. The fact that the $\mathbf{B}$-field remains unchanged with respect to any gauge choice is, in fact, due to a symmetry that results in a very important law: the conservation of charge. Such symmetries provide a foundation for the fundamental theories of nature.

For a current density that *is not changing* with time, it is convenient to choose a gauge in which $\vec{\nabla} \cdot \mathbf{A} = 0$. It is convenient because we also have $\vec{\nabla} \cdot \mathbf{J} = 0$, for the electrical current density (from the conservation of charge and electro-neutrality in conductors). For a simple case, we can choose $\mathbf{J} = J_z(x, y)\hat{\mathbf{k}}$ and $\mathbf{A} = A_z(x, y)\hat{\mathbf{k}}$ (which indeed both satisfy $\vec{\nabla} \cdot \mathbf{J} = 0$ and $\vec{\nabla} \cdot \mathbf{A} = 0$). In this case we simply obtain from Eq. (III.3.113) the following result:

$$\frac{\partial^2 A_z(x, y)}{\partial x^2} + \frac{\partial^2 A_z(x, y)}{\partial y^2} = \mu_0 J_z(x, y)$$

$$\text{with} \quad \mathbf{B}(x, y) = \frac{\partial A_z(x, y)}{\partial y}\hat{\mathbf{i}} - \frac{\partial A_z(x, y)}{\partial x}\hat{\mathbf{j}} \tag{III.3.114}$$

The l.h.s of Eq. (III.3.114) is our first example of a partial differential equation, which we may solve to find $A_z(x, y)$, for a given $J_z(x, y)$, and so find $\mathbf{B}(x, y)$.

**Example III.3.11 (Physical Application ▶▶).** We can also apply the Gauss divergence theorem to electric fields. Using Eq. (III.3.106) we can write

$$\int_V \vec{\nabla} \cdot \mathbf{E}(\mathbf{r}) d\mathbf{r} = \frac{Q}{\varepsilon_0} = \frac{1}{\varepsilon_0} \int_V \rho(\mathbf{r}) d\mathbf{r} \tag{III.3.115}$$

where $\rho(\mathbf{r})$ is the charge density. Thus, we obtain

$$\vec{\nabla} \cdot \mathbf{E}(\mathbf{r}) = \frac{\rho(\mathbf{r})}{\varepsilon_0} \tag{III.3.116}$$

This is the famous Poisson's equation. A particularly convenient way of recasting it is to write the electric field in terms of the electrostatic potential so that $\mathbf{E}(\mathbf{r}) = -\vec{\nabla}\varphi(\mathbf{r})$. In terms of the potential this yields

$$-\vec{\nabla}^2\phi(\mathbf{r}) = \frac{\rho(\mathbf{r})}{\varepsilon_0} \tag{III.3.117}$$

Note that we have a similar equation for gravity where

$$-\nabla^2 V_g(\mathbf{r}) = 4\pi G \rho_M(\mathbf{r}) \qquad \text{(III.3.118)}$$

where $V_g(\mathbf{r})$ is the gravitational potential, $G$ is the gravitational constant, and $\rho_M(\mathbf{r})$ is the mass density. Given a charge density $\rho(\mathbf{r})$, or mass density $\rho_M(\mathbf{r})$, we can solve either the partial differential equations Eq. (III.3.117) or (III.3.118) to obtain the electric or gravitational field, respectively.

In these last three chapters, thanks to your patience, we have come a long way. We have learnt how to describe surfaces in terms vectors and their curvature, how to integrate over surfaces. We have learnt about the differential operators of vector calculus, extending our concept of the line integral to vectors and in 3D, Stokes' and Gauss's theorems, and finally we came across partial differential equations. The solution of these we'll start dealing with in Chapter 6 of Part III of Vol. 2. In the next chapter we move on to learning about differential eigenvalue equations, requisite for solving partial differential equations.

## 3.4.  Key Points

- A 3D vector field is defined as $\mathbf{F}(\mathbf{r}) = F_x(\mathbf{r})\hat{\mathbf{i}} + F_y(\mathbf{r})\hat{\mathbf{j}} + F_z(\mathbf{r})\hat{\mathbf{k}}$, where the components $F_x(\mathbf{r}) = F_x(x, y, z)$, $F_y(\mathbf{r}) = F_y(x, y, z)$ and $F_z(\mathbf{r}) = F_z(x, y, z)$ are functions of the three coordinates $x, y,$ and $z$.
- We may take the divergence of a vector field by taking the dot product of the gradient operator $\vec{\nabla} = \hat{\mathbf{i}}\partial/\partial x + \hat{\mathbf{j}}\partial/\partial y + \hat{\mathbf{k}}\partial/\partial z$ with it, such that $\vec{\nabla} \cdot \mathbf{F}(\mathbf{r}) = \partial F_x/\partial x + \partial F_y/\partial y + \partial F_z/\partial z$.
- $\Delta\Phi = \vec{\nabla} \cdot \mathbf{F}(\mathbf{r})\Delta x\Delta y\Delta z$ is the net flux flowing out of a cuboid with infinitesimal dimensions. Physically, flux may be a current $I$ (number of particles flowing through a surface per second), where $\mathbf{F}(\mathbf{r}) = \mathbf{J}(\mathbf{r})$ is the current density.
- If a current represents matter which is not converted or destroyed, the continuity equation $\partial c(\mathbf{r}; t)/\partial t = \vec{\nabla} \cdot \mathbf{J}(\mathbf{r}; t)$ applies, where $c(\mathbf{r}; t) = c(x, y, z, t)$ is the concentration at a point $\mathbf{r}$, and $\mathbf{J}(r; t)$ is

the current density (which may change with time $t$) flowing into and out of that point.

- We may take the curl of a vector field by taking its cross product with the gradient operator $\vec{\nabla} = \hat{\mathbf{i}}\partial/\partial x + \hat{\mathbf{j}}\partial/\partial y + \hat{\mathbf{k}}\partial/\partial z$ such that $\vec{\nabla} \times \mathbf{F} = (\partial F_z/\partial y - \partial F_y/\partial z)\hat{\mathbf{i}} + (\partial F_x/\partial z - \partial F_z/\partial x)\hat{\mathbf{j}} + (\partial F_y/\partial x - \partial F_x/\partial y)\hat{\mathbf{k}}$.

- The interpretation of the curl is less straightforward than the divergence. The $x, y$ and $z$ components of the curl measure the amount of circulation (or vorticity) of the vector field around infinitesimal circles of radius $\Delta r$ about the $x, y$ and $z$ axes respectively.

- We may also take the tensorial product of the gradient operator $\vec{\nabla} = \hat{\mathbf{i}}\partial/\partial x + \hat{\mathbf{j}}\partial/\partial y + \hat{\mathbf{k}}\partial/\partial z$ with a vector field; namely, $\vec{\nabla} \otimes \mathbf{F}(\mathbf{r})$. This can be represented by a matrix as given by Eq. (III.3.31).

- We can apply a Laplacian operator $\vec{\nabla}^2 = \partial^2/\partial x^2 + \partial^2/\partial y^2 + \partial^2/\partial z^2$ to a vector field such that $\vec{\nabla}^2 \mathbf{F}(\mathbf{r}) = \vec{\nabla}^2 F_x(\mathbf{r})\hat{\mathbf{i}} + \vec{\nabla}^2 F_y(\mathbf{r})\hat{\mathbf{j}} + \vec{\nabla}^2 F_y(\mathbf{r})\hat{\mathbf{k}}$. This is equivalent to $\vec{\nabla} \cdot (\vec{\nabla} \otimes \mathbf{F}(\mathbf{r}))$.

- We can extend the concept of line integration into 3D using vector fields, as given by Eq. (III.3.37). Such line integrals can be evaluated by the use of Eq. (III.3.39).

- If we can write $\mathbf{F}(\mathbf{r}) = \vec{\nabla}\phi(\mathbf{r})$, the value of the line integral of $\mathbf{F}(\mathbf{r})$ does not depend on the path taken between two points $\mathbf{r}_1$ and $\mathbf{r}_2$, and it evaluates to $\phi(\mathbf{r}_2) - \phi(\mathbf{r}_1)$. In this case $\mathbf{F}(\mathbf{r})$ is called a conservative field.

- Using the surface integral defined by Eq. (III.3.78) we may compute the net flux over a surface $S$ from a vector field $\mathbf{F}(\mathbf{r})$.

- In electrostatics, the net flux of the electrostatic field (flux units: $Vm^{-1} \times m^2$) through a closed surface is always the charge it encloses divided by the vacuum dielectric constant $\varepsilon_0$. This is known as Gauss's law (Eq. (III.3.79)).

- Two useful theorems in vector integral calculus are Stokes' and Gauss's theorem. The former relates the line integral of a closed curve $C$ bounding a surface $S$, to a surface integral over that surface, given by Eq. (III.3.93) (this can be thought of a generalization of Green's theorem to surfaces which are not flat). The latter relates the surface integral over a closed surface $S$ to a volume integral over the volume $V$ enclosing that surface, given by Eq. (III.3.106).

- Using Gauss's theorem we may obtain Poisson's equation for electrostatic potential, namely $-\vec{\nabla}^2\varphi(\mathbf{r}) = \rho(\mathbf{r})/\varepsilon_0$. This is an example of a partial differential equation, the solution to which we will consider later.

# Chapter 4

# Differential Eigenvalue Equations and Special Functions

## 4.1. Differential Equations Handled as Matrix Equations: Sturm–Liouville Theory Explained and Unleashed

*Homogeneous differential equations in the form of eigenvalue equations*

Let's consider first homogeneous linear second order differential equations on a function of one variable, $f(x)$, defined in some domain $a \leq x \leq b$. Such equations take the form $r(x)f''(x) + q(x)f'(x) + \tilde{p}(x)f(x) = 0$, where $f'(x)$ and $f''(x)$ are the first and second derivatives with respect to $x$, and $r$, $q$, and $\tilde{p}$ are $x$-dependent coefficients. Such equations can be presented in the form of an eigenvalue equation (see Part II, Vol. 1). Indeed, we are free to introduce $p(x) = \tilde{p}(x) + E$, and then rewrite our equation as

$$p(x)f(x) + q(x)\frac{df(x)}{dx} + r(x)\frac{d^2 f(x)}{dx^2} = \hat{H}f(x) = Ef(x) \quad \text{(III.4.1)}$$

where $\hat{H}$ denoted a *differential operator*:

$$\hat{H} = p(x) + q(x)\frac{d}{dx} + r(x)\frac{d^2}{dx^2} \quad \text{(III.4.2)}$$

In Eq. (III.4.1), $E$ becomes the eigenvalue of a particular differential operator $\hat{H}$ (we'll find out what this means below). We should stress that $p(x)$, $q(x)$ and $r(x)$ can generally be any functions of $x$. Any homologous linear second order differential equation will satisfy Eq. (III.4.1). Writing such equations in an eigenvalue form is particularly useful for their solutions, as we shall later see, as well as being a standard for applications in quantum mechanics. Setting this approach to solving homogeneous equations, it will be easy to proceed to solving linear inhomogeneous equations of the form $\hat{H}f(x) = g(x)$.

We warn that the material that follows, boxed as being *optional*, is rather technical, albeit still by no means rigorous (this would have made things far worse...we're not in the business of providing you with rigorous, too technical, proofs). But we think that grasping its main ideas will be insightful for understanding a lot of what will be going on in this book. You may want to skip this in the first reading of Part III, and move on to the later, more practical, sections. But in any case, before tackling this, the readers should be confident in matrix/linear algebra, eigenvalue equations and diagonalization procedure (as presented in Chapter 4, Part II, Vol. 1).

---

*Optional: Linear homogeneous differential equations as matrix equations.* To simplify things in what follows, we'll for the moment consider $a = 0$ so that $f(x)$ is defined in the domain $0 \leq x \leq b$. Now, Eq. (III.4.1) is, actually, the same as the matrix eigenvalue equations that were encountered in Part II, Vol. 1, if we were to consider the function $f(x)$ as a vector and represent $\hat{H}$ as matrix. You may shout, what a nonsense, how can we represent a scalar function as a vector? But be patient, and here is what we mean.

First, let's divide the $x$-axis into a series of discrete points $x_j = j\varepsilon$, where $\varepsilon$ is a spacing between points that we take to be infinitesimally small, to form a continuum (see Fig. III.4.1). At each of these points we can define a value of the function $f_j = f(x_j)$, and then present this information in a form of an infinite column vector

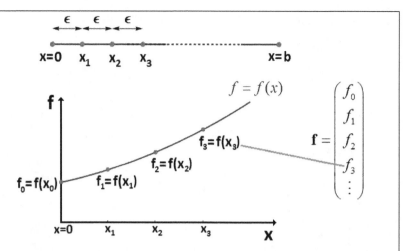

**Fig. III.4.1** Here, we show how a function $f(x)$, here defined in the domain $0 < x \leq b$ can be represented by a matrix. We divide the input variable into a series of points $\{0, x_1, x_2 + \cdots\}$ with a separation between adjacent points of $\varepsilon$, such that $x_j = j\varepsilon$. We then evaluate the function at each of these points $f_j = f(x_j)$. Each of these values forms entries of the vector, as shown in the figure.

$$\mathbf{f} = \begin{pmatrix} f_0 \\ f_1 \\ f_2 \\ f_3 \\ \vdots \end{pmatrix} \tag{III.4.3}$$

Also, we can similarly define values $p_j = p(x_j)$, and of course we can write $p_j f_j = p(x_j) f(x_j)$. Therefore, the operation of multiplying a function $f(x)$ by another function $p(x)$ can be viewed as multiplying Eq. (III.4.3) by a diagonal matrix such that

$$\mathbf{p.f} = \begin{pmatrix} p_0 & 0 & 0 & 0 & \cdots \\ 0 & p_1 & 0 & 0 & \cdots \\ 0 & 0 & p_2 & 0 & \cdots \\ 0 & 0 & 0 & p_3 & \cdots \\ \vdots & \vdots & \vdots & \vdots & \ddots \end{pmatrix} \begin{pmatrix} f_0 \\ f_1 \\ f_2 \\ f_3 \\ \vdots \end{pmatrix} = \begin{pmatrix} p_0 f_0 \\ p_1 f_1 \\ p_2 f_2 \\ p_3 f_3 \\ \vdots \end{pmatrix} \tag{III.4.4}$$

Now, what about differentiation? Can we represent this operation as a matrix? Yes, we can. First we can write

$$f'(x_i) = \frac{df(x_i)}{dx_i} = \frac{f(x_j + \varepsilon) - f(x_j - \varepsilon)}{2\varepsilon} = \frac{f_{j+1} - f_{j-1}}{2\varepsilon} \quad \text{(III.4.5)}$$

The function $f'(x_i)$ can also be represented as a vector, so we can represent the operation of differentiation as multiplying Eq. (III.4.3) with a matrix $\hat{\mathbf{D}}$ such that

$$\hat{\mathbf{D}}.\mathbf{f} = \begin{pmatrix} 0 & \dfrac{1}{2\varepsilon} & 0 & 0 & \cdots \\ -\dfrac{1}{2\varepsilon} & 0 & \dfrac{1}{2\varepsilon} & 0 & \cdots \\ 0 & -\dfrac{1}{2\varepsilon} & 0 & \dfrac{1}{2\varepsilon} & \cdots \\ 0 & 0 & -\dfrac{1}{2\varepsilon} & 0 & \cdots \\ \vdots & \vdots & \vdots & \vdots & \ddots \end{pmatrix} \begin{pmatrix} f_0 \\ f_1 \\ f_2 \\ f_3 \\ \vdots \end{pmatrix} = \begin{pmatrix} f_1/2\varepsilon \\ (f_2 - f_0)/2\varepsilon \\ (f_3 - f_1)/2\varepsilon \\ (f_4 - f_2)/2\varepsilon \\ \vdots \end{pmatrix}$$

$$\text{(III.4.6)}$$

The second derivative of a function is arrived at by simply considering the product $\hat{\mathbf{D}}.\hat{\mathbf{D}}.\mathbf{f}$. Thus, we can represent Eq. (III.4.1) as matrix equation

$$[\mathbf{p} + \mathbf{q}.\hat{\mathbf{D}} + \mathbf{r}.\hat{\mathbf{D}}.\hat{\mathbf{D}}]\mathbf{f} = \mathbf{Hf} = E\mathbf{f} \quad \text{(III.4.7)}$$

Here, $\mathbf{q}$ and $\mathbf{r}$ are given by the diagonal matrices

$$\mathbf{q} = \begin{pmatrix} q_0 & 0 & 0 & 0 & \cdots \\ 0 & q_1 & 0 & 0 & \cdots \\ 0 & 0 & q_2 & 0 & \cdots \\ 0 & 0 & 0 & q_3 & \cdots \\ \vdots & \vdots & \vdots & \vdots & \ddots \end{pmatrix} \qquad \mathbf{r} = \begin{pmatrix} r_0 & 0 & 0 & 0 & \cdots \\ 0 & r_1 & 0 & 0 & \cdots \\ 0 & 0 & r_2 & 0 & \cdots \\ 0 & 0 & 0 & r_3 & \cdots \\ \vdots & \vdots & \vdots & \vdots & \ddots \end{pmatrix}$$

$$\text{(III.4.8)}$$

where $q_j = q(x_j)$ and $r_j = r(x_j)$; and **H** is the matrix specific to the differential operator $\hat{H}$. Indeed, the reader should recognize Eq. (III.4.7), as a matrix eigenvalue equation, which we came across in Chapter 3 of Part II (Vol. 1) Thus, the functions that solve Eq. (III.4.1) are then represented by the eigenvectors of Eq. (III.4.7).

In fact such an eigenvalue equation, for a second order differential equation, is doubly degenerate; i.e. two eigenvectors/functions yield the same eigenvalue—we won't prove this. Notwithstanding, the reader should be aware that two different functions should solve Eq. (III.4.1) (with $E$ fixed), as it is a second order differential equation, and the general solution should be written as a linear combination of them (recall our experience of solving second order differential equations in Chapter 6 of Part II (Vol. 1)). We'll write these two degenerate eigenvectors as $\mathbf{f}_k^{(1)}$ and $\mathbf{f}_k^{(2)}$, which yield a specific eigenvalue $E_k$, where $k = 1, 2, 3 \ldots$. Thus, for the eigenvalue equations, we may write

$$\mathbf{H}.\mathbf{f}_k^{(1)} = \mathbf{H} \begin{pmatrix} f_{0,k}^{(1)} \\ f_{1,k}^{(1)} \\ f_{2,k}^{(1)} \\ f_{3,k}^{(1)} \\ \vdots \end{pmatrix} = E_k \begin{pmatrix} f_{0,k}^{(1)} \\ f_{1,k}^{(1)} \\ f_{2,k}^{(1)} \\ f_{3,k}^{(1)} \\ \vdots \end{pmatrix} = E_k \mathbf{f}_k^{(1)} \qquad \text{(III.4.9)}$$

$$\mathbf{H}.\mathbf{f}_k^{(2)} = \mathbf{H} \begin{pmatrix} f_{0,k}^{(2)} \\ f_{1,k}^{(2)} \\ f_{2,k}^{(2)} \\ f_{3,k}^{(2)} \\ \vdots \end{pmatrix} = E_k \begin{pmatrix} f_{0,k}^{(2)} \\ f_{1,k}^{(2)} \\ f_{2,k}^{(2)} \\ f_{3,k}^{(2)} \\ \vdots \end{pmatrix} = E_k \mathbf{f}_k^{(2)} \qquad \text{(III.4.10)}$$

Obviously, there should be an infinite number of eigenvalues (though these can be repeated), as the matrix **H** is infinite.

## *The diagonalization of the matrix* **H**

Now, importantly, we can diagonalize the matrix **H** so that (as we learnt from Chapter 4 of Part I, Vol. 1)

$$\mathbf{R}^{-1}\mathbf{H}\mathbf{R} = \mathbf{H}_D \quad \text{and} \quad \mathbf{R}\mathbf{H}_D\mathbf{R}^{-1} = \mathbf{H} \tag{III.4.11}$$

As before, the diagonalizing matrix can be constructed from the eigenvectors in the same way. In this case, we can write for the matrix **R**:

$$\mathbf{R} = (\dots \mathbf{f}_3^{(1)}, \mathbf{f}_2^{(1)}, \mathbf{f}_1^{(1)}, \mathbf{f}_1^{(2)}, \mathbf{f}_2^{(2)}, \mathbf{f}_3^{(2)} \dots)$$

$$= \begin{pmatrix} \cdots & f_{0,3}^{(1)} & f_{0,2}^{(1)} & f_{0,1}^{(1)} & f_{0,1}^{(2)} & f_{0,2}^{(2)} & \cdots \\ \cdots & f_{1,3}^{(1)} & f_{1,2}^{(1)} & f_{1,1}^{(1)} & f_{1,1}^{(2)} & f_{1,2}^{(2)} & \cdots \\ \cdots & f_{2,3}^{(1)} & f_{2,2}^{(1)} & f_{2,1}^{(1)} & f_{2,1}^{(2)} & f_{2,2}^{(2)} & \cdots \\ \cdots & f_{3.3}^{(1)} & f_{3,2}^{(1)} & f_{3,1}^{(1)} & f_{3,1}^{(2)} & f_{3,2}^{(2)} & \cdots \\ \cdot\cdot & \vdots & \vdots & \ddots & \vdots & \vdots & \cdot\cdot \end{pmatrix} \tag{III.4.12}$$

Then, the diagonal matrix **H**$_D$ is given by

$$\mathbf{H}_D = \begin{pmatrix} \ddots & \vdots & \vdots & \vdots & \vdots & \vdots & \cdot\cdot \\ \cdots & E_3 & 0 & 0 & 0 & 0 & \cdots \\ \cdots & 0 & E_2 & 0 & 0 & 0 & \cdots \\ \cdots & 0 & 0 & E_1 & 0 & 0 & \cdots \\ \cdots & 0 & 0 & 0 & E_1 & 0 & \cdots \\ \cdots & 0 & 0 & 0 & 0 & E_2 & \cdots \\ \cdot\cdot & \vdots & \vdots & \vdots & \vdots & \vdots & \ddots \end{pmatrix} \tag{III.4.13}$$

Note that we are free to arrange the eigenvectors in any way we choose, in constructing **R**, provided that we order the entries in **H**$_D$ in the same way: each eigenvalue appearing in the same column position as its corresponding eigenvector.

## The inverse of the matrix H

We can express the inverse matrix of $\mathbf{H}$ as

$$\mathbf{H}^{-1} = \mathbf{R}\mathbf{H}_D^{-1}\mathbf{R}^{-1} \tag{III.4.14}$$

and $\mathbf{H}_D^{-1}$ is given as

$$\mathbf{H}_D^{-1} = \begin{pmatrix} \ddots & \vdots & \vdots & \vdots & \vdots & \vdots & \iddots \\ \cdots & 1/E_3 & 0 & 0 & 0 & 0 & \cdots \\ \cdots & 0 & 1/E_2 & 0 & 0 & 0 & \cdots \\ \cdots & 0 & 0 & 1/E_1 & 0 & 0 & \cdots \\ \cdots & 0 & 0 & 0 & 1/E_1 & 0 & \cdots \\ \cdots & 0 & 0 & 0 & 0 & 1/E_2 & \cdots \\ \iddots & \vdots & \vdots & \vdots & \vdots & \vdots & \ddots \end{pmatrix} \tag{III.4.15}$$

You can check this by multiplying it with $\mathbf{H}$ given by Eq. (III.4.11), with $\mathbf{R}.\mathbf{R}^{-1} = \mathbf{I}$, the identity matrix. As we'll see below, this inverse matrix actually represents Green's function we first encountered in Chapter 6 of Part II (Vol. 1).

## Linear inhomogeneous differential equations as matrix equations and their solution

Now, suppose that we want to solve the *inhomogeneous* equation, i.e. the one in which the r.h.s does not depend on $f$:

$$p(x)f(x) + q(x)\frac{df(x)}{dx} + r(x)\frac{d^2f(x)}{dx^2} = \hat{H}f(x) = g(x) \tag{III.4.16}$$

This is equivalent to solving the matrix equation

$$[\mathbf{p} + \mathbf{q}.\hat{\mathbf{D}} + \mathbf{r}.\hat{\mathbf{D}}.\hat{\mathbf{D}}]\mathbf{f} = \mathbf{H}\mathbf{f} = \mathbf{g} \tag{III.4.17}$$

where $\mathbf{g}$ is the vector representation of the function $g(x)$.
Simply, the formal solution to Eq. (III.4.17) (and thus Eq. (III.4.16)) can be written in the matrix form as

$$\mathbf{f} = \mathbf{H}^{-1}\mathbf{g} = \mathbf{R}\mathbf{H}_D^{-1}\mathbf{R}^{-1}\mathbf{g} \tag{III.4.18}$$

To illustrate all of what we discussed so far, we'll now consider an example.

**Example III.4.1 (Maths Practice ♫♪).** Let's consider a eigenvalue-differential equation with constant coefficients such that

$$-b\frac{d^2 f}{dx^2} + ic\frac{df}{dx} = Ef \qquad \text{(III.4.19)}$$

Here, we'll consider the eigenfunctions $f$ are defined in the range $-\infty \leq x \leq \infty$ and no specific boundary conditions apply. The eigenfunctions for this equation are very simple:

$$f_k(x) = \exp(-ikx) \qquad \text{(III.4.20)}$$

where we'll consider $k$ as being real, as well as $E$. Substituting Eq. (III.4.20) into Eq. (III.4.19) we obtain an quadratic equation on the eigenvalues:

$$bk^2 + ck = E \qquad \text{(III.4.21)}$$

We note that $k$ can be negative or positive and there are two values of $k$ for every value of $E$:

$$k_\pm = \frac{-c \pm \sqrt{c^2 + 4bE}}{2b} \qquad \text{(III.4.22)}$$

Thus, for every eigenvalue $E$ we have two functions $f_k^+(x) = \exp(-ik_+x)$ and $f_k^-(x) = \exp(-ik_-x)$. Note that to satisfy Eq. (III.4.19) we can choose $k$ to be any (real) value, and therefore the values of $E$ form a continuous range.

It's worth asking what the matrix **R** is in this case. To do this we'll split $k$ into discrete values $k = k_l = l\Delta$ (each value of $l$, an integer, corresponding to a choice of $E$); but with $\Delta$ chosen to be infinitesimally small. We consider the range values from $l = -\infty$ to $l = \infty$, so that for all the $k$ values considered, we include both the $k_-$ and $k_+$ eigenfunctions. Strictly speaking, when we come to consider limiting cases, $x$ should extend from $-\pi/\Delta$ to $\pi/\Delta$ when we divide up $k$ into discrete values. Again, we may divide $x$ into discrete points such that $x = x_j = j\varepsilon$, where again $\varepsilon$ is infinitesimally small, and $j$

runs from $j = -\pi/(\Delta\varepsilon)$ to $j = \pi/(\Delta\varepsilon)$. The eigenvector for the value $k_l$ is then given by

$$\mathbf{f}_{k_l} = \frac{\Delta}{2\pi} \begin{pmatrix} \vdots \\ \exp(ix_2 k_l) \\ \exp(ix_1 k_l) \\ 1 \\ \exp(-ix_1 k_l) \\ \exp(-ix_2 k_l) \\ \vdots \end{pmatrix} = \frac{\Delta}{2\pi} \begin{pmatrix} \vdots \\ \exp(2i\Delta\varepsilon l) \\ \exp(i\Delta\varepsilon l) \\ 1 \\ \exp(-i\Delta\varepsilon l) \\ \exp(-2i\Delta\varepsilon l) \\ \vdots \end{pmatrix} \tag{III.4.23}$$

In this case, the matrix $\mathbf{R}$ is constructed from Eq. (III.4.23), to give

$$\mathbf{R} = \frac{\Delta}{2\pi} \begin{pmatrix} \ddots & \vdots & \vdots & \vdots & \vdots & \vdots & \iddots \\ \cdots & \exp(ix_2 k_2) & exp(ix_2 k_1) & 1 & \exp(-ix_2 k_1) & \exp(-ix_2 k_2) & \cdots \\ \cdots & \exp(ix_1 k_2) & \exp(ix_1 k_1) & 1 & \exp(-ix_1 k_1) & \exp(-ix_1 k_2) & \cdots \\ \cdots & 1 & 1 & 1 & 1 & 1 & \cdots \\ \cdots & \exp(-ix_1 k_2) & \exp(-ix_1 k_1) & 1 & \exp(ix_1 k_1) & \exp(ix_1 k_2) & \cdots \\ \cdots & \exp(-ix_2 k_2) & \exp(-ix_2 k_2) & 1 & \exp(ix_2 k_1) & \exp(ix_2 k_2) & \cdots \\ \iddots & \vdots & \vdots & \vdots & \vdots & \vdots & \ddots \end{pmatrix}$$

$$\tag{III.4.24}$$

Now, $\mathbf{R}^{-1} = (2\pi\varepsilon/\Delta)\mathbf{R}^\dagger$ and $\mathbf{H}_D$ is given by

$$\mathbf{H}_D = \begin{pmatrix} \ddots & \vdots & \vdots & \vdots & \vdots & \vdots & \iddots \\ \cdots & bk_2^2 - ck_2 & 0 & 0 & 0 & 0 & \cdots \\ \cdots & 0 & bk_1^2 - ck_1 & 0 & 0 & 0 & \cdots \\ \cdots & 0 & 0 & 0 & 0 & 0 & \cdots \\ \cdots & 0 & 0 & 0 & bk_1^2 + ck_1 & 0 & \cdots \\ \cdots & 0 & 0 & 0 & 0 & bk_2^2 + ck_2 & \cdots \\ \iddots & \vdots & \vdots & \vdots & \vdots & \vdots & \ddots \end{pmatrix}$$

$$\tag{III.4.25}$$

Note that the form of the inverse matrix can be deduced from the fact that

$$\lim_{\substack{\Delta \to 0 \\ \varepsilon \to 0}} \varepsilon \Delta \sum_{j=-\pi/\Delta\varepsilon}^{\pi/\Delta\varepsilon} \exp(i x_j (k_n - k_m))$$

$$= \lim_{\Delta \to 0} \Delta \int_{-\pi/\Delta}^{\pi/\Delta} \exp(i x (k_n - k_m)) dx = 2\pi \delta_{n,m} \qquad \text{(III.4.26)}$$

We remind the reader that the Kronecker delta is defined as $\delta_{n,n} = 1$ and $\delta_{n,m} = 0$ when $n \neq m$. As $k_n - k_m = (n - m)\Delta$, it is easy to see where the r.h.s of Eq. (III.4.26) comes from.

Now, let's consider what happens if we act with the matrix $\mathbf{R}^{-1}$ on a vector $\mathbf{g}$ representing a function $g(x)$. We get

$$\tilde{\mathbf{g}} = \mathbf{R}^{-1}\mathbf{g} = \begin{pmatrix} \tilde{g}_1 \\ \tilde{g}_2 \\ \tilde{g}_3 \\ \tilde{g}_4 \\ \vdots \end{pmatrix} \quad \text{and} \quad \tilde{g}_l = \varepsilon \sum_j \exp(x_j k_l) g_j \qquad \text{(III.4.27)}$$

If we let $\varepsilon \to 0$ and $\Delta \to 0$, the multiplication of $\mathbf{R}$ with $\mathbf{g}$ is indeed equivalent to a Fourier transform of $g(x)$, namely

$$\tilde{g}(k) = \int_{-\infty}^{\infty} \exp(ikx) g(x) dx \qquad \text{(III.4.28)}$$

Thus, we find out that Fourier transforms diagonalize the matrix that represents the differential operator

$$\hat{H} = -b\frac{d^2}{dx^2} + ic\frac{d}{dx} \qquad \text{(III.4.29)}$$

Next, let us consider the solution to the inhomogeneous equation

$$-b\frac{d^2 f}{dx^2} + ic\frac{df}{dx} = g(x) \qquad \text{(III.4.30)}$$

which can be written formally (from Eq. (III.4.18)) as

$$\mathbf{f} = \mathbf{H}^{-1}\mathbf{g} = \mathbf{R}\mathbf{H}_D^{-1}\tilde{\mathbf{g}} \qquad \text{(III.4.31)}$$

where

$$
\mathbf{H}_D^{-1} = \begin{pmatrix}
\ddots & \vdots & \vdots & \vdots & \vdots & \vdots & \reflectbox{$\ddots$} \\
\cdots & \dfrac{1}{bk_2^2 - ck_2} & 0 & 0 & 0 & 0 & \cdots \\
\cdots & 0 & \dfrac{1}{bk_1^2 - ck_1} & 0 & 0 & 0 & \cdots \\
\cdots & 0 & 0 & \infty & 0 & 0 & \cdots \\
\cdots & 0 & 0 & 0 & \dfrac{1}{bk_1^2 + ck_1} & 0 & \cdots \\
\cdots & 0 & 0 & 0 & 0 & \dfrac{1}{bk_2^2 + ck_2} & \cdots \\
\reflectbox{$\ddots$} & \vdots & \vdots & \vdots & \vdots & \vdots & \ddots
\end{pmatrix}
$$

$$(\text{III.4.32})$$

Equation (III.4.31) is then seen to be equivalent to writing, in practice, the solution

$$
f(x) = \frac{1}{2\pi} \int_{-\infty}^{\infty} \frac{\tilde{g}(k)\exp(-ikx)dk}{bk^2 + ck} \tag{III.4.33}
$$

In this case, if $\tilde{g}(0) \neq 0$, particular care needs to be taken in integrating over the $1/k$ pole (though it still leads to a finite result, provided $\tilde{g}(0)$ is finite), which we'll discuss how to do later.

### The Sturm–Liouville form

To finish this section, let us discuss the theory created by Sturm and Liouville (brilliant XIX Century French mathematicians) for solving linear differential equations. In the cases where it works, for the homologous equations we deal with the eigenfunctions defined in the domain $0 \leq x \leq b$ that satisfy the generalized boundary conditions

Jacques Charles
François Sturm
(1803–1855)

$$
\alpha_1 f(0) + \alpha_2 \frac{df(x)}{dx}\bigg|_{x=0} = 0 \quad \beta_1 f(b) + \beta_2 \frac{df(x)}{dx}\bigg|_{x=b} = 0 \quad (\text{III.4.34})
$$

Note that the real constants $\alpha_1$, $\alpha_2$, $\beta_1$ and $\beta_2$ may have any values subject to a particular boundary value problem. Note that once we have set the boundary conditions, Eq. (III.4.34), our homologous linear differential equation, specified by Eq. (III.4.1), will

Joseph Liouville
(1809–1882)

have only one eigenfunction for every eigenvalue, $f_k(x)$ (or eigenvector $\mathbf{f}_k$) which is itself a linear combination of two original eigenfunctions, or the representative eigenvectors, i.e. $\mathbf{f}_k = A_k \mathbf{f}_k^{(1)} + B_k \mathbf{f}_k^{(2)}$ that satisfy Eq. (III.4.34). The rotation matrix $\mathbf{R}$ and $\mathbf{H}_D$ are then constructed to be

$$\mathbf{R} = (\ldots \mathbf{f}_{k-2}, \mathbf{f}_{k-1}, \mathbf{f}_k, \mathbf{f}_{k+1}, \mathbf{f}_{k+2}, \ldots)$$

$$\mathbf{H}_D = \begin{pmatrix} \ddots & \vdots & \vdots & \vdots & \vdots & \vdots & \iddots \\ \cdots & E_{k-2} & 0 & 0 & 0 & 0 & \cdots \\ \cdots & 0 & E_{k-1} & 0 & 0 & 0 & \cdots \\ \cdots & 0 & 0 & E_k & 0 & 0 & \cdots \\ \cdots & 0 & 0 & 0 & E_{k+1} & 0 & \cdots \\ \cdots & 0 & 0 & 0 & 0 & E_{k+2} & \cdots \\ \iddots & \vdots & \vdots & \vdots & \vdots & \vdots & \ddots \end{pmatrix} \quad \text{(III.4.35)}$$

Now, importantly, we can write Eq. (III.4.1) in the Sturm–Liouville form:

$$\frac{d}{dx}\left[ R(x)\frac{df(x)}{dx} \right] + S(x)f(x) = \hat{H}_{sl}f(x) = \lambda \omega(x)f(x) \quad \text{(III.4.36)}$$

where we have that

$$p(x) = S(x)/\omega(x) \quad q(x) = R'(x)/\omega(x) \quad r(x) = R(x)/\omega(x)$$

$$\text{(III.4.37)}$$

We will not discuss here how to make the correct, convenient choice of $R(x)$, $S(x)$ and $\omega(x)$, but we'll just assume that we can do it in general. The advantage of writing our equation this way is that the matrix that represents $\hat{H}_{sl}$, $\mathbf{H}_{sl}$ is Hermitian, i.e. $\mathbf{H}_{sl} = \mathbf{H}_{sl}^\dagger$

(provided that the boundary conditions, Eq. (III.4.34), are met). In this case the diagonalization matrix, $\mathbf{R}_{sl}$ will be unitary, so that $\mathbf{R}_{sl}^{-1} = \mathbf{R}_{sl}^{\dagger}$, and the eigenvectors of $\mathbf{H}_{sl}$ are orthogonal. In terms of matrices, Eq. (III.4.36) is equivalent to

$$\mathbf{H}_{sl}.\mathbf{f} = \boldsymbol{\omega}.\mathbf{f} \tag{III.4.38}$$

where

$$\boldsymbol{\omega} = \begin{pmatrix} \omega_1 & 0 & 0 & 0 & \cdots \\ 0 & \omega_2 & 0 & 0 & \cdots \\ 0 & 0 & \omega_3 & 0 & \cdots \\ 0 & 0 & 0 & \omega_4 & \cdots \\ \vdots & \vdots & \vdots & \vdots & \ddots \end{pmatrix} \tag{III.4.39}$$

and $\omega_j = \omega(x_j)$.

## A generalized orthogonality condition

Now, let's do something a bit more concrete, to see exactly in real terms what benefit we get from converting our equation into the format of Eq. (III.4.36). First, let us formulate an orthoganlity condition between the eigenfunctions (vectors) $f_k(x)$, in integral form. We have already considered the one for $f_k(x) = \exp(ikx)$, given by Eq. (III.4.26), as is used in Fourier transforms. Now, let's generalize it to any eigenfunction satisfying an equation of the Sturm–Louiville form, Eq. (III.4.36). This is a useful tool to develop, as it facilitates particular transformations to solve inhomogeneous differential equations, as well as being able to represent any function as a series of eigenfunctions that satisfy Eq. (III.4.36). To do this, we consider the integral

$$I = \int_0^b dx [f_2^*(x) \hat{H}_{sl}(x) f_1(x) - f_1(x) \hat{H}_{sl}(x) f_2^*(x)]$$

$$= (\lambda_1 - \lambda_2) \int_0^b dx \omega(x) f_1(x) f_2^*(x) \tag{III.4.40}$$

Let us evaluate the l.h.s of Eq. (III.4.40). First, using the Sturm–Liouville form (Eq. (III.4.36)) we can write this integral as

$$I = \int_0^b f_2^*(x) \frac{d}{dx}\left[R(x)\frac{df_1(x)}{dx}\right] dx$$

$$- \int_0^b f_1(x) \frac{d}{dx}\left[R(x)\frac{df_2^*(x)}{dx}\right] dx \qquad (III.4.41)$$

Integration of Eq. (III.4.41) by parts yields

$$I = R(b)\left[f_2^*(b)\frac{df_1(x)}{dx}\bigg|_{x=b} - f_1(b)\frac{df_2^*(x)}{dx}\bigg|_{x=b}\right]$$

$$- R(0)\left[f_2^*(0)\frac{df_1(x)}{dx}\bigg|_{x=0} - f_1(0)\frac{df_2^*(x)}{dx}\bigg|_{x=0}\right] \qquad (III.4.42)$$

Now, using the boundary conditions, which both eigenfunctions must satisfy, (Eq. (III.4.34)), it is possible to show that $I = 0$, since $f_1'(0) = -\alpha_1 f_1(0)/\alpha_2$, $f_2'^*(0) = -\alpha_1 f_2^*(0)/\alpha_2$, $f_1'(b) = -\beta_1 f_1(a)/\beta_2$ and $f_2'^*(b) = -\beta_1 f_2^*(b)/\beta_2$. Thus, the l.h.s of Eq. (III.4.40) vanishes, and

$$(\lambda_1 - \lambda_2)\int_0^b dx\,\omega(x) f_1(x) f_2^*(x) = 0 \qquad (III.4.43)$$

So either we must have that both functions have the same eigenvalue or we must have that

$$0 = \int_0^b dx\,\omega(x) f_1(x) f_2^*(x) \qquad (III.4.44)$$

This can be thought of as a generalized orthogonality condition for the functions that solve Eq. (III.4.1).

Actually, we can show all of this, with just one line, in a more abstract way, by using simply the Hermitian properties of $\mathbf{H}_{sl}$ and the dot product

$$\mathbf{f}_2^*.\mathbf{H}_{sl}.\mathbf{f}_1 = \lambda_1 \mathbf{f}_2^*.\boldsymbol{\omega}.\mathbf{f}_1 = \mathbf{f}_2^*.\mathbf{H}_{sl}^\dagger.\mathbf{f}_1 = \lambda_2 \mathbf{f}_2^*.\boldsymbol{\omega}.\mathbf{f}_1 \qquad (III.4.45)$$

where $\mathbf{f}_2^*$ and $\mathbf{f}_1$ are the vector representations of $f_1(x)$ and $f_2^*(x)$, respectively. Again, in Eq. (III.4.45), we get $\mathbf{f}_j^*.\boldsymbol{\omega}.\mathbf{f}_k = 0$, unless $\lambda_j = \lambda_k$.

## Utilizing Sturm–Liouville theory

What we have shown just above is equivalent to showing in the matrix representation that $\tilde{\boldsymbol{\omega}} = \mathbf{R}^\dagger \boldsymbol{\omega} \mathbf{R}$, where $\tilde{\boldsymbol{\omega}}$ is a *diagonal* matrix (which is a consequence of $\mathbf{f}_j^* . \boldsymbol{\omega} . \mathbf{f}_k = 0$, unless $j = k$). It follows that $\mathbf{I} = \tilde{\boldsymbol{\omega}}^{-1} \mathbf{R}^\dagger \boldsymbol{\omega} \mathbf{R}$; this means that we must indeed have $\mathbf{R}^{-1} = \tilde{\boldsymbol{\omega}}^{-1} \mathbf{R}^\dagger \boldsymbol{\omega}$. This is very useful result for solving inhomogeneous equations, because we can now write Eq. (III.4.18) as

$$\mathbf{f} = \mathbf{R} \mathbf{H}_D^{-1} \tilde{\boldsymbol{\omega}}^{-1} \mathbf{R}^\dagger \boldsymbol{\omega} \mathbf{g} = \mathbf{R} \mathbf{H}_D^{-1} \tilde{\boldsymbol{\omega}}^{-1} \tilde{\mathbf{g}} \qquad \text{(III.4.46)}$$

where now $\tilde{\mathbf{g}} = \mathbf{R}^\dagger \boldsymbol{\omega} \mathbf{g}$. Let's make Eq. (III.4.46) more concrete in terms of functions. First we can write a transform (where for generality we have replaced 0 with $a$; the same boundary conditions still apply at that point)

$$\tilde{g}_k = S[g(x)] = \int_a^b f_k^*(x) \omega(x) g(x) dx \qquad \text{(III.4.47)}$$

which is equivalent to $\tilde{\mathbf{g}} = \mathbf{R}^\dagger \boldsymbol{\omega} \mathbf{g}$. Secondly, Eq. (III.4.46) is equivalent to writing the inverse transform

$$f(x) = S^{-1}[\tilde{g}_k / E_k] = \sum_k \frac{f_k(x) \tilde{g}_k}{E_k \tilde{\omega}_k} \qquad \text{(III.4.48)}$$

Both Eqs. (III.4.47) and (III.4.48) constitute the solution to an inhomogeneous equation, Eq. (III.4.16), provided that $f(x)$ is subject to the same boundary condition constraint, Eq. (III.4.34), as the eigenfunctions used for the transform and inverse transform. However, this is not much of a problem, as we can write the full solution to Eq. (III.4.16) as $f_T(x) = f(x) + f_H(x)$, where $f_H(x)$ satisfies the homogenous equation (Eq. (III.4.16) with $g(x) = 0$). Now, $f_H(x)$ can be chosen for $f_T(x)$ to satisfy boundary conditions other than Eq. (III.4.34).

This is a practically important result! Indeed, we see that there exists a generalized set of special transformations $S[g(x)]$ (which indeed includes, in particular, the Fourier transform) and their inverses, $S^{-1}[\tilde{g}_k]$, which act in a similar way to inverse Fourier transforms and Fourier series. As we shall see, these have uses other than just finding the solution to inhomogeneous differential equations.

> Finding the eigenvalues and vectors of a differential eigenvalue equation, and so any special transformations, is usually a bit of a higher 'pilotage' than simple operations with the Fourier transforms. We'll discuss further solvable examples in the next sections and practical examples for such more complicated transforms.

## 4.2. Series Solutions and Special Polynomials

### *Back to Earth*

Although we think that our optional theoretical digression was insightful, to get a better understanding of linear differential equations and how to solve them, let's now return to more practical issues. We'll consider some classic eigenvalue equations that arise in quantum mechanics as well as in classical physics that are solved by particular combinations of functions we already know.

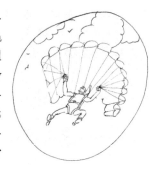

**Example III.4.2 (Physical Application ▶▶).** Here, we'll consider the Schrödinger equation for a simple harmonic oscillator. This is a classic problem in quantum mechanics: it describes the small oscillation (or vibration) of chemical bonds, understanding of which is absolutely crucial for the basics of molecular spectroscopy. At the same time, this is a brilliant example of how to handle eigenvalue equations (those who have done some quantum mechanics might be already familiar with the solutions, but still this will be a useful exercise in how to solve such equations from scratch).

To approximate the vibrations of a system of two vibrating atoms in a molecule (as shown in Fig. III.4.2), the equation reads as

$$-\frac{\hbar^2}{2\mu}\frac{d^2\psi(x)}{dx^2} + \frac{kx^2}{2}\psi(x) = \hat{H}\psi(x) = E\psi(x) \qquad \text{(III.4.49)}$$

The derivative term in the differential operator $\hat{H}$ is associated with the kinetic energy of the relative motion between two masses on a spring, which can be considered as the motion of a single particle

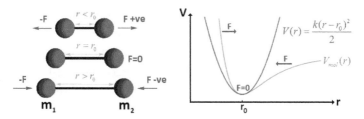

**Fig. III.4.2** We consider a vibrating molecule consisting of two atoms of mass $m_1$ and $m_2$. When the molecule compresses to a bond length smaller than the equilibrium bond length $r_0$, there is a force on the atoms pulling them apart. When the molecule is extended to a bond length larger than the equilibrium length, there is a force pushing the atoms together. As bond length moves away from $r_0$ the potential energy increases. On the right, we plot the potential energy as a function of bond length $r$, $V_{mol}$. When the vibrations are small, we may approximate the potential energy by the parabolic function $V(r) = k(r - r_0)^2/2$, where the 'spring constant' is given by $k = d^2 V_{mol}(r)/dr^2|_{r=r_0}$. This is called the harmonic approximation. Such a potential energy is shown as the red curve, whereas the exact form, which becomes unharmonic at large deviations from the equilibrium position (rising less steeply than the parabola at bond length larger than the equilibrium one, and steeper than the parabola at short distances) is depicted by the green curve. In the text we solve the Schrödinger equation with a harmonic form of potential energy, to obtain wave functions and energy levels (accounting for anharmonicity leads to more complicated equations and solutions that one can find in specialized textbooks).

with reduced mass $\mu = m_1 m_2/(m_1 + m_2)$, where $m_1$ and $m_2$ are the masses of the two atoms. The term that depends on $kx^2/2$ is the elastic potential energy for a spring obeying Hooke's law (describing the bond), where $x = r - r_0$ is the extension about the equilibrium spring (bond) length, $r_0$ and $k$ is the spring constant. The eigenvalues $E$ represent the allowed energy levels that the vibrating atoms can take. We will now show how to solve Eq. (III.4.49).

The first thing to do is to rescale Eq. (III.4.49) to make it read in a neat form. We introduce the variable change $s = x\sqrt{\mu\omega/\hbar}$, where $s$ is a dimensionless length variable and we have a parameter, the classical vibrational frequency $\omega = \sqrt{k/\mu}$. Following the lessons of Chapter 10 of Part I, Vol. 1, let us check that it is indeed dimensionless: $\left[\frac{\mu\omega}{\hbar}\right] = \left[\frac{M}{tMVL}\right] = \left[\frac{1}{L^2}\right]$, so that $[s] = [x\sqrt{\mu\omega/\hbar}] = \left[\frac{L}{L}\right]$ is dimensionless.

We can then write our equation with all terms dimensionless, and thereby without too many symbols distracting our attention:

$$\frac{d^2 u}{ds^2} = (E_R - s^2)u \qquad\qquad \text{(III.4.50)}$$

where $u(s) = \psi(x)$ and $E_R = 2E/\hbar\omega$. Before presenting the systematic method for solving this equation, let's do some analysis.

Let's look for special values of $E_R$, where we can reduce the order of the differential equation Eq. (III.4.50), recollecting Chapter 5 of Part II, Vol. 1. Generally, any linear second order differential equation

$$\frac{d^2u(s)}{ds^2} + P(s)\frac{du(s)}{ds} + Q(s)u(s) = 0 \qquad \text{(III.4.51)}$$

if we can find functions $p(s)$ and $q(s)$, that satisfy

$$P(s) = q(s) + p(s) \quad Q(s) = p(s)q(s) + \frac{dq(s)}{ds} \qquad \text{(III.4.52)}$$

can be reduced to two first order ones:

$$p(s)v(s) + \frac{dv(s)}{ds} = 0 \quad \text{and} \quad v(s) = \frac{du(s)}{ds} + q(s)u(s) \qquad \text{(III.4.53)}$$

In the case of Eq. (III.4.50), $P(s) = 0$ and $Q(s) = E_R - s^2$, so that $q(s) = -p(s)$ such that

$$E_R - s^2 = -q(s)^2 + \frac{dq(s)}{ds} \qquad \text{(III.4.54)}$$

Let's begin by setting $E_R = 1$ in Eq. (III.4.54), and seeing what solution we obtain. In this case, $q(s) = s$ would be the solution to Eq. (III.4.54) (we could set $E_R = -1$, with $q(s) = -s$, but we can rule this out as unphysical, as the total energy of our system must remain positive; so we restrict ourselves to $E_R > 0$). Thus, in this case, we may solve the linear equations

$$-sv(s) + \frac{dv(s)}{ds} = 0 \quad \text{and} \quad v(s) = \frac{du(s)}{ds} + su(s) \qquad \text{(III.4.55)}$$

It is easy to see either by inspection, or by separation of variables, that the solution to the first equation is $v(s) = A\exp(s^2/2)$. Thus, we are left with solving the equation

$$A\exp(s^2/2) = \frac{du(s)}{ds} + su(s) \qquad \text{(III.4.56)}$$

This equation we know how to solve from Chapter 5, Part II, Vol. 1. We multiply both sides by an integrating factor $f(s) = \exp(s^2/2)$ to

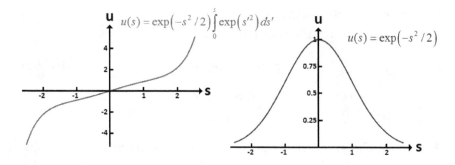

**Fig. III.4.3** Plots of the two functions appearing in Eq. (III.4.58).

obtain

$$A \exp(s^2) = \frac{d}{ds}[\exp s^2/2u(s)] \tag{III.4.57}$$

Integrating both sides of Eq. (III.4.57) with respect to $s$ we find

$$u(s) = A \exp(-s^2/2) \int_0^s \exp(s'^2)ds' + B \exp(-s^2/2) \tag{III.4.58}$$

We plot the two functions that make up the general solution in Fig. III.4.3. Now, as opposed to doing more blind maths, it's time to think some physics; good physical scientists with a solid grasp of maths should be able to interchange freely between the two!

In Eq. (III.4.49), $\psi(x)$ describes a wave function for which we require

$$1 = \int_{-\infty}^{\infty} P(x) = \int_{-\infty}^{\infty} \psi^*(x)\psi(x)dx \tag{III.4.59}$$

as $P(x) = \psi^*(x)\psi(x)$ is the probability distribution of finding an extension $x$ between the two atoms.

Looking at the plots in Fig. III.4.3, it is clear that this means we must have that $A = 0$ (as the first term in Eq. (III.4.58) diverges at large $x$; we can discard this solution, as we need to integrate from $-\infty$ to $\infty$). Still, note that sometimes the discarded solution is needed for Eq. (III.4.50), if the wave function should be restricted to a finite range or some more complicated potential for which the harmonic potential forms only a part. But we won't deal with this here. Thus, we have the solution for $E_R = 1$ that reads as

$$u(s) = B \exp(-s^2/2) \tag{III.4.60}$$

Both $E_R = 1$ and Eq. (III.4.60) are indeed physically valid solutions to Eq. (III.4.50).

**Exercise.** Equation (III.4.54) has another, slightly less obvious solution for $E_R = 3$, which is $q(x) = (x - 1/x)$. Show that it is indeed a solution, and hence find the general solution to Eq. (III.4.50), using Eq. (III.4.53), for that eigenvalue. Demonstrate that the physical solution is $u(s) = B s \exp(-s^2/2)$.

**Example III.4.2 (Continued).** It is tricky to solve Eq. (III.4.54) just by trying different values of $E_R$ for which the physical constraint, Eq. (III.4.54), is satisfied, so we now turn to the general method of solution. This relies of the fact that we have a clue on how to solve Eq. (III.4.50). Our experience, so far, suggests that we should look for solutions in the form

$$u(s) = \mathcal{H}(s) \exp(-s^2/2) \tag{III.4.61}$$

Substituting into Eq. (III.4.50), we find the following equation for $\mathcal{H}(s)$:

$$\frac{d^2 \mathcal{H}(s)}{ds^2} - 2s\frac{d\mathcal{H}(s)}{ds} + (E_R - 1)\mathcal{H}(s) = 0 \tag{III.4.62}$$

Indeed, $\mathcal{H}(s) = 1$ and $\mathcal{H}(s) = s$ satisfy Eq. (III.4.62), with $E_R = 1$ and $E_R = 3$, respectively. From this, it seems a good idea to try two types of power series solutions to find all eigenfunctions and permitted values for energy. So, we write two possible trial solutions:

$$\mathcal{H}_{even}(s) = a_0 + a_2 s^2 + a_3 s^4 + \cdots \tag{III.4.63}$$

$$\mathcal{H}_{odd}(s) = a_0 s + a_2 s^3 + a_3 s^5 + \cdots \tag{III.4.64}$$

These represent even and odd functions, respectively; and only such functions are allowed, as the probability distribution must be even, which the symmetry of the problem dictates. Let's start by substituting Eq. (III.4.63) into Eq. (III.4.62)

$$\frac{d^2}{ds^2}[a_0 + a_2 s^2 + a_3 s^4 + \cdots + a_{2p+2} s^{2p+2} + \cdots]$$

$$- 2s\frac{d}{ds}[a_0 + a_2 s^2 + a_3 s^4 + \cdots + a_{2p} s^{2p} + \cdots]$$

$$+ (E_R - 1)[a_0 + a_2 s^2 + a_3 s^4 + \cdots + a_{2p} s^{2p} + \cdots] = 0$$

$$\Rightarrow [2a_2 + 4 \times 3a_3 s^2 + \cdots + (2p+2)(2p+1)a_{2p+2} s^{2p} + \cdots]$$

$$- 2[2a_2 s^2 + 4a_3 s^4 + \cdots + 2na_{2p} s^{2p} + \cdots]$$

$$+ (E_R - 1)[a_0 + a_2 s^2 + a_3 s^4 + \cdots + a_{2p} s^{2p} + \cdots] = 0$$

$$\text{(III.4.65)}$$

Now, Eq. (III.4.65) must be satisfied for all the coefficients of $s$, meaning that each set of terms on the l.h.s of Eq. (III.4.65), multiplied by the same power of $s$ must sum to zero. This leads to relationships between the coefficients of the power series (called recursion relationships):

$$2a_2 + (E_R - 1)a_0 = 0$$

$$(2p+2)(2p+1)a_{2p+2} + (E_R - 4p - 1)a_{2p} = 0 \quad \text{(III.4.66)}$$

where $p$ is an integer. We can do the same thing substituting Eq. (III.4.64) into Eq. (III.4.62). In this case we obtain the recursion relations

$$6a_2 + (E_R - 3)a_0 = 0$$

$$(2p+3)(2p+2)a_{2p+3} + (E_R - 4p - 3)a_{2p+1} = 0 \quad \text{(III.4.67)}$$

Let's consider the (physical) case in which the series terminates. In this case we require the values $E_R = 1 + 4m$ (where $m$ is any positive integer or zero), for the even functions, so that the series expansion terminates at $p = m$. When we substitute $E_R = 1 + 4m$ into Eq. (III.4.66) with $p = m$ (where $p$ must be an integer as we originally defined it) we see that equation reduces to $(2m+2)(2m+1)a_{2m+2} = 0$. Thus, all terms for which $p > m$ simply vanish. Similarly if we choose

$E_R = 3 + 4m$, for the odd functions, this series expansion will again terminate at $p = m$.

Now, the above choices for $E_R$, are the only physical choices; if we don't choose eigenvalues that will terminate the series in Eqs. (III.4.66) and (III.4.67), we have a problem with the solution, $u(s)$, diverging at $|s| \to \infty$. To see this, let's look at both recursion relations when $p$ is very large. In this case, Eqs. (III.4.66) and (III.4.67) reduce to

$$a_{2p+2} \approx \frac{1}{n} a_{2p} \quad \text{and} \quad a_{2p+3} \approx \frac{1}{n} a_{2p+1} \qquad \text{(III.4.68)}$$

This suggests that at large $s$ we can write (as only the large $p$ terms in the series will matter)

$$\tilde{H}_1(s)/a_0 \approx \sum_{p=0}^{\infty} \frac{s^{2p}}{p!} = \exp(s^2) \quad \text{and}$$

$$\tilde{H}_2(s)/a_0 \approx \sum_{p=0}^{\infty} \frac{s^{2p+1}}{p!} = s \exp(s^2) \qquad \text{(III.4.69)}$$

We see that the problem is that when $s$ is large both $u(s) = \tilde{H}_1(s) \exp(-s^2/2)$ and $u(s) = \tilde{H}_2(s) \exp(-s^2/2)$ will blow up (!!) as $\exp(-s^2/2)$. As we require the normalization condition, Eq. (III.4.59), to hold, these solutions cannot physically exist as wave-functions, unless we have a physical situation where we can restrict the allowed range of $\psi(x)$ somehow, as discussed before; but, we are not doing this here.

We see that both odd and even allowed polynomial functions have energy eigenvalues $E_R = 1 + 2n$, where $n$ is zero or any positive integer. This can be seen because the allowed choices of $E_R$ were $E_R = 1 + 4m$ (for even functions) and $E_R = 3 + 4m$ (for odd functions), where $m$ is any zero or positive integer. These two sets of values are covered by the expression $E_R = 1 + 2n$. Let's obtain a polynomial solution for $\mathcal{H}(s)$. We consider first a couple of specific eigenvalues. For instance, when $E_R = 5$, we have that $a_2 = -2a_0$.

Therefore, our solution is

$$\mathcal{H}(s) = \mathcal{H}_{even}(s) = a_0(1 - 2s^2) \qquad \text{(III.4.70)}$$

and for $E_R = 7$ we have $a_3 = -2/3a_1$ so that

$$\mathcal{H}(s) = \mathcal{H}_{odd}(s) = a_0(s - 2s^3/3) \qquad \text{(III.4.71)}$$

Pafnuty Chebyshev
(1821–1894)

In the polynomial solutions generated by Eqs. (III.4.66) and (III.4.67), with $E_R = 1 + 2n$, the convention is to choose $a_0$, which is arbitrary, for each solution so that coefficient of the highest power of $s$ in the polynomial is $2^m$. These special polynomials that solve Eq. (III.4.62) are a called *Hermite polynomials*. Laplace introduced the notion of them, but not in the standard form that we present below, so they do not bear his name.

A slight digression: strictly speaking the mathematician Chebyshev (often considered a founding father of Russian mathematics) should be credited for the development of them in 1859. So, you may be asking why are they not called Chebyshev polynomials. Well, at that time he didn't get credit for them, and they were 'rediscovered' by Charles Hermite in 1865. And in any case a different set of polynomials already bears his name. All of this happened long before the birth of quantum mechanics, but we should point out that Hermite polynomials have many other applications in probability and numerical analysis, as well as abstract maths—often something that has a completely different application finds its way into another branch of physical sciences.

For the solution to Eq. (III.4.50), we have

$$u_n(s) = H_n(s)\exp(-s^2/2) \quad \text{with} \quad E_R = 1 + 2n \qquad \text{(III.4.72)}$$

where $H_n(s)$ stands for a Hermite polynomial of order $n$. The first few Hermite polynomials are given by

$$H_0(s) = 1, \quad H_1(s) = 2s, \quad H_2(s) = -2 + 4s^2,$$
$$H_3(s) = -12s + 8s^3, \quad H_4(s) = 12 - 48s^2 + 16s^4 \qquad \text{(III.4.73)}$$

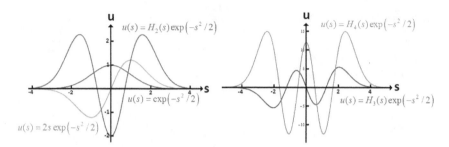

**Fig. III.4.4** Plots of the functions which correspond to physical solutions of the Schrödinger equation that satisfy Eq. (III.4.50), i.e. $H_n(s) \exp(-s^2/2)$, shown for the first four quantum numbers $n$. We see that each solution crosses the $s$-axis $n$ times. These crossing points are quite interesting physically, as they correspond to points where the probability density of finding the molecule with those particular bond lengths is zero—forbidden configurations. Note that points where the functions are at their maximal and minimal values correspond to most probable bond lengths.

In Fig. III.4.4, we plot the solutions for $u$ given by Eq. (III.4.73).

Higher-order Hermite polynomials can be generated using the following recursion relation:

$$H_{n+1}(s) = 2sH_n(s) - 2nH_{n-1}(s) \tag{III.4.74}$$

The eigenfunctions are orthogonal, and orthogonality is given by

$$\int_{-\infty}^{\infty} u_n(s)u_l(s)ds = \sqrt{\pi}2^n n!\delta_{l\cdot n} \tag{III.4.75}$$

**Exercise.** Hermite polynomials can also be generated through the expression

$$H_n(s) = (-1)^n \exp(s^2)\frac{d^n}{ds^n}\exp(-s^2) \tag{III.4.76}$$

Use this formula and integrating by parts to show that Eq. (III.4.75) is true.

**Example III.4.2 (Continued).** Scaling back into length $x$, and enforcing the orthogonality condition, Eq. (III.4.59), we finally arrive at

$$\psi(x) = \psi_n(x) = (n!2^n)^{1/2} \left(\frac{\mu\omega}{\hbar\pi}\right)^{1/4} H_n(x\sqrt{\mu\omega/\hbar}) \exp\left(-\frac{\mu\omega x^2}{2\hbar}\right)$$

$$\text{(III.4.77)}$$

with $E = E_n = \hbar\omega(n + 1/2)$, which represents the famous law of quantization of harmonic oscillator energies.

**Example III.4.3 (Maths Practice ♫♪).** Here, we'll consider an example which comes up all the time in the solution of partial differential equations in spherical geometry (as we shall see later in Chapter 7 of Part III, Vol. 2). Consider the following eigenvalue equation written in terms of the polar angle $\theta$:

$$-\frac{1}{\sin\theta}\frac{d}{d\theta}\left[\sin\theta\frac{d\Theta(\theta)}{d\theta}\right] + \frac{m^2\Theta(\theta)}{\sin^2\theta} = E\Theta(\theta) \qquad \text{(III.4.78)}$$

where $m$ is an integer. In quantum mechanics, Eq. (III.4.78) in fact arises in the solution of the 3D Schrödinger equation in spherical polar coordinates (for any form of potential energy not depending on $\theta$ or $\phi$), where $\Theta(\theta)$ is part of the 3D wave function $\psi(r, \theta, \phi) = R(r)\Theta(\theta)\Phi(\phi)$. In solving Eq. (III.4.78), we'll make no boundary condition choices except the condition that $\Theta(\theta)$ should remain finite-valued within the domain $0 \leq \theta \leq \pi$. This condition actually puts a big restriction on the solutions that we may have.

We start by making the variable substitution $z = \cos\theta$, noting that

$$\frac{d}{d\theta} = \frac{dz}{d\theta}\frac{d}{dz} = -\sin\theta\frac{d}{dz} = -\sqrt{1 - z^2}\frac{d}{dz} \qquad \text{(III.4.79)}$$

Using Eq. (III.4.79), Eq. (III.4.78) can be rewritten in the form called *Legendre's equation*:

$$\frac{d}{dz}\left[(1 - z^2)\frac{dP(z)}{dz}\right] + \left(E - \frac{m^2}{1 - z^2}\right)P(z) = 0$$

$$\text{(III.4.80)}$$

Adriene-Marie
Legendre
(1752–1833)

where the functions $P(z) = P(\cos\theta) = \Theta(\theta)$ are defined only within the domain $-1 \leq z \leq 1$. First, we'll consider the simplest case of Eq. (III.4.80), where $m = 0$, which then simplifies to

$$\frac{d}{dz}\left[(1 - z^2)\frac{dP(z)}{dz}\right] + EP(z) = 0 \qquad (\text{III.4.81})$$

To solve Eq. (III.4.81), we can write the following power series expansion:

$$P(z) = a_o + a_1z + a_2z^2 + a_3z^3 + \cdots \qquad (\text{III.4.82})$$

Importantly, we will require that $P(1)$ and $P(-1)$ are finite, and that will put some restrictions on the choice of eigenvalues, $E$, as we shall see. We first substitute Eq. (III.4.82) into Eq. (III.4.81) so that

$$(1 - z^2)\frac{d^2}{dz^2}[a_0 + a_1z + a_2z^2 + a_3z^3 + \cdots]$$
$$- 2z\frac{d}{dz}[a_0 + a_1z + a_2z^2 + a_3z^3 + \cdots]$$
$$= -E[a_0 + a_1z + a_2z^2 + a_3z^3 + \cdots]$$
$$\Rightarrow (1 - z^2)[2a_2 + 6a_3z + \cdots + n(n - 1)a_nz^{n-2} + \cdots]$$
$$- 2[a_1z + 2a_2z^2 + 3a_3z^3 + \cdots + na_nz^n + \cdots]$$
$$= -E[a_0 + a_1z + a_2z^2 + a_3z^3 + \cdots + a_nz^n + \cdots] \qquad (\text{III.4.83})$$

From Eq. (III.4.83) it unambiguously follow that

$$(n + 1)(n + 2)a_{n+2} = (n(n - 1) + 2n - E)a_n = (n(n + 1) - E)a_n$$
$$(\text{III.4.84})$$

Importantly, if we choose $E = l(l+1)$ where $l$ is a positive integer; we see from Eq. (III.4.84) that the series terminates at $n = l$ provided that it contains either only odd or even powers of $z$. If we do not make these choices, we have the problem that $P(1) \to \infty$; the power series diverges at $z = 1$. In this case, the solution cannot physically describe the polar angle part of the wave function of a particle. To see how this problem comes about, let's again consider when $n$ is

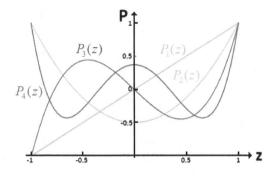

**Fig. III.4.5** The graphs of the first four Legendre polynomials $P_l(z)$ defined in the domain $-1 \leq z \leq 1$. We see that each Legendre polynomial has $1 - 1$ turning points. Also, we have $P_1(z) = z$, i.e. simply a straight line.

very large. In this case the recursion relation is approximated by

$$a_{n+2} \approx a_n \qquad (III.4.85)$$

Since $\lim_{n \to \infty}(a_{n+2}/a_n) = 1$, the series diverges at $z = 1$ (as well as at $z = -1$).

Using Eq. (III.4.84) with $E = l(l+1)$, we obtain polynomial solutions. For instance, for $l = 0$, $l = 1$ and $l = 2$, we have the solutions

$$P_{l=0}(z) = a_0, \quad P_{l=1}(z) = za_0 \quad \text{and} \quad P_{l=2}(z) = (1 - 3z^2)a_0 \qquad (III.4.86)$$

For each solution, the choice of $a_0$ is arbitrary and we may choose it so that our solutions are the Legendre polynomials $P_l(z)$, which are generated by

$$P_l(z) = \frac{1}{2^l l!} \frac{d^l}{dz^l}(z^2 - 1)^l \qquad (III.4.87)$$

In particular we have that $P_0(z) = 1$, $P_1(z) = z$, $P_2(z) = (3z^2 - 1)/2$. We plot some of the Legendre polynomials in Fig. III.4.5.

**Exercise.** Substitute Eq. (III.4.87) into the Legendre equation given in Eq. (III.4.87), and show that with $E = l(l + 1)$ it is indeed a solution.

**Example III.4.3 (Continued).** Now, let's consider the solution to Eq. (III.4.80) when $m \neq 0$. One particular way of solving

Eq. (III.4.80) is a series solution with $E = l(l+1)$. The series solution, for a particular positive value of $m$, actually sums to the following:

$$P_l^m(z) = (-1)^m(1-z^2)^{m/2}\frac{d^m P_l(z)}{dz^m} \qquad \text{(III.4.88)}$$

For negative values of $m$ we have that $P_l^{-m}(z) = P_l^m(z)$. The functions $P_l^m(z)$ are called the *associated Legendre functions*.

To show by substitution that Eq. (III.4.88) is a solution of Eq. (III.4.80) is a bit tricky, but it is still not rocket science. The trick is to differentiate Eq. (III.4.81) $m$ times to start with, so that

$$\frac{d^{m+1}}{dz^{m+1}}\left[(1-z^2)\frac{dP(z)}{dz}\right] + E\frac{d^m P(z)}{dz^m} = 0 \qquad \text{(III.4.89)}$$

Next, we utilize the Leibniz formula for multiple differentiation of a product, which states

$$\frac{d^m(g(x)h(x))}{dx^m} = \sum_{j=1}^{m}\frac{m!}{(m-j)!j!}\frac{d^j g(x)}{dx^j}\frac{d^{m-j}h(x)}{dx^{m-j}} \qquad \text{(III.4.90)}$$

Thus, through Eq. (III.4.90), we can write

$$(1-z^2)\frac{d^{m+2}P_l(z)}{dz^{m+2}} - 2z(m+1)\frac{d^{m+1}P_l(z)}{dz^{m+1}}$$

$$+ (E - m(m+1))\frac{d^m P_l(z)}{dz^m} = 0 \qquad \text{(III.4.91)}$$

Next, we can multiply Eq. (III.4.91) by $(1-z^2)^{m/2}$ so that

$$(1-z^2)^{m/2+1}\frac{d^{m+2}P_l(z)}{dz^{m+2}} - 2z(m+1)(1-z^2)^{m/2}\frac{d^{m+1}P_l(z)}{dz^{m+1}}$$

$$+ (E - m(m+1))(1-z^2)^{m/2}\frac{d^m P_l(z)}{dz^m} = 0 \qquad \text{(III.4.92)}$$

**Exercise.** By substituting Eq. (III.4.88) into Eq. (III.4.80), obtain Eq. (III.4.92), thereby showing that it satisfies Eq. (III.4.80).

**Example III.4.3 (Continued).** We plot some associated Legendre functions in Fig. III.4.6.

Now, we should realize that we've considered only one set of functions that solve Eq. (III.4.78). As Eq. (III.4.78) is a second

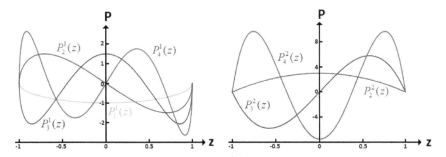

**Fig. III.4.6** The first four associated Legendre functions defined by Eq. (III.4.87). We note that if we keep $l$, but increase the absolute value of $m$, the number of turning points diminishes. Note that, to always require a non-zero function, we require $m \leq l$.

order differential equation, we should write as a general solution $P(z) = A P_l^m(z) + B Q_l^m(z)$, where $Q_l(z)$ and $Q_z^m(z)$, are called *Legendre polynomials of the second kind* and *associated Legendre functions of the second kind*, respectively. The reason why we don't need to include them in our solution is that they are singular at the points $z = 1$ and $z = -1$. However, they can come up in solutions where we have boundary conditions that restrict the range of $\theta$ over which $\Theta(\theta)$ is defined (such boundary conditions can also relax the requirement that $E = l(l+1)$). Those functions are defined in the following way

$$Q_l(x) = \frac{P_l(x)}{2} \ln\left(\frac{1+x}{1-x}\right) - \sum_{m=1}^{l} \frac{1}{m} P_{m-1}(x) P_{l-m}(x) \quad \text{(III.4.93)}$$

$$Q_l^m(z) = (-1)^m (1-z^2)^{m/2} \frac{d^m Q_l(z)}{dz^m} \quad \text{(III.4.94)}$$

We plot some of these functions in Fig. III.4.7.

Considering problems where $Q_l(z)$, $Q_z^m(z)$ occur would lead us too far; we'd better stop this example here. We simply wanted to show the reader what the second solution to Eq. (III.4.78) is.

**Example III.4.4 (Physical Application ▶▶).** As the last example in this section, we'll consider the Schrödinger equation for the radial part of the wave function in the hydrogen atom, where the radial position of the electron is described by Fig. III.4.8. This is

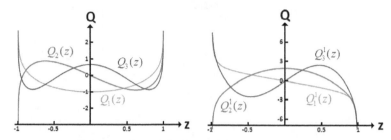

**Fig. III.4.7** Plots of some of the Legendre polynomials of the second kind, $Q_l(z)$, and some of the their associated Legendre functions $Q_l^1(z)$. One should clearly note that these functions diverge at the points $z = 1$ and $z = -1$ either to $\infty$ or $-\infty$.

**Fig. III.4.8** An illustration of a possible electron position relative to the proton nucleus, at a distance $r$ away, in a hydrogen atom (Of course, the electron can never be 'fixed' in that position, or in any other position!).

given by

$$-\frac{\hbar^2}{2m_e}\left(\frac{d^2}{dr^2} + \frac{2}{r}\frac{d}{dr}\right)R(r) - \frac{e^2}{4\pi\varepsilon_0 r}R(r)$$

$$+\frac{\hbar^2 l(l+1)}{2m_e r^2}R(r) = \hat{H}R(r) = ER(r) \qquad \text{(III.4.95)}$$

Before solving it, we'll briefly discuss the terms in the differential operator $\hat{H}$ (the Hamiltonian). The first term is associated with the kinetic energy of the electron in the radial direction; the second term is the electrical potential energy of the electron at a given position from the nucleus (just a proton, for hydrogen atom); and the last term is it's rotational kinetic energy that depends on the positive integer (including zero) $l$. Each value of $l$ corresponds to an electron state with a particular value of magnitude of angular momentum. In Eq. (III.4.95) we have used the mass of the electron, $m_e$. Strictly speaking we should use the reduced mass of the proton-electron system, and $r$ should be the distance away from the centre of mass. However, since the mass of the proton is so much larger, using $m_e$ and measuring $r$ from nucleus is actually a very good approximation.

We'll solve Eq. (III.4.95) for bound states, where we consider $E < 0$. First, to bring the equation to a more compact form, we rescale it in terms of dimensionless quantities:

$$s = \frac{(-8m_e E)^{1/2} r}{\hbar} \qquad \lambda = \frac{e^2}{4\pi\varepsilon_0 \hbar}\left(-\frac{m_e}{2E}\right)^{1/2} \qquad \text{(III.4.96)}$$

**Exercise.** Using the logic described in Chapter 10, Part I, Vol. 1 show that these variables are dimensionless.

**Example III.4.4 (Continued).** We then rewrite our Schrödinger equation as

$$\frac{d^2\tilde{R}(s)}{ds^2} + \frac{2}{s}\frac{d\tilde{R}(s)}{ds} + \left(\frac{\lambda}{s} - \frac{1}{4} - \frac{l(l+1)}{s^2}\right)\tilde{R}(s) = 0 \qquad \text{(III.4.97)}$$

where $R(r) = \tilde{R}(s)$. Let's first examine Eq. (III.4.97) when $s$ is large. In this case we can start by making obvious approximations to Eq. (III.4.97). Indeed, for $s \gg 1$,

$$\frac{d^2\tilde{R}(s)}{ds^2} + \frac{2}{s}\frac{d\tilde{R}(s)}{ds} - \frac{1}{4}\tilde{R}(s) \approx 0 \qquad \text{(III.4.98)}$$

but being even more 'daring', we can also neglect the middle term of Eq. (III.4.98); and actually we have justification for doing so. Note, first, that $\exp(-s/2)$ solves the further simplified equation

$$\frac{d^2\tilde{R}(s)}{ds^2} - \frac{1}{4}\tilde{R}(s) = 0 \qquad \text{(III.4.99)}$$

Hence, we can estimate the omitted middle term of Eq. (III.4.98) as $-1/s\exp(-s/2)$, so it can indeed be neglected in the limit of large $s$. Of course, $\exp(s/2)$ also solves Eq. (III.4.99), but that solution can be discarded: it diverges at large $s$ (large distances) and is unphysical; if you are not convinced—it does not satisfy the normalization

condition,

$$\int_0^\infty r^2 R(r)^2 dr = 1 \qquad \text{(III.4.100)}$$

where $r^2 R(r)^2$ is the probability distribution of finding the electron at distance $r$ away from the nucleus.

This analysis suggests that we should make the substitution

$$\tilde{R}(s) = u(s)\exp(-s/2) \qquad \text{(III.4.101)}$$

and focus on the function $u(s)$. Note that we can write $\ln \tilde{R}(s) = \ln u(s) - s/2$, and we expect for large $s$, $\ln \tilde{R}(s) \approx -s/2$. On substitution of Eq. (III.4.101) into Eq. (III.4.97), we find the equation on $u$ to be

$$\frac{d^2 u(s)}{ds^2} - \left(1 - \frac{2}{s}\right)\frac{du(s)}{ds} + \left[\frac{\lambda - 1}{s} - \frac{l(l+1)}{s^2}\right]u(s) = 0 \quad \text{(III.4.102)}$$

In this equation, the coefficients in front of the second and third terms diverge when $s = 0$; this is called a singular point. It is still does not necessarily mean that $u$ must diverge at this point, as terms in Eq. (III.4.102) have different signs and may compensate each other. However, because of the divergence of coefficients a regular series

Ferdinand Georg
Frobenius
(1849–1917)

solution of the kind we used in the two previous examples will not work. Thus, we need to use a different approach, one first proposed by German mathematician Frobenius. To understand how this works in this case (we will talk more about the Frobenius method after this example), let us first examine Eq. (III.4.102) close to when $s = 0$. Here, we need keep only the most singular terms in the coefficients, and so approximate it by

$$\frac{d^2 u(s)}{ds^2} + \frac{2}{s}\frac{du(s)}{ds} - \frac{l(l+1)}{s^2}u(s) \approx 0 \qquad \text{(III.4.103)}$$

The form of Eq. (III.4.103) clearly suggests a leading order solution that is a power of $s$, i.e. $u(s) \approx u_0 s^\gamma$. We simply find the value of $\gamma$ by substituting this form into Eq. (III.4.103) so that we obtain

$$\gamma(\gamma - 1) + 2\gamma - l(1 + l) = 0 \Rightarrow \gamma(\gamma + 1) = l(l + 1) \quad \text{(III.4.104)}$$

There are two solutions to Eq. (III.4.104), but only one satisfies the normalization condition, Eq. (III.4.100), and that is $\gamma = l$. The next step is to write, instead of a regular power series expansion, the series

$$u(s) = s^\gamma \sum_{n=0}^{\infty} a_n s^n \quad \text{(III.4.105)}$$

Substitution of Eq. (III.4.105) into Eq. (III.4.102) with $\gamma = l$ yields

$$\sum_{n=0}^{\infty} [(n + 1)n a_{n+1} + (n + 1)a_{n+1}(2l + 2)$$
$$+ (\lambda - l - 1 - n)a_n]s^{n-1} = 0 \quad \text{(III.4.106)}$$

As all the coefficients in front of each $s^{n-1}$-term in Eq. (III.4.106) must vanish, we obtain the recursion relationship

$$(n + 1)(n + 2l + 2)a_{n+1} = (n + l + 1 - \lambda)a_n \quad \text{(III.4.107)}$$

Here, it is the same old story as in the previous examples: if we choose $\lambda = \tilde{n} + l + 1$, where $\tilde{n}$ is some integer, the series terminates at $n = \tilde{n}$ and things will be fine. Here, we will get a certain type of finite polynomial again, which we discuss later. However, if we don't make this choice, we will get an unacceptable solution for the electron wave function. Indeed, the series then would be dominated at large $s$ by the large $n$ terms in the series. For large $n$, we get

$$n a_{n+1} \approx a_n \quad \text{(III.4.108)}$$

The solution to this equation is $a_n \approx 1/n!$ so that at large $s$

$$u(s) \approx \sum_{n=0}^{\infty} \frac{a_0 s^n}{n!} = \exp(s) \qquad \text{(III.4.109)}$$

This would give $\tilde{R}(s) \approx \exp(s/2)$, the solution to Eq. (III.4.97) for large $s$. This solution clearly does not satisfy the normalization condition, Eq. (III.4.100), as it is diverging when $s \to \infty$. Thus such solutions and eigenvalues must be discarded.

All in all, we now have convinced ourselves that $\lambda = n_l = \tilde{n} + l + 1$, and its value must be positive because $\tilde{n}$ and $l$ are both positive numbers. This creates the restriction that $l + 1 \leq n_l$ for allowed solutions of Eq. (III.4.102). The polynomials that solve Eq. (III.4.102) with $\lambda = n_l = \tilde{n} + l + 1$ are called *Laguerre* and *Associated Laguerre* polynomials, which are denoted by $L_q(s) = L_q^{p=0}(s)$ and $L_q^p(s)$, respectively. Associated Laguerre polynomials are sometimes called *Sonin Polynomials*, after the Russian mathematician who first derived them, Nikolay Sonin. These are conveniently (in one particular definition of them) given by a closed form expression

Edmund Laguerre
(1849–1915)

Nikolay Sonin
(1849–1915)

$$L_q^p(s) = \sum_{j=0}^{q} (-1)^j \frac{(q+p)!}{(q-j)!(j+p)!} \frac{x^j}{j!} \qquad \text{(III.4.110)}$$

We plot some of these functions in Fig. III.4.9.

The general solution to Eq. (III.4.102) with the physical constraint, Eq. (III.4.100), is given by

$$u(s) = -As^l L_{n_l-l-1}^{2l+l}(s) \qquad \text{(III.4.111)}$$

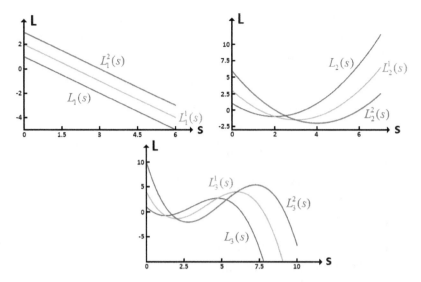

**Fig. III.4.9** Plots of the various Laguerre polynomials $L_q^p(s)$. We see that the index $q$ controls the order of the polynomial and its number of turning points, while one of the main effects of changing the index $p$ is to shift the curve described by the polynomial to the right along the $s$-axis.

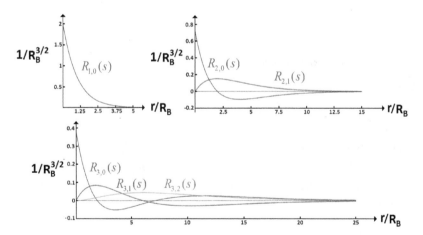

**Fig. III.4.10** Plots of the radial part of the wave function $R_{n,l}(r)$ for the first three values of $n$ and corresponding to allowed values of $l$; the distance r is measured in Bohr radii. Keeping $l$ fixed and increasing $n$, the number of nodes (points where the wave function is zero) increases. As the radial probability distribution function for finding the electron is $r^2 R_{n,l}(r)^2$, these points correspond to 'forbidden' distances away from the proton, where the electron cannot be found. Also in line with the wave function, at large values of $r$, the probability density diminishes exponentially with increasing distance—all in all, the electron is localized around the nucleus.

Combining this with Eqs. (III.4.101), (III.4.96) and (III.4.100), thus determining the normalization constant $A$, the solution to Eq. (III.4.95) is found to be

$$R_{n_l,l}(r) = \left(\frac{2}{n_l R_B}\right)^{3/2} \left(\frac{(n_l - l - 1)!}{2n(n_l + l)!}\right)^{1/2} \left(\frac{2r}{n_l R_B}\right)^l$$

$$\times L_{n_1-l-1}^{2l+1}\left(\frac{2r}{n_l R_B}\right) \exp\left(-\frac{r}{n_l R_B}\right) \qquad \text{(III.4.112)}$$

where $R_B = 4\pi\varepsilon_0 \hbar^2/me^2$ (this is in SI units, in Gaussian units we set $4\pi\varepsilon_0 = 1$) is the Bohr radius. We plot of few of these wave functions in Fig. III.4.10.

**Exercise.** To get accustomed to the use of Laguerre-Sonin polynomial, sketch or plot—knowing the form of the wave functions (Fig. III.4.10) and Eq. (III.4.112)—the probability distributions for finding the electron $f_{n,l}(r) = r^2 R_{n,l}(r)^2$ for the first three values of $n$. By considering the normalization condition, Eq. (III.4.100), show that the wave functions are correctly normalized for these values of $n$.

### *Power series solutions*

We'll round off this section with a more general discussion about the power series solution of linear differential equations and the *Frobenius method*. Suppose, in general, we want to find a power series of a linear homogenous differential equation, i.e. Eq. (III.4.51), about the point $s = 0$, where both $Q(s)$ and $P(s)$ have power series expansions, though they can be singular at $s = 0$. We'll suppose that we can write Eq. (III.4.51) for small $s$ as

$$\frac{d^2 u(s)}{ds^2} + \frac{1}{s}(p_0 + p_1 s + p_2 s^2 + \cdots)\frac{du(s)}{ds}$$

$$+ \frac{1}{s^2}(q_0 + q_1 s + q_2 s^2 + \cdots)u(s) = 0 \qquad \text{(III.4.113)}$$

The form of Eq. (III.4.113) suggests that for very small $s$, $u(s) \sim s^r$, where $r$ can be positive or negative. Then, following the method of

Frobenius, the solution can be written as

$$u(s) = s^r \sum_{k=0}^{\infty} a_k s^k \qquad \text{(III.4.114)}$$

Substitution of Eq. (III.4.113) into Eq. (III.4.114) yields

$$\sum_{k=0}^{\infty} (k+r)(k+r-1)a_k s^{k+r-2}$$

$$+ (p_0 + p_1 s + p_2 s^2 + \cdots) \sum_{n=0}^{\infty} (k+r)a_k s^{r+k-2}$$

$$+ (q_0 + q_1 s + q_2 s^2 + \cdots) \sum_{k=0}^{\infty} a_k s^{k+r-2} = 0 \qquad \text{(III.4.115)}$$

The lowest power of $s$ in Eq. (III.4.115) is given by $s^{r-2}$, and the coefficients of this power obey

$$r(r-1)a_0 s^{r-2} + r p_0 a_0 s^{r-2} + q_0 a_0 s^{r-2} = 0$$

$$\Rightarrow (r^2 - r) + r p_0 + q_0 = 0 \qquad \text{(III.4.116)}$$

Thus, from Eq. (III.4.116) we can determine $r$. It is given by

$$r = r_{\pm} = \frac{1 - p_0 \pm \sqrt{(1-p_0)^2 - 4q_0}}{2} \qquad \text{(III.4.117)}$$

Note that for a real power series solution we require that $(1-p_0)^2 - 4q_0 \geq 0$. Provided that $r_+$ and $r_-$ are not separated by an integer or zero, two power series solutions of the form of Eq. (III.4.114), with $r = r_+$ and $r = r_-$, will form two linearly independent solutions to Eq. (III.4.51).

Comparing higher powers of $s$ will then determine the coefficients $a_1$, and so on. For instance, comparing each of the $s^{r-1}$ coefficients in Eq. (III.4.115) yields a relationship between $a_1$ and $a_0$:

$$(1+r)r a_1 + p_0(1+r)a_1 + p_1 r a_0 + q_0 a_1 + q_1 a_1 = 0 \qquad \text{(III.4.118)}$$

By comparing the $s^r$ coefficients the $a_2$ can be determined, etc.

There are complications when $r_+ - r_- = n$, where $n$ is an integer, or the roots are repeated. The two roots in this case do not generate separate functions; we'll show this when we come to consider Bessel functions in the next section. In this case one solution is still Eq. (III.4.114), with $r = r_+$, i.e.

$$u_1(s) = s^{r_+} \sum_{k=0}^{\infty} a_k s^k \qquad\qquad (\text{III}.4.119)$$

with the $a_k$ coefficients still determined through Eq. (III.4.115). However, to ensure a linearly independent solution, we must instead write for the second solution

$$u_2(s) = C u_1(s) \ln s + s^{r_-} \sum_{k=0}^{\infty} b_k s^k \qquad\qquad (\text{III}.4.120)$$

which we substitute back into Eq. (III.4.113) to determine both $C$ and the $b_k$. We won't prove this; we just leave this as a recipe.

Matters simplify if $p_0 = q_0 = q_1 = 0$. In this case we simply have $r = 0$, and we have the simple power series solution. In this case substituting Eq. (III.4.114) into Eq. (III.4.113) yields

$$\sum_{k=2}^{\infty} k(k-1)a_k s^{k-2} + (p_1 + p_2 s + \cdots) \sum_{n=1}^{\infty} k a_k s^{k-1}$$

$$+ (q_2 + q_3 s \ldots) \sum_{k=0}^{\infty} a_k s^k = 0 \qquad\qquad (\text{III}.4.121)$$

The coefficients multiplying $s^0$ should then satisfy

$$2a_2 + p_1 a_1 + q_2 a_0 = 0 \qquad\qquad (\text{III}.4.122)$$

Here, the choices of both $a_0$ and $a_1$ are arbitrary, and they define two linearly independent solutions (one with $a_0 = 1$ and $a_1 = 0$, the other with $a_0 = 0$ and $a_1 = 1$). A linear combination of these two solutions forms the general solution. Higher order coefficients are determined by comparing $s$, $s^2$ coefficients, and so on, in Eq. (III.4.121).

We can also look to find a power series for the solution of a homogenous linear differential equation about some other point $s = s_0$. The simplest way to do this is to make a substitution

$s = s_0 + x$ and rewrite Eq. (III.4.51) so that

$$\frac{d^2\tilde{u}(x)}{dx^2} + P(x + s_0)\frac{d\tilde{u}(x)}{dx} + Q(x + s_0) = 0 \qquad \text{(III.4.123)}$$

where $\tilde{u}(x) = u(s_0 + x)$. We then expand $P(x + s_0)$ and $Q(x + s_0)$ in powers of $x$ about the point $s_0$, and use a power series solution, or the method of Frobenius where appropriate, to solve Eq. (III.4.123) close to $s_0$.

Representing a solution $u(s)$ as a power series in $s$ can sometimes be of limited use, as the range of value of $s$ that we want to look at may require many terms in the series. In other cases the power series may not even be valid for an arbitrary number of terms within the range of $s$ of interest (this is certainly the case if $P(s)$ and $Q(s)$ become singular at some other point than the point we expand about in the range of interest, as we saw it in the case of Legendre's equation without the condition $E = l(l+1)$). Sometimes, as we have seen in Examples III.4.2 and III.4.4, the trick is first to try to deduce the behaviour of large $s$, i.e. find a solution $u_\infty(s)$ for which $\ln u(s) \to \ln u_\infty(s)$ when $s \to \infty$. Then, to substitute $u(s) = h(s)u_\infty(s)$ and find an equation on $h(s)$. The power series expansion of $h(s)$ is then likely to work better and sometimes (in the cases of physical interest) it will terminate as a finite order polynomial, yielding a closed form solution.

There are other techniques for solving linear differential equations that are available to us, which are approximation techniques. We will discuss some of these in Chapter 6 of Part III, Vol. 2, as well as those for handling non-linear differential equations.

## 4.3. Introducing Bessel Functions

In this section we examine a special but very important class of linear differential equations, called Bessel's equations, introduced by famous Prussian mathematician W.F. Bessel, known also for his works in astronomy and geodesy. These are solved by the so-called *Bessel functions*, new functions that we will discuss in some detail.

### Bessel's equation

This famous equation reads:

$$x^2 \frac{d^2 y}{dx^2} + x \frac{dy}{dx} + (x^2 - \alpha^2) y = 0 \qquad \text{(III.4.124)}$$

It comes up in a lot of physical problems. For instance, Bessel's equation may arise from the following eigenvalue equation, which comes up in the solution of partial differential equations:

Friedrich Wilhelm
Bessel
(1784–1846)

$$-\frac{d^2 \tilde{y}(x)}{dx^2} - \frac{1}{x} \frac{d\tilde{y}(x)}{dx} + \frac{\alpha^2 \tilde{y}(x)}{x^2} = E\tilde{y}(x) \qquad \text{(III.4.125)}$$

Indeed, by setting $\tilde{y}(x) = y(x\sqrt{E})$ we get Eq. (III.4.124). It is also useful to note that Eq. (III.4.125) can be written in the Sturm–Liouville form (see Eq. (III.4.36)):

$$-\frac{d}{dx}\left[ x \frac{d\tilde{y}(x)}{dx} \right] + \frac{\alpha^2 \tilde{y}(x)}{x} = Ex\tilde{y}(x) \qquad \text{(III.4.126)}$$

Bessel's equation (Eq. (III.4.124)) can be solved close to $x = 0$ by series expansion, using the method of Frobenius. This is the series expansion of a new type of function called the Bessel function, which solves Bessel's equation.

### Bessel functions of the first kind

The Bessel function of order $\alpha$ has the power series expansion, obtained by applying the method of Frobenius to Eq. (III.4.124), of

$$J_\alpha(x) = \sum_{m=0}^{\infty} \frac{(-1)^m}{m!\,\Gamma(m + \alpha + 1)} \left(\frac{x}{2}\right)^{2m+\alpha} \qquad \text{(III.4.127)}$$

Here, we have the Gamma function $\Gamma(x)$, which we introduced in Chapter 7 of Part II, Vol. 1.

For the values of $\alpha = n + 1/2$, where $n$ is an integer, Bessel functions are simply related to trigonometric functions. They can be generated by the formulae

$$J_{n+1/2}(x) = (-1)^n x^{n+1/2} \sqrt{\frac{2}{\pi}} \left[\frac{1}{x}\frac{d}{dx}\right]^n \left(\frac{\sin x}{x}\right)$$

$$J_{-n+1/2}(x) = x^{n+1/2} \sqrt{\frac{2}{\pi}} \left[\frac{1}{x}\frac{d}{dx}\right]^n \left(\frac{\cos x}{x}\right)$$

(III.4.128)

so that, for example,

$$J_{1/2}(x) = \sqrt{\frac{2}{\pi x}} \sin x, \quad J_{-1/2}(x) = \sqrt{\frac{2}{\pi x}} \cos x$$

$$J_{3/2}(x) = \sqrt{\frac{2}{\pi x}} \left(\frac{\sin x}{x} - \cos x\right)$$

(III.4.129)

$$J_{-3/2}(x) = -\sqrt{\frac{2}{\pi x}} \left(\frac{\cos x}{x} + \sin x\right)$$

In particular, these functions come up in the solution of a partial differential equation called the Helmholtz equation in spherical geometry (to be discussed later). These are sometime written in terms of *spherical Bessel functions*, which are defined as

$$j_n(x) = \sqrt{\frac{\pi}{2x}} J_{n+1/2}(x)$$

(III.4.130)

We plot some spherical Bessel functions in Fig. III.4.11.

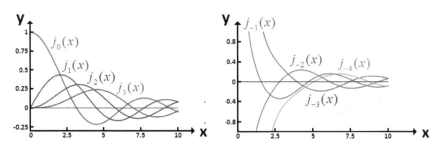

**Fig. III.4.11** Plots of curves described by some of the spherical Bessel functions, $y = j_n(x)$, for both positive and negative $n$. Plots for the positive values are shown on the left as well as for $n = 0$. We see that $j_0(0) = 1$, whereas $j_n(0) = 0$ for $n > 0$. Plots for the negative values of $n$ are shown on the right; they diverge at $x = 0$ as $\propto 1/x^n$. All curves cross the line $y = 0$ an infinite number of times, and the amplitude of their oscillations slowly diminishes with increasing $x$.

Both $J_\alpha(x)$ and $J_{-\alpha}(x)$ are solutions to Eq. (III.4.124). For non-integer values of $\alpha$, $J_\alpha(x)$ and $J_{-\alpha}(x)$ are (linearly) independent of each other, and so we can write the general solution as

$$y(x) = AJ_\alpha(x) + BJ_{-\alpha} \qquad \text{(III.4.131)}$$

### Bessel functions of the second kind

When $\alpha = n$, an integer, $J_n(x)$ and $J_{-n}(x)$ are not independent solutions. An indication of this is that, in Eq. (III.4.116), we find that $r_+ = n$ and $r_- = -n$ for Bessel's equation; the roots are separated by an integer. We can check that $J_n(x)$ and $J_{-n}(x)$ are related by considering the series expansion for $J_{-n}(x)$. From Eq. (III.4.127) this reads as

$$J_{-n}(x) = \sum_{m=0}^{\infty} \frac{(-1)^m}{m!\,\Gamma(m - n + 1)} \left(\frac{x}{2}\right)^{2m-n} \qquad \text{(III.4.132)}$$

Now, we recollect from our discussion of the Gamma function that $\Gamma(m - n + 1) = \infty$ for $m \leq n - 1$; this means that these terms in the series vanish! Then

$$J_{-n}(x) = \sum_{m=n}^{\infty} \frac{(-1)^m}{m!\,\Gamma(m - n + 1)} \left(\frac{x}{2}\right)^{2m-n}$$
$$= \sum_{l=0}^{\infty} \frac{(-1)^{l+n}}{(l + n)!\,\Gamma(l + 1)} \left(\frac{x}{2}\right)^{2l+n} \qquad \text{(III.4.133)}$$

Now, we note that $\Gamma(l + 1) = l!$ and $\Gamma(l + n + 1) = (l + n)!$ so that

$$J_{-n}(x) = \sum_{l=0}^{\infty} \frac{(-1)^{l+n}}{(l)!\,\Gamma(l + m + 1)} \left(\frac{x}{2}\right)^{2l+n} = (-1)^n J_n(x) \qquad \text{(III.4.134)}$$

Indeed $J_{-n}(x)$ and $J_n(x)$ differ by, at most, only a sign, and so are *not linearly* independent. The way out is to construct a series solution of the form that we encountered in Eq. (III.4.120). This defines what is called a Bessel function of the second kind, or a *Neumann* function.

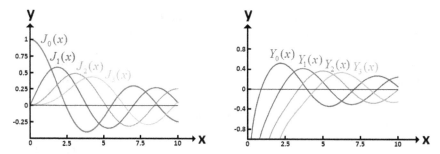

**Fig. III.4.12** Plots of the cylindrical Bessel functions, $J_n(x)$ and $Y_n(x)$, where $n$ is an integer. On the left, we show Bessel functions of the first kind, for the first four values of $n$. We see that $J_n(0) = 0$ for $n \geq 1$, and $J_0(0) = 1$; and that they have multiple (infinite) values of $x$ where $J_n(x) = 0$, the Bessel function zeros. On the right, we show Bessel functions of the second kind, again for the first four values of $n$. These functions look similar to $J_n(x)$, but with one important difference: $Y_n(x)$ diverges at $x = 0$. For $Y_0(x)$, the dominant behaviour close to $x = 0$ is $\sim \ln x$ and for $Y_n(x)$ with $n \geq 1$ it is $\sim 1/x^n$. At large values of $x$, the amplitudes of all the Bessel functions decrease $\propto 1/\sqrt{x}$ (see Eq. (III.4.138)).

The series expansion is unwieldy and we do not give it here (if you like, you can look it up in a specialist textbook or derive it yourself). A far more convenient and compact definition of these functions is (they can be defined for any value of $\alpha$):

$$Y_\alpha(x) = \frac{J_\alpha(x)\cos(\alpha\pi) - J_{-\alpha}(x)}{\sin(\alpha\pi)} \qquad \text{(III.4.135)}$$

For integer values $n$ they are defined as the $\alpha \to n$ limit of Eq. (III.4.135). The general solution to Eq. (III.4.124) for $\alpha = n$ is given as

$$y_n(x) = AJ_n(x) + BY_n(x) \qquad \text{(III.4.136)}$$

We show plots of both $J_n(x)$ and $Y_n(x)$ in Fig. III.4.12 for integer values. Note that both functions are oscillatory and they have an infinite number of zero values $x = x_j$, where $J_n(x_j) = 0$ and $Y_n(x_j) = 0$.

## Useful recursion relations and asymptotic forms for Bessel functions

The two types of Bessel functions satisfy the following recursion relations:

$$\frac{2\alpha}{x} J_\alpha(x) = J_{\alpha-1}(x) + J_{\alpha+1}(x)$$

$$\frac{2\alpha}{x} Y_\alpha(x) = Y_{\alpha-1}(x) + Y_{\alpha+1}(x) \tag{III.4.137}$$

$$2\frac{dJ_\alpha(x)}{dx} = J_{\alpha-1}(x) - J_{\alpha+1}(x)$$

$$2\frac{dY_\alpha(x)}{dx} = Y_{\alpha-1}(x) - Y_{\alpha+1}(x) \tag{III.4.138}$$

Also useful are the approximate forms for the Bessel functions at large $x$ ($\alpha$ fixed):

$$J_\alpha(x) \approx \sqrt{\frac{2}{\pi x}} \cos(x - \alpha\pi/2 - \pi/4)$$

$$Y_\alpha(x) \approx \sqrt{\frac{2}{\pi x}} \sin(x - \alpha\pi/2 - \pi/4) \tag{III.4.139}$$

Note that both expressions in Eq. (III.4.139) are not approximations but exact expressions when $\alpha = 1/2$.

**Example III.4.5 (Maths Practice ♪♪).** Suppose we want to solve

$$x^2\frac{d^2y}{dx^2} + x\frac{dy}{dx} + (x^2 - 1)y = 0 \tag{III.4.140}$$

with the boundary condition that $y(1) = 0$ and $y'(1) = b$. This is Bessel's equation with $\alpha = 1$, and thus we immediately write the general solution as

$$y(x) = AJ_1(x) + BY_1(x) \tag{III.4.141}$$

The boundary conditions then allow us to determine $A$ and $B$ through

$$0 = AJ_1(1) + BY_1(1) \quad b = AJ_1'(1) + BY_1'(1) \tag{III.4.142}$$

Thus, we find from Eq. (III.4.142) that

$$A = \frac{bY_1(1)}{Y_1(1)J_1'(1) - J_1(1)Y_1'(1)} \qquad B = -\frac{bJ_1(1)}{Y_1(1)J_1'(1) - J_1(1)Y_1'(1)}$$

$$\text{(III.4.143)}$$

**Exercise.** Through the definition of $Y_n(x)$, Eq. (III.4.135) and Bessel's equation (Eq. (III.4.124)), and by considering leading order terms in the power series expansion Eq. (III.4.127), prove that

$$Y_n(x)J_n'(x) - J_n(x)Y_n'(x) = 2/(\pi x) \qquad \text{(III.4.144)}$$

**Example III.4.5 (Continued).** Thus we have for our solution to the boundary value problem that

$$y(x) = \frac{\pi b(Y_1(1)J_1(x) + J_1(1)Y_1(x))}{2} \qquad \text{(III.4.145)}$$

### *Bessel-Fourier series*

Bessel functions of integer order come up when solving partial differential equations in cylindrical geometry. In fact, the Bessel functions of integer order are cylindrical (or circular) analogues of $\sin x$ and $\cos x$. To start to understand this analogy, suppose we have a function $f(x, y)$, defined over a rectangle ($0 \leq x \leq a$ and $0 \leq y \leq b$). We know that we can represent this as a 2D Fourier series:

$$f(x, y) = \sum_{n=-\infty}^{\infty} a_{n,m} \exp\left( i \left( \frac{2\pi nx}{a} + \frac{2\pi my}{b} \right) \right) \qquad \text{(III.4.146)}$$

where

$$a_{n,m} = \frac{1}{ab} \int_0^a dx \int_0^b dy f(x, y) \exp\left( -i \left( \frac{2\pi nx}{a} + \frac{2\pi my}{b} \right) \right)$$

$$\equiv \int_A f(\mathbf{R}) R_{n,m}(\mathbf{R}) d\mathbf{R} \qquad \text{(III.4.147)}$$

Here, $A$ is the area of the rectangle, and the $R_{n,m}(\mathbf{R}) = [1/(ab)] \exp(-i(2\pi nx/a + 2\pi my/b))$ function can be referred to as *rectangular harmonics*. Now, what about if we have a function defined over the domain of a circle? Can we write some kind of analogous

Fourier series? For circles it is convenient to use polar coordinates, $\phi$ and $R$, and write our function $g(R, \phi)$ over a circle of radius $a$. The answer to the raised question is yes, and that series will involve... Bessel functions. Such series expansions are very useful when we come to consider the solution of partial differential equations in cylindrical geometry.

To illustrate such a series let's consider for simplicity the case for the circle $g(a, \phi) = 0$. Then, we can represent the function $g(R, \phi)$ as a Fourier series in $\phi$:

$$g(R, \phi) = \sum_{n=-\infty}^{\infty} G_n(R) \exp(in\phi) \tag{III.4.148}$$

as the function must be periodic such that $g(R, \phi) = g(R, \phi + 2\pi)$ to remain single-valued, where we require that $G_n(a) = 0$. Now, if we want to expand $G_n(R)$ in terms of Bessel functions we would write

$$g(R, \phi) = \sum_{n=-\infty}^{\infty} G_n(R) \exp(in\phi) = \sum_{n=-\infty}^{\infty} \sum_{j=1}^{\infty} b_{n,j} J_n(k_{n,j} R) \exp(in\phi)$$

$$\tag{III.4.149}$$

To fulfil our imposed boundary condition $G_n(a) = 0$, these Bessel functions $J_n(k_{n,j} R)$ must vanish at $R = a$, so that $J_n(k_{n,j} a) = 0$. Thus, $k_{n,j}$ must be such that $k_{n,j} a$ are zeros of the Bessel function of order $n$. For each Bessel function of index $n$ we will start counting $j$ from those that correspond to the lowest values of $k_{n,j} a$. Note that at $R = 0$, we can already see from Eq. (III.4.127) (for integer values) that $J_n(0) = 0$ for $n > 0$, and $J_0'(0) = 0$ (see Fig III.4.12). Now, since Bessel functions satisfy a Sturm–Liouville equation, Eq. (III.4.126), and we have appropriate boundary conditions on them (Eq. (III.4.34)), we can write an orthogonality condition for the Bessel functions:

$$\int_0^a J_n(k_{n,j} R) J_n(k_{n,k} R) R dR = \frac{\delta_{j,k}}{2} a^2 J_{|n|+1}(ak_{n.j})^2 \tag{III.4.150}$$

For the orthogonality condition, we require $j = k$ for the integral to be non zero. This arises from the Sturm–Liouville form,

Eq. (III.4.126): for a discussion on how this is derived see the boxed part of Section 4.1. For $j = k$, we can derive Eq. (III.4.150), by first considering the indefinite integral (you can find it in tables of integrals):

$$\int J_n(kR)J_n(kR)RdR = \frac{R^2}{2}[J_n(kR)^2 - J_{n-1}(kR)J_{n+1}(kR)]$$

$$(\text{III.4.151})$$

Note also that using $J_{-n}(x) = (-1)^n J_n(x)$, we can write

$$J_n(kR)^2 = J_{-n}(kR)^2 = J_{|n|}(kR)^2$$

$$J_{n-1}(kR)J_{n+1}(kR) = J_{1-n}(kR)J_{-n-1}(kR) = J_{|n|-1}(kR)J_{|n|+1}(kR)$$

$$(\text{III.4.152})$$

Now to validate Eq. (III.4.150), we first use that fact that for $k = k_{n,j}$, $J_n(k_{n,j}a) = 0$. Due to the recursion relation $\frac{2n}{x}J_n(x) = J_{n-1}(x) + J_{n+1}(x)$ it follows that $J_{|n|-1}(k_{n,j}a) = -J_{|n|+1}(k_{n,j}a)$, and so from Eq. (III.4.151) we obtain Eq. (III.4.150).

We can use Eq. (III.4.150) to calculate the coefficients $b_{n,j}$ in the sum of Eq. (III.4.149), so that

$$\frac{1}{2\pi}\int_0^{2\pi}\int_0^a g(R,\phi)J_n(k_{n,k}R)e^{-in\phi}RdRd\phi = \frac{a^2}{2}J_{|n|+1}(ak_{n,j})^2 b_{n,j}$$

$$(\text{III.4.153})$$

or alternatively, introducing the *cylindrical harmonics*, $C_{n,j}(\mathbf{R}) = \frac{J_n(k_{n,k}R)e^{-in\phi}}{\pi a^2 J_{n+1}(ak_{n,j})^2}$, we may write

$$\int_A g(\mathbf{R})C_{n,j}(\mathbf{R})d\mathbf{R} = b_{n,j} \qquad (\text{III.4.154})$$

where $A$ is over the area of a circle.

All these derivations may seem a bit abstract, but they serve a purpose. With the help of such series we can solve a wide class of physical problems that have a cylindrical symmetry. We demonstrate its power later, when we get to partial differential equations. Below we will show two maths practice examples demonstrating the routine of finding the coefficients of such series.

**Example III.4.6 (Maths Practice ♫♪).** Let us take some simple function the behaviour of which we know, defined within the domain of a circle of radius $a$, to show how a Bessel-Fourier series (Eq. (III.4.149)) works. For instance, consider the function which, in polar coordinates, reads $g(R, \phi) = (R - a) \cos \phi$. Let's calculate the coefficients $b_{n,j}$ of the series and see how the latter represents this function. Because $g(a, \phi) = 0$, we can apply the orthogonality condition Eq. (III.4.150). Using Eq. (III.4.153), we can write

$$\int_0^{2\pi} \int_0^a (R - a) \cos \phi J_n(k_{n,j}R)e^{-in\phi} RdRd\phi = \frac{a^2}{2} J_{n+1}(ak_{n,j})^2 b_{n,j}$$

(III.4.155)

The condition $g(a, \phi) = 0$ also imposes a constraint on the values $k_{n,j}$: according to Eq. (III.4.149) they must satisfy

$$J_n(k_{n,j}a) = 0 \qquad (III.4.156)$$

As $\cos \phi = (\exp(i\phi) + \exp(-i\phi))/2$, after integration over $\phi$ in the l.h.s of Eq. (III.4.155) only $n = 1$ and $n = -1$ will give non-zero result; thus all other $b$-coefficients are zero. That means that

$$b_{1,j} = \frac{1}{a^2 J_2(ak_{1,j})^2} \int_0^a (R - a)J_1(k_{1,j}R)RdR$$

$$b_{-1,j} = \frac{1}{a^2 J_2(ak_{-1,j})^2} \int_0^a (R - a)J_{-1}(k_{-1,j}R)RdR$$

(III.4.157)

Don't worry about any zeros of $J_2(ak_{1,j})$, because $J_2(ak_{1,j}) \neq J_1(ak_{1,j}) = 0$, $J_2(ak_{-1,j}) \neq J_{-1}(ak_{-1,j}) = 0$. Now, simply because $J_{-1}(x) = -J_1(x)$, as learnt previously in this section, we have that $k_{-1,j} = k_{1,j}$, and so $b_{-1,j} = -b_{1,j}$. We can also rescale the first integral in Eq. (III.4.157) by $a$ so that

$$I = \int_0^1 (x - 1)J_1(x_{1,j}x)xdx = J_2(x_{1,j})^2 b_{1,j}/a \qquad (III.4.158)$$

where the $x_{1,k}$ are the zeros of $J_1(x)$ where $J_1(x_{1,k}) = 0$. Let's consider the first five zeros of $J_1(x)$ given in a table along with the associated values of $I$ and $J_2(x_{1,j})^2$ that have been calculated numerically, once and for all.

**Table III.4.1** We show the values of the first five zeros of $J_1(x)$, not including $x = 0$. Using these values we compute the integral given by Eq. (III.4.158) and $J_2(x_{1.j})^2$. These table entries are used to calculate the coefficients $A_{1,j}$ in the Bessel-Fourier series of the function $(R - a)\cos\phi$.

| $j$ | $x_{1,j}$ | $I$ | $J_2(x_{1.j})^2$ |
|-----|-----------|-----------|-------------------|
| 1 | 3.832 | −0.07439 | 0.1622 |
| 2 | 7.016 | −0.01949 | 0.09007 |
| 3 | 10.173 | −0.009893 | 0.06235 |
| 4 | 13.324 | −0.005542 | 0.04768 |
| 5 | 16.471 | −0.003730 | 0.03860 |

The data in Table III.4.1 helps us to determine the first five term in the series solution for $(R - a)\cos\phi$. We have, included also three more terms of that series (find them yourself as an exercise).

In addition, we have included three more

$$(R - a)\cos\phi \approx -a(e^{i\phi} + e^{-\phi})(0.917 J_1(x_{1,1}R/a) + 0.433 J_1(x_{1,2}R/a)$$
$$+ 0.317 J_1(x_{1,3}R/a) + 0.233 J_1(x_{1,4}R/a)$$
$$+ 0.193 J_1(x_{1,5}R/a) + 0.159 J_1(x_{1,6}R/a)$$
$$+ 0.139 J_1(x_{1,7}R/a) + 0.121 J_1(x_{1,8}R/a))$$

$$(III.4.159)$$

**Exercise.** In the same manner as above, find the coefficients in the series of Eq. (III.4.159) for these extra three terms. For this you need to compute first the Bessel function zeros $x_{1,6}$, $x_{1,7}$, and $x_{1,8}$.

**Example III.4.6 (Continued).** We plot the exact function and series in Fig. III.4.13.

It is important to realize that the series fails as we move to $R = 0$. This is because $J_1(x)$ is always zero at $x = 0$, and therefore the r.h.s of Eq. (III.4.159) vanishes at $R = 0$, whereas the value of the function does not: $(R - a)|_{R=0} = -a$. To capture smaller and smaller $R$ we need more and more terms in the series. If we had an infinite number of terms, we would still have discontinuity at $R = 0$. Note also that we can consider functions where $g(a, \phi) \neq 0$; however, as we require

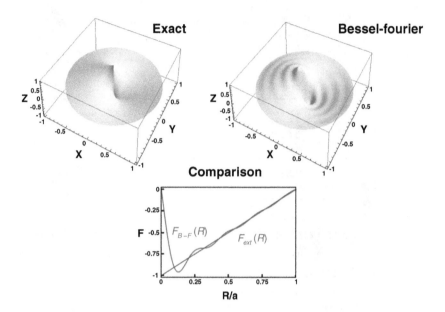

**Fig. III.4.13**   In these plots we compare the exact function $(R-a)\cos\phi$ with the result given by the eight first terms given by Eq. (III.4.158) in the Bessel-Fourier expansion. The top left plot is a surface plot of the function $(R-a)\cos\phi$, and in the top right is a similar plot of the Bessel-Fourier-series expansion, Eq. (III.4.158), (both plotted for $a = 1$, $R \leq 1$). We see that series expansion captures most of the features; however, discrepancies are slight oscillations (ripples) around the surface represented by the function and the peak and the trough are not so sharp as for the exact function. In the bottom plot we compare the exact radial part of the function $F_{ext}(R) = (R-a)$ with that computed by the Bessel-Fourier series $F_{B-F}(R)$, given by Eq. (III.4.158). We see that agreement is very good for $R/a > 0.5$. Agreement between the series expansion and the exact function can be improved by including more and more terms in the series (although this will not not help at $R$ very close to 0, read the text).

that the Bessel functions be zero at the values $R = a$, we'll have the same problem in being not able to have a series that accurately describes that point. Note that we had the same kind of problem with Fourier series in Chapter 2 of Part II, Vol. 1, but we already have seen its utility. We chose this example also to highlight the problem; some caution is always needed with series representations of functions. Nevertheless, they are very useful, especially when it comes to the solution of partial differential equations.

**Example III.4.7 (Maths Practice ♫♪).** A classic example of a function of both $R$ and $\phi$, in which the dependence on these variables is inseparable, is $f(R, \varphi) = \exp(ik_r R \cos \phi)$, appearing, for example, in the theory of scattering phenomena. This can be written in terms of a series of Bessel functions (in the domain $0 \leq x \leq \infty$) as

$$\exp(ik_r R \cos \phi) = \sum_{n=-\infty}^{\infty} (i)^n J_n(k_r R) \exp(in\phi) \qquad \text{(III.4.160)}$$

Such a series requires that the following integrals must hold (due to the orthogonality conditions of both $\exp(in\phi)$ and $J_n(k'_r R)$):

$$\frac{1}{2\pi} \int_0^{2\pi} \exp(ik_r R \cos \phi - in\phi) = (i)^n J_n(k_r R) \qquad \text{(III.4.161)}$$

$$\frac{1}{2\pi} \int_0^{\infty} RdR \int_0^{2\pi} J_n(k'_r R) \exp(ik_r R \cos \phi - in\phi) = \frac{(i)^n \delta(k - k')}{k}$$

$$\text{(III.4.162)}$$

The series given by Eq. (III.4.160), as well as the integrals, are very useful when considering 2D Fourier transforms in polar coordinates.

### Hankel transforms

We can extend $a$ to infinity in our Bessel series defined by Eq. (III.4.149). In this case, the spacing between the $k_{n.j}$ becomes infinitesimally small and we have a continuum of values. Here, we can present any smooth function of the radial coordinate $H(R)$ as

$$H(R) = \int_0^{\infty} \tilde{H}(k) J_n(kR) k\, dk \qquad \text{(III.4.163)}$$

When $a \to \infty$, the orthogonality condition (Eq. (III.4.150)) becomes

$$\int_0^{\infty} J_n(k'R) J_n(kR) R\, dR = \frac{\delta(k - k')}{k} \qquad \text{(III.4.164)}$$

This allows us to write the Bessel function analogue of the Fourier transform (called a Hankel transform, named so after a mathematician about whom we will speak more in Chapter 8 of this part of the book).

$$\int_0^\infty H(R) R J_n(kR) dR = \tilde{H}(k) \qquad \text{(III.4.165)}$$

and Eq. (III.4.163) is the inverse transform. We'll use these transforms to solve the inhomogeneous Bessel equation in the next section.

## *Modified Bessel functions and the modified Bessel equation*

Now are there any Bessel function analogues to the exponential functions $\exp(-\alpha x)$ and $\exp(\alpha x)$, and if so what equation do they solve? We recollect that $\cos(\alpha x)$ and $\sin(\alpha x)$ solve

$$\frac{dy(x)}{dx^2} = -\alpha^2 y(x) \qquad \text{(III.4.166)}$$

Now if we make the change $x \to ix$, we indeed get the equation that the exponential functions solve. Therefore, let's make the same change ($x \to ix$) to Bessel's equation, whereupon we obtain

$$x^2 \frac{d^2 y}{dx^2} + x \frac{dy}{dx} - (x^2 + \alpha^2) y = 0 \qquad \text{(III.4.167)}$$

This is called the *Modified Bessel equation* and is solved by modified Bessel functions which are the cylindrical analogues of exponential functions. There are two types of them. The first type are called *Modified Bessel Functions of the first kind*, which play an analogous role in the cylindrical geometry to $\exp(\alpha x)$ (more precisely, the hyperbolic functions $\cosh(\alpha x)$ and $\sinh(\alpha x)$, as we'll see). For positive $x$ they exponentially increase. Such functions are denoted by $I_\alpha(x)$. In the same way as $J_\alpha(x)$, these functions can be represented as a power series, so that

$$I_\alpha(x) = \sum_{m=0}^\infty \frac{1}{m! \Gamma(m + \alpha + 1)} \left(\frac{x}{2}\right)^{2m+\alpha} \qquad \text{(III.4.168)}$$

When $\alpha$ is non-integer we get the general solution to Eq. (III.4.167) in the form

$$y(x) = AI_\alpha(x) + BI_{-\alpha}(x) \qquad \text{(III.4.169)}$$

When $\alpha = n + 1/2$ the modified Bessel functions can be expressed in terms of $\cosh x$ and $\sinh x$, and are generated from the formulae

$$I_{n+1/2}(x) = \sqrt{\frac{2x}{\pi}} x^n \left[\frac{1}{x}\frac{d}{dx}\right]^n \frac{\sinh x}{x}$$
$$I_{-n-1/2}(x) = \sqrt{\frac{2x}{\pi}} x^n \left[\frac{1}{x}\frac{d}{dx}\right]^n \frac{\cosh x}{x} \qquad \text{(III.4.170)}$$

Again, these functions crop up when we solve partial differential equations in spherical geometry. We can define *modified spherical Bessel functions* as $i_n(x) = \sqrt{\frac{\pi}{2x}} I_{n+1/2}(x)$.

When $\alpha = n$, where $n$ is integer, again $I_n(x)$ and $I_{-n}(x)$ are no longer independent of each other, and we have that $I_n(x) = I_{-n}(x)$. For a second independent solution, we need to introduce modified Bessel functions of the second kind. These are defined as

$$K_\alpha(x) = \frac{\pi}{2} \frac{I_{-\alpha}(x) - I_\alpha(x)}{\sin(\alpha\pi)} \qquad \text{(III.4.171)}$$

We can also define modified spherical Bessel functions of the second kind as $k_n(x) = \sqrt{\frac{2}{\pi x}} K_{n+1/2}(x)$. The modified Bessel functions satisfy the recursion relationships

$$\frac{2\alpha}{x} I_\alpha(x) = I_{\alpha-1}(x) - I_{\alpha+1}(x)$$

$$-\frac{2\alpha}{x} K_\alpha(x) = K_{\alpha-1}(x) - K_{\alpha+1}(x) \qquad \text{(III.4.172)}$$

$$2\frac{dI_\alpha(x)}{dx} = I_{\alpha-1}(x) + I_{\alpha+1}(x)$$

$$-2\frac{dK_\alpha(x)}{dx} = K_{\alpha-1}(x) + K_{\alpha+1}(x) \qquad \text{(III.4.173)}$$

We plot the Bessel functions $I_n(x)$ and $K_n(x)$ in Fig. III.4.14 for some values of $n$. We see that the $I_n(x)$ are monotonically increasing functions with $x$, whereas the $K_n(x)$ are monotonically decreasing.

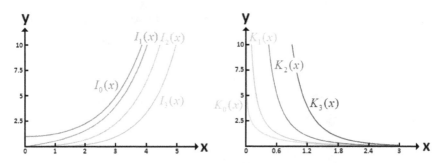

**Fig. III.4.14**   Plots of the modified cylindrical Bessel functions (shown for indicated integer values of $n$). These functions are monotonic. On the left we show modified Bessel functions of the first kind, which at large values of $x$ increase exponentially with increasing $x$. On the right we show modified Bessel functions of the second kind. These, at large values, diminish exponentially with increasing $x$. The modified Bessel functions $K_n(x)$ are also singular at $x = 0$. For $n = 0$, the dominant behaviour close to $x = 0$ is $K_0(x) \sim \ln x$, for $n \neq 0$, $\underbrace{K_n(x)}_{x \to 0} \sim 1/x^n$.

Also useful are the approximations of the modified Bessel functions for large $x$ (regardless of order):

$$I_\alpha(x) \approx \frac{e^x}{\sqrt{2\pi x}}, \quad K_\alpha(x) \approx \sqrt{\frac{\pi}{2x}} e^{-x} \qquad \text{(III.4.174)}$$

**Example III.4.8 (Physical Application ▶▶).**
Here, we consider a physical example of using Bessel functions, which does not rely on partial differential equations (we have yet to cover these), but uses vectors. This example comes from the theory of elasticity. We'll look at a vertically standing elastic rod and calculate the critical length at which it starts to buckle. Supposedly way back in the days of yore, long before the days of Pharaoh Tut, there were those unfortu-

nate builders of the Tower of Babel, who didn't have the handy knowledge that we're going impart to you now, through applying some maths you've learnt. They built too high: the tower buckled and collapsed. If you buy that ancient tall story, had ancient Baby-lonians/Sumerians grasped what we hope you will grasp, perhaps we'd all be speaking some dialect of ancient Sumerian. ... Unlike in

English, there are 12 noun Cases, and the Sumerian number system had base 60, not 10...so then again, perhaps, it's not such a bad thing they didn't. So, how tall a tower can you build and still be sure it will not buckle?

Recall that we have already talked about the elastic energy due to bending which for a narrow rod has the form

$$E_{el} = \frac{B}{2} \int_0^L \left( \frac{d\hat{t}(s)}{ds} \right)^2 ds \qquad \text{(III.4.175)}$$

Here, $B$ is the bending elastic modulus of the rod, $\hat{t}$ is the tangent vector of the line describing the rod and $s$ is the unit arc length. We will describe the rod by the curve $\mathbf{r}(s) = x(s)\hat{i} + y(s)\hat{j}$, as shown in Fig. III.4.15.

Thus, we'll consider the rod standing in $x-y$ plane, where $y$ will be the vertical direction. As we learnt, the tangent vector is then given by

$$\hat{t}(s) = \frac{dx(s)}{ds}\hat{i} + \frac{dy(s)}{ds}\hat{j} = \sin\theta(s)\hat{i} + \cos\theta(s)\hat{j} \qquad \text{(III.4.176)}$$

parametrized through angle $\theta(s)$ that it makes with the $y$-axis. Thus, we find that we can write Eq. (III.4.175) as

$$E_{el} = \frac{B}{2} \int_0^L \left( \frac{d\theta(s)}{ds} \right)^2 ds \qquad \text{(III.4.177)}$$

Now, let's suppose we have some moment acting on the rod which is a function of the angle $\theta$ along the rod $M(\theta)$. Then it must be the case that the work done by this moment (which is got by integrating it over theta) should be equal to the elastic energy. Thus, we must have that

$$\frac{B}{2} \int_0^L \left( \frac{d\theta(s)}{ds} \right)^2 ds = \int_0^{\theta(L)} M(\theta)d\theta = \int_0^L \tilde{M}(s)\frac{d\theta}{ds}ds \qquad \text{(III.4.178)}$$

This tells us that we must have

$$B\frac{d\theta}{ds} = \tilde{M}(s) \qquad \text{(III.4.179)}$$

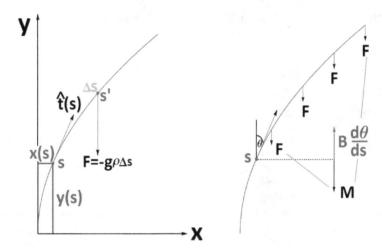

**Fig. III.4.15** We show schematic pictures of the mechanics of a vertical rod bent away from its equilibrium position. On the left we show how the positions of the rod can be characterized by two coordinates that depend on $s$, such that the position vector is $\mathbf{r}(s) = x(s)\hat{\mathbf{i}} + y(s)\hat{\mathbf{j}}$. At each position on the rod we may define a tangent vector $\hat{\mathbf{t}}(s) = d\mathbf{r}(s)/ds$; its rate of change can be used to characterize the degree of bending. Above the point $s$, we may consider a small segment of length $\Delta s$ centred at point $s'$ along the rod. The mass of this segment $\rho\Delta s$ (where $\rho$ is rod's linear mass density) exerts a gravitational force $F = -g\rho\Delta s$ downwards. On the right, we show how the orientation of the tangent vector of the rod can be parametrized through an angle $\theta(s)$, the one it makes with the y-axis. Furthermore, the degree of bending can be quantified through the derivative $d\theta(s)/ds$. We consider the gravitational force, acting on each segment, acting at points $s'$ along the rod. Each segment gives a contribution $\Delta\mathbf{M} = (\mathbf{r}(s') - \mathbf{r}(s)) \times \mathbf{F}$, to the moment $M$ about the point $s$. For the rod to remain in equilibrium, this must be balanced by the internal moment $Bd\theta(s)/ds$, which comes from elastic resistance to bending. From this equilibrium, we can write down a differential equation that describes the shape of the rod, as discussed in the text.

The l.h.s of Eq. (III.4.179) is the moment from elastic forces that balance the applied moment $\tilde{M}(s)$, keeping the rod in equilibrium. Now, if we suppose that the rod is already bent, then gravity supplies the moment $\tilde{M}(s)$. Let's calculate it. The rod is fixed at its base, so that $\theta(0) = 0$ and the highest part of the rod is at $s = L$. Each small section of the rod has mass $\rho\Delta s$ (here we denoted by $\rho$ the mass density of the rod multiplied by its cross-section area; this may be called *linear mass density*), so the force due to gravity of that mass is $\mathbf{F} = -g\rho\Delta s\hat{\mathbf{j}}$. Now, let's consider a point $s$ along the rod, and above that another point $s'$. At the point, $s'$ the force of gravity is acting and wants to bend the rod; it tries to rotate the tangent vector of the rod about point $s$. Thus, gravity acting at the point $s'$ contributes a moment at the point $s$, which we can calculate

$$\Delta \tilde{\mathbf{M}} = (\mathbf{r}(s') - \mathbf{r}(s)) \times \mathbf{F} = -(\mathbf{r}(s') - \mathbf{r}(s)) \times g\rho\Delta s\hat{\mathbf{k}}$$

$$= \begin{vmatrix} \hat{\mathbf{i}} & \hat{\mathbf{j}} & \hat{\mathbf{k}} \\ x(s) - x(s') & y(s) - y(s') & 0 \\ 0 & -g\rho\Delta s & 0 \end{vmatrix} = g\rho(x(s') - x(s))\Delta s\hat{\mathbf{k}}$$

$$\text{(III.4.180)}$$

The situation is outlined in Fig. III.4.15. Summing up all such contributions, the magnitude of the total moment about the point at $s$ from points above $s$, due to gravity is

$$\tilde{M}(s) = g\rho \int_s^L (x(s') - x(s))ds \qquad \text{(III.4.181)}$$

At every point $s$, this is balanced by the elastic moment through Eq. (III.4.179). Thus, combining Eqs. (III.4.179) and (III.4.181), we have that

$$B\frac{d\theta}{ds} = g\rho \int_s^L (x(s') - x(s))ds' \qquad \text{(III.4.182)}$$

We can differentiate both sides of Eq. (III.4.182) with respect to $s$ so that

$$B\frac{d^2\theta}{ds^2} = g\rho(x(s') - x(s))|_{s'=s} - \frac{dx(s)}{ds}g\rho \int_s^L ds'$$

$$\Rightarrow B\frac{d^2\theta}{ds^2} = \frac{dx(s)}{ds}g\rho(s - L) \qquad \text{(III.4.183)}$$

$$\Rightarrow \frac{d^2\theta(s)}{ds^2} = \frac{g\rho}{B}\sin\theta(s)(s - L)$$

Note that in writing the bottom line, we have used the definition of the tangent vector $\hat{\mathbf{t}}(s)$, Eq. (III.4.176). Solving Eq. (III.4.183) in general is hard, as it is non-linear and it is more complicated than the sine-Gordon equation we considered in Chapter 6 of Part II, Vol. 1.

However, here, we'll consider small bending such that $\sin\theta(s) \approx \theta(s)$. This allows us to write

$$\frac{d^2\tilde\theta(t)}{dt^2} + t\tilde\theta(t) = 0 \qquad (\text{III}.4.184)$$

where, to work neatly (as always!), we have introduced the dimensionless variable $t = (g\rho/B)^{1/3}(L - s)$ and $\tilde\theta(t) = \theta(s)$. The general solution to Eq. (III.4.184) can be written in terms of Bessel functions:

$$\tilde\theta(t) = At^{1/2}J_{1/3}(2t^{3/2}/3) + Ct^{1/2}J_{-1/3}(2t^{3/2}/3) \qquad (\text{III}.4.185)$$

**Exercise.** Check that Eq. (III.4.185) is indeed the solution to Eq. (III.4.184).

**Example III.4.8 (Continued).** Now, at equilibrium at the free end of the rod ($s = L$) there is no gravitational moment as there is no mass above this point (you can see this from Eq. (III.4.182)). This means (from Eq. (III.4.179)) that we require

$$\left.\frac{d\tilde\theta(t)}{dt}\right|_{t=0} = 0 \qquad (\text{III}.4.186)$$

We can consider the small $t$ behaviour of both Bessel functions using the series expansion. For small values of arguments, from Eq. (III.4.127) we get

$$J_{1/3}(x) \approx \frac{x^{1/3}}{2^{1/3}\Gamma(4/3)}\left(1 - \frac{3x^2}{16}\right)$$

$$J_{-1/3}(x) \approx \frac{2^{1/3}}{\Gamma(2/3)x^{1/3}}\left(1 - \frac{3x^2}{8}\right) \qquad (\text{III}.4.187)$$

From here it is easy to see (and you should check it yourself) that to satisfy the condition Eq. (III.4.186) we require that in Eq. (III.4.185) $A = 0$. This is because only the gradient of the second function in Eq. (III.4.187) is zero at the point $t = 0$. Thus, the solution reads as

$$\tilde\theta(t) = Ct^{1/2}J_{-1/3}(2t^{3/2}/3) \qquad (\text{III}.4.188)$$

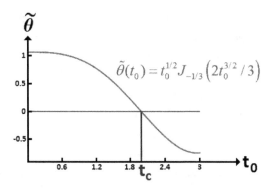

**Fig. III.4.16** We plot the function $\tilde{\theta}(t_0) = t_0^{1/2} J_{-1/3}(2t_0^{3/2}/3)$ which is obtained as a solution to the linearized equation, Eq. (III.4.180), at the fixed end of the rod. Through the boundary condition that $\theta = 0$ at $s = 0$, the value $t_0 = t_c$ at which of $\tilde{\theta}(t_0)$ function crosses zero to a critical length of the rod $L_c$. Below this value, the coefficient $C$ that multiplies this function is required to be zero and the rod stays straight. At $L_c$, $C$ can take any value and the rod is unstable to bending. To predict the shape of the bent rod for $L \geq L_c$ the solution of the full non-linear equation, Eq. (III.4.184) is required.

Now, we recollect that also we must have $\theta(s = 0) = 0$. This means that at the rescaled position $t = t_0 = (g\rho/B)^{1/3}L$ (which corresponds to $s = 0$) we require that

$$Ct_0^{1/2} J_{-1/3}(2t_0^{3/2}/3) = 0 \;\Rightarrow\; J_{-1/3}\left(2t_0^{3/2}/3\right) = 0 \text{ or } C = 0$$

$$(\text{III.4.189})$$

Let's plot $J_{-1/3}(2t_0^{3/2}/3)$. We do so in Fig. III.4.16. If $L$ is too small, we see that the condition $J_{-1/3}(2t_0^{3/2}/3) = 0$ is never satisfied, and we must have $C = 0$, which corresponds to a straight rod. However, there is a critical length at which $J_{-1/3}(2t_0^{3/2}/3) = 0$ is satisfied. Above this length, $\theta(s)$ is no longer required to be zero. At this point, the rod is unstable to bending, as we have no restriction on $C$ and so the degree of bending is unlimited (actually, what limits the size of the bending are the non-linear terms in $\theta(s)$ that we neglected in Eq. (III.4.183) when taking small angle approximation). The critical length at which buckling can occur (with the slightest of pushes) is

$$2t_0^{3/2}/3 \approx 1.87 \;\Rightarrow\; (g\rho/B)^{1/2} L_c^{3/2} \approx 2.81 \;\Rightarrow\; L_c \approx 1.99(B/g\rho)^{1/3}$$

$$(\text{III.4.190})$$

The result makes sense in two ways: (i) The stiffer the rod (the larger $B$), the longer would be the critical height; and (ii) The critical height would be shorter on Jupiter, where $g$ is 9 times larger than on Earth, or if the rod was denser. But, does the 1/3 exponent make sense? We should point out that by using back of the envelope... what we theorists like to call very short calculations...dimensional analysis, which we discussed at the end Part I, in Vol. 1, we would have immediately got $L_c \sim (B/g\rho)^{1/3}$, verifying the exponent (how else can you get from $B/g\rho$ a quantity of the dimensionality of length?). The above presented sophisticated analysis, however, has got us the numerical factor $\approx 2$, which is important, as it 'allows you to build a two times taller tower than if 'restricted' by a scaling estimate'.

**Example III.4.9 (Physical Application ▶▶).** Here, we look at an example that uses modified Bessel functions. In the previous chapter we introduced Poisson's equation and the field of a uniformly charged cylinder. Now, let us suppose that the cylinder of constant charge density, of radius $b$, is immersed in electrolyte solution (as illustrated in Fig. III.4.17) and we want to calculate the electrostatic potential. Such a cylinder might model a large, rod-like, weakly

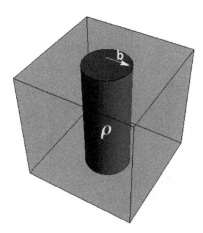

**Fig. III.4.17** A cylinder of radius $b$ and uniform charge density $\rho$ submerged in electrolyte solution.

charged macro-ion in a solution, and we want to estimate the electric potential $\varphi$ about it. Here, we'll suppose for simplicity that the dielectric constants inside and outside are both $\varepsilon$. But the charge density still depends only on the radial coordinate $R$ as well as the potential. Now because of the cylindrical symmetry of the system, the electrostatic potential and its minus derivative—electric field—depend only on $R$, and the Laplacian term $\vec{\nabla}^2\varphi$ (which we encountered in Chapter 3 of Part III, Vol. 2) simplifies to

$$\vec{\nabla}^2\varphi(R) = \frac{1}{R}\frac{d}{dR}\left[R\frac{d\varphi(R)}{dR}\right] \tag{III.4.191}$$

**Exercise.** Show that Eq. (III.4.191) holds by writing $\varphi(R) = \varphi(\sqrt{x^2 + y^2})$ and using the form of the Laplacian operator that we introduced in Eq. (III.3.17).

**Example III.4.9 (Continued).** Thus, we have for Poisson's equation in SI units (Eq. III.3.114) (don't worry those who hate to use them in electrostatics, as much as we do—we will soon go dimensionless):

$$\frac{1}{R}\frac{d}{dR}\left[R\frac{d\varphi(R)}{dR}\right] = -\frac{\rho(R)}{\varepsilon\varepsilon_0} \tag{III.4.192}$$

where $\varepsilon_0$ is the dielectric constant of vacuum. The charge density $\rho(R)$ comprises two contributions, one from the charge density of the cylinder, $\rho$, and the other from the ions in the electrolyte, $\rho_{ion}(R)$. For the electrolytic solution in which of positive and negative ions are present in equal proportions (called a 1:1 electrolyte), the charge density, in the what is called the mean-field approximation, can be given by a combination of two Boltzmann exponentials

$$\rho_{ion}(R) = Zec_0\left[\exp\left(-\frac{eZ\varphi(R)}{k_BT}\right) - \exp\left(\frac{eZ\varphi(R)}{k_BT}\right)\right] \tag{III.4.193}$$

Here, $\varphi(R)$ is the self-consistent electrostatic potential caused by charges of all ions and the charges of the rod. The exponentials reflect the competition between the electrostatic force on the cations and anions, respectively, and the entropy due to thermal fluctuations.

Here, $e$ is the fundamental charge, and $Z$ is the magnitude of the valence of the ions. For simplicity we considered here the case of a 1:1 electrolyte, in which both type of ions must have equal and opposite charge, and this is reflected in the form of Eq. (III.4.193). The parameter $c_0$ stands for the concentrations of both positive and negative ions in the bulk (at $R \to \infty$), which must be equal to conserve electro-neutrality of the bulk. Combining the last two equations we get a non-liner differential equation on $\varphi(R)$. It is too hard to deal with it, so, we limit the following discussion to the case when $eZ\varphi(R)/k_B T \ll 1$, so that we can expand the exponentials to get

$$\frac{\rho_{ion}(R)}{\varepsilon\varepsilon_0} = -\frac{2e^2 Z^2 c_0 \varphi(R)}{\varepsilon\varepsilon_0 k_B T} = -\frac{1}{R_D^2}\varphi(R) \qquad (\text{III.4.194})$$

where $R_D$ is the Debye screening length, which we considered in one of the examples of Chapter 10, Part I, Vol. 1. Thus we can write down equations for $\varphi(R)$ inside and outside the cylinder:

$$\frac{1}{R}\frac{d}{dR}\left[R\frac{d\varphi(R)}{dR}\right] = -\frac{\rho}{\varepsilon\varepsilon_0} \quad 0 \le R \le b \qquad (\text{III.4.195})$$

$$\frac{1}{R}\frac{d}{dR}\left[R\frac{d\varphi(R)}{dR}\right] = \frac{1}{R_D^2}\varphi(R) \quad b \le R \le \infty \qquad (\text{III.4.196})$$

Let's first solve for the potential in the cylinder, for which $R \le b$, where we have that

$$\frac{1}{R}\frac{d}{dR}\left[R\frac{d\varphi(R)}{dR}\right] = -\frac{\rho}{\varepsilon\varepsilon_0} \;\Rightarrow\; R\frac{d\varphi(R)}{dR} = -\frac{\rho R^2}{2\varepsilon\varepsilon_0}$$

$$\Rightarrow \frac{d\varphi(R)}{dR} = -\frac{\rho R}{2\varepsilon\varepsilon_0} \;\Rightarrow\; \varphi(R) = -\frac{\rho R^2}{4\varepsilon\varepsilon_0} + \varphi_0 \qquad (\text{III.4.197})$$

When solving for $\varphi(R)$ we have the requirement (due to Gauss's law) that the field is zero at the centre of the cylinder, so there is no constant of integration the first time we integrate up in Eq. (III.4.197) but $\varphi_0$ has to be kept for now as a constant of the second integration. Now, we require that both the electric field and potentials are continuous across the surface of the cylinder. This means, firstly (from Eq. (III.4.197)) that we have the boundary condition on the potential outside the cylinder as

$$\frac{d\varphi(R)}{dR}\bigg|_{R=b} = -\frac{\sigma}{\varepsilon\varepsilon_0} \qquad \text{(III.4.198)}$$

where the effective surface charge density is $\sigma = \rho b/2$. This makes sense, as the total charge of the cylinder is $Q = 2\pi bh\sigma$, where $h$ would be the height of the cylinder (where we assume $h \gg R_D$, to neglect any edge effects), and since also $Q = \pi b^2 h\rho = V\rho$, where $V$ is the volume of the cylinder. Indeed, Eq. (III.4.198) can also be derived from Gauss's law by constructing a cylinder through which the flux passes through close to the cylinder surface (as we considered in Example III.3.8).

Secondly, we require that the electric field is zero in the bulk, at $R = \infty$, so that

$$\frac{d\varphi(R)}{dR}\bigg|_{R=\infty} = 0 \qquad \text{(III.4.199)}$$

To solve Eq. (III.4.196), subject to the boundary conditions Eq. (III.4.198) and Eq. (III.4.199), we first introduce a dimensionless variable $s = R/R_D$ such that $\tilde{\varphi}(s) = \varphi(R)$, so that

$$\frac{1}{s}\frac{d}{ds}\left[s\frac{d\tilde{\varphi}(s)}{ds}\right] = \tilde{\varphi}(s) \Rightarrow s^2\frac{d^2\tilde{\varphi}(s)}{ds^2} + s\frac{d\tilde{\varphi}(s)}{ds} - s^2\tilde{\varphi}(s) = 0$$

$$\text{(III.4.200)}$$

Equation (III.4.200) is the modified Bessel equation, Eq. (III.4.167), with $\alpha = 0$. Thus, it is easy to see that the general solution should read as

$$\varphi(R) = AI_0(R/R_D) + BK_0(R/R_D) \qquad \text{(III.4.201)}$$

To satisfy Eq. (III.4.199), we require that $A = 0$, and to satisfy Eq. (III.4.198) we have that

$$\frac{B}{R_D}\frac{dK_0(s)}{ds}\bigg|_{s=b/R_b} = -\frac{B}{R_D}K_1(R/R_D) = -\frac{\sigma}{\varepsilon\varepsilon_0}$$

$$\Rightarrow B = \frac{R_B\sigma}{\varepsilon\varepsilon_0 K_1(R/R_B)} \qquad \text{(III.4.202)}$$

Note that you can show that the derivative of $\frac{dK_0(s)}{ds}$ is $-K_1(s)$, through $K_{-1}(s) = K_1(s)$ and the recursion relation given in

Eq. (III.4.173). Thus we get for the potential outside the cylinder:

$$\varphi(R) = \frac{\sigma R_D K_0(R/R_D)}{\varepsilon \varepsilon_0 K_1(b/R_D)} \tag{III.4.203}$$

Now, the potential must also be continuous across the cylinder surface for us to be able to determine $\varphi_0$:

$$\varphi(b) = \frac{\rho b R_D K_0(b/R_D)}{2\varepsilon \varepsilon_0 K_1(b/R_D)} = -\frac{\rho b^2}{4\varepsilon \varepsilon_0} + \varphi_0 \tag{III.4.204}$$

Thus, the potential inside the cylinder is given by

$$\varphi(R) = \frac{(b^2 - R^2)}{4} \frac{\rho}{\varepsilon \varepsilon_0} + \frac{\rho b R_D K_0(b/R_D)}{2\varepsilon \varepsilon_0 K_1(b/R_D)} \tag{III.4.205}$$

We plot the potential and magnitude of the electric field, $E = -d\varphi/dR$ in Fig. III.4.18.

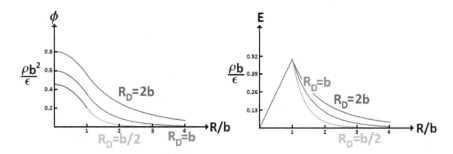

**Fig. III.4.18** We show both the electric potential and field strength inside and outside the cylinder. The electrostatic potential and electric field are given respectively in the units $\rho b^2 / \in$ and $\rho b^2 / \in$ (where $\in \equiv \varepsilon \varepsilon_0$). Generally the Debye length depends on the concentration of electrolyte, decreasing in inverse proportion to the square root of the concentration of ions; we show the curves for three particular values of $R_D$. On the left, the highest curve is taken for example to be equal to the diameter of cylinder $R_D = 2b$, whereas the middle curve a value $R_D = b$ is used, and for the bottom curve $R_D = b/2$. The red and purple parts of the curves correspond to the solution inside the cylinder, whereas the rest corresponds to the solution outside. On the right, we see that inside the cylinder the solutions of all three Debye screening lengths produce the same electric field strength, shown in red. This is physically correct, as we have not allowed electrolyte ions to penetrate the cylinder so there is no screening. The field strength, here, can be obtained by considering Gauss's law (discussed in Chapter 3 of Part III of Vol. 2) with the cylindrical surface of integration inside the charged cylinder. Outside the cylinder, the top curve is the field strength calculated with screening length $R_D = 2b$, the middle one is with $R_D = b$, and the bottom one with $R_D = b/2$.

Bessel equations are the last types of homogenous ordinary linear differential equation with non-constant coefficients that we'll discuss; more types of equations with their solutions can be found in other mathematics textbooks (along with further discussion). To conclude this chapter we'll now consider solving inhomogeneous linear differential equations.

## 4.4.   Solving Inhomogeneous Linear Differential Equations

In Section 4.1 we had quite an abstract discussion about solving inhomogeneous equations. Here, we'll be a bit more specific and present some examples on how to actually do this. In Chapter 5, Part II, Vol. 1 we were already solving the inhomogeneous equation in which the coefficients $p(x)$, $q(x)$ and $r(x)$, in Eq. (III.4.16), were all constant—which is the easiest case, of course! Here, we'll look at more complicated examples.

### The inhomogeneous Bessel equation

Here, we'll see how to solve the inhomogeneous Bessel equation. This equation can come up when we solve Poisson's equation or a wave equation with an inhomogeneous driving term in cylindrical (or spherical) geometry. In general, such equations have the form

$$-\frac{d^2y(x)}{dx^2} - \frac{1}{x}\frac{dy(x)}{dx} + \frac{\alpha^2 y(x)}{x^2} + m^2 y(x) = f(x) \qquad \text{(III.4.206)}$$

where we want to solve for $y(x)$ in the domain $0 \leq x \leq \infty$. To solve Eq. (III.4.206) we use the direct and inverse Hankel transforms introduced in previous subsection. Using the definition of the inverse Hankel transform, Eq. (III.4.163), we can write

$$-\left[\frac{d^2}{dx^2} + \frac{1}{x}\frac{d}{dx} - \frac{\alpha^2}{x^2} - m^2\right]\int_0^\infty k_r \tilde{y}(k_r)J_\alpha(k_r x)dk_r$$
$$= \int_0^\infty k_r \tilde{f}(k_r)J_\alpha(k_r x)dk_r \qquad \text{(III.4.207)}$$

where $\tilde{y}(k_r)$ and $\tilde{f}(k_r)$ are the Hankel transformed functions of $y(x)$ and $f(x)$. Now, we can bring the differential operator inside the integral on the l.h.s of Eq. (III.4.207), and use the eigenvalue equation, Eq. (III.4.125) (which was solved by the eigenfunctions $J_\alpha(\sqrt{E}R)$, where in this case $k_r = \sqrt{E}$), to write

$$\int_0^\infty k_r(k_r^2 + m^2)\tilde{y}(k_r)J_\alpha(k_r x)dk_r = \int_0^\infty k_r\tilde{f}(k_r)J_\alpha(k_r x)dk_r$$

(III.4.208)

To satisfy this, we need the integrands of the two integrals to be equal, and hence

$$\tilde{y}(k_r) = \frac{\tilde{f}(k_r)}{k_r^2 + m^2}$$

(III.4.209)

Thus, the solution to the inhomogeneous equation can be written as

$$y(x) = \int_0^\infty \frac{k_r\tilde{f}(k_r)}{k_r^2 + m^2}J_\alpha(k_r x)dk_r$$

(III.4.210)

This is a particular solution of this equation, which obviously exists. As the solution of any linear differential equation can be presented as the sum of the general solution of the homologous equation and any particular solution of the inhomogeneous equation, a complete general solution to Eq. (III.4.206) may be written as

$$y_T(x) = y(x) + AK_\alpha(mx) + BI_\alpha(mx)$$

(III.4.211)

where $A$ and $B$ are chosen to satisfy any boundary conditions imposed on the full solution $y_T(R)$.

**Exercise.** Show that both $K_\alpha(mR)$ and $I_\alpha(mx)$ are solutions to homogenous equation of Eq. (III.4.206) and hence that a complete general solution can indeed be written as Eq. (III.4.211).

**Example III.4.10 (Maths Practice ♫♪).** Transparent, fully analytical solutions of important physical problems related to solving inhomogeneous Bessel equation rely on being able to perform the integral (III.4.210) by complex contour integration, an important subject that we will consider later in this book. So here, we'll demonstrate such solution on what may seem (and it really is!) a rather abstract, but still instructive example of the form of $f(x)$ in Eq. (III.4.206),

$$-\frac{d^2y(x)}{dx^2} - \frac{1}{x}\frac{dy(x)}{dx} + \frac{y(x)}{x^2} + m^2y(x) = \beta J_1(kx) \qquad \text{(III.4.212)}$$

with the boundary condition that $y(1) = 0$ and $y'(1) = 0$ (our excuse for considering such a crazy form of $f(x)$ is that we will not have to use complex contour integration, and will obtain analytical results straightaway).

We first find the particular solution to the inhomogeneous equation. In this case we write (using Hankel transforms)

$$-\left[\frac{d}{dx^2} + \frac{1}{x}\frac{d}{dx} - \frac{1}{x^2} - m^2\right]\int_0^\infty k_r\tilde{y}(k_r)J_1(k_rx)dk_r$$

$$= \beta\int_0^\infty \delta(k_r - k)J_1(k_rx)dk_r = \int_0^\infty \tilde{f}(k_r)J_1(k_rx)k_rdk_r \qquad \text{(III.4.213)}$$

Thus, we have that, from Eq. (III.4.209),

$$\tilde{y}(k_r) = \frac{\beta}{k_r}\frac{\delta(k_r - k)}{k_r^2 + m^2} \qquad \text{(III.4.214)}$$

So, we obtain for the inhomogeneous solution

$$y_I(x) = \beta\int_0^\infty \frac{\delta(k_r - k)}{k_r^2 + m^2}J_1(k_rx)dk_r = \frac{\beta J_1(kx)}{k^2 + m^2} \qquad \text{(III.4.215)}$$

The total solution then reads as

$$y(x) = \frac{\beta J_1(Kx)}{k^2 + m^2} + AK_1(mx) + BI_1(mx) \qquad \text{(III.4.216)}$$

We then find $A$ and $B$ through the boundary conditions. We obtain the following equations from the boundary conditions:

$$-\frac{\beta J_1(k)}{k^2 + m^2} = AK_1(m) + BI_1(m)$$

$$-\frac{\beta k J_1'(k)}{k^2 + m^2} = AmK_1'(m) + BmI_1'(m) \qquad \text{(III.4.217)}$$

Solving Eq. (III.4.217) we find

$$A = \left[\frac{\beta(mI_1'(m)J_1(k) - kI_1(m)J_1'(k))}{k^2 + m^2}\right]$$

$$\times [mK_1'(m)I_1(m) - mK_1(m)I_1'(m)]^{-1} \qquad (\text{III.4.218})$$

and

$$B = -\left[\frac{\beta(mK_1'(m)J_1(k) - kK_1(m)J_1'(k))}{k^2 + m^2}\right]$$

$$\times [mK_1'(m)I_1(m) - mK_1(m)I_1'(m)]^{-1} \qquad (\text{III.4.219})$$

We can simplify these solutions using the equality $I_n(x)\frac{dK_n(x)}{dx} - K_n(x)\frac{dI_n(x)}{dx} = -\frac{1}{x}$. Therefore, we have as a full solution to the boundary value problem

$$y(x) = \frac{\beta}{k^2 + m^2}\{J_1(kx) - [mI_1'(m)J_1(k) - kI_1(m)J_1'(k)]K_1(mx)$$

$$+ [mK_1'(m)J_1(k) - kK_1(m)J_1'(k)]I_1(mx)\}$$

$$(\text{III.4.220})$$

For your amusement, in Fig. III.4.19 we plot some solutions given by Eq. (III.4.220) for various values of the parameters $k$ and $m$.

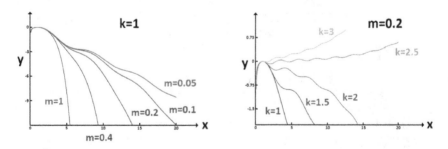

**Fig. III.4.19** Plots of the function $y_T(x)/\beta$, where $y_T(x)$ is given by the r.h.s. of Eq. (III.4.216), for various values of both $k$ and $m$. On the l.h.s we keep $k$ fixed at 1 and vary $m$. We see that the gradient becomes more oscillatory as we decrease $m$, as $J_1(kx)$ becomes more important in the solution. On the r.h.s we fix $m = 0.2$ and vary $k$. Increasing the value of $k$ increases the amount of oscillation in the gradient and makes the solutions more positive at large $x$. We see on the plots that we have correctly satisfied the boundary condition $y(1) = y'(1) = 0$.

**Exercise.** Solve Eq. (III.4.212) subject to the boundary condition $\tilde{y}(0) = 0$ and $\lim_{x \to \infty} y(x) = 0$. **Hint:** You don't need to involve the modified Bessel functions in the solution for these boundary conditions. **Think:** Why not?

### The inhomogeneous Legendre equation

Next, we will look at solutions to the inhomogeneous Legendre equation that reads as

$$-\frac{1}{\sin\theta}\frac{d}{d\theta}\left[\sin\theta\frac{d\Theta(\theta)}{d\theta}\right] + \frac{m^2\Theta(\theta)}{\sin^2\theta} = G(\cos\theta) \qquad \text{(III.4.221)}$$

Such an equation can arise, for example, from solving Poisson's equation for the electrostatic potential in spherical coordinates with a charge density that depends only on the polar angle $0 \le \theta \le \pi$ (in that case $m = 0$). Below, however, we will consider a more general case of this equation, when $m$ is any positive integer. We'll suppose that $G(\cos\theta)$ is a single-valued non-singular function at the points $\theta = 0$ and $\theta = \pi$. In addition as a boundary condition we'll also suppose that $\Theta(\theta)$ remains finite at $\theta = 0$ and $\theta = \pi$, a typical physical requirement for many applications. Then, we can write both $G(\cos\theta)$ and $\Theta(\theta)$ in terms of associated Legendre polynomials (for positive $m$), so that

$$G(\cos\theta) = \sum_{l=m}^{\infty} \tilde{G}_l P_l^m(\cos\theta) \quad \Theta(\theta) = \sum_{l=m}^{\infty} \tilde{\Theta}_l P_l^m(\cos\theta) \quad \text{(III.4.222)}$$

Note the these sums start at $l = m$, as the associated Legendre functions do not exist for $l < m$. These functions form a complete set, so that any smooth non-singular functions $G(\cos\theta)$ and $\Theta(\theta)$ can be represented in such a way (rather like Fourier and Fourier-Bessel series). Substituting Eq. (III.4.222) into Eq. (III.4.221) we obtain

$$\sum_{l=m}^{\infty} \tilde{\Theta}_l \left[ -\frac{1}{\sin\theta}\frac{d}{d\theta}\left[\sin\theta\frac{dP_l^m(\cos\theta)}{d\theta}\right] + \frac{m^2 P_l^m(\cos\theta)}{\sin^2\theta} \right]$$

$$= \sum_{l=m}^{\infty} \tilde{G}_l P_l^m(\cos\theta) \qquad \text{(III.4.223)}$$

Now the l.h.s of Eq. (III.4.223) satisfies the eigenvalue equation Eq. (III.4.78) with $E = l(l+1)$. Thus, we can write

$$\sum_{l=m}^{\infty} \tilde{\Theta}_l l(l+1) P_l^m(\cos\theta) = \sum_{l=m}^{\infty} \tilde{G}_l P_l^m(\cos\theta) = G(\cos\theta) \quad \text{(III.4.224)}$$

Because the associated Legendre polynomials are orthogonal, so that

$$\int_{-1}^{1} P_l^m(z) P_{l'}^m(z) dz = \int_0^{\pi} P_l^m(\cos\theta) P_{l'}^m(\cos\theta) \sin\theta d\theta$$

$$= \frac{2\delta_{l,l'}}{2l+1} \cdot \frac{(l+m)!}{(l-m)!} \quad \text{(III.4.225)}$$

we are able to define a Legendre transform to determine $\tilde{G}_l$ and so $\tilde{\Theta}_l$:

$$\tilde{G}_l = \frac{2l+1}{2} \frac{(l-m)!}{(l+m)!} \int_0^{\pi} G(\cos\theta) P_l^m(\cos\theta) \sin\theta d\theta,$$

$$\text{(III.4.226)}$$

$$\tilde{\Theta}_l = \frac{\tilde{G}_l}{l(l+1)}$$

Last of all, when considering the full solution to Eq. (III.4.221), we should consider a solution to the homogenous equation

$$-\frac{1}{\sin\theta} \frac{d}{d\theta} \left[ \sin\theta \frac{d\Theta(\theta)}{d\theta} \right] + \frac{m^2 \Theta(\theta)}{\sin^2\theta} = 0 \quad \text{(III.4.227)}$$

which is

$$\Theta_H(\theta) = A \left( \frac{1+\cos\theta}{1-\cos\theta} \right)^{m/2} + B \left( \frac{1-\cos\theta}{1+\cos\theta} \right)^{m/2} \quad \text{when } m > 0$$

$$\text{(III.4.228)}$$

and

$$\Theta_H(\theta) = A \ln\left( \frac{1+\cos\theta}{1-\cos\theta} \right) + B \quad \text{when } m = 0 \quad \text{(III.4.229)}$$

**Exercise.** Check that Eq. (III.4.228) and Eq. (III.4.229) are solutions to Eq. (III.4.227). **Hint:** Make the substitution $z = \cos\theta$.

For $m \neq 0$ the solutions to the homogenous equation are singular at the points $\theta = 0$ and $\theta = \pi$. The solution for the case of $m = 0$ is singular at $\theta = 0$. Thus for $m \neq 0$ we have to assume that $A = B = 0$, i.e. add nothing to the particular solution of the inhomogeneous equation. For the case of $m = 0$, we must adopt $A = 0$, whereas the constant $B$ can take any value. Thus, the full solution to Eq. (III.4.221), satisfying the requirement that $\Theta$ is *not singular* reads as

$$\Theta_m(\cos\theta) = \begin{cases} \displaystyle\sum_{l=m}^{\infty} \frac{\tilde{G}_l}{l(l+1)} P_l^m(\cos\theta), & m \neq 0 \\ \tilde{G}_0 + B, & m = 0 \end{cases} \tag{III.4.230}$$

**Example III.4.11 (Maths Practice ♪♪).** Let's obtain the solution for a specific case of the inhomogeneous Legendre equation, which is

$$-\frac{1}{\sin\theta}\frac{d}{d\theta}\left[\sin\theta\frac{d\Theta(\theta)}{d\theta}\right] = \cos^2\theta \tag{III.4.231}$$

In this case we have the form of Eq. (III.4.221) with $m = 0$ and $G(\cos\theta) = \cos^2\theta$. To solve it, firstly, we find the Legendre transformation of $\cos^2\theta$. To do this faster, we can use a bit of common sense. Since $P_0(\cos\theta) = 1$ and $P_2(\cos\theta) = (1/2)(3\cos^2\theta - 1)$ (see Section 4.2), we know that we are able to write

$$\cos^2\theta = \frac{2}{3}P_2(\cos\theta) + \frac{1}{3}P_0(\cos\theta) \tag{III.4.232}$$

This suggests that $\tilde{G}_l = (2/3)\delta_{l,2} + (1/3)\delta_{l,0}$ in Eq. (III.4.226), and we have for the general solution to Eq. (III.4.231)

$$\Theta_m(\cos\theta) = B + A\ln\left(\frac{1 + \cos\theta}{1 - \cos\theta}\right) + \frac{1}{9}P_2(\cos\theta) \tag{III.4.233}$$

Here, we adsorbed $P_0(\cos\theta) \equiv 1$ into the constant $B$ arising from the homogenous solution to Eq. (III.4.231), which will be chosen to satisfy any boundary condition. If we require $\Theta_m(\cos\theta)$ to be non-singular at the points $\theta = 0$ and $\theta = \pi$, we must have $A = 0$. Furthermore, if we require, as a boundary condition, $\Theta_m(1) = 0$, we obtain

$$\Theta_m(\cos\theta) = \frac{1}{9}\left(P_2(\cos\theta) - 1\right) = \frac{1}{6}\left(\cos^2\theta - 1\right) \qquad \text{(III.4.234)}$$

**Exercise.** Verify that Eq. (III.4.234) is a solution to Eq. (III.4.231) by direct substitution.

### General remarks about solving inhomogeneous linear differential equations

We'll conclude this section with a few general statements. Firstly, we can write the solution, $f(x)$, to the equation

$$p(x)f(x) + q(x)\frac{df(x)}{dx} + r(x)\frac{d^2f(x)}{dx^2} = g(x) \quad \text{for } a \le x \le b$$
$$\text{(III.4.235)}$$

in another way:

$$f(x) = \int_a^b G(x, x')g(x')dx' \qquad \text{(III.4.236)}$$

The function $G(x, x')$ is the 1D Green's function. It satisfies the equation

$$p(x)G(x, x') + q(x)\frac{dG(x, x')}{dx} + r(x)\frac{d^2G(x, x')}{dx^2} = \delta(x - x')$$
$$\text{(III.4.237)}$$

This is the same equation as Eq. (III.4.235), but with $g(x) = \delta(x-x')$. From Eq. (III.4.236) it follows that this form of $g(x)$ will give us $f(x) = G(x, x')$.

Now, if we multiply both sides by $g(x')$ and integrate over $x'$ we can indeed show that, in general, Eq. (III.4.236) is a solution of

Eq. (III.4.235). Green's function can then be written [this is obtained by considering Eqs. (III.4.47) and (III.4.48) with $g(x) = \delta(x - x')$] as

$$G(x, x') = \sum_k \frac{f_k^*(x) f_k(x') \omega(x')}{E_k \tilde{\omega}_k} \qquad (\text{III.4.238})$$

where the functions $f_k(x)$ satisfy the eigenvalue equation (Eq. (III.4.1)) with eigenvalues $E_k$, which has the same coefficients $p(x)$, $q(x)$ and $r(x)$ as Eq. (III.4.235). The functions $\omega(x')$ and $\tilde{\omega}_k$ are defined through the orthogonality relationship

$$\int_a^b f_k(x) f_{k'}(x) \omega(x) dx = \delta_{k,k'} \tilde{\omega}_k \qquad (\text{III.4.239})$$

For example, in the case of Bessel's equation (with $a = 0$, $b = \infty$), we have $f_k(x) = J_n(kx)$. Here, we have $\omega(x) = kx$, as required by the Sturm–Liouville form, Eq. (III.4.126) (see the boxed part of first section of this chapter), and from the integral on the l.h.s we obtain $\tilde{\omega}_k = 1/k$.

Let's suppose $f_k(x)$ satisfies the 1D Schrödinger equation we have talked about. Here the functions $f_k(x)$ are orthogonal. This equation is already in Sturm–Liouville form (Eq. (III.4.36)) where $\omega(x) = 1$, and such eigenfunctions can be normalized such that $\tilde{\omega}_k = 1$.

When Eq. (III.4.235) is subject to boundary value constraints that the functions $f_k(x)$ don't satisfy, we must write the solution as a sum of homogenous and non-homogenous parts, and then choose the unknown coefficients of the general solution of the homogenous equation to satisfy the boundary conditions. This is just as we did in the examples which considered the inhomogeneous Bessel equation and Legendre equation, although we didn't explicitly invoke Green's function there. However, sometimes it is better to find a Green's function that already satisfies the boundary conditions; this can be useful if we want to prepare a result for a general form of $g(x)$.

It will be easier to demonstrate practical applications of the techniques that we've learnt in this section after we introduce complex contour integration. This additional technique is required for the solution of inhomogeneous equations in physical situations.

## 4.5. Key Points

- We can consider *homogeneous* ordinary linear differential equations as *eigenvalue equations* where $\hat{H}f(x) = Ef(x)$. Here, $\hat{H}$ is a differential operator acting on an *eigenfunction* resulting in the same eigenfunction but multiplied by the *eigenvalue*, $E$.
- Simple differential operators are $1, d/dx$ and $d^2/dx^2$; when these act on a function they take its zeroth, first and second order derivatives, respectively. Any second order linear differential operator can be written as $\hat{H}(x) = p(x) + q(x)d/dx + r(x)d^2/dx^2$.
- For a particular differential operator, there are an infinite set of eigenfunctions $f_n(x)$, each with its characteristic eigenvalue $E_n$.
- The eigenfunctions $f_n(x)$ obey an orthogonality relationship $\int_a^b \omega(x)f_n^*(x)f_m(x)dx = \tilde{\omega}_n\delta_{n,m}$, where $\omega(x)$ is a particular function associated with the particular type of eigenvalue equation. When $\hat{H}$ in that equation, which the eigenfunctions $f_n(x)$ satisfy, is of the form $\hat{H} = a\frac{d^2}{dx^2} + v(x)$, we have that $\omega(x) = 1$. For other differential operators, $\omega(x)$ is found by converting the eigenvalue equation into its Sturm–Liouville form (Eq. (III.4.36)).
- Differential eigenvalue equations may be viewed as *matrix equations*, where the eigenfunctions are represented by an *eigenvectors*—a columnar matrix with components $f_n(x)$. Each entry is an output value from particular input value (see Fig. III.4.1). The differential operator, $\hat{H}$ can also be viewed as a matrix.
- Fourier transforms and other integral transforms can also be viewed as matrices. These diagonalize the matrix representing a particular differential operator, transforming the differential operator into a function.
- Some classical examples of eigenvalue equations are the Schrödinger equation on a simple harmonic oscillator, Legendre's equation and the radial equation for a hydrogen atom. How they are solved is discussed in Section 4.2.
- For a solution (like Legendre's equation) defined in a finite domain we may attempt a series solution $f(x) = \Sigma_{n=0}^{\infty}a_nx^n$. Substitution of such an ansatz (a presumed answer) yields a recursion relationship

from which one can relate $a_n$ to $a_{n-1}$ and $a_{n-2}$, and so on. For a homogeneous second order differential equation, two of the coefficients are arbitrary, leading to two independent functions that solve it. For instance, one solution is got by choosing $a_1 = 0$, the other by choosing $a = 0$.

- Such equations (for instance the harmonic oscillator or the radial Schrödinger equations), where the solution are defined in an infinite domain (e.g from 0 to $\infty$), can be solved by first inferring the dominant behaviour of the solution $f(x)$ at $|x| \to \infty$. This can be done by neglecting terms in the differential equation less important at $|x| \to \infty$, yielding a solution $f_\infty(x)$. We may then write our solution as $f(x) = u(x)f_\infty(x)$ and find the resulting differential equation on $u(x)$. For $u(x)$ again we may attempt a series solution, i.e. $u(x) = \sum_{n=0}^\infty b_n x^n$.

- In certain cases, like the radial equation, the leading order term in a series is not a constant, it is rather $x^s$ where $s \neq 0$. In these cases we must use the method of Frobenius (outlined at the end of Section 4.2) to generate a power series solution.

- Power series solutions can have limited applicability, if the convergence of the series is slow, and it cannot be summed to an already known function. Sometimes, the approximation techniques that we will discuss in Chapter 6 of Part III of Vol. 2 work better.

- There is a useful set of functions called *Bessel functions*, $J_n(x)$ and $Y_n(x)$, which are the solutions of the *Bessel's equation*. Plots of these functions for integer values are given in Fig. III.4.12. Also useful are the *modified Bessel functions* $I_n(x)$ and $K_n(x)$, which are solutions to the *modified Bessel equation* (Bessel's equation with $x \to ix$). Plots of these functions for integer values are given in Fig. III.4.14.

- Bessel functions and modified Bessel functions appear in physical problems in spherical and cylindrical geometry, for instance the solution of partial differential equations (Chapter 7 of Part III, Vol. 2). The former can be considered analogous to $\sin x$ and $\cos x$ and the latter to $\sinh x$ and $\cosh x$, which all appear in the same problem in rectangular geometry.

- In general, the solution to a linear *inhomogeneous* equations $\hat{H}f(x) = g(x)$ defined in the domain $a \leq x \leq b$ is given by $f(x) = \Sigma_k \frac{f_k(x)\tilde{g}_k}{E_k\tilde{\omega}_k}$. Here, $f_k(x)$ and $E_k$ are the eigenfunctions of $\hat{H}$, and $\omega(x)$ and $\tilde{\omega}_k$ are specific to the orthogonality requirement between the $f_k(x)$ (see above), whereas $\tilde{g}_k$ is the integral transform of the eigenfunction of the homogeneous equation, defined as $\tilde{g}_k = \int_a^b f_k^*(x)\omega(x)g(x)dx$.

- One useful integral transform, analogous to the Fourier transform, is the *Hankel transform*. The Hankel transform of a function $f(R)$ is defined as $\tilde{f}(k) = \int_0^\infty RJ_n(kR)f(R)dR$, and the inverse Hankel transform is $f(k) = \int_0^\infty kJ_n(kR)\tilde{f}(k)dk$; in this case $\omega(R) = R$ and $\tilde{\omega}_k = 1/k$. The Hankel transform can be efficiently used to solve the inhomogeneous Bessel equation, Eq. (III.4.206), as shown in Section 4.4.

## Chapter 5

# Coupled Linear Second Order Differential Equations and Higher Order Equations

### 5.1. Finite Systems of Coupled Linear Homogeneous Differential Equations with Constant Coefficients

Suppose your physical problem has to be described by a system of interconnected linear differential equations. To give you a first lesson of how to solve such systems of equations and what may come out of it, we will consider in this book the cases when they are simple enough to contain only constant coefficients. We'll start by looking at a physical example where they come up.

**Example III.5.1 (Physical Application ▶▶).** Consider a setup where we have two coupled simple harmonic oscillators. This could be a situation where we have two springs attached to each other. We'll suppose that one end of the first spring, with spring constant $k_1$, is attached to the wall, i.e. is fixed, but the other one is attached to mass $m_1$; the second spring, with spring constant $k_2$, is then attached to first spring at that mass; a second mass $m_2$ is attached at its other end. For mathematical simplicity, we'll choose the springs to have the same natural (equilibrium) length, $l$, although when oscillating the springs may acquire different extensions. We furthermore consider

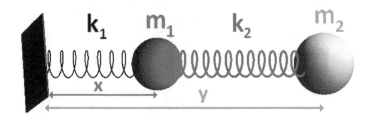

**Fig. III.5.1** A system of two masses connected to two springs, for which we want to solve the equations of motion. The left end of the first spring, with spring constant $k_1$ is fixed. At the other end of it sits mass $m_1$ with a displacement away from the fixed end of $x(t)$. Attached also to this mass is a second spring, with its own spring constant, $k_2$, which is in turn attached to a second mass $m_2$. This second mass has displacement away from the fixed end of $y(t)$.

the springs to be oriented in the same direction (lying on one line), i.e. we will exclude the possibility of their bending.

Such a situation is illustrated in Fig. III.5.1, where we denoted by $x$ the position of mass $m_1$ and by $y$ be the position of mass $m_2$.

There are two forces acting on mass $m_1$, one from each of the springs $m_1$, with the total force,

$$F_1 = -k_1(x - l) + k_2(y - x - l) \qquad \text{(III.5.1)}$$

The force on mass $m_2$ will be

$$F_2 = -k_2(y - x - l) \qquad \text{(III.5.2)}$$

Thus, we can write through Newton's second law the following equations of motion:

$$m_1 \frac{d^2 x}{dt^2} = -k_1(x - l) + k_2(y - x - l), \quad m_2 \frac{d^2 y}{dt^2} = -k_2(y - x - l)$$
$$\text{(III.5.3)}$$

First, it is handy to define new variables $\bar{x} = x - l$ and $\bar{y} = y - 2l$ so that

$$m_1 \frac{d^2 \bar{x}}{dt^2} = (-k_1 - k_2)\bar{x} + k_2 \bar{y}, \quad m_2 \frac{d^2 \bar{y}}{dt^2} = k_2 \bar{x} - k_2 \bar{y} \qquad \text{(III.5.4)}$$

How do we solve the system of Eq. (III.5.4)? One way to do so is in the same way as we did for a single equation with constant coefficients.

We make the substitutions

$$\bar{x} = x_0 \exp(i\omega t) \quad \text{and} \quad \bar{y} = y_0 \exp(i\omega t) \tag{III.5.5}$$

and look for solutions for $\omega$, and the correct relationship between $x_0$ and $y_0$ for each of those frequencies of vibration. Note that in doing such substitutions we implicitly assume that $\bar{x}(t)$ and $\bar{y}(t)$ will oscillate with the same frequencies; if you relaxed this assumption to let the frequencies be different, you should see that the resulting equations could not be satisfied at an arbitrarily value of time $t$. Furthermore, such a situation can be physically expected in the stationary regime of oscillations of these springs, because when one spring moves, the other moves too. In few lines below we will see, however, how these two springs sort out their relationships — how they will oscillate together in a concerted fashion, each accommodating *two vibrational modes*.

Substitution of Eq. (III.5.5) yields

$$m_1\omega^2 x_0 = (k_1 + k_2)x_0 - k_2 y_0, \quad m_2\omega^2 y_0 = -k_2 x_0 + k_2 y_0 \tag{III.5.6}$$

We can write Eq. (III.5.6) in matrix form:

$$\begin{pmatrix} (k_1 + k_2)/m_1 & -k_2/m_1 \\ -k_2/m_2 & k_2/m_2 \end{pmatrix} \begin{pmatrix} x_0 \\ y_0 \end{pmatrix} = \omega^2 \begin{pmatrix} x_0 \\ y_0 \end{pmatrix} = \lambda \begin{pmatrix} x_0 \\ y_0 \end{pmatrix} \tag{III.5.7}$$

Equation (III.5.7) is simply an eigenvalue equation of the form we met in Chapter 4, Part II, Vol. 1 — consult it, if you have forgotten how to solve it. We can obtain values for $\lambda = \omega^2$ by solving the secular equation. This yields

$$\begin{vmatrix} \frac{k_1+k_2}{m_1} - \lambda & -\frac{k_2}{m_1} \\ -\frac{k_2}{m_2} & \frac{k_2}{m_2} - \lambda \end{vmatrix} = \left[ \frac{k_1 + k_2}{m_1} - \lambda \right] \left( \frac{k_2}{m_2} - \lambda \right) - \frac{k_2^2}{m_1 m_2} = 0$$

$$\Rightarrow \lambda^2 - \lambda \left( \frac{k_2}{m_2} + \frac{k_1 + k_2}{m_1} \right) \lambda + \frac{k_1 k_2}{m_1 m_2} = 0 \tag{III.5.8}$$

The solution of this quadratic equation on $\lambda$ gives

$$\lambda = \lambda_{\pm} = \frac{1}{2} \left( \frac{k_2}{m_2} + \frac{k_1 + k_2}{m_1} \pm \sqrt{\left[ \frac{k_2}{m_2} + \frac{k_1 + k_2}{m_1} \right]^2 - \frac{4 k_1 k_2}{m_1 m_2}} \right)$$

(III.5.9)

For these solutions for $\lambda$, we find from Eq. (III.5.7) the relationship between $y_0$ and $x_0$: we have that either

$$y_0 = \frac{m_1}{k_2} \left( \frac{k_1 + k_2}{m_1} - \lambda_+ \right) x_0 \ \text{ or } \ y_0 = \frac{m_1}{k_2} \left( \frac{k_1 + k_2}{m_1} - \lambda_- \right) x_0$$

(III.5.10)

For $\omega$ we obtain four roots: $\omega_1 = \sqrt{\lambda_+}$, $\omega_2 = -\sqrt{\lambda_+} (= -\omega_1)$, $\omega_3 = \sqrt{\lambda_-}$, and $\omega_4 = -\sqrt{\lambda_-} (= -\omega_3)$. Noting the relationships between $x_0$ and $y_0$ (Eq. (III.5.10)), the general solution to Eq. (III.5.4) must be constructed as

$$\bar{x}(t) = A e^{i\omega_1 t} + A^* e^{-i\omega_1 t} + B e^{i\omega_3 t} + B^* e^{-i\omega_3 t} \qquad \text{(III.5.11)}$$

$$\bar{y}(t) = \frac{m_1}{k_2} \left( \frac{k_1 + k_2}{m_1} - \lambda_+ \right) \left( A e^{i\omega_1 t} + A^* e^{-i\omega_1 t} \right)$$

$$+ \frac{m_1}{k_2} \left( \frac{k_1 + k_2}{m_1} - \lambda_- \right) \left( B e^{i\omega_3 t} + B^* e^{-i\omega_3 t} \right) \quad \text{(III.5.12)}$$

The coefficients $A, B$ and their complex conjugates $A^*$ and $B^*$, are chosen to satisfy four 'boundary conditions'; since we are dealing here with time dependence these could be conditions imposed at certain points of time, for example initial conditions on $y(t)$ and $x(t)$ or its derivatives. The obtained solutions are commonly referred to as the *normal modes of vibration*, each with its characteristic frequency $\omega_1$ and $\omega_2$.

Alternatively, we may recast Eqs. (III.5.11) and (III.5.12) in a more transparent form by introducing phases. This will help us to visualize how these two spring-connected beads manage to move, each having the same two different frequencies of vibration, Namely,

$$\bar{x}(t) = A_1 \cos(w_1 t + \phi_1) + A_3 \cos(w_3 t + \phi_3) \qquad \text{(III.5.13)}$$

$$\bar{y}(t) = \frac{m_1}{k_2} \left( \frac{k_1 + k_2}{m_1} - \lambda_+ \right) A_1 \cos(w_1 t + \phi_1)$$

$$+ \frac{m_1}{k_2} \left( \frac{k_1 + k_2}{m_1} - \lambda_- \right) A_3 \cos(w_3 t + \phi_3) \qquad \text{(III.5.14)}$$

If we choose from our initial conditions that both springs start from rest, i.e. $\frac{d\bar{x}}{dt}\big|_{t=0} = \frac{d\bar{y}}{dt}\big|_{t=0} = 0$, then $\phi_1 = \phi_3 = 0$, and the initial displacements $\bar{x}(0)$ and $\bar{y}(0)$ will give us equations on $A_1$ and $A_3$ values.

**Exercise.** Show that Eqs. (III.5.13) and (III.5.14) do follow from Eqs. (III.5.11) and (III.5.12). Use Eqs. (III.5.13) and (III.5.14) to find the specific solution to Eq. (III.5.4) with the boundary conditions $\bar{x}'(0) = \bar{y}'(0) = \bar{y}(0) = 0$, $\bar{x}(0) = a$. Plot $\bar{y}(t)$ and $\bar{x}(t)$ in units of $a$ to see how the two eigenmodes manifest themselves in the resulting vibration of this system (for simplicity take both masses and spring constants to be the same).

**Example III.5.1 (Continued).** To better understand how this coupled system actually 'implements' its eigen vibration modes, it's worth looking at a couple of limiting cases. The first one to consider, as in the exercise above, is when both spring constants and masses are equal, so that $k = k_1 = k_2$ and $m = m_1 = m_2$. In this case, Eqs. (III.5.9) and (III.5.10) yield

$$\lambda = \lambda_\pm = \frac{k}{2m} \left( 3 \pm \sqrt{5} \right) \qquad \text{(III.5.15)}$$

$$y_0 = \left( 2 - \lambda_+ \frac{m}{k} \right) x_0 = -\frac{1}{2} \left( \sqrt{5} - 1 \right) x_0$$

$$y_0 = \left( 2 - \lambda_- \frac{m}{k} \right) x_0 = \frac{1}{2} \left( \sqrt{5} + 1 \right) x_0 \qquad \text{(III.5.16)}$$

We see that in the normal mode of vibration with the lower frequency $w = \sqrt{\frac{3-\sqrt{5}}{2} \cdot \frac{k}{m}}$ the masses oscillate in phase, and for the higher frequency mode $w = \sqrt{\frac{3+\sqrt{5}}{2} \cdot \frac{k}{m}}$ the masses vibrate out of

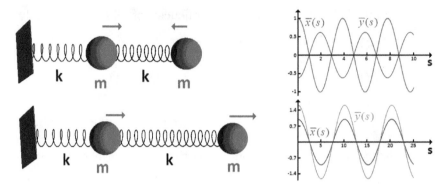

**Fig. III.5.2** We illustrate normal modes of vibration for our system, when both spring constants and masses are the same. In one mode, the two masses vibrate in opposite directions to each other (top left). In a second mode the masses move in the same direction (bottom left). On the right, we plot the displacements $\bar{x}(s) = x(s) - l$ and $\bar{y}(s) = y(s) - 2l$, away from the equilibrium configuration of the springs, for both modes of vibration. Here the variable $s$ stands for dimensionless time, $s = t\sqrt{k/m}$; the amplitude of $\bar{x}(s)$ is normalized to 1. The mode in which the masses move in opposite directions has a higher vibrational frequency than the one where the masses move in the same direction. Also we see that for the former, the amplitude of $\bar{y}(s)$ is smaller than $\bar{x}(s)$; but for the latter the amplitude of $\bar{y}(s)$ is larger.

phase with a phase difference $\phi = \pi$. Remarkably, the ratio of the two frequencies is universal, $\sqrt{\frac{3+\sqrt{5}}{3-\sqrt{5}}}$, i.e. the frequency of the anti-symmetric mode of motion of the two beads is 2.62 times higher than that of the symmetric one. These modes of vibration are illustrated in Fig. III.5.2.

Now, what happens when we make $m_1 \gg m_2$ (Fig. III.5.3)? In this case, Eq. (III.5.9) simplifies to

$$\lambda_+ \approx k_2/m_2 \quad \text{and} \quad \lambda_- \approx k_1/m_1 \qquad \text{(III.5.17)}$$

**Exercise.** Show that Eq. (III.5.17) follows from Eq. (III.5.9) in the limiting case $m_1 \gg m_2$.

**Example III.5.1 (Continued).** Thus, in this second limiting case, we have either the characteristic frequency of the mass $m_1$ with its spring of spring constant $k_1$, or the characteristic frequency of the mass $m_2$ with its spring of spring constant $k_2$. This suggests that effectively, in each case, one mass vibrates on one spring, but the

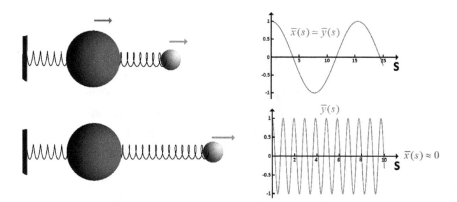

**Fig. III.5.3** An illustration of the normal modes when $m_1 \gg m_2$. In one mode (top of diagram) the masses move with the same velocity so that the spring that connects $m_2$ and $m_1$ effectively stays at its equilibrium length $l$. This means that $\bar{y}(s) \gg \bar{x}(s)$, where $s = t\sqrt{k_1/m_1}$. In the second mode (bottom of diagram) the mass $m_1$ effectively remains stationary while mass $m_2$ vibrates with the frequency $\approx \sqrt{k_2/m_2}$. We plot the displacement $\bar{y}(s)$ on the right, amplitude normalized to 1. We see when $m_1 \gg m_2$, when $m_1$ is stationary, the frequency of vibration is much larger than for the mode when the bigger mass, $m_1$ is moving with $m_2$. In both plots on the right $s = t\sqrt{k/m_1}$, where we have displayed the case when the spring constants are the same, $k_1 = k_2 = k$, and only the masses are dramatically different.

other spring is extended by not an appreciable amount. When $\lambda = \lambda_-$, we find from Eq. (III.5.10) that

$$y_0 \approx x_0 \qquad \text{(III.5.18)}$$

So indeed only the first spring, of spring constant $k_1$, which connects the heavy bead to the wall is extending appreciably, the second much lighter mass's movement is locked with that of the much heavier one. When $\lambda = \lambda_+$ we find from Eq. (III.5.10) that

$$x_0 \approx -\frac{m_2}{m_1}y_0 \quad \Rightarrow \quad |y_0| \gg |x_0| \qquad \text{(III.5.19)}$$

This means that in this mode of vibration, the heavier mass is effectively stationary and only the spring with the lighter mass is extending appreciably. The normal modes of this limiting case are visualized in Fig. III.5.3.

The results of this analysis make physical sense, and help us to believe that Eqs. (III.5.9) and (III.5.10) are correct.

**Example III.5.2 (Physical Application ▶▶).** Nothing is forever! People say that diamonds are, but we are not sure about that, and still less about the Universe. But the spring, if you stretch or compress it against its equilibrium length, sooner or later will stop vibrating, due to internal or external friction. Let's extend our previous example with the masses of different sizes by adding a dissipation term (friction) into the equations of motion with different damping coefficients ($\gamma_1$ and $\gamma_2$). In this case, the equations of motion (Eq. (III.5.4)), would read

$$m_1 \frac{d^2\bar{x}}{dt^2} + \gamma_1 \frac{d\bar{x}}{dt} = (-k_1 - k_2)\bar{x} + k_2\bar{y}, \quad m_2 \frac{d^2\bar{y}}{dt^2} + \gamma_2 \frac{d\bar{y}}{dt} = k_2\bar{x} - k_2\bar{y}$$

$$\text{(III.5.20)}$$

Note that the dimensionality of the damping coefficient is $[M/t]$, so that it is different from the dimensionality of frequency, $[1/t]$.

To solve these equations we will, again, make the substitutions of Eq. (III.5.5) and write the result in matrix form,

$$\begin{pmatrix} \dfrac{k_1 + k_2}{m_1} + i\dfrac{\omega\gamma_1}{m_1} & -\dfrac{k_2}{m_1} \\ -\dfrac{k_2}{m_2} & \dfrac{k_2}{m_2} + i\dfrac{\omega\gamma_2}{m_2} \end{pmatrix} \begin{pmatrix} x_0 \\ y_0 \end{pmatrix} = \omega^2 \begin{pmatrix} x_0 \\ y_0 \end{pmatrix} = \lambda \begin{pmatrix} x_0 \\ y_0 \end{pmatrix}$$

$$\text{(III.5.21)}$$

This is again a matrix eigenvalue equation, with eigenvalues $\lambda = \omega^2$. Thus, we find $\omega$ by solving the secular equation

$$\begin{vmatrix} k_1 + k_2 + i\omega\gamma_1 - \omega^2 m_1 & -k_2 \\ -k_2 & k_2 + i\omega\gamma_2 - \omega^2 m_2 \end{vmatrix} = 0 \qquad \text{(III.5.22)}$$

This would yield a rather complicated quartic equation, for which there would be no simple solution. To avoid complicated algebra, we will not be considering the general case here. However, let's examine the special case where all the masses, spring constants and damping factors are the same, i.e. $m = m_1 = m_2$, $\gamma = \gamma_1 = \gamma_2$ and $k = k_1 = k_2$, when all equations become simple, but we can still learn useful lessons from them. In this case, one can show (as an exercise, we leave the reader to do this, solving the resulting quartic equation) that

$$m\omega^2 - i\gamma\omega = \frac{k}{2}\left(3 \pm \sqrt{5}\right) \tag{III.5.23}$$

It is also not hard to show that we have the same relationships between $x_0$ and $y_0$ as there were without damping. From Eq. (III.5.23) we obtain for $\omega$:

$$\omega_{1,\pm} = \frac{i\gamma}{2m} \pm \sqrt{\frac{k\left(3 + \sqrt{5}\right)}{2m} - \frac{\gamma^2}{4m^2}}$$

$$\omega_{2,\pm} = \frac{i\gamma}{2m} \pm \sqrt{\frac{k\left(3 - \sqrt{5}\right)}{2m} - \frac{\gamma^2}{4m^2}} \tag{III.5.24}$$

Both normal modes of oscillation are now damped and decay with time. We plot $\bar{x}(t)$ and $\bar{y}(t)$ for each pure mode of vibration in Fig. III.5.4.

We see that the two modes have different values of *critical damping*, which is the value of damping above which oscillations disappear (recall Example II.6.11, Part II, Chapter 6, Vol. 1). These can be found by equalizing the expressions under the square roots to zero, giving for the asymmetric, higher frequency mode $\gamma_{cr} = \sqrt{2mk(3 + \sqrt{5})}$, whereas for the symmetric lower frequency mode it is $\gamma_{cr} = \sqrt{2mk(3 - \sqrt{5})}$. All in all, the normal mode which has the smaller oscillation frequency when $\gamma = 0$ has a lower value of critical damping.

The general solution can be written as

$$\bar{x}(t) = e^{-\delta t}\left(Ae^{i\omega_1 t} + A^* e^{-i\omega_1 t}\right) + e^{-\delta t}\left(Be^{i\omega_2 t} + B^* e^{-i\omega_2 t}\right) \tag{III.5.25}$$

$$\bar{y}(t) = -\frac{e^{-\delta t}}{2}\left(\sqrt{5} + 1\right)\left(Ae^{i\omega_1 t} + A^* e^{-i\omega_1 t}\right)$$

$$+ \frac{e^{-\delta t}}{2}\left(\sqrt{5} - 1\right)\left(Be^{i\omega_2 t} + B^* e^{-i\omega_2 t}\right) \tag{III.5.26}$$

where $\delta = \frac{\gamma}{2m}$, $\omega_1 = \sqrt{\frac{k(3+\sqrt{5})}{2m} - \delta^2}$ and $\omega_2 = \sqrt{\frac{k(3-\sqrt{5})}{2m} - \delta^2}$

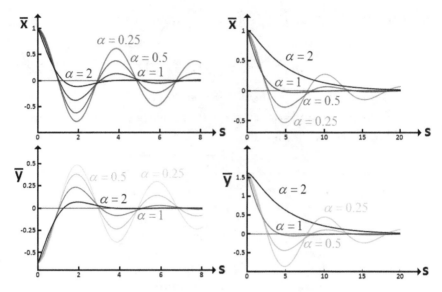

**Fig. III.5.4** We plot the displacements as a function of dimensionless time $s = t\sqrt{k/m}$ for the normal modes of vibration (for the case when the damping factors, spring constants and masses of the two springs and beads are all the same), with damping. The curves correspond to various values of the dimensionless parameter that controls the amount of damping $\alpha = \gamma/\sqrt{mk}$. On the left, we show plots for the normal mode in which the masses move in opposite directions. As we increase $\alpha$ the displacements decay away more quickly. At the highest value of $\alpha$ considered, $\alpha = 2$, we are still below the value for critical damping for this mode $\alpha_c = 3.24$ (you should check this value!), and we can see slight, but heavily damped oscillations. On the right we show plots for the normal mode where the masses move in the same direction. In this case when we reach a value of $\alpha = 2$ we have exceeded the value for critical damping $\alpha_c = 1.24$ (check again!) and we have pure exponential decay. For the other values of $\alpha$ considered we have oscillations which are damped more as we increase $\alpha$ towards the critical value.

## *Solving systems of homogeneous linear equations with constant coefficients in general*

In principle, we can solve any system of $n$ homogeneous linear differential equations of the form

$$\frac{d^2 y_j}{dt^2} + \sum_{s=1}^{n} b_{j,s} \frac{dy_s}{dt} + \sum_{s=1}^{n} c_{j,s} y_s = 0 \qquad \text{(III.5.27)}$$

Making the substitutions $y_j(t) = y_{0,j} \exp(-i\omega t)$, we get a linear eigenvalue equation on $k$,

$$-\omega^2 y_{0,j} - i\omega \sum_{s=1}^{n} b_{j,s} y_{0,s} + \sum_{s=1}^{n} c_{j,s} y_{0,s} = 0 \qquad \text{(III.5.28)}$$

that we can rewrite in the matrix form as

$$\mathbf{H} . \mathbf{y}_0 = 0 \qquad \text{(III.5.29)}$$

where $\mathbf{y}_0 = \begin{pmatrix} y_{0,1} \\ y_{0,2} \\ \vdots \\ y_{0,j} \\ \vdots \end{pmatrix}$ and matrix $\mathbf{H}$ is given by

$\mathbf{H}=$

$$\begin{pmatrix} c_{1,1}-i\omega b_{1,1}-\omega^2 & c_{1,2}-i\omega b_{1,2} & \cdots & c_{1,n-1}-i\omega b_{1,n-1} & c_{1,n}-i\omega b_{1,n} \\ c_{2,1}-i\omega b_{2,1} & c_{2,2}-i\omega b_{2,2}-\omega^2 & \cdots & c_{2,n-1}-i\omega b_{2,n-1} & c_{2,n}-i\omega b_{2,n} \\ \vdots & \vdots & \ddots & \vdots & \vdots \\ c_{n-1,1}-i\omega b_{n-1,1} & c_{n-1,2}-i\omega b_{n-1,2} & \cdots & c_{n-1,n-1}-i\omega b_{n-1,n-1}-\omega^2 & c_{n-1,n}-i\omega b_{n-1,n} \\ c_{n,1}-i\omega b_{n,1} & c_{n,2}-i\omega b_{n,2} & \cdots & c_{n,n-1}-i\omega b_{n,n-1} & c_{n,n}-i\omega b_{n,n}-\omega^2 \end{pmatrix}$$

$$\text{(III.5.30)}$$

As always, the normal modes are found through finding the $k$ values from the requirement that $\det[\mathbf{H}] = 0$. This condition leads to a polynomial equation with $2n$ roots, real and complex. Here, we'll suppose that the coefficients, $c_{j,s}$ and $b_{j,s}$ are chosen so that there are no degeneracies (repeated eigenvalues) as this rather complicates matters. Any complex roots correspond to damped solutions. Then, as usual, for each thus allowed value of $\omega$ ($\omega_m$ that satisfies $\det[\mathbf{H}] = 0$), we will be solving for the $y_{0,j}$ amplitudes in terms of $y_{0,1}$ to find the eigenvectors $\mathbf{y}_0^{(m)}$. The general solution can then be written as a linear combination of the $2n$ eigenvectors $\mathbf{y}_0^{(m)}$ such that

$$\mathbf{y}(t) = \sum_{m=1}^{2n} \mathbf{y}_0^{(m)} \exp(-i\omega_m t) \text{ where}$$

$$\mathbf{y}(t) = \begin{pmatrix} y_1(t) \\ y_2(t) \\ \vdots \\ y_3(t) \\ \vdots \end{pmatrix} \text{ and } \mathbf{y}_0^{(m)} = \begin{pmatrix} y_{0,1}^{(m)} \\ y_{0,2}^{(m)} \\ \vdots \\ y_{0,j}^{(m)} \\ \vdots \end{pmatrix} \tag{III.5.31}$$

Of course, in the general case finding such eigenvalues and amplitudes will be a task for the computer, so we humbly stop here...

## 5.2. Inhomogeneous Coupled Linear Differential Equations with Constant Coefficients

We can also solve coupled inhomogeneous equations of the form

$$\frac{d^2 y_j(t)}{dt^2} + \sum_{s=1}^{n} b_{j,s} \frac{dy_k(t)}{dt} + \sum_{s=1}^{n} c_{j,s} y_s(t) = f_j(t) \tag{III.5.32}$$

As before, when we dealt with single equations, we make a Fourier transforms of both $f_j(t)$ and $y_j(t)$ so that we can write

$$-\omega^2 \tilde{y}_j(\omega) - i\omega \sum_{s=1}^{n} b_{j,s} \tilde{y}_s(\omega) + \sum_{s=1}^{n} c_{j,s} \tilde{y}_s(\omega) = \tilde{f}_j(\omega) \tag{III.5.33}$$

where

$$\tilde{y}_j(\omega) = \int_{-\infty}^{\infty} y_j(t) \exp(i\omega t) dt \quad \tilde{f}_j(\omega) = \int_{-\infty}^{\infty} f_j(t) \exp(i\omega t) dt$$

$$\tag{III.5.34}$$

The solution to Eq. (III.5.33) is

$$\tilde{\mathbf{y}}(\omega) = \mathbf{H}^{-1}.\tilde{\mathbf{f}}(\omega) \text{ where } \tilde{\mathbf{f}}(\omega) = \begin{pmatrix} \tilde{f}_1(\omega) \\ \tilde{f}_2(\omega) \\ \vdots \\ \tilde{f}_n(\omega) \end{pmatrix} \text{ and } \tilde{\mathbf{y}}(k) = \begin{pmatrix} \tilde{y}_1(\omega) \\ \tilde{y}_2(\omega) \\ \vdots \\ \tilde{y}_n(\omega) \end{pmatrix}$$

(III.5.35)

Where $\mathbf{H}^{-1}$ is the inverse of the matrix $\mathbf{H}$ defined by Eq. (III.5.30). Then we can obtain $y_j(t)$ by the inverse Fourier transform

$$y_j(t) = \frac{1}{2\pi} \int_{-\infty}^{\infty} \tilde{y}_j(\omega) \exp(-i\omega t) d\omega \tag{III.5.36}$$

**Example III.5.3 (Physical Application ►►).** We return to our system of two masses on springs (with pairs of damping coefficients, masses, and spring constants, again taken for simplicity to be the same: $\gamma = \gamma_1 = \gamma_2$, $m = m_1 = m_2$, and $k = k_1 = k_2$) and suppose that mass $m_2$ (the one at the end of the two springs) is loaded with charge $Q$. We then apply an electric field, periodic in time, $E(t) = E(t + T)$ where $T$ is a period, with the time dependence of the triangular sawtooth form

$$E(t) = E_0 \cdot \begin{cases} t/T, & 0 \le t < T/2 \\ 1 - t/T, & T/2 \le t < T \end{cases} \tag{III.5.37}$$

along the direction of the spring, which will act on this charge. Assuming that the acceleration of the charged mass is small enough to neglect electromagnetic effects, the equations of motion for our double-spring system will get modified to

$$m\frac{d^2\bar{x}}{dt^2} + \gamma\frac{d\bar{x}}{dt} = -2k\bar{x} + k\bar{y}, \quad m\frac{d^2\bar{y}}{dt^2} + \gamma\frac{d\bar{y}}{dt} = k\bar{x} - k\bar{y} + QE(t)$$

(III.5.38)

As well as illustrating how to solve coupled inhomogeneous equations, this example is a good application of Fourier series. A useful lesson: *when faced with an inhomogeneous term which is periodic, it usually makes sense to write it as a Fourier series.* In a moment, we'll see why that is so.

We express $E(t)$ as a Fourier series such that

$$E(t) = E_0 \sum_{n=-\infty}^{\infty} c_n \exp\left(-\frac{2\pi i n t}{T}\right) \tag{III.5.39}$$

Before evaluating the $c_n$ coefficients, let's see why it makes sense to employ a Fourier series. Each item is denoted by subscript $n$, which represents the contribution of each frequency mode $\omega_n = 2\pi n/T$. Once we substitute Eq. (III.5.39) into Eq. (III.5.38) we obtain equations on $\bar{x}_n(t)$ and $\bar{y}_n(t)$, the components of each mode:

$$m\frac{d^2\bar{x}_n}{dt^2} + \gamma\frac{d\bar{x}_n}{dt} = -2k\bar{x}_n + k\bar{y}_n$$

$$m\frac{d^2\bar{y}_n}{dt^2} + \gamma\frac{d\bar{y}_n}{dt} = k\bar{x}_n - k\bar{y}_n + QE_0c_n \exp\left(-\frac{2\pi i n t}{T}\right) \tag{III.5.40}$$

As discussed above, we now take the Fourier transforms of both equations. This yields

$$(-m\omega^2 + i\gamma\omega)\tilde{x}_n(\omega) = -2k\tilde{x}_n(\omega) + k\tilde{y}_n(\omega) \tag{III.5.41}$$

$$(-m\omega^2 + i\gamma\omega)\tilde{y}_n(\omega) = k\tilde{x}_n(\omega) - k\tilde{y}_n(\omega) + 2\pi QE_0c_n\delta\left(\omega - \frac{2\pi n}{T}\right) \tag{III.5.42}$$

where $\tilde{x}_n(\omega)$, $\tilde{y}_n(\omega)$ and the delta function are given by

$$\tilde{x}_n(\omega) = \int_{-\infty}^{\infty} \bar{x}_n(t) \exp\left(i\omega t\right) dt, \quad \tilde{y}_n(\omega) = \int_{-\infty}^{\infty} \bar{y}_n(t) \exp\left(i\omega t\right) dt$$

$$2\pi\delta\left(\omega - \frac{2\pi n}{T}\right) = \int_{-\infty}^{\infty} \exp\left(i\left(\omega - \frac{2\pi n}{T}\right)t\right) dt \tag{III.5.43}$$

One way of dealing with solving Eqs. (III.5.41) and (III.5.42) would be to eliminate either $\tilde{x}(\omega)$ or $\tilde{y}(\omega)$, which is simple to do for a system of two equations. This would have been, however, more cumbersome if we had a larger system of equations (like those discussed in Part II, Vol. 1). Thus, to refresh your matrix-algebra skills, we'll solve things the matrix way. In the matrix format Eqs. (III.5.41) and (III.5.42) read as

$$\mathbf{H}(\omega) \begin{pmatrix} \tilde{x}_n(\omega) \\ \tilde{y}_n(\omega) \end{pmatrix} = 2\pi Q E_0 c_n \delta \left( \omega - \frac{2\pi n}{T} \right) \begin{pmatrix} 0 \\ 1 \end{pmatrix} \qquad \text{(III.5.44)}$$

where

$$\mathbf{H}(\omega) = \begin{pmatrix} -m\omega^2 + i\gamma\omega + 2k & -k \\ -k & -m\omega^2 + i\gamma\omega + k \end{pmatrix} \qquad \text{(III.5.45)}$$

We can then write the solution of Eq. (III.5.44) through the inverse matrix $\mathbf{H}^{-1}(\omega)$

$$\begin{pmatrix} \tilde{x}_n(\omega) \\ \tilde{y}_n(\omega) \end{pmatrix} = 2\pi Q E_0 c_n \delta \left( \omega - \frac{2\pi n}{T} \right) \mathbf{H}^{-1}(\omega) \begin{pmatrix} 0 \\ 1 \end{pmatrix} \qquad \text{(III.5.46)}$$

To find the inverse matrix we use the recipe given in Chapter 4, Part II, Vol. 1, and hence

$$\mathbf{H}^{-1}(\omega) = \frac{1}{(m\omega^2 - i\gamma\omega)^2 - 3k(m\omega^2 - i\gamma\omega) + k^2}$$
$$\times \begin{pmatrix} -m\omega^2 + i\gamma\omega + k & k \\ k & -m\omega^2 + i\gamma\omega + 2k \end{pmatrix} \qquad \text{(III.5.47)}$$

Therefore, we can write a particular solution of Eqs. (III.5.41) and (III.5.42) as

$$\begin{pmatrix} \tilde{x}_n(\omega) \\ \tilde{y}_n(\omega) \end{pmatrix} = 2\pi Q E_0 c_n \delta \left( \omega - \frac{2\pi n}{T} \right) \frac{1}{R(\omega)^2 - 3kR(\omega) + k^2}$$
$$\times \begin{pmatrix} k \\ -R(\omega) + 2k \end{pmatrix} \qquad \text{(III.5.48)}$$

where $R(\omega) = m\omega^2 - i\gamma\omega$. We then inverse Fourier transform back to find

$$\begin{pmatrix} \tilde{x}_n(t) \\ \tilde{y}_n(t) \end{pmatrix} = Q E_0 c_n \frac{\exp\left( \frac{2\pi nit}{T} \right)}{R\left( \frac{2\pi n}{T} \right)^2 - 3kR\left( \frac{2\pi n}{T} \right) + k^2} \begin{pmatrix} k \\ R\left( \frac{2\pi n}{T} \right) + 2k \end{pmatrix}$$
$$\text{(III.5.49)}$$

The full solution of our equations, Eq. (III.5.38), is then obtained by taking the sum of the modes $\bar{x}_n(t)$ and $\bar{y}_n(t)$, which is

$$\begin{pmatrix} \bar{x}(t) \\ \bar{y}(t) \end{pmatrix} = QE_0 \sum_{n=-\infty}^{\infty} \begin{pmatrix} \bar{x}_n(t) \\ \bar{y}_n(t) \end{pmatrix} \qquad \text{(III.5.50)}$$

One advantage in presenting the solution in a form of a series is often that we don't necessarily need all the terms in the series. This applies particularly to limiting cases which we'll explore. Furthermore, and most importantly, by writing $E(t)$ as a Fourier series we can express the solution as a sum of terms, each having analytical expressions.

Our last task is to find the $c_n$ coefficients by performing the Fourier integrals (see Part II, Vol. 1). Recollect that these should read as

$$c_n = -\frac{1}{T} \int_0^T \frac{E(t)}{E_0} \exp\left(\frac{2\pi int}{T}\right) \qquad \text{(III.5.51)}$$

Splitting the integral in two parts,

$$c_n = \frac{1}{T^2} \int_0^{T/2} t \exp\left(\frac{2\pi int}{T}\right) dt + \frac{1}{T^2} \int_{T/2}^T (T - t) \exp\left(\frac{2\pi int}{T}\right) dt$$
$$\text{(III.5.52)}$$

These integrals evaluate to

$$c_n = \frac{1}{2\pi^2 n^2} \left[(-1)^n - 1\right] \quad \text{when } n \neq 0$$

$$c_0 = \frac{1}{4} \quad \text{for } n = 0 \qquad \text{(III.5.53)}$$

**Exercise.** Show that Eq. (III.5.52) evaluates to Eq. (III.5.53). **Think:** By considering $n$ being a continuous variable in Eq. (III.5.52) perform the integral and show in the limit $n \to 0$ that one indeed obtains $c_0 = 1/4$.

**Example III.5.3 (Continued).** The full solutions, presented in dimensionless form, $\bar{x}(t)k/QE_0$ and $\bar{y}(t)k/QE_0$, are well approximated by retaining only the $n = -1, 0, 1$ modes in the sum, for the

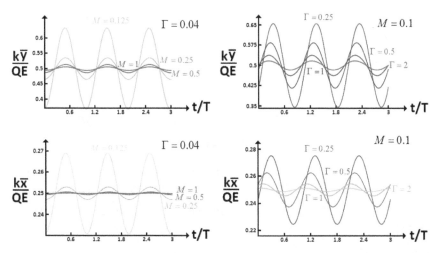

**Fig. III.5.5** We plot the inhomogeneous solution to Eq. (III.5.38) with Eq. (III.5.37), as a function of $t/T$ for various values of dimensionless parameters $M = m/(T^2k)$ and $\Gamma = \gamma/(Tk)$. For the parameter range considered, keeping only the $n = -1, 0, 1$ modes is a good approximation. The dominant frequency of the oscillations is thus $\omega = 2\pi/T$, where $T$ is the period of oscillation of the external field. One general observation is that the average values of both $\bar{x}(t)$ and $\bar{y}(t)$ are not zero. This is explained by the fact that the time-averaged value of the electric field is not zero, but in fact is equal to $E_0/4$, so the average equilibrium positions are no longer at $\bar{x} = \bar{y} = 0$. Also, we see that $k\bar{x}/QE_0$ oscillates around $\bar{x} = 0.25$, while $k\bar{y}/QE_0$ oscillates around $\bar{y} = 0.5$. This is as it should be, because at equilibrium the extensions of both springs are the same and have the value $0.25QE_0/k$, due to the average electric force $QE_0/4$ acting on $m_2$. We also note that the response of the uncharged mass (with displacement $\bar{x}(t)$) is much less than that of the other mass (with displacement $\bar{y}(t)$), to which the external force (electric field) is applied. On the l.h.s plots, we keep the parameter that controls damping $\Gamma$ fixed at the value of $\Gamma = 0.04$ and vary $M$. As we increase $M$, we see that the amplitude of the oscillations diminishes, but their phase remains constant. On the r.h.s plots we fix $M = 0.1$ and vary $\Gamma$. We see that, as we increase $\Gamma$, we also diminish the amplitude of the oscillations, but also we affect the phase. Furthermore, with increasing $\Gamma$ the oscillations become more out of phase with the oscillating electric field. This is because, for resonance in general, even for a single spring, increased damping causes a greater lag in response.

range of dimensionless parameters $M = m/(T^2k)$ and $\Gamma = \gamma/(Tk)$ specified in Fig. III.5.5.

Furthermore, for $m/T \gg \gamma$ we may simplify the result even further, keeping only $n = -1, 0, 1$ modes:

$$\begin{pmatrix} \bar{x}(t) \\ \bar{y}(t) \end{pmatrix} \approx -\frac{QE_0T^4 \cos\left(\frac{2\pi t}{T}\right)}{8\pi^6} \begin{pmatrix} k \\ \frac{4\pi^2 m}{T^2} \end{pmatrix} + \frac{QE_0}{4k} \begin{pmatrix} 1 \\ 2 \end{pmatrix} \qquad \text{(III.5.54)}$$

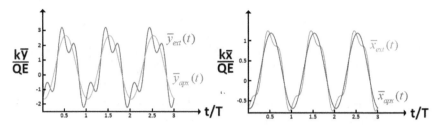

**Fig. III.5.6** Here, we plot the displacements $\bar{x}$ and $\bar{y}$ for the parameters $M = 0.0073684$ and $\Gamma = 0.0025$. The value $M$ has been chosen so that the frequency $\omega = 6\pi/T$, for the $n = -3, 3$ modes, lies extremely close to the natural frequency of the system $\sqrt{k/2m}(\sqrt{5}+3)^{1/2}$ (this is why we haven't rounded the value of $M$ up!). In this case, the $n = -1, 0, 1$ mode approximation, described by $\bar{x}_{apx}(t)$ and $\bar{y}_{apx}(t)$, does not work well compared with the full summation (exact solutions), $\bar{x}_{ext}(t)$ and $\bar{y}_{ext}(t)$. This is because the contribution from the $n = 3, -3$ modes is significantly enhanced by resonance (discussed in Part II, Vol. 1). If we were to decrease $M$ and $\Gamma$ further, we would be more likely to find more disagreement between the approximation and the exact solution. But this discrepancy does not increase monotonically with the reduction of $M$. Notably, if $M$ is chosen so that one of the modes of the oscillating electric field (for which $n \geq 3$) is close to one of the natural normal mode frequencies of this system, the disagreement will be most pronounced. What to do in such cases is described in the text.

Let's try to understand why we can do this. Our system of springs and masses cannot respond suddenly to the sawtooth varying electric force, due to the inertia of the masses on the springs and damping. Instead the motion follows the dominant lowest frequency of this force, which comes from $n = 1, -1$ as well as the static $n = 0$ component, unless we reduce both the masses and damping coefficient.

Indeed, when $m/(T^2 k) \ll 1$ and $\gamma/(Tk) \ll 1$, the approximation of retaining only the $n = -1, 0, 1$ modes starts to fail. In Fig. III.5.6 we show when it starts to happen, and the $n = 3, -3$ modes in the sum, Eq. (III.5.50), become important.

Also, when $\gamma$ is very small, there may be a particular mode $n = n_c$ which lies very close to the natural frequency. Its value is such that

$$n_c \approx \pm \frac{\omega_0 T}{2\pi} \sqrt{\frac{k}{2m}\left(3 \pm \sqrt{5}\right)} \qquad (\text{III.5.55})$$

In this case, the solution will be dominated by this mode $n = n_c$, and other terms in the series, Eq. (III.5.50), can be neglected. In Fig. III.5.7 we plot cases where $n_c = \pm 3$, where this mode is right at the resonant frequency. Here, we plot the approximate solution

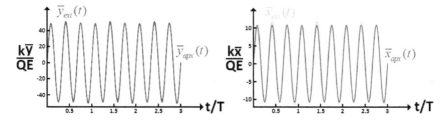

**Fig. III.5.7** Here, we plot the displacements for the exact solution, $\bar{x}_{ext}$ and $\bar{y}_{ext}$ as well as for an approximation where only the $n = 3, -3$ modes are retained, $\bar{x}_{apx}$ and $\bar{y}_{apx}$. The curves are plotted here for the value of $M = (3 + \sqrt{5})/(72\pi^2)$, which corresponds to modes $n_c = \pm 3$ having exactly one of the resonant frequencies, and we have chosen $\Gamma = 0.00005$ to be sufficiently small for such resonance to matter. We see that just retaining the modes $n = 3, -3$ works very well, when compared with the full sum over all modes. It will work even better as one reduces $\Gamma$ further.

keeping only the $n = 3$ mode and the exact solution that keeps all modes in the sum.

We see in this case the $n = 3, -3$ clearly dominate and just considering these modes in the sum is an excellent approximation, when those modes are close to the natural frequency of vibration $\omega = \sqrt{3 + \sqrt{5}}\sqrt{k/2m}$. Looking at what we see in Fig. III.5.7 may seem rather surprising: we had a sawtoothed variation in an electric field that had period $T$, but we see that $\bar{x}(t)$ and $\bar{y}(t)$ are vibrating at period $T/3$, which for the parameter value $M = (3 + \sqrt{5})/(72\pi^2)$, is at the resonant frequency of just one of the normal modes! However, it is not the overall period of the changing electric field that matters, it's the modes at resonant frequency that contribute to the sawtooth variation that matters when damping is small. This is a classic illustration of a resonance.

## 5.3.  Infinite Systems of Equations: Sound Waves

**Example III.5.4 (Physical Application ▶▶).** Here, we consider a long chain of masses with springs between them, as illustrated in Fig. III.5.8. We'll suppose for simplicity that all the masses, spring constants, and equilibrium length of each spring are the same, and the motion is undamped.

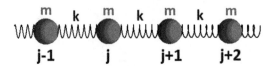

**Fig. III.5.8** An infinite chain of masses and springs all with the same mass $m$ and spring constants $k$. We solve the equations of motion for such a system in the text.

The force on mass $j$ is

$$F = -k(x_j - x_{j+1} - l) - k(x_j - x_{j-1} - l) \qquad \text{(III.5.56)}$$

Firstly, we may write $\delta x_j = x_j - \langle x \rangle = x_j - jl$, where $\langle x \rangle = jl$ is the mean (or equilibrium) displacement; here $l$ is an input parameter: the equilibrium length of each spring. Thus, the equations of motion, for each of the relative displacements $\delta x_j$, are

$$m\frac{d^2\delta x_j}{dt^2} = -k(\delta x_j - \delta x_{j+1}) - k(\delta x_j - \delta x_{j-1}) \qquad \text{(III.5.57)}$$

If the number of masses and springs is very large, solving Eq. (III.5.57) by the method outlined by Eqs. (III.5.27)–(III.5.31) will be very complicated. However, we can do a trick: we can introduce what is sometimes called a discrete Fourier transform. This method is exact in limit where the number of springs is infinite. For such a transform, we may write

$$\delta x_j = \frac{1}{2\pi} \int_{-\pi/x_0}^{\pi/x_0} \delta\tilde{x}(q) \exp(iqjl) dq \qquad \text{(III.5.58)}$$

Equation (III.5.58) comes about by first considering a continuous function $\delta x(\bar{x})$ that represents the displacements $\delta x_j$ as functions of the equilibrium displacement along the chain, $\bar{x}$. Introducing $\bar{x} \equiv jl$, we may write a Fourier representation for $\delta x_j = \delta x(jl)$. This reads as

$$\delta x(jl) = \frac{1}{2\pi} \int_{-\infty}^{\infty} \delta\tilde{x}(q) \exp(iqjl) dq \qquad \text{(III.5.59)}$$

Now, we realize that $\delta\tilde{x}(q) = \delta\tilde{x}(q + 2\pi/l)$ , i.e. it is a periodic function, since

$$\frac{1}{2\pi}\int_{-\infty}^{\infty}\delta\tilde{x}(q + 2\pi/l)\exp(iqjl)dq = \frac{1}{2\pi}\int_{-\infty}^{\infty}\delta\tilde{x}(q)\exp(iqjl + i2\pi j)dq$$

$$= \frac{1}{2\pi}\int_{-\infty}^{\infty}\delta\tilde{x}(q)\exp(iqjl)dq$$

$$(\text{III.5.60})$$

This means that we should restrict the range of integration, as we do in Eq. (III.5.58) to obtain a one-to-one mapping between $\delta x_j$ and $\delta\tilde{x}(q)$ for every combination of the $\{\delta x_j\}$. Using Eq. (III.5.58), we can rewrite Eq. (III.5.57) as

$$\int_{-\pi/l}^{\pi/l} m\frac{d^2\delta\tilde{x}(q,t)}{dt^2}e^{iqjl}dq$$

$$= \int_{-\pi/l}^{\pi/l}\left[-k\delta\tilde{x}(q,t)(1 - e^{iql}) - k\delta\tilde{x}(q,t)(1 - e^{iql})\right]e^{iqjl}dq$$

$$\Rightarrow m\frac{d^2\delta\tilde{x}(q,t)}{dt^2} = -2k\delta\tilde{x}(q,t)(1 - \cos(ql)) \qquad (\text{III.5.61})$$

By making the discrete Fourier transform, we've managed to decouple all the equations for all the masses given by Eq. (III.5.57). Now, we simply make the substitution $\delta\tilde{x}(q,t) = \delta\hat{x}_0(q)e^{-i\omega t}$ to find

$$m\omega^2\hat{x}_0(q) = 2k\hat{x}_0(q)(1 - \cos(ql)) \Rightarrow \omega = \pm\sqrt{\frac{2k}{m}}\cdot(1 - \cos(ql))^{1/2}$$

$$(\text{III.5.62})$$

The general solution to Eq. (III.5.57) can then be written as

$$\delta x_j(t) = \int_{-\pi/l}^{\pi/l}\delta\hat{x}_0^+(q)e^{i\left(qjl - \sqrt{\frac{2k}{m}(1-\cos(ql))^{1/2}}t\right)}dq$$

$$+ \int_{-\pi/l}^{\pi/l}\delta\hat{x}_0^-(q)e^{i\left(qjl + \sqrt{\frac{2k}{m}(1-\cos(ql))^{1/2}}t\right)} \qquad (\text{III.5.63})$$

The two functions $\delta \hat{x}_0^+ (q)$ and $\delta \hat{x}_0^- (q)$ appear here because we should write $\delta \tilde{x}(q, t) = \delta \hat{x}_0^+ (q) e^{-i\sqrt{\frac{2k}{m}(1-\cos(ql))}t} + \delta \hat{x}_0^- (q) e^{+i\sqrt{\frac{2k}{m}(1-\cos(ql))}t}$ to take into account that there are two roots for $\omega$ for every value of $q$. These functions may be determined through a set of initial conditions, $\delta x_j(0)$ and $\delta x'_j(0)$ for our system of springs, through the discrete inverse transforms

$$\sum_{j=-\infty}^{\infty} \delta x_j(0) e^{-ijql} = \frac{2\pi}{l} \left( \delta \hat{x}_0^+ (q) + \delta \hat{x}_0^- (q) \right) \tag{III.5.64}$$

$$\sum_{j=-\infty}^{\infty} \delta x'_j(0) e^{-ijql} = \frac{2\pi}{l} \left( -i\sqrt{\frac{2k}{m}[1 - \cos(ql)]}\, \delta \hat{x}_0^+ (q) \right.$$

$$\left. +i\sqrt{\frac{2k}{m}[1 - \cos(ql)]}\, \delta \hat{x}_0^- (q) \right) \tag{III.5.65}$$

**Exercise.** Show that the inverse discrete transforms given by Eq. (III.5.64) and (III.5.65) do indeed take these forms. **Hint:** Consider the time derivative of Eq. (III.5.63) and that

$$\sum_{j=-\infty}^{\infty} e^{ijq'l} e^{-ijql} = 2\pi \delta \left( (q - q')l \right) = \frac{2\pi}{l} \delta(q - q') \tag{III.5.66}$$

**Example III.5.4 (Continued).** What does Eq. (III.5.63) describe? Nothing else than a supposition of travelling longitudinal waves, along the springs, with different wave numbers $q = 2\pi/\lambda$, which travel with speed $v = \sqrt{\frac{2k}{m}} \frac{\sqrt{1-\cos(ql)}}{|q|}$. We plot a 'dimensionless velocity' $v_R = v\sqrt{m/k/l}$ as a function of $ql$ in Fig. III.5.9.

The relationship between $\omega$ and $k$ for waves is called a *dispersion relationship*: waves with different wave lengths have different speeds and when travelling through a medium they separate out or disperse.

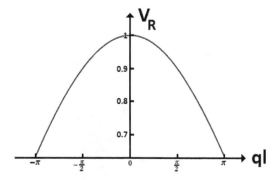

**Fig. III.5.9** We plot $v_R = v/l\sqrt{k/m}$ as function of $ql$. Note that as $\|ql\|$ increases the speed of the waves diminishes.

One thing to note is that if $ql \ll 1$ the wave's speed is not affected by the average spacing between the springs, $l$. In this limit, replacing $\cos(ql)$ with $1 - q^2l^2/2$, we get $v = l\sqrt{k/m} = \sqrt{K/\rho}$, where $\rho = m/l$ is the linear mass density of the chain (note, here, we suppose the springs to be massless), and $K = kl$, where $K$ is a modulus of stiffness that does not depend on spring length. What we have come up with in this fairly simple example is a description of 1D sound waves! We'll discuss the continuum version of Eq. (III.5.57), 1D wave equation later, when dealing with partial differential equations.

As a step closer to real systems we'll extend our discussion of coupled differential equations for a 3D lattice, to describe soundwaves in solids in classical mechanics. Since this is a bit complicated, so we 'boxed' this discussion as optional, for curious readers.

---

*Optional: Classical vibrations for a 3D cubic lattice.* We can extend the above described approach to a 3D cubic lattice, which may represent some types of crystal. Classically (we'll not consider quantum mechanical effects here), sound waves for such solids can be modelled by considering a 3D array of masses on springs. The springs represent bonds or interactions between the atoms making up the crystal, where the equilibrium spring length

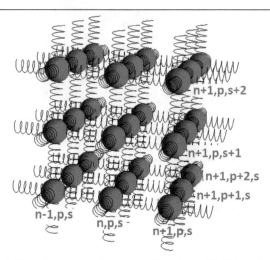

**Fig. III.5.10** A 3D cubic array of masses and springs. All the spring constants are equal to $k$ and all the masses are $m$. We have labelled some of the masses to show how the $\{n, p, s\}$ labelling works. For instance, if we move from one atom to a neighbouring one in the positive $x$-direction, the index $n$ increases by 1.

corresponds to the equilibrium spacing between atoms in the crystal. Each mass is given three integer indices $\{n, p, s\}$ describing its position in the lattice. The situation is illustrated in Fig. III.5.10.

The position of the centres of each mass can then be described as

$$\mathbf{r}_{n,p,s} = \mathbf{r}_{n,p,s}^{(0)} + \delta\mathbf{r}_{n,p,s} \qquad \text{(III.5.67)}$$

where $\mathbf{r}_{n,p,s}^{(0)} = l(n\hat{\mathbf{i}} + p\hat{\mathbf{j}} + s\hat{\mathbf{k}})$ is the position of all the masses when all the springs are at their equilibrium length $l$. The force on each mass comes from the six springs that surround it, which for the mass labelled by $\{n, p, s\}$ is

$$\mathbf{F}_{n,p,s} = k(\delta\mathbf{r}_{n+1,p,s} - \delta\mathbf{r}_{n,p,s}) + k(\delta\mathbf{r}_{n-1,p,s} - \delta\mathbf{r}_{n,p,s})$$

$$+ k(\delta\mathbf{r}_{n,p+1,s} - \delta\mathbf{r}_{n,p,s}) + k(\delta\mathbf{r}_{n,p-1,s} - \delta\mathbf{r}_{n,p,s}) \qquad \text{(III.5.68)}$$

$$+ k(\delta\mathbf{r}_{n,p,s+1} - \delta\mathbf{r}_{n,p,s}) + k(\delta\mathbf{r}_{n,p,s-1} - \delta\mathbf{r}_{n,p,s})$$

The equation of motion for each mass is given by

$$m\frac{d^2\mathbf{r}_{n,p,s}}{dt^2} = \mathbf{F}_{n,p,s} \tag{III.5.69}$$

We can solve this system of equations by writing the 3D version of the discrete Fourier transform, which reads

$$\delta\mathbf{r}_{n,p,s}(t) = \frac{1}{2\pi}\int_{-\pi/l}^{\pi/l}\int_{-\pi/l}^{\pi/l}\int_{-\pi/l}^{\pi/l}\delta\tilde{\mathbf{r}}(\mathbf{q},t)e^{i(q_x n + q_y p + q_z s)l}\,dq_x dq_y dq_z$$

$$\tag{III.5.70}$$

With the use of this definition, Eqs. (III.5.68) and (III.5.69) result in a single, decoupled equation on a Fourier component of $\delta\tilde{\mathbf{r}}(\mathbf{q},t)$:

$$m\frac{d^2\delta\tilde{\mathbf{r}}(\mathbf{q},t)}{dt^2} = -2k\delta\tilde{\mathbf{r}}(\mathbf{q},t)\left[(1-\cos(q_x l)) + (1-\cos(q_y l))\right.$$

$$\left. +(1-\cos(q_z l))\right] \tag{III.5.71}$$

Equation (III.5.71) is solved by writing

$$\delta\tilde{\mathbf{r}}(\mathbf{q},t) = \delta\hat{\mathbf{r}}_0^+(\mathbf{q})e^{-i\sqrt{\frac{2k}{m}[3-\cos(q_x l)-\cos(q_y l)-\cos(q_z l)]}\,t}$$

$$+\delta\hat{\mathbf{r}}_0^-(\mathbf{q})e^{i\sqrt{\frac{2k}{m}[3-\cos(q_x l)-\cos(q_y l)-\cos(q_z l)]}\,t}$$

$$\tag{III.5.72}$$

Thus, substituting Eq. (III.5.72) into Eq. (III.5.70) one obtains the general solution. The functions $\delta\hat{\mathbf{r}}_0^+(\mathbf{q})$ and $\delta\hat{\mathbf{r}}_0^-(\mathbf{q})$ may be determined through the initial conditions on $\delta\mathbf{r}_{n,p,s}(0)$ and $\delta\mathbf{r}'_{n,p,s}(0)$ by considering the discrete form of the inverse Fourier transform

$$\sum_{n=-\infty}^{\infty}\sum_{p=-\infty}^{\infty}\sum_{s=-\infty}^{\infty}\delta\mathbf{r}_{n,p,s}(t)e^{-il(nq_x + pq_y + sq_z)} = \left(\frac{2\pi}{l}\right)^3\delta\tilde{\mathbf{r}}(\mathbf{q},t)$$

$$\tag{III.5.73}$$

and its derivative with respect to time.

**Exercise.** Show that for an initial condition $\delta\mathbf{r}_{n,p,s}(0) = f_n$ and $\delta\mathbf{r}'_{n,p,s}(0) = g_n$, the general solution reduces to that obtained in Example III.5.4 for a 1D chain. What physically does this solution in 3D describe?

---

### Limiting case for long wave lengths

Let's examine the limiting case when $q_x l \ll 1$, $q_y l \ll 1$ and $q_z l \ll 1$. Here, Eq. (III.5.72) reduces to

$$\delta \tilde{\mathbf{r}}(\mathbf{q}, t) = \delta \hat{\mathbf{r}}_0^+(\mathbf{q}) \exp\left(-il\,|\mathbf{q}|\,t\sqrt{k/m}\right)$$

$$+\, \delta \hat{\mathbf{r}}_0^-(\mathbf{q}) \exp\left(il\,|\mathbf{q}|\,t\sqrt{k/m}\right)$$

where $|\mathbf{q}| = \sqrt{q_x^2 + q_y^2 + q_z^2}$. This implies that the frequency of a wave with wavelength $\lambda = 2\pi/|\mathbf{q}|$ is given by $\omega = |\mathbf{q}|\sqrt{l^2 k/m} = |\mathbf{q}|\sqrt{k/(l\rho)} = |\mathbf{q}|\sqrt{Y/\rho}$ (the *dispersion relation*), where $\rho = m/l^3$ is density of the solid and $Y = k/l$ is its Young's modulus or stiffness (measured in Newtons per meter squared). Thus, the speed of sound in a solid (for long wave lengths) is $v = \omega/|\mathbf{q}| = \sqrt{Y/\rho}$.

---

### Remarks about solving coupled linear differential equations in general

We have talked about solving systems of coupled linear differential equations with constant coefficients. It is possible to extend some of these ideas to the solution of more complicated coupled linear differential equations, through the appropriate choice of functions for the solution, or the correct integral transforms in the inhomogeneous equations. For instance, we might want to solve two coupled inhomogeneous Bessel equations such as

$$-\frac{d^2 x(s)}{ds^2} - \frac{1}{s}\frac{dx(s)}{ds} + \frac{\alpha^2 x(s)}{s^2} + ax(s) + by(s) = f(x)$$

$$\tag{III.5.74}$$

$$-\frac{d^2 y(s)}{ds^2} - \frac{1}{s}\frac{dy(s)}{ds} + \frac{\alpha^2 y(s)}{s^2} + dy(s) + cx(s) = g(x)$$

$$\tag{III.5.75}$$

The inhomogeneous solution of this system of equations can be solved through Hankel transforms:

$$\int_0^\infty sx(s)J_\alpha(k_r s)ds = \tilde{x}(k_r), \quad \int_0^\infty sy(s)J_\alpha(k_r s)ds = \tilde{y}(k_r)$$

$$\text{(III.5.76)}$$

$$\int_0^\infty sf(s)J_\alpha(k_r s)ds = \tilde{f}(k_r), \quad \int_0^\infty sg(s)J_\alpha(k_r s)ds = \tilde{g}(k_r)$$

$$\text{(III.5.77)}$$

In this case, these Hankel transforms allow us to write the matrix equation

$$\begin{pmatrix} k_r^2 + a & b \\ c & k_r^2 + d \end{pmatrix} \begin{pmatrix} \tilde{x}(k_r) \\ \tilde{y}(k_r) \end{pmatrix} = \begin{pmatrix} \tilde{f}(k_r) \\ \tilde{g}(k_r) \end{pmatrix} \qquad \text{(III.5.78)}$$

We can solve this matrix equation by finding the inverse matrix, so obtaining $\tilde{x}(k_r)$ and $\tilde{y}(k_r)$, and then performing the inverse Hankel transform back to get the inhomogeneous solution for both $x(s)$ and $y(s)$.

**Exercise.** By solving Eq. (III.5.78), write the solutions for $x(s)$ and $y(s)$ as integrals over $k_r$ by considering the inverse Hankel transforms of both $\tilde{x}(k_r)$ and $\tilde{y}(k_r)$.

## 5.4. Differential Equations of Arbitrary Order

### *Some general remarks*

Before finishing this chapter, let us briefly discuss differential equations of arbitrary order. We have only a few points to say here, and in a moment you will see why (for more details, we refer the reader to more specialized texts). Already, we've come across 4th order equations and higher, but in disguise, as we solve coupled second differential equations. In Example III.5.1, we could first differentiate the equation for $\bar{x}''(t)$ (in Eq. (III.5.4)) an extra two times with respect to $t$. Then from the resulting equation we could use Eq. (III.5.4) to eliminate both $\bar{y}(t)$ and $\bar{y}''(t)$. This would yield the following 4th order differential equation

$$\frac{d^4\bar{x}}{dt^4} + \left[\frac{k_1 + k_2}{m_1} + \frac{k_2}{m_2}\right]\frac{d^2\bar{x}}{dt^2} + \frac{k_1 k_2}{m_1 m_2} = 0 \qquad \text{(III.5.79)}$$

Now, let's consider the case of a homogeneous $n$th order linear differential equation, of the form

$$b_n\frac{d^n y}{dt^n} + b_{n-1}\frac{d^{n-1}y}{dt^{n-1}} + \cdots + b_2\frac{d^2 y}{dt^2} + b_1\frac{dy}{dt} + b_0 = 0 \qquad \text{(III.5.80)}$$

The general solution takes the form

$$y(t) = \sum_{m=1}^{n} A_m \exp(i\omega_m t) \qquad \text{(III.5.81)}$$

where the $\omega_m$ are roots of the polynomial equation

$$b_n(i)^n \omega^n + b_{n-1}(i)^{n-1}\omega^{n-1} + \cdots - \omega^2 b_2 + i\omega b_1 + b_0 = 0 \quad \text{(III.5.82)}$$

**Exercise.** Solve the polynomial equation for $\omega$, Eq. (III.5.82), for the case considered in Eq. (III.5.79) and show that one indeed obtains Eq. (III.5.9), as we found it by matrix methods.

To obtain a specific solution we would need to specify $n$ boundary conditions to find all of the coefficients.

To solve the inhomogeneous equation in the domain $-\infty \le x \le \infty$

$$b_n\frac{d^n y}{dx^n} + b_{n-1}\frac{d^{n-1}y}{dx^{n-1}} + \cdots + b_2\frac{d^2 y}{dx^2} + b_1\frac{dy}{dx} + b_0 = f(x) \quad \text{(III.5.83)}$$

(we do not specify here what the function $y$ and variable $x$ correspond to) we would use the Fourier transforms

$$\int_{-\infty}^{\infty} f(x)\exp(iqx)dx = \tilde{f}(q), \qquad \int_{-\infty}^{\infty} y(x)\exp(iqx)dx = \tilde{y}(q)$$
$$\text{(III.5.84)}$$

**Exercise.** Using Eq. (III.5.84) write the inhomogeneous solution to Eq. (III.5.83) as an integral over $q$ with $\tilde{f}(q)$ in the integrand.

In physical problems you may encounter a general $n$th order linear differential equation with variable coefficients,

$$a_n(x)\frac{d^n y}{dx^n} + a_{n-1}(x)\frac{d^{n-1}y}{dx^{n-1}} + \cdots + a_2(x)\frac{d^2 y}{dx^2}$$

$$+ a_1(x)\frac{dy}{dx} + a_0(x) = f(x) \tag{III.5.85}$$

Beyond second order these may be much more complicated to solve. But, in any case, it is a general rule that to fully specify a unique (specific) solution to Eq. (III.5.85) we do indeed need $n$ boundary conditions.

We devoted this chapter to exact solutions of differential equations. In the next chapter we proceed to approximate methods for solving such equations.

## 5.5. Key Points

- We can solve a system of $n$ coupled linear homogeneous differential equations with constant coefficients by writing each of the solutions as $y_j(x) = \tilde{y}_j \exp(iqx), j = 1, 2, \ldots, n$. We then have a system of linear algebraic homogeneous equations in terms of the $\tilde{y}_j$. These can be written as a matrix equation $\mathbf{M}.\mathbf{y} = 0$, where $\mathbf{y}$ is a column vector containing the entries $\tilde{y}_j$. As we found out before, in Part II, Vol. 1, these equations are solved by requiring $\det[\mathbf{M}] = 0$.
- We can solve a system of $n$ coupled linear inhomogeneous differential equations by performing Fourier transforms on solutions $\tilde{y}_j(q) = \int_{-\infty}^{\infty} y_j(x) \exp(iqx)dx$ and inhomogeneous (driving) terms $\tilde{f}_j(q) = \int_{-\infty}^{\infty} f_j(x) \exp(iqx)dx, j = 1, 2, \ldots, n$. We then obtain a system of linear algebraic inhomogeneous equations in terms of the $\tilde{y}_j(q)$ and $\tilde{f}_j(q)$. These can be written as a matrix equation $\mathbf{M}.\mathbf{y} = \mathbf{f}$, where $\mathbf{y}$ and $\mathbf{f}$ are column vectors containing the entries $\tilde{y}_j(q)$ and $\tilde{f}_j(q)$. The solutions $y_j(x)$ are obtained through $\mathbf{f} = \mathbf{M}^{-1}\mathbf{y}$ and the subsequent inverse Fourier transforming back.
- An infinite system of linear homogeneous equations with constant coefficients that can be written as a single equation, relating

$y_j(x)$ to $y_{i+1}(x) \ldots y_{j+m}(x)$ and $y_{j+1}(x) \ldots y_{j+m}(x)$, where $m$ is a finite integer, can be solved through the discrete Fourier transform $y(x_n) = \frac{1}{2\pi} \int_{-\pi/l}^{\pi/l} \tilde{y}(q) \exp(iqnl) dq$ where $x_n = nl$ (see Example III.5.4).

- An $n$th order homogeneous linear differential equation with constant coefficients can be solved through the substitution $y(t) = \exp(i\omega t)$, resulting in a polynomial equation of the form $b_n(i)^n \omega^n + b_{n-1}(i)^{n-1}\omega^{n-1} + \cdots - \omega^2 b_2 + i\omega b_1 + b_0 = 0$. The general solution is $y(t) = A_1 \exp(i\omega_1 t) + A_2 \exp(i\omega_2 t) + \cdots + A_n \exp(i\omega_n t)$, if the roots $\omega_n$ of the polynomial equation are all different.

- Any $n$th order inhomogeneous linear differential equation *with constant coefficients* may be solved by Fourier transform.

- In general, to have a unique solution to an $n$th order differential equation we need to specify $n$ boundary conditions.

## Chapter 6

# Approximation Techniques for Ordinary Differential Equations

## 6.1. Perturbation Theory for Linear Differential Equations

### *Perturbation theory for homogeneous linear differential equations*

Let's consider a homogeneous linear differential equation of the form

$$\frac{d^2y(x)}{dx^2} + P(x)\frac{dy(x)}{dx} + Q(x)y(x) = 0 \qquad \text{(III.6.1)}$$

Now, suppose that $P(x)$ and $Q(x)$ are complicated, and we do not know how to solve Eq. (III.6.1) exactly analytically, but we can solve another equation

$$\frac{d^2y_0(x)}{dx^2} + P_0(x)\frac{dy_0(x)}{dx} + Q_0(x)y_0(x) = 0 \qquad \text{(III.6.2)}$$

If $P(x)$ and $Q(x)$ are not too different from $P_0(x)$ and $Q_0(x)$, we can attempt a perturbation expansion about the solution to Eq. (III.6.2) to approximate the solution to Eq. (III.6.1).

First, we may write $P(x) - P_0(x) = \alpha\Delta P(x)$ and $Q(x) - Q_0(x) = \beta\Delta Q(x)$. Here, $\alpha$ and $\beta$ are perturbation parameters that control the size of the differences between Eqs. (III.6.1) and (III.6.2).

We can then rewrite Eq. (III.6.1) as

$$\frac{d^2y}{dx^2} + (P_0(x) + \alpha\Delta P(x))\frac{dy}{dx} + (Q_0(x) + \beta\Delta Q(x)) = 0 \quad \text{(III.6.3)}$$

Next is to write the solution as a power series in both $\alpha$ and $\beta$:

$$y(x) = y^{(0,0)}(x) + \alpha y^{(1,0)}(x) + \beta y^{(0,1)}(x) + \alpha^2 y^{(2,0)}(x)$$
$$+ \alpha\beta y^{(1,1)}(x) + \beta^2 y^{(0,2)}(x) + \cdots \quad \text{(III.6.4)}$$

where $y^{(0,0)}(x) = y_0(x)$, and substitute Eq. (III.6.4) into Eq. (III.6.3) so that

$$\frac{d^2 y^{(0,0)}(x)}{dx^2} + P_0(x)\frac{dy^{(0,0)}(x)}{dx} + Q_0(x)y^{(0,0)}(x)$$

$$+ \alpha\left[\frac{d^2 y^{(1,0)}(x)}{dx^2} + P_0(x)\frac{dy^{(1,0)}(x)}{dx} + \Delta P(x)\frac{dy^{(0,0)}(x)}{dx} + Q_0(x)y^{(1,0)}(x)\right]$$

$$+ \beta\left[\frac{d^2 y^{(0,1)}(x)}{dx^2} + P_0(x)\frac{dy^{(0,1)}(x)}{dx} + Q_0(x)y^{(0,1)}(x) + \Delta Q(x)y^{(0,0)}(x)\right]$$

$$+ \alpha^2\left[\frac{d^2 y^{(2,0)}(x)}{dx^2} + P_0(x)\frac{dy^{(2,0)}(x)}{dx} + \Delta P(x)\frac{dy^{(1,0)}(x)}{dx} + Q_0(x)y^{(2,0)}(x)\right]$$

$$+ \beta^2\left[\frac{d^2 y^{(0,2)}(x)}{dx^2} + P_0(x)\frac{dy^{(0,2)}(x)}{dx} + Q_0(x)y^{(0,2)}(x) + \Delta Q(x)y^{(0,1)}(x)\right]$$

$$+ \alpha\beta\left[\frac{d^2 y^{(1,1)}(x)}{dx^2} + P_0(x)\frac{dy^{(1,1)}(x)}{dx} + \Delta P(x)\frac{dy^{(0,1)}(x)}{dx}\right.$$

$$\left. + Q_0(x)y^{(1,1)}(x) + \Delta Q(x)y^{(1,0)}(x)\right] + \cdots = 0 \quad \text{(III.6.5)}$$

Now, we must have that the coefficients of *each* power in the expansion, $\alpha^s\beta^r$ sum to zero. This requirement gives us a hierarchy of equations. The zeroth order term $y^{(0,0)}(x) = y_0(x)$ simply satisfies Eq. (III.6.2). Both $y^{(1,0)}(x)$ and $y^{(0,1)}(x)$ satisfy the inhomogeneous equations

$$\frac{d^2 y^{(1,0)}(x)}{dx^2} + P_0(x)\frac{dy^{(1,0)}(x)}{dx} + Q_0(x)y^{(1,0)}(x)$$

$$= -\Delta P(x)\frac{dy^{(0,0)}(x)}{dx} \tag{III.6.6}$$

$$\frac{d^2 y^{(0,1)}(x)}{dx^2} + P_0(x)\frac{dy^{(0,1)}(x)}{dx} + Q_0(x)y^{(0,1)}(x)$$

$$= -\Delta Q(x)y^{(0,0)}(x) \tag{III.6.7}$$

Successive orders in the perturbation expansion may be calculated through the resulting inhomogeneous equations:

$$\frac{d^2 y^{(n,m)}(x)}{dx^2} + P_0(x)\frac{dy^{(n,m)}(x)}{dx} + Q_0(x)y^{(n,m)}(x)$$

$$= -\Delta Q(x)y^{(n,m-1)}(x) - \Delta P(x)\frac{dy^{(n-1,m)}(x)}{dx} \tag{III.6.8}$$

**Example III.6.1 (Physical Application ▶▶).** The first example we have chosen (picked for the simplicity of the solution that does not necessarily require any knowledge of complex contour integration, yet to be covered, for the evaluation of the correction), may look contrived. But in fact it is not, as such systems have relevance in problems involving generation of electrical signal from mechanical motion. Let's look at an LC circuit that contains a capacitor and an inductor, sketched in Fig. III.6.1, with capacitance $C(t)$ changing with time (for simplicity we'll neglect the resistance of the wires). We will consider a special kind of $C(t)$ variation.

Suppose there is an elastic insulating material between plates of the capacitor. At $t < 0$ the distance between the plates is at equilibrium value for this material, $d_0$. We will suppose that at some time $t < 0$ the capacitor has been charged to charge $Q = Q_0$ on the plates and then it is disconnected from a power source and connected to the inductor of inductance $L$. At $t = 0$, when we suppose that all the charge $Q_0$ is on the capacitor (not flowing yet through the inductor), we suddenly pull the plates apart, stretching the material, to a distance $d_0 + d_1$. We assume for this material that the plates of

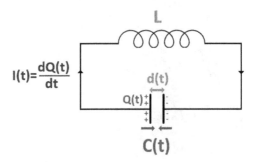

**Fig. III.6.1** Schematic picture of an LC circuit, where the distance between the capacitor plates is diminishing from a stretched value, so that the capacitance, $C(t)$, is increasing as a function of time. The problem formulated in this example is to solve a differential equation for the charge on the capacitor, $Q(t)$, using perturbation theory.

the capacitor are performing overdamped harmonic motion, returning to $d_0$ without oscillations, i.e. the distance between the two plates varies as $d(t) = d_0 + d_1 e^{-\frac{t}{\tau}}\theta(t)$, where $\tau$ is the relaxation time, and $\theta(t)$ is a Heaviside step-function (recall Section 6.6, Part I, Vol. 1) that ensures that the plates are set in motion right after time $t = 0$. Also we'll assume that the motion between the plates is slow enough that we can neglect complicating electromagnetic effects. The initial conditions (at $t = 0$) are $I = dQ/dt = 0$ and $Q = Q_0$.

For this circuit the charge on the capacitor satisfies the following homogeneous equation:

$$L\frac{d^2Q}{dt^2} + \frac{Q}{C(t)} = 0 \qquad\qquad (\text{III.6.9})$$

The capacitance of a plane dielectric capacitor is related to the distance between the two plates $C(t) = \varepsilon\varepsilon_0 A/d(t)$. In this expression $A$ is the area of the plates, $\varepsilon$ is the dielectric constant of the medium between them, which for simplicity will be assumed to be independent of the variation of $d(t)$, and $\varepsilon_0$ is the dielectric constant of vacuum (to make engineering students happy, we use here SI units). This standard expression allows us to rewrite (III.6.9) as

$$L\frac{d^2Q}{dt^2} + [\beta_0 + \beta_1\theta(t)e^{-\frac{t}{\tau}}]Q = 0 \qquad\qquad (\text{III.6.10})$$

where $\beta_0 = \frac{d_0}{A\varepsilon\varepsilon_0}$ and $\beta_1 = \frac{d_1}{A\varepsilon\varepsilon_0}$.

Well, we know how to solve Eq. (III.6.10) when $\beta_1 = 0$, as it is the homogeneous equation for a harmonic oscillator. So we could approximate a solution of this equation by treating the term $\beta_1 \theta(t) e^{-\frac{t}{\tau}}$ as a perturbation. Note that if we are going to consider only a small number of terms in the perturbation series, such an approach will be justified, strictly speaking, when the initial stretching $d_1 \ll d_0$; but, in fact, this approximation may appear to work even when $d_1 < d_0$. As in our general perturbation scheme considered just before this example, we can choose $\alpha = 0$, $\beta = \beta_1$ and $\Delta Q(x) = \theta(t) e^{-\frac{t}{\tau}}$. Thus, we may write the power series

$$Q(t) = Q^{(0)}(t) + \beta_1 Q^{(1)}(t) + (\beta_1)^2 Q^{(2)}(t) + \cdots \qquad \text{(III.6.11)}$$

As the leading order solution is simply the solution to the harmonic oscillator equation, it is given by

$$Q^{(0)}(t) = B \cos(\omega_0 t + \phi_0) \qquad \text{(III.6.12)}$$

where $\omega_0 = \sqrt{\beta_0/L}$. Here, $B$ and $\phi_0$ are constants to be determined from the initial conditions. Had we had $\beta_1 = 0$, we would have $B = Q_0$ and $\phi_0 = 0$. Hence, a perturbation expansion for $\phi_0$ will start from the term first order in $\beta_1$, i.e. $\phi_0 = \beta_1 \phi_0^{(1)} + (\beta_1)^2 \phi_0^{(2)} + \cdots$, where the $\phi_0^{(n)}$ are chosen to satisfy the boundary conditions to whatever order in perturbation theory we care to approximate the solution.

The equation for the next to leading order solution reads

$$L \frac{d^2 Q^{(1)}}{dt^2} + \beta_0 Q^{(1)} = -B e^{-\frac{t}{\tau}} \cos(\omega_0 t) \, \theta(t) \qquad \text{(III.6.13)}$$

Note that on the r.h.s of Eq. (III.6.13), we had to put $\phi_0 = 0$, as here we are looking for the correction which is first order in $\beta_1$. However, we still need to include $\phi_0 \approx \beta_1 \phi_0^{(1)}$ in Eq. (III.6.12).

Equation (III.6.13) is an inhomogeneous linear differential equation with constant coefficients and one way it can be solved is by Fourier transforms. However, to get the result this way requires complex contour integration, which we have yet to cover. But fortunately, for this equation form it is not hard to find the Green's function $\bar{G}(t - t')$). It's given by $\bar{G}(t - t') = \frac{\sin(\omega_0 |t - t'|)}{2\omega_0}$.

**Exercise.** Show that $\sin(\omega_0|t - t'|)$ satisfies the equation

$$\left[\frac{d}{dt^2} + \omega_0^2\right] \sin(\omega_0|t - t'|) = 2\omega_0 \delta(t - t') \qquad \text{(III.6.14)}$$

Thus, show that we can write

$$Q^{(1)}(t) = -\frac{B}{2\beta_0} \int_0^\infty \sin\left(|\omega_0 t - x|\right) e^{-px} \cos\left(x\right) dx \qquad \text{(III.6.15)}$$

where we have the dimensionless parameter $p = \frac{1}{\tau \omega_0}$. **Hint:** You may find Eq. (III.4.236) (with $G(t, t') = \bar{G}(t - t')$ and $b = -a = \infty$) useful for writing Eq. (III.6.15) as a solution to Eq. (III.6.13).

**Exercise.** Show that Eq. (III.6.15) evaluates to

$$Q^{(1)}(t) = \frac{B}{2\beta_0} \left[ -\frac{1}{2} \left( \frac{1 - 2e^{-\frac{t}{\tau}}}{p} + \frac{p\left(1 + 2e^{-\frac{t}{\tau}}\right)}{p^2 + 4} \right) \times \sin(\omega_0 t) \right.$$

$$\left. + \frac{1 - 2e^{-\frac{t}{\tau}}}{p^2 + 4} \cos\left(\omega_0 t\right) \right] \qquad \text{(III.6.16)}$$

for $t > 0$. **Hint:** Write $\sin(|\omega_0 t - x|)$ and $\cos(x)$ in terms of $\exp(i|\omega_0 t - x|)$, $\exp(-i|\omega_0 t - x|)$ and $\exp(ix)$ and $\exp(-ix)$; also, split integration into two parts: from 0 to $\omega_0 t$ and $\omega_0 t$ to $\infty$.

**Example III.6.1 (Continued).** The perturbation expansion effectively generates a power series in dimensionless ratio $\beta_1/\beta_0$, and the approximation of including just the leading order term correction works well for $\beta_1/\beta_0 \ll 1$. In Fig. III.6.2 we plot the correction $\frac{\beta_0}{B} Q^{(1)}(t)$ for various values of $p$.

Now, let's apply our initial conditions that at $t = 0$, $Q = Q_0$ and $I = 0$. To satisfy the initial condition at $t = 0$, for small deviation of $\phi_0$ from zero, we may write

$$Q^{(0)}(t) \approx B \cos\left(\omega_0 t\right) + \beta_1 B \phi_0^{(1)} \sin(\omega_0 t) \qquad \text{(III.6.17)}$$

**Exercise.** Show that the approximate solution $Q(t) \approx Q^{(0)}(t) + \beta_1 Q^{(1)}(t)$ at $t = 0$ yields $Q(0) \approx B\left(1 - \frac{\beta_1}{2\beta_0(4+p^2)}\right)$ so that

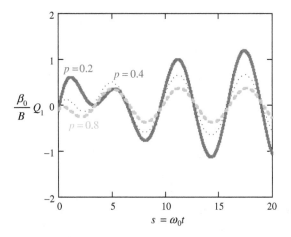

**Fig. III.6.2** The correction $Q^{(1)}\beta_0/B$ shown for indicated values of the parameter $p$ as a function of dimensionless time $s$. We see that decreasing $p$ increases the size of the correction. This is because the perturbation persists $\beta_1 e^{-\frac{t}{\tau}}$ for longer times. As $s$ and $p$ become larger, the correction becomes more sinusoidal.

$$B \approx \frac{Q_0}{1 - \frac{\beta_1}{2\beta_0(4+p^2)}} \tag{III.6.18}$$

and for no initial current (i.e. $I(0) = 0$) we must have that

$$\phi_0 \approx \beta_1 \phi_0^{(1)} = -\left(\frac{\beta_1}{4\beta_0}\right)\left(\frac{1}{p} + \frac{p}{p^2+4}\right) \tag{III.6.19}$$

Finally, using Eqs. (III.6.16), (III.6.17) and (III.6.19), we can write our full approximate solution for the charge, as a function of time, as

$$Q(t) \approx B\left[\cos(\omega_0 t) + \frac{\beta_1}{\beta_0}\tilde{Q}^{(1)}(t)\right] \tag{III.6.20}$$

$$\tilde{Q}^{(1)}(t) = \frac{1}{4}\left\{2\frac{1-2e^{-\frac{t}{\tau}}}{p^2+4}\cos(\omega_0 t)\right.$$

$$\left. -\left(\frac{1-2e^{-\frac{t}{\tau}}}{p} + \frac{p\left(1+2e^{-\frac{t}{\tau}}\right)}{p^2+4}\right) \times \sin(\omega_0 t)\right\} \tag{III.6.21}$$

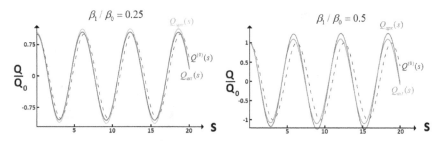

**Fig. III.6.3** Here, we plot the exact numerical solution to Eq. (III.6.10), $Q_{ext}(s)$ against both the leading order ($\beta_1 = 0$) solution $Q^{(0)}(s)$, and an approximate perturbation theory solution that includes the first order correction, $Q_{apx}(t) = B\cos(\omega_0 t) + B\beta_1\tilde{Q}^{(1)}(t)/\beta_0$ as functions of dimensionless time, $s = \omega_0 t$. The solutions have applied the initial ($t = 0$) conditions $Q(t = 0) = Q_0$ and $\frac{dQ}{dt}\big|_{t=0} = 0$. Note that $Q^{(0)}(s)$ would be the solution with the distance between the plates of the capacitor remaining constant at $d_0$ with respect to time, i.e. for $\beta_1 = 0$. In both panels, each corresponding to indicated value of $\beta_1/\beta_0$, the curves are plotted for $p = 1$. In both panels we see that $Q_{apx}(s)$ does a better job at approximating the solution $Q_{ext}(s)$ than simply neglecting $\beta_1$. For the value $\beta_1/\beta_0 = 0.25$ (l.h.s), agreement between the approximation and the exact solution is excellent, for the value $\beta_1/\beta_0 = 0.5$ (r.h.s) it is still not too bad.

where $B$ is given by Eq. (III.6.18). Strictly speaking, we should have expanded Eq. (III.6.18) for small $\beta_1/\beta_2$, but we leave it as it is because Eq. (III.6.18) exactly satisfies the initial conditions.

In Fig. III.6.3 we test the obtained approximate solution against the exact numerical solution of our basic equation, showing the results as a function of $S = \omega_0 t$.

Let us first discuss what we physically expect for the qualitative differences between the full solution to Eq. (III.6.10) and when $d_1 = 0$ ($\beta_1 = 0$), and see if our solutions to this problem make any sense.

The frequency of oscillation of $Q(t)$ for a normal LC circuit is given by $\omega = \frac{1}{\sqrt{LC}}$, so as the capacitance goes down we expect the frequency of oscillation to increase. Since increasing the distance between the two plates decreases the capacitance, we might reasonably expect that initially the frequency of oscillation in the charge goes up, but then should decrease as the plates return to their original position. However, due to the initial change in frequency, we would now expect that the phase would be different from the case when $d_1 = 0$. This is indeed what we see in Fig. III.6.3, when we compare $Q^{(0)}(t)$ with the perturbed solution.

Now, let's discuss how well the perturbation theory approximation works when compared to the exact numerical solution. We see that agreement is excellent for $\beta_1/\beta_0 = 0.25$ and is still pretty reasonable for $\beta_1/\beta_0 = 0.5$. If we make $p$ smaller or $\beta_1/\beta_0$ larger the accuracy of the approximation will become worse. Accuracy can be improved by including the next to leading order correction $Q^{(2)}(s)$ although, one should point out, because of the relatively complicated form of $Q^{(1)}(s)$ this result is expected to be quite cumbersome. In perturbation theory, in most cases, as we go to higher orders in the expansion parameter things can get very complicated and, on a practical level, to have a useful formula one should consider the next to leading order correction, at most. In the case when $p$ is large, it is better to try another approach—the WKB approximation that we'll discuss in the next section of this chapter. On the other hand, in some other situations, certain terms in the perturbation expansion follow a series which one can sum up, resulting in a much better approximation.

### *Perturbation expansion for differential eigenvalue equations*

The perturbation approach is often used for solving eigenvalue equations, in particular in quantum mechanics. Here, we apply it to equation

$$r(x)\frac{d^2 f_j(x)}{dx^2} + q(x)\frac{df_j(x)}{dx} + p(x)f_j(x) = \hat{H}f_j(x) = \lambda_j f_j(x)$$

(III.6.22)

where $f_j(x)$ is the eigenfunction of differential operator $\hat{H}$ corresponding to eigenvalue $\lambda_j$. Let's try to solve this based on knowing how to solve another eigenvalue equation which is close to Eq. (III.6.22) in the form

$$r_0(x)\frac{d^2 f_j^{(0)}(x)}{dx^2} + q_0(x)\frac{df_j^{(0)}(x)}{dx} + p_0(x)f_j^{(0)}(x)$$
$$= \hat{H}_0 f_j^{(0)}(x) = \lambda_j^{(0)} f_j^{(0)}(x)$$

(III.6.23)

First, we write Eq. (III.6.22) as

$$\hat{H}f_j(x) = (\hat{H}_0 + \alpha\Delta\hat{H})f_j(x) = \lambda_j f_j(x) \qquad \text{(III.6.24)}$$

Here, $\alpha\Delta\hat{H}$ is the perturbation of the operator $\hat{H}$, which distinguishes the latter from $\hat{H}_0$.

We can present the eigenfunctions that we seek to find as a sum over a complete set (or basis) of eigenfunctions $f_n^{(0)}(x)$,

$$f_j(x) = \sum_n \tilde{f}_{j,n} f_n^{(0)}(x) \qquad \text{(III.6.25)}$$

as we could for any other function $g(x)$. Notably when $\alpha \to 0$, we can write

$$f_j(x) \Rightarrow f_j^{(0)}(x) \qquad \text{(III.6.26)}$$

that is, the amplitude $\tilde{f}_{j,n} \Rightarrow \delta_{jn}$. Crucially, we'll assume that this proceeds smoothly as we diminish $\alpha$. We can therefore choose to rewrite Eq. (III.6.25) as

$$f_j(x) = f_j^{(0)}(x) + \sum_{n \neq j} \tilde{f}_{j,n} f_n^{(0)}(x) \qquad \text{(III.6.27)}$$

The essence of such a perturbation expansion is that we can write the coefficients and the eigenvalues as power series in the smallness of parameter $\alpha$ such that

$$\tilde{f}_{j,n} = \alpha\tilde{f}_{j,n}^{(1)} + \alpha^2\tilde{f}_{j,n}^{(2)} + \cdots = \sum_{m=1}^{\infty} \alpha^m \tilde{f}_{j,n}^{(m)} \qquad \text{(III.6.28)}$$

$$\lambda_j = \lambda_j^{(0)} + \alpha\lambda_j^{(1)} + \alpha^2\lambda_j^{(2)} = \sum_{m=0}^{\infty} \alpha^m \lambda_j^{(m)} \qquad \text{(III.6.29)}$$

Substitution of Eqs. (III.6.27), (III.6.28) and (III.6.29) into the eigenvalue equation Eq. (III.6.24) yields

$$\left(\hat{H}_0 + \alpha\Delta\hat{H}\right)\left(f_j^{(0)}(x) + \sum_{n\neq j}(\alpha\tilde{f}_{j,n}^{(1)} + \alpha^2\tilde{f}_{j,n}^{(2)} + \cdots)f_n^{(0)}(x)\right)$$

$$= \left(\lambda_j^{(0)} + \alpha\lambda_j^{(1)} + \alpha^2\lambda_j^{(2)} + \cdots\right)$$

$$\times \left(f_j^{(0)}(x) + \sum_{n\neq j}(\alpha\tilde{f}_{j,n}^{(1)} + \alpha^2\tilde{f}_{j,n}^{(2)} + \cdots)f_n^{(0)}(x)\right) \quad \text{(III.6.30)}$$

Now, Eq. (III.6.30) needs to be satisfied identically for each power of $\alpha$. Comparing $\alpha^0$ coefficients, we simply get back Eq. (III.6.23). Comparing the $\alpha$ and $\alpha^2$ coefficients, and using the fact that $\hat{H}_0 f_n^{(0)}(x) = \lambda_n^{(0)}$, yields the equations

$$\sum_{n\neq j}\tilde{f}_{j,n}^{(1)}\lambda_n^{(0)}f_n^{(0)}(x) + \Delta\hat{H}f_j^{(0)}(x) = \lambda_j^{(0)}\sum_{n\neq j}\tilde{f}_{j,n}^{(1)}f_n^{(0)}(x) + \lambda_j^{(1)}f_j^{(0)}(x)$$

$$\text{(III.6.31)}$$

$$\sum_{n\neq j}\tilde{f}_{j,n}^{(2)}\lambda_n^{(0)}f_n^{(0)}(x) + \Delta\hat{H}\sum_{n\neq j}\tilde{f}_{j,n}^{(1)}f_n^{(0)}(x)$$

$$= \lambda_j^{(0)}\sum_{n\neq j}\tilde{f}_{j,n}^{(2)}f_n^{(0)}(x) + \lambda_j^{(1)}\sum_{n\neq j}\tilde{f}_{j,n}^{(1)}f_n^{(0)}(x) + \lambda_j^{(2)}f_j^{(0)}(x)$$

$$\text{(III.6.32)}$$

Now, we multiply both sides of Eqs. (III.6.31) and (III.6.32) by the complex conjugate eigenfunction $f_m^{*(0)}(x)$ and the special function $\omega(x)$ which appears in the Sturm–Liouville (Eq. (III.4.236)) form of Eq. (III.6.23) and integrate over $x$. By doing this, we can exploit the orthogonality condition that states

$$\int_a^b \omega(x)f_m^{*(0)}(x)f_n^{(0)}(x)dx = \delta_{n,m}\tilde{\omega}_n \quad \text{(III.6.33)}$$

By using Eq. (III.6.33) we are supposing that $f_m^{(0)}(x)$ satisfies the boundary conditions given by Eq. (III.4.236) (but generalized to be at the points $a$ and $b$, not 0 and $b$); this also underlies being first able to write Eq. (III.6.25). Thus, if $m = j$, we may find expressions for the eigenvalue corrections. The first two corrections read as

$$\lambda_j^{(1)} = \frac{M_{j,j}}{\tilde{\omega}_j}, \quad \lambda_j^{(2)} = \frac{1}{\tilde{\omega}_j} \sum_{n \neq j} \tilde{f}_{j,n}^{(1)} M_{j,n} \tag{III.6.34}$$

where the matrix elements are given by

$$M_{j,n} = \int_a^b w(x) f_j^{*(0)}(x) \Delta \hat{H} f_n^{(0)}(x) dx \tag{III.6.35}$$

or more explicitly they can be calculated through

$$M_{j,n} = \frac{1}{\alpha} \int_a^b w(x) f_j^{*(0)}(x) \left[ (r(x) - r_0(x)) \frac{d^2}{dx^2} \right.$$

$$+ (q(x) - q_0(x)) \frac{d}{dx} + (p(x) - p_0(x)) \bigg]$$

$$\times f_n^{(0)}(x) dx \tag{III.6.36}$$

If $m \neq j$, we find expressions for the coefficients

$$\tilde{f}_{j,m}^{(1)} = \frac{M_{m,j}}{(\lambda_j^{(0)} - \lambda_m^{(0)}) \tilde{\omega}_m} \tag{III.6.37}$$

$$\tilde{f}_{j,m}^{(2)} = \frac{1}{\tilde{\omega}_m (\lambda_j^{(0)} - \lambda_m^{(0)})} \sum_{n \neq j} \tilde{f}_{j,n}^{(1)} M_{m,n} - \frac{\lambda_j^{(1)} \tilde{f}_{j,m}^{(1)}}{(\lambda_j^{(0)} - \lambda_m^{(0)})} \tag{III.6.38}$$

Combining Eqs. (III.6.34), (III.6.37), and (III.6.38) we can solve for $\tilde{f}_{j,n}^{(1)}$, $\tilde{f}_{j,n}^{(2)}$, $\lambda_j^{(1)}$ and $\lambda_j^{(2)}$ to obtain

$$f_j(x) = f_j^{(0)}(x) + \alpha \sum_{m \neq j} \frac{f_m^{(0)}(x) M_{m,j}}{(\lambda_j^{(0)} - \lambda_m^{(0)}) \tilde{\omega}_m}$$

$$+ \alpha^2 \sum_{n \neq j} \sum_{m \neq j} \frac{f_m^{(0)}(x) M_{m,n} M_{n,j}}{\tilde{\omega}_n \tilde{\omega}_m (\lambda_j^{(0)} - \lambda_n^{(0)})(\lambda_j^{(0)} - \lambda_m^{(0)})} + \cdots \tag{III.6.39}$$

$$\lambda_j = \lambda_j^{(0)} + \frac{\alpha M_{j,j}}{\tilde{\omega}_j} + \frac{\alpha^2}{\tilde{\omega}_j} \sum_{n \neq j} \frac{M_{j,n} M_{n,j}}{(\lambda_j^{(0)} - \lambda_n^{(0)}) \tilde{\omega}_n} + \cdots \tag{III.6.40}$$

Note that we have a problem with the expansions if eigenvalues are degenerate, such that there are values $\lambda_j^{(0)} = \lambda_n^{(0)}$ for $n \neq j$. In this case the perturbation expansion that we have used breaks down and we have to use a different type of expansion, arising from what is called degenerate perturbation theory, but we won't cover this here.

**Example III.6.2 (Physical Application ▶▶).** Here, we'll take an example from quantum mechanics. We consider the eigenvalue equation

$$-\frac{\hbar^2}{2m}\frac{d\psi(x)}{dx} + (kx^2 - \alpha x^3)\psi(x) = E\psi(x) \qquad \text{(III.6.41)}$$

This is the 1D Schrödinger equation for a particle with position-dependent potential energy $kx^2 - \alpha x^3$. The form of Eq. (III.6.41) is physically relevant when we talk about the vibration of atoms in a diatomic molecule. Here, the term $-\alpha x^3$ represents a next to leading order correction to the harmonic approximation, which comes from expanding the full interaction potential describing the bonding about the equilibrium bond length.

If $\alpha = 0$, we have the simple harmonic oscillator equation that we already considered in Example III.4.3, the solution of which was expressed in terms of Hermite polynomials, $H_n(x)$. Thus, we can generate a perturbation expansion which is valid for small values of $\alpha$ (or small vibrational energies, as we shall see) if we choose that

$$\hat{H}^{(0)} = -\frac{\hbar^2}{2m}\frac{d}{dx} + kx^2 \quad \text{and} \quad \Delta H = -x^3 \qquad \text{(III.6.42)}$$

where $\alpha$ here is the expansion parameter. To build this expansion, we follow step-by-step the formalism of the beginning of this sub-section.

Equation (III.6.41) is already in Sturm–Liouville form (the eigenfunctions of a complete Schrödinger equation always are, which has physical significance that we won't discuss here), so simply $\omega(x) = 1$. Let's compute the correction to the ground state $(j = 0)$. First, we need to calculate the matrix elements $M_{m,0}$; to do this we'll use the mathematical result

$$\int_{-\infty}^{\infty} e^{-x^2} H_n(x) H_m(x) dx = \sqrt{\pi} 2^n n! \delta_{n,m} \tag{III.6.43}$$

Also, we'll make use of exact expressions for Hermite polynomials to write

$$x^3 = \frac{1}{8}(H_3(x) + 6H_1(x)) \tag{III.6.44}$$

Combining Eqs. (III.6.43) and (III.6.44), we get

$$I_{n,0} = \int_{-\infty}^{\infty} e^{-x^2} x^3 H_n(x) dx = \frac{3\sqrt{\pi}}{2}(4\delta_{n,3} + \delta_{n,1}) \tag{III.6.45}$$

As you will see in a moment, Eq. (III.6.45) will help us to calculate the matrix elements $M_{n,0}$. First, we have for the eigenfunctions of the simple harmonic oscillator (see Example III.4.2)

$$\psi_n^{(0)} = \frac{1}{\sqrt{2^n n!}} \left(\frac{m\omega}{\hbar\pi}\right)^{1/4} H_n(x\sqrt{m\omega/\hbar}) \exp\left(-\frac{m\omega x^2}{2\hbar}\right) \tag{III.6.46}$$

It then follows that (recollecting that $H_0(x) = 1$)

$$M_{n,0} = -\frac{1}{\sqrt{2^n n!}} \left(\frac{m\omega}{\hbar\pi}\right)^{1/2} \int_{-\infty}^{\infty} x^3 H_n(x\sqrt{m\omega/\hbar}) \exp\left(-\frac{m\omega x^2}{\hbar}\right) dx \tag{III.6.47}$$

Next, we can rescale $z = x\sqrt{m\omega/\hbar}$ to obtain

$$M_{n,0} = -\frac{1}{\sqrt{2^n n!}} \frac{1}{\sqrt{\pi}} \left(\frac{\hbar}{m\omega}\right)^2 \int_{-\infty}^{\infty} z^3 H_n(z) \exp\left(-z^2\right) dz$$

$$\Rightarrow \quad M_{n,0} = -\frac{3}{2}\left(\frac{\hbar}{m\omega}\right)^2 \left(\frac{\delta_{n,3}}{\sqrt{3}} + \frac{\delta_{n,1}}{\sqrt{2}}\right) \tag{III.6.48}$$

Since $E_n^{(0)} = \hbar\omega(n + 1/2)$ are the energy eigenvalues of the simple harmonic oscillator, we may rewrite Eqs. (III.6.39) and (III.6.40) as

$$\psi_j(x) = \psi_j^{(0)}(x) + \frac{\alpha}{\omega\hbar} \sum_{n\neq j} \frac{M_{j,n}\psi_n^{(0)}(x)}{(j-m)} +$$

$$\left(\frac{\alpha}{\omega\hbar}\right)^2 \sum_{n\neq j}\sum_{m\neq j} \frac{M_{n,m}M_{m,j}\psi_n^{(0)}(x)}{(j-n)(j-m)} + \cdots \qquad \text{(III.6.49)}$$

$$E_j = E_j^{(0)} + \frac{\alpha^2}{\hbar\omega} \sum_{n\neq j} \frac{M_{j,n}M_{n,j}}{(j-n)} + \cdots \qquad \text{(III.6.50)}$$

Note that $\tilde{\omega}_j = 1$ here, as the eigenfunctions are normalized. Note also that in this case, we always have $M_{j,j} = 0$ (**Think:** Why?), a fact that we have used in writing Eq. (III.6.50). Let's be more explicit for $j = 0$, the ground state, utilizing the obtained expressions for $M_{0,n}$ and $M_{n,0}$, and look at the corrections of leading order in $\alpha$. These are

$$\psi_0(x) = \psi_0^{(0)}(x) + \frac{3}{2}\frac{\alpha\hbar}{m^2\omega^3} \left(\frac{\psi_1^{(0)}(x)}{\sqrt{2}} + \frac{\psi_3^{(0)}(x)}{3\sqrt{3}}\right) \qquad \text{(III.6.51)}$$

$$E_0 = \hbar\omega\left(\frac{1}{2} - \frac{11}{8}\left(\frac{\alpha\hbar}{m^2\omega^3}\right)^2 + \cdots\right) \qquad \text{(III.6.52)}$$

For these results to be valid, we must require that $\alpha\hbar/m^2\omega^3 \ll 1$ i.e. the calculated correction to the harmonic oscillator 'zero point energy' is indeed small. In Fig. III.6.4, we plot the rescaled wave function $\bar{\psi}_0^{(0)}(z) = (\hbar\pi/m\omega)^{1/4}\psi_0^{(0)}(x)$, the perturbed wave function $\bar{\psi}_0(z) = (\hbar\pi/m\omega)^{1/4}\psi_0(x)$, and the difference between the two, $\Delta\psi(z) = \bar{\psi}_0(z) - \bar{\psi}_0^{(0)}(z)$—all as functions of $z = x\sqrt{m\omega/\hbar}$.

A subtlety, here, is that the potential described by Eq. (III.6.41) as it stands is, in fact, unstable! This is because the potential energy, $V(x) = kx^2 - \alpha x^3$, goes to infinitely negative values as we take $x \to \infty$. This means that technically speaking the true ground state for such potential used to demonstrate the machinery of the

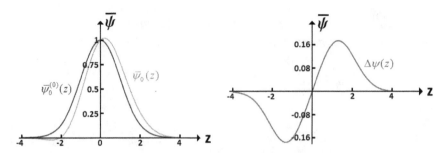

**Fig. III.6.4** In the left panel we compare the solution for the ground state wave-function of the unperturbed Schrödinger equation for a harmonic oscillator $\bar{\psi}_0^{(0)}(z) = (\hbar\pi/m\omega)^{1/4}\psi_0^{(0)}(x)$ with $\bar{\psi}_0(z)$, which is the solution obtained via the first order by perturbation theory in the smallness of parameter $\alpha$ (exactly, when $\alpha\hbar/m^2\omega^3 \ll 1$. The $x$-axis is presented in dimensionless units $z = x\sqrt{m\omega/\hbar}$). $\bar{\psi}(z)$ accounts for adding the small anharmonic term $-\alpha x^3$ to the potential energy of the system, as shown in Eq. (III.6.49). The main effect of the perturbation is to shift the wave function to the right. Physically, this makes sense, as the energy is lowered for positive values of $z$ and increased for negative values, so the average distance between two atoms (which the wave function may describe) will be a bit larger. In the right panel we plot the difference $\Delta\psi(z) = \bar{\psi}_0(z) - \bar{\psi}_0^{(0)}(z)$ between the perturbed and unperturbed wavefunctions. In both plots a value of $\alpha\hbar/m^3\omega^3 = 0.2$ was used.

perturbation theory, does not correspond to the value $j = 0$, rather it is one of infinite negative energy localized at $x = \infty$. This problem would be resolved if we had added an $x^4$ term (with a positive sign!) in Eq. (III.6.41); that would have stabilized the potential, as a next order correction to the harmonic approximation.

**Exercise.** Calculate the matrix elements $M_{l,m}$ for two different per-turbations: $\Delta H = -x^3$ and $\Delta H = x^3$ and the simple harmonic oscillator eigenfunctions. Use these to calculate the leading order correction to the first excited state ($j = 1$). **Hint:** You'll need to write $x^4$ in terms of Hermite polynomials $H_4(x)$, $H_2(x)$ and $H_0(x)$; expressions for these can be found in Eq. (III.4.236).

**Exercise.** Consider the perturbation $\beta x^4$ to the simple harmonic oscillator equation and recalculate the corrections to the ground state.

You may be disappointed that we spend so much time on rather cumbersome manipulation of equations, which even when legitimate gives only small corrections. But these help us to understand the trends of more subtle effects, as well as those that can often be experimentally detectable. We'll deal with perturbation expansions of inhomogeneous differential equations in the last part of this book, where you will see their power in full, but this sub-section lays the basis.

## 6.2. The Elegant WKB Approximation

### *The basic method*

Another method of approximating linear differential equations is the famous WKB approximation. This approximation method was developed in the mid-1920s independently by three theoretical physicists—Wentzel, Kramers, and Brillouin—in attempting to solve the Schrödinger equation in quantum mechanics, but its application is not just restricted to that. Here, we'll generalize things a little more to a wider class of linear differential equations.

Gregor Wentzel
(1898–1978)

We start a demonstration of this method by considering a specific eigenvalue equation of the form

$$\frac{d^2y(x)}{dx^2} + \frac{1/4 - \alpha^2}{x^2}y(x) = -Ey(x) \quad \text{(III.6.53)}$$

Hendrik Kramers
(1894–1952)

At the moment, we'll suppose $E > 0$. Note that Bessel's equation and the simple harmonic oscillator equation are particular examples of Eq. (III.6.53).

Let us rescale the eigenfunctions of such an equation so that $\tilde{y}^+(s) = y(x)$, where $s = \sqrt{E}x$. We can then write

Leon Brillouin
(1889–1969)

$$\frac{d^2\tilde{y}^+(s)}{ds^2} + \frac{1/4 - \alpha^2}{s^2}\tilde{y}^+(s) = -\tilde{y}^+(s) \quad \text{(III.6.54)}$$

The general solution to Eq. (III.6.54), for $E > 0$ can be written as

$$\tilde{y}_\alpha^+(s) = A^+ s^{1/2} J_\alpha(s) + B^+ s^{1/2} Y_\alpha(s)$$

**Exercise.** Show that this equation is indeed the general solution of Eq. (III.6.54) (you'll need to consider Bessel's equation), and also verify that when $\alpha = 1/2$, we get $\tilde{y}(s) = A\sin(s) + B\cos(s)$.

Utilizing this knowledge we try to approximate a more general differential equation:

$$-\frac{d^2 f(x)}{dx^2} - \frac{b}{x}\frac{df(x)}{dx} + \frac{c}{x^2} f(x) = E(x)f(x) \qquad \text{(III.6.55)}$$

This type of equation may emerge when solving partial differential equations in rectangular, cylindrical or spherical geometry. After making the substitution $f(x) = x^{-b/2}g(x)$, we can write

$$\frac{d^2 g(x)}{dx^2} + \frac{1/4 - \alpha^2}{x^2} g(x) = -E(x)g(x) \qquad \text{(III.6.56)}$$

where $\alpha^2 = \frac{1}{4} + \frac{b^2}{4} - \frac{b}{2} + c$. Note we could generally have a situation when $\alpha$ is imaginary, but here we'll consider $\alpha^2 > 0$ (for cases of physical interest). This equation looks rather like Eq. (III.6.53), but with $E(x)$ instead of constant $E$; this form we'll exploit. We'll first deal with $E(x) > 0$.

The WKB approximation to the solution of Eq. (III.6.55) relies on $E(x)$ *varying slowly* (and being not too small). Let us try an approximation to $g(x)$ of the form

$$g(x) = \psi(x)\tilde{y}_\alpha^+ \left(\int_0^x \sqrt{E(x')}dx'\right) \qquad \text{(III.6.57)}$$

[Note that the dimensionality of $E(x)$ is $[x^{-2}]$, so that $\int_0^x \sqrt{E(x')}dx'$, the argument of the newly introduced function $\tilde{y}_\alpha^+$, is dimensionless.] We will now substitute this form into Eq. (III.6.56) and see what we can then neglect. Differentiating Eq. (III.6.57) with respect to $x$, we find (in intermediate equations below, we will for brevity of notations temporarily omit the $+$ superscript of $\tilde{y}$-function, but restore it in the final result)

$$\frac{dg(x)}{dx} = \frac{d\psi(x)}{dx}\tilde{y}_\alpha\left(\int_0^x \sqrt{E(x')}dx'\right)$$
$$+ \psi(x)\sqrt{E(x)}\tilde{y}'_\alpha\left(\int_0^x \sqrt{E(x')}dx'\right) \qquad \text{(III.6.58)}$$

and

$$\frac{d^2g(x)}{dx^2} = \frac{d\psi^2(x)}{dx^2}\tilde{y}_\alpha\left(\int_0^x \sqrt{E(x')}dx'\right)$$
$$+ 2\frac{d\psi(x)}{dx}\sqrt{E(x)}\tilde{y}'_\alpha\left(\int_0^x \sqrt{E(x')}dx'\right)$$
$$+ \frac{\psi(x)}{2\sqrt{E(x)}}\frac{dE(x)}{dx}\tilde{y}'_\alpha\left(\int_0^x \sqrt{E(x')}dx'\right)$$
$$+ \psi(x)E(x)\tilde{y}''_\alpha\left(\int_0^x \sqrt{E(x')}dx'\right) \qquad \text{(III.6.59)}$$

Substituting Eqs. (III.6.57), and (III.6.59) into Eq. (III.6.55) and using the fact that the functions $\tilde{y}''_\alpha(s), \tilde{y}'_\alpha(s)$ and $\tilde{y}_\alpha(s)$ satisfy Eq. (III.6.54), as well as $g(x)$ satisfying Eq. (III.6.56), we arrive at an equation on $\psi(x)$:

$$\frac{d^2\psi}{dx^2} + 2\sqrt{E(x)}\frac{d\psi}{dx} + \frac{\psi(x)}{2\sqrt{E(x)}}\frac{dE(x)}{dx} = 0 \qquad \text{(III.6.60)}$$

Now *the key point of the WKB approximation is to neglect the double derivative term*, so that

$$2\sqrt{E(x)}\frac{d\psi(x)}{dx} + \frac{\psi(x)}{2\sqrt{E(x)}}\frac{dE(x)}{dx} \approx 0 \qquad \text{(III.6.61)}$$

The solution to Eq. (III.6.61) can then be easily obtained by separation of variables, which reads as

$$\ln \psi(x) - \ln C = -\frac{1}{4} \int_0^x \frac{dE(x)}{dx} \frac{dx}{E(x)} = -\frac{1}{4} \ln \left( \frac{E(x)}{E(0)} \right) \quad \text{(III.6.62)}$$

Thus, we can write our approximate solution for $g(x)$ as

$$g(x) \approx (E(x))^{-1/4} \tilde{y}_\alpha^+ \left( \int_0^x \sqrt{E(x')} dx' \right) \quad \text{(III.6.63)}$$

Note that we have absorbed the constants $C$ and $E(0)$ appearing in Eq. (III.6.62) into the constants $A$ and $B$ standing in the general form for $\tilde{y}_\alpha^+$.

This result is obtained under the condition that we could neglect the second derivative in Eq. (III.6.60). This is justified if $dE(x)/dx \ll E(x)^{3/2}$ and $d^2E(x)/dx^2 \ll E(x)^2$ (again, everything is consistent here dimensionally, since the dimensionality of $E(x)$ is $1/x^2$).

**Example III.6.3 (Maths Practice ♫♪).** Let's consider approximating the solution to the following equation

$$-\frac{d^2u(x)}{dx^2} - \frac{1}{x}\frac{du(x)}{dx} + \frac{\gamma u(x)}{x^2} = D^2(2 + \sin(kx))^2 u(x) \quad \text{(III.6.64)}$$

If $D$ is large but $k$ is sufficiently small such that $E(x)$ varies slowly, and we may try to apply the WKB approximation. With the substitution $u(x) = \frac{g(x)}{\sqrt{x}}$, one gets for $g(x)$,

$$\frac{d^2g(x)}{dx^2} + \frac{1/4 - \gamma}{x^2} g(x) = -D^2(2 + \sin(kx))^2 g(x) \quad \text{(III.6.65)}$$

Then, the WKB approximation yields the approximate solution (from Eq. (III.6.57) and (III.6.63), with $\alpha = \sqrt{\gamma}$):

$$g(x) \approx \frac{\tilde{y}_{\sqrt{\gamma}} \left( \frac{D}{k} \int_0^x (2 + \sin kx') dx' \right)}{\sqrt{D (2 + \sin kx)}}$$

$$\Rightarrow g(x) \approx \tilde{y}_{\sqrt{\gamma}} \left( \frac{D}{k} \frac{(2kx + 1 - \cos kx)}{\sqrt{D (2 + \sin kx)}} \right)$$

$$\Rightarrow u(x) \approx \sqrt{\frac{1 \frac{(2kx+1-\cos kx)}{x}}{2 + \sin kx}} \left[ AJ_{\sqrt{\gamma}} \left( \frac{D}{k} (2kx + 1 - \cos kx) \right) \right.$$

$$\left. + BY_{\sqrt{\gamma}} \left( \frac{D}{k} (2kx + 1 - \cos kx) \right) \right]$$

$$\text{(III.6.66)}$$

Let's look at the specific case where $\gamma = 1/4$, Eq. (III.6.66) simplifies to

$$\bar{u}(x) = \bar{y}(s; p) \approx \frac{A \sin \left( 2s + \frac{1 - \cos ps}{p} \right) + B \cos \left( 2s + \frac{1 - \cos ps}{p} \right)}{\sqrt{s (2 + \sin ps)}}$$

$$\text{(III.6.67)}$$

where, to tidy things up, we have written our approximate solution as function of $s = Dx$ with parameter $p = k/D$. In particular, we'll consider a specific solution that satisfies the boundary conditions $\bar{u}(s = 0) = 0$ and $\lim_{s \to 0} \{ s^{1/2} \bar{y}'(s) \} = 1$.

**Exercise.** To satisfy these boundary conditions show that the constants have to take values $B = 0$ and $A = 2^{1/2}$. **Hint:** As you'll want to consider the small $s$ behaviour of Eq. (III.6.67), you can simplify your analysis by straightaway taking $\sin x \approx x$ and $\cos x \approx 1 - x^2/2$.

**Example III.6.3 (Continued).** Let's test how good our approximation to this boundary value problem is, taking Eq. (III.6.67) with $B = 0$ and $A = 2^{1/2}$ and comparing the approximate result for the $g$-function, written in rescaled variables,

$$\bar{g}_{apx}(s) = s^{1/2} \bar{y}_{apx}(s) = (1 + 0.5 \sin ps)^{-1/2}$$

$$\times \sin \left( 2s + \frac{(1 - \cos ps)}{p} \right)$$

$$\text{(III.6.68)}$$

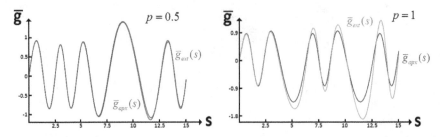

**Fig. III.6.5** We plot the exact numerical solution of Eq. (III.6.9), $\bar{g}_{ext}(s)$ (with boundary condition $\bar{g}_{ext}(0) = 0$ and $\bar{g}'_{ext}(0) = 1$) against the WKB approximation result $\bar{g}_{apx}(s)$, given by Eq. (III.6.68), for two values of the parameter $p$, as functions of the rescaled variable $s = Dx$. On the l.h.s we see that for a value of $p = 0.5$, the WKB approximation is excellent; and for $p = 1$, though still fair, it already beginning break down. Agreement between $\bar{g}_{ext}(s)$ and $\bar{g}_{apx}(s)$ becomes progressively worse as we increase $p$. We may attempt to improve the accuracy of approximation at $p = 1$ by treating $\psi''(x)$, in Eq. (III.6.60), as a perturbation within a WKB theory. We won't consider it here, but leave it as a problem. Note that the solutions $\bar{g}_{ext}(s)$ and $\bar{g}_{apx}(s)$ relate to the solution to Eq. (III.6.64) with $\gamma = 1/4$, where $y(x) = x^{-1/2}\bar{g}(Dx)$.

with the numerical solution of rescaled Eq. (III.6.65) with $\gamma = 1/4$, namely

$$-\frac{d^2\bar{g}(s)}{ds^2} = (2 + \sin ps)^2 \bar{g}(s) \qquad \text{(III.6.69)}$$

with the corresponding boundary conditions $\bar{g}(0) = 0$, $\bar{g}'(0) = 1$. In Fig. III.6.5, we see that the WKB approximation to this equation works well up to moderate values, $p = 0.5$, but starts to break down at $p = 1$. This demonstrates how useful the WKB approximation could be.

## *WKB approximation for negative $E(x)$*

What happens if $E(x)$ is negative? The solution to Eq. (III.6.53) with negative values of $E$ is

$$\tilde{y}_\alpha^-(s) = A^- s^{1/2} I_\alpha(s) + B^- s^{1/2} K_\alpha(s) \qquad \text{(III.6.70)}$$

It is not oscillating any more. Thus, if we reconsider Eq. (III.6.56) for $E(x) < 0$, the WKB approximation will yield for $g(x)$

$$g(x) = (-E(x))^{-1/4} \tilde{y}_\alpha^- \left( \int_0^x \sqrt{-E(x')} dx' \right) \qquad \text{(III.6.71)}$$

**Example III.6.4 (Physical Application ▶▶).** We can actually extend these ideas to another set of equations. We'll illustrate this with a funny physical example. Let's consider the motion of an object, which has mass $m$ and drag $\gamma$, whose potential energy depends on distance and time in a crazy way: $V(x,t) = \frac{k(t)}{2}(b^2 - x^2)$. Is this that crazy? We might suppose that $V(x,t)$ is the gravitational potential energy on a parabolic surface surface as a function of horizontal position, and the surface itself moves upwards in the vertical direction with respect to time—you might envisage that for a short time this might describe a surfer surfing down a growing wave (what the surfer is doing in such situation is crazy enough, but at least this may happen).

For the sake of simplicity, we'll suppose that this gradient of the force $F(t) = k(t)x$ changes with time linearly and slowly, as $k(t) = a_0(1 + \lambda t)$. The equation of motion for the $x$ position of the particle (or our brave surfer) then reads as

$$m\frac{d^2x}{dt^2} + \gamma\frac{dx}{dt} = k(t)x \tag{III.6.72}$$

where $k(t)$ is positive. How can we use the WKB approximation in this case? The answer is that we start, again, by thinking what we would do if $k(t)$ was constant. In this case we would substitute into Eq. (III.6.72) $x(t) = A\exp(\kappa t)$ and solve for $\kappa$, (see Chapter 6, Part II, Vol. 1). This suggests that the WKB way to approximate the solution of Eq. (III.6.72) is to start by writing $x(t)$ in the form

$$x(t) = \psi(t)\exp\left(\int_0^t \kappa(t')dt'\right) \tag{III.6.73}$$

**Exercise.** Show that by substituting Eq. (III.6.73) into Eq. (III.6.72), we are left with two equations on $\psi(t)$ and $\kappa(t)$ that read as

$$m\kappa(t)^2 + \gamma\kappa(t) = k(t) \tag{III.6.74}$$

$$\frac{d\psi^2(t)}{dt^2} + \left(\frac{\gamma}{m} + 2\kappa(t)\right)\frac{d\psi(t)}{dt} + \psi(t)\frac{d\kappa(t)}{dt} = 0 \tag{III.6.75}$$

**Example III.6.4 (Continued).** Equation (III.6.74) is simply what we would have had with $k(t)$ constant with respect to time. It yields two roots $\kappa(t) = \kappa_\pm(t) = \frac{-\gamma \pm \sqrt{\gamma^2 + 4mk(t)}}{2m}$. Next, following the WKB procedure we can neglect $\frac{d\psi^2(t)}{dt^2}$ in Eq. (III.6.75). Intuitively we could expect this is to be acceptable when $\lambda$ is small, i.e. when $\kappa(t)$ varies with time slowly, but what the quantitative criterium this implies we can learn by the substitution of the approximate WKB result that we will obtain for $\psi(t)$ into this equation, and then ask ourselves when the second derivative can indeed be neglected. This we will obtain in a moment.

**Exercise.** Show that when we neglect $\frac{d\psi^2(t)}{dt^2}$ the solution to Eq. (III.6.75) reads as

$$\psi(t) = D\left|\kappa(t) + \frac{\gamma}{2m}\right|^{-1/2} \tag{III.6.76}$$

**Hint:** Rearrange the resulting equation so that $\psi(t)$ terms are moved to one side of the equation and all $\kappa(t)$ terms on the other. **Think:** Why do require the modulus sign in Eq. (III.6.76)?

**Example III.6.4 (Continued).** Hence, it is easy to check that the criterium for neglecting $\frac{d\psi^2(t)}{dt^2}$ is

$$\frac{1}{m}\frac{dk(t)}{dt} \ll \left(\frac{\gamma^2}{m^2} + 4\frac{k(t)}{m}\right)^{3/2}$$

For our simple example of $k(t) = a_0(1 + \lambda t)$ this condition simplifies to $\frac{1}{m}\lambda a_0 \ll (\frac{\gamma^2}{m^2} + 4\frac{a_0}{m})^{3/2}$, so that $\lambda$ must be small to satisfy $\lambda \ll \frac{(\gamma^2 + 4a_0 m)^{3/2}}{a_0 m^2}$.

Next, using the fact that we have two roots to Eq. (III.6.74), we can write an approximate general solution as

$$x(t) = \frac{\exp\left(-\frac{\gamma t}{2m}\right)}{(\gamma^2 + 4mk(t))^{1/4}} \left[ A \exp\left(\frac{1}{2m} \int_0^t \sqrt{\gamma^2 + 4mk(t)}\, dt'\right) \right.$$

$$\left. + B \exp\left(-\frac{1}{2m} \int_0^t \sqrt{\gamma^2 + 4mk(t)}\, dt'\right) \right]$$

$$\text{(III.6.77)}$$

Now, let's suppose that $x = 0$ at $t = 0$. It this case it must be $A = -B$, so that

$$x(t) = \frac{2A \exp\left(-\frac{\gamma t}{2m}\right)}{(\gamma^2 + 4mk(t))^{1/4}} \sinh\left(\frac{1}{2m} \int_0^t \sqrt{\gamma^2 + 4mk(t)}\, dt'\right)$$

$$\text{(III.6.78)}$$

For the form $k(t) = a_0(1 + \lambda t)$, we can perform the integration inside the sinh function, yielding

$$x(t) \approx \frac{2A}{(\gamma^2 + 4ma_0(1 + \lambda t))^{1/4}} \sinh\left(\frac{U(t)}{12m^2 a_0 \lambda}\right) \qquad \text{(III.6.79)}$$

where

$$U(t) = \left(\gamma^2 + 4ma_0(1 + \lambda t)\right)^{3/2} - \left(\gamma^2 + 4ma_0\right)^{3/2} \qquad \text{(III.6.80)}$$

Differentiating with respect to time, the velocity $v(t) = dx/dt$ is given by the expression

$$v(t) \approx A \exp\left(-\frac{\gamma t}{2m}\right) \left\{ \frac{(\gamma^2 + 4ma_0(1 + \lambda t))^{1/4}}{m} \cosh\left(\frac{U(t)}{12m^2 a_0 \lambda}\right) \right.$$

$$-\frac{2}{(\gamma^2 + 4ma_0(1 + \lambda t))^{1/4}}$$

$$\times \left[ \frac{\gamma}{2m} + \frac{ma_0 \lambda}{(\gamma^2 + 4ma_0(1 + \lambda t))^{1/4}} \right]$$

$$\left. \times \sinh\left(\frac{U(t)}{12m^2 a_0 \lambda}\right) \right\} \qquad \text{(III.6.81)}$$

Note that we have not yet specified the initial condition for the speed at $t = 0$. Let us specify it as

$$v(t = 0) = v_0 = x_0 \left( \frac{\gamma^2}{m^2} + \frac{4a_0}{m} \right)^{1/2} \qquad \text{(III.6.82)}$$

where $x_0$ is some parameter of dimensionality of length (do not mix it up with $x(t = 0) = 0$).

It follows, then, that $A = x_0(\gamma^2 + 4ma_0)^{1/4}$. This reformatting allows us to write things in a neater fashion, where the parameter $x_0$ controls the initial speed of the object; everything that follows could be rewritten back in terms of $v_0$.

**Exercise.** Show that the solution subject to initial condition Eq. (III.6.82) and $x(t = 0) = 0$ can be written as

$$y(s) = x(t)/x_0 \approx \frac{2}{\left( 1 + \frac{4\alpha\beta s}{1+4\alpha} \right)^{1/4}} e^{-s} \sinh \left( \frac{\bar{U}(s)}{6\alpha\beta} \right) \qquad \text{(III.6.83)}$$

where

$$\bar{U}(s) = (1 + 4\alpha(1 + \beta s))^{3/2} - (1 + 4\alpha)^{3/2} \qquad \text{(III.6.84)}$$

where $\alpha = \frac{ma_0}{\gamma^2}$, $\beta = 2\lambda m/\gamma$ and $s = \gamma t/(2m)$. Also show that the WKB approximation will work best when $\beta \ll (1 + \alpha)^{3/2}/\alpha$. **Hint:** This can be obtained through $\frac{1}{m}\lambda a_0 \ll \left( \frac{\gamma^2}{m^2} + 4\frac{a_0}{m} \right)^{3/2}$.

**Example III.6.4 (Continued).** The initial condition, Eq. (III.6.82), then becomes in terms of the rescaled variables $y'(0) = (1 + 4\alpha)^{1/2}$. It's worth a quick comment here: we have been a bit sloppy in terms of what we told you was good practice. We proceeded in the way above to be more illustrative.

The best way of approaching things, however, would have been by rescaling all variables and parameters up front, so that Eq. (III.6.72) would read

$$\frac{1}{4} \frac{d^2 y(s)}{ds^2} + \frac{1}{2} \frac{dy(s)}{ds} = \alpha(1 + \beta s)y(s) \qquad \text{(III.6.85)}$$

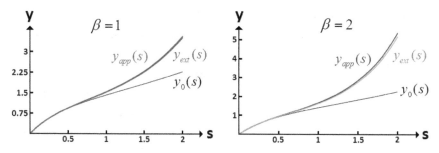

**Fig. III.6.6**  We compare the WKB result $y_{app}(s)$ (given by Eq. (III.6.83)) against the exact solution $y_{ext}(s)$ (of Eq. (III.6.85)), as well as with the constant $k$ the solution of $\beta = 0$, $y_0(s)$. In both panels we consider a parameter value of $\alpha = 0.25$; the panels are plotted for indicated values of $\beta$. In the left (right) hand panel, the red (purple) and blue (green) curves correspond to approximate and exact solutions, respectively. We see that at large values of $s$, $y_0(s)$ really does not work in describing the solution, but $y_{app}(s)$ does a very good job indeed. However, as we increase $\beta$, $y_{app}(s)$ doesn't do quite so well in approximating the solution to Eq. (III.6.85).

**Exercise.** Show that Eq. (III.6.85) can be obtained from Eq. (III.6.72) by suitable rescaling as described above. Also, derive Eq. (III.6.83) from Eq. (III.6.85), using the WKB approximation.

In Fig. III.6.6, we compare a numerical solution of Eq. (III.6.85) with that of our WKB approximation (Eq. (III.6.83)), as well as showing what happens when we simply set $\beta = 0$ (i.e. the potential doesn't change in time).

When we compare the solution for $\beta = 0$ with those solutions where $\beta = 1$ and $\beta = 2$ we see some important qualitative differences. When $\beta = 0$ the external force acting on the object is $F = a_0 x$, and since we have chosen $\alpha = 0.25$ we are sitting in a regime where damping is significant. In this case, the exponential factors $\exp(-s)$ and $\exp\left(\frac{\bar{U}(s)}{6\alpha\beta}\right)$ almost cancel each other out and it looks like we have reached some terminal velocity for $s > 1$. In fact, this motion is a slow exponential increase and the object will start to speed up at larger $s$.

We also see that for very small values of $s$ it is sufficient to neglect $\beta$ in our solution, but at larger values this clearly does not work. Indeed, the $\beta = 0$ solution fails to capture a large acceleration due to the time dependence of $k(t)$ that starts kicking in at around $s \approx 1$.

In Fig. III.6.6 we see that the WKB approximation does a good to excellent job when compared with the exact solution for the parameter values considered. If we were to make $\alpha$ smaller and $\beta$ larger we would find that the WKB approximation would work less well, but still this is pretty good going. Now, let's move on to some further, even more useful, elaborations to the WKB method.

### Turning points in the WKB approximation

Let's reconsider Eq. (III.6.56) with $\alpha = 1/2$, for simplicity. How do we find appropriate boundary conditions to match the two approximate solutions ($E(x) > 0$ and $E(x) < 0$) when $E(x)$ changes sign at a point $x_1$ (called a turning point)? The problem is that when $E(x) \to 0$ both WKB approximate solutions diverge to infinity (see Eqs. (III.6.63) and (III.6.71)), and the assumption used in generating them is no longer valid. Thus, we cannot use the simple method of matching the two WKB approximate formulas through their continuity and continuity of their derivatives. A more sophisticated method is needed.

We start by considering the following equation

$$\frac{d^2 \tilde{g}_T(s)}{ds^2} = -se_1 \tilde{g}_T(s) \qquad \text{(III.6.86)}$$

where $\tilde{g}_T(s) = g_T(x)$ and $s = x - x_1$; $e_1$ is a bit (sorry!) exotic notation for a constant: $e_1 \equiv \frac{dE}{dx}\big|_{x=0}$. This equation is a valid approximation of Eq. (III.6.56) (when $\alpha = 1/2$) close to the value $x_1$, where $E(x)$ changes sign, and we can write $E(x) \approx e_1(x - x_1)$.

In fact, we have already solved Eq. (III.6.86) for positive $s$ in Example III.4.8. As we are going to consider the $s < 0$ domain as well, its general solution can be written in a more convenient way, as

$$\underbrace{\tilde{g}(s)}_{s>0} = A_T s^{1/2} \left( J_{-1/3}(2e_1^{1/2} s^{3/2}/3) + J_{1/3}\left(2e_1^{1/2} s^{3/2}/3\right) \right)/3$$

$$+ C_T s^{1/2} \left( J_{-1/3}\left(2e_1^{1/2} s^{3/2}/3\right) - J_{1/3}\left(2e_1^{1/2} s^{3/2}/3\right) \right)/3$$

$$\text{(III.6.87)}$$

and

$$\underbrace{\tilde{g}(s)}_{s<0} = A_T |s|^{1/2} \pi^{-1} K_{1/3} \left(2e_1^{1/2} |s|^{3/2} /3\right) /3 + C_T |s|^{1/2}$$

$$\times \left(I_{1/3}\left(2e_1^{1/2} |s|^{3/2} /3\right) + I_{-1/3}\left(2e_1^{1/2} |s|^{3/2} /3\right)\right)/3 \quad \text{(III.6.88)}$$

You should check this by substituting Eqs. (III.6.87) and (III.6.88) into Eq. (III.6.86) (recollect Example III.4.8). We wrote Eqs. (III.6.87) and (III.6.88) in this particular way, because it will be easier to express them in terms of what are called Airy functions. These have useful asymptotic forms at $s \gg 1$ which we'll want to exploit. The Airy functions are defined (with $z \geq 0$) as:

$$Ai(z) \equiv z^{1/2}\pi^{-1}K_{1/3}\left(2z^{3/2}/3\right)/3$$

$$Ai(-z) \equiv z^{1/2}\pi^{-1}\left(J_{1/3}\left(2z^{3/2}/3\right) + J_{-1/3}\left(2z^{3/2}/3\right)\right)/3$$

$$Bi(z) = z^{1/2}\left(I_{1/3}\left(2z^{3/2}/3\right) + I_{-1/3}\left(2z^{3/2}/3\right)\right)/3$$

$$Bi(-z) = z^{1/2}\left(J_{-1/3}\left(2z^{3/2}/3\right) - J_{1/3}\left(2z^{3/2}/3\right)\right)/3 \quad \text{(III.6.89)}$$

We plot the Airy functions in Fig. III.6.7.

By plotting these functions we already see that they have some interesting and potentially useful features. On the r.h.s (positive $s$) these functions behave monotonically, whereas on the l.h.s (negative $s$), they are oscillatory. We are going to exploit these features.

For $|s| \gg 1$, the solutions (Eqs. (III.6.87) and (III.6.88)) simplify to their asymptotic forms

$$\tilde{g}(s) \approx \frac{e_1^{-1/12} s^{-1/4}}{\sqrt{\pi}} \left[A_T \sin\left(\frac{2}{3}e_1^{1/2}s^{3/2} + \frac{\pi}{4}\right)\right.$$

$$\left. +C_T \cos\left(\frac{2}{3}e_1^{1/2}s^{3/2} + \frac{\pi}{4}\right)\right], \quad s > 0 \quad \text{(III.6.90)}$$

$$\tilde{g}(s) \approx \frac{e_1^{-1/12}}{2\sqrt{\pi}}(-s)^{-1/4}\left[A_T \exp(-2e_1^{1/2}(-s)^{3/2}/3)\right.$$

$$\left. + 2C_T \exp(2e_1^{1/2}(-s)^{3/2}/3)\right], \quad s < 0 \quad \text{(III.6.91)}$$

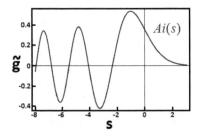 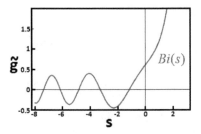

**Fig. III.6.7** Plots of the Airy functions, $\tilde{g} = Ai(s)$ and $\tilde{g} = Bi(s)$. The exact solutions Eq. (III.6.74) can be written as linear combinations of these two functions. These functions have oscillatory behaviour for $s < 0$. However, for $s > 0$ they have either exponential decay in the case of $Ai(s)$ (left panel) or exponential growth in the case of $Bi(s)$ (right panel). As discussed in the text, a particularly useful feature is that the WKB approximations of Eq. (II.6.74) correspond to the asymptotic forms of $Ai(e_1^{1/3}s)$ and $Bi(e_1^{1/3}s)$, which the Airy functions tend to when $s \gg 1$ and $s \ll 1$. We can use this feature to build approximate solutions where $E(x)$ in Eq. (III.6.55) changes sign. Some art lovers might appreciate the Mondrian crosses (more entertainment on Mondrian later!) in each of these plots; they denote $s = 0$ and $\tilde{g} = 0$, to orientate the viewer.

We can now go back to consider the WKB solutions to Eq. (III.6.86) with $E(x) = e_1(x - x_1)$:

$$g_T(x) \approx e_1^{-1/4}(x - x_1)^{-1/4}[A^+ \sin(2e_1^{1/2}(x - x_1)^{3/2}/3)$$

$$+ B^+ \cos(2e_1^{1/2}(x - x_1)^{3/2}/3)], \quad x - x_1 > 0 \quad \text{(III.6.92)}$$

$$g_T(x) \approx e_1^{-1/4}(x_1 - x)^{-1/4}[A^- \exp(-2e_1^{1/2}(x_1 - x)^{3/2}/3)$$

$$+ B^- \exp(2e_1^{1/2}(x_1 - x)^{3/2}/3)], \quad x - x_1 < 0 \quad \text{(III.6.93)}$$

It seems magical that both Eqs. (III.6.90) and (III.6.91) are of the same form as the WKB approximations, Eq. (III.6.92) and (III.6.93). However, this is due to the asymptotic forms having the same structure as dictated by Eq. (III.6.57), with prefactor $\psi(x) = (x - x_1)^{-1/4}$ (or $\psi(x) = (x_1 - x)^{-1/4}$ for Eq. (III.6.71)). We are going to utilize this, but first we need to *convince you of why this is useful*.

Let's consider the WKB approximations, Eq. (III.6.63) and (III.6.71) (with $\alpha = 1/2$), and the full function $E(x)$ that has a zero at $x_1$. Close to this turning point, we may expand out $E(x) = e_1(x - x_1) + e_2(x - x_1)^2 + \cdots$. Then, near $x_1$, Eqs. (III.6.92) and (III.6.93) are the dominant singular contributions to the WKB approximation, obtained by keeping only the first term in the Taylor expansion of $E(x)$; but close to the turning point it is due to these terms the WKB approximation breaks down. In the WKB approximation corrections coming from $e_2(x - x_1)^2$, and higher order terms in the Taylor expansion, are not singular at the turning point and are harmless there, so we don't have to worry about them. The ingenious part is, then, if we choose the coefficients so that Eqs. (III.6.90) and (III.6.91) are exactly the same as Eqs. (III.6.92) and (III.6.93), we can deal with the problematic WKB contribution that comes from $e_1(x - x_1)$. By matching asymptotic behaviours of the Airy functions to the WKB approximation, we can use the Airy functions to interpolate across the turning point between the two WKB approximations on either side of it.

So, let's do it. Matching Eqs. (III.6.92) and (III.6.93) with Eqs. (III.6.90) and (III.6.91), respectively, we require that

$$
\begin{aligned}
A_T &= \sqrt{\frac{\pi}{2}} e_1^{-1/6}(A^+ + B^+) = 2\sqrt{\pi} e_1^{-1/6} A^- \\
C_T &= \sqrt{\frac{\pi}{2}} e_1^{-1/6}(B^+ - A^+) = \sqrt{\pi} e_1^{-1/6} B^-
\end{aligned}
\tag{III.6.94}
$$

Thereby, we can also apply WKB approximation to Eq. (III.6.55) with $E(x)$ changing sign through Eqs. (III.6.63), (III.6.70) and (III.6.71) and the linking conditions, Eqs. (III.6.94).

This is a clever construction, and we hope that you can appreciate the beauty of it. Wenzel, Kramers and Brillouin were indeed very smart people, who have made classical contributions to different areas of physics. This mathematical diversion for them was motivated by struggling with Schrödinger's equation, when solving physical problems.

**Example III.6.5 (Physical Application ▶▶).** As we have mentioned, first applications of the WKB approximation go back to quantum mechanics. However, we reiterate that it can be applied to other problems where one needs to solve differential eigenvalue or homogeneous equations. In certain instances, we can also use it to approximate Green's functions needed to solve tricky homogeneous or inhomogeneous equations.

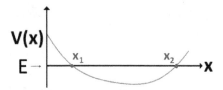

**Fig. III.6.8**  Schematic plot showing the type of potential, $V(x)$, which we deal with in Example III.6.5. It has two turning points, $x_1$ and $x_2$, for energy eigenvalue $E$ at which $E - V(x) = 0$. Between $x_1$ and $x_2$, that difference is positive.

However, let's not get ahead of ourselves. In this example, we apply the WKB-method to approximate a bound state in 1D quantum mechanics without specifying the form of the potential well, the qualitative shape of which is sketched in Fig. III.6.8. We first write down a Schrödinger equation that describes the wave function of a particle of mass $m$, with total energy $E$ and potential energy $V(x)$:

$$-\frac{\hbar^2}{2m}\frac{d^2\psi(x)}{dx^2} + V(x)\psi(x) = E\psi(x) \qquad (\text{III.6.95})$$

Now, we'll suppose it is the case that $E - V(x) > 0$ when $x_1 < x < x_2$, and beyond that interval of $x$, $E - V(x) < 0$. The latter means that kinetic energy is negative, which in classical mechanics is strictly forbidden, but in quantum mechanics this rule can be violated to a certain extent. Also, we suppose that $\psi(x)$ is defined in the whole domain $-\infty < x < \infty$. The points $x_1$ and $x_2$, depicted in Fig. III.6.8, are called the *turning points*. **Think:** In classical mechanics, why these are called turning points?

In this particular case, the general forms Eq. (III.6.55), Eqs. (III.6.63) and (III.6.71) (where $\alpha = 1/2$ in this case), can be written as

$$\underbrace{\psi(x)}_{x_1<x<x_2} = \frac{1}{\left\{\frac{2m}{\hbar^2}(E - V(x))\right\}^{\frac{1}{4}}}$$

$$\times \left[ A^+ \cos\left(\frac{\sqrt{2m}}{\hbar}\int_{x_1}^{x}\sqrt{E - V(x)}dx'\right) \right.$$

$$\left. + B^+ \sin\left(\frac{\sqrt{2m}}{\hbar}\int_{x_1}^{x}\sqrt{E - V(x)}dx'\right) \right] \qquad (\text{III.6.96})$$

$$\underbrace{\psi(x)}_{x<x_1} = A_1^- \frac{\exp^{\frac{\sqrt{2m}}{\hbar} \int_{x_1}^x \sqrt{V(x)-E}dx'}}{\left\{\frac{2m}{\hbar^2}(V(x)-E)\right\}^{\frac{1}{4}}} \tag{III.6.97}$$

$$\underbrace{\psi(x)}_{x>x_2} = A_2^- \frac{\exp^{-\frac{\sqrt{2m}}{\hbar} \int_{x_2}^x \sqrt{V(x)-E}dx'}}{\left\{\frac{2m}{\hbar^2}(V(x)-E)\right\}^{\frac{1}{4}}} \tag{III.6.98}$$

In writing both, Eqs. (III.6.97) and (III.6.98) we have neglected one of the solutions that appears in both Eqs. (III.6.70) and (III.6.71). This is because we require that $\psi(x) \to 0$ when $|x| \to \infty$, as quantum mechanics dictates: indeed, a particle is not likely to tunnel far into the classically forbidden region.

In quantum mechanics, the (basic) WKB approximation is applicable when the potential varies smoothly; more precisely when the following conditions are satisfied:

$$\left|\frac{dV(x)}{dx}\right| \ll |V(x) - E|^{3/2} \frac{\sqrt{m}}{\hbar} \qquad \left|\frac{dV^2(x)}{dx^2}\right| \ll |V(x) - E|^2 \frac{m}{\hbar^2}$$
$$\tag{III.6.99}$$

We could write the WBK result in the classically allowed region as

$$\underbrace{\psi(x)}_{x_1<x<x_2} = \frac{1}{\left\{\frac{2m}{\hbar^2}(E - V(x))\right\}^{\frac{1}{4}}}$$

$$\times \left[\tilde{A}^+ \cos\left(\frac{\sqrt{2m}}{\hbar} \int_x^{x_2} \sqrt{E - V(x)}dx'\right)\right.$$

$$\left. + \tilde{B}^+ \sin\left(\frac{\sqrt{2m}}{\hbar} \int_x^{x_2} \sqrt{E - V(x)}dx'\right)\right] \tag{III.6.100}$$

which is another convenient form for us to write, alternative to Eq. (III.6.96), when matching solutions.

**Exercise.** Show that Eqs. (III.6.96) and (III.6.100) are equivalent and relate $A^+$ and $B^+$ to $\tilde{A}^+$ and $\tilde{B}^+$, respectively. **Hint:** You may want to use the trivial fact that

$$\int_a^b f(x)dx = -\int_b^a f(x)dx \tag{III.6.101}$$

In the case we're considering, near the two turning points $x_1$ and $x_2$ (where the WKB approximation on its own breaks down) we can approximate $\frac{2m}{\hbar^2}(E - V(x)) \approx e_1^{(1)}(x - x_1)$, and $\frac{2m}{\hbar^2}(E - V(x)) \approx e_1^{(2)}(x_2 - x)$, respectively. Here, $e_1^{(1)}$ and $e_1^{(2)}$ are the coefficients of first order terms in the Taylor expansion near the corresponding turning points. Thus, we may write

$$\frac{d^2\psi(x)}{dx^2} = -(x - x_1)e_1^{(1)}\psi(x), \quad \frac{d^2\psi(x)}{dx^2} = -(x_2 - x)e_1^{(2)}\psi(x)$$

(III.6.102)

These are solved in a similar manner to Eq. (III.6.86), as described before, but we also require that $\psi(x) \to 0$ when $|x| \to \infty$. So the solutions to these equations can be written in terms of the Airy functions (defined in Eq. (III.6.89)),

$$\psi(x) = C_T Ai\left(\left(e_1^{(1)}\right)^{1/3}(x_1 - x)\right)$$

$$\psi(x) = A_T Ai\left(\left(e_1^{(2)}\right)^{1/3}(x - x_2)\right)$$

(III.6.103)

where $C_T$ and $A_T$ are constants. Following the WKB procedure, *far from the turning points*, we use the asymptotic forms of the Airy functions which are (for $j = 1, 2$):

$$\psi(x) \approx \frac{C_T\left(e_1^{(j)}\right)^{-1/12}\left((-1)^j(x_j - x)\right)^{-1/4}}{\sqrt{\pi}}$$

$$\times \sin\left(\frac{2}{3}\left(e_1^{(j)}\right)^{1/2}\left((-1)^j(x_j - x)\right)^{3/2} + \frac{\pi}{4}\right)$$

(III.6.104)

within the classically allowed region, and

$$\psi(x) \approx \frac{C_T\left(e_1^{(j)}\right)^{-1/12}\left((-1)^j(x - x_j)\right)^{-1/4}}{2\sqrt{\pi}}$$

$$\times \exp\left(-\frac{2\left(e_1^{(j)}\right)^{1/2}\left((-1)^j(x - x_j)\right)^{3/2}}{3}\right)$$

(III.6.105)

within the classically forbidden region.

Then, we substitute $e_1^{(1)}(x - x_1)$ for $\sqrt{2m}(E - V(x))/\hbar$ into Eqs. (III.6.96) and (III.6.97) to obtain formulae by performing the integration. Next, we choose the coefficients $(A^+, B^+ \text{ and } A_1^-)$ in the formulae so that they are exactly the same as the asymptotic solutions given by Eqs. (III.6.104) and (III.6.105) (with $j = 1$). Similarly, we substitute $\sqrt{2m}(E - V(x))/\hbar$ with $e_1^{(2)}(x_2 - x)$ in Eqs. (III.6.97) and (III.6.98), perform the integration, and match the coefficients with the asymptotic solutions again given by Eqs. (III.6.104) and (III.6.105) (with $j = 2$). Doing so, we obtain the following relationships between coefficients:

$$A^+ = B^+ \quad C_T = \sqrt{2\pi}(e_1^{(1)})^{-1/6}A^+ = 2\sqrt{\pi}(e_1^{(1)})^{-1/6}A_1^-$$
$$\tilde{A}^+ = \tilde{B}^+ \quad A_T = \sqrt{2\pi}(e_1^{(2)})^{-1/6}\tilde{A}^+ = 2\sqrt{\pi}(e_1^{(2)})^{-1/6}A_2^- \quad \text{(III.6.106)}$$

Now, importantly, the wave function $\psi(x)$ in the region $x_1 < x < x_2$ must satisfy

$$\psi(x) = 2A_1^- \left( \sqrt{\frac{2m}{\hbar}} \sqrt{E - V(x)} \right)^{-1/2}$$

$$\times \sin \left( \sqrt{\frac{2m}{\hbar}} \int_{x_1}^{x} \sqrt{E - V(x)}dx' + \frac{\pi}{4} \right)$$

$$= 2A_2^- \left( \sqrt{\frac{2m}{\hbar}} \sqrt{E - V(x)} \right)^{-1/2}$$

$$\times \sin \left( \sqrt{\frac{2m}{\hbar}} \int_{x}^{x_2} \sqrt{E - V(x)}dx' + \frac{\pi}{4} \right) \quad \text{(III.6.107)}$$

This is derived using Eqs. (III.6.96), (III.6.100) and (III.6.106). Thus, following from Eq. (III.6.107), we have a requirement for all values of $x$, between the two turning points, which reads as

$$\sin \left( \frac{\sqrt{2m}}{\hbar} \int_{x_2}^{x} \sqrt{E - V(x)}dx' + \frac{\sqrt{2m}}{\hbar} \int_{x_1}^{x_2} \sqrt{E - V(x)}dx' + \frac{\pi}{4} \right)$$

$$= -A_2^-/A_1^- \sin \left( \frac{\sqrt{2m}}{\hbar} \int_{x_2}^{x} \sqrt{E - V(x)}dx' - \frac{\pi}{4} \right) \quad \text{(III.6.108)}$$

For Eq. (III.6.108) to be satisfied for all values of $x$, the following condition must be fulfilled:

$$\frac{\sqrt{2m}}{\hbar} \int_{x_1}^{x_2} \sqrt{E - V(x)}\, dx' = \pi(n + 1/2) \qquad \text{(III.6.109)}$$

This is a crucial result! Eq. (III.6.109) is called the Sommerfeld–Wilson quantization condition, which determines the only allowed values $E_n$ of $E$ for the bound states of Eq. (III.6.95). Of course, mathematically this is a complicated integral condition, but for some forms of $V(x)$, as we will see in an example below, it gives an analytical result for the spectrum of $E_n$ values. Substitution of Eq. (III.6.109) (without need to solve it for $E$) into Eq. (III.6.108) leads to

$$A_1^-(-1)^n = A_2^- \qquad \text{(III.6.110)}$$

Thus, the WKB wave-function in the whole region (with $A \equiv A_1^-$ to simplify the notation) reads as

$$\psi(x) = A \left( \frac{\sqrt{2m}}{\hbar} \sqrt{V(x) - E_n} \right)^{-\frac{1}{2}}$$

$$\times \exp\left( -\frac{\sqrt{2m}}{\hbar} \int_x^{x_1} \sqrt{V(x) - E_n}\, dx' \right) \qquad x \ll x_1$$

$$\psi(x) = 2A\sqrt{\pi}(e_1^{(1)})^{-1/6} Ai\left( (e_1^{(1)})^{1/3}(x_1 - x) \right) \qquad x \approx x_1$$

$$\psi(x) = 2A \left( \sqrt{\frac{2m}{\hbar}} \sqrt{E_n - V(x)} \right)^{-\frac{1}{2}}$$

$$\times \sin\left( \sqrt{\frac{2m}{\hbar}} \int_{x_1}^x \sqrt{E_n - V(x)}\, dx' + \frac{\pi}{4} \right) \qquad x_1 \ll x \ll x_2$$

$$\psi(x) = (-1)^n 2A\sqrt{\pi} \left( e_1^{(2)} \right)^{-1/6} Ai\left( \left( e_1^{(2)} \right)^{1/3}(x - x_2) \right) \qquad x \approx x_2$$

$$\psi(x) = (-1)^n A \left( \frac{\sqrt{2m}}{\hbar} \sqrt{V(x) - E_n} \right)^{-\frac{1}{2}}$$

$$\times \exp\left( -\frac{\sqrt{2m}}{\hbar} \int_{x_2}^{x} \sqrt{V(x) - E_n} dx' \right) \quad x_2 \ll x \quad \text{(III.6.111)}$$

The Airy functions can be used in the regions where the WKB approximations starts to diverge as $|x - x_1|^{-1/4}$ and $|x - x_2|^{-1/4}$, where clearly the WKB approximation on its own is breaking down. The constant $A$ may be determined by the normalization condition

$$1 = \int_{-\infty}^{\infty} \psi(x)^2 dx \quad \text{(III.6.112)}$$

For a specified form of $V(x)$, one first does the integrals inside the formulae given in Eq. (III.6.111), secondly one has to evaluate the integral in Eq. (III.6.112). In practice, this has to be done numerically, except for in the simplest of cases where an analytical formula for $A$ might be obtained.

### The Sommerfeld–Wilson quantization condition

Interestingly, the 'quantization condition', in the form of Eq. (III.6.109) (as well as the forms of the approximate eigenfunctions, Eq. (III.6.111)), can be applied beyond quantum mechanics. It can be applied to the WKB approximation of any differential eigenvalue (or homogeneous) equation of the form of Eq. (III.6.56), with $\alpha = 1/2$, for a region where $E(x) < 0$ in (III.6.57) for both $-\infty < x < x_1$ and $x_2 < x < \infty$, and $E(x) > 0$ for $x_1 < x < x_2$ [on notation: in Example III.6.5, $E(x) = E - V(x)$]. This is provided that we have the boundary condition for the solution $g(x) \to 0$ when $|x| \to \infty$. For

Arnold Sommerfeld
(1888–1951)

William Wilson
(1975–1965)

this general case, we have

$$\int_{x_1}^{x_2} \sqrt{E(x)}dx' = \pi(n + 1/2) \tag{III.6.113}$$

**Example III.6.6 (Maths Practice ♫♪).** Let's approximate the allowed eigenvalues $\lambda_n$ of the following equation:

$$\frac{d^2 f_n(s)}{ds^2} - B|s|^\gamma f_n(s) = -\lambda_n f_n(s) \tag{III.6.114}$$

where we require that $f(s) \to 0$ when $|s| \to \infty$. First, we need to find the two turning points. These are the $s$ values where $\lambda_n - B|s|^\gamma = 0$; thus $s_1 = -(\lambda_n/B)^{1/\gamma}$ and $s_2 = (\lambda_n/B)^{1/\gamma}$, and between these two points $E(s) = \lambda_n - B|s|^\gamma$ remains positive. The general quantization condition (Eq. (III.6.113)) then yields

$$\sqrt{\lambda_n} \int_{-(\lambda_n/B)^{1/\gamma}}^{(\lambda_n/B)^{1/\gamma}} ds \sqrt{1 - B|s|^\gamma/\lambda_n} = \pi(n + 1/2) \tag{III.6.115}$$

We can rescale $u = (B/\lambda_n)^{1/\gamma}s$ to obtain

$$\sqrt{\lambda_n}(\lambda_n/B)^{1/\gamma} \int_{-1}^{1} \sqrt{1 - |u|^\gamma}du = \pi(n + 1/2) \tag{III.6.116}$$

The integral evaluates to

$$\int_{-1}^{1} \sqrt{1 - |u|^\gamma}ds = \sqrt{\pi}\frac{\Gamma(1 + 1/\gamma)}{\Gamma(3/2 + 1/\gamma)} \tag{III.6.117}$$

where $\Gamma(x)$ is the Euler Gamma function. Thus, the allowed eigenvalues are approximately given by

$$\lambda_n = B^{2/(\gamma+2)} \left(\sqrt{\pi}(n + 1/2)\frac{\Gamma(3/2 + 1/\gamma)}{\Gamma(1 + 1/\gamma)}\right)^{2\gamma/(\gamma+2)} \tag{III.6.118}$$

**Example III.6.7 (Physical Application ►►).** Particularly interesting in the previous example are the values $\gamma = 2$, $B = 1$, as this corresponds to the rescaled Schrödinger equation for the simple harmonic oscillator (Eq. (III.4.236)) where $s = x\sqrt{\mu\omega/\hbar}$ and $\lambda_n = 2E_n/(\hbar\omega)$. Here, we'll see how well the WKB approximation for 'bound states' works, as we already know the exact solution. First, we set $\gamma = 2$, $B = 1$ in Eq. (III.6.118). Recalling that $\Gamma(2) = 1$ and $\Gamma(3/2) = \sqrt{\pi}/2$, we get

$$E_n = \hbar\omega(n + 1/2) \tag{III.6.119}$$

Remarkably, this is the same result as the exact law of quantization of harmonic oscillator! Not bad at all for what supposed to be an approximation.

Will it work as well for eigenfunctions? To answer this question, let us find what the WKB approximation gives for them, and then plot them against the exact solution. We first note some results for indefinite integrals which we encounter when doing this:

$$\int \sqrt{1 - u^2}\, du = \frac{1}{2}\left[ u\sqrt{1 - u^2} + \arcsin(u) \right] \tag{III.6.120}$$

$$\int \sqrt{u^2 - 1}\, du = \frac{1}{2}\left[ u\sqrt{1 - u^2} - \operatorname{sgn}(u)\ln(u\operatorname{sgn}(u) + \sqrt{u^2 - 1}) \right] \tag{III.6.121}$$

This allows us to write for the WKB's Eq. (III.6.111)):

$$\psi(s) = \frac{A}{2\left(s^2 - 2n - 1\right)^{1/4}}$$

$$\times \exp\left\{ \frac{2n+1}{2}\left[ \frac{s}{2n+1}\sqrt{s^2 - 2n - 1} \right.\right.$$

$$\left.\left. + \ln\left( -\frac{s}{\sqrt{2n+1}} + \sqrt{\frac{s^2}{2n+1} - 1} \right) \right] \right\}$$

$$s < -\sqrt{2n+1} \tag{III.6.122}$$

$$\psi(s) = \frac{A}{2\left(2n + 1 - s^2\right)^{1/4}}$$

$$\times \cos\left\{ \frac{2n+1}{2}\left[ \frac{s\sqrt{2n+1-s^2}}{2n+1} \right.\right.$$

$$\left.\left. + \arcsin\left( \frac{s}{\sqrt{2n+1}} \right) \right] + n\frac{\pi}{2} \right\}$$

$$-\sqrt{2n+1} \le s \le \sqrt{2n+1} \tag{III.6.123}$$

$$\psi(s) = A\frac{(-1)^n}{2\,(s^2 - 2n - 1)^{1/4}}$$

$$\times \exp\left\{-\frac{2n+1}{2}\left[\frac{s}{2n+1}\sqrt{s^2 - 2n - 1}\right.\right.$$

$$\left.\left.- \ln\left(\frac{s}{\sqrt{2n+1}} + \sqrt{\frac{s^2}{2n+1} - 1}\right)\right]\right\}$$

$$s > \sqrt{2n+1} \qquad\qquad\qquad\qquad\qquad (\text{III.6.124})$$

Near the turning points, $e_1^{(1)} = e_1^{(2)} = 2\sqrt{2n+1}$, so that the solutions near 'turning points' are given by

$$\psi(s) = 2^{-1/6}A\sqrt{\pi}(2n+1)^{-1/2}\text{Ai}(-2^{1/3}(2n+1)^{1/6}(\sqrt{2n+1}+s))$$
$$(\text{III.6.125})$$

$$\psi(s) = 2^{-1/6}A\sqrt{\pi}(2n+1)^{-1/2}\text{Ai}(-2^{1/3}(2n+1)^{1/6}(s - \sqrt{2n+1}))$$
$$(\text{III.6.126})$$

We now plot the functions given by Eqs. (III.6.122)–(III.6.126) for $n = 0, 1, 2$ and 10 and, in Fig. III.6.9, see how these results agree with the exact solution (Eq. (III.4.236)).

Note that, to make our life easier, we cheated a bit: we have not determined the constant $A$ by the normalization condition (Eq. (III.6.112)), as strictly speaking we should have done. As we are testing the WKB approximation against the exact solution, we instead, for simplicity, fix it so that at one point the exact and approximate solutions are the same. For even values of $n$ exact solution and approximate solution are chosen to coincide at $s = 0$; whereas for odd values of $n$, the derivatives of the two solutions are matched at this point. If the approximation is a good one, there would be little discrepancy between the value of $A$ calculated this way and the one obtained through normalization constraint without knowing the exact solution.

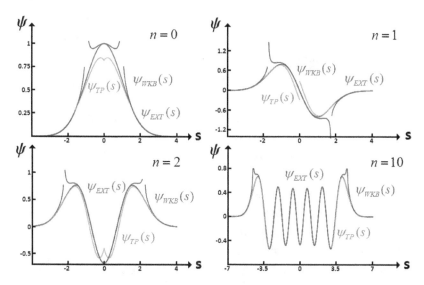

**Fig. III.6.9** Here, we compare the exact wavefunctions of the simple harmonic oscillator, given by Eq. (III.4.236), with the WKB approximation, Eq. (III.6.109), also including the 'turning-point solutions' given by Eqs. (III.6.110) and (III.6.111). We show it for $n = 0, 1, 2$ and 10 as functions of the rescaled coordinate variable $s = x\sqrt{m\omega/\hbar}$. The normalization constant of the wave functions has been chosen so that the values of the WKB and exact calculation match at $s = 0$ for even values of $n$, and for $n = 1$ so that their derivatives match at $s = 0$, as opposed to calculating them through Eq. (III.6.112). In each panel, we plot both $\psi_{EXT}(s)$, the exact solution (shown in red), and $\psi_{WKB}(s)$ the WKB approximation (shown in blue) over the whole range of s-values considered (with the exception of close to the turning point singularities for the WKB approximation). The turning point solutions $\psi_{TP}(s)$ (shown in green) are plotted over a limited range of values close to the turning points. The WKB approximation works worst for $n = 0$, and best for $n = 10$. For $n = 0$ it is, unfortunately, impossible to construct a continuous interpolation across the turning points at $s = 1, -1$, that would have joined the corresponding wings at $|s| > 1$ with those at $|s| < 1$; those wings do not 'meet'. Still, even for $n = 0$ the turning point solutions seem to provide a decent interpolation across the turning points, and results in a very accurate approximation. As we increase $n$ further, the WKB approximation becomes even more accurate.

At $n = 0$ the WKB approximation works well in places away from the turning points, where we see the singularities, which are artefacts of the WKB approximation. We can use the turning point solutions to go across the turning points, but unfortunately a continuous interpolation, in this case, between these solutions and the WKB functions is not possible. However, we see that the WKB approximation gets progressively better as we increase $n$. At $n = 10$ it is incredibly accurate, especially if we use the turning point solutions. With these we can see that we can interpolate over the artefact singularity, leading to a very accurate approximation. The reason why the WKB approximation works so well at large values of $n$ is because the wave function is oscillating much more rapidly compared to the rate of change of the potential energy.

Here, we have tested the WKB approximation against an exact solution that we knew. Of course, the main value of the WKB technique is its ability to solve the cases where exact solutions are unknown, or for arbitrary form of potential energy; this is what it is for! It can be used, for example, for better approximations of the potential energy of chemical bonds, for which we do not have exact solutions—although the integrals we have to deal with may become more cumbersome.

**Exercise.** Find approximate eigenfunctions for Eq. (III.6.114) with $\gamma = 1$, using the WKB approximation and also finding the turning point solutions. You'll need to use Eq. (III.6.118) to find the allowed eigenvalues.

## 6.3.    Approximate Solutions of Differential Equations When Coefficients at High Derivatives are Small

This is a branch of applied mathematics related to the notions of 'singular perturbation techniques', 'boundary layer problems', and 'matching asymptotic expressions'. There are classical books about it (see e.g. J.D. Cole, *Perturbation Methods in Applied Mathematics*, Blaidell, Waltham, 1968; W. Eckhaus, *Matched Asymptotic Expansion and Singular Perturbations*, Academic Press, New York, 1973; D.R. Smith, *Singular Perturbation Theory*, Cambridge Univ. Press, Cambridge, 1985). Here, we will not be able to go into any details of these techniques, and instead refer the reader to specialized books, but we will just give an idea of what it is all about and show a couple of examples, when they work and when they do not.

### Recap: Exact solution for dissipating oscillator and its limiting cases

Let us go back for a moment and recall the solution for a decaying linear oscillator motion. We studied those types of equations and their solutions in Part II, Vol. 1, in Section 6.2, Example II.6.11. But here it will be convenient for us to prepare the solution for different initial conditions than in the exercise (II.6.194). We want to have this exact solution at hand and fully understood, as we will be comparing with it the results obtained by approximate methods and of more complicated equations.

Consider equation of simple 1D harmonic oscillatory motion of a mass $m$ on a spring of stiffness $k$ fixed at one end:

$$m\frac{d^2x}{dt^2} + \gamma\frac{dx}{dt} + kx = 0 \qquad \text{(III.6.127)}$$

where $x$ stands for the deviation from the equilibrium position, and $\gamma$ is the dissipation coefficient. This second order equation needs two 'boundary' conditions. We consider the initial conditions

$$x(0) = 0, \quad \left.\frac{dx}{dt}\right|_{t=0} = v \qquad \text{(III.6.128)}$$

i.e. at the beginning the spring was at equilibrium but the mass momentarily receives a momentum $mv$.

Following the same steps as described in the Section II.6.2 of Vol. 1, we get the solution satisfying these initial conditions (derive this as an exercise) as

$$x(t) = \frac{v}{\omega}e^{-\beta t}\frac{\sin\left(\omega\sqrt{1 - \frac{\beta^2}{\omega^2}}\,t\right)}{\sqrt{1 - \frac{\beta^2}{\omega^2}}} \qquad \text{(III.6.129)}$$

where

$$\omega = \sqrt{\frac{k}{m}}, \quad \beta = \frac{\gamma}{2m} \qquad \text{(III.6.130)}$$

Both of these parameters, the 'frequency' of oscillations, $\omega$, and their 'decay decrement', $\beta$, have dimensionality of inverse time. The names given to them become clear from the limiting case of weak dissipation. Let us first prepare expressions for all important limiting cases.

1. Weak dissipation, $\omega \gg \beta$. Neglecting the second term under the square root, we get simply

$$x(t) = \frac{v}{\omega}e^{-\beta t}\sin(\omega t) \qquad \text{(III.6.131)}$$

2. Strong dissipation $\omega \ll \beta$

First, when $\omega < \beta$, if the argument under the square root becomes negative, taking the $\sqrt{-1} = i$ out of the square root, and recalling the relationships between the trigonometric and hyperbolic functions, in this case $\sin(iu) = i\sinh(u)$, we rewrite (III.6.129) as

$$x(t) = \frac{v}{\omega}e^{-\beta t}\frac{\sinh\left(\omega\sqrt{\frac{\beta^2}{\omega^2} - 1}\, t\right)}{\sqrt{\frac{\beta^2}{\omega^2} - 1}} = \frac{v}{\beta}e^{-\beta t}\frac{\sinh\left(\beta\sqrt{1 - \frac{\omega^2}{\beta^2}}\, t\right)}{\sqrt{1 - \frac{\omega^2}{\beta^2}}}$$

$$\text{(III.6.132)}$$

Using the definition of the sinh function and expanding the r.h.s expression in the smallness of $\frac{\omega^2}{\beta^2}$, we obtain

$$x(t) \approx \frac{v}{2\beta}e^{-\beta t}\left\{ e^{\beta\left(1 - \frac{\omega^2}{2\beta^2}\right)t} - e^{-\beta\left(1 - \frac{\omega^2}{2\beta^2}\right)t} \right\}$$

$$= \frac{v}{2\beta}\left\{ e^{-\frac{\omega^2}{2\beta}t} - e^{-\left(2\beta - \frac{\omega^2}{2\beta}\right)t} \right\} \qquad \text{(III.6.133)}$$

The first term in this expression is slowly decaying, and the second one decays faster but we need to keep both terms, to make sure that the first of the initial conditions is exactly satisfied, and the second one will be approximately satisfied (to the accuracy of the small terms $\sim \frac{\omega^2}{2\beta^2}$)—check this yourself.

3. The formula for the border case, $\omega = \beta$, is obtained by replacement from Eq. (III.6.129) by approximating in it the sinh-function, $\sinh\left(\beta\sqrt{1 - \frac{\omega^2}{\beta^2}}\, t\right) \approx \left(\beta\sqrt{1 - \frac{\omega^2}{\beta^2}}\, t\right)$, getting thereby

$$x(t) = vte^{-\beta t} \qquad \text{(III.6.134)}$$

### The case of strong dissipation: Matching asymptotic solutions

Now, let us come to subject of this section. Look at Eq. (III.6.127), and ask: if the rigidity of the spring and the dissipation are the dominating effects, but not the inertia (i.e. both $\gamma$ and $k$ are large enough and $m$ is relatively small), can we just get the result (III.6.133) by neglecting the first term straightaway in Eq. (III.6.127)?

And the answer is—not at all, because the equation will become first order and it will not be able to satisfy *two* initial conditions. For this you need the method of matching the asymptotic solutions.

A regular way to illustrate this method is to first rewrite Eq. (III.6.127) by dividing it by $m$ and writing $\frac{d^2x}{dt^2} + 2\beta\frac{dx}{dt} + \omega^2 x = 0$. Then by introducing dimensionless time $s = \beta t$, we can rewrite this equation as

$$\left(\frac{\beta}{\omega}\right)^2 \frac{d^2x}{ds^2} + 2\left(\frac{\beta}{\omega}\right)^2 \frac{dx}{ds} + x = 0 \qquad \text{(III.6.135)}$$

Since we are considering the case when $(\beta/\omega) \gg 1$ (and the more so $(\beta/\omega)^2 \gg 1$), it is obvious that at $s \sim 1$ the first two terms in this equation will be dominating. In simple physical terms this will mean that at small times deviation from equilibrium is still small but the acceleration and velocity, which the first two terms represent, respectively, are not. Then one can apply the so-called *singular perturbation technique,* which will get you the answer to any accuracy of parameter $(\omega/\beta)^2 \ll 1$. We will not be presenting this technique here, but will show the basic result, using the key underlying idea of the method: introduction of the so-called *internal* and *external* solutions in two regions—small and long time.

In the internal region, $s < 1$, we can keep only the first two terms in this equation, $\frac{d^2x_{int}}{ds^2} + 2\frac{dx_{int}}{ds} \approx 0$, the solution of which is obtained by first integration, $\frac{dx_{int}}{ds} + 2x_{int} = C$, where $C$ is a constant of integration. Using the initial condition, Eq. (III.6.128), we immediately see that $C = v/\beta$. Next, using separation of variables, we get $\frac{dx_{int}}{\frac{v}{\beta} - 2x_{int}} = ds \Rightarrow \int_0^{x_{int}} \frac{dx'}{\frac{v}{\beta} - 2x'} = s$; specifying the limits of integration in this form we have already used the condition that at the initial time moment the deviation from equilibrium is zero. Performing integration we obtain

$$\ln\left|x_{int} - \frac{v}{2\beta}\right| - \ln\left|-\frac{v}{2\beta}\right| = -2s$$

so that exponentializing this equality we finally get

$$x_{int} = \frac{v}{2\beta}\left(1 - e^{-2s}\right) \tag{III.6.136}$$

In the external region, Eq. (III.6.134) is simplified to $2\left(\frac{\beta}{\omega}\right)^2 \frac{dx_{ext}}{ds} + x_{ext} = 0$, i.e. the mass still moves but acceleration has been fully damped by dissipation (e.g. friction). The solution of this equation is again obtained by separation of variables, giving us

$$x_{ext} = A e^{-\left(\frac{\omega^2}{2\beta^2}\right)s} \tag{III.6.137}$$

where $A$ is the integration constant. Now how do we determine it? The idea is to match the two solutions:

$$x_{int}(s \to \infty) = x_{ext}(s \to 0) \tag{III.6.138}$$

Applying this condition to expressions (III.6.136) and (III.6.137) we find that

$$A = \frac{v}{2\beta} \tag{III.6.139}$$

Now, we can combine the two asymptotic solutions into one interpolation formula $x = \frac{v}{2\beta}(1 - e^{-2s})e^{-\left(\frac{\omega^2}{2\beta^2}\right)s}$, which, going back to dimensional time, reads

$$x(t) = \frac{v}{2\beta}\left(e^{-\left(\frac{\omega^2}{2\beta}\right)t} - e^{-\left(2\beta + \frac{\omega^2}{2\beta}\right)t}\right) \tag{III.6.140}$$

If we compare this result and the approximate form given by Eq. (III.6.133) to the exact solution, Eq. (III.6.132), we see that two approximate formulas look almost the same, but (III.6.140) is even better, because it not only gives zero at time zero, but recovers

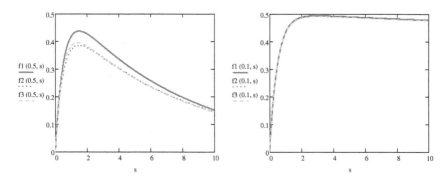

**Fig. III.6.10**   We compare three functions $f_1(p, s) = e^{-s}\sinh(\sqrt{1-p^2}\,s)/\sqrt{1-p^2}$, $f_2(p, s) = \frac{1}{2}\{e^{-\frac{p^2}{2}s} - e^{-(2-\frac{p^2}{2})s}\}$ and $f_3(p, s) = \frac{1}{2}\left(1 - e^{-2s}\right)e^{-\frac{p^2}{2}s}$, to which the exact solution, the asymptotic form of the exact solution, and the 'matching-asymptotic-expansions' results are, respectively, proportional to—all with the same factor $(v/\beta)]$, with $p = \omega/\beta$. Both approximate expressions should be valid at $p \ll 1$. As we see from the left panel, the approximate forms look fair but are not too accurate for $p = 0.5$. For $p = 0.1$, however, they are already indistinguishable from each other and from the exact solution, as seen in the right panel.

exactly the value of velocity, $v$, at time zero (check this by differentiation of Eq. (III.6.140)). In Fig. III.6.10 we compare the three results.

## Considering the same problem with variable coefficients

So far, we just reproduced known results, showing how well the matching scheme could work. Let us, using this method, get something for which we do not have an exact solution. We consider an equation,

$$m\frac{d^2x}{dt^2} + \gamma(t)\frac{dx}{dt} + k(t)x = 0 \qquad (\text{III.6.141})$$

with the same initial conditions, but in which dissipation coefficient $\gamma(t)$ and the spring stiffness $k(t)$ change with time (but never vanish); we consider the mass constant, although one can imagine cases when it is leaking out or evaporating—you can play around with it yourself!

First we rewrite it again in the form

$$\frac{d^2x}{dt^2} + \beta(t)\frac{dx}{dt} + \omega^2(t)x = 0 \tag{III.6.142}$$

where

$$[\omega(t)]^2 = \sqrt{\frac{k(t)}{m}}, \quad \beta(t) = \frac{\gamma(t)}{2m} \tag{III.6.143}$$

If thus defined $\omega(t)$ and $\beta(t)$ are smooth and nonvanishing functions, and we can, going along the same steps as in the previous subsection, obtain an interpolation formula:

$$x = \frac{v}{2\beta(0)}\left(1 - e^{-2\beta(0)t}\right) e^{-\int \left(\frac{\omega(t)^2}{2\beta(t)}\right)t} \tag{III.6.144}$$

To illustrate this result let us consider the case when dissipation does not change but the spring gets tired; specifically

$$\gamma(t) = \gamma = \text{const}, \quad k(t) = k_\infty + \frac{k - k_\infty}{1 + (t/\tau)^2} \tag{III.6.145}$$

(i.e. the spring constant smoothly varies between the initial value, $k$, and the finite, smaller value $k_\infty \le k$, with the relaxation time $\tau$). Combining Eqs. (III.6.144), (III.6.143), (III.6.145), and performing the integration we get

$$x = v\frac{m}{\gamma}\left(1 - e^{-\frac{\gamma}{m}t}\right)\exp\left\{-\frac{k}{\gamma}\left[\frac{k_\infty}{k}t + \left(1 - \frac{k_\infty}{k}\right)\tau\arctan\left(\frac{t}{\tau}\right)\right]\right\} \tag{III.6.146}$$

When $k_\infty = k$, this results reduces to the one obtained in the previous subsection. When $k_\infty < k$; the amplitude will decay slower at long times. For $k_\infty = 0$, it will never return to $x = 0$, but saturate at $x = v\frac{m}{\gamma}e^{-\frac{\pi}{2}\frac{k\tau}{\gamma}}$. The larger will be the ratio $\frac{k\tau}{\gamma}$, the closer this level will be to zero, which makes perfect sense. Indeed, at large $\tau$ the decay of the stiffness will be slower and the system will have time to return to smaller value of $x$. The same trend will take place with smaller dissipation, i.e. smaller values of $\gamma$, or larger initial stiffness of the spring, $k$.

For the criteria of the method described in this section or higher order corrections we refer the readers to the specialized books cited above.

## 6.4. Another Case to Learn from: Nondissipating Oscillator with the Spring Getting Tired

Here, we will present just one example, trying to use intuitive methods first and then get the result by a more regular approach.

**Example III.6.8 (Physical Application ▶▶).** Consider the equation of an ideal oscillator (no dissipation), in which the stiffness of a spring decays with time to zero, so that $\frac{d^2x}{dt^2} + \frac{\omega^2}{1+(t/\tau)^2}x = 0$, with the same initial conditions as given by Eq. (III.6.128). We can rewrite this differential equation in a more convenient form:

$$\left(1 + \frac{t^2}{\tau^2}\right)\frac{d^2x}{dt^2} + \omega^2 x = 0 \qquad \text{(III.6.147)}$$

Introducing new time variable $\omega t \equiv \xi$, we can rewrite it as

$$\left(1 + p^2\xi^2\right)\frac{d^2x}{d\xi^2} + x = 0, \quad p = \frac{1}{\omega\tau} \qquad \text{(III.6.148)}$$

Below we consider the case when $p < 1$, i.e. the decay time of the stiffness of the spring is longer then the inverse frequency.

It looks obvious that when $\xi \gg p^{-1}$ (i.e. $t \gg \tau$) this equation can be simplified to

$$p^2\xi^2\frac{d^2x}{d\xi^2} + x \simeq 0 \qquad \text{(III.6.149)}$$

whereas in the opposite limit $\xi \ll p^{-1}$ (i.e. $t \ll \tau$) it is rather

$$\frac{d^2x}{d\xi^2} + x \simeq 0 \qquad \text{(III.6.150)}$$

The general solution of Eq. (III.6.149) is found in the form of

$$x \propto \xi^\lambda \qquad \text{(III.6.151)}$$

with characteristic equation on $\lambda$:

$$p^2\lambda(\lambda - 1) + 1 \simeq 0 \qquad\qquad (\text{III.6.152})$$

i.e. $p^2\lambda^2 - p^2\lambda + 1 \simeq 0 \Rightarrow \lambda = \frac{p^2 \pm \sqrt{p^4 - 4p^2}}{2p^2} = \frac{1}{2}\{1 \pm \sqrt{1 - \frac{4}{p^2}}\} = \frac{1}{2}\{1 \pm i\sqrt{\frac{4}{p^2} - 1}\}$. Thus two linear independent solutions of this equation are: $x = C_+ \exp\{[\frac{1}{2} + \frac{i}{2}\sqrt{\frac{4}{p^2} - 1}]\ln\xi\}$, $C_- \exp\{[\frac{1}{2} - \frac{i}{2}\sqrt{\frac{4}{p^2} - 1}]\ln\xi\}$, where $C_+$ and $C_-$ are free constants. It will be more convenient to rewrite this result in terms of real quantities, so that the two possible solutions will read as

$$x = A_1\sqrt{\xi}\cos\left\{\sqrt{4p^{-2} - 1} \cdot \ln\xi\right\}$$
$$x = A_2\sqrt{\xi}\sin\left\{\sqrt{4p^{-2} - 1} \cdot \ln\xi\right\} \qquad (\text{III.6.153})$$

where $A_1$ and $A_2$ are new free constants.

The possible solutions of the 'small-times' equation, Eq. (III.6.150), is, of course,

$$x = \tilde{A}_1\cos\xi \quad x = \tilde{A}_2\sin\xi \qquad\qquad (\text{III.6.154})$$

and in view of the specified initial conditions (Eq. (III.6.128)) the sine option should be chosen with its constant specified:

$$x = \frac{v}{\omega}\sin\xi \qquad\qquad (\text{III.6.155})$$

But there is no way to match Eqs. (III.6.155) and (III.6.153) through the free constants, or even by a reasonable interpolation formula. We thus consider an approach similar to WKB, which will be legitimate, however, only in the limit of $p \ll 1$.

To perform this analysis, let us consider generally an equation

$$\frac{d^2x}{dt^2} + \omega^2 f^2(t)x = 0 \qquad\qquad (\text{III.6.156})$$

with the initial conditions given by Eq. (III.6.128).

In the example considered above

$$f(t) = \frac{1}{\sqrt{1 + (t/\tau)^2}} \tag{III.6.157}$$

But let us keep the form of $f(t)$ general for a moment, just assuming that it is a smooth function varying slowly over the period $T = \frac{2\pi}{\omega}$.

Keeping in mind the slow rate of variation of $f(t)$, we would look for the solution of Eq. (III.6.156) in the form of functions

$$x_{\pm}(t) = z(t)e^{\pm i\omega \int_0^t f(t)dt} \tag{III.6.158}$$

The exponential function describes the fast oscillation, whereas $z(t)$ should be the slow-varying modulating factor. Substituting this form into Eq. (III.6.156), we obtain

$$\frac{1}{\omega}\frac{d^2z}{dt^2} \pm 2if(t)\frac{dz}{dt} \pm i\frac{df}{dt}z = 0 \tag{III.6.159}$$

As $z(t)$ is expected to be a slowly varying function, we can expect the first term in this equation be small, and let's neglect it. We then get for $z(t)$ an approximate equation,

$$2f(t)\frac{dz}{dt} + \frac{df}{dt}z \simeq 0 \Rightarrow \frac{1}{z}\frac{dz}{dt} = -\frac{1}{2f(t)}\frac{df}{dt}$$

$$\Rightarrow \frac{d\ln z}{dt} = -\frac{1}{2}\frac{d\ln f}{dt}$$

$$\Rightarrow \ln z = \ln\frac{1}{\sqrt{f}} + A$$

where $A$ is a constant of integration, and hence

$$z(t) \simeq \frac{C}{\sqrt{f(t)}} \tag{III.6.160}$$

Combining Eqs. (III.6.160) and (III.6.158), we obtain that the general solutions of Eq. (III.6.156) can be approximated by

$$x_{\pm}(t) \simeq \frac{C_{\pm}}{\sqrt{f(t)}}e^{\pm i\omega \int_0^t f(t)dt} \tag{III.6.161}$$

To satisfy the initial conditions we need to take the sinusoidal combinations of these functions to obtain

$$x(t) \simeq \frac{v}{\omega} \frac{\sin\left\{\omega \int_0^t f(t)dt\right\}}{\sqrt{f(t)}} \tag{III.6.162}$$

It is a beautiful equation, but let us see—what will it give us in the case of our example, when $f(t)$ is given by Eq. III.6.157)? Since $\int_0^t \frac{1}{\sqrt{1+(t/\tau)^2}}dt = \tau\{\ln \frac{t+\sqrt{\tau^2+t^2}}{\tau}\}$, we get

$$x(t) \simeq \frac{v}{\omega} \sqrt[4]{1+\left(\frac{t}{\tau}\right)^2} \sin\left\{\omega\tau \ln\left[\frac{t}{\tau} + \sqrt{1+\left(\frac{t}{\tau}\right)^2}\right]\right\} \tag{III.6.163}$$

Interesting function! The graph of it is shown in Fig. III.6.11. The physics of the result is tricky. Differentiating Eq. (III.6.163) over time, i.e. calculating the velocity of oscillation as a function of time,

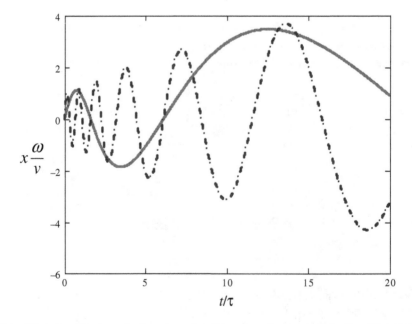

Fig. III.6.11 We show the functions describing the solution given by Eq. (III.6.163). solid curve corresponds to $p = (\omega\tau)^{-1} = 0.4$ (the case at the border of rigorous applicability of the equation) and dash-dotted curve to $p = 0.1$. As expected, with softening of the spring with time, amplitude of oscillations gets larger and larger and their actual period increases.

you will see that the velocity will also oscillate, but with an amplitude slowly going to zero, in contrast to the amplitude of $x(t)$ which
does not converge. Does it make sense? Thus, if the spring constant
vanishes the masses connected by the spring can go anywhere far
from each other but with a slower and slower speed. Does it violate
determinism of classical mechanics? Not at all: for any time t you can
calculate the distance between the masses, i.e. you will know how far
the spring will expand. But it seems from this solution that if we
wait long enough, it could go to another galaxy. Of course, because
the velocity goes to zero this will happen only in the next 100 million
years. Does it look strange? Is that this approximate solution does
not work at long times?

Another limiting example helps to relax these concerns: the solution is OK, just blame it on the model described by Eq. (III.6.147).
Indeed, imagine for a moment that instead of continuously decreasing spring constant we had abruptly cut the spring. The two masses
in the oscillator would then continue going away from each other
by inertia, with the velocities they acquired at the time moment
when the spring was chopped. Indeed, instead of a parabolic potential energy, we get it momentarily flat. In the considered case of
the oscillator with the 'tiring' spring, that flattening proceeds but
gradually, so there is nothing paradoxical in this solution. Of course,
this model will not describe the reality (i) the spring length, however tired the spring is, cannot expand to infinity; (ii) if the spring
is getting tired, one must include dissipation into the equation.

The main mathematical lesson for us is that the equations given
in Eq. (III.6.153) at small $p$ are no way near to the elegant solution,
given by Eq. (III.6.163), although it exhibits similar trends, such as
the increase of the amplitude and the period of oscillations with time,
as long as the spring gets softer and softer. Whereas Eq. (III.6.163)
does reproduce the sinusoidal solution of Eq. (III.6.155) at small
times, because $\frac{t}{\tau} \ll 1, \ln[\frac{t}{\tau} + \sqrt{1 + (\frac{t}{\tau})^2}] \approx \ln[1 + \frac{t}{\tau}] \approx \frac{t}{\tau}$, the solution does not evolve into the sinusoidal form of Eq. (III.6.153); it
rather becomes $x(t) \simeq \frac{v}{\omega} \sqrt[2]{\frac{t}{\tau}} \sin \{\omega\tau \ln [\frac{2t}{\tau}]\}$. In other words, matching asymptotic expansion without a regular procedure for it, is, gen-
erally, unsafe.

## 6.5.    Key Points

- We introduced to you some popular methods for approximate solution of ordinary differential equations with variable coefficients, including the eigenvalue equations.
- For differential equations of the form $\hat{H}f(x) = 0$ and $\hat{H}f(x) = Ef(x)$ that lie close to equations $\hat{H}_0 f_0(x) = 0$ and $\hat{H}_0 f_0(x) = E_0 f_0(x)$, respectively, which have analytical solutions (where $\hat{H}$ and $\hat{H}_0$ are differential operators), we may generate a perturbation expansion by considering, $f(x) = f_0(x) + \alpha f_1(x) + \alpha^2 f_2(x) + \cdots$ where $\hat{H} - \hat{H}_0 = \alpha \Delta \hat{H}$.
- General expressions for the perturbation expansion for second order linear homogeneous differential equations are given by Eqs. (III.6.4), (III.6.6), (III.6.7) and (III.6.8).
- General expressions for the perturbation expansion for second order differential eigenvalue equations are given by Eqs. (III.6.39) and (III.6.40).
- Perturbation theory works well provided that the size of $\alpha \Delta \hat{H}$ is not too large. If it is small enough we may approximate the exact solution with $\psi(x) \approx \psi_0(x) + \alpha \psi_1(x)$.
- As we calculate successive corrections, $\psi_n(x)$, we may obtain more accurate results. However, the results for perturbation theory start to become rapidly more cumbersome, and their derivation more laborious, as we increase $n$. There may be a point where calculating the next correction $\psi_{n+1}(x)$ is not worthwhile.
- Another approximation technique for (linear) differential equations is the WKB approximation. Equations that have the form $y''(x) + by'(x)/x + cy(x) = E(x)y(x)$ may be approximated this way. This is done by first reducing them to the form $g''(x) + (1/4 - \alpha^2)g(x)/x = E(x)g(x)$, where $y(x) = x^{-b/2}g(x)$ and $\alpha^2 = 1/4 + b^2/4 - b/2 + c$.
- The WKB approximation relies on $E(x)$ not varying too fast and too much and on it not becoming too small (in the basic WKB approximation).

- For $E(x) > 0$, the WKB approximate solution is, $g(x) \approx (E(x))^{-1/4}\tilde{y}_\alpha^+ \left( \int_0^x \sqrt{E(x')}dx \right)$ where $\tilde{y}_\alpha^+(s) = A^+ s^{1/2} J_\alpha(s) + B^+ s^{1/2} Y_\alpha(s)$.

- For $E(x) < 0$, the WKB approximate solution is $g(x) \approx (-E(x))^{-1/4}\tilde{y}_\alpha^- \left( \int_0^x \sqrt{-E(x')}dx' \right)$, where $\tilde{y}_\alpha^-(s) = A^- s^{1/2} I_\alpha(s) + B^- s^{1/2} K_\alpha(s)$.

- Near the turning points, where $E(x) = 0$, WKB approximation does not work. We have specified the procedure that helps to approximate the solution near these points.

- A situation where $E(x)$ has two turning points, $x_1$ and $x_2$, such that $E(x) < 0$ for both $x < x_1$ and $x_2 < x$ results in the Sommerfeld–Wilson quantization condition: $\int_{x_1}^{x_2} \sqrt{E(x)}dx' = \pi(n + 1/2)$.

- There are many other methods for the approximate solving of differential equations of equations with variable coefficients for which we refer the reader to specialized literature. Some examples of these approaches have been considered in the last two subsections of this chapter, with a demonstration of how using them without sufficient care may lead to incorrect/inaccurate results.

# Chapter 7

# Homogeneous Partial Differential Equations

## 7.1. Delving into Laplace's Equation

### Laplace's equation in 2D Cartesian coordinates

In Chapter 3 of Part III, Vol. 2 we introduced the Laplacian operator, defined as $\vec{\nabla}^2 = \vec{\nabla}.\vec{\nabla}$, the divergence of the gradient of a function. Now Laplace's equation is defined as

$$\vec{\nabla}^2 \phi(\mathbf{r}) = 0 \tag{III.7.1}$$

Functions that satisfy Eq. (III.7.1) are called *harmonic*. Understanding how to solve Eq. (III.7.1) will teach us a lot about solving homogeneous partial differential equations. Let's start by considering the simplest case of the 2D Laplace equation in Cartesian coordinates. This can be written as

$$\vec{\nabla}^2 f(x, y) = \frac{\partial^2 f(x, y)}{\partial x^2} + \frac{\partial^2 f(x, y)}{\partial y^2} = 0 \tag{III.7.2}$$

Let's first look at obtaining a general solution of this equation. To do this, we use what is referred to as the separation of variables technique. Namely, we look at a specific solution to Eq. (III.7.2) where we may write

$$f(x, y) = X(x)Y(y) \tag{III.7.3}$$

Substitution of Eq. (III.7.3) into Eq. (III.7.2) yields

$$\frac{d^2X(x)}{dx^2}Y(y) = -X(x)\frac{d^2Y(y)}{dy^2} \Rightarrow \frac{1}{X(x)}\frac{d^2X(x)}{dx^2} = -\frac{1}{Y(y)}\frac{d^2Y(y)}{dy^2}$$

(III.7.4)

We see that what is on the l.h.s on the last line of Eq. (III.7.4) depends only on $x$, and what is on the r.h.s of it depends only on $y$. This implies that that r.h.s should be equal to a constant with respect to $x$, the l.h.s of it equalling a constant with respect to $y$, and this must be obviously the *same* constant!

Thus, we get

$$\frac{1}{X(x)}\frac{d^2X(x)}{dx^2} = k^2$$

(III.7.5)

where $k^2$ is an arbitrary constant that does not depend on either $x$ or $y$. Whether $k^2$ should be positive or negative (in principle, $k$ can be imaginary) depends on the boundary conditions, but here we'll suppose it is positive. We should already know how to solve Eq. (III.7.5) (see Chapter 6, Part II, Vol. 1, for details). Its solution is

$$X(x) = A_x \exp(kx) + B_x \exp(-kx)$$

(III.7.6)

From Eq. (III.7.4), we similarly have

$$\frac{1}{Y(y)}\frac{d^2Y(y)}{dy^2} = -k^2$$

(III.7.7)

and a real solution to this equation is

$$Y(y) = A_y \exp(iky) + A_y^* \exp(-iky)$$

(III.7.8)

So, we may write a solution to Eq. (III.7.2) of the form

$$f_k(x, y) = (A_x \exp(kx) + B_x \exp(-kx))$$
$$\times (A_y \exp(-iky) + A_y^* \exp(iky))$$

(III.7.9)

Now an important thing to realize is that $f(x, y) = f_k(x, y) + f_{k'}(x, y)$ where $k \neq k'$ is also a solution to Eq. (III.7.2). Also, the constants, $A_x$, $B_x$, $A_y$ and $A_y^*$ can be different for each choice of $k$. Thus, we are led to writing down a general solution of the following form:

$$f(x, y) = \sum_k \left( \tilde{A}_k \exp(kx) + \tilde{B}_k \exp(-kx) \right) \exp(-iky) \quad \text{(III.7.10)}$$

where we require that $\tilde{B}_{-k} = \tilde{A}_k^*$ and $\tilde{A}_{-k} = \tilde{B}_k^*$ to guarantee that the solution is real. To find both $\tilde{A}_k$ and $\tilde{B}_k$ we need to specify boundary conditions. As we will see, they will define not only the constants, but also the allowed $k$-values (so far we shyly avoided speaking about them!).

### A simple rectangular boundary value problem

In a system with a rectangular boundary, a supposition of linear solutions obtained by separation of variables in Cartesian coordinates allows us to formulate the full solution in a tractable form. These are when $f(x, y)$ is defined within a rectangle where $0 < x < x_1$ and $0 < y < y_1$. For the solution to be uniquely determined, we need to specify functions along the perimeter of the rectangle. For instance, let us consider the case when

$$f(0, y) = g_A(y) \quad f(x_1, y) = g_B(y) \quad f(x, 0) = f(x, y_1) = 0$$
$$\text{(III.7.11)}$$

as illustrated in Fig. III.7.1. In this case, we would obtain for the solution

$$f(x, y) = \sum_{n=1}^{\infty} \left( \tilde{A}_n \exp(k_n x) + \tilde{B}_n \exp(-k_n x) \right) \sin(k_n y) \quad \text{(III.7.12)}$$

where the values of $k$ must be discretized to $k_n = n\pi/y_1$ and we must have only combinations of $\exp(ik_n x) - \exp(-ik_n x)$ to satisfy $f(x, 0) = f(x, y_1) = 0$, and this would be possible only for such values of $k$. To satisfy the other two boundary conditions in Eq. (III.7.11)

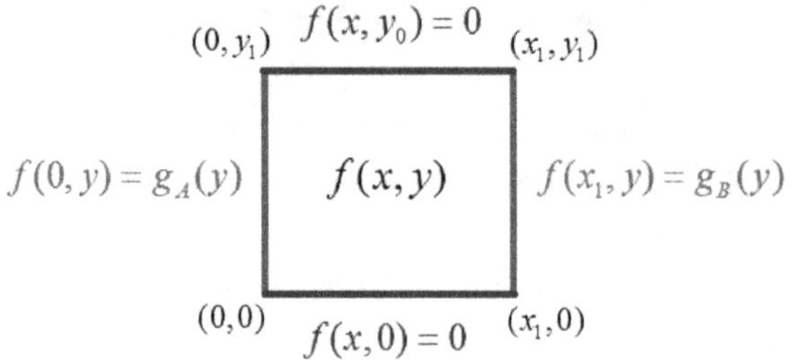

**Fig. III.7.1**  An illustration of a simple rectangular boundary value problem. In this example, on two sides of the rectangle, shown at the top and bottom, the output value of the function is held at zero. On the other two sides, we specify $f(0, y) = g_A(y)$ and $f(x_1, y) = g_B(y)$.

we require

$$g_A(y) = \sum_{n=1}^{\infty} \left( \tilde{A}_n + \tilde{B}_n \right) \sin(k_n y) \tag{III.7.13}$$

$$g_B(y) = \sum_{n=1}^{\infty} \left( \tilde{A}_n \exp(k_n x_1) + \tilde{B}_n \exp(-k_n x_1) \right) \sin(k_n y) \tag{III.7.14}$$

So, here we are, job done! Indeed, Eqs. (III.7.13) and (III.7.14) are Fourier series, so we know that we can obtain the coefficients through the integrals:

$$\begin{cases} \tilde{A}_n + \tilde{B}_n = \dfrac{2}{y_1} \displaystyle\int_0^{y_1} g_A(y) \sin(k_n y) dy \\[3mm] \tilde{A}_n \exp(k_n x_1) + \tilde{B}_n \exp(-k_n x_1) = \dfrac{2}{y_1} \displaystyle\int_0^{y_1} g_B(y) \sin(k_n y) dy \end{cases} \tag{III.7.15}$$

We thus have a system of two elementary linear equation, from which you can easily find $\tilde{A}_n$ and $\tilde{B}_n$.

**Exercise.** Write the solution for the coefficients $\tilde{A}_n$ and $\tilde{B}_n$ in determinants and then get a simple expression for them in the particular case when $g_A(y) = 0$ and $g_B(y) = 1$.

**Example III.7.1 (Maths Practice ♫♪).** Let's consider the solution given by Eq. (III.7.12) within a rectangular region, $0 < x < x_1$ and $0 < y < y_1$, with the boundary conditions $f(x_1, y) = ay(y - y_1)$ and $f(0, y) = f(x, 0) = f(x, y_1) = 0$. Two of those conditions,

$f(x, 0) = f(x, y_1) = 0$ have already been used in writing the solution in the form of Eq. (III.7.12), in which $k_n = n\pi/y_1$. So we need to focus on the other two conditions, $f(0, y) = 0$ and $f(x_1, y) = ay(y - y_1)$. These translate into

$$0 = \sum_{n=1}^{\infty} \left( \tilde{A}_n + \tilde{B}_n \right) \sin(k_n y) \tag{III.7.16}$$

$$ay(y - y_1) = \sum_{n=1}^{\infty} \left( \tilde{A}_n \exp(k_n x_1) + \tilde{B}_n \exp(-k_n x_1) \right) \sin(k_n y) \tag{III.7.17}$$

Equation (III.7.16) requires that $\tilde{A}_n = -\tilde{B}_n$. Thus, we're left with

$$ay(y - y_1) = 2 \sum_{n=1}^{\infty} \tilde{A}_n \sinh(k_n x_1) \sin(k_n y) \tag{III.7.18}$$

This is a Fourier series where the coefficients are calculated as

$$\tilde{A}_n \sinh(k_n x_1) = \frac{1}{y_1} \int_0^{y_1} ay(y - y_1) \sin(k_n y) dy \tag{III.7.19}$$

With $k_n = n\pi/y_1$, evaluation of the integral on the r.h.s yields

$$\tilde{A}_n = \frac{2ay_1^2}{\pi^3} \frac{(-1)^n - 1}{n^3} \frac{1}{\sinh(k_n x_1)} \tag{III.7.20}$$

Thus, we obtain the solution

$$f(x, y) = \frac{4ay_1^2}{\pi^3} \sum_{n=1}^{\infty} \frac{(-1)^n - 1}{n^3} \frac{\sinh(k_n x)}{\sinh(k_n x_1)} \sin(k_n y) \tag{III.7.21}$$

Evaluation of the sum is rather tricky, though the sum can be performed numerically in no time; in the last section of this part of the book, we'll show how one may attempt to evaluate such sums analytically. We plot the result numerically as a function of $x/x_1$ and $y/y_1$ in Fig. III.7.2 for two values of the aspect ratio $\eta = x_1/y_1$.

Let's look at two limiting cases where we may readily obtain closed form analytical expressions for the sums. In the first one, where

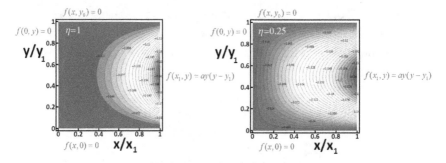

**Fig. III.7.2**   We plot solution, $f(x,y)/a$ to Laplace's equation with specified rectangular boundary conditions. Namely, we consider a rectangle of dimensions $x_1$ and $y_1$, where we specify on three sides of the rectangle that the output value of the function is zero. On the side where $x = x_1$ we specify the boundary condition $f(x_1, y) = ay(y - y_1)$. We look at two values of the aspect ratio parameter $\eta = x_1/y_1$. In the left panel we show the result for a square; $\eta = 1$. We see that for a large section, shown in the darkest shade of blue, that $f(x,y) \simeq 0$; the solution decays quickly away from the $x = x_1$ edge. The right panel displays the case of $\eta = 0.25$ which corresponds to the rectangle being narrower in the x-direction. Here the solution decays away less, as we move from the $x = x_1$ edge. If we make $\eta$ small enough we recover the limiting case discussed in the text.

$x_1 \ll y_1$ and $y \ll y_1$, we may take $k_n x$ to be small as the dominant terms in the sum are for small $n$. In this case, we can write

$$\frac{\sinh(k_n x)}{\sinh(k_n x_1)} \approx \frac{x}{x_1} \tag{III.7.22}$$

to obtain

$$f(x,y) \approx \frac{4ay_1^2}{\pi^3} \frac{x}{x_1} \sum_{n=1}^{\infty} \frac{(-1)^n - 1}{n^3} \sin(k_n y)$$

$$\Rightarrow f(x,y) \approx \frac{ayx(y - y_1)}{x_1} \approx -\frac{ay_1}{x_1} xy \tag{III.7.23}$$

The second line of Eq. (III.7.23) is obtained by realizing that the sum is simply a Fourier series of $y(y - y_1)$, as in Eqs. (III.7.18) and (III.7.20), as well as requiring $y \ll y_1$. Indeed, for the particular case of $y_1 = \infty$, this becomes the solution to Laplace's equation for a *semi-infinite slab* $(0 < y < \infty)$ of finite width $x_1$ that satisfies the boundary conditions $f(x,0) = f(0,y) = 0$ and $f(x_1, y) = -ay_1 y$

(you can check this by substituting the r.h.s of Eq. (III.7.23) back into Eq. (III.7.2)).

Let us also examine the opposite case when $x_1, x \gg y_1$, $x_1 - x \gg y_1$. Here, we may approximate in the sum $\sinh(k_n x)/\sinh(k_n x_1) \approx \exp(k_n(x - x_1))$ so that

$$f(x, y) \approx \frac{4ay_1^2}{\pi^3} \sum_{n=1}^{\infty} \frac{(-1)^n - 1}{n^3} e^{k_n(x - x_1)} \sin(k_n y)$$

$$\approx -\frac{8ay_1^2}{\pi^3} e^{-\frac{\pi(x_1 - x)}{y_1}} \sin\left(\frac{\pi y}{y_1}\right) \tag{III.7.24}$$

where, due to the exponentially decaying factors, we have neglected all terms in the sum except for the leading ($n = 1$) term.

### Boundary value problem for an infinite slab

Let us consider an infinite slab where $-\infty \leq y \leq \infty$ and $0 < x < x_1$, as illustrated in Fig. III.7.3.

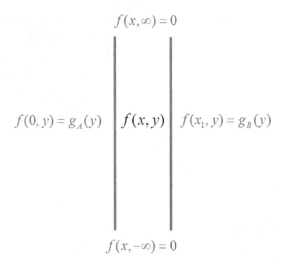

$$f(x, \infty) = 0$$

$$f(0, y) = g_A(y) \qquad f(x, y) \qquad f(x_1, y) = g_B(y)$$

$$f(x, -\infty) = 0$$

**Fig. III.7.3** We illustrate a generic boundary value problem for an infinite slab. On the two sides of the slab shown in red we specify the boundary conditions $f(0, y) = g_A(y)$ and $f(x_1, y) = g_B(y)$, as well as requiring that $f(x, y) \to 0$ when $\|y\| \to \infty$.

If we have that $f(x, y) \to 0$ when $|y| \to \infty$ the solution, Eq. (III.7.10), may be written as an integral:

$$f(x, y) = \frac{1}{2\pi} \int_{-\infty}^{\infty} \left( \tilde{A}(k) \exp(kx) + \tilde{B}(k) \exp(-kx) \right) \exp(-iky) dk$$

(III.7.25)

On the sides of the slab we may apply the boundary conditions $f(0, y) = g_A(y)$ and $f(x_1, y) = g_B(y)$ (note that we require $g_A(y) \to 0$ and $g_B(y) \to 0$, when $|y| \to \infty$). Then, the coefficients $\tilde{A}(k)$ and $\tilde{B}(k)$ are determined through the Fourier transforms

$$\tilde{A}(k) + \tilde{B}(k) = \int_{-\infty}^{\infty} g_A(y) \exp(iky) dy$$

$$\tilde{A}(k) \exp(kx_1) + \tilde{B}(k) \exp(-kx_1) = \int_{-\infty}^{\infty} g_B(y) \exp(iky) dy$$

(III.7.26)

In the next example, we'll show one way in which we can relax the requirements $g_1(y) \to 0$ and $g_2(y) \to 0$, when $|y| \to \infty$ for an infinite slab.

**Example III.7.2 (Physical Application ▶▶).** A classic example where Laplace's equation occurs is the steady-state of heat-flow. Here, steady-state means that the temperature $T$, however distributed in space, remains constant with time. $T$ satisfies a 2D Laplace equation, if we suppose that the temperature is uniform in the $z$-direction (see Fig. III.7.4). Namely, we have that

$$\nabla^2 T(x, y) = 0 \qquad (III.7.27)$$

Let's consider a wall infinite in the $y$-direction, so that $-\infty < y < \infty$, and the slab thickness is $x_1$, so that in the $x$-direction it is confined: $0 < x < x_1$. Suppose that along the front side of the wall, at $x = 0$, we can maintain a temperature that can be approximated by $T_1(y) = T_0 + \alpha \exp(-\lambda y^2)$. Here, we might imagine that for $x < 0$ (outside of the wall) we have a large volume that, for the most part, stays constant at $T_0$. But the front side has been heated for a significant time by a long heating-element (or some other continuous heat

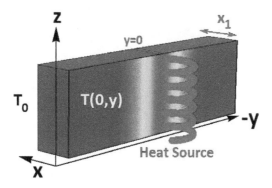

**Fig. III.7.4** The boundary problem considered in Example III.7.2. Here, we suppose that there is a heat source on one side of a wall, and we have a steady state solution to the heat flow equation (Laplace's equation). We suppose that we can model the temperature on one side of an infinite wall of thickness $x_1$, as $T(0, y) = T_0 + \alpha \exp(-\lambda y^2)$ and on the other side it is uniform temperature $T_0$. Note that here (for convenience in sketching) we have reversed the $y$-axis, so to be in the $y$-direction.

source) lying outside, parallel to the $z$-axis. On the other side, at $x = x_1$, we have $T_2(y) = T_0$, which assumes that is there is another large volume for $x > x_1$ that also stays at a roughly constant temperature. For the steady-state solution to work we must have relatively high thermal conductivity in the outside regions compared with the inside of the wall. We illustrate the situation in Fig III.7.4.

Thus, the boundary conditions on Eq. (III.7.27) are $T(0, y) = T_0 + \alpha \exp(-\lambda y^2)$, $T(x_1, y) = T_0$ and we'll require that $T(x, y) - T_0 \to 0$ when $|y| \to \infty$. We can use Eq. (III.7.25) to write the general solution as

$$T(x, y) = T_0 + \frac{1}{2\pi} \int_{-\infty}^{\infty} \left( \tilde{A}(k) \exp(kx) + \tilde{B}(k) \exp(-kx) \right)$$
$$\times \exp(-iky) dk \tag{III.7.28}$$

Now, the boundary conditions on either side of the wall yield

$$\alpha \exp(-\lambda y^2) = \frac{1}{2\pi} \int_{-\infty}^{\infty} \left( \tilde{A}(k) + \tilde{B}(k) \right) \exp(-iky) dk \tag{III.7.29}$$

$$0 = \frac{1}{2\pi} \int_{-\infty}^{\infty} \left( \tilde{A}(k) \exp(kx_1) + \tilde{B}(k) \exp(-kx_1) \right) \exp(iky) dk$$

$$(\text{III.7.30})$$

From Eq. (III.7.30), we require that for non-zero $\tilde{A}(k)$ and $\tilde{B}(k)$

$$\tilde{B}(k) = -\tilde{A}(k) \exp(2kx_1) \qquad (\text{III.7.31})$$

Through the Fourier transform of Eq. (III.7.29), we have that

$$\alpha \int_{-\infty}^{\infty} \exp(-\lambda y^2) \exp(iky) dy = \tilde{A}(k) + \tilde{B}(k)$$

$$= \tilde{A}(k) \left( 1 - \exp(2kx_1) \right) \qquad (\text{III.7.32})$$

Thus to find $\tilde{A}(k)$ and $\tilde{B}(k)$, we need to perform the integral in Eq. (III.7.32):

$$I(k) = \alpha \int_{-\infty}^{\infty} \exp(-\lambda y^2) \exp(iky) dy$$

$$= \alpha \int_{-\infty}^{\infty} \exp \left( -\lambda(y + ik/2\lambda)^2 \right) \exp(-k^2/4\lambda) dy$$

$$\Rightarrow I(k) = \alpha \exp(-k^2/4\lambda) \int_{-\infty}^{\infty} \exp(-\lambda y^2) dy \qquad (\text{III.7.33})$$

In the last step in Eq. (III.7.33) we've assumed that we can integrate along the real axis when we have made a change of variables which is a complex number. Though it's true for Gaussian functions, *it is not generally true for any function*. We'll justify this procedure, and learn more about when we can use it, when we come to integration in the complex plane in the next chapter. But here we can evaluate the remaining integral on the bottom line of Eq. (III.7.33) to be

$$\int_{-\infty}^{\infty} \exp(-\lambda y^2) dy = \frac{1}{\sqrt{\lambda}} \int_{-\infty}^{\infty} \exp(-\lambda y^2) d(\sqrt{\lambda} y)$$

$$= \frac{1}{\sqrt{\lambda}} \underbrace{\int_{-\infty}^{\infty} \exp(-t^2) dt}_{\sqrt{\pi}} = \sqrt{\frac{\pi}{\lambda}}$$

(we have discussed already **In Chapter 1 of this Part (Example III.1.11)** how the result for such integral is obtained). Hence,

$$I(k) = \alpha \sqrt{\frac{\pi}{\lambda}} \exp(-k^2/4\lambda) \tag{III.7.34}$$

Then, we get

$$\tilde{A}(k) = \alpha \sqrt{\frac{\pi}{\lambda}} \frac{\exp(-k^2/4\lambda)}{(1 - \exp(2kx_1))} \quad \text{and}$$

$$\tilde{B}(k) = \alpha \sqrt{\frac{\pi}{\lambda}} \frac{\exp(2kx_1)\exp(-k^2/4\lambda)}{(1 - \exp(2kx_1))} \tag{III.7.35}$$

Thus, our full solution can be written as

$$T(x,y) = T_0 + \frac{\alpha}{2\sqrt{\pi\lambda}} \int_{-\infty}^{\infty} \frac{\exp(kx) - \exp(k(2x_1 - x))}{(1 - \exp(2kx_1))}$$

$$\times \exp(-k^2/4\lambda)\exp(iky)dk$$

$$\Rightarrow T(x,y) = T_0 + \frac{\alpha}{\sqrt{\pi\lambda}} \int_{0}^{\infty} \frac{\sinh(k(x_1 - x))}{\sinh(kx_1)} \exp\left(-k^2/4\lambda\right)\cos(ky)dk$$

$$\tag{III.7.36}$$

For a general case, we cannot get a closed form analytical expression for the integral in Eq. (III.7.36). But we can get an expression for it when the wall is very thin. (We can also get an expression when the wall is very thick, but this is rather complicated, involving error functions; so that we will consider only the thin wall case). But first let us illustrate the general solution.

Although it is not the best outcome, the integral in Eq. (III.7.36) already defines $T(x,y)$ as a function. We can numerically evaluate it for any values of $x$ and $y$. It useful to first rescale the integration variable $k \to k/x_1$ to reduce the number of parameters; a useful measure in any numerical calculations. In Fig. III.7.5, we plot $\Delta f(x,y) = (f(x,y) - T_0)/\alpha$ in terms of $x/x_1$ and $y/x_1$, for a few values of the dimensionless parameter $\gamma_R = \lambda x_1^2$.

Now let us enjoy considering the limiting case. If the wall is thin, $x_1$ and $x$ will be small. Because of the Gaussian appearing in the integrand in Eq. (III.7.36), only $k$-values for which $|k| < 2\sqrt{\lambda}$ contribute

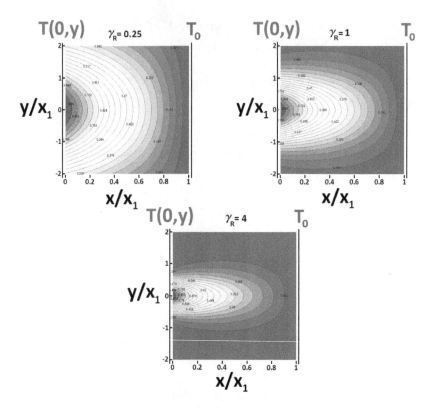

**Fig. III.7.5** A contour plot of $\Delta f(x, y) = (f(x, y) - T_0)/\alpha$, where $f(x, y)$ is the solution to Laplace's equation for the boundary-value problem specified in Example III.7.2, in terms of the rescaled lengths $x/x_1$ and $y/x_1$. The dimensionless parameter that controls the relative decay of $\Delta f(x, y)$ is $\gamma_R = \lambda x_1^2$. We plot $\Delta f(x, y)$ for three values of this parameter, $\gamma_R = 0.25$, $\gamma_R = 1$ and $\gamma_R = 4$. We see that as $\gamma_R$ increases, the region where $\Delta f(x, y)$ is sizable (shown in light blues, whites, yellows and reds) diminishes and the contour plot becomes more tear-dropped.

significantly to the integral. Then, we can approximate

$$\frac{\sinh(k(x_1 - x))}{\sinh(kx_1)} \approx \frac{x_{1'} - x}{x_1} \qquad \text{(III.7.37)}$$

provided that $x_1\sqrt{\lambda} \ll 1$. This allows us to write

$$T(x, y) \approx T_0 + \frac{\alpha}{2\sqrt{\pi\lambda}} \frac{x - x_1}{x_1} \int_{-\infty}^{\infty} \exp(-k^2/4\lambda) \exp(iky) dk$$

$$\Rightarrow T(x, y) \approx T_0 + \frac{x_1 - x}{x_1} \alpha \exp(-\lambda y^2) \qquad \text{(III.7.38)}$$

In obtaining the last line we have dealt with the same form of integral as we did in Eq. (III.7.33). Note that $T_0 + \alpha(x_1 - x)/x_1$ is the solution to Laplace's equation that we would have had, had we chosen boundary conditions $T(0, y) = T_0 + \alpha$ and $T(x_1, y) = T_0$ (you should check this as an exercise). This describes a *uniform temperature gradient* in the $x$-direction, when the wall is thin enough to keep the original temperature distribution $\alpha \exp(-\lambda y^2)$ in the $y$-direction unchanged. Thus, Eq. (III.7.38) makes sense. If we want corrections to Eq. (III.7.38), where the temperature distribution in the $y$-direction starts to change as we move along $x$, we need to expand out the sinh-functions in Eq. (III.7.37) to the next to the leading order.

**Exercise.** Find the next to leading order correction to Eq. (III.7.38).

---

***Optional: The rectangular boundary value problem in more depth.*** We start this more sophisticated discussion by noting that Eq. (III.7.10) is *not the complete general solution*. Indeed, we could have chosen $k^2$ to be negative (for imaginary $k$). To take account of both possibilities, we should write a general solution of the form

$$f(x, y) = f_1(x, y) + f_2(x, y) \tag{III.7.39}$$

$$f_1(x, y) = \sum_k \left( \tilde{A}_k \exp(kx) + \tilde{B}_k \exp(-kx) \right) \exp(iky) \tag{III.7.40}$$

$$f_2(x, y) = \sum_k \left( \tilde{C}_k \exp(ky) + \tilde{D}_k \exp(-ky) \right) \exp(ikx) \tag{III.7.41}$$

To fully determine all the coefficients, we now specify

$$f(0, y) = g_A(y) \quad f(x_1, y) = g_B(y) \tag{III.7.42}$$

$$f(x, 0) = h_A(x) \quad f(x, y_1) = h_B(x) \tag{III.7.43}$$

on all sides of the rectangle. Let's see how we can satisfy all these boundary conditions, with one concession to make things relatively simple: we'll suppose that we have at the corners $f(0, 0) = f(0, y_1) = f(x_1, 0) = f(x_1, y_1) = 0$. Now the trick is to suppose that $f_2(x, y)$ satisfies the boundary condition, Eq. (III.7.43), while $f_1(x, y)$ satisfies boundary condition Eq. (III.7.42). Then, to satisfy all these boundary conditions we choose that

$$f_2(0, y) = f_2(x_1, y) = 0 \quad f_1(x, 0) = f_1(x, y_1) = 0 \qquad \text{(III.7.44)}$$

From Eq. (III.7.44) it follows that

$$f_1(x, y) = \sum_{n=1}^{\infty} \left( \tilde{A}_n \exp(k_n^{(1)} x) + \tilde{B}_n \exp(-k_n^{(1)} x) \right) \sin(k_n^{(1)} y)$$

$$\text{(III.7.45)}$$

$$f_2(x, y) = \sum_{n=1}^{\infty} \left( \tilde{A}_n \exp(k_n^{(2)} y) + \tilde{B}_n \exp(-k_n^{(2)} y) \right) \sin(k_n^{(2)} x)$$

$$\text{(III.7.46)}$$

where

$$k_n^{(1)} = \frac{\pi n}{y_1} \quad \text{and} \quad k_n^{(2)} = \frac{\pi n}{y_2} \qquad \text{(III.7.47)}$$

The coefficients $\tilde{A}_n$, $\tilde{B}_n$, $\tilde{C}_k$ and $\tilde{D}_n$ are determined through the following Fourier series:

$$g_1(y) = \sum_{n=1}^{\infty} \left( \tilde{A}_n + \tilde{B}_n \right) \sin(k_n^{(1)} y)$$

$$g_2(y) = \sum_{n=1}^{\infty} (\tilde{A}_n \exp(k_n^{(1)} x_1) + \tilde{B}_n \exp(-k_n^{(1)} x_1)) \sin(k_n^{(1)} y)$$

$$\text{(III.7.48)}$$

$$h_1(x) = \sum_{n=1}^{\infty} \left( \tilde{C}_n + \tilde{D}_n \right) \sin(k_n^{(2)} x)$$

$$h_2(x) = \sum_{n=1}^{\infty} (\tilde{C}_n \exp(k_n^{(2)} y_1) + \tilde{D}_n \exp(-k_n^{(2)} y_1)) \sin(k_n^{(2)} x)$$

$$\text{(III.7.49)}$$

If we specified that $f(0, 0) \neq f(0, y_1) \neq f(x_1, 0) \neq f(x_1, y_1) \neq 0$, we would need to consider an additional solution $f_3(x, y) = E + Fx + Gy + Hxy$. This function, as well as $f_1(x, y)$ and $f_2(x, y)$, satisfies

Laplace's equation. The extra solution, $f_3(x, y)$, corresponds to the choice of $k = 0$ in Eq. (III.7.5). The constants $E$, $F$, $G$ and $H$ can be chosen to give different values of $f(x, y)$ at the corners of our rectangle to satisfy those specified boundary conditions. As the details in this case are more complicated and the solution gets more cumbersome, we'll refrain from doing this (energetic readers might want to examine this for themselves).

Note that if $h_A(x) = h_B(x) = g_A(y) = g_B(y) = 0$ the solution to the Laplace equation is, in fact, the trivial one $f(x, y) = 0$.

Generally, we are not restricted to defining at the boundaries just *values of the function*. One can also fix values of the partial derivatives, along the sides. For instance, we could replace one of the boundary conditions on our rectangle with

$$\left. \frac{\partial f(x, y)}{\partial y} \right|_{y=0} = h_1(x) \qquad \text{(III.7.50)}$$

We'll discuss such boundary condition choices, as in Eq. (III.7.50), in some later examples.

---

***Optional: The uniqueness of a solution to the 2D Laplace equation.*** In general, let us suppose that the domain, in which the solution to Laplace's equation $f(x, y)$ is defined, is bordered by the closed curve $\mathbf{r}(s) = x(s)\hat{\mathbf{i}} + y(s)\hat{\mathbf{j}}$, as shown in Fig. III.7.6.

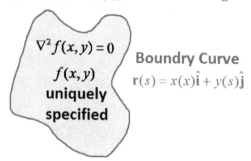

**Fig. III.7.6** If we can specify a boundary $\mathbf{r}(s) = x(s)\hat{\mathbf{i}} + y(s)\hat{\mathbf{j}}$ about an area such that we can fix $\tilde{f}(s) = f(x(s), y(s))$, then the solution to Laplace's equation $f(x, y)$ is uniquely defined.

If we know the boundary value function $\tilde{f}(s) = f(x(s), y(s))$ we can uniquely determine $f(x, y)$ (we won't prove this, but hopefully the discussion beforehand, based on common sense reasoning, will suggest to you that this is the case). In a way, this should not come as too much of a surprise, for the solutions to second order ordinary differential equations can be thought as being defined over a domain on a 1D line, where we could specify two points on that line to determine a unique solution. This suggests that the solution to the 2D Laplace equation—a second order partial differential equation defined over domain on a 2D surface spanned by $x$ and $y$—should indeed be uniquely determined by a closed curve on that surface.

Obviously, if the function $f(x, y)$ is smooth and continuous in the bounded region, $f(x, y) \equiv 0$ is the solution to $\nabla^2 f(x, y) = 0$ everywhere within bounded region, if at the boundary $\tilde{f}(s) = 0$.

### The 3D Laplace equation in Cartesian coordinates

Now, let's move on to the 3D. Let's first consider it in Cartesian coordinates

$$\vec{\nabla}^2 \phi(\mathbf{r}) = \frac{\partial^2 \phi(x, y, z)}{\partial x^2} + \frac{\partial \phi(x, y, z)}{\partial y^2} + \frac{\partial \phi(x, y, z)}{\partial z^2} = 0 \quad \text{(III.7.51)}$$

Again, we perform the standard routine of separation of variables by writing

$$\phi(\mathbf{r}) = X(x)Y(y)Z(z) \quad \text{(III.7.52)}$$

On substitution of Eq. (III.7.52) into Eq. (III.7.51) we obtain

$$\frac{1}{X(x)} \frac{d^2 X(x)}{dx^2} + \frac{1}{Y(y)} \frac{d^2 Y(y)}{dy^2} + \frac{1}{Z(z)} \frac{d^2 Z(z)}{dz^2} = 0 \quad \text{(III.7.53)}$$

To satisfy Eq. (III.7.53), for the three independent variables $(x, y, z)$, we require that

$$\frac{1}{X(x)}\frac{d^2X(x)}{dx^2} = p^2 \quad \frac{1}{Y(y)}\frac{d^2Y(y)}{dy^2} = q^2 \quad \frac{1}{Z(z)}\frac{d^2Z(z)}{dz^2} = r^2$$

$$\text{(III.7.54)}$$

and

$$p^2 + q^2 + r^2 = 0 \qquad \text{(III.7.55)}$$

Let's understand why this is so. Suppose we write $Z''(z)/Z(z) = r^2$, we can think of this relationship as defining a new 'variable' $r^2$. Then, we may define $\chi(x,y) = X''(x)/X(x) + Y''(y)/Y(y)$ and so Eq. (III.7.53) yields $-r^2 = \chi(x,y)$. This equation is simply a description of a surface where we are free to vary $x$ and $y$ to choose $-r^2$. Again, nothing stops us from writing $Y''(z)/Y(z) = q^2$, as a change of variables. Thus, substituting this into Eq. (III.7.53) requires that $X''(z)/X'(z) = p^2 = -r^2 - q^2$.

We can select three possibilities that can satisfy real boundary conditions (chosen so that we can define 2D Fourier series at each of the faces of the cuboid region that we would want to solve Eq. (III.7.51) within). These are

$$1. \ k_x^2 + k_y^2 = r^2 > 0, -k_x^2 = p^2 < 0 \text{ and } -k_y^2 = q^2 < 0$$

$$2. \ k_y^2 + k_z^2 = p^2 > 0, -k_y^2 = q^2 < 0 \text{ and } -k_z^2 = r^2 < 0$$

$$3. \ k_x^2 + k_z^2 = q^2 > 0, -k_x^2 = p^2 < 0 \text{ and } -k_z^2 = r^2 < 0$$

$$\text{(III.7.56)}$$

For simplicity, here we'll consider only the first of these possibilities with the relevant boundary conditions; an extended discussion is presented as optional material in a subsequent box.

### Simple cuboid boundary condition

We first want to solve Eq. (III.7.51) in a cuboid region specified by $0 \le x \le x_1$, $0 \le y \le y_1$ and $0 \le z \le z_1$, as illustrated in Fig. III.7.7.

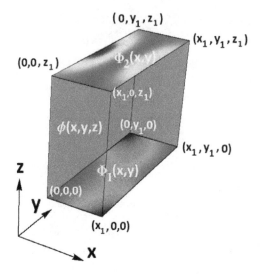

**Fig. III.7.7**  An illustration of boundary conditions for the cuboid region defined for $0 \leq x \leq x_1, 0 \leq y \leq y_1$ and $0 \leq z \leq z_1$. On two of the faces we show density plots for some arbitrary choice of the boundary functions which have been selected for illustrative purposes (red for high values and blue for low values). The boundary functions $\Phi_1(x,y) = \phi(x,y,0)$ and $\Phi_2(x,y) = \phi(x,y,z_1)$ can be, in fact, quite general. For the simple cuboid boundary conditions discussed in the text the function $\phi(x,y,z)$ is set equal to zero on the rest of the faces (see Eq. III.7.57).

We'll start by considering the simplest case with the boundary conditions on four sides of the cuboid (a more general case is presented in the box below):

$$\phi(0,y,z) = \phi(x_1,y,z) = \phi(x,0,z) = \phi(x,y_1,z) = 0 \qquad \text{(III.7.57)}$$

These conditions can be enforced through $X(0) = X(x_1) = 0$ and $Y(0) = Y(y_1) = 0$. Then, then solutions to Eq. (III.7.54) with $k_x^2 + k_y^2 = r^2$, $-k_x^2 = p^2$ and $-k_y^2 = q^2$, that satisfy these conditions are

$$X(x) = C \sin(k_n x) \quad Y(y) = D \sin(k_m y)$$

$$Z(z) = A \exp\left(z\sqrt{k_n^2 + k_m^2}\right) + B \exp\left(-z\sqrt{k_n^2 + k_m^2}\right)$$

$$\text{(III.7.58)}$$

where we have had to discretize $k_x$ and $k_y$ so that $k_x = k_n = \pi n/x_1$ and $k_y = k_m = \pi m/y_2$ for $X(x)$ and $Y(y)$ to satisfy the conditions

$X(0) = X(x_1) = 0$ and $Y(0) = Y(y_1) = 0$. Again, a sum of two solutions of the form of Eqs. (III.7.52) and (III.7.58), each term in the series with different values of $n$ and $m$, can satisfy Eq. (III.7.51). The upshot is that we should write a 'general' solution of the form

$$\phi(x, y, z) = \sum_{n=1}^{\infty} \sum_{m=1}^{\infty} \left( A_{n,m} \exp\left( z\sqrt{k_n^2 + k_m^2} \right) \right.$$

$$\left. + B_{n,m} \exp\left( -z\sqrt{k_n^2 + k_m^2} \right) \right) \sin(k_n x) \sin(k_m y)$$

$$(\text{III.7.59})$$

To determine the constants $A_{n,m}$ and $B_{n,m}$ we need to specify functions $\Phi_1(x, y)$ and $\Phi_2(x, y)$ at the two other faces of the cuboid such that

$$\phi(x, y, 0) = \Phi_1(x, y), \quad \phi(x, y, z_1) = \Phi_2(x, y) \qquad (\text{III.7.60})$$

Eqs. (III.7.59) and (III.7.60) combined yield the 2D Fourier series

$$\Phi_1(x, y) = \sum_{n=1}^{\infty} \sum_{m=1}^{\infty} (A_{n,m} + B_{n,m}) \sin(k_n x) \sin(k_m y) \qquad (\text{III.7.61})$$

$$\Phi_2(x, y) = \sum_{n=1}^{\infty} \sum_{m=1}^{\infty} \left( A_{n,m} \exp\left( z_1 \sqrt{k_n^2 + k_m^2} \right) \right.$$

$$\left. + B_{n,m} \exp\left( -z_1 \sqrt{k_n^2 + k_m^2} \right) \right) \sin(k_n x) \sin(k_m y)$$

$$(\text{III.7.62})$$

Hence, we have

$$A_{n,m} + B_{n,m} = \frac{4}{x_1 y_1} \int_0^{x_1} dx \int_0^{y_1} dy \Phi_1(x, y) \sin(k_n x) \sin(k_m y)$$

$$(\text{III.7.63})$$

$$A_{n,m} \exp\left( z_1 \sqrt{k_n^2 + k_m^2} \right) + B_{n,m} \exp\left( -z_1 \sqrt{k_n^2 + k_m^2} \right)$$

$$= \frac{4}{x_1 y_1} \int_0^{x_1} dx \int_0^{y_1} dy \Phi_2(x, y) \sin(k_n x) \sin(k_m y) \qquad (\text{III.7.64})$$

Thus, we have arrived at a system of two linear equations on $A_{n,m}$ and $B_{n,m}$. Solving them, we obtain

$$A_{n,m} = \frac{2}{x_1 y_1 \sinh\left(z_1\sqrt{k_n^2 + k_m^2}\right)} \int_0^{x_1} dx \int_0^{y_1} dy$$

$$\times \left[\Phi_2(x,y) - \exp\left(-z_1\sqrt{k_n^2 + k_m^2}\right)\Phi_1(x,y)\right]\sin(k_n x)\sin(k_m y)$$

$$\text{(III.7.65)}$$

$$B_{n,m} = \frac{2}{x_1 y_1 \sinh\left(z_1\sqrt{k_n^2 + k_m^2}\right)} \int_0^{x_1} dx \int_0^{y_1} dy$$

$$\times \left[\exp\left(z_1\sqrt{k_n^2 + k_m^2}\right)\Phi_1(x,y) - \Phi_2(x,y)\right]\sin(k_n x)\sin(k_m y)$$

$$\text{(III.7.66)}$$

**Exercise.** Suppose we have the cuboid boundary conditions

$$\phi(0,y,z) = \phi(x_1,y,z) = \phi(x,0,z) = \phi(x,y_1,z) = 0,$$
$$\phi(x,y,0) = \Phi_1(x,y) = 0$$

and

$$\phi(x,y,z_1) = \Phi_2(x,y) = ax(x-x_1)y(y-y_1)$$

Show that the solution to Laplace's equation (Eq. (III.7.51)) with these conditions can be written as the following Fourier series:

$$\phi(x,y,z) = a\left(\frac{2y_1^2}{\pi^3}\right)\left(\frac{2x_1^2}{\pi^3}\right)\sum_{n=1}^{\infty}\sum_{m=1}^{\infty}\frac{[(-1)^n - 1][(-1)^m - 1]}{m^3 n^3}$$

$$\times \frac{\sinh\left(z\sqrt{k_n^2 + k_m^2}\right)}{\sinh\left(z_1\sqrt{k_n^2 + k_m^2}\right)}\sin(k_n x)\sin(k_m y)$$

**Hint:** You can do this by using Eq. (III.7.65) and (III.7.66) with $\Phi_1(x,y) = 0$ and $\Phi_2(x,y) = ax(x-x_1)y(y-y_1)$. How you obtain the result is similar to Example III.7.1.

## Infinite wall boundary condition

Here, we'll also look at the boundary value problem for an infinite wall where $0 \leq x \leq x_1$ and $-\infty < y < \infty$ and $-\infty < z < \infty$, where we suppose that $\phi(x, y, z) \to 0$ when either $|y| \to \infty$ or $|z| \to \infty$. (**Think:** Which of the three cases in Eq. (III.7.56) does this correspond to?). At the two faces we specify $\phi(0, y, z) = \Phi_1(y, z)$ and $\phi(x_0, y, z) = \Phi_2(y, z)$ (note that both $\Phi_1(y, z) \to 0$ and $\Phi_2(y, z) \to 0$, when either $|y| \to \infty$ or $|z| \to \infty$). In this case the solution is written as (a 3D generalization from the 2D case)

$$\phi(x, y, z) = \frac{1}{2\pi} \int_{-\infty}^{\infty} dk_y \int_{-\infty}^{\infty} dk_z \left( \tilde{A}(k_y, k_z) \exp\left( x\sqrt{k_y^2 + k_z^2} \right) \right.$$
$$\left. + \tilde{B}(k_y, k_z) \exp\left( -x\sqrt{k_y^2 + k_z^2} \right) \right) \exp(-i(k_y y + k_z z))$$

$$\text{(III.7.67)}$$

where (show it yourself!)

$$\tilde{A}(k_y, k_z) = \frac{1}{2\sinh\left(x_0\sqrt{k_y^2 + k_z^2}\right)} \int_{-\infty}^{\infty} dy \int_{-\infty}^{\infty} dz$$
$$\times \left( \Phi_2(y, z) - \exp\left( -x_0\sqrt{k_y^2 + k_z^2} \right) \Phi_1(y, z) \right)$$
$$\times \exp(i(k_y y + k_z z))$$

$$\text{(III.7.68)}$$

$$\tilde{B}(k_y, k_z) = \frac{1}{2\sinh\left(x_0\sqrt{k_y^2 + k_z^2}\right)} \int_{-\infty}^{\infty} dy \int_{-\infty}^{\infty} dz$$
$$\times \left( \exp\left( x_0\sqrt{k_y^2 + k_z^2} \right) \Phi_1(y, z) - \Phi_2(y, z) \right)$$
$$\times \exp(i(k_y y + k_z z))$$

$$\text{(III.7.69)}$$

**Exercise.** Show that Eq. (III.7.67) satisfies Laplace's equation (Eq. (III.7.51)).

**Exercise.** Now suppose we return to our infinite wall in Example III.7.2 and replace our boundary conditions with $T(0, y, z) = T_0 + \alpha \exp(-\lambda(y^2 + z^2))$ and $T(x_1, y, z) = T_0$. Find an expression for the solution of the 3D Laplace equation as an integral over $k_y$ and $k_z$. **Hint:** You'll need to start with

$$T(x,y,z) = T_0 + \frac{1}{2\pi}\int_{-\infty}^{\infty} dk_y \int_{-\infty}^{\infty} dk_z \left(\tilde{A}(k_y,k_z)\exp\left(x\sqrt{k_y^2+k_z^2}\right)\right.$$

$$+\tilde{B}(k_y,k_z)\exp\left(-x\sqrt{k_y^2+k_z^2}\right)\right)\exp(-i(k_y y + k_z z))$$

$$\text{(III.7.70)}$$

and determine the coefficients through Eqs. (III.7.68) and (III.7.69); also look at Example III.7.2 to see how to do some of the integrals. Explore the limiting case when the wall is very thin.

---

*Optional: The cuboid boundary value problem in more depth.* Let's consider a more general solution to Laplace's equation $\phi(\mathbf{r})$ defined in a cuboid; $0 \le x \le x_1$, $0 \le y \le y_1$ and $0 \le z \le z_1$. One way to fully specify a unique solution to Laplace's equation would be to specify functions of two variables at each of the six faces, so that

$$\phi(x,y,0) = \Phi_1(x,y), \quad \phi(x,y,z_1) = \Phi_2(x,y), \quad \phi(x,0,z) = \Phi_3(x,z)$$

$$\phi(x,y_1,z) = \Phi_4(x,z), \quad \phi(0,y,z) = \Phi_5(y,z), \quad \phi(x_1,y,z) = \Phi_6(y,z)$$

$$\text{(III.7.71)}$$

To satisfy all of these conditions (in Eq. (III.7.71)) we'll need to use all three possibilities in Eq. (III.7.71). Thus, we need to write a general solution as

$$\phi(x,y,z) = \phi_1(x,y,z) + \phi_2(x,y,z) + \phi_3(x,y,z) \qquad \text{(III.7.72)}$$

with

$$\phi_1(x,y,z) = \sum_{n=1}^{\infty}\sum_{m=1}^{\infty}\left(A_{n,m}\exp\left(z\sqrt{k_n^2+k_m^2}\right)\right.$$

$$+ B_{n,m}\exp\left(-z\sqrt{k_n^2+k_m^2}\right)\right)\sin(k_n x)\sin(k_m y)$$

$$\text{(III.7.73)}$$

$$\phi_2(x, y, z) = \sum_{n=1}^{\infty} \sum_{l=1}^{\infty} \left( C_{n,l} \exp\left( y\sqrt{k_n^2 + k_l^2} \right) \right.$$

$$\left. + D_{n,l} \exp\left( -y\sqrt{k_n^2 + k_l^2} \right) \right) \sin(k_n x) \sin(k_l z)$$

$$\text{(III.7.74)}$$

$$\phi_3(x, y, z) = \sum_{m=1}^{\infty} \sum_{l=1}^{\infty} \left( E_{m,l} \exp\left( x\sqrt{k_m^2 + k_l^2} \right) \right.$$

$$\left. + F_{m,l} \exp\left( -x\sqrt{k_m^2 + k_l^2} \right) \right) \sin(k_m y) \sin(k_l z)$$

$$\text{(III.7.75)}$$

where we have assumed that at the edges of the cuboid $\phi(x, y, z) = 0$. The six sets of coefficients, $A_{n,m}$, $B_{n,m}$, $C_{n,l}$, $D_{n,l}$, $E_{m,l}$, and $F_{m,l}$ can then be determined by employing 2D Fourier series at each of the faces of the cuboid, using Eq. (III.7.71). For instance, to determine $C_{n,l}$ and $D_{n,l}$, we have that

$$\Phi_3(x, z) = \sum_{n=1}^{\infty} \sum_{l=1}^{\infty} (C_{n,l} + D_{n,l}) \sin(k_n x) \sin(k_l z) \qquad \text{(III.7.76)}$$

$$\Phi_4(x, z) = \sum_{n=1}^{\infty} \sum_{l=1}^{\infty} \left( C_{n,l} \exp\left( y_1 \sqrt{k_n^2 + k_l^2} \right) \right.$$

$$\left. + D_{n,l} \exp\left( -y_1 \sqrt{k_n^2 + k_l^2} \right) \right) \sin(k_n x) \sin(k_l z)$$

$$\text{(III.7.77)}$$

Thus, we can start to generalize the method of finding a unique solution to any boundary condition specified at the surface of a cuboid.

### Laplace's equation in cylindrical polar coordinates

We'll now consider the solution of Laplace's equation within a cylinder (or outside a cylinder, or between two coaxial cylinders). Here, we use cylindrical polar coordinates, such that $x = R\cos\phi$, $y = R\sin\phi$

and $z$. In these coordinates Laplace's equation reads as

$$\vec{\nabla}^2 \psi(R, \phi, z) = \frac{1}{R} \frac{\partial}{\partial R} \left( R \frac{\partial \psi(R, \phi, z)}{\partial R} \right)$$

$$+ \frac{1}{R^2} \frac{\partial^2 \psi(R, \phi, z)}{\partial \phi^2} + \frac{\partial^2 \psi(R, \phi, z)}{\partial z^2} = 0$$

(III.7.78)

where we'll deal with a region for which $0 \leq \phi < 2\pi$, so that we'll already require that $\psi(R, \phi, z) = \psi(R, \phi + 2\pi, z)$. Yet again, we can use separation of variables to write $\psi(R, \phi, z) = \rho(R)\Phi(\phi)Z(z)$ so that Eq. (III.7.78) becomes

$$\frac{1}{R\rho(R)} \frac{d}{dR} \left( R \frac{d\rho(R)}{dR} \right) + \frac{1}{R^2 \Phi(\phi)} \frac{d^2\Phi(\phi)}{d\phi^2} + \frac{1}{Z(z)} \frac{d^2 Z(z)}{dz^2} = 0$$

(III.7.79)

Now, using the same arguments as in the above sections, we can write

$$\frac{1}{Z(z)} \frac{d^2 Z(z)}{dz^2} = k^2$$

(III.7.80)

Note that, as usual, we have a choice for $k^2$ to be negative or positive so either

$$Z(z) = A \exp(ik_z z) + A^* \exp(-ik_z z) \text{ for } -k_z^2 = k^2 < 0$$

(III.7.81)

$$Z(z) = B + Cz \text{ for } k^2 = 0$$

(III.7.82)

or

$$Z(z) = D \exp(k_z z) + E \exp(-k_z z) \text{ for } k_z^2 = k^2 > 0$$

(III.7.83)

From Eq. (III.7.79) it then follows that

$$\frac{R}{\rho(R)} \frac{d}{dR} \left( R \frac{d\rho(R)}{dR} \right) + R^2 k^2 = -\frac{1}{\Phi(\phi)} \frac{d^2\Phi(\phi)}{d^2\phi} = n^2$$

(III.7.84)

Note that for $\Phi(\phi)$ we have the periodicity requirement $\Phi(\phi) = \Phi(\phi + 2\pi)$, so $n$ must be a real positive integer, such that

$$\Phi(\phi) = \exp(in\phi) = \exp(in(\phi + 2\pi))$$

(III.7.85)

Then, finally, from Eq. (III.7.84), we obtain

$$-\frac{d^2\rho(R)}{dR^2} - \frac{1}{R}\frac{d\rho(R)}{dR} + \frac{n^2\rho(R)}{R^2} = k^2\rho(R) \qquad \text{(III.7.86)}$$

This is precisely the eigenvalue equation (Eq. (III.4.125)) that we discussed in relation with Bessel functions in Chapter 4 of Part III, Vol. 2. Its general solutions are

$$\rho(R) = A_r J_n(k_z R) + B_r Y_n(k_z R) \quad k_z^2 = k^2 > 0 \quad \text{(III.7.87)}$$

$$\rho(R) = A_r R^n + B_r R^{-n} \qquad\qquad k^2 = 0 \qquad \text{(III.7.88)}$$

$$\rho(R) = A_r I_n(k_z R) + B_r K_n(k_z R) \quad -k_z^2 = k^2 < 0$$

$$\text{(III.7.89)}$$

## *Boundary conditions for Laplace's equation inside a cylinder*

Here, we'll consider the general solution inside a cylinder (where $0 \leq z \leq H$ and $0 \leq R \leq a$), as illustrated in Fig. III.7.8. In this case we require that the solutions *must not be singular* at $R = 0$, and so require $B_r = 0$ in Eqs. (III.7.87), (III.7.88) and (III.7.89). We should note that for a layer inside two coaxial regions generally, both

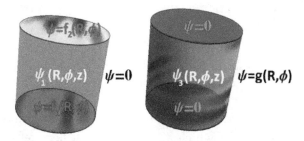

**Fig. III.7.8** A schematic illustration of two types of simple cylindrical boundary conditions. On the left, we set $\psi(a, \varphi, z) = 0$ at the surface of the cylinder of radius $a$ (shown as uniform blue) and at the top and the bottom of the cylinder we fix $\psi(R, \phi, 0) = f_1(R, \phi)$ and $\psi(R, \phi, H) = f_2(R, \phi)$, respectively (shown by the 'density plots' for an arbitrary choice of these functions). We label this solution on the left as $\psi(R, \phi, z) = \psi_1(R, \phi, z)$. On the right, we set $\psi = 0$ (shown again as uniform blue) at both ends of the cylinder ($z = 0$ and $z = H$) and we specify $\psi(b, \phi, z) = g(R, \phi)$ (illustrated by a density plot about the cylinder) at the cylinder surface. In such a case we label $\psi(R, \phi, z) = \psi_3(R, \phi, z)$.

coefficients will be non-zero; and outside a cylinder in infinite space, physical solutions will correspond to $A_r = 0$. However, as we have to discuss many other things, we won't consider these possibilities (we leave the reader to work these out, or refer to another textbook.

A full general solution, inside a cylinder, may contain three possible contributions, $\psi_1(R, \phi, z)$, $\psi_2(R, \phi, z)$ and $\psi_3(R, \phi, z)$ (corresponding to the choices $k^2 > 0$, $k^2 = 0$ and $k^2 < 0$), which are expressed as

$$\psi_1(R, \phi, z) = \sum_{m=1}^{\infty} \sum_{n=-\infty}^{\infty} (A_{n,m} \exp(\kappa_{m,n} z)$$

$$+ B_{n,m} \exp(-\kappa_{m,n} z)) J_n(\kappa_{m,n} R) \exp(in\phi)$$

(III.7.90)

$$\psi_2(R, \phi, z) = \sum_{n=-\infty}^{\infty} \exp(in\phi)(C_n + D_n z) R^{|n|} \qquad \text{(III.7.91)}$$

$$\psi_3(R, \phi, z) = \sum_{m=1}^{\infty} \sum_{n=-\infty}^{\infty} A_n I_n(k_m R) \exp(in\phi)$$

$$\times \sin(k_m z) \text{ where } k_m = 2\pi m/H \qquad \text{(III.7.92)}$$

To keep things simple, we'll consider separately the cases when the first or the second solution is applicable. If we have a boundary condition that $\psi(a, \phi, z) = 0$ or that

$$\left. \frac{\partial \psi(R, \phi, z)}{\partial R} \right|_{R=a} = 0 \qquad \text{(III.7.93)}$$

we only require $\psi(x, y, z) = \psi_1(x, y, z)$. For $\psi_1(a, \phi, z) = 0$, the values of $\kappa_{n.m}$ are determined through the zeros of the Bessel function such that $J_n(\kappa_{m,n} a) = 0$, for which $\kappa_{n.m} \neq 0$ (the $m$ here refers to which zero of the Bessel function we have). If Eq. (III.7.93) is satisfied, their values are determined through positions of the Bessel function maxima and minima, such that $J'_n(\kappa_{m,n} a) = 0$.

If we have the boundary condition $\psi(R, \phi, 0) = \psi(R, \phi, H) = 0$, we would require that $\psi(x, y, z) = \psi_3(x, y, z)$. More complicated boundary conditions may require combinations of $\psi_1(x, y, z)$, $\psi_2(R, \phi, z)$

and $\psi_3(R, \phi, z)$, in a similar fashion to what was discussed for generalized cuboid boundary conditions in one of the optional boxes. For instance, if we wanted to fix, strictly, a non-zero value at the two edges of the cylinder we would use $\psi_2(R, \phi, z)$. However, note that the list of functions given by Eqs. (III.7.90), (III.7.91) and (III.7.92) is not exhaustive, although it covers a wide variety of situations. One particular case where we would need something different is if we were to require that both $z$-partial derivatives of $\psi(x, y, z)$ were zero at both ends of the cylinder. In this case, we should replace $\sin(k_m z)$, in the sum in Eq. (III.7.92), with $\cos(k_m z)$ (note that this would also be the case for the cuboid) and use this as our solution. Let's now look at an example of a simple boundary value problem for cylinders.

**Example III.7.3 (Physical Application ▶▶).** Here, we'll consider the steady-state diffusion equation, for the concentration $c(\mathbf{r})$ of a molecule dissolved in solution, in a cylindrical pipe. Let's first understand what the steady-state diffusion equation is. Fick's law states (we introduced the 1D case of this law in Chapter 10, Part I, Vol. 1) that the current density (the number of such molecules diffusing per unit area per unit time) is given by

$$\mathbf{J}(\mathbf{r}) = -D\vec{\nabla}c(\mathbf{r}) \qquad \text{(III.7.94)}$$

where $D$ is the diffusion constant (which determines the ability of species to diffusion). In Chapter 3 of Part III, Vol. 2, we discussed the continuity equation in relation to the divergence of a vector field, Eq. (III.3.10). From Eq. (III.3.10), we see that if the concentration is not changing with respect to time, i.e. is in a steady state, we require $\vec{\nabla}.\mathbf{J}(\mathbf{r}) = 0$. Thus, the concentration, as a function of position, must satisfy Laplace's equation such that

$$\vec{\nabla}^2 c(\mathbf{r}) = 0 \qquad \text{(III.7.95)}$$

Now, let us suppose that we have an impenetrable cylindrical tube of radius $a$, our pipe, connected to two reservoirs, as depicted in Fig. III.7.9. At one end of the tube we'll suppose that we know the concentration to be $c(R, \phi, H) = c^{(0)}(R) = \beta \exp(-(R/a)^2)$ and at

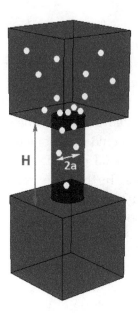

**Fig. III.7.9** A schematic picture of the setup in which a pipe of diameter $2a$ and height $H$, open to two large reservoirs. Here, we want to find the concentration of a particular species of molecule in the pipe. At the top end of the pipe the concentration of molecules is supposed to be known, specified as $c(R, \phi, H) = c^{(0)}(R)$ where $c^{(0)}(R) = \beta \exp(-(R/a)^2)$. To assume a steady state solution the concentration at the bottom of the pipe is assumed to be close to zero (molecules quickly disappear in that reservoir, or there is a 'demon' killing them as its entrance).

the other end the concentration is effectively $c(R, \phi, 0) = 0$, thus assuming that the molecules are able to quickly diffuse out into the second reservoir.

To see what other boundary conditions we need, let's consider the form of the diffusion current in cylindrical coordinates, in which we may write the gradient operator as

$$\vec{\nabla} = \hat{\mathbf{R}} \frac{\partial}{\partial R} + \frac{\hat{\boldsymbol{\phi}}}{R} \frac{\partial}{\partial \phi} + \hat{\mathbf{z}} \frac{\partial}{\partial z} \qquad \text{(III.7.96)}$$

where $\hat{\mathbf{R}}$ is a radial unit vector pointing away from the axis of the cylinder and $\hat{\mathbf{z}}$ points in the direction of the $z$-axis. The unit vectors are all perpendicular to each other (orthogonal) such that $\hat{\mathbf{R}} \cdot \hat{\boldsymbol{\phi}} = \hat{\mathbf{z}} \cdot \hat{\boldsymbol{\phi}} = \hat{\mathbf{R}} \cdot \hat{\mathbf{z}} = 0$. From Fick's law, the component of the current flowing in the radial direction (through a cylinder of radius $R$ in the tube) is

$$J_R(R) = \hat{\mathbf{R}}.\mathbf{J}(R) = -D\frac{\partial c(\mathbf{r})}{\partial R} \tag{III.7.97}$$

Since the tube is impenetrable, no current can flow out of it through the cylinder wall—that is, the current in the radial direction at $R = a$ must be zero. Thus, we require the boundary condition

$$\left.\frac{\partial c(R, \phi, z)}{\partial R}\right|_{R=a} = 0 \tag{III.7.98}$$

Now, Eq. (III.7.98) suggests that we should use the solution given by Eq. (III.7.90) to describe the concentration flowing through the pipe, so we write

$$c(R, \phi, z) = \sum_{m=0}^{\infty} \sum_{n=-\infty}^{\infty} (A_{n,m} \exp(\kappa_{m,n} z) + B_{n,m} \exp(-\kappa_{m,n} z))$$
$$\times J_n(\kappa_{m,n} R) \exp(in\phi) \tag{III.7.99}$$

where we require that $J_n'(\kappa_{m,n} a) = 0$, where for convenience count $m$ from 0 to $\infty$. Now, since there is no $\phi$-dependence in either of the two boundary conditions at the bottom and top ends of the pipe ($z = 0$ and $z = H$) we need to retain only the $n = 0$ terms in the series, Eq. (III.7.99). So we get:

$$c(R, \phi, z) = \sum_{m=0}^{\infty} (A_{0,m} \exp(\kappa_{m,0} z) + B_{0,m} \exp(-\kappa_{m,0} z)) J_0(\kappa_{m,0} R) \tag{III.7.100}$$

The boundary conditions at the bottom and the top of the pipe, require, respectively

$$0 = c(R, \phi, 0) = \sum_{m=0}^{\infty} (A_{0,m} + B_{0,m}) J_0(\kappa_{m,0} R) \tag{III.7.101}$$

$$\beta \exp(-(R/a)^2) = c(R, \phi, H) = \sum_{m=0}^{\infty} (A_{0,m} \exp(\kappa_{m,0} H)$$
$$+ B_{0,m} \exp(-\kappa_{m,0} H)) J_0(\kappa_{m,0} R) \tag{III.7.102}$$

Both conditions are expressed as Bessel-Fourier series, which we came across in Chapter 4 of Part III, Vol. 2. From, Eq. (III.7.101) we require that $A_{0,m} = -B_{0,m}$, so that we have

$$\beta \exp(-(R/a)^2) = 2 \sum_{m=0}^{\infty} A_{0,m} \sinh(\kappa_{m,0}H) J_0(\kappa_{m,0}R) \quad \text{(III.7.103)}$$

Now in this case where $J_0'(\kappa_{m,0}a) = 0$, the Bessel functions satisfy an orthogonality relationship, which reads slightly differently from the one for $J_0(\kappa_{m,0}a) = 0$ given by Eq. (III.4.150) in Chapter 4 of Part III, Vol. 2, namely–

$$\int_0^a RJ_0(\kappa_{m,0}R)J_0(\kappa_{m',0}R) = \frac{\delta_{m,m'}a^2}{2} J_0(\kappa_{m,0}a)^2 \quad \text{(III.7.104)}$$

In this case, the value of the integral when $m = m'$ is found through Eq. (III.4.151) of Chapter 4 of Part III, Vol. 2, with $J_0'(\kappa_{m,0}a) = 0$. [due to the properties of the Bessel functions (check handbooks or look at their definitions in this book) $J_1 = -J' = J_{-1}$, and thus both $J_1(k_{m,0}a)$ and $J_1(k_{m,0}a)$ appearing in the r.h.s of Eq. (III.7.104) must vanish]

Equation (III.7.104) allows us to compute the coefficients $A_{0,m}$ through the integral

$$I_m = \frac{1}{a^2} \int_0^a R \exp\left(-(R/a)^2\right) J_0(\kappa_{m,0}R)dR$$

$$= \int_0^1 s \exp(-s^2) J_0(x_{m,0}s)ds \quad \text{(III.7.105)}$$

A relationship between $I_m$ and $A_{0,m}$ can be obtained by multiplying both sides of Eq. (III.7.103) by $RJ_0(\kappa_{m,0}R)$ and integrating with respect to $R$ from 0 to $a$. Then we may exploit Eq. (III.7.104) to yield

$$\beta a^2 I_m = A_{0,m}a^2 \sinh(x_{m,0}H/a)J_0(x_{m,0})^2$$

$$\Rightarrow A_{0,m} = \frac{\beta I_m}{\sinh(x_{m,0}H/a)J_0(x_{m,0})^2} \quad \text{(III.7.106)}$$

The positions of the maxima and minima, $x_{m,0}$, integrals $I_m$, and $\tilde{A}_{0,m} = \frac{1}{\beta}A_{0,m} \sinh\left(\frac{x_{m,0}H}{a}\right)$ are simply numbers and their first five values are given in Table III.7.1.

**Table III.7.1** In the left column we show the values of the first five turning points (maxima and minima) of $J_0(x)$, $x_{m,0}$. The middle column displays the corresponding values of the integral $I_m$ (defined by Eq. (III.7.105)) calculated for the value of $x_{m,0}$, and the right column shows the coefficients $\tilde{A}_{0,m}$ (defined in the text) that appear in the final result, Eq. (III.7.107).

| $x_{m,0}$ | $I_m$ | $\tilde{A}_{0,m}$ |
|---|---|---|
| 0 | 0.3161 | 0.3161 |
| 3.8317 | 0.03565 | 0.2198 |
| 7.0156 | -0.004867 | -0.05404 |
| 10.1735 | 0.001846 | 0.02961 |
| 13.3237 | -0.0009259 | -0.01948 |

For those who have doubts, we show in Fig. III.7.11 how well the Bessel-Fourier series of the function $\beta \exp\{-(R/a)^2\}$, Eq. (III.7.103), reproduces its exact form with just those first five terms!

But accounting for all the terms, we can write the particular solution of this boundary value problem as

$$c(R, \phi, z) \simeq \frac{0.632\beta z}{H} + 2\beta \sum_{m=1}^{\infty} \tilde{A}_{0,m} \frac{\sinh(x_{m,0}z/a)}{\sinh(x_{m,0}H/a)} J_0(x_{m,0}R/a)$$

$$(\text{III.7.107})$$

Note that that the $m = 0$ term simplifies to the first term of Eq. (III.7.107), because $2\tilde{A}_{0,0} = 0.632$, so we have written explicitly as the first term; in writing it we also used that

$$\lim_{k \to 0} \frac{\sinh(kz)}{\sinh(kH)} = \frac{z}{H} \qquad (\text{III.7.108})$$

We plot the solution in Fig III.7.10 along a cross section (taken to be at $y = 0$ so that $R = |x|$) of the pipe by considering the first five terms of the series, in terms of the rescaled lengths $x/a$ and $z/H$ for three different values of the parameter $\Gamma = H/a$ that controls the aspect ratio of the pipe. This is a reasonable approximation to the full series solution, as five terms represent well $\beta \exp\{-(R/a)^2\}$, (see Fig. III.7.11).

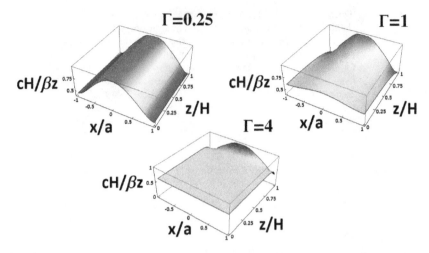

**Fig. III.7.10**   We show plots of $c(R, \phi, z)H/\beta z$ (retaining only terms up to $m = 4$ in the formula for the concentration Eq. (III.7.105)) along the tube for a cross-section taken in the $x$-$z$ plane. For the particular boundary conditions, the concentration does not depend on the polar angle $\phi$, but only on the distance $R$ away from the $z$-axis and the height $z$. Thus, the concentration profile in the $y$-$z$ plane is the same as in the $x$-$z$ plane (shown here), or any other plane bisected by the $z$-axis. Note that in all of these concentrations, the gradient in the $x$-direction is zero at the wall of the pipe ($R = a$), to satisfy the impermeable boundary condition (Eq. (III.7.96)). The three plots shown correspond to different values of parameter $\Gamma = H/a$ that controls the relative dimensions of the pipe, its aspect ratio. When $\Gamma = 0.25$, so that the pipe is wider than it is long, the concentration significantly varies as a function of $R$ until well inside the pipe along the $z$-axis, with a minimum value at the pipe wall (see top left). Also, as we increase $\Gamma$ (see plots for $\Gamma = 1$ and $\Gamma = 4$ — top right and bottom respectively) the relative proportion of the pipe over which the concentration stays roughly constant with respect to $R$, increases.

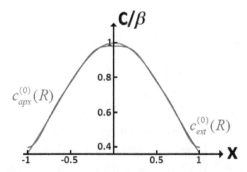

**Fig. III.7.11**   We compare the exact boundary condition function $c_{ext}^{(0)}(R) = \beta \exp(-(R/a)^2)$ with an approximate function $c_{apx}^{(0)}(R)$ where we retain terms in the series given in Eq. (III.7.101) up to $m = 4$. This truncation of the number of terms has been used in approximating the full concentration profile in Fig. III.7.10.

From Eq. (III.7.107) we can explore the most interesting limiting case. When the pipe is very long such that $H \gg a$ and when $z < H$, we get

$$c(R, \phi, z) \approx \frac{0.632\beta z}{H} \text{ and } \mathbf{J}(R, \phi, z) \approx -\frac{0.632\beta}{H}\hat{\mathbf{z}} \qquad \text{(III.7.109)}$$

This result makes sense perfect sense: it describes a uniform current flowing from larger values of $z$ to smaller values along the direction of the concentration gradient. The trend towards this limiting case is clearly seen in Fig. III.7.10.

## Laplace's equation in spherical polar coordinates

Last of all, before we go on to consider other partial differential equations (PDEs), we will look at the 3D Laplace equation in spherical polar coordinates:

$$\frac{1}{r^2}\frac{\partial}{\partial r}\left(r^2 \frac{\partial f(r, \theta, \phi)}{\partial r}\right) + \frac{1}{r^2 \sin^2 \theta}\frac{\partial^2 f(r, \theta, \phi)}{\partial \phi^2}$$

$$+ \frac{1}{r^2 \sin \theta}\frac{\partial}{\partial \theta}\left(\sin \theta \frac{\partial f(r, \theta, \phi)}{\partial \theta}\right) = 0 \qquad \text{(III.7.110)}$$

For simplicity, we'll define Eq. (III.7.110) inside a sphere or shell, also with the requirement that $f(r, \theta, \phi)$ is never singular at any point within the sphere. Therefore, we must have that $f(r, \theta, \phi) = f(r, \theta, \phi+2\pi)$. What we do next? You know it: separation of variables! This produces

$$f(r, \theta, \phi) = \rho(r)\Theta(\theta)\Phi(\phi) \qquad \text{(III.7.111)}$$

so that we may write

$$\frac{\sin^2 \theta}{\rho(r)}\frac{d}{dr}\left(r^2 \frac{d\rho(r)}{dr}\right) + \frac{\sin \theta}{\Theta(\theta)}\frac{d}{d\theta}\left(\sin \theta \frac{d\Theta(\theta)}{d\theta}\right)$$

$$= n^2 = -\frac{1}{\Phi(\phi)}\frac{d^2\Phi(\phi)}{d\phi^2} \qquad \text{(III.7.112)}$$

where again we have the requirement that $n$ is integer (which comes from) $f(r, \theta, \phi) = f(r, \theta, \phi + 2\pi)$, and $\Phi(\phi) = \exp(in\phi)$. Then, From Eq. (III.7.112) we may write

$$\frac{1}{\rho(r)} \frac{d}{dr} \left( r^2 \frac{d\rho(r)}{dr} \right) = \lambda = \frac{n^2}{\sin^2 \theta} - \frac{1}{\Theta(\theta) \sin \theta} \frac{d}{d\theta} \left( \sin \theta \frac{d\Theta(\theta)}{d\theta} \right)$$

(III.7.113)

Thus, we have the remaining equations

$$\frac{d}{dr} \left( r^2 \frac{d\rho(r)}{dr} \right) = \lambda \rho(r) - \frac{1}{\sin \theta} \frac{d}{d\theta} \left( \sin \theta \frac{d\Theta(\theta)}{d\theta} \right)$$

$$+ \frac{n^2 \Theta(\theta)}{\sin^2 \theta} = \lambda \Theta(\theta)$$

(III.7.114)

Now, on the r.h.s of Eq. (III.7.114), we have the Legendre equation. As we found out in Chapter 4, we require $\lambda = l(l+1)$ where $l$ is an integer, for $\Theta(\theta)$ to remain finite, and the Legendre equation is then solved by $\Theta(\theta) = P_l^m(\cos \theta)$, the associated Legendre functions. The equation on the l.h.s of Eq. (III.7.114) can be written as

$$r^2 \frac{d^2 \rho(r)}{dr^2} + 2r \frac{\partial \rho(r)}{\partial r} = l(l+1)\rho(r)$$

(III.7.115)

This equation can simply be solved by the substitution $\rho(r) = r^\gamma$ which yields

$$\gamma(\gamma - 1) + 2\gamma = l(l+1)$$

(III.7.116)

Thus, we can have either $\gamma = l$ or $\gamma = -1 - l$. Hence, for $\rho(r)$ we have the general solution to Eq. (III.7.115) of the form

$$\rho(r) = A_r r^l + B_r r^{-l-1}$$

(III.7.117)

### Boundary value problem inside a sphere

Here, we'll consider $f(r, \theta, \phi)$ as defined inside a sphere centred at the origin, as illustrated in Fig. III.7.12. Since we require the solution to be finite at $r = 0$, must set $B_r = 0$ in Eq. (III.7.117). Combining the

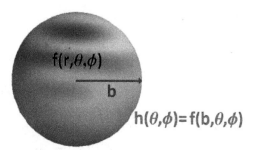

**Fig. III.7.12** This is not a picture of a gas-giant planet Jupiter, but a diagram illustrating a possible spherical boundary condition for a function $f(r, \theta, z)$ satisfying Laplace's equation, defined within the sphere. Here, we fix $h(\theta, \phi) = f(b, \theta, \phi)$ at the surface of the sphere (the darkness represent a density plot of an arbitrary choice of $h(\theta, \phi)$). We could also fix a partial derivative of $f(r, \theta, z)\theta$ at the sphere surface. This is considered in Example III.7.4.)

solutions $\rho(r) = r^l$, $\Theta(\theta) = P_l^m(\cos\theta)$ and $\Phi(\phi) = \exp(in\phi)$ together we can construct a 'general' solution

$$f(r, \theta, \phi) = \sum_{l=0}^{\infty} \sum_{|m| \leq l} A_{l,m} Y_l^m(\theta, \phi) r^l \tag{III.7.118}$$

Here, the $Y_l^m(\theta, \phi)$ are the spherical harmonics, which are given by

$$Y_l^m(\theta, \phi) = \left[ \frac{2l + 1}{4\pi} \frac{(l - m)!}{(l + m)!} \right]^{1/2} (-1)^m \exp(im\phi) P_l^m(\cos\theta) \tag{III.7.119}$$

We can determine the coefficients $A_{l,m}$ through a boundary condition at the surface of the sphere (of radius $b$),

$$f(b, \theta, \phi) = h(\theta, \phi) = \sum_{l=0}^{\infty} \sum_{|m| \leq l} A_{l,m} Y_l^m(\theta, \phi) b^l \tag{III.7.120}$$

or alternatively we can impose a boundary condition on the partial derivatives at the surface of the sphere, as we'll consider in the next example. The spherical harmonics satisfy the orthogonality relation

$$\int_0^{2\pi} d\phi \int_0^{\pi} \sin\theta d\theta Y_l^{*m}(\theta,\phi) Y_{l'}^{m'}(\theta,\phi) = \delta_{l,l'}\delta_{m,m'} \qquad (\text{III.7.121})$$

Thus, we can compute the coefficients $A_{l,m}$ in the solution through the integral (spherical harmonic transform)

$$b^{-l} \int_0^{2\pi} d\phi \int_0^{\pi} \sin\theta d\theta h(\theta,\phi) Y_l^m(\theta,\phi) = A_{l,m} \qquad (\text{III.7.122})$$

**Example III.7.4 (Physical Application ▶▶).** Let us consider the temperature of a ball centred at the origin covered by an insulating band across its equator, as pictured in Fig III.7.13. We'll suppose that heat is flowing uniformly into the top of the sphere and heat is leaving uniformly from the bottom, and inside the ball the temperature has reached a steady state. We want to find the temperature inside the ball.

To do this, you need to know that the temperature flow of is governed by Fick's law. This states that the current density of the flow of temperature—the amount of heat energy divided by the specific heat

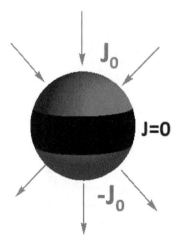

**Fig. III.7.13** A ball with a insulating band across it (shown in black). Here, we assume a uniform heat flow density of $J_0$ into the top of the sphere (shown in red) and a uniform heat flow out of the bottom of the sphere (shown in blue) of $J_0$ ($-J_0$ into the sphere).

capacity flowing per unit area per unit time (called, sometimes, a 'temperature current')—is given by

$$J(\mathbf{r}) = -D_T \vec{\nabla} T(\mathbf{r}) \tag{III.7.123}$$

where the coefficient $D_T$ is called the thermal diffusivity (for those interested in details of heat transfer: it is defined by the ratio of the thermal conductivity over the specific heat capacity per unit volume). Heat flow is governed by Fick's law because it is a diffusive-random process. In the steady state, the same amount of heat must be flowing into the ball as that which is leaving into the surroundings. This means that

$$J_s = D_T \left.\frac{\partial T}{\partial r}\right|_{r=b} \tag{III.7.124}$$

where $J_s$ is the heat-current density flowing (note the sign!) from the surroundings *into* the ball that itself has thermal diffusivity $D_T$.

To model our system we will assume that

$$J_s(\theta) = J_0 \qquad 0 \le \theta < \pi/2 - \alpha$$

$$J_s(\theta) = 0 \qquad \pi/2 - \alpha \le \theta \le \pi/2 + \alpha$$

$$J_s(\theta) = -J_0 \qquad \pi/2 + \alpha < \theta < \pi \tag{III.7.125}$$

where $2\alpha$ is the angular width of the isolating strip. Inside the ball $T$ satisfies the Laplace equation, $\vec{\nabla}^2 T = 0$ (for a steady state), so from differentiating the 'general' solution, Eq. (III.7.118), with respect to $r$, and using Eq. (III.7.124), we have

$$J_s(\theta) = \sum_{l=1}^{\infty} \sum_{|m| \le l} l A_{l,m} Y_l^m(\theta, \phi) b^{l-1} \tag{III.7.126}$$

But you would not expect any azimuthal dependence of the result on $\phi$. Do not worry, there will not be. In this case, the $A_{l,m}$ coefficients are given by the integral

$$b^{1-l} \int_0^{2\pi} d\phi \int_0^{\pi} \sin\theta d\theta J_s(\theta) Y_l^m(\theta, \phi) = l A_{l,m} \tag{III.7.127}$$

The integral on the l.h.s of Eq. (III.7.127) is non-zero only when $m = 0$ and hence, Eqs. (III.7.119) and (III.7.125) into Eq. (III.7.127), we have that

$$l^{-1}b^{1-l}\pi^{1/2}(2l+1)^{1/2}J_0 I_l(\alpha) = A_{l,0} \qquad \text{(III.7.128)}$$

where

$$I_l(\alpha) = \int_0^{\pi/2-\alpha} \sin\theta P_l(\cos\theta)d\theta - \int_{\pi/2+\alpha}^{\pi} \sin\theta P_l(\cos\theta)d\theta$$

$$\text{(III.7.129)}$$

Changing variables we can rewrite Eq. (III.7.129) as

$$I_l(\alpha) = \int_{\cos(\pi/2+\alpha)}^{1} P_l(z)dz - \int_{-1}^{\cos(\pi/2-\alpha)} P_l(z)dz \qquad \text{(III.7.130)}$$

To evaluate these integrals we need to look up the following result:

$$\int P_l(z)dz = \frac{1}{(2l+1)}[P_{l+1}(z) - P_{l-1}(z)] \qquad \text{(III.7.131)}$$

Thus, we can write

$$(2l+1)I_l(\alpha) \equiv G_l(\alpha) \equiv P_{l+1}(\cos(\pi/2-\alpha)) + P_{l+1}(\cos(\pi/2+\alpha))$$

$$- P_{l-1}(\cos(\pi/2-\alpha)) - P_{l-1}(\cos(\pi/2+\alpha))$$

$$\text{(III.7.132)}$$

where we have used the fact that $P_n(1) = 1$ and $P_n(-1) = (-1)^n$. The full solution is then expressed as

$$T(\theta,r) = T_0 + \sum_{l=1}^{\infty} \frac{rG_l(\alpha)}{2l}P_l(\cos\theta)\left(\frac{r}{b}\right)^{l-1} \qquad \text{(III.7.133)}$$

We plot $\Delta T(\theta,r) = (T(\theta,r) - T_0)/b$ as a function $r$ and $\theta$ in Fig. III.7.14, for a few values of $\alpha$.

**Exercise.** Find the solutions to Laplace's equation $\vec{\nabla}^2 T = 0$ with boundary condition Eq. (III.7.124), when $J_s = J_0(3\cos^2\theta - 2\cos\theta - 1)$ and $J_s = J_0(3\cos\theta\sin\theta + \sin\theta)\sin\phi$. **Hint:** You need to look at the forms of the spherical harmonics and try to write the $J_s$ in terms of them.

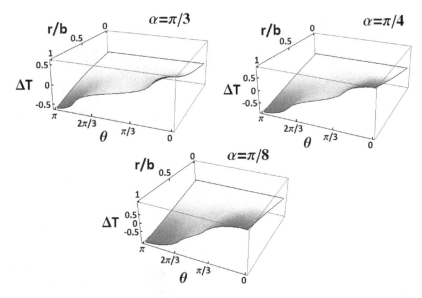

**Fig. III.7.14** We show plots of the function $\Delta T(\theta, r) = (T(\theta, r) - T_0)/b$ for three values of the parameter $\alpha$ (the angular width of the insulating band). Where the colour is red (and yellow) the temperature is hotter than $T_0$, the average temperature of the ball; and where the colour is blue, the temperature is cooler. In the 'hot' area, heat is entering into the sphere, whereas the cool region is where the heat is leaving. One can see that on the surface of the sphere (corresponding to $r/b = 1$) the radial derivative is zero within the insulating strip; in the blue area it is negative (a condition for the heat to flow out), and in the red area positive (a condition for heat to flow in). At the centre of the sphere (corresponding to $r/b = 0$), the temperature is always at $T_0$. As we increase the parameter $\alpha$, which is the angle from the equator, away up and down from the insulation disk, we see more pronounced variation in $\Delta T(\theta, r)$.

**Exercise.** Consider a steady-state solution when

$$J_s(\theta) = J_0 \quad 0 \leq \theta < \pi/2 - \alpha$$

$$J_s(\theta) = 0 \quad \pi/2 - \alpha \leq \theta \leq \pi/2 + \beta$$

$$J_s(\theta) = -J_1 \quad \pi/2 + \beta < \theta < \pi \qquad \text{(III.7.134)}$$

where $J_s(\theta)$ is given by Eq. (III.7.124). What is the requirement on the relation between $\alpha$ and $\beta$? Physically interpret this requirement in terms of temperature and net heat flow. **Hint:** You need to consider the Gauss divergence theorem for the integral $\oint \mathbf{J}_s.d\mathbf{s}$ (this is the net heat flow into the ball at its surface), as well as the steady-state requirement that $\vec{\nabla}.\mathbf{J} = 0$.

*Optional: The uniqueness of the solution to the 3D Laplace equation.* In general, we can get a unique solution of a 3D Laplace equation in a region of space. If the region or volume is bounded by a closed surface $\mathbf{r}(s,t) = x(s,t)\hat{\mathbf{i}} + y(s,t)\hat{\mathbf{j}} + z(s,t)\hat{\mathbf{k}}$ on which we can specify a function $\Phi(s,t) = \phi(x(s,t), y(s,t), z(s,t))$, the solution is unique . Such a boundary condition is commonly referred to as *Dirichlet problem.* Also, if $\Phi(s,t) = \phi(x(s,t), y(s,t), z(s,t)) = 0$, then the solution to $\nabla^2\phi(x,y,z) = 0$ is $\phi(x,y,z) = 0$ (you can test this using the coordinate systems we have considered). If we specify $\Phi(s,t)$ equal to some partial derivative $\phi(x,y,z)$ at the surface, the solution (if it can be obtained) may be unique up to a constant. This is the case when we define a partial deriva-

Peter Dirichlet
(1805–1859)

Carl Neumann
(1798–1895)

tive in the direction of the normal to be a function of $s$ and $t$; such a boundary condition is referred to as a *Neumann problem.* We already considered such a boundary condition in Example III.7.4.

This finishes our discussion about how to solve Laplace's equation in our introduction to partial differential equations. Tired of it or not, in the next section, you will be able to apply what we've learnt to solving the linearized Poisson-Boltzmann equation and the Helmholtz equation.

## 7.2. The Linearized Poisson-Boltzmann and Helmholtz Equations

### *The form of both equations*

The linearized Poisson Boltzmann and Helmholtz equations are of the form

$$\vec{\nabla}^2\psi(\mathbf{r}) - \kappa^2\psi(\mathbf{r}) = 0 \tag{III.7.135}$$

When $\kappa^2 > 0$ we have what can be referred to as either a linearized Poisson-Boltzmann equation or Debye equation (as the latter also appears in the theory of electrolytes) and when $\kappa^2 < 0$ we have what is generally referred to as a Helmholtz equation. We'll consider two examples of how to solve such equations and how they come about.

**Example III.7.5 (Physical Application ▶▶).** Here, we'll calculate the electrostatic potential from a charged macromolecule that may be approximated as a long cylinder of radius $a$ with a surface charge density $\sigma(\phi, z)$ in *cylindrical polar coordinates*, where $z$ is the coordinate along the axis of the cylinder, and $\phi$ is the azimuthal angle. We'll suppose that the interior of that cylinder has low relative dielectric constant $\varepsilon_r^{(1)} \approx 2$ (determined just by the electronic polarizability of the molecule), while the outside is surrounded by water which will be for simplicity characterized by its large relative macroscopic dielectric constant, $\varepsilon_r^{(2)}$ ($\approx 80$); the full dielectric constant in SI units in each case is given by $\varepsilon^{(j)} = \varepsilon_r^{(j)} \varepsilon_0$, where $\varepsilon_0$ is the dielectric constant of the vacuum (in our much loved Gaussian units the latter is just $=1$, but then in the next equation a $4\pi$ factor will appear in the r.h.s). In addition to this we will consider that in the water, we have salt — a fully dissociated electrolyte of singly-charged cations and anions. We want to calculate the electrostatic potential, $\varphi(\mathbf{r})$, that such a molecule will cause outside itself, in the solution.

If $e\varphi(\mathbf{r})/k_B T \ll 1$, we may use Eq. (III.7.135) to approximate the equation on $\varphi(\mathbf{r})$ (replacing $\psi(\mathbf{r})$ with $\varphi(\mathbf{r})$), with $\kappa = 1/R_D$ where $R_D$ is the Debye screening length (which we discussed in Chapter 10, Part I, Vol. 1). Also, when $\varepsilon_r^{(2)} \gg \varepsilon_r^{(1)}$ we may use the boundary condition

$$\left. \frac{\partial \varphi(R, \phi, z)}{\partial R} \right|_{R=a} = -\frac{\sigma(\phi, z)}{\varepsilon^{(1)}} \qquad \text{(III.7.136)}$$

Because away from the charge source, in the bulk of electrolyte, the electric field due to our molecules is totally screened by ionic atmosphere (we will see in a moment that it will be exponentially small), the electrostatic potential will not vary there, and we may take it for zero, adopting thereby the second boundary condition: $\varphi(\infty, \phi, z) = 0$.

We start then by using the separation of variables technique and substitute $\varphi(R, \phi, z) = \rho(R)\Phi(\phi)Z(z)$ into Eq. (III.7.135). Recollecting that the Laplace operator in cylindrical coordinates reads $\Delta = \vec{\nabla}^2 = \frac{1}{R}\left(R\frac{\partial}{\partial R}\right) + \frac{1}{R^2}\frac{\partial^2}{\partial \phi^2} + \frac{\partial^2}{\partial z^2}$, this yields

$$\frac{1}{R}\frac{d}{dR}\left(R\frac{d\rho(R)}{dR}\right) + \frac{1}{R^2\Phi(\phi)}\frac{d^2\Phi(\phi)}{d^2\phi} + \frac{1}{Z(z)}\frac{dZ(z)}{dz} = \kappa^2$$

This equation is only different from the Laplace one in that $\kappa \neq 0$. The periodicity requirement that $\Phi(\phi) = \Phi(\phi + 2\pi)$ means, again, that we must write $\Phi(\phi) = \exp(in\phi)$. Substituting this directly into Eq. (III.7.135) yields

$$\frac{1}{R}\frac{d}{dR}\left(R\frac{d\rho(R)}{dR}\right) + \frac{n^2}{R^2} = \kappa^2 - \frac{1}{Z(z)}\frac{dZ(z)}{dz} \qquad (III.7.137)$$

As the l.h.s depends only on $R$ and the r.h.s depends only on $z$, we must choose that $\frac{1}{Z(z)}\frac{dZ(z)}{dz}$ must be equal to some constant, independent of $R$ or $z$, so that

$$-\frac{1}{Z(z)}\frac{dZ(z)}{dz} = k_z^2 \qquad (III.7.138)$$

$$\frac{1}{R}\frac{d}{dR}\left(R\frac{d\rho(R)}{dR}\right) + \frac{n^2}{R^2} = \kappa^2 + k_z^2 \qquad (III.7.139)$$

with $k_z^2 > 0$. Equation (III.7.139) is simply Eq. (III.7.86), but with $k^2$ replaced by $-(k^2 + k_z^2)$. Secondly, the solution to Eq. (III.7.139) that we require (to satisfy $\varphi(\mathbf{r}) \to 0$ as $R \to \infty$) is

$$\rho(R) = AK_n\left(\sqrt{\kappa^2 + k_z^2}R\right) \qquad (III.7.140)$$

For the general solution to Eq. (III.7.138) we write

$$Z(z) = B\exp(ik_z z) + C\exp(-ik_z z) \qquad (III.7.141)$$

where $B$ and $C$ are free constants yet to be determined by the boundary conditions.

As before, when we were considering Laplace's equation, the general solution can be built of a linear supposition of solutions, each with different $k_z$ and $n$. If we suppose that the molecule is infinitely long, $k_z$ is free to take a continuum of values from $-\infty$ to $\infty$. Thus, for that case, we may write down the general solution

$$\varphi(R, \phi, z) = \frac{1}{2\pi} \sum_{n=-\infty}^{\infty} \int_{-\infty}^{\infty} dk_z A_n(k_z) K_n\left(\sqrt{\kappa^2 + k_z^2}R\right)$$

$$\times \exp(-in\phi)\exp(-ik_z z) \qquad \text{(III.7.142)}$$

Using our surface boundary condition we are able to find the coefficients $A_n(k_z)$, which uniquely determine $\varphi(R, \phi, z)$. To this we first write

$$-\frac{\sigma(\phi, z)}{\varepsilon^{(1)}} = \frac{\partial\varphi(R, \phi, z)}{\partial R}\bigg|_{R=a} = \frac{1}{2\pi}\sum_{n=-\infty}^{\infty}\int_{-\infty}^{\infty} dk_z A_n(k_z)\sqrt{\kappa^2 + k_z^2}$$

$$\times K_n'\left(\sqrt{\kappa^2 + k_z^2}\, a\right)\exp(-in\phi)\exp(-ik_z z) \quad \text{(III.7.143)}$$

If we define a double Fourier integral (transform) for the surface charge density

$$\tilde{\sigma}(n, k_z) = \frac{1}{2\pi}\int_0^{2\pi} d\phi \int_{-\infty}^{\infty} dz\sigma(\phi, z)\exp(in\phi)\exp(ik_z z) \quad \text{(III.7.144)}$$

it is not hard to see that we need (using the properties of Fourier series and transforms that we discussed in Part II, Vol. 1)

$$A_n(k_z) = -\frac{\tilde{\sigma}(n, k_z)}{\varepsilon^{(1)}\sqrt{\kappa^2 + k_z^2}K_n'\left(\sqrt{\kappa^2 + k_z^2}a\right)} \qquad \text{(III.7.145)}$$

to satisfy the surface boundary condition.

Let's now suppose that we have a molecule for which the charge distribution has helical symmetry so that charge density remains the same when $\phi \to \phi + g\Delta z$ and $z \to z + \Delta z$, describing a corkscrew movement. In this case we may write $\sigma(\phi, z) = \bar{\sigma}(\phi - gz)$, describing a helical charge density with helical pitch $H = 2\pi/g$. From this it follows that

$$\tilde{\sigma}(n, k_z) = \frac{1}{2\pi} \int_0^{2\pi} d\phi \int_{-\infty}^{\infty} dz \bar{\sigma}(\phi - gz) \exp(in\phi) \exp(ik_z z)$$

$$\Rightarrow \tilde{\sigma}(n, k_z) = \frac{1}{2\pi} \int_0^{2\pi} d\phi \int_{-\infty}^{\infty} dz \bar{\sigma}(\phi) \exp(in\phi) \exp(i(k_z + gn)z)$$

$$\Rightarrow \tilde{\sigma}(n, k_z) = \delta(k_z + ng) \int_0^{2\pi} d\phi \bar{\sigma}(\phi) \exp(in\phi) \equiv \zeta_n \delta(k_z + ng)$$

$$\text{(III.7.146)}$$

Thus, we obtain

$$\varphi(R, \phi, z) \equiv \bar{\varphi}(R, \phi - gz) = -\frac{1}{2\pi\varepsilon^{(1)}} \sum_{n=-\infty}^{\infty} \frac{\zeta_n K_n(\kappa_n R)}{\kappa_n K_n'(\kappa_n R)}$$

$$\times \exp(-in(\phi - gz))$$

$$= -\frac{1}{2\pi\varepsilon^{(1)}} \left\{ \frac{\zeta_0 K_0(\kappa_n R)}{\kappa_0 K_0'(\kappa_n R)} + 2 \sum_1^{\infty} \frac{\zeta_n K_n(\kappa_n R)}{\kappa_n K_n'(\kappa_n R)} \right.$$

$$\left. \times \cos(n(\phi - gz)) \right\} \qquad \text{(III.7.147)}$$

We see that the potential is also invariant under the combined transformation $\phi \to \phi + g\Delta z$ and $z \to z + \Delta z$.

**Exercise.** Consider a new form $\sigma(\phi, z) = \frac{\sigma_0}{2}(\delta(\phi - gz) + \delta(\phi + \pi - gz))$, which represents a charge density of two equally spaced helical strands with the same uniform charge. What symmetries does this charge distribution possess? Calculate $\zeta_n$ and the electrostatic potential. For what values of $n$ is there no contribution? Show that the electrostatic potential possesses the same symmetries. What values of $n$ must we select to give non-zero contributions for four equally spaced strands with the same uniform charge?

**Exercise.** Suppose, instead, that we want to calculate the electric potential in a 1:1 electrolyte solution inside a cylindrical pore of radius $a$, where both Eqs. (III.7.135) and (III.7.136) apply. At the

$$\psi(\mathbf{a},\theta,\phi)=0$$

**Fig. III.7.15** An Schematic illustration of a particle (represented by a small red sphere) of mass of $m$ confined to a spherical cavity of radius $a$ with impenetrable walls (so that outside the sphere, the particle would have infinite energy), shown in black.

centre of the pore the potential remains finite. **Think:** What restriction does this place on the choice of modified Bessel functions satisfying Eq. (III.7.139)? Find the solution for $\varphi(R,\phi,z)$ in terms of $\tilde{\phi}(n,k_z)$ given by Eq. (III.7.144).

**Example III.7.6 (Physical Application ▶▶).** Here, we'll take an example from quantum mechanics and consider a particle of mass $m$ in a spherical cavity of radius $a$, as illustrated in Fig. III.7.15. (If the particle is an electron, this example describes a type of what is called a *quantum dot*.) This is a simple example of the 3D Schrödinger equation, the general solution of which we will discuss last of all. The 3D Schrödinger equation in this case is

$$-\frac{\hbar^2}{2m}\vec{\nabla}^2\psi(\mathbf{r}) = E\psi(\mathbf{r}) \qquad (\text{III.7.148})$$

where $\psi(\mathbf{r})$ is the wave function of the particle and $E$ is its energy. Note that Eq. (III.7.148) is simply the Helmholtz equation (Eq. (III.7.135)) with $-k^2 = 2mE/\hbar^2$. We require that $\psi(a,\theta,\phi) = 0$, so that the cavity is a kind of infinite wall potential well. Now, if $\psi(\mathbf{r})$ satisfied the Laplace equation we would have to require that $\psi(\mathbf{r}) = 0$ everywhere, but as $\psi(\mathbf{r})$ satisfies the Helmholtz equation we can in fact have $\psi(\mathbf{r}) \neq 0$. However, $E$ is only allowed to be certain values for $\psi(\mathbf{r}) \neq 0$, permitted eigenvalues of energy. Here, we will find some of these energy eigenvalues and wave functions for those values.

Again, we can make a separation of variables $\psi(\mathbf{r}) = \rho(r)\Theta(\theta)\Phi(\phi)$; in this case Eq. (III.7.112) becomes modified to

$$\frac{\sin^2\theta}{\rho(r)}\frac{\partial}{\partial r}\left(r^2\frac{\partial\rho(r)}{\partial r}\right) + \frac{\sin\theta}{\Theta(\theta)}\frac{\partial}{\partial\theta}\left(\sin\theta\frac{\partial\Theta(\theta)}{\partial\theta}\right)$$

$$+\frac{1}{\Phi(\phi)}\frac{d^2\Phi(\phi)}{d\phi^2} + k^2\sin^2\theta = 0 \qquad\qquad \text{(III.7.149)}$$

where $k^2 = -\kappa^2 = 2mE/\hbar^2$. And again, to satisfy the periodicity condition we require $\Phi(\phi) = \exp(in\phi)$, where $n$ is an integer, so that we may write

$$\frac{\sin^2\theta}{\rho(r)}\frac{\partial}{\partial r}\left(r^2\frac{\partial\rho(r)}{\partial r}\right) + \frac{\sin\theta}{\Theta(\theta)}\frac{\partial}{\partial\theta}\left(\sin\theta\frac{\partial\Theta(\theta)}{\partial\theta}\right)$$

$$-m^2 + k^2 r^2 \sin^2\theta = 0$$

$$\Rightarrow \frac{1}{\rho(r)}\frac{\partial}{\partial r}\left(r^2\frac{\partial\rho(r)}{\partial r}\right) + k^2 r^2 = \lambda$$

$$= -\frac{1}{\Theta(\theta)\sin\theta}\frac{\partial}{\partial\theta}\left(\sin\theta\frac{\partial\Theta(\theta)}{\partial\theta}\right) + \frac{m^2}{\sin^2\theta} \quad \text{(III.7.150)}$$

As before, for the solutions $\Theta(\theta)$ to be non-singular, we require $\lambda = l(l+1)$ with $l$ zero or a positive integer and $\Theta(\theta) = P_l^m(\cos\theta)$. What is different from solving the Laplace equation is that the equation on $\rho(r)$ now reads as

$$-\frac{d^2\rho(r)}{dr^2} - \frac{2}{r}\frac{d\rho(r)}{dr} + \frac{l(l+1)}{r^2}\rho(r) = k^2\rho(r) \qquad \text{(III.7.151)}$$

If we write $\rho(r) = (kr)^{-1/2}u(r)$, it can be shown that $u(r)$ satisfies Bessel's equation with $\alpha = l + 1/2$ (we leave this as an exercise). Thus, we can write the solutions to Eq. (III.7.151) as spherical Bessel functions

$$\rho(r) = j_l(kr) \equiv \left(\frac{\pi}{2kr}\right)^{-1/2} J_{l+1/2}(kr) \qquad\qquad \text{(III.7.152)}$$

Note that we discard the solution to Bessel's equation involving $Y_{l+1/2}(kr)$ as we require the wave function and its derivatives not

**Table III.7.2** We present some of the first few allowed values of $k_{n,l}a$ that must satisfy the constraint $j_l(k_{n,l}a) = 0$. Here the $n$ value corresponds to a particular zero of $j_l(x)$ that satisfies $j_l(x) = 0$. Note here that we count here $n = 1$ as the first zero for which $k_{n,l}a \neq 0$. Each value in the table belongs to a column, where all values within it have the same $l$ value, specified at the top; and the values in a single row all have the same $n$ value given in the first column.

| $ak_{n,l}$ | $l = 0$ | $l = 1$ | $l = 2$ | $l = 3$ |
|---|---|---|---|---|
| $n = 1$ | $\pi$ | 4.49 | 5.76 | 6.99 |
| $n = 2$ | $2\pi$ | 7.73 | 9.10 | 10.42 |
| $n = 3$ | $3\pi$ | 10.90 | 12.32 | 13.70 |
| $n = 4$ | $4\pi$ | 14.07 | 15.51 | 16.92 |

to become singular. Now, the requirement that $\psi(b, \theta, \phi) = 0$ can be met if we require that $j_l(ka) = 0$; thus $k = \sqrt{2mE}/\hbar$ becomes quantized for values $k = k_{n,l}$ (here $n$ is the order of the Bessel function zero at $r = a$) that satisfy $j_l(k_{n,l}a) = 0$; the values of $ak_{n,l}$ are simply numbers. The first few of them are given in Table III.7.2.

In Fig. III.7.16 we plot the radial functions for some of the first few quantum numbers. In Fig. III.7.17, we show the energy spectra of the first few energy levels. The full set of eigenfunctions for each energy eigenvalue $E_{n,l} = \hbar^2 k_{n,l}^2/2m$ is thus given by

$$\psi_{n,l,m}(r, \theta, \phi) = C_{n,l} Y_l^{(m)}(\theta, \phi) j_l(\kappa_{n,l} r) \qquad \text{(III.7.153)}$$

Here, $C_{n,l}$ is the normalization constant and it can be found by requiring that

$$\int_0^a r^2 dr \int_0^{2\pi} d\phi \int_0^{\pi} \sin\theta d\theta \psi_{n,l,m}(r, \theta, \phi) \psi_{n,lm}^*(r, \theta, \phi) = 1 \qquad \text{(III.7.154)}$$

**Exercise.** Use Eq. (III.4.150) to find an orthogonality condition for the spherical Bessel functions. **Hint:** Think about writing the $r$-integral in Eq. (III.7.154) in terms of Bessel functions. Use Eq. (III.7.154) to find the value $C_{n,l}$.

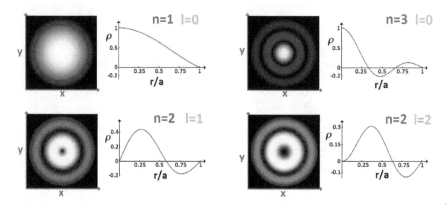

**Fig. III.7.16**  We show plots of the radial wave functions for a spherical cavity. We have picked as examples four different sets of $n$ and $l$ values. For each value, on the left, we present a density plot of the magnitude of the wave function (in the $x-y$ plane bisecting the spherical cavity at the equator) and, on the right, a plot of the un-normalized wave function in terms of $r/a$ (the value 1 is the wall of the cavity). In the density plots the dark areas correspond to nodes, where the likelihood of finding the particle is very small, whereas the brighter bands correspond to maxima and minima in the wave function where one is most likely to find the particle. The value $n$ determines the number of nodes, or dark fringes; when we increase $n$ by one, we get one extra fringe. When $l = 0$ there is central maxima, the particle is most likely to be at the centre of the cavity. This totally changes for the states with $l \neq 0$: here it is forbidden for a particle to be exactly at the centre of the cavity, as seen by the dark spot at the centres of the bottom two density plots. A Physical explanation behind this is that value $l$ corresponds to the magnitude of angular momentum that a particle has. It may seem that the particles with a quantum state of nonzero $l$ behaves as if it is, classically orbiting around the centre of the cavity. In terms of classical physics, this would be maintained by a central force pulling the particle inwards with the acceleration acting towards the centre of the cavity. In fact, this force would actually not be present in classical physics as the potential energy is constant inside the sphere, so it is a 'violation' of the classical equations of motion! All in all this analogy is not good. Furthermore, at the centre of the cavity, the 'acceleration' required to keep particles in such orbits becomes infinite thus this region should be forbidden. As we increase $l$ we see that the dark central region grows. An increase in angular momentum would increase this additional acceleration that would be required to hold a particle in an orbit close to the centre. This decreases the likelihood that a particle will be found there, due to the increased additional acceleration needed, violating classical mechanics even more! A general wisdom of quantum mechanics: a particle is least likely to be in regions where classical mechanics is most violated.

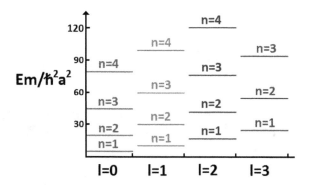

**Fig. III.7.17** Here, we show the energy spectra for a particle in a spherical cavity. The allowed energies are related to the allowed wave number values $k_{n,l}$ through $E_{n,l} = \hbar^2 k_{n,l}^2 / 2m$. We see two trends. First, as we increase $n$ the energy dramatically increases; for $l = 0$ that energy is proportional to $n^2$. Second, as we increase $l$ keeping $n$ fixed we increase the allowed energy.

## 7.3. Thinking of Time: Wave and Diffusion Equations

Let us move away from steady states and consider time-dependent processes in different spatial dimensions.

### The 1D wave equation

One important partial differential equation in the physical sciences is the wave equation. To find out what it is and one of the ways it comes about, let's go back to Example. III.5.4, where we had a chain of springs. First, we can rewrite Eq. (III.5.56) as

$$\rho \frac{\partial^2 \delta x(jl, t)}{\partial t^2} = K \frac{\delta x((j+1)l, t) - 2\delta x(jl, t) + \delta x((j-1)l, t)}{l^2}$$

(III.7.155)

Here, $l$ is the equilibrium length of each spring (we consider them to be the same in the chain), and we have defined $\delta x(jl, t) = \delta x_j(t)$, with linear density $\rho = m/l$, bulk modulus, $\rho = m/l$ and $K = lk$ where $k$ is the spring constant). Now, what happens if we take $l \to 0$, increasing the number of springs so that we can replace $jl$ with a continuum of $x$ values? First, we define a function of two variables $\psi(x, t)$ such that

$$\lim_{l \to 0} \{\delta x(jl, t)\} = \psi(x, t) \qquad \text{(III.7.156)}$$

Next, we note that

$$\lim_{x_0 \to 0} \left\{ \frac{\delta x((j+1)l, t) - 2\delta x(jl, t) + \delta x((j-1)l, t)}{l^2} \right\} = \frac{\partial^2 \psi(x, t)}{\partial x^2}$$
$$\text{(III.7.157)}$$

Therefore, using Eqs. (III.7.156) and (III.7.157), we may write Eq. (III.7.155) in the limit $l \to 0$ as

$$\frac{\partial^2 \psi(x, t)}{\partial t^2} = v^2 \frac{\partial^2 \psi(x, t)}{\partial x^2} \qquad \text{(III.7.158)}$$

This is the 1D *wave equation*, which we have derived for 1D longitudinal sound waves in Chapter 5 of Part III, Vol. 2. Here, $v$ is the speed of the wave, and for the considered chain, $v = \sqrt{K/\rho}$. It can also be derived for transverse waves on strings, where the speed of the wave is $v = \sqrt{T/\rho}$, where $T$ is the tension in the string. As we have many other subjects to cover, we won't give this derivation; it can be found in any general university level physics textbook.

Now, we already know the solution to Eq. (III.5.56) for an infinite 1D system. Thus, we can obtain the solution to Eq. (III.7.158) by considering the $l \to 0$ limit of Eq. (III.5.62):

$$\psi(x, t) = \int_{-\infty}^{\infty} A^+(q) \exp(iqx - ivqt) dq$$
$$+ \int_{-\infty}^{\infty} A^-(q) \exp(iqx + ivqt) dq \qquad \text{(III.7.159)}$$

where $\omega = qv$. The solution to the wave equation is the linear supposition of waves travelling in either of two directions along a 1D line with speed $v$; both $\pm v$ are solutions to the wave equation (Eq. (III.7.158)). The two coefficients $A^+(q)$ and $A^-(q)$ can be determined by any initial condition (specifying $\psi(x, 0)$ at $t = 0$) and any other boundary condition specifying $\psi(x, t)$ at a particular time.

**Exercise.** By substitution of Eq. (III.7.159) into Eq. (III.7.158) show that the former is the solution to the wave equation.

**Example III.7.7 (Physical Application ▶▶).** Let's look at a 1D travelling wave, moving from one medium to another, where the speed of the wave is different in each medium. (For simplicity we ignore any effects of dissipation, which would make the problem more complicated.) This could be a linear, coherent wave front of sound passing from one material to another with the interface perpendicular to the direction of propagation, as shown in Fig. III.7.18. In one medium (region 1 that the wave initially travels in) we have a wave speed $v_1$, and for the other medium (region 2) it is $v_2$.

When travelling from one medium to another, part of the wave gets reflected back. Here, we will consider a sinusoidal wave of a single frequency. Therefore, in region 1 (defined by $x < 0$) we can write as a solution (from Eq. (III.7.159) with a single frequency)

$$\psi_1(x,t) = I \cos\left(k_1 x - \omega_1 t + \phi\right) + R \cos\left(k_1 x + \omega_1 t - \phi\right) \quad \text{(III.7.160)}$$

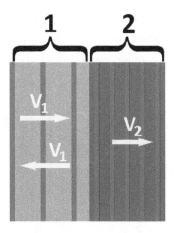

**Fig. III.7.18** This is schematic representation of a sound wave (the crests are represented with red lines) travelling with speed $v_1$ in a medium (labelled 1, denoted by light blue) in a direction perpendicular to the interface with another medium (labelled 2). Part of the wave enters the other medium travelling with a speed $v_2$, while a proportion of the wave gets reflected to travel away from the interface with speed $v_1$. In medium 2 the distance between the crests is shorter than in medium 1, which means that in this illustration $v_2 < v_1$ as the wave frequency must remain the same in both media (see text).

where $\omega_1/k_1 = v_1$. And in region 2 (for which $x > 0$) we can write (where we suppose that the wave is travelling in one direction)

$$\psi_2(x, t) = T \cos(k_2 x - \omega_2 t + \phi) \qquad \text{(III.7.161)}$$

where $\omega_2/k_2 = v_2$. Now, across the interface we have boundary conditions of the continuity of the amplitudes and their derivatives

$$\psi_1(0, t) = \psi_2(0, t) \text{ and } \left.\frac{\partial \psi_1(x, t)}{\partial x}\right|_{x=0} = \left.\frac{\partial \psi_2(x, t)}{\partial x}\right|_{x=0} \qquad \text{(III.7.162)}$$

The first condition is rather obvious as we should have a continuous displacement at the interface. The second condition is required, as otherwise the second derivative of $\psi(x, t)$ would become singular at this point, resulting in an infinite restoring force acting on the wave (in the continuum, the restoring force is $\propto d^2\psi/dx^2$, which is the r.h.s of Eq. (III.7.155) in the discrete case of masses on springs).

Applying the first of these boundary conditions, we require that

$$I \cos(-\omega_1 t + \phi) + R \cos(\omega_1 t - \phi) = T \cos(-\omega_2 t + \phi) \quad \text{(III.7.163)}$$

For Eq. (III.7.163) to be satisfied, at all times, we must require that $\omega_1 = \omega_2$: i.e. the frequencies of the waves in the two media must be the same. The second boundary condition yields

$$k_1 I \sin(-\omega_1 t + \phi) + k_1 R \sin(\omega_1 t - \phi) = k_2 T \cos(-\omega_1 t + \phi)$$
$$\text{(III.7.164)}$$

Thus, from Eqs. (III.7.163) and (III.7.164) we find that

$$I + R = T \quad k_1 I - k_1 R = k_2 T \qquad \text{(III.7.165)}$$

From Eq. (III.7.165) we can solve for both $T$ and $I$ in terms of the incident amplitude:

$$R = \frac{(k_1 - k_2)I}{(k_1 + k_2)}, \quad T = \frac{2k_1 I}{k_1 + k_2} \qquad \text{(III.7.166)}$$

Now, since $\omega = \omega_1 = \omega_2$, we can write $k_1 = \omega/v_1$ and $k_2 = \omega/v_2$. Thus, we may rewrite Eq. (III.7.166) in terms of the wave speeds as

$$R = \frac{(v_2 - v_1)I}{(v_1 + v_2)}, \quad T = \frac{2v_2 I}{v_1 + v_2} \qquad \text{(III.7.167)}$$

Note that if $v_2 \ll v_1$, the wave is almost totally reflected back; if $v_1 \approx v_2$, very little is reflected and most of the wave is transmitted. For sound waves, the change in velocity between media would be associated with a change in the density of the material, $\rho$, or its stiffness, $K$. Note that because of solids being much stiffer, soundwaves travel much faster in them than in liquids or gasses. Thus, our analysis tells us that for sound reaching an interface between a solid and a gas from the solid side, a lot of the sound will get reflected back into the solid. This is indeed what is observed.

### The wave equation in higher dimensions

We can generalize the wave equation to 2D or 3D (which may describe more generally the propagation of sound, or one of the vector components of the electric or magnetic field of electromagnetic radiation), where it reads as

$$\frac{\partial^2 \psi(\mathbf{r}, t)}{\partial t^2} = v^2 \vec{\nabla}^2 \psi(\mathbf{r}, t) \tag{III.7.168}$$

We can then generalize Eq. (III.7.159), to write the general solution to Eq. (III.7.168) in infinite space, which reads as

$$\psi(\mathbf{r},t) = \int A^+(\mathbf{q}) \exp(i\mathbf{q}.\mathbf{r} - i\omega t) d\mathbf{q} + \int A^-(\mathbf{q}) \exp(i\mathbf{q}.\mathbf{r} + i\omega t) d\mathbf{q} \tag{III.7.169}$$

where $\omega / |\mathbf{q}| = v$.

**Exercise.** Verify by writing $\mathbf{q}.\mathbf{r} = q_x x + q_y y + q_z z$ and using $\nabla^2 = \partial^2/\partial x^2 + \partial^2/\partial y^2 + \partial^2/\partial z^2$ that Eq. (III.7.169) is indeed a solution to Eq. (III.7.168).

In other applications for which we have spatial boundary conditions (for instance, standing waves in a spherical cavity), when considering waves of a particular frequency $\omega$ one should write $\psi(\mathbf{r}, t) = \tilde{\psi}_1(\mathbf{r}, \omega) \exp(i\omega t) + \tilde{\psi}_2(\mathbf{r}, \omega) \exp(-i\omega t)$. Substituting this into Eq. (III.7.168), we find that both $\tilde{\psi}_1(\mathbf{r}, \omega)$ and $\tilde{\psi}_2(\mathbf{r}, \omega)$ satisfy the Helmholtz equation with $\kappa^2 = k^2 = \omega^2/v^2$, which can be dealt

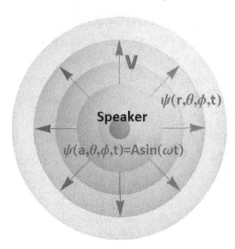

**Fig. III.7.19** A schematic illustration of uniform spherical sound waves being emitted from a speaker with frequency $\omega$. We suppose that at a spherical surface close to the speaker the sound is emitted uniformly with displacement $\psi(a, \theta, \phi, t) = A \sin(\omega t)$. Using this as a boundary condition and the fact that we must have the waves travelling outwards we are able to find the form of $\psi(r, \theta, \phi, t)$.

with using the separation of variables technique, as we discussed previously. If required, we can construct a more general solution from a supposition (sum) of solutions with different frequencies.

**Example III.7.8 (Physical Application ▶▶).** Let's consider a speaker emitting sound of one frequency. We'll suppose that at $r = a$, i.e. at the surface of a sphere centred about the speaker, we have $\psi(a, \theta, \phi, t) = A \sin \omega t$, as illustrated in Fig. III.7.19. We want to find the displacement $\psi(r, \theta, \phi, t)$ away from that sphere. Due to the considered sound being of just one frequency, the solution to the wave equation, written as a linear supposition of two independent solutions with the same frequency $\omega$, reads

$$\psi(\mathbf{r}, t) = \frac{1}{2i} \left( \tilde{\psi}_1(\mathbf{r}, \omega) \exp(i\omega t) + \tilde{\psi}_2(\mathbf{r}, \omega) \exp(-i\omega t) \right) \quad \text{(III.7.170)}$$

It is convenient, though not necessary, so to write Eq. (III.7.170), to take from both $\tilde{\psi}_1(\mathbf{r}, \omega)$ and $\tilde{\psi}_2(\mathbf{r}, \omega)$ a factor $1/2i$. Substitution of Eq. (III.7.170) into the wave equation, Eq. (III.7.168), yields the Helmholtz equations

$$\vec{\nabla}^2 \tilde{\psi}_1(\mathbf{r}, \omega) + k^2 \tilde{\psi}_1(\mathbf{r}, \omega) = 0$$
$$\vec{\nabla}^2 \tilde{\psi}_2(\mathbf{r}, \omega) + k^2 \tilde{\psi}_2(\mathbf{r}, \omega) = 0 \qquad \text{(III.7.171)}$$

where $k = v/\omega$. Since we now have some experience in solving partial differential equations, let's think first, instead of doing blind maths straightaway, about the symmetry of the problem. Namely, our boundary condition at the surface of the sphere about the speaker does not depend on $\theta$ or $\phi$. This means that we can automatically infer that $\tilde{\psi}_1(r, \theta, \phi, \omega) = \Psi_1(r)$ and $\tilde{\psi}_2(r, \theta, \phi, \omega) = \Psi_2(r)$. Also, as we require the solution to be real, we require $\Psi_2(r) = -\Psi_1^*(r)$. Thus, we can write Eq. (III.7.171), using the form of the Laplace operator in spherical polar coordinates, i.e. the l.h.s of Eq. (III.7.110) as

$$\frac{1}{r^2} \frac{d}{dr} \left( r^2 \frac{d\Psi_1(r)}{dr} \right) + k^2 \Psi_1(r) = 0$$

$$\frac{1}{r^2} \frac{d}{dr} \left( r^2 \frac{d\Psi_2(r)}{dr} \right) + k^2 \Psi_2(r) = 0 \qquad \text{(III.7.172)}$$

These equations are of the same type as Eq. (III.7.151) but with $l = 0$; thus their general solutions can be written as

$$\Psi_1(r) = (\pi/2r)^{1/2} (A_1 J_{1/2}(kr) + B_1 J_{-1/2}(kr))$$
$$= (A_1 \sin(kr) + B_1 \cos(kr))/r \qquad \text{(III.7.173)}$$
$$\Psi_2(r) = (\pi/2r)^{1/2} (A_2 J_{1/2}(kr) + B_2 J_{-1/2}(kr))$$
$$= (A_2 \sin(kr) + B_2 \cos(kr))/r \qquad \text{(III.7.174)}$$

However, it is in fact more convenient to write these solutions as

$$\Psi_1(r) = (\tilde{A} \exp(ikr) + \tilde{B} \exp(-ikr))/r$$
$$\Psi_2(r) = -(\tilde{B}^* \exp(ikr) + \tilde{A}^* \exp(-ikr))/r \qquad \text{(III.7.175)}$$

where we have exploited the fact that $\Psi_2(r) = -\Psi_1^*(r)$. Substituting Eq. (III.7.175) into Eq. (III.7.170) we find that

$$\psi(\mathbf{r}, t) = \frac{1}{2ir} \left( \tilde{A} \exp(ikr + i\omega t) - \tilde{A}^* \exp(-ikr - i\omega t) \right)$$

$$+ \frac{1}{2ir} \left( \tilde{B} \exp(-ikr + i\omega t) - \tilde{B}^* \exp(ikr - i\omega t) \right)$$

$$= \frac{r_A}{r} \sin(kr + \omega t + \phi_A) + \frac{r_B}{r} \sin(-kr + \omega t + \phi_B)$$

$$\text{(III.7.176)}$$

where in the last line we have used $\tilde{A} = r_A \exp(i\phi_A)$, $\tilde{B} = r_B \exp(i\phi_B)$, $\tilde{A}^* = r_A \exp(-i\phi_A)$ and $\tilde{B}^* = r_B \exp(-i\phi_B)$. Now, $\sin(kr+\omega t+\phi_A)$ represents a wave travelling inwards while $\sin(-kr+\omega t + \phi_B)$ represents a wave travelling outwards. As we are dealing with a speaker and not a receiver we can immediately set $r_A = 0$. In order to satisfy the remaining boundary condition we require

$$A \sin \omega t = \frac{r_B}{a} \sin(-ka + \omega t + \phi_B) \tag{III.7.177}$$

For Eq. (III.7.177) to hold we require that $\phi_B = ka$ and $r_B = aA$. Thus, in the end we obtain a very simple solution of this wave equation,

$$\psi(\mathbf{r}, t) = A \left( \frac{a}{r} \right) \sin(k(a - r) + \omega t) \tag{III.7.178}$$

which prescribes that away from the source the sound amplitude decays at a rate inversely proportional to the distance.

### The diffusion equation

The diffusion equation describes the concentration of particles that have completely random motion. Recalling the continuity equation (Eq. (III.3.10) from Chapter 3 of Part III, Vol. 2) we may write the diffusion equation as

$$\frac{\partial c(\mathbf{r}, t)}{\partial t} = -\vec{\nabla} . J(\mathbf{r}, t) = D\vec{\nabla}^2 c(\mathbf{r}, t) \tag{III.7.179}$$

where we have combined it with Fick's law in 3D, namely $\mathbf{J}(\mathbf{r}) = -D\nabla c(\mathbf{r})$.

The flow of temperature is governed by an equation of the same form, where the diffusion constant $D$ for particles is replaced with the thermal diffusivity $D_T$ of the medium that the heat is flowing through.

In fact, for a long time, people actually thought that heat was the diffusion of some type of substance called caloric, before they found out that really it was the transfer of random kinetic energy. Nevertheless, heat flow obeys the diffusion equation due to the completely random motion of particles—but this is not due to particle diffusion, but energy randomly transferred through their collisions. A legend tells us that Albert Einstein had a deal with his mathematician friend M. Grossmann, to teach him differential geometry and tensor cal-

Marcel Grossmann
(1878–1936)

culus, which Einstein needed to know for his general relativity theory. Einstein benefited from Grossmann's classes a lot, later even publishing papers with Grossman, in 1913 and 1914, on general relativity—his rare non-solo papers. In exchange Einstein taught him principles of physics. After Einstein's classes in thermodynamics, which Einstein himself admired almost as much as he loved Euclid geometry—for their beauty and rigour, Grossman admitted that he did learn something from Einstein in return. Namely, Grossman confessed that earlier he felt uncomfortable seating down onto the chair heated by the person who was using that chair before. But after Einstein's explanation, Grossmann got to terms with sitting on such chairs, having understood that no substance belonging to the previous guy will flow into any part of his own body!

One way we can solve Eq. (III.7.179) is by the separation of variables method by writing

$$c(\mathbf{r}, t) = T(t)C(\mathbf{r})$$

Substitution of this product into Eq. (III.7.179) yields

$$\frac{1}{T(t)}\frac{\partial T(t)}{\partial t} = -Dk^2 = D\frac{\nabla^2 C(\mathbf{r})}{C(\mathbf{r})} \qquad \text{(III.7.180)}$$

Note that in this case we must require $k^2 \geq 0$, otherwise we will have the unphysical situation of backwards diffusion, and this would violate the second law of thermodynamics. Also, one should be aware that $k = 0$ corresponds to the steady-state solution, which we learnt satisfies Laplace's equation. Thus, we have that

$$\frac{\partial T(t)}{\partial t} = -Dk^2 T(t) \quad \nabla^2 C(\mathbf{r}) + k^2 C(\mathbf{r}) = 0 \qquad \text{(III.7.181)}$$

The solution to the equation on the l.h.s of Eq. (III.7.181) is simple, and we obtain

$$T(t) = A\exp(-Dk^2 t) \qquad \text{(III.7.182)}$$

The r.h.s of Eq. (III.7.181) is the Helmholtz equation that we dealt with previously. We can write a general solution to Eq. (III.7.179) in terms of Eq. (III.7.182) and the solution to the Helmholtz equation, $C_k(\mathbf{r})$, for different values of $k$. This reads as

$$c(\mathbf{r}, t) = \sum_k C_k(\mathbf{r})\exp(-Dk^2 t) \qquad \text{(III.7.183)}$$

We can find a specific solution to Eq. (III.7.179) by specifying appropriate boundary conditions, and so finding the $C_k(\mathbf{r})$, which we will look to do in a couple of examples. Also, note that at $t \to \infty$ only the $k = 0$ steady-state solution survives: the diffusion moves towards a steady state where $\nabla.\mathbf{J} = 0$. Note that this does not necessarily imply thermodynamic equilibrium (maximal entropy) where $\mathbf{J} = 0$, simply only that the concentration must satisfy Laplace's equation.

**Example III.7.9 (Physical Application ▶▶).** Let us consider a small drop of soluble dye that we have squirted into water. We will suppose that the water is calm, with no advection (movement of the water) going on. Thus, we can use the diffusion equation to predict how the dye may spread out into the water, if we know an initial concentration. At $t = 0$, we'll assume that the dye is confined within small spherical region of radius $a$ with a uniform concentration $c_0$. Thus, the concentration may be described, at $t = 0$, as

$$c(r, \theta, \phi, 0) = c_0 \theta(a - r) \qquad \text{(III.7.184)}$$

where $\theta(a - r)$ is the Heaviside step function we discussed in Chapter 6, Part I, Vol. 1; by definition, of course, $r \geq 0$. First we can write the solution $c(\mathbf{r}, t) = c(r, \theta, \phi, t)$ in terms of Eq. (III.7.183). Thus, at $t = 0$ we must have (from Eq. (III.7.183)) that

$$c_0 \theta(r - a) = \sum_k C_k(\mathbf{r}) \qquad \text{(III.7.185)}$$

Now, let's consider $C_k(\mathbf{r})$, the solutions to the Helmholtz equation. First because of the symmetry of the problem we may again immediately suppose that $C_k(\mathbf{r})$ depends only on $r$ in spherical polar coordinates so that $C_k(\mathbf{r}) = \rho_k(r)$, where $\rho_k(r)$ satisfies the equation

$$\frac{1}{r^2} \frac{d}{dr} \left( r^2 \frac{d\rho_k(r)}{dr} \right) + k^2 \rho_k(r) = 0 \qquad \text{(III.7.186)}$$

This is solved (as we found out before) by the general solution

$$\rho_k(r) = \frac{1}{r} \left( A \sin(kr) + B \cos(kr) \right) \qquad \text{(III.7.187)}$$

Now, as we are solving the diffusion equation for the concentration over all space, we require that the solution remains finite at $r = 0$, so we must set $B = 0$. As the space is effectively infinite, we are unrestricted in our choice of $k$. This means that we can rewrite Eqs. (III.7.183) and (III.7.185) as (using Eq. (III.7.187))

$$c(\mathbf{r}, t) = \frac{1}{r} \int_0^\infty A(k) \sin(kr) \exp(-Dk^2 t) dk \qquad \text{(III.7.188)}$$

$$c_0 \theta(r - a) = \frac{1}{r} \int_0^\infty A(k) \sin(kr) dk \qquad \text{(III.7.189)}$$

Using the fact that

$$\int_0^\infty \sin(k'r) \sin(kr) dr = \frac{\pi \delta(k - k')}{2} \qquad \text{(III.7.190)}$$

(Eq. (III.7.190) can be obtained by considering orthogonality conditions for Bessel functions of order 1/2, Eq. (III.4.150) and their definition, Eq. (III.4.128)) we can write from Eqs. (III.7.189) and (III.7.190)

$$A(k) = \frac{2c_0}{\pi} \int_0^\infty \theta(r-a) r \sin(kr) dr = \frac{2c_0}{\pi} \int_0^a r \sin(kr) dr$$

(III.7.191)

We won't perform the integral in Eq. (III.7.191), though it is easy enough, but rather substitute it back into the solution. The reason for doing this is that it will make evaluation of the $k$-integral in Eq. (III.7.188) more straightforward. Thus, we obtain for the solution

$$c(\mathbf{r},t) = \frac{c_0}{\pi r} \int_0^a r' dr' \int_{-\infty}^\infty dk \sin(kr) \sin(kr') \exp\left(-Dk^2 t\right)$$

(III.7.192)

We can then evaluate the $k$-integral by first writing (using double angle identities)

$$c(\mathbf{r},t) = \frac{c_0}{2\pi r} \int_0^a r' dr' \int_{-\infty}^\infty dk \left[\cos(k(r-r')) - \cos(k(r+r'))\right]$$
$$\times \exp\left(-Dk^2 t\right)$$

$$\Rightarrow c(\mathbf{r},t) = \frac{c_0}{2\pi r} \mathrm{Re} \int_0^a r' dr' \int_{-\infty}^\infty dk \left[\exp(ik(r-r')) - \exp(ik(r+r'))\right]$$
$$\times \exp\left(-Dk^2 t\right)$$

(III.7.193)

We can then complete the squares such that

$$c(\mathbf{r},t) = \frac{c_0}{2\pi r} \mathrm{Re} \int_0^a r' dr' \int_{-\infty}^\infty dk$$
$$\times \left[\exp\left(-Dt\left(k - i\frac{(r-r')}{2Dt}\right)^2\right) \exp\left(-\frac{(r-r')^2}{4Dt}\right)\right.$$
$$\left. - \exp\left(-Dt\left(k - i\frac{(r+r')}{2Dt}\right)^2\right) \exp\left(-\frac{(r+r')^2}{4Dt}\right)\right]$$

(III.7.194)

We'll suppose again (yet, without justification—that will eventually come in the next chapter, but for now just believe us!) that the integrals can then be written as

$$c(\mathbf{r}, t) = \frac{c_0}{2\pi r} \int_0^a r' dr' \int_{-\infty}^{\infty} dk \exp\left(-Dtk^2\right)$$

$$\times \left[\exp\left(-\frac{(r-r')^2}{4Dt}\right) - \exp\left(-\frac{(r+r')^2}{4Dt}\right)\right]$$

$$(\text{III.7.195})$$

It then follows, on evaluation of the $k$-integrals that

$$c(\mathbf{r}, t) = \frac{c_0}{2\pi r} \sqrt{\frac{\pi}{Dt}} \int_0^a r' dr' \left[\exp\left(-\frac{(r-r')^2}{4Dt}\right) - \exp\left(-\frac{(r+r')^2}{4Dt}\right)\right]$$

$$\Rightarrow c(\mathbf{r}, t) = \frac{c_0}{2\pi r} \sqrt{\frac{\pi}{Dt}} \int_{-a}^a r' dr' \exp\left(-\frac{(r-r')^2}{4Dt}\right) \qquad (\text{III.7.196})$$

We can evaluate the final remaining integral in terms of the error function and Gaussian functions; but before doing this, it is worth looking at a limiting case, to make sense of our result, when the time taken for the drop to diffuse is long. When $t \gg a^2/D$, the Gaussian in the integrand of Eq. (III.7.196) is dominated by $r \gg r'$. Thus, in this case, we can approximate

$$c(\mathbf{r}, t) \approx \frac{c_0}{2} \sqrt{\frac{1}{\pi Dt}} \int_0^a dr' \frac{r'^2}{Dt} \exp\left(-\frac{r^2}{4Dt}\right)$$

$$= \frac{c_0 a^3}{6Dt} \sqrt{\frac{1}{\pi Dt}} \exp\left(-\frac{r^2}{4Dt}\right)$$

$$= \frac{N}{8\pi Dt} \sqrt{\frac{1}{\pi Dt}} \exp\left(-\frac{r^2}{4Dt}\right) \qquad (\text{III.7.197})$$

where $N = 4\pi c_0 a^3/3$ is the total number of particles within the droplet.

To try to understand what Eq. (III.7.197) describes, let's consider a different initial condition, one that starts with the concentration all condensed at the origin:

$$c(r, \theta, \phi, 0) = \frac{N}{4\pi r^2} \delta(r) \qquad \text{(III.7.198)}$$

You can check that Eq. (III.7.198) is indeed what you would have with all the $N$ particles fixed at the point $r = 0$, by integrating Eq. (III.7.198) over all space. We would have that

$$\frac{N}{4\pi r^2} \delta(r) = \frac{1}{r} \int_0^\infty A(k) \sin(kr) dk \qquad \text{(III.7.199)}$$

From Eqs. (III.7.190) and (III.7.199), one can show that $A(k) = Nk/2\pi^2$, and substituting this into Eq. (III.7.188), we get

$$c(\mathbf{r}, t) = \frac{N}{2\pi^2 r} \int_0^\infty k \sin(kr) \exp(-Dk^2 t) dk$$

$$= -\frac{N}{2\pi^2 r} \frac{\partial}{\partial r} \int_0^\infty \cos(kr) \exp(-Dk^2 t) dk$$

$$\Rightarrow c(\mathbf{r}, t) = -\frac{N}{4\pi^2 r} \text{Re} \frac{\partial}{\partial r} \int_{-\infty}^\infty \exp(-Dt(k - ir/(2Dt))^2) dk$$

$$\times \exp\left(-r^2/(4Dt)\right)$$

$$\Rightarrow c(\mathbf{r}, t) = -\frac{N}{4\pi^2 r} \sqrt{\frac{\pi}{Dt}} \frac{\partial}{\partial r} \left[ \exp\left(-r^2/(4Dt)\right) \right]$$

$$= \frac{N}{8\pi Dt} \sqrt{\frac{1}{\pi Dt}} \exp\left(-r^2/(4Dt)\right) \qquad \text{(III.7.200)}$$

So we do indeed get the same result as given by Eq. (III.7.197). Thus, we see that over long timescales, $t \gg a^2/D$, where $a$ is the size of droplet, we can approximate the diffusion by considering the dye concentrated at a single point. This makes sense, as what determines the *length scales* we need to consider for the problem at a particular time is $\sqrt{Dt}$. Thus, at longer times we need to consider the concentration over larger length scales, as the drop spreads out. If we look

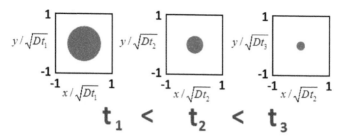

$$t_1 \quad < \quad t_2 \quad < \quad t_3$$

**Fig. III.7.20** Dear reader, this is not a conceptualist postmodern art study in walking away from a Japanese insert flag, but an illustration of how an extended object becomes more point-like as we change the scale of $x$ and $y$ through $\sqrt{Dt}$. As we make $t$ larger we go to larger length scales and the circle becomes more point-like. In terms of diffusion, this means that at long diffusion times where large length scales matter, an extended initial concentration profile can be approximately described as a concentration at a single point. An idea of scale can be every important in making approximation in physical systems, as we discussed in Chapter 10, Part I, Vol. 1.

at larger length scales, our initial drop (at $t = 0$) will start to look more and more like a point, as illustrated in Fig. III.7.20.

Now, let's go back and obtain a full evaluation of the integral in Eq. (III.7.196). First we tidy up and rescale the integral

$$c(\mathbf{r}, t) = \frac{2Dtc_0}{\pi r} \sqrt{\frac{\pi}{Dt}} \int_{-(a+r)/(2\sqrt{Dt})}^{(a-r)/(2\sqrt{Dt})} (s' + s) \exp\left(-s'^2\right) ds' \tag{III.7.201}$$

where $2\sqrt{Dt}s = r$ and $2\sqrt{Dt}s' = r'$. In Chapter 6, Part I, Vol. 1, we introduced the error function. Using the integral that defines it (see Chapter 6, Part I), presented previously, we can evaluate the integral in Eq. (III.7.201) and obtain

$$c(\mathbf{r}, t) = \frac{c_0}{r} \sqrt{\frac{Dt}{\pi}} \left[ \exp\left(-\frac{(a+r)^2}{4Dt}\right) - \exp\left(-\frac{(a-r)^2}{4Dt}\right) \right]$$
$$+ \frac{c_0}{2} \left[ \text{erf}\left(\frac{a-r}{2\sqrt{Dt}}\right) + \text{erf}\left(\frac{a+r}{2\sqrt{Dt}}\right) \right] \tag{III.7.202}$$

Let's consider the limiting case, opposite to the one we considered before, namely the behaviour at short times, when $\sqrt{Dt}/a \ll 1$.

We recall one of the definitions of the theta function as a limit of a theta-like function (given in Chapter 6, Part I, Vol. 1)

$$\theta(x) = \frac{1}{2} \lim_{\sigma \to 0} \left\{ 1 + \mathrm{erf}\left(\frac{x}{\sigma}\right) \right\} \qquad (\text{III.7.203})$$

and the fact that always $r + a > 0$ so that second error function in Eq. (III.7.202) tends to 1, when $t \to 0$. Thus, in the limit $t \to 0$, we recover the correct limiting case, the initial condition $c(r, \theta, \phi, 0) = c_0 \theta(a - r)$.

In Fig. III.7.21 we display plots of the obtained full solution in the $x{-}y$ plane in terms of $x/a$ and $y/a$ at various values of the dimensionless 'diffusion time' $t_R = tD/a^2$, showing how it evolves

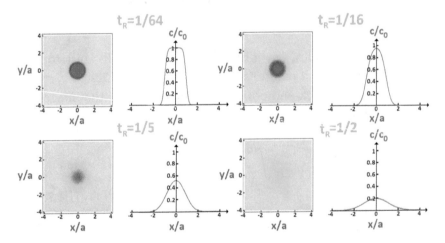

**Fig. III.7.21**   We show a sequence of pictures in time ('snapshots') generated in the $x{-}y$ plane of our diffusing drop of dye which we've made red, scaled by $a$ so that the initial radius of the droplet is 1. On the right of these pictures, we plot the concentration profile as a function of $x$ at $y = 0$ (along the $x$-axis). We start at a very short diffusion time $t_R = 1/64$, top left, after the drop has been introduced. At this point, the concentration profile is uniform across the centre of the droplet, but already on the droplet edges the dye is starting to spread out. At a slightly longer time $t_R = 1/16$, top right, across the centre of the dye the concentration is no longer uniform; the concentration profile starts to assume a more Gaussian character. At $t_R = 1/5$, bottom left, the drop is starting to look far more diffuse, looking at the concentration profile we see that the maximum concentration has decreased significantly and has spread out. At time $t_R = 1/2$, bottom right, it looks almost as if there is no dye visible; however, a concentration profile is still seen in the accompanying graph, even more spread out and with much lower maximum concentration at the centre. For longer times it will spread out even more and maximum concentration will get even lower.

with time. Note that due to spherical symmetry any other cross-section passing through the origin will give the same plot, for instance in the $y-z$ plane.

**Exercise.** Expand out Eq. (III.7.202) in the limit $\sqrt{Dt}/a \gg 1$ and check that the leading order term does indeed correspond to Eq. (III.7.200), as it should.

### Checking your result

A useful way of checking whether you have obtained the correct solution to the diffusion equation is to integrate the concentration over all space. This must remain a constant with respect to time in the diffusion process. Why? Physically, this integral gives the total number of particles which must be conserved (in the absence of chemical reactions). Mathematically, solutions with the initial condition $C_{\text{int}}(\mathbf{r})$ must obey the limiting behaviour $\mathbf{J}(\mathbf{r},t) = -D\vec{\nabla}c(\mathbf{r},t) \to 0$ when $|\mathbf{r}| \to \infty$. With this property it is easy to show from the diffusion equation that

$$\frac{\partial}{\partial t} \int c(\mathbf{r},t)d\mathbf{r} = 0 \qquad\qquad (\text{III.7.204})$$

by Gauss's divergence theorem, as illustrated through doing one of the exercises below.

**Exercise.** Integrate the first two sides of the diffusion equation, Eq. (III.7.179), over a sphere of radius $a$, making use of Gauss's divergence theorem (see Chapter 3). Express the r.h.s as an integral of $\mathbf{J}$ over the surface of the sphere and the l.h.s as $dN_a/dt$ where $N_a$ is the total number of diffusing particles enclosed by the sphere. This expression should make physical sense, because the surface integral gives the flux, the number flowing outside the sphere per unit time. Now taking into account that in the limit of $a \to \infty$, $\mathbf{J}(\mathbf{r},t) = -D\vec{\nabla}c(\mathbf{r},t) \to 0$, show that diffusion equation ensures that the total number of particles, $N_\infty$, remains constant.

**Exercise.** Mathematically check that the solution described by Eq. (III.7.202) does satisfy the requirement $\mathbf{J}(\mathbf{r},t) = -D\vec{\nabla}c(\mathbf{r},t) \to 0$

when $|\mathbf{r}| \to \infty$. Thus, the integral $\int_0^\infty r^2 c(r,t) dr$ must remain constant, as shown by doing the previous exercise. Show numerically, or otherwise, that Eq. (III.7.202) satisfies this second requirement, to check that it indeed is the correct solution (we ourselves actually did this to check that we indeed had the right answer to Example III.7.9; you should so as well!).

**Example III.7.10 (Physical Application ▶▶).** Here, we'll consider *a single particle* that undergoes random (Brownian) motion. The probability distribution $P(\mathbf{r},t)$ of finding the particle is governed by a diffusion equation. We have that

$$\int P(\mathbf{r},t) d\mathbf{r} = 1, \qquad \frac{\partial P(\mathbf{r},t)}{\partial t} = -D\vec{\nabla} P(\mathbf{r},t) \qquad \text{(III.7.205)}$$

Let's consider the case where we have to solve this diffusion equation in a limited region, again in a pipe with impermeable walls that we'll take to be of radius $b$ and very long. Initially, at time $t$, we know the position of the particle was at a point $(0,0,0)$ inside the pipe. Our initial condition is thus

$$P(\mathbf{r},0) = \frac{1}{2\pi R}\delta(z)\delta(R) \qquad \text{(III.7.206)}$$

where $z$ is the coordinate directed along the axis of the cylinder, and $R$ is the radial coordinate. This case is illustrated in Fig. III.7.22. You should check that Eq. (III.7.205) indeed satisfies the normalization condition (l.h.s of Eq. (III.7.205)) and realize that this must correspond to the position of a particle known with certainty to lie at $(0,0,0)$. The boundary condition at the pipe boundary is determined by no flow of the particles through it (we refer the reader to Example III.7.4 for a justification of it); namely

$$\left.\frac{\partial P(R,\phi,z,t)}{\partial R}\right|_{R=b} = 0 \qquad \text{(III.7.207)}$$

We follow the separation of variables technique by writing $P(\mathbf{r},t) = T(t)C_k(\mathbf{r})$, and so are able to write a general solution in the form of Eq. (III.7.183). The next part of the procedure is to find solutions to the Helmholtz equation, $C_k(\mathbf{r})$, in cylindrical geometry subject to

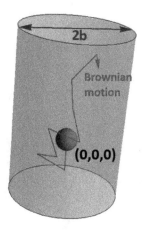

**Fig. III.7.22** In Example III.7.10 we consider a particle performing Brownian (random) motion in an impenetrable tube. We suppose that the particle starts at the position $(0,0,0)$ right at the centre of the tube. The probability density of finding the particle at successive times is described through a diffusion equation (Eq. (III.7.205)). The impenetrable walls place a spatial boundary condition on the diffusion, Eq. (III.7.207).

the boundary condition given by Eq. (III.7.207), as well as requiring $C_k(\mathbf{r}) \to 0$ when $z \to \infty$ and $C_k(\mathbf{r})$ being finite at $R = 0$. To find the $C_k(\mathbf{r})$ for cylindrical geometry we use the separation of variables method $C_k(\mathbf{r}) = \rho(R)\Phi(\phi)Z(z)$ in Eq. (III.7.181). We find that

$$\rho(R) = A_R J_n\left(R\sqrt{k^2 - k_z^2}\right) \quad \Phi(\phi) = A_\phi \exp(in\phi)$$

$$Z(z) = A_z \exp(ik_z z) \tag{III.7.208}$$

**Exercise.** Through the separation of variables technique, modify Eqs. (III.7.79)–(III.7.89) so that they apply to the Helmholtz equation. Then show that the solutions in Eq. (III.7.208) do indeed hold.

**Example III.7.10 (Continued).** However, because of the symmetry of the problem, we need only consider $n = 0$, as the solution will not depend on $\phi$. To satisfy the boundary condition requirement that $J_0'\left(b\sqrt{k^2 - k_z^2}\right) = 0$, we require $b\sqrt{k^2 - k_z^2} = x_{m,0}$, where $x_{m,0}$ are the positions of the maximum and minimum values of $J_0(x)$ (some of the first few values of $x_{m,0}$ were tabulated previously in Table III.7.1).

Thus, we can write

$$C_k(\mathbf{r}) = \sum_m A_m(k_z) J_0\left(\frac{R x_{m,0}}{b}\right) \exp(ik_z z) \qquad \text{(III.7.209)}$$

Using then Eqs. (III.7.183) and (III.7.209), we obtain

$$P(R, z, t) = \int_{-\infty}^{\infty} dk_z \sum_m A_m(k_z) J_0\left(\frac{R x_{m,0}}{b}\right) \exp(ik_z z) \exp(-D k_z^2 t)$$

$$\times \exp\left(-\frac{D(x_{m,0})^2 t}{b^2}\right) \qquad \text{(III.7.210)}$$

$$P(R, z, 0) = \frac{\delta(z)\delta(R)}{2\pi R} = \int_{=\infty}^{\infty} dk_z \sum_m A_m(k_z) J_0\left(\frac{R x_{m,0}}{b}\right) \exp(ik_z z)$$

$$\text{(III.7.211)}$$

using the fact that we can write $k^2 = k_z^2 + (x_{m,0}/b)^2$. The values of $A_m(k_z)$ are found by first multiplying Eq. (III.7.211) by $\exp(-ik_z' z)/2\pi$, and by integrating over $z$, so that

$$\frac{\delta(R)}{4\pi^2 R} = \sum_m A_m(k_z) J_0\left(\frac{R x_{m,0}}{b}\right) \qquad \text{(III.7.212)}$$

where we have used $\int_{-\infty}^{\infty} \delta(z) \exp(-ik_z' z) dz = 1$ and $\int_{-\infty}^{\infty} \exp(i(k_z - k_z')z) dz = 2\pi\delta(k_z - k_z')$. Next we can use the orthogonality condition for the Bessel functions, Eq. (III.7.104), to write

$$\frac{1}{4\pi^2} \int_0^R \delta(R) J_0\left(\frac{R x_{m,0}}{b}\right) dR = \frac{A_m(k_z) b^2 J_0(x_{m,0})^2}{2}$$

$$\Rightarrow A_m(k_z) = \frac{1}{2\pi^2 b^2} \frac{1}{J_0(x_{m,0})^2} \qquad \text{(III.7.213)}$$

So, $A_m$ does not actually depend on $k_z$. Thus, the full diffusion equation solution reads as

$$P(R, z, t) = \frac{1}{2\pi^2 b^2} \int_{-\infty}^{\infty} dk_z \sum_{m=0}^{\infty} J_0\left(\frac{R x_{m,0}}{b}\right) \frac{\exp(ik_z z) \exp(-D t k_z^2)}{J_0(x_{m,0})^2}$$

$$\times \exp\left(-\frac{D(x_{m,0})^2 t}{b^2}\right) \qquad \text{(III.7.214)}$$

We can immediately perform the integral over $k_z$, as we did in the previous example. This yields

$$P(R, z, t) = \frac{1}{2\pi b^2} \sqrt{\frac{1}{\pi Dt}} \exp\left(-\frac{z^2}{4Dt}\right) \sum_{m=0}^{\infty} J_0\left(\frac{Rx_{m,0}}{b}\right)$$

$$\times \frac{1}{J_0(x_{m,0})^2} \exp\left(-\frac{D(x_{m,0})^2 t}{b^2}\right) \qquad \text{(III.7.215)}$$

By introducing the rescaled variables $R/b$, $z/b$ and $t_R = tD/b^2$, we plot the approximate solution $\bar{P}(R, z, t) = b^3 P(R, z, t)$ in Fig. III.7.23, keeping terms up to $m = 4$ in Eq. (III.7.215), which is valid up to very short diffusion times and the values of $t_R$ shown.

**Exercise.** Show numerically for the diffusion times $t_R = 0.05$, $t_R = 0.15$ and $t_R = 0.45$ that the ratios between the $m = 5$ and the $m = 4$ terms in Eq. (III.7.215) are tiny. By thinking about the Bessel function zeros, demonstrate, without numerical calculation of it, that the ratio between the $m = 6$ and $m = 5$ terms will be considerably smaller. What does this tell you about the convergence of the series and the accuracy of the approximation used to generate the plots in Fig. III.7.23?

**Example III.7.10 (Continued).** Let's look at the limiting behaviour of Eq. (III.7.215). For large diffusion times for which $t \gg b^2/D$ things are simple: all terms in the sum for which $m \neq 0$ become negligible and we are just left with the $m = 0$ term. We note that $x_{0,0} = 0$, as $J_0(x)$ has a maximum at $x = 0$ and $J_0(0) = 1$, and so obtain

$$P(R, z, t) \approx \frac{1}{2\pi b^2} \sqrt{\frac{1}{\pi Dt}} \exp\left(-\frac{z^2}{4Dt}\right) \qquad \text{(III.7.216)}$$

This is simply the solution to the 1D diffusion equation along the axis of the tube with an initial condition $P(z, t) \approx \delta(z)/2\pi b^2$. This describes the situation of it being equally likely to find the particle at any position along the cross-section of the tube perpendicular to the $z$-axis. This is indeed what we start to see in the last of our plots in Fig. III.7.23.

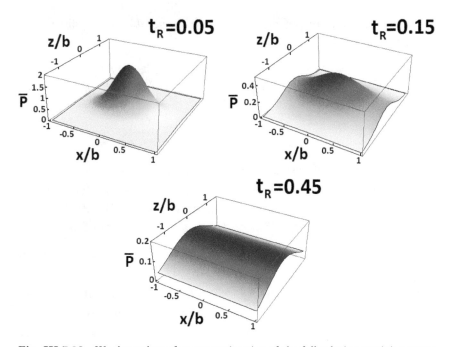

**Fig. III.7.23** We show plots of an approximation of the full solution retaining terms in the series in Eq. (III.7.221) up to $m = 4$ for different values of reduced, dimensionless time $t_R$, defined in the text, values $t_R = 0.05$, $t_R = 0.15$ and $t_R = 0.45$, as a function of $x/b$, across the tube [with $y = 0$, $R = \sqrt{x^2 + y^2}$], and $z/b$, along the tube. This approximation is excellent for all the values of $t_R$ considered. We leave it as an exercise (see in the text) to show that the $m = 5$ term is negligible and that the series converges rapidly for the values of $t_R$ considered. In the plot for $t_R = 0.05$ (top left plot), $\bar{P}$ is more or less what we would have had if there was no tube and there was unrestricted diffusion; the probability density is yet to be significant at the wall of the tube ($x/b = -1$ and $x/b = 1$). When we get to the value $t_R = 0.15$, $\bar{P}$ is starting to look markedly different than a diffusing particle in free space. At the wall, the gradient of $\bar{P}$ is zero in the x-direction, as is required by the boundary condition of an impermeable tube, as it is now 'condensing' near the axis (close to $z = 0$), $\bar{P}$ is significant at the wall— there is a good chance of finding the particle close to the wall. When we get to $t_R = 0.45$, there is little dependence of $\bar{P}$ on $x$ (and $R$), and now $\bar{P}$ is effectively described by a 1D diffusion equation (see the analysis in the text).

**Exercise.** Show that Eq. (III.7.216) is indeed a solution to the 1D diffusion equation

$$\frac{\partial P}{\partial t} = D \frac{\partial^2 P}{\partial z^2} \qquad (\text{III.7.217})$$

**Example III.7.10 (Continued).** To find the limiting case for small diffusion times for which $t \ll b^2/D$ is rather tricky. The analysis of this limiting case is therefore left boxed as optional material, which the reader might want to skip on a first read. However, the physically transparent result of this analysis is given just after the boxed material.

---

***Optional: Some tricky analysis.*** For small times $t \ll b^2/D$, the sum in Eq. (III.7.215) is dominated by small $R$ values due to the oscillatory behaviour of $J_0(Rx_{m,0}/b)$, and accordingly very large values of $x_{m,0}$, as $\exp(-D(x_{m,0})^2 t/b^2)$ in Eq. (III.7.215) remains small. Thus, we can use the asymptotic forms of the Bessel functions to write

$$J_0(x_{m,0}) \approx \sqrt{2/(\pi x_{m,0})} \cos\left(x_{m,0} - \pi/4\right) \qquad \text{(III.7.218)}$$

Also, from inspection of Eq. (III.7.218), we see that for large values $x_{m,0} \approx m\pi - 3\pi/4$. Thus, we can write

$$J_0(x_{m,0})^{-2} \approx \frac{\pi x_{m,0}}{2} \left(\cos\left(m\pi - \pi\right)\right)^{-2} = \frac{\pi x_{m,0}}{2} \qquad \text{(III.7.219)}$$

Now, we can use Eq. (III.7.219) to approximate Eq. (III.7.215) by

$$P(R, z, t) \approx \frac{1}{4\pi b R} \sqrt{\frac{1}{\pi D t}} \exp\left(-\frac{z^2}{4Dt}\right) \sum_m \tilde{k}_{m,0} J_0\left(\tilde{k}_{m,0}\right)$$

$$\times \exp\left(-D(\tilde{k}_{m,0})^2 t/R^2\right) \qquad \text{(III.7.220)}$$

where $\tilde{k}_{m,0} = R x_{m,0}/b$. Then, for small $R$, we can approximate the sum over $m$ by an integral over $\tilde{k}_{m,0}$:

$$P(R, z, t) \approx \frac{1}{4R^2} \sqrt{\frac{1}{\pi D t}} \exp\left(-\frac{z^2}{4Dt}\right) \int_0^\infty \tilde{k} J_0\left(\tilde{k}\right)$$

$$\times \exp\left(-D\tilde{k}^2 t/R^2\right) d\tilde{k}$$

$$\Rightarrow \quad P(R, z, t) \approx \frac{1}{4} \sqrt{\frac{1}{\pi D t}} \exp\left(-\frac{z^2}{4Dt}\right) \int_0^\infty k J_0\left(kR\right)$$

$$\times \exp\left(-Dk^2 t\right) dk \qquad \text{(III.7.221)}$$

We can do the integral in Eq. (III.7.221) by a trick that is worth learning. In Example III.4.7, we learnt that we could write

---

$$J_0\left(kR\right) = \frac{1}{2\pi}\int_0^{2\pi} \exp(ikR\cos\theta_k)d\theta_k \qquad \text{(III.7.222)}$$

In fact, due to the periodicity of the integrand, we can modify Eq. (III.7.222) to be

$$J_0\left(kR\right) = \frac{1}{2\pi}\int_0^{2\pi} \exp(ikR\cos(\theta_k - \phi_R))d\theta_k \qquad \text{(III.7.223)}$$

We can use Eq. (III.7.223) to write

$$P(R,z,t) \approx \frac{1}{8\pi}\sqrt{\frac{1}{\pi Dt}}\exp\left(-\frac{z^2}{4Dt}\right)\int_0^{2\pi} d\theta_k \int_0^\infty k\, dk$$
$$\times \exp(ikR\cos(\theta_k - \phi))\exp\left(-Dk^2 t\right)$$
$$\text{(III.7.224)}$$

Now, we use the identity $kR\cos(\theta_k - \theta_R) \equiv \mathbf{k.R}$, where $\mathbf{k} = k_x\hat{\mathbf{i}} + k_y\hat{\mathbf{j}} = k(\cos(\theta_k)\hat{\mathbf{i}} + \sin(\theta_k)\hat{\mathbf{j}})$ and $\mathbf{R} = x\hat{\mathbf{i}} + y\hat{\mathbf{j}} = R\left(\cos(\phi_R)\hat{\mathbf{i}} + \sin(\phi_R)\hat{\mathbf{j}}\right)$ so that

$$P(R,z,t) \approx \frac{1}{8\pi}\sqrt{\frac{1}{\pi Dt}}\exp\left(-\frac{z^2}{4Dt}\right)$$
$$\times \int \exp\left(-Dk^2 t\right)\exp(i\mathbf{k.R})d\mathbf{k} \qquad \text{(III.7.225)}$$

Eq. (III.7.225) is a 2D Fourier transform. Most importantly, we realize that we can evaluate it in the Cartesian coordinates $k_x = k\cos\theta_k$ and $k_y = k\sin\theta_k$. In these coordinates Eq. (III.7.225) reads as

$$P(R,z,t) \approx \frac{1}{8\pi}\sqrt{\frac{1}{\pi Dt}}\exp\left(-\frac{z^2}{4Dt}\right)\int_{-\infty}^\infty dk_x \int_{-\infty}^\infty dk_y$$
$$\times \exp\left(-D(k_x^2 + k_y^2)t\right)\exp(i(k_x x + k_y y))$$
$$\text{(III.7.226)}$$

The double integral here factorizes into the product of two independent integrals—one over $k_x$, another over $k_y$. Evaluation of those integral is elementary and gives the result shown below, outside this box.

The probability distribution in the limit $t \ll b^2/D$ reads as

$$P(R, z, t) \approx \frac{1}{8} \left( \frac{1}{\pi Dt} \right)^{3/2} \exp \left( -\frac{z^2 + y^2 + x^2}{4Dt} \right)$$

$$= \frac{1}{8} \left( \frac{1}{\pi Dt} \right)^{3/2} \exp \left( -\frac{z^2 + R^2}{4Dt} \right) \qquad \text{(III.7.227)}$$

Equation (III.7.227) is exactly what we would have for the probability distribution of a particle diffusing in space from the point $(0, 0, 0)$ in free space (no boundaries). This makes sense, as at small times, when the probability distribution is very narrow, there is little chance that the particle has reached the sides of the pipe, which will affect its motion. In other words, the random walker will yet know nothing about the sad fact that they are confined in a tube. A note, not important for this 'mathematical' example in which the randomly moving particle was point-like: in reality when the size of the random walker is not much smaller than the tube radius, the walker will learn about the confinement much earlier. This effect can be easily handled mathematically, but we now will not go into it.

We conclude this example by again emphasizing an important lesson that you should have learnt in working through it: scale matters! Here, the time, diffusion constant and width of the tube sets the scale of the problem. When we consider a wide tube, small diffusion constant and relatively short diffusion time, so that $t \ll b^2/D$ we can forget about the walls of the tube. Conversely when $t \gg b^2/D$ (and even for $t \sim b^2/D$) we can treat the diffusion as 1D diffusion problem in $z$. Only for intermediate values (the range depending on how accurate we need the solution to be) do we need to consider the series in Eq. (III.7.215).

### *The propagator method for solving the diffusion equation*

As opposed to writing a general solution of the form of Eq. (III.7.183) and solving the Helmholtz equation, if we have an initial condition $c_{\text{int}}(\mathbf{r}) = c(\mathbf{r}, t = 0)$ a powerful way of solving the diffusion equation (Eq. (III.7.179)) is through the propagator or diffusion-kernel method. The key to this method is first to consider the following initial condition:

$$c_{\text{int}}(\mathbf{r}) = c(\mathbf{r}, t = 0) = \delta(\mathbf{r} - \mathbf{r}') \tag{III.7.228}$$

For this initial condition, the solution to the diffusion equation (in infinite space) reads as

$$c_p(\mathbf{r} - \mathbf{r}', t) = \frac{1}{8} \left( \frac{1}{\pi D t} \right)^{3/2} \exp\left( -\frac{(\mathbf{r} - \mathbf{r}')^2}{4Dt} \right) \tag{III.7.229}$$

Equation (III.7.229) is called the diffusion kernel or *propagator*. We can use it to solve for any initial condition $c_{\text{int}}(\mathbf{r}) = c(\mathbf{r}, t = 0)$ by first writing

$$c_{\text{int}}(\mathbf{r}) = \int \delta(\mathbf{r} - \mathbf{r}') c_{\text{int}}(\mathbf{r}') d\mathbf{r}' \tag{III.7.230}$$

Now the key is to realize that the solution to Eq. (III.7.179) can be written as the supposition of solutions with initial conditions $\delta(\mathbf{r} - \mathbf{r}')$, each with a different value of $\mathbf{r}'$ (Eq. (III.7.229)). Thus, we can write the solution to the diffusion equation as

$$c(\mathbf{r}, t) = \int c_p(\mathbf{r} - \mathbf{r}', t) c_{\text{int}}(\mathbf{r}') d\mathbf{r}' = \frac{1}{8} \left( \frac{1}{\pi D t} \right)^{3/2}$$

$$\times \int c_{\text{int}}(\mathbf{r}') \exp\left( -\frac{(\mathbf{r} - \mathbf{r}')^2}{4Dt} \right) d\mathbf{r}' \tag{III.7.231}$$

Note that this method can also be applied to situations where the diffusion is controlled by spatial boundary conditions, provided that the form of propagator $c_p(\mathbf{r}, \mathbf{r}', t)$ (the concentration with condition $c_p(\mathbf{r}, \mathbf{r}', t = 0) = \delta(\mathbf{r} - \mathbf{r}')$) is known and is known to satisfy the particular boundary value problem.

**Exercise.** Use the initial condition $c_{\text{int}}(\mathbf{r}) = c_0 \theta(a - r)$ in Eq. (III.7.231) to show that the solution $c(\mathbf{r}, t)$ can indeed be written as Eq. (III.7.196). **Hint:** You will need to write Eq. (III.7.231) in spherical polar coordinates (i.e. $r$, $\theta$ and $\phi$) and choose $\mathbf{r}$ so that $\mathbf{r}.\mathbf{r}' = rr' \cos\theta$.

## 7.4. Separation of Variables for Other Partial Differential Equations

*Separation of variables for one family of partial differential equations*

Let us consider linear partial differential equations of the form

$$\nabla^2 \psi(\mathbf{r}) + \lambda(\mathbf{r})\psi(\mathbf{r}) = 0 \qquad \text{(III.7.232)}$$

More complicated forms of Eq. (III.7.232) than $\lambda(\mathbf{r})$ being simply constant with respect to $\mathbf{r}$ (which we have already considered) may arise for the Schrödinger equation of a particle where $\lambda(\mathbf{r}) = 2m(E - V(\mathbf{r}))/\hbar$, with $E$ being the particle energy eigenvalue and $V(\mathbf{r})$ its potential energy; and they are known to arise in other areas of physics (for instance in the solutions of the Fokker-Planck equation). In Cartesian coordinates, the separation of variables method (i.e. $\psi(\mathbf{r}) = X(x)Y(y)Z(z)$) will be of use only if we can write $\lambda(\mathbf{r}) = \lambda_x(x) + \lambda_y(y) + \lambda_z(z)$. For cylindrical polars we would require $\lambda(\mathbf{r}) = \lambda_R(R) + \frac{1}{R^2}\lambda_\phi(\phi) + \lambda_z(z)$ and for spherical polars we would require $\lambda(\mathbf{r}) = \lambda_R(r) + \frac{1}{r^2 \sin^2 \theta}\lambda_\phi(\phi) + \frac{1}{r^2}\lambda_\theta(\theta)$ to be able to make use of the separation of variables method. Alas, for other forms of $\lambda(\mathbf{r})$ the separation of variables technique is useless!

**Example III.7.11 (Physical Application ▶▶).** As a final example in this section, we'll look at solving the full 3D time-independent Schrödinger equation for motion in a central potential $V(\mathbf{r}) = V_{ct}(r = \|\mathbf{r}\|)$. This reads as

$$-\frac{\hbar^2}{2m}\nabla^2 \psi(\mathbf{r}) + V_{ct}(r)\psi(\mathbf{r}) = E\psi(\mathbf{r}) \qquad \text{(III.7.233)}$$

In this case, the Schrödinger equation is separable in spherical polar coordinates. Thus, using the form of Laplace operator in spherical coordinates, and writing $\phi(\mathbf{r}) = R(r)\Phi(\phi)\Theta(\theta)$, we obtain

$$\frac{\sin^2 \theta}{R(r)}\frac{d}{dr}\left(r^2\frac{dR(r)}{dr}\right) + \frac{\sin \theta}{\Theta(\theta)}\frac{d}{d\theta}\left(\sin \theta\frac{d\Theta(\theta)}{d\theta}\right)$$

$$+ \frac{1}{\Phi(\phi)}\frac{d^2\Phi(\phi)}{d\phi^2} + \frac{2m}{\hbar^2}r^2 \sin^2 \theta(E - V(r)) = 0 \quad \text{(III.7.234)}$$

As $\Psi(r, \theta, \phi) = \Psi(r, \theta, \phi + 2\pi)$, we must have that $\Phi(\phi) = \exp(im_l\phi)$ (where $m_l$ is an integer) so that

$$\frac{1}{R(r)} \frac{d}{dr}\left(r^2 \frac{dR(r)}{dr}\right) + \frac{2m}{\hbar^2} r^2 (E - V(r)) = \lambda$$

$$= -\frac{1}{\Theta(\theta)\sin(\theta)} \frac{d}{d\theta}\left(\sin\theta \frac{d\Theta(\theta)}{d(\theta)}\right) + \frac{m_l^2}{\sin^2\theta} \quad \text{(III.7.235)}$$

Again, as in previous cases, the r.h.s of Eq. (III.7.235) is Legendre's equation for which (for $\Theta(\theta)$ to remain finite at $\theta = 0$ and $\theta = \pi$) we require $\lambda = l(l+1)$ and the solutions are the associated Legendre functions $P_l^{m_l}(\cos\theta)$. Thus, we are finally left with the radial equation

$$\frac{1}{r^1}\frac{d}{dr}\left(r^2\frac{dR(r)}{dr}\right) + \frac{2m}{\hbar}(E - V(r))R(r) - \frac{l(l+1)}{r^2}R(r) = 0$$

$$\text{(III.7.236)}$$

For the hydrogen atom, the potential energy of the electron is electrostatic, and if we write it in Gaussian units,

$$V = -\frac{e^2}{r} \quad \text{(III.7.237)}$$

We have already solved Eq. (III.7.236) for this case (i.e. with Eq. (III.7.237)) in Example III.3.4. In general, the full solution for the hydrogen atom can be written as

$$\psi(r, \theta, \phi) = \psi_{n,l,m}(r, \theta, \phi) = R_{n,l}(r) Y_n^{m_l}(\theta, \phi) \quad E = E_n = -\frac{me^4}{2\hbar^2 n^2}$$

$$\text{(III.7.238)}$$

where $Y_n^{m_l}$ are the spherical harmonic functions defined by Eq. (III.7.119) and $R_{n,l}(r)$, the radial functions, are given by (as we saw before)

$$R_{n,l}(r) = \left(\frac{2}{n_l R_B}\right)^{3/2} \left(\frac{(n_l - l - 1)!}{2n(n_l + l)!}\right)^{1/2} \left(\frac{2r}{n_1 R_B}\right)^l$$

$$\times L_{n_l - l - 1}^{2l+1}\left(\frac{2r}{n_l R_B}\right) \exp\left(-\frac{r}{n_l R_B}\right) \quad \text{(III.7.239)}$$

where $L_q^p(s)$ are the previously discussed Laguerre polynomials.

## Separation of variables in general

Generally separation of variables, $\psi(\mathbf{r}) = u(x_1)v(x_2)w(x_3)$, works for any partial differential equation in a 3D coordinate system specified by three independent variables $x_1, x_2$ and $x_3$, if we are able to first write the resulting equation in the form

$$H_1\left(\frac{d^2u}{dx_1^2}, \frac{du}{dx_1}, x_1\right) = H_2\left(\frac{d^2v}{dx_2^2}, \frac{dv}{dx_2}, x_2\right) = H_3\left(\frac{d^2w}{dx_3^2}, \frac{dw}{dx_3}, x_3\right) = k$$

(III.7.240)

where the functions $H_1, H_2$, and $H_3$ depend on what partial differential equation we are considering. If we cannot write the equation in this form, the separation of variables method does not work.

We may still be able to tackle equations that we cannot separate through an appropriate change of variables (this is equivalent to finding a coordinate system where the partial differential equation is separable). Also we may be able to reduce the equation to separable form through a particular choice of 3D integral transforms, writing $\psi(\mathbf{r})$ in a different way, or through other methods, but we won't go into them here. For more discussion on this, we refer the reader to more specialist texts.

## 7.5. Key Points

- Many physical laws are encoded in partial differential equations. Thus, in deriving many predictive physical formulas we must solve such equations.
- One of the simplest second order partial differential equations we can deal with is Laplace's equation $\nabla^2\phi = 0$.
- The 2D and 3D Laplace equations in Cartesian coordinates may be solved by the separation of variables method, where we write $\phi(x, y) = X(x)Y(y)$ and $\phi(x, y, z) = X(x)Y(y)Z(z)$. This method allows us to obtain differential equations for $X(x)$, $Y(y)$ and $Z(z)$ (see Eqs. (III.7.4), (III.7.5), (III.7.7), (III.7.53), (III.7.54) and (III.7.55)).

- For a 2D-boundary value problem of the form $\phi(x,0) = \phi(x,y_1) = 0$ in a rectangular region $0 \le x \le x_1$ and $0 \le y \le y_1$, the specific solutions $X(x)$ and $Y(y)$ are $X_n(x) = A_n \exp(-\pi n x/y_1) + B_n \exp(\pi n x/y_1)$ and $Y_n(y) = C_n \sin(\pi n y/y_1)$. The most general solution to such a boundary value problem is $\phi(x,y) = \sum_{n=1}^{\infty} X_n(x) Y_n(y)$. We can specify the coefficients $A_n C_n$ and $B_n C_n$, through the Fourier series of boundary value functions $\phi(0,y) = g_A(y)$ and $\phi(x_1,y) = g_B(y)$.

- For boundary problems where $\phi(x,\infty) = \phi(x,-\infty) = 0$ and $0 \le x \le x_1$ and $-\infty \le y \le \infty$, the most general solution is $\phi(x,y) = \frac{1}{2\pi} \int_{-\infty}^{\infty} X(x,k) \exp(-iky) dk$ where $X(x,k) = A(k) \exp(kx) + B(k) \exp(-kx)$. The coefficients $A(k)$ and $B(k)$ may be determined through Fourier transforms of the boundary condition functions $\phi(0,y) = g_A(y)$ and $\phi(x_1,y) = g_B(y)$.

- Boundary value problems for the 3D Laplace equation in Cartesian geometry are dealt with in a manner similar to those involving the 2D Laplace equation. For instance, when we specify $\phi(x,0,z) = \phi(x,y_1,z) = \phi(0,y,z) = \phi(x_1,y_1,z) = 0$ in the cuboid region $0 < x < x_1$, $0 < y < y_1$ and $0 \le z \le z_1$, the most general solution that we may write satisfying these conditions is $\phi(x,y) = \sum_{n=1}^{\infty} \sum_{n=1}^{\infty} X_n(x) Y_m(y) Z_{n,m}(Z)$. Here, the functions are $X_n(x) = \sin(n\pi x/x_1)$, $Y_m(y) = \sin(m\pi y/y_1)$, and $Z_{n,m}(y) = C_{n,m} \exp((n^2\pi^2/x_1^2 + m^2\pi^2/y_1^2)^{1/2}z) + D_{n,m} \exp(-(n^2\pi^2/x_1^2 + m^2\pi^2/y_1^2)^{1/2}z)$. To specify the coefficients $C_{n,m}$ and $D_{n,m}$ we consider the 2D Fourier series of the boundary condition functions $\phi(x,y,0) = \Phi_1(x,y)$ and $\phi(x,y,z_1) = \Phi_2(x,y)$.

- When we deal with problems of cylindrical or spherical symmetry, we must write a partial differential equation that we want to solve in cylindrical polars $\{R, \phi, z\}$ or spherical polars $\{r, \theta, \phi\}$, respectively. These forms, for Laplace's equation, are given by Eqs. (III.7.78) and (III.7.110).

- The 3D Laplace equation written in both cylindrical and spherical polars is separable. This means that in order to solve $\nabla^2 \varphi = 0$, we may write $\varphi(R, \phi, z) = \rho(R)\Phi(\phi)Z(z)$, in cylindrical polars, and $\varphi(r, \theta, \phi) = \rho(r)\Phi(\phi)\Theta(\theta)$, in spherical coordinates.

- When we deal with an entire cylindrical regions, where $0 \leq \phi \leq 2\pi$, or entire spherical regions, for which $0 \leq \phi \leq 2\pi$ and $0 \leq \theta \leq \pi$, we have the important constraints that $\Phi(\phi) = \Phi(\phi + 2\pi)$ and $\Theta(\theta)$ must remain finite, and these impose crucial limitations on the allowed solutions.

- For Laplace's equation in cylindrical polar coordinates, with the solution defined in the full domain of $0 \leq \phi \leq 2\pi$, we have three types of solution:

  1. $\Phi(\phi) = \exp(in\phi), Z(z) = A_k \exp(Kz) + B_k \exp(-Kz),$
     $\rho(R) = C_{n,k}J_n(kR) + D_{n,k}Y_n(kR)$
  2. $\Phi(\phi) = \exp(in\phi), Z(z) = A + Bz, \rho(R) = C_nR^n + D_nR^{-n}$
  3. $\Phi(\phi) = \exp(in\phi), Z(z) = A_k \exp(ikz) + B_k \exp(-ikz),$
     $\rho(R) = C_{n,k}I_n(kR) + D_{n,k}K_n(kR)$

  where $n$ is an integer. To satisfy a particular boundary value problem we pick one, or more, of these solutions (see examples and text of Section 7.1).

- For Laplace's equation in spherical polar coordinates, for which the solution is defined in the domain $0 \leq \phi \leq 2\pi$ and $0 \leq \theta \leq \pi$, its general form is $\varphi(r, \theta, \phi) = \Sigma_{l=0}^{\infty}\Sigma_{m=-1}^{l}(A_rr^l + B_rr^{-l-1})Y_l^m(\theta, \phi)$. We set $B_r = 0$ if the solution is defined within a sphere about the origin. We set $A_l = 0$ if the solution is defined outside a sphere about the origin and extends to $r = \infty$.

- Examples of other partial differential equations that we examine in the text are: the Helmholtz equation $\nabla^2\varphi + \kappa^2\varphi = 0$; the wave equation $v^2\nabla^2\psi = \delta^2\psi/\delta t^2$; and the diffusion equation $\delta\psi/\delta t = D\nabla^2\psi$.

- The Helmholtz, wave equation and diffusion equation may all be dealt with by employing the separation of variables technique in Cartesian, cylindrical polar and polar coordinates subject to the system's geometry.

- For the time-dependent equations, we may first write for the solution $\psi(\mathbf{r}, t) = T(t)\varphi(\mathbf{r})$, where $\varphi(\mathbf{r})$ satisfies the Helmholtz equation. The method of solution of the Helmholtz equation is very similar to that of Laplace's equation (see examples in Section. 7.2 for details).
- A unique solution of the diffusion equation $\partial c/\partial t = D\nabla^2 c$ can be solved by specifying an initial concentration $c(\mathbf{r}, 0) = c_0(\mathbf{r})$.
- One powerful way of writing a solution to the diffusion equation is the propagator method. Here, we may write the concentration profile as $c(\mathbf{r}, t) = \int c_p(\mathbf{r}, \mathbf{r}'; t)c_0(\mathbf{r}')d\mathbf{r}$, where the propagator $c_p(\mathbf{r}, \mathbf{r}'; t)$ is the solution to the diffusion equation with initial condition $c_p(\mathbf{r}, \mathbf{r}'; 0) = \delta(\mathbf{r} - \mathbf{r}')$.
- In 3D free space, the propagator is $c_p(\mathbf{r} - \mathbf{r}', t) = \frac{1}{8}\left(\frac{1}{\pi Dt}\right)^{3/2}$ $\exp\left(-\frac{(\mathbf{r}-\mathbf{r}')^2}{4Dt}\right)$.

# Chapter 8

# Complex Contour Integration

*The theory of functions of complex variables* is one of the most beautiful branches of mathematics, with multiple physical applications. There are wonderful textbooks about it, and so we don't intend to go into any great depth here. Nevertheless, in this chapter we will give a brief overview of contour integration which helps not only  obtain exact formulae for the functions defined as integrals but also to get compact approximate, asymptotic expressions for them, when full analytical expressions are not possible.

Most of the functions we deal with in ordinarily life and research emerge from exploring the relationships between real quantities. We used to think about these functions as defined on the *real axis*. When the theory of functions of complex variables was born, its genius revelation was that a great deal of what a function can do for you (e.g, how it, standing in the integrand, will determine the integral) is determined 'on the complex plane', by analytical continuation of the function to the complex values of its argument(s). The physical meaning of the argument(s) usually (not always) makes sense only on the real axis, so attributing the argument a complex component sounds like a fiction, but the truth is that the fate of the function is determined by its properties on the complex plane. The complex plane, as we have seen in previous chapters on complex numbers and variables, also establishes an important connection between oscillatory functions and monotonically decaying or increasing functions, for instance $\cos(x)$ and $\cosh(y)$ are really manifestations of the same function $\cos(z)$ (with $z = x + iy$) defined across the whole complex plane. We will exploit this very important link in this chapter.

Before pushing on further with the business of this book we thought that we should entertain you a little. Mystics, and dabblers in hokum and hocus-pocus might feel more of a connection of complex numbers with our lives at our scale in the universe, such as our *real* life and fate is determined by something that is 'written somewhere in the skies—in heaven' (our complex plane!); but this may dangerously bring us to astrology, or other wonders. As scientists and users of applied mathematics, we will consider these not as nice metaphors, but focus in this chapter on practical consequences of the beautiful relations between the functions defined for their real arguments and their analytical continuations to the complex plane.

We complete this digression by saying something about one of the most important figures in this branch of mathematics: there were great founding fathers and heroes of it, though in the 19th century, perhaps the most important was Cauchy. His theorems about the analytical continuation of the function into the complex plane and

Augustin-Louis Cauchy
(1789–1857)

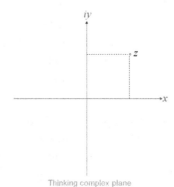

Thinking complex plane

the theory of residues have 'changed the world', making possible new methods of analysis. On a biographical note, Cauchy's father was a high official in the Paris police before the French revolution. Cauchy senior, and his family, fled Paris while Robespierre was causing people's heads to roll. Like his father, Cauchy was promoted as an important official in the later, Napoleonic era, working under Laplace (who was also in the French government—how unashamedly meritocratic it was!), and also trained as a successful engineer before becoming a mathematical great. We'll be looking at some of what Cauchy did.

Now back to business; we promised there will be no rigorous proofs of theorems. However, we'll make sense of the techniques we are going to use, to be able to use them properly, and with understanding. Before we do anything, we'll need to know how to integrate in the complex plane, dealing with notion of contour integration. Next, we will need to formulate Cauchy's integral theorem. A simple non-rigorous proof of it will be given in a box, helpful but not critical for getting a grasp of this topic. In the main text, we give an example to show how this theorem works. From this we will come, by considering an example where the integral theorem doesn't work (integration around a pole), to an even more important theorem for the evaluation of integrals: the *residue theorem*.

## 8.1. The Basics of Contour Integration

### *What is complex contour integration?*

So let's first talk about contour integrals of functions of complex variables $f(z)$. As we discussed previously, in Chapter 8 of Part II (Vol. 1), these are essentially functions of two variables, $x$ and $y$, that need to be connected through $z = x + iy$. *Contour integrals* are simply line integrals along paths in the complex plane; a particular path is denoted by a letter (usually $C$, which labels the *contour*) at the bottom of the integral sign. As we talked about when discussing functions of two variables, in Chapter 1 of Part II (Vol. 1), a particular path can be described through another (real) variable $t$, where

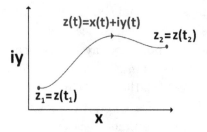

**Fig. III.8.1** An illustration of a contour integration path in the complex plane. The path is specified through a function $z(t) = x(t) + iy(t)$ of real variable $t$ that traces a curve in the complex plane. End points are specified by values $t = t_1$ and $t = t_2$ that deliver through the functional dependence of $x(t)$ and $y(t)$ the complex values of $z_1 = z(t_1)$ and $z_2 = z(t_2)$.

both $x$ and $y$ are functions of $t$, i.e. $x(t)$ and $y(t)$. Integration along a particular path in the complex plane, is then given by

$$\int_C f(z)dz = \int_{t_1}^{t_2} f(z(t))\frac{dz(t)}{dt}dt$$

$$= \int_{t_1}^{t_2} f(x(t) + iy(t))\left(\frac{dx(t)}{dt} + i\frac{dy(t)}{dt}\right)dt \quad \text{(III.8.1)}$$

where the end points are given by $z_1 = x(t_1) + iy(t_1)$ and $z_2 = x(t_2) + iy(t_2)$. A possible path of integration is drawn in Fig. III.8.1 to illustrate this.

Note that if we change the direction along the path, this is simply swapping $t_2$ with $t_1$. If we choose a contour or path that goes in the opposite direction to $C$ with the same end points, which we will denote by $C'$ we have

$$\int_{C'} f(z)dz = -\int_C f(z)dz \quad \text{(III.8.2)}$$

**Example III.8.1 (Maths Practice ♫♪).** To start introducing you to contour integration, we begin with a rather unconventional example. This comes about by revisiting Eq. (I.3.67) of Chapter 3 of Part II (Vol. 1), where we encountered the integral

$$I = \int_{-1}^{2} \frac{dx}{(x-1)} \quad \text{(III.8.3)}$$

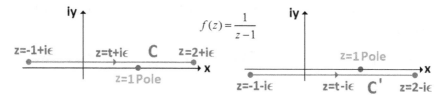

**Fig. III.8.2** Here, we show two ways we can define the integral $I$ given by Eq. (III.8.3). In one case, shown on the left, we take an integration contour $C$ that sits just above the pole in $f(z) = (z-1)^{-1}$, specified by $z(t) = t + i\varepsilon$, going from end points $z = -1 + i\varepsilon$ to $z = 2 + i\varepsilon$. This defines the integral $I_\varepsilon$ which can be denoted by $I^+$ (given by Eq. III.8.6), when $\varepsilon \to 0$. This is one possible definition of $I$. In a second case the path of integration $C'$ sits below the pole, in the complex plane. This path is specified by $z(t) = t - i\varepsilon$ with end points $z = -1 - i\varepsilon$ and $z = 2 - i\varepsilon$. This defines the integral $I_{-\varepsilon}$ named $I^-$ (given by Eq. (III.8.9)), when $\varepsilon \to 0$, on second definition of $I$.

As we discussed previously, this integral wasn't well defined; especially so with respect to the logarithms of negative numbers that come up in the answer. To define the integral, the trick is to consider a contour of integration in the complex plane slightly off the real axis of the function $f(z) = 1(z-1)$. We specify a path $C$ given by $x = t$ and $y = i\varepsilon$ (illustrated in Fig. III.8.2), where $-1 \le t \le 2$ so that

$$I_\varepsilon = \int_C \frac{dz}{z-1} = \int_{-1}^{2} dt \frac{1}{t + i\varepsilon - 1} = \ln(1 + i\varepsilon) - \ln(-2 + i\varepsilon) \quad \text{(III.8.4)}$$

Using the fact that $\ln z = \ln|z| + i\theta$, where $|z|$ and $\theta$ are the modulus and argument of $z$, we can rewrite Eq. (III.8.4) as

$$I_\varepsilon = \ln(\sqrt{1 + \varepsilon^2}) - \ln(\sqrt{4 + \varepsilon^2}) + i\arctan(\varepsilon) - i\arctan(-\varepsilon/2) + i\pi \tag{III.8.5}$$

Now, we could define our original integral (Eq. (III.8.3)) as

$$I = I^+ = \lim_{\varepsilon \to 0} I_\varepsilon = i\pi - \ln 2 \tag{III.8.6}$$

However, this definition is not the only one we could make. We could also consider a path $C'$ given by $x = t$ and $y = -i\varepsilon$, where $-1 \le t \le 2$. In this case we would write

$$I_{\varepsilon-} = \int_{C'} \frac{dz}{z-1} = \int_{-1}^{2} dt \frac{1}{t - i\varepsilon - 1} = \ln(1 - i\varepsilon) - \ln(-2 - i\varepsilon) \tag{III.8.7}$$

This second logarithm is on the other side of the branch cut of $\ln z$, and thus we should write

$$I_{-\varepsilon} = \ln(\sqrt{1+\varepsilon^2}) - \ln(\sqrt{4+\varepsilon^2}) + i\arctan(\varepsilon) - i\arctan(\varepsilon/2) - i\pi$$

$$\text{(III.8.8)}$$

This leads to an alternative definition of $I$, which reads as

$$I = I^- = \lim_{\varepsilon \to 0} I_{-\varepsilon} = -i\pi - \ln 2 \qquad \text{(III.8.9)}$$

In addition, we could also define what is called the principle (or real) part of Eq. (III.8.3)

$$I = \frac{1}{2}(I^+ + I^-) = -\ln 2 \qquad \text{(III.8.10)}$$

We see that there are three ways we can define $I$. Nevertheless, as long as we are clear which way we want to present it, we can make the integral be well defined. Sometimes, how we choose to define such integrals can be very important for the physics that they describe. Never let yourself relax in this respect!

**Example III.8.2 (Maths Practice ♪♪).** Let's evaluate the contour integral

$$I = \int_C \frac{1-z^2}{1+z^2} \frac{dz}{z} \qquad \text{(III.8.11)}$$

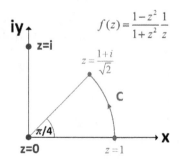

**Fig. III.8.3** We sketch the path of integration considered in Example III.8.2, which is part of unit circle with end points $z = 1$ and $z = (1 + i)/\sqrt{2}$. Thick dots show two of the poles of the integrand $f(z) = (1 - z^2)(1 + z^2)^{-1}z^{-1}$ in this quadrant.

along the path $C$, shown in Fig. III.8.3, specified by $x = \cos\theta$ and $y = \sin\theta$, where $\theta$ is the polar angle, with $0 \le \theta \le \pi/4$. In this case we may write $z = \exp(i\theta) = \cos\theta + i\sin\theta$. Using Eq. (III.8.1), we can thus rewrite Eq. (III.8.11) as

$$I = \int_0^{\pi/4} \left(\frac{1 - \exp(2i\theta)}{1 + \exp(2i\theta)}\right) \frac{1}{\exp i\theta} \frac{d(\exp i\theta)}{d\theta} d\theta$$

$$= i \int_0^{\pi/4} \left(\frac{\exp(-i\theta) - \exp(i\theta)}{\exp(-i\theta) + \exp(i\theta)}\right) d\theta$$

$$\Rightarrow \quad I = 2 \int_0^{\pi/4} \frac{\sin\theta}{\cos\theta} d\theta = 2[-\ln(\cos\theta)]_0^{\pi/4}$$

$$= -2\ln(\cos(\pi/4)) = 2\ln\sqrt{2} = \ln 2 \qquad (III.8.12)$$

## *Cauchy's integral theorem*

Now imagine that we have two paths that start and finish at the same end points $z_1$ and $z_2$ in the complex plane. We may describe one path through the variable $t$ and the other through another variable $t'$, so that for one path we have $z = z_1(t) = x_1(t) + iy_1(t)$ and for the other we have $z = z_2(t') = x_2(t') + iy_2(t')$. Along the paths, we may have that $t_1 \le t \le t_2$ and $t_1' \le t' \le t_2'$, and generally for values of $t$ and $t'$ (except for the end points, $z_1$ and $z_2$) we have $z_1(t) \ne z_2(t')$. We'll denote these two paths as $C_1$ and $C_2$, as illustrated in Fig. III.8.4.

Cauchy's integral theorem states that as long as the function $f(z)$ is completely analytic between two paths—which is another way of saying that the derivative of the function with respect to the complex variable $z$, i.e. $f'(z) = df(z)/dz$, must be finite and well defined, as well as $f(z)$ being finite, throughout that region—then,

$$\int_{C_1} f(z)dz = \int_{C_2} f(z)dz \qquad (III.8.13)$$

This means that for Eq. (III.8.13) to hold, there must be no cuts or poles (such features were discussed in Chapter 8, Part II, Vol. 1) within the area bounded by the two paths.

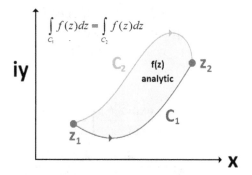

**Fig. III.8.4** This figure illustrates one formulation of Cauchy's integral theorem. Two paths in the complex plane $C_1$ (shown in blue) and $C_2$ (shown in green) start from the same end point $z_1$ and terminate at the same end point $z_2$ (shown in red). If the integrand $f(z)$ is analytic in the region between them (yellow) then the contour integrals over those paths evaluate to the same value, i.e. Eq. (III.8.13) holds.

The theorem is usually stated in a different way, where we consider a closed path. A closed path is one which starts at a point in the complex plane $z(t_1)$ and returns back to the same point. Such a closed path can be constructed by considering a path in the opposite direction to $C_2$, which we denote by $C_2'$, going from $t_2$ to $t_1$. We then have a closed path where we go along $C_1$ and return by $C_2'$, and this whole close loop path which we'll denote by $C$, as illustrated in Fig. III.8.5:

$$\oint_C f(z)dz = \int_{C_1} f(z)dz + \int_{C_{2'}} f(z)dz = \int_{C_1} f(z)dz - \int_{C_2} f(z)dz$$
$$(\text{III.8.14})$$

Here the symbol (circle) in the integral sign denotes a closed path and in what direction we move about it. Also, we should point out that any closed path (as long $f(z)$ remains analytic along its path) the starting position on the closed path doesn't matter. **Think:** This is not actually hard to prove, but how would you show this?

Now, if Eq. (III.8.13) holds, then it must follow that

$$\oint_C f(z)dz = 0 \qquad\qquad (\text{III.8.15})$$

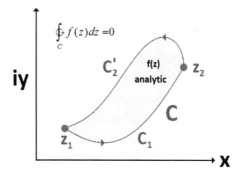

**Fig. III.8.5** An illustration of how reversing the path $C_2$ (considered Fig. III.8.4) to $C_2'$, leads of the classic formulation of Cauchy's integral theorem for closed paths. Namely, that the integral of a complex function $f(z)$ around a closed path is zero if $f(z)$ is analytic over the whole region contained within that path. In the case illustrated, the closed path of integration is labelled by $C$ which comprises the paths $C_1$ and $C_2'$.

This is the standard version of Cauchy's integral theorem. For an interested reader, a short non-rigorous proof is provided in the box below (but it is optional and can be skipped at first reading). It requires knowledge of Green's theorem and the Cauchy–Riemann equations for analytic functions of a complex variable, which were discussed in Part II (Vol. 1). However, we will put Eq. (III.8.15) to the test by considering a simple example below. In cases where Cauchy's integral theorem holds, the direction around the closed path doesn't matter. However, we'll show that when we integrate around poles the direction of integration does actually matter (in this case, as we shall show, Eq. (III.8.15) doesn't apply).

---

*Optional: A simple proof of Cauchy's integral theorem.* We can give a simple non-rigorous proof of Eq. (III.8.15) by considering the l.h.s of the equation as a line integral around a closed path:

$$\oint_C f(z)dz = \oint_C [P(x,y)dx + Q(x,y)dy] \qquad \text{(III.8.16)}$$

where $P(x,y) = f(x+iy)$ and $Q(x,y) = if(x+iy)$. Application of Green's theorem (see Chapter 1 of Part III, Vol. 2) then yields

$$\oint_C f(z)dz = \int_A \left[ \frac{\partial Q(x,y)}{\partial x} - \frac{\partial P(x,y)}{\partial y} \right] dxdy$$

$$= \int_A \left[ i\frac{\partial f(x+iy)}{\partial x} - \frac{\partial f(x+iy)}{\partial y} \right] dxdy \qquad \text{(III.8.17)}$$

where $A$ is the area enclosed by the contour. The requirement for $f(z)$ to be analytic throughout the whole of $A$ (see the box in Chapter 8, Part II, Vol. 1) is that

$$\frac{\partial f(x+iy)}{\partial x} = -i\frac{\partial f(x+iy)}{\partial y} \qquad \text{(III.8.18)}$$

everywhere in $A$. Thus, if this holds, the area integral vanishes and we have Eq. (III.8.15).

**Example III.8.3 (Maths Practice ♫♪).** Here, we demonstrate that Cauchy's integral works for an analytical function of a complex variable. Let's consider a simple function $f(z) = z^2$, which is analytical over the entire complex plane, and consider a closed path around (i) a unit square and (ii) unit circle, both centred at the origin of the complex plane.

Let's consider the first path about the square (which we'll denote by $S$). To specify the path, we start at the point $z_1 = 1/2 - i/2$ and move to the point $z_2 = 1/2 + i/2$, along the path $C_1$ specified by $x = 1/2$ and $y = t_1$ where $-1/2 \leq t_1 \leq 1/2$. Next, we move from $z_2 = 1/2 + i/2$ to $z_3 = -1/2 + i/2$ along path $C_2$, specified by $x = t_2$ and $y = 1/2$ where $-1/2 \leq t_2 \leq 1/2$. Then, we move from $z_3 = -1/2 + i/2$ to $z_4 = -1/2 - i/2$ along path $C_3$ specified by $x = -1/2$ and $y = t_3$ where $-1/2 \leq t_3 \leq 1/2$. Finally, we return to point $z_1$ from point $z_4$ along the path $C_4$ specified by $x = t_4$ and $y = -1/2$, where $-1/2 \leq t_4 \leq 1/2$. The full path is illustrated in Fig. III.8.6. Thus, putting everything

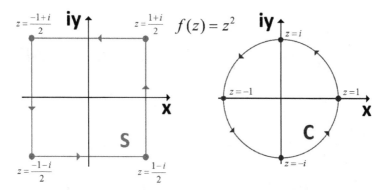

**Fig. III.8.6** We put Cauchy's integral theorem to the test by considering contour integrals over closed paths of the completely analytic function $f(z) = z^2$. One path is a square with vertices $z = (1 + i)/2$, $z = (1 - i)/2$, $z = -(1 + i)/2$, and $z = -(1 - i)/2$ (left panel). The second path is a unit circle ($|z| = 1$), centred at the origin (right panel). The contour integrals over both paths evaluate to zero as shown in Example III.8.3.

together, we have

$$\oint_S z^2 dz = \int_{C_1} z^2 dz + \int_{C_2} z^2 dz + \int_{C_3} z^2 dz + \int_{C_4} z^2 dz$$

$$\Rightarrow \quad \oint_S z^2 dz = \int_{-1/2}^{1/2} z(t_1)^2 \frac{dz(t_1)}{dt_1} dt_1 + \int_{1/2}^{-1/2} z(t_2)^2 \frac{dz(t_2)}{dt_2} dt_2$$

$$+ \int_{1/2}^{-1/2} z(t_3)^2 \frac{dz(t_3)}{dt_3} dt_3 + \int_{-1/2}^{1/2} z(t_4)^2 \frac{dz(t_4)}{dt_4} dt_4$$

$$\Rightarrow \quad \oint_S z^2 dz = \int_{-1/2}^{1/2} z(t_1)^2 \frac{dz(t_1)}{dt_1} dt_1 - \int_{-1/2}^{1/2} z(t_3)^2 \frac{dz(t_3)}{dt_3} dt_3$$

$$- \int_{-1/2}^{1/2} z(t_2)^2 \frac{dz(t_2)}{dt_2} dt_2 + \int_{-1/2}^{1/2} z(t_4)^2 \frac{dz(t_4)}{dt_4} dt_4$$

$$(\text{III.8.19})$$

To calculate these four integrals, we need to write the explicit dependence of the integrands on the variables of integration. These are:

$$\frac{dz(t_1)}{dt_1} = \frac{d}{dt_1}(1/2 + it_1) = i \qquad z(t_1)^2 = (1/4 - t_1^2) + it_1 \qquad \text{(III.8.20)}$$

$$\frac{dz(t_2)}{dt_2} = \frac{d}{dt_2}(t_2 + i/2) = 1 \qquad z(t_2)^2 = (t_2^2 - 1/4) + it_2 \qquad \text{(III.8.21)}$$

$$\frac{dz_3(t)}{dt_3} = \frac{d}{dt_3}(1/2 + it_3) = i \qquad z(t_3)^2 = (1/4 - t_3^2) + it_3 \qquad \text{(III.8.22)}$$

$$\frac{dz(t_4)}{dt_4} = \frac{d}{dt_4}(t_4 - i/2) = 1 \qquad z(t_4)^2 = (t_4^2 - 1/4) - it_4 \qquad \text{(III.8.23)}$$

Hence, we have

$$\oint_S z^2 dz = i \int_{-1/2}^{1/2} [(1/4 - t_1^2) + it_1]dt_1 - i \int_{-1/2}^{1/2} [(1/4 - t_3^2) + it_3]dt_3$$

$$+ \int_{-1/2}^{1/2} [(t_4^2 - 1/4) - it_4]dt_4 - \int_{-1/2}^{1/2} [(t_2^2 - 1/4) + it_2]dt_2 = 0$$

$$\text{(III.8.24)}$$

Thus, we see that in this case Eq. (III.8.15) works.

Now, let's see if the same happens when we consider integrating $f(z) = z^2$ about a unit circle. Here, it is convenient to write $z$ in polars so that $z = |z|\exp(i\theta)$. This path we'll denote by $C$ (also shown in Fig. III.8.6), and we'll specify the contour or path as $z = \exp(i\theta)$. Here, returning back to the same point is going from $\theta = 0$ to $\theta = 2\pi$, so we have

$$\oint_C z^2 dz = \int_0^{2\pi} z(\theta)^2 \frac{dz(\theta)}{d\theta} d\theta \qquad \text{(III.8.25)}$$

Now, $z(\theta)^2 = \exp(2i\theta)$ and we have that $dz(\theta)/d\theta = i\exp(i\theta)$. Substituting these in the integral we get

$$\oint_C z^2 dz = i \int_0^{2\pi} \exp(3i\theta)d\theta$$

$$= i \int_0^{2\pi} \cos(3i\theta)d\theta - \int_0^{2\pi} \sin(3i\theta)d\theta = 0 \qquad \text{(III.8.26)}$$

So again Eq. (III.8.15) works and it works for any closed path—as it should, if the function is analytical inside the contour.

**Example III.8.4 (Maths Practice ♫♪).** Let's now put Cauchy's integral theorem to good use and reconsider the evaluation of a type of integral that has come up many times in our physical examples:

$$I = \int_{-\infty}^{\infty} \exp(-a(x + i\gamma)^2)dx \qquad (III.8.27)$$

First we can write Eq. (III.8.27) as a contour integral:

$$I = -\int_{C_1} \exp(-az^2)dz \qquad (III.8.28)$$

where the path of integration $C_1$ is specified along $z = x + i\gamma$ from $x = \infty$ to $x = -\infty$. Now, let us consider the closed path shown in Fig. III.8.7 around a box with vertices $z_1 = -a$, $z_2 = a$, $z_3 = a + i\gamma$ and $z_4 = -a + i\gamma$. The function $f(z) = \exp(-\alpha z^2)$ is entirely analytic in the complex plane; thus Cauchy's integral theorem tells us that

$$0 = \oint_C \exp(-az^2)dz$$
$$= \int_{C_1} \exp(-az^2)dz + \int_{C_2} \exp(-az^2)dz + \int_{C_3} \exp(-az^2)dz$$
$$+ \int_{C_4} \exp(-az^2)dz \qquad (III.8.29)$$

In the limit $a \to \infty$, the contributions from contours $C_2$ and $C_4$ at $x = \infty$ and $x = -\infty$ are infinitesimally small as $\exp(-ax^2) \to 0$. The path of integration $C_3$ is specified along $z = x$ from $x = \infty$ to $x = -\infty$, so that

$$\int_{C_3} \exp(-az^2)dz = \int_{-\infty}^{\infty} \exp(-ax^2)dx \qquad (III.8.30)$$

Thus, we have justified that

$$I = \int_{-\infty}^{\infty} \exp(-a(x + i\gamma)^2)dx = \int_{-\infty}^{\infty} \exp(-ax^2)dx = \sqrt{\frac{\pi}{a}} \qquad (III.8.31)$$

We used this result, Eq. (III.8.31), in some of the physical examples we considered before.

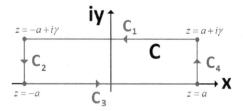

**Fig. III.8.7**   In the evaluation of the integral given by Eq. (III.8.27), we consider this closed rectangular contour $C$, with vertices $z = a$, $z = -a$, $z = a + i\gamma$ and $z = -a + i\gamma$. The function $f(z) = \exp(-az^2)$ is completely analytical, so the integral of it around this path evaluates to zero. We can divide the path $C$ into four separate paths: $C_1$, $C_2$, $C_3$ and $C_4$. In the limit where $a \to \infty$, the contributions shown in blue vanish ($C_2$ and $C_4$) and the contributions shown in red survive ($C_1$ and $C_3$). In this limit the integral along $C_1$ becomes the integral we want to evaluate multiplied by a minus sign. The integral along $C_3$ becomes the standard integral of the Gaussian function $f(x) = \exp(-ax^2)$ between $x = -\infty$ and $x = \infty$, leading us to Eq. (III.8.31).

**Example III.8.5 (Maths Practice ♫♪).** Let's again put Cauchy's integral theorem to good use to evaluate another integral, which will be also the basis of the *stationary phase approximation* and *the saddle point approximation* in the method of steepest descent for approximating integrals, that we will consider soon—all under the slogan of this chapter: 'going into the complex plane!' This example is itself a good illustration of how one can use the complex plane and the Cauchy integral theorem to deal with rapidly oscillating functions by converting them into monotonically decaying functions, by a trick that is termed '*rotation of the contour of integration*'. Let us consider the integral

$$I = \int_{-\infty}^{\infty} \cos(-\alpha t^2)\,dt \qquad\qquad (\text{III.8.32})$$

If we were to plot the integrand as a function of $t$ on the real axis ($t = x$), we would see that this function oscillates faster and faster as $x$ increases. How do we determine such an integral? First, the convergence of this integral is conditional, and we will discuss this in a moment, after we evaluate this integral using complex analysis, by first converting the integrand into a Gaussian. Indeed, as subject to Euler's equation $\cos(u) = \text{Re}[\exp(iu)]$, we can rewrite Eq. (III.8.32) as

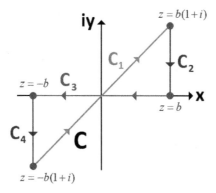

**Fig. III.8.8** A closed path that we can use to evaluate the integral, Eq. (III.8.32). The total path $C$, consists of $C_1$ along the path $z(t) = t \exp(i\pi/4)$ from $z = -b(1+i)$ to $z = b(1+i)$, followed by path $C_2$ that goes to $z = b$, then path $C_3$ that crosses over $C_1$ going to $z = -b$. Finally, the path is closed by $C_4$ that goes from $z = -b$ to $z = -b(1+i)$. By Cauchy's integral theorem the integral of $f(z) = \exp(\alpha z^2)$ over $C$ is zero. In the limit where we take $b \to \infty$, the integral along the contour $C_1$ can be related back to the original integral through Eq. (III.8.34), whereas, in this limit, the integral along $C_3$ becomes the integral of $f(x) = \exp(-\alpha x^2)$ with limits of integration $x = -\infty$ and $x = \infty$. The contributions from $C_2$ and $C_4$, just about vanish in this limit.

$$I = \text{Re} \int_{-\infty}^{\infty} \exp(-i\alpha t^2)dt \qquad (III.8.33)$$

Next, we can express the integral $I$ in terms of a contour integral $f(z) = \exp(-\alpha z^2)$ along the path $C_1$ specified by $z = x(t) + iy(t) = t \exp(i\pi/4)$ from $t = -b$ to $t = b$, in the limit $b \to \infty$ (see Fig. III.8.8), so that Eq. (III.8.33) becomes

$$I = \text{Re}\left[\exp(-i\pi/4) \lim_{b\to\infty} \int_{C_1} \exp(-\alpha z^2)dz\right] \qquad (III.8.34)$$

**Exercise.** Show that Eq. (III.8.34) does indeed hold by applying Eq. (III.8.1) to the path $z(t) = t \exp(i\pi/4)$.

Now, we can construct a closed path, $C$, as shown in Fig. III.8.8. Again, to such a path we can apply Cauchy's integral theorem, although of a slightly complicated topology. To see why Cauchy's integral theorem is applicable, one should consider the interception between $C_1$ and $C_3$ path sections. This point may be thought of as the junction between two simple triangular closed paths for the integral; along each of them the integral is zero. The total closed path

$C$ is the sum of these two closed paths, and so the integral around it must also be zero. Thus, we may write

$$0 = \oint_C \exp(-\alpha z^2)dz$$

$$= \int_{C_1} \exp(-\alpha z^2)dz + \int_{C_2} \exp(-\alpha z^2)dz + \int_{C_3} \exp(-\alpha z^2)dz$$

$$+ \int_{C_4} \exp(-\alpha z^2)dz \qquad\qquad\qquad\qquad \text{(III.8.35)}$$

The contributions from $C_2$ and $C_4$ vanish when $b \to \infty$.

**Technical point.** Strictly speaking, the convergence of the integral Eq. (III.8.32) is conditional: we met this concept in Part I (Vol. 1)]. This is the same for other integrands that contain purely oscillatory functions. However, what the integral Eq. (III.8.33) exactly convergences to can be specified in a consistent manner. This may be done by specifying $\cos(\alpha t^2)\exp(-\varepsilon t^2)$ for our integrand in Eq. (III.8.32) instead of $\cos(\alpha t^2)$, where $\varepsilon$ is an infinitesimally small real positive number. This infinitesimally small number converges the integral and so causes both $C_2$ and $C_4$ to vanish.

The contribution $C_3$ is then along $z = x$ from $x = \infty$ to $x = -\infty$. Thus, we may write

$$I = \int_{-\infty}^{\infty} \cos(\alpha t^2)dt = \text{Re}\left[\exp(-i\pi/4) \int_{-\infty}^{\infty} \exp(-\alpha t^2)dt\right]$$

$$\Rightarrow \quad I = \frac{1}{\sqrt{2}} \int_{-\infty}^{\infty} \exp(-\alpha t^2)dt = \sqrt{\frac{\pi}{2\alpha}} \qquad\qquad \text{(III.8.36)}$$

Why does $I$ diminish with $\alpha$? First we could have rescaled the original integral with $s = \sqrt{\alpha}t$, $I = \int_{-\infty}^{\infty} \cos(\alpha t^2)dt = \frac{1}{\sqrt{\alpha}}\int_{-\infty}^{\infty} \cos(s^2)ds$. The non-trivial point here is that $\int_{-\infty}^{\infty} \cos(s^2)ds$ converges conditionally, and when we converge it in the systematic way described above, it becomes just the number $\sqrt{\pi/2}$. But this is a formal answer. In common sense terms, larger values of $\alpha$ make the integrand oscillating faster, actually—faster and faster with increase of $t$ (because of $t^2$ in the argument), so that the integral 'accumulates' less.

In general, we can use similar constructions to evaluate much nicer integrals of the form, which converge unconditionally:

$$I = \int_{-\infty}^{\infty} \cos(\alpha t^2)\exp(-\beta t^2)dt \quad I = \int_{-\infty}^{\infty} \sin(\alpha t^2)\exp(-\beta t^2)dt$$

$$\text{where } \beta \geq 0 \qquad \text{(III.8.37)}$$

The reader may start to appreciate that complex contour integration is a very powerful tool, as we can use it to convert integrals over troublesome oscillating functions (which could be difficult to calculate numerically to sufficient accuracy) into ones with easy-to-handle decaying functions.

**Exercise.** Evaluate

$$I = \int_{-\infty}^{\infty} \sin(\alpha t^2)\exp(-\alpha t^2)dt \qquad \text{(III.8.38)}$$

by considering $f(z) = \exp(-\sqrt{2}\alpha z^2)$ and an appropriate closed integration contour.

### The complex anti-derivative and definite integration over complex numbers

Let's consider two functions of a complex variable $f(z)$ and $F(z)$ which are analytic at all points in some region of the complex plane, for which we may write

$$\frac{dF(z)}{dz} = f(z) \qquad \text{(III.8.39)}$$

Then, we can define a complex 'anti-derivative' (the integral) such that

$$F(z) = \int f(z)dz \qquad \text{(III.8.40)}$$

This allows us to define the definite integration of analytic functions in the complex plane:

$$\int_{z_1}^{z_2} f(z)dz = F(z_2) - F(z_1) \qquad \text{(III.8.41)}$$

Equation (III.8.41) holds provided that, within that region, Eq. (III.8.39) is well defined, and the contour integral between the points $z_2$ and $z_1$ is independent of path. Indeed, the latter must be so for $f(z)$ to be analytic. We won't be proving these statements; you may, if you like, find a proof somewhere else, but all this is rather obvious.

**Example III.8.6 (Maths Practice ♪♪).** The definite integral

$$I = \int_0^{\alpha+i\beta} z \exp(-z^2)dz \qquad \text{(III.8.42)}$$

is well defined over the entire complex plane, as the function $f(z) = z \exp(-z^2)$ is entirely analytic. Here, we have that

$$-\frac{1}{2}\frac{d}{dz}[\exp(-z)^2] = z \exp(-z^2) \qquad \text{(III.8.43)}$$

Therefore, according to Eq. (III.8.41)

$$2I = 1 - \exp(-(\alpha+i\beta)^2) = 1 - \exp(\beta^2 - \alpha^2 - 2i\alpha\beta) \qquad \text{(III.8.44)}$$

**Example III.8.7 (Maths Practice ♪♪).** We now give an important and classic demonstration of a situation where Cauchy's integral theorem fails, if only to highlight that in these kinds of cases we will need to use another theorem. Consider a contour integral of a function $f(z)$ about a point $z = 0$ where $f(z)$ has a first order pole. Here, we'll just consider a path around a unit circle: $z = \exp(i\theta)$, $0 \le \theta \le 2\pi$, and consider the simplest example of such function: $f(z) = 1/z$. The path of integration is shown in Fig. III.8.9—a unit radius circle around a pole at $z = 0$. Before we proceed, it is important to realize that both $f(0)$ and $f'(0)$ are infinite, and so the function is not analytic at the point $z = 0$—this is the reason why, as will see, Cauchy's integral theorem will not work here. But we want to see the difference; for this, let's evaluate the integral and see what we get for an answer. The contour integral evaluates as

$$\oint_C \frac{1}{z}dz = \int_0^{2\pi} \frac{1}{z(\theta)}\frac{dz(\theta)}{d\theta}d\theta = i\int_0^{2\pi} d\theta = 2\pi i \qquad \text{(III.8.45)}$$

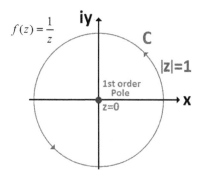

$$f(z) = \frac{1}{z}$$

**Fig. III.8.9** This an illustration of a situation where Cauchy's integral theorem doesn't work. We consider integrating the function $f(z) = 1/z$ about a unit circle (for which $|z| = 1$). The closed path of integration is drawn. Within the centre of the unit circle at $z = 0$ there lies a pole, where $f(z)$ is infinite and therefore not analytic. This is the reason why Cauchy's theorem breaks down.

Here, the *direction we circulate matters*; if we go in the opposite direction we get

$$\oint_{C'} \frac{1}{z}dz = \int_{2\pi}^{0} \frac{1}{z(\theta)} \frac{dz(\theta)}{d\theta}d\theta = -2\pi i \qquad (III.8.46)$$

Actually, this value of $2\pi i$ doesn't depend on what path we take around the pole, only its sign depends on the direction we circulate—clockwise or anticlockwise. To see why this is so, just look at Fig. III.8.10. Using Eq. (III.8.13), we can deform any closed path provided that the contour crosses no poles or cuts when we deform it. It also doesn't matter where the pole is, provided that our contour goes about that pole.

## Contour integration around poles

Now, let's consider what happens when $f(z) = 1/z^{2+n}$, where the power is an integer with $n \geq 0$. Evaluating the integral about the same unit circle, in fact we have

$$\oint_{C} \frac{1}{z^{2+n}}dz = \int_{0}^{2\pi} \frac{1}{z(\theta)^{2+n}} \frac{dz(\theta)}{d\theta}d\theta = i \int_{0}^{2\pi} \exp(-i(1+n)\theta)d\theta = 0$$
$$(III.8.47)$$

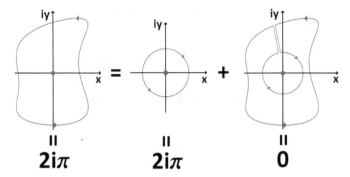

**Fig. III.8.10**   In this figure we show how $1/z$ integrated over an arbitrary closed contour evaluates to $2\pi i$, if that contour encloses the pole at $z = 0$. The arbitrary integration contour shown on the left can be decomposed into the integration over a unit circle (middle graph), which we evaluated to be $2\pi i$ plus a second closed contour, shown on the right. Note that this is indeed so, because the contour about the inner circle of the contour, on the right-most graph, cancels out the unit circle integration shown in the middle graph. Also, in the contour displayed on the right-most graph, the two connecting lines (shown in green) between the inner circle and the outer path, may also be deformed to lie on top of each other and cancel each other out. Then by Cauchy's integral theorem, integration of $1/z$ over the right-most closed contour yields zero.

in spite of these functions not satisfying the analytic requirement normally needed for Cauchy's integral theorem to apply. So, what is going on? Let's try to understand it, and for this we will need to do some auxiliary investigations.

Let us consider $f(z) = g(z)/(z - z_0)$, where $g(z_0)$ is finite and analytic, but $f(z)$ 'blows up' at $z = z_0$. The contour integral of $f(z)$ about closed path $C$ about the pole at $z = z_0$ can be written as

$$\oint_C \frac{g(z)}{z - z_0}\, dz \qquad\qquad (III.8.48)$$

The trick in evaluating Eq. (III.8.48) is to deform the contour to a path which is very small circle about $z = z_0$ of radius $\varepsilon$. Now in this limit, $\lim_{\varepsilon \to 0} g(z) = g(z_0)$. Therefore, we can write

$$\oint_C \frac{g(z)}{z - z_0}\, dz = g(z_0) \oint_C \frac{1}{z - z_0}\, dz = 2\pi i g(z_0) \qquad (III.8.49)$$

Now, let's consider

$$\oint_C \frac{g(z)}{(z - z_0)^2}\, dz \qquad\qquad (III.8.50)$$

This does not generally evaluate to zero. Special care has to be taken now, close to $z_0$; we need to use the Taylor series for analytic functions and write

$$g(z) = g(z_0) + g'(z_0)(z - z_0) + \frac{g''(z_0)}{2}(z - z_0)^2 + \cdots \qquad \text{(III.8.51)}$$

Substituting this Taylor series, Eq. (III.8.51), into Eq. (III.8.50) we obtain

$$\oint_C \frac{g(z)}{(z - z_0)^2} dz = g(z_0) \oint_C \frac{1}{(z - z_0)^2} dz + g'(z_0) \oint_C \frac{1}{(z - z_0)} dz$$
$$+ \frac{g''(z_0)}{2} \oint_C dz + \cdots \qquad \text{(III.8.52)}$$

Now, all terms in this series vanish, except for the one that depends on $g'(z_0)$, so that we obtain

$$\oint_C \frac{g(z)}{(z - z_0)^2} dz = g'(z_0) \oint_C \frac{1}{(z - z_0)} dz + \cdots = 2\pi i g'(z_0) \qquad \text{(III.8.53)}$$

So it is equal to 0 if $g(z) = g_0$, where $g_0$ is a constant.

Last, but not least, using similar arguments as Eqs. (III.8.50)–(III.8.53), we can show that

$$\oint_C = \frac{g(z)}{(z - z_0)^m} dz = 2\pi i \frac{1}{(m - 1)!} \lim_{z \to z_0} \frac{d^{m-1} g(z)}{dz^{m-1}} \qquad \text{(III.8.54)}$$

**Exercise.** Derive Eq. (III.8.54) by considering the complex Taylor series Eq. (III.8.51).

What we have obtained in Eq. (III.8.54) is much more important than just a rather formal integral result, an 'example'. In fact, all of this is crucial for understanding one of the most important laws of the complex analysis: the *residue theorem*. We present this later, in Section 6.2, as well as examples of applications of this powerful theorem. But before doing so, we need to formulate and explain a few further useful rules.

### Rules for the deformation of contours

It is worth summarizing some rules about how we can manipulate integration over contours, which we'll simply state, and you should

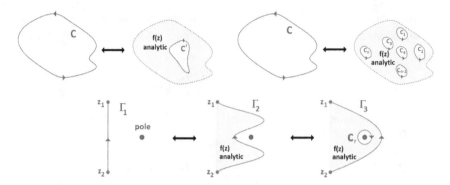

**Fig. III.8.11** An illustration of some of the basic rules for deforming and manipulating closed contours, the black arrows indicate that we may go either way. We show in the top left corner that we may deform one closed contour $C$ into another closed contour $C'$, or as shown on top r.h.s deform and divide the contour into many contours $C_1, C_2, \ldots, C_n$, provided that $f(z)$ is analytic in the yellow shaded regions, i.e. no poles or cuts. We may deform a contour over a pole, here shown for a path $\Gamma_1$ provided that we include a residual contour around the pole. The integral of $f(z)$ over $\Gamma_1$ must equal the sum of the integrations over $\Gamma_3$ and $C_r$.

learn. Understanding them relies on common sense, based on what we have already discussed. With Cauchy's integral theorem, non-rigorous proofs of rules 2 and 3 are rather simple, but this book is not about proofs, and we leave it up to the reader to justify these rules (looking at Fig. III.8.10 you should already have a good idea how to do so). These rules are summarized in Fig. III.8.11, so you can just look at it, but we also list them below.

1. We may deform the contour of integration $C_1$ between two points $z_1$ and $z_2$ into another contour $C_2$, *between the same two points*, provided that the integrand $f(z)$ is *analytic everywhere between these two curves.*

2. We may deform a closed path of integration $C$ into another closed path of integration $C'$, provided that the integrand $f(z)$ is *analytic everywhere between these two curves.* So that we may write

$$\oint_C f(z)dz = \oint_{C'} f(z)dz \qquad \text{(III.8.55)}$$

3. We may divide up a closed path of integration $C$ into many other smaller closed paths $C_1, C_2, \ldots, C_n$ provided that $f(z)$ remains

analytic in the region between these close curves and path $C$. The integral over $C$ becomes the sum of all the integrals over the smaller contours, i.e.

$$\oint_C f(z)dz = \sum_{j=1}^{n} \oint_{C_j} f(z)dz \qquad \text{(III.8.56)}$$

4. A contour of integration may not be deformed, with the integral maintaining the same value, by cutting through a pole or branch cut. Unless, in the case of the pole, we leave behind a residual closed contour around the pole. The value of the integral is conserved if we do this. This is because these are places at which $f(z)$ *is not analytic.*

## 8.2.   The Residue Theorem and Its Applications

### *The golden theorem of contour integration*

The *residue theorem*, introduced by Cauchy in 1825, reads as follows. Let a function $f(z)$ be analytical on the complex plane of its argument $z$, except at particular points $a_k$ where it has poles of order $m_k$; label $k$ distinguishes the pole, and the integer $m$ indicates its *order* (recall the definition of the pole order in Chapter 8, Part II, Vol. 1)]. We will suppose that there are $n$ such points. Then, for integration of such function over a contour, as shown in Fig. III.8.12, where the path of integration should be taken anti-clockwise

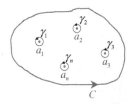

**Fig. III.8.12**   This figure illustrates the residue theorem which we may use to evaluate a closed contour containing poles at multiple points $a_1, a_2 \ldots a_n$. Using one of our rules for deforming contours, the integral of $f(z)$ about the path $C$ is the same a the sum of the integrals about closed paths $\gamma_1, \gamma_2 \ldots \gamma_n$ as in Eq. (III.8.58).

(going the other direction changes the sign of the integral), the following equality holds:

$$\oint_C f(z)dz = 2\pi i \sum_{k=1}^{n} \text{res}\{f(z)\}_{z=a_k} \qquad \text{(III.8.57)}$$

or, equivalently,

$$\oint_C f(z)dz = 2\pi i \sum_{k=1}^{n} \oint_{\gamma_k} f(z)dz \qquad \text{(III.8.58)}$$

where the $\gamma_k$ are contours about individual poles (Fig. III.8.12).

The symbol "res" stands for a *residue* of a pole. Its value depends on the order $m$ and its mathematical definition is

$$\text{res}\{f(a_k)\}_{z=a_k} = \frac{1}{(m_k-1)!} \lim_{z \to a_k} \frac{d^{m_k-1}}{dz^{m_k-1}} \{(z-a_k)^m f(z)\}$$

$$\text{(III.8.59)}$$

Let's try to understand why this theorem should hold. Note that we can write $f(z) = g_k(z)/(z-a_k)^{m_k}$, where $g_k(z)$ is a function that is finite at a point $z = a_k$. Then, each integral (along paths $\gamma_k$ about points $z = a_k$) can be evaluated in similar fashion as Eq. (III.8.54). From this it is not hard to see that if Eq. (III.8.58) holds (which is appropriately deforming the closed contour through one of our rules discussed previously), then Eq. (III.8.57) should also hold.

The particular case of a first order pole (i.e. $m = 1$) looks particularly simple. In this case Eq. (III.8.59) reduces to

$$\text{res}\{f(z)\}_{z=a_k} = \lim_{z \to a_k} \{(z-a_k)f(z)\} \qquad \text{(III.8.60)}$$

Let us consider function $f(z)$ having only first order poles; we'll derive a useful formula. We may represent $f(z)$ as a fraction $f(z) = \phi(z)/\psi(z)$, where $\psi(z)$ and $\phi(z)$ are analytic over the entire complex plane. In order to have the first order poles of $f(z)$ at $a_k$, $\psi(z)$ must have first order zeros at $a_k$, and $\phi(z)$ remain finite and non-zero. Then, for $z$ close to $a_k$, we can write $\psi(z) \approx \psi'(a_k)(z-a_k)$, and thus

$\frac{\psi(z)}{(z-a_k)} = f'(a_k)$, so that for each first order residue

$$\text{res}\,\{f(z)\}_{z=a_k} = \lim_{z \to a_k} \left\{ \frac{\phi(z)}{\psi(z)}(z - a_k) \right\} = \lim_{z \to a_k} \frac{\phi(z)}{\frac{\psi(z)}{(z-a_k)}} = \frac{\phi(a_k)}{\frac{d\psi(z)}{dz}\Big|_{z=a_k}}$$

$$(\text{III.8.61})$$

It is straightforward to extend the representation $f(z) = \phi(z)/\psi(z)$ to $f(z)$ having any order poles. Simply, we require $\psi(z)$ to have an $m$ order of zero where $f(z)$ has an $m$ order pole.

**Exercise.** Extend the residue formula given by Eq. (III.8.61) to a second order pole, where $\psi(z)$ has a second order zero. **Hint:** Think about the form of a Taylor series in $\psi(z)$ about a second order zero.

**Example III.8.8 (Maths Practice ♫♪).** Let's consider using the residue theorem to calculate a contour integral for an arbitrary closed path

$$I(1) = \oint_C f(z)dz = \oint_C \frac{1}{z^2 + 1} dz \qquad (\text{III.8.62})$$

This integrand has two poles that lie at $z = \pm i$ and we can write

$$I(1) = \oint_C \frac{dz}{(z + i)(z - i)} \qquad (\text{III.8.63})$$

Now, from Eq. (III.8.57), the integral in Eq. (III.8.63) could evaluate to 0, $2\pi i \,\text{res}\{f(z)\}_{z=i} + 2\pi i \,\text{res}\{f(z)\}_{z=-i}$, or $2\pi i \,\text{res}\{f(z)\}_{z=i}$ and $2\pi i \,\text{res}\{f(z)\}_{z=-i}$, depending on whether we choose the closed path about no poles, about both poles, or either pole, respectively, as shown in and described in the caption to Fig. III.8.13.

Let's now calculate the two residues

$$\text{res}\{f(z)\}_{z=i} = \lim_{z \to i} \left\{ \frac{(z - i)}{(z - i)(z + i)} \right\} = \lim_{z \to i} \left\{ \frac{1}{(z + i)} \right\} = \frac{1}{2i}$$

$$(\text{III.8.64})$$

$$\text{res}\{f(z)\}_{z=-i} = \lim_{z \to -i} \left\{ \frac{(z + i)}{(z - i)(z + i)} \right\} = \lim_{z \to -i} \left\{ \frac{1}{(z - i)} \right\} = -\frac{1}{2i}$$

$$(\text{III.8.65})$$

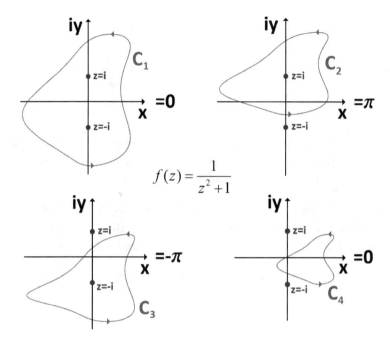

**Fig. III.8.13**  We show four possible contours, with the integrand $f(z) = (z^2 + 1)^{-1}$, over a closed path integration. When (top left corner) we have a contour $C_1$ that encloses both poles the integral evaluates to zero, as that is the sum of the two residues. Note that in this case this is purely coincidental (!) for our choice of $f(z)$; it does not hold generally for functions with two poles at $z = i$ and $z = -i$. When we have a contour that encloses the $z = i$ pole in the anti-clockwise direction (top right corner, $C_2$) the contour integral evaluates to $\pi$. Conversely, when the integration contour encloses the pole at $z = -i$ in an anti-clockwise direction (bottom left corner, $C_3$) the contour integral evaluates to $-\pi$. Lastly, when the integration contour contains no poles (bottom right corner, $C_4$), the contour integral of course evaluates to zero.

When we enclose only one pole, we get

$$I(1) = \oint_C \frac{dz}{(z+i)(z-i)} = \pi \qquad \text{if } C \text{ encloses only the } z = i \text{ pole}$$

$$\text{(III.8.66)}$$

$$I(1) = \oint_C \frac{dz}{(z+i)(z-i)} = -\pi \qquad \text{if } C \text{ encloses only the } z = -i \text{ pole}$$

$$\text{(III.8.67)}$$

If we enclose both poles or no poles at all we get $I(1) = 0$.

**Example III.8.9 (Maths Practice ♫♪).** Let's put what we've learnt to practice and consider the real integral

$$I(b) = \int_{-\infty}^{\infty} \frac{dk}{k^2 + b^2} \qquad \text{(III.8.68)}$$

First working neatly, we can first rescale $k = bx$, so that we get

$$I(b) = \frac{1}{b} \int_{-\infty}^{\infty} \frac{dx}{x^2 + 1} = \frac{1}{b} I(1) \qquad \text{(III.8.69)}$$

Now, we can evaluate the remaining integral in two ways; we'll show that they arrive at the same result. One way is to not bother about the complex plane and simply make a substitution, $x = \tan\theta$, in which case the integral becomes

$$I(b) = \frac{1}{b} \int_{-\pi/2}^{\pi/2} d\theta = \frac{\pi}{b} \qquad \text{(III.8.70)}$$

This is of course the simplest way, but now we'll do the same evaluation using residues in the complex plane, just to learn the technique. Pushing you to use here the residue theorem has only pedagogical value, but we'll extend this in the next example to something of physical importance, where we can only evaluate the integral by complex integration.

The idea is to consider an integration contour which is a semicircle in either the upper or lower half of the complex plane enclosing one of the poles, as is illustrated in Fig. III.8.14. Here, we'll consider the top half and enclose the $z = i$ pole. Then, using the residue theorem we get

$$I(1) = \oint_C \frac{dz}{z^2 + 1} = \pi \qquad \text{(III.8.71)}$$

Now, we can write

$$I(1) = \int_{C_1} \frac{dz}{z^2 + 1} + \int_{C_2} \frac{dz}{z^2 + 1} \qquad \text{(III.8.72)}$$

Here, contour $C_1$ lies along the real axis so that $z = x$, and $-R \geq x \geq R$ so that

$$\int_{C_1} \frac{dz}{z^2 + 1} = \int_{-R}^{R} \frac{dx}{x^2 + 1} \qquad \text{(III.8.73)}$$

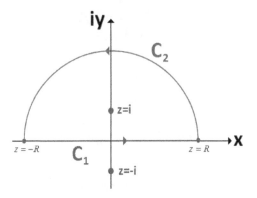

**Fig. III.8.14** The closed integration contour used to evaluate the integral given by Eq. (III.8.69). It comprises two parts. The first is $C_1$ (shown in purple) that goes along the real axis from $z = -R$ to $z = R$. The second is $C_2$ (shown in red) which is the path around a circle of radius $R$, starting at $z = R$ and terminating at $z = -R$. At $|R| \to \infty$ the integration over $C_1$ will return the integral that we want to evaluate, whereas the integral over $C_2$ will vanish.

For the other path $C_2$ we have that $z(\theta) = R \exp i\theta$, where $0 \le \theta\pi$ so that we have

$$\int_{C_2} \frac{dz}{z^2 + 1} = -\int_0^\pi \frac{iR \exp(i\theta)d\theta}{R^2 \exp(2i\theta) + 1} \qquad \text{(III.8.74)}$$

Now, let's consider when $R \to \infty$:

$$\lim_{R \to \infty} \int_{C_1} \frac{dz}{z^2 + 1} = \int_{-\infty}^{\infty} \frac{dx}{x^2 + 1} \quad \text{and}$$

$$\lim_{R \to \infty} \int_{C_2} \frac{dz}{z^2 + 1} = -\lim_{R \to \infty} \frac{i}{R} \int_0^\pi \exp(-i\theta)d\theta = 0 \qquad \text{(III.8.75)}$$

Thus, we get

$$I(b) = \frac{1}{b} \int_{-\infty}^{\infty} \frac{dx}{x^2 + 1} = \frac{\pi}{b} \qquad \text{(III.8.76)}$$

In the considered case, we had a choice; we could have enclosed the pole $z = -i$, by considering a path along the real axis combined with a semicircle in the bottom half of the plane. The answer, of course, would have been exactly the same.

**Exercise.** Make a closed contour about $z = -i$ considering the paths $z = x$, with $-R \le x \le R$ and $z(\theta) = R \exp(i\theta)$, where $-\pi \le \theta \le 0$.

The value of the integral counter-clockwise around this path is given by Eq. (III.8.67). Show that when $R \to \infty$ the result is again Eq. (III.8.76).

**Example III.8.10 (Maths Practice ♫♪).** Let's do contour integration to evaluate the integral

$$I(s) = \int_{-\infty}^{\infty} \frac{1}{x^2 + 1} \exp(-isx) dx \qquad (III.8.77)$$

We can do this integral in the way that we did for $s = 0$, as covered in the previous example. We look at closed paths $C^+$ and $C^-$ made of a straight line along the real axis and semi-circle, encompassing one of the poles at $z = \pm i$, respectively. However, a little more care is needed here in the choice of integration contour: bottom half or top half? Let us consider the paths, $C_2^+$ and $C_2^-$, on the semicircles of radius $R$, in the upper and lower half of the complex plane, respectively. For the path $C_2^+$, $z(\theta) = R\exp(i\theta)$ where $0 \le \theta \le \pi$, we have for $R \gg 1$

$$\int_{C_2^+} \frac{1}{z^2 + 1} \exp(isz) dz \approx \frac{i}{R} \int_0^\pi \exp(-i\theta) \exp(-Rs(\sin\theta - i\cos\theta)) d\theta$$

$$(III.8.78)$$

Now, if $s > 0$, when $R \to \infty$ the contribution over $C_2^+$ becomes infinitesimally small, which is what we want! However, this is not the case when $s < 0$; in this case the exponential term $\exp(-Rs\sin\theta)$ in the integrand grows as $R$ increases and the contribution over $C_2^+$ diverge, instead of vanishing. Now, let us consider $C_2^-$ in the lower half of the plane, with $z(\theta) = R\exp(i\theta)$ where $-\pi \le \theta \le 0$. Here, we have for $R \gg 1$

$$\int_{C_2^-} \frac{1}{z^2 + 1} \exp(isz) dz$$

$$\approx \frac{i}{R} \int_{-\pi}^0 \exp(-i\theta) \exp(-Rs(\sin\theta - i\cos\theta)) d\theta$$

$$\approx \frac{i}{R} \int_0^\pi \exp(i\theta) \exp(Rs(\sin\theta + i\cos\theta)) d\theta \qquad (III.8.79)$$

We see that the opposite is true; when $s < 0$, the integral in Eq. (III.8.79) becomes infinitesimally small when $R \to \infty$, whereas for $s > 0$ the integral blows up.

Thus, we see that for $s > 0$, we need to close our contour of integration in the top half of the plane (contour $C$), and for $s < 0$ we need to close it in the bottom half (contour $C'$). Part of the art of using complex contour integration is choosing an appropriate contour, by understanding how integrands behave as functions in the complex plane in various limits.

Let us proceed: for $s > 0$, we should enclose the $z = i$ pole, using contour $C$, and thus we can write (in the limit $R \to \infty$)

$$I = \oint_C \frac{1}{z^2 + 1} \exp(isz)dz = \pi \exp(-s) \qquad \text{(III.8.80)}$$

where we have calculated the residue of the integrand at $z = i$. For $s < 0$, we should enclose the $z = -i$ pole using contour $C'$. In this case, we get

$$I \oint_{C'} \frac{1}{z^2 + 1} \exp(isz)dz = \pi \exp(s) \qquad \text{(III.8.81)}$$

Note that in the derivation of Eq. (III.8.81) closing the contour into the lower semi-plane makes the contour running not anti-clockwise, but clockwise. Thus using the residue theorem equation we were obliged to take the residue with the negative sign. Therefore, combining Eqs. (III.8.80) and (III.8.81) our integral evaluates to

$$I(s) = \pi \exp(-|s|) \qquad \text{(III.8.82)}$$

**Example III.8.11 (Maths Practice ♫♪).** When considering Fourier transforms in Chapter 2 of Part I (Vol. 1), we came across an integral of the form

$$I = \int_{-\infty}^{\infty} \frac{\sin(px)}{x} dx \quad p > 0 \qquad \text{(III.8.83)}$$

It is sometimes, unjustly, referred to as the Laplace–Dirichlet integral, but it was actually first evaluated in 1781 by Euler. Simple rescaling $s = px$ shows us that the integral doesn't, in fact, depend on $p$:

$$I = \int_{-\infty}^{\infty} \frac{\sin(x)}{x}dx = \frac{1}{i}\int_{-\infty}^{\infty} \frac{\exp(ix)}{x}dx \qquad \text{(III.8.84)}$$

Here, we have used the fact that $\exp(ix) = \cos x + i\sin x$, and the integrand $\cos x/x$ is an odd function and so integrates to zero.

Looking at the form (III.8.84), we may introduce a function of a complex variable $z$, $f(z) = e^{iz}/(iz)$, and a contour of integration for it, shown in Fig. III.8.15. Integration over this full contour is zero, because this function $f(z)$ has no singularities inside of it. Thus, we can write

$$\int_{-R}^{-r} f(z)dz + \int_{C_r} f(z)dz + \int_{r}^{R} f(z)dz + \int_{C_R} f(z)dz = 0$$

$$\text{(III.8.85)}$$

The contour $C_R$ which is specified by $z(\theta) = R\exp(i\theta)$ for which $0 \leq \theta \leq \pi$, and so

$$\int_{C_R} f(z)dz = \int_{0}^{\pi} \exp(-R(\sin\theta + i\cos\theta))d\theta \qquad \text{(III.8.86)}$$

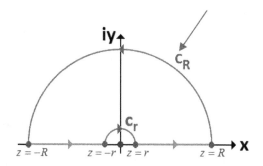

**Fig. III.8.15** We display a closed integration contour that can be used to evaluate the integral given by Eq. (III.8.82). This contour comprises four parts. Two of these are integration along the real axis from $-R$ to $-r$, and from $r$ to $R$. Then there is integration around two semi-circles centred about the pole at $z = 0$ (mid point) of radius $r$ and $R$, paths $c_r$ and $C_R$, respectively.

The contour $c_r$ is specified by $z(\theta) = r \exp(i\theta)$ for which $0 \le \theta \le \pi$. Therefore,

$$\int_{c_r} f(z)dz = -\int_0^\pi d\theta \exp\{-r(\sin\theta + i\cos\theta)\} \qquad \text{(III.8.87)}$$

Note that the minus sign is because we are going from $\theta = \pi$ to $\theta = 0$, i.e. clockwise, but not anti-clockwise. Next, we can take the limits $R \to \infty$ and $r \to 0$ of the l.h.s of Eq. (III.8.85). We note that

$$\lim_{\substack{R \to \infty \\ r \to 0}} \left\{ \int_{-R}^{-r} f(z)dz + \int_r^R f(z)dz \right\} = \int_{-\infty}^\infty \frac{\exp(ix)}{ix} dx = I$$

$$\text{(III.8.88)}$$

$$\lim_{R \to \infty} \int_{c_R} f(z)dz = 0 \quad \lim_{r \to 0} \int_{c_r} f(z)dz = -\int_0^\pi d\theta = -\pi$$

$$\text{(III.8.89)}$$

Combining Eqs. (III.8.85), (III.8.88) and (III.8.89), we obtain

$$I = \pi \qquad \text{(III.8.90)}$$

**Exercise.** Use the same method of scaling and complex integration to evaluate

$$I = \int_{-\infty}^\infty \frac{1 - \cos(px)}{x^2} dx \qquad \text{(III.8.91)}$$

**Example III.8.12 (Maths Practice ♫♪).** We can use complex contour integration to evaluate integrals of the class

$$I = \int_0^{2\pi} F(\cos\theta, \sin\theta)d\theta \qquad \text{(III.8.92)}$$

where $F(\cos\theta, \sin\theta)$ is a rational function of both $\cos\theta$ and $\sin\theta$ with no singularities for $0 \le \theta \le 2\pi$. As a simple case, let's consider

$$I = \int_0^{2\pi} \frac{d\theta}{b + \cos\theta} \quad \text{where } b > 1 \qquad \text{(III.8.93)}$$

The trick in evaluating Eq. (III.8.93) is to make it an integral that is got from a closed contour integral by specifying the path around a unit circle, as was shown in Fig. III.8.9. This path is $z(\theta) = \exp(i\theta)$, for $0 \le \theta \le 2\pi$. Therefore, we can write

$$\cos\theta = \frac{1}{2}\left(z + \frac{1}{z}\right) \quad \text{and} \quad dz = i\exp(i\theta)d\theta = izd\theta \qquad \text{(III.8.94)}$$

Thus, we can transform Eq. (III.8.93) into

$$I = \frac{2}{i} \oint_C \frac{dz}{2bz + z^2 + 1} \qquad \text{(III.8.95)}$$

We can then evaluate the integral in Eq. (III.8.95) by the residue theorem via looking for the zeros of $z^2 + 2bz + 1$. These are given by (through application of the quadratic roots formula)

$$z = z_\pm = -b \pm \sqrt{b^2 - 1} \qquad \text{(III.8.96)}$$

Since $b > 1$, the only pole that matters is the one at $z_+$, as this is the only one enclosed by the integration contour. By considering the residue of the integrand at the point $z_+$, we have that

$$I = \frac{2}{i} \oint_C \frac{dz}{2bz + z^2 + 1}$$
$$= 4\pi \lim_{z \to z_+} \left\{ \frac{(z - z_+)}{(z - z_+)(z - z_-)} \right\} = \frac{4\pi}{(z_+ - z_-)}$$
$$\Rightarrow \quad I = \frac{2\pi}{\sqrt{b^2 - 1}} \qquad \text{(III.8.97)}$$

Let's test this result on limiting cases. We see that when $b \to 1$ the integral tends to infinity. This singularity is expected, because at $b = 1$, the denominator $b + \cos\theta$ in our initial integral becomes zero at $\theta = \pi$. Furthermore, when we 'went complex plane' the pole in that case settles on the path of integration, at $\theta = \pi$. In the opposite limit, when $b \gg 1$ we get $I = 2\pi/b$, which should be expected as the integrand is practically constant, being very close to $b$. So the obtained result makes perfect sense.

For the general case, we may perform the same trick and write for the integral of $F(\cos\theta, \sin\theta)$, i.e. of any rational function of $\cos\theta$ and $\sin\theta$:

$$I = \int_0^{2\pi} F(\cos\theta, \sin\theta)\,d\theta$$

$$= \frac{1}{i}\oint_{|z|=1} F\left(\frac{1}{2}\left(z+\frac{1}{z}\right), \frac{1}{2i}\left(z-\frac{1}{z}\right)\right)\frac{dz}{z} \qquad \text{(III.8.98)}$$

where we have specified the path around the unit circle as $|z| = 1$. Thus, we can evaluate real integrals of rational functions of trig functions (l.h.s of Eq. (III.8.98)) from the residue theorem (Eq. (III.8.57)), calculating the residues of the poles enclosed within the path $|z| = 1$.

**Exercise.** Evaluate

$$I = \int_0^{2\pi} \frac{d\theta}{(b+\cos\theta)^2} \quad \text{with } b > 1 \qquad \text{(III.8.99)}$$

**Example III.8.13 (Maths Practice ♫♪).** Let us evaluate the integral

$$I(a) = \int_{-\infty}^{\infty} dx\,\frac{e^{ax}}{1+e^x}, \quad 0 < a < 1 \qquad \text{(III.8.100)}$$

We take $f(z) = e^{az}(1+e^z)^{-1}$ and the integration contour, which surrounds a pole at $z = i\pi$. This contour is illustrated in Fig. III.8.16.

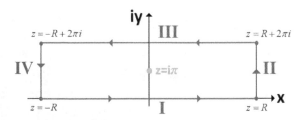

**Fig. III.8.16**   We show an integration contour that can be used in the limit $R \to \infty$ to evaluate the integral given by Eq. (III.8.93). It comprises four sections, $I$, $II$, $III$ and $IV$ about a rectangle of vertices $z = -R$, $z = R$, $z = R+2\pi i$ and $z = -R+2\pi i$, enclosing the pole $z = i\pi$. In the limit $R \to \infty$, the contribution from sections $II$ and $IV$ (shown in blue) disappears.

Splitting integration over the indicated four parts of this contour, we may write

$$\int_I f(z)dz + \int_{II} f(z)dz + \int_{III} f(z)dz + \int_{IV} f(z)dz$$
$$= 2\pi i \operatorname{res}\{f(z)\}_{z=i\pi} \qquad \text{(III.8.101)}$$

The residue evaluates to be

$$2\pi i \operatorname{res} f(i\pi) = 2\pi i \lim_{z \to i\pi} \left\{ \frac{e^{az}(z - i\pi)}{1 + e^z} \right\}$$
$$= 2\pi i \lim_{z \to i\pi} \left\{ \frac{e^{az}(z - i\pi)}{e^{i\pi}(z - i\pi)} \right\} = -2\pi i e^{ai\pi} \qquad \text{(III.8.102)}$$

So we have established what the sum of the four integrals should be. Now, let us see what these integrals are. Firstly, contribution I is evaluated along the path $z = x$ for $-R \leq x \leq R$ so that

$$\int_I f(z)dz = \int_{-R}^R \frac{e^{ax}}{1 + e^x}dx \qquad \text{(III.8.103)}$$

Contribution III is evaluated along the path $z = x+2\pi i$ for $-R \leq x \leq R$ so that

$$\int_{III} f(z)dz = -e^{2i\pi a} \int_{-R}^R \frac{e^{ax}}{1 + e^{x+2\pi i}}dx = -e^{2i\pi a} \int_I f(z)dz \qquad \text{(III.8.104)}$$

Contribution II is evaluated along the path $z = +R + iy$ for $0 \leq y \leq 2\pi$ and IV along the path $z = -R + iy$ for $0 \leq y \leq 2\pi$. Therefore, we have

$$\int_{II} f(z)dz = i \int_0^{2\pi} \frac{e^{aiy}e^{aR}}{1 + e^{iy}e^{+R}}dy$$
$$\int_{IV} f(z)dz = -i \int_0^{2\pi} \frac{e^{aiy}e^{-aR}}{1 + e^{iy}e^{-R}}dy \qquad \text{(III.8.105)}$$

If we take the limit $R \to \infty$ we can see that as $0 < a < 1$ the contributions from II and IV will vanish, and we obtain from Eqs. (III.8.101), (III.8.102), (III.8.103) and (III.8.104)

$$I(a)(1 - e^{a2\pi i}) = -2\pi i e^{a\pi i}$$

$$\Rightarrow \quad I(a) = -2\pi i \frac{e^{a\pi i}}{1 - e^{a2\pi i}} = -2\pi i \frac{1}{e^{-a\pi i} - e^{a\pi i}}$$

$$= \frac{\pi}{\frac{e^{a\pi i} - e^{-a\pi i}}{2i}} = \frac{\pi}{\sin(a\pi)} \tag{III.8.106}$$

Let us try to check and understand this result. We first plot the graph of this function in Fig. III.8.17. The obtained expression for $I(a)$, Eq. (III.8.106), does exactly what is expected from it, which is easier to trace if we convert the integral with integration range from $-\infty$ to $\infty$ to one from 0 to $\infty$. Thus we obtain

$$I(a) = \int_{-\infty}^{\infty} dx \frac{e^{ax}}{1 + e^x} = \int_{0}^{\infty} dx \frac{\cosh(ax) + \cosh((a-1)x)}{2[\cosh(x/2)]^2} \tag{III.8.107}$$

The integrand in the last form of the integral, as a function of $a$, is symmetric about $a = 1/2$, and so is the integral $I(a)$. The integral must diverge at both $a \to 0$ and $a \to 1$, because at large $x$ in each

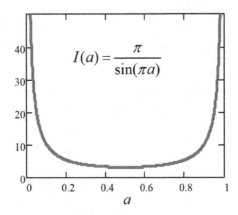

Fig. III.8.17   The graph of the answer obtained for $I(a)$.

of these limits, the integrand becomes a constant (=1). Both of these divergences in $a$ are seen in the graph shown in Fig. III.8.17.

Let us examine the singular nature of our integral Eq. (III.8.100) at small $a$ and check whether our result, Eq. (III.8.106), corresponds with this. When $a$ is small, the main contribution to the integral given by Eq. (III.8.100) comes from large negative values of $x$ for which $1/(1 + e^x) \approx 1$. Thus we may approximate

$$I(a) \approx \int_{-\infty}^{0} e^{ax} dx = \frac{1}{a} \int_{0}^{\infty} e^{-y} dy = \frac{1}{a} \qquad \text{(III.8.108)}$$

Now, let's check our result, Eq. (III.8.106), at small $a$:

$$I(a) = \frac{\pi}{\sin(a\pi)} \approx \frac{\pi}{a\pi} = \frac{1}{a} \qquad \text{(III.8.109)}$$

**Exercise.** Perform the same analysis for $a$ close to 1. **Hint:** You should note that in this limit the main contribution to the integral gives the domain of large values of $x$, where the integrand tends to $e^{-(1-a)x}$. Calculate the integral to show that it is then becomes equal to $\frac{1}{1-a}$. Then check it against a Taylor expansion of $\frac{\pi}{\sin(a\pi)}$ near to $a = 1$; not much to check, because as you will see, the first term of that series turns out to be $\frac{1}{1-a}$.

**Exercise.** At $a = 1/2$ from Eq. (III.8.107) we have that

$$I(1/2) = \int_{0}^{\infty} dx \frac{1}{\cosh(x/2)} \qquad \text{(III.8.110)}$$

Show by evaluating the integral in Eq. (III.8.110) without any contour integration, just by making the substitution $e^{x/2} = t$, that one gets the same result as substituting $a = 1/2$ into Eq. (III.8.106).

Thus, the expected behaviour perfectly matches the contour integration result at both ends of the interval of $a$, as well as at the middle value, so we may be confident in our result—but if you doubt it, go ahead and check it against numerical integration! But the result for this integral looks amazingly beautiful (the result often attributed to Euler; we did not check it, but this great mind seems seldom to be absent from the pages of this book).

## 8.3.   Contour Integration Around Branch Cuts

In Chapter 8, Part II , Vol. 1, we introduced branch cuts. Now, we'll consider evaluating integrals by taking contours about them.

### *Integration about the branch cut of* $\ln z$

To start off with, let us consider the following integral

$$I = \int_0^\infty f(x)dx \qquad\qquad (\text{III.8.111})$$

where $f(x)$ is some non-even function which is analytic in the complex plane except at poles, depicted in Fig. III.8.18 as $a_i$-points $(i = 1, 2, \ldots n)$. There is a way to evaluate this integral by considering

$$F = \oint_C f(-z) \ln z \, dz \qquad\qquad (\text{III.8.112})$$

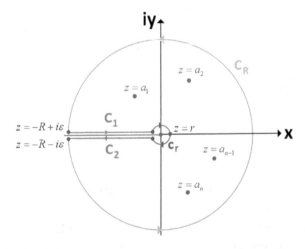

**Fig. III.8.18**   An illustration of keyhole integration around a cut. Functions such as $\ln z$ and $z^\alpha$, where $-1 < \alpha < 1$ have a branch cut along the real axis extending from $z = 0$ to $z = -\infty$. Such a cut is shown as a red line, which appears for the functions $f(-z) \ln z$ and $f(-z)z^\alpha$, where $f(z)$ is a completely analytic function except at the poles $a_k$. The complete contour of integration encloses these poles. To evaluate integrals of $f(x)$ and $f(x)x^\alpha$ from 0 to $\infty$, we may use an integration contour $C$ comprising the contributions $C_R$, $C_1$, $C_2$ and $c_r$. The paths $C_1$ and $C_2$ are shown in aquamarine, whereas the paths $C_R$ and $c_r$ are shown as green and blue curves, respectively. Here, $\varepsilon$ is taken to be infinitesimally small.

Here, $C_R$ is taken about a circle of radius $R$ connecting with a contour taken about the branch cut from $x = -R$ to $x = 0$. The full contour is shown in Fig. III.8.18. We can evaluate the integral in Eq. (III.8.112), using the residue theorem, as

$$\oint_C f(-z) \ln z\, dz = 2\pi i \sum_{k=1}^{n} \text{res}\left\{f(-z)\ln(z)\right\}_{z=a_k} \qquad \text{(III.8.113)}$$

Next we suppose that function $f(-z)$, in the limit of $R \to \infty$, behaves as $f(R\exp i\theta) R \ln R \to 0$ (where $-\pi \le \theta \le \pi$), so that the integral about the big circle with path $C_R$ (shown in Fig. III.8.18 in green) vanishes when $R \to \infty$. Thus, we are left with the contour about the cut which can be written as three contributions

$$F = \int_{C_1} f(-z) \ln z\, dz + \int_{C_2} f(-z) \ln z\, dz + \int_{C_r} f(-z) \ln z\, dz$$
$$\text{(III.8.114)}$$

Here, with $\varepsilon$ taken to be small, $C_1$ is specified by $z = x + i\varepsilon$ with $-\infty \le x \le 0$ and $C_2$ is specified by $z = x - i\varepsilon$ with $\infty \le x \le 0$ running in the opposite direction to $C_1$; and $c_r$ may be specified by $z = r \exp i\theta$, where $-\pi \le \theta \le \pi$. We find that

$$\int_{C_1} f(-z) \ln z\, dz = \int_{-\infty}^{0} f(-x - i\varepsilon) \ln(x + i\varepsilon)\, dx$$

$$\int_{C_2} f(-z) \ln z\, dz = - \int_{-\infty}^{0} f(-x + i\varepsilon) \ln(x - i\varepsilon)\, dx \qquad \text{(III.8.115)}$$

and

$$\int_{c_r} f(-z) \ln z\, dz = -ir \int_{-\pi}^{\pi} f(-r \exp(i\theta)) \ln(r \exp(i\theta)) \exp(i\theta)\, d\theta$$
$$\text{(III.8.116)}$$

Now, for small $\varepsilon$, on either side of the cut, recalling Eq. (III.8.117) and the overall analysis in Section 8.2 of Part II of in Vol. 1, we have

$$\ln(x + i\varepsilon) \approx \ln|x| + i\pi \qquad \ln(x - i\varepsilon) \approx \ln|x| - i\pi \qquad \text{(III.8.117)}$$

Thus, we can write

$$\int_{C_1} f(-z) \ln z \, dz + \int_{C_2} f(-z) \ln z \, dz$$

$$= \int_{-\infty}^{0} (f(-x - i\varepsilon) - f(-x + i\varepsilon)) \ln |x| dx$$

$$+ i\pi \int_{-\infty}^{0} (f(-x - i\varepsilon) + f(-x + i\varepsilon)) dx \qquad \text{(III.8.118)}$$

In the limit $r \to 0$, the contribution from Eq. (III.8.116) should vanish, and putting $\varepsilon = 0$ in Eq. (III.8.118) we obtain

$$F = 2\pi i \int_{-\infty}^{0} f(-x) dx = 2\pi i \int_{0}^{\infty} f(x) dx$$

$$= 2\pi i \sum_{k=1}^{n} \text{res}\{f(-z) \ln(z)\}_{z=a_k}$$

$$\Rightarrow \quad \int_{0}^{\infty} f(x) dx = \sum_{k=1}^{n} \text{res}\{f(-z) \ln(z)\}_{z=a_k} \qquad \text{(III.8.119)}$$

**Example III.8.14 (Maths Practice ♩♪).** Let's demonstrate this technique, using for an example the integral

$$I(a, b, c) = \int_{0}^{\infty} f(x) dx = \int_{0}^{\infty} \frac{x + c}{(x + a)(x^2 + b^2)} dx \qquad \text{(III.8.120)}$$

Note that this integral could be evaluated by partial fractions, but this would be rather time consuming, so we will 'treat ourselves' by using the techniques we have just learned. Firstly, we need to check that this integral satisfies the formulated criteria. For large $x$ the integrand decreases $\sim 1/x^2$, and thus

$$f(R \exp i\theta) R \ln(R) \approx \ln(R) \exp(-4i\theta)/R \qquad \text{(III.8.121)}$$

Hence, $f(R \exp i\theta) R \ln(R) \to 0$ when $R \to \infty$. Secondly, we may write

$$f(-z) = \frac{c - z}{(a - z)(z^2 + b^2)} \qquad \text{(III.8.122)}$$

which is analytic everywhere except at the poles $z = a$ and $z = \pm ib$. Therefore, we can indeed evaluate the integral using the contour integral, Eq. (III.8.112), and so the formula Eq. (III.8.119). Hence,

$$I(a,b,c) = \lim_{z \to a} \left\{ \frac{(-z+c)(z-a)}{(a-z)(z^2+b^2)} \ln z \right\}$$

$$+ \lim_{z \to -ib} \left\{ \frac{(-z+c)(z+ib)}{(a-z)(z^2+b^2)} \ln z \right\}$$

$$+ \lim_{z \to ib} \left\{ \frac{(-z+c)(z-ib)}{(a-z)(z^2+b^2)} \ln z \right\} \qquad \text{(III.8.123)}$$

We may evaluate each of these limits in turn. First

$$\lim_{z \to a} \left\{ \frac{(-z+c)(z-a)}{(a-z)(z^2+b^2)} \ln z \right\} = \frac{a-c}{(a^2+b^2)} \ln a \qquad \text{(III.8.124)}$$

Secondly, we have that

$$\lim_{z \to ib} \left\{ \frac{(c-z)(z-ib)}{(a-z)(z^2+b^2)} \ln z \right\} = \frac{1}{2ib} \frac{(c-ib)}{(a-ib)} \left( \ln b + \frac{i\pi}{2} \right)$$

$$= -\frac{i}{2b} \frac{(c-ib)(a+ib)}{a^2+b^2} \left( \ln b + \frac{i\pi}{2} \right)$$

$$\text{(III.8.125)}$$

Lastly, we get

$$\lim_{z \to -ib} \left\{ \frac{(c-z)(z+ib)}{(a-z)(z^2+b^2)} \ln z \right\} = -\frac{1}{2ib} \frac{(c+ib)}{(a+ib)} \left( \ln b - \frac{i\pi}{2} \right)$$

$$= \frac{i}{2b} \frac{(c+ib)(a-ib)}{a^2+b^2} \left( \ln b - \frac{i\pi}{2} \right)$$

$$\text{(III.8.126)}$$

Combining Eqs. (III.8.123), (III.8.124), (III.8.125) and (III.8.126) we obtain expectedly a real result (if imaginary terms did not cancel out, that would have meant we have made an error somewhere!):

$$I(a,b,c) = \frac{a-c}{a^2+b^2} \ln\left(\frac{a}{b}\right) + \frac{\pi}{2b} \frac{ac+b^2}{a^2+b^2} \qquad \text{(III.8.127)}$$

To check this result, let's look at a particular case when $a = c$. In this case the integral, Eq. (III.8.120), reduces to the known integral

$$I(a, b, a) = \int_0^\infty \frac{1}{(x^2 + b^2)} dx = \frac{\pi}{2b} \qquad \text{(III.8.128)}$$

This is indeed what we get from Eq. (III.8.127).

What about when $a = b$ and $c = 0$? In this case we deal again with another well-known integral,

$$I(b, b, 0) = \int_0^\infty \frac{x}{(x + b)(x^2 + b^2)} dx = \frac{1}{b} \int_0^\infty \frac{s}{(s + 1)(s^2 + 1)} ds = \frac{\pi}{4b} \qquad \text{(III.8.129)}$$

because $\int_0^\infty \frac{s}{(s+1)(s^2+1)} ds = \frac{\pi}{4}$ (check this numerically or derive it yourself), whereas in this particular case our Eq. (III.8.127), $I(a = b, b, c = 0)$, gives the same result.

### Integration about the branch cut of $z^\alpha$

We can also look to evaluate integrals of the form

$$I = \int_0^\infty x^\alpha f(x) dx \qquad \text{(III.8.130)}$$

For this integral to be finite, we have a restriction on $\alpha$ such that $f(x)x^{\alpha+1} \to 0$ both when $x \to 0$ and $x \to \infty$. We may use the same 'keyhole' integration contour, $C$, as was considered previously in writing Eq. (III.8.112), shown in Fig. III.8.18. Here the cut remains at the same place (although the discontinuity is different). This allows us through the residue theorem to write

$$F = \oint_C f(-z)z^\alpha dz = 2\pi i \sum_{k=1}^n \text{res}\{f(-z)z^\alpha\}_{z=a_k} \qquad \text{(III.8.131)}$$

where again $f(z)$ is analytic everywhere, except at poles. Provided that, for the contour about the circle of radius $R$, $R^{\alpha+1} f(R\exp i\theta) \to 0$ when $R \to \infty$, we may safely neglect this contribution. So that again we have

$$F = \int_{C_1} f(-z)z^\alpha \, dz + \int_{C_2} f(-z)z^\alpha \, dz + \int_{c_r} f(-z)z^\alpha \, dz \quad \text{(III.8.132)}$$

where $C_1$, $C_2$ and $c_r$ are the contours shown in Fig. III.8.18. On evaluation we have

$$\int_{C_1} f(-z)z^\alpha \, dz = \int_{-\infty}^{0} f(-x - i\varepsilon)(x + i\varepsilon)^\alpha dx$$

$$\int_{C_2} f(-z)z^\alpha \, dz = -\int_{-\infty}^{0} f(-x + i\varepsilon)(x - i\varepsilon)^\alpha dx \quad \text{(III.8.133)}$$

Next, taking the $\varepsilon \to 0$ limit, we get

$$\lim_{\varepsilon \to 0}(x + i\varepsilon)^\alpha = (|x| \exp(i\pi))^\alpha = |x|^\alpha \exp(i\pi\alpha) \quad \text{(III.8.134)}$$

$$\lim_{\varepsilon \to 0}(x - i\varepsilon)^\alpha = (|x| \exp(-i\pi))^\alpha = |x|^\alpha \exp(-i\pi\alpha) \quad \text{(III.8.135)}$$

Thus, in this case we obtain

$$\int_{C_1} f(-z)z^\alpha \, dz + \int_{C_2} f(-z)z^\alpha \, dz$$

$$= (\exp(i\pi\alpha) - \exp(-i\pi\alpha)) \int_{-\infty}^{0} f(-x)|x|^\alpha dx \quad \text{(III.8.136)}$$

Again, we require that the contribution from $c_r$ vanishes, which is true provided that $r^{\alpha+1}f(r\exp i\theta) \to 0$ when $r \to 0$. Thus, combining Eqs. (III.8.131), (III.8.132) and (III.8.136), we may write

$$2i\sin(\pi\alpha) \int_0^\infty f(x)x^\alpha dx = 2i\pi \sum_{k=1}^{n} \text{res}\{f(-z)z^\alpha\}_{z=a_k}$$

$$\Rightarrow \quad \int_0^\infty f(x)x^\alpha dx = \frac{\pi}{\sin(\pi\alpha)} \sum_{k=1}^{n} \text{res}\{f(-z)z^\alpha\}_{z=a_k}$$

$$\text{(III.8.137)}$$

**Example III.8.15 (Maths Practice ♫♪).** Using this technique, we'll evaluate the integral

$$I(b, \alpha) = \int_0^\infty \frac{x^\alpha}{x^2 + b^2} dx \qquad \text{(III.8.138)}$$

First, let us rescale the integral with $x = by$ so that

$$I(b, \alpha) = \frac{1}{b^{1-\alpha}} \int_0^\infty \frac{x^\alpha}{x^2 + 1} dx \qquad \text{(III.8.139)}$$

The integral in Eq. (III.8.139) satisfies the criteria for using Eq. (III.8.131) to evaluate it, where here $f(-z) = 1/(z^2 + 1)$. So we can use the formula given in Eq. (III.8.137) to write

$$I(b, \alpha) = \frac{\pi}{b^{1-\alpha} \sin(\pi\alpha)} \left[ \lim_{z \to i} \left( \frac{z - i}{z^2 + 1} z^\alpha \right) + \lim_{z \to -i} \left( \frac{z + i}{z^2 + 1} z^\alpha \right) \right]$$

$$\Rightarrow \quad I(b, \alpha) = \frac{\pi}{2ib^{1-\alpha} \sin(\alpha\pi)} \left[ \exp\left( \frac{i\pi\alpha}{2} \right) - \exp\left( -\frac{i\pi\alpha}{2} \right) \right]$$

$$= \frac{\pi}{b^{1-\alpha} \sin(\alpha\pi)} \sin\left( \frac{\alpha\pi}{2} \right) = \frac{\pi}{2b^{1-\alpha}} \sec\left( \frac{\alpha\pi}{2} \right)$$

$$\text{(III.8.140)}$$

Let's look at some limiting cases. What happens when $\alpha \to 1$? We may write then $\cos\left( \frac{\alpha\pi}{2} \right) \approx \frac{\pi}{2}(1 - \alpha)$ so that

$$I(b, \alpha) \approx \frac{1}{(1 - \alpha)}, \quad \alpha \approx 1 \qquad \text{(III.8.141)}$$

Clearly, when $|\alpha| = 1$ the integral diverges, as it should. When $\alpha = 0$, $\cos\left( \frac{\alpha\pi}{2} \right) \approx 1$, so we obtain $I(b, 1) = \frac{\pi}{2b}$. This is what we expect, as we found that $2I(b, 0) = \pi/b$ in Example III.8.9.

Of course, we are not limited to just integrals of the forms of Eq. (III.8.111) and (III.8.130). There are many other ways we can exploit integration about cuts. We'll finish this section with a slightly more complicated application.

## 8.4. Dealing with Integrals of Rapidly Oscillating Functions: Stationary Phase Approximation and the Method of Steepest Descent

We introduce you now to approximate methods of analysis of integrals containing rapidly oscillating functions in the integrands.

### *Stationary phase approximation*

Let us suppose that we have to deal with an integral of the form

$$I = \int_a^b f(x) \exp(i\lambda\omega(x)) dx \qquad (III.8.142)$$

where $\lambda$ is very large and $f(x)$ is a much slower-varying function, or possibly even a constant. Now in regions where the rate of change of $\omega(x)$, and its magnitude, are large, the integrand will oscillate rapidly and effectively its contribution to the integral will be negligible. Thus, any dominant contribution to the integral will be at points where the magnitude of $\omega(x)$ is minimal. In what follows, we'll suppose that $\omega(x)$ has only one turning point at $x = x_0$. Thus, in a way analogous to the Laplace method where we dealt with integrals on the real axis (Part II, Vol. 1), we can expand the integrand out around $x = x_0$. This yields (to leading order in the expansion)

$$I \approx \exp(i\lambda\omega(x_0)) \int_a^b f(x) \exp\left(\frac{i\lambda\omega''(x_0)(x-x_0)^2}{2} + \cdots\right) dx$$

$$\approx f(x_0) \exp(i\lambda\omega(x_0)) \int_{-\infty}^{\infty} \exp\left(\frac{i\lambda\omega''(x_0)y^2}{2}\right) dy \qquad (III.8.143)$$

In Example III.8.4 we considered how to perform an integral of the form of the bottom line of Eq. (III.8.143). It evaluates to

$$\int_{-\infty}^{\infty} \exp\left(\frac{i\lambda\omega''(x_0)y^2}{2}\right) dy = \exp\left(\frac{i\pi}{4}\right) \sqrt{\frac{2\pi}{\lambda\omega''(x_0)}} \qquad (III.8.144)$$

Combining Eqs. (III.8.143) and (III.8.144) leads to

$$I \approx f(x_0) \exp\left(i\left[\lambda\omega(x_0) + \frac{\pi}{4}\right]\right) \sqrt{\frac{2\pi}{\lambda\omega''(x_0)}} \qquad (III.8.145)$$

So far, these were trivial manipulations. But the power of the method based on them, is that we can use them to approximate oscillatory integrals on the real axis. The next example shows how it works.

**Example III.8.16 (Maths Practice ♪♪).** Let us use the stationary phase approximation to estimate the integral

$$I(\lambda, c) = \int_a^b \frac{1}{x^2 + 1} \cos(\lambda(x^4 - 2c^2 x^2)) dx$$

$$\text{where } \lambda \gg 1 \text{ and } b > 0, \ a > 0 \qquad \text{(III.8.146)}$$

First we can write this integral in a similar form to Eq. (III.8.187):

$$I(\lambda, c) = \text{Re} \int_a^b \frac{1}{x^2 + 1} \exp(i\lambda\omega(x))) dx \qquad \text{(III.8.147)}$$

where $\omega(x) = x^4 - 2c^2 x^2$. Next, we find the point $x_0$ in the integration range where $d\omega/dx = 0$:

$$4x_0^3 - 4c^2 x_0 = 0 \quad \Rightarrow \quad x_0 = c \qquad \text{(III.8.148)}$$

Note that $x = 0$ is also a turning point, as it is a trivial root of the l.h.s function in Eq. (III.8.148), but it falls out of the range of integration as long as $a > 0$. We compute both $\omega(x_0)$ and the second derivative $\omega''(x_0)$

$$\omega(x_0) = \omega(c) = -c^4 \quad \text{and} \quad \omega''(x_0) = 12x_0^2 - 4c^2 = 8c^2$$

$$\text{(III.8.149)}$$

Then, we use the formula, Eq. (III.8.145), to obtain

$$I(\lambda, c) \approx \frac{1}{2c} \frac{1}{c^2 + 1} \sqrt{\frac{\pi}{\lambda}} \text{Re}[\exp(i(-c^4\lambda + \pi/4))]$$

$$= \frac{1}{2c} \frac{1}{c^2 + 1} \sqrt{\frac{\pi}{\lambda}} \cos(c^4\lambda - \pi/4) \qquad \text{(III.8.150)}$$

Note that this result is only valid when $\lambda \gg 1$, and if $c$ is not too close to $a$ or $b$, as both of the latter parameters have disappeared from the answer.

In Fig. III.8.19, we compare an exact numerical evaluation of Eq. (III.8.146) with that of the stationary phase approximation. We

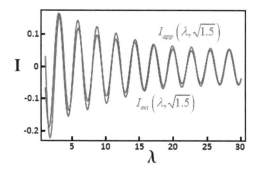

**Fig. III.8.19** We compare the exact numerical evaluation of Eq. (III.8.146) with its stationary phase approximation, given by Eq. (III.8.150). Here, a parameter value of $c^2 = 1.5$ has been used. The limits of integration in the exact integral are $a = 1$ and $b = 3$. The red curve corresponds to the approximation and the blue curve to the exact evaluation of the integral.

see already for $\lambda > 2$ the approximation is reasonable. We can do better by considering higher order corrections to Eq. (III.8.150) that arise from considering Taylor series about $x = c$.

**Exercise.** Apply the stationary phase approximation to the integral considered in Eq. (III.8.146), when $a < 0 < b < c$. Note here that the sign of $\omega''(x_0)$ does not matter in applying Eq. (III.8.145). **Hint:** To get the answer write $\sqrt{\omega''(x)} = \exp(i\pi/2)\sqrt{-\omega''(x)}$, and note that your turning point of interest will be the one at $x = 0$.

### Saddle point approximation

Now, we'll combine the stationary phase approximation with Laplace's method and the Cauchy integral theorem to create a powerful tool for approximating integrals: the famous *steepest descent* or *saddle point* method. Let us consider, first, an integral of the form

$$I = \int_a^b \exp(\lambda(u_0(x) + iv_0(x)))dx \qquad \text{(III.8.151)}$$

where $u_0(x) \equiv u(x, y = 0)$, $v_0(x) \equiv v(x, y = 0)$, and, again, $\lambda \gg 1$. For this case, as we'll discuss, the estimate of $I$ is dominated by turning points in $\omega(z) = u(x, y) + iv(x, y)$, which may not lie on the real axis. We will choose a path in the complex plane that maximizes the steepest descent of $\omega(z)$, i.e. use an idea similar to the

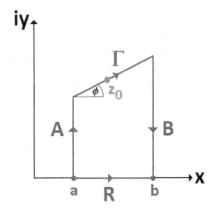

**Fig. III.8.20** We illustrate the deformation of a contour used to generate the saddle point approximation. We start by considering the integral of $f(z) = \exp(\lambda\omega(z))$ along the real axis between the points $a$ and $b$, which is represented by the path $R$. We may deform the contour into that which comprises three paths $A$, $B$ and $\Gamma$, as shown in the figure, provided that $f(z)$ remains analytic between the old path and the new one. The path $\Gamma$ is described by $z - z_0 = t\exp(i\phi)$, where $z_0$ is a saddle point for which the $z$-derivative of $\omega(z)$ vanishes, which the path of integration now passes through. When $\lambda \gg 1$, we can neglect the contributions from $A$ and $B$, shown in blue.

Laplace method. To do this, we will need to use Cauchy's integral theorem to deform the contour that corresponds to the integration in Eq. (III.8.151).

We can first express $I$ as a complex contour integral over the path $R$ along the real axis, starting at $z = x = a$ and terminating at $z = x = b$ (as in Fig. III.8.20). Then

$$I = \int_R \exp(\lambda\omega(z))dz \qquad (III.8.152)$$

Now the key to the saddle point approximation is to realize that the integral is estimated by a dominant contribution, at the turning points in both $u(x,y)$ and $v(x,y)$ along a chosen path of integration $z(t) = x(t) + iy(t)$, for which

$$\frac{du(x(t),y(t))}{dt} = 0 \qquad \frac{dv(x(t),y(t))}{dt} = 0 \qquad (III.8.153)$$

If there are multiple turning points we choose the one at $z_0$ where $|\omega(z_0)|$ is largest. Note that this will work well, unless the turning points are too close to each other on the complex plain, or the values of $\omega(z)$ at those turning points are too close. But we will stay away from this more complicated case, referring the interested reader to more specialized literature.

We won't prove it, but Eq. (III.8.153) can then be satisfied if the integration contour passes through a point for which

$$\left.\frac{d\omega(z)}{dz}\right|_{z=z_0} = 0 \qquad\qquad \text{(III.8.154)}$$

These points correspond to saddle points in both $u(x, y)$ and $v(x, y)$, as functions of $x$ and $y$ (see Chapter 8, Part II, Vol. 1). For this reason this whole approach is often called the *saddle point approximation*.

To take advantage of a saddle point in the complex plane, we deform the contour $R$ to the one that we label as $\Gamma$, which passes through the point $z_0$ as a straight line, as shown in Fig. III.8.20. To obtain the steepest descent of $\omega(z)$ away from $\omega(z_0)$, we will choose the path to be at an angle $\phi$ to the real axis (see Fig. III.8.20), as we'll see below. This works provided that $f(z) = \exp(\lambda\omega(z))$ is analytic over the region where we want to deform the contour. **Technical point.** In fact, we can deal with poles that lie in the way of the deformation, but we have to modify our integration contour by splitting away closed contours about the poles, as illustrated in Fig. III.8.11, to move the contour past them. This can lead to a more complicated result; however, if the poles lie far away from $z_0$ it may be possible to ignore them. From the deformation of the contour, we have that

$$I = \int_R \exp(\lambda\omega(z))dz \approx \int_\Gamma \exp(\lambda\omega(z))dz \qquad\qquad \text{(III.8.155)}$$

An approximate sign must be written, and not just in view of the technical remark above when we have poles 'messing around'; there is also another strict reason to do this. When writing Eq. (III.8.155) we have neglected the contributions from paths $A$ and $B$ from the end points $a$ and $b$ (shown in Fig. III.3.21). But this is legitimate, as those contributions will be small when $\lambda \gg 1$. Then, the rest is straightforward. We expand $w(z)$, which must be analytic in $z$, in a power series about $w(z_0)$. Thus, we can write

$$I = \exp(\lambda w(z_0)) \int_\Gamma \exp\left(\frac{\lambda w''(z_0)(z-z_0)^2}{2}\right)$$

$$\times \left(1 + \frac{\lambda w^{(3)}(z_0)(z-z_0)^3}{3!} + \frac{\lambda w^{(4)}(z_0)(z-z_0)^4}{4!} + \cdots\right) dz$$

$$\text{(III.8.156)}$$

where $w^{(3)}(z_0)$ and $w^{(4)}(z_0)$ are third and fourth order derivatives of $w(z)$ at the point $z = z_0$, respectively. The straight line that passes through $z_0$, i.e. the path $\Gamma$, is then specified by $z(t) - z_0 = t \exp\{i\phi\}$. We may also write $w''(z_0)$ in polar form

$$w''(z_0) = |w''(z_0)| \exp(i(\theta(z_0))) \qquad \text{(III.8.157)}$$

We are free to choose $\Gamma$ so that $w''(z_0)(z-z_0)^2$ is real and negative (so that our expansion is convergent, and the resultant integrals easy to deal with); thus, we find that we must choose $\phi = (\pi - \theta(z_0))/2$. Making this choice for the integration contour $\Gamma$, as $\lambda \gg 1$, we can safely shift the limits of integration over $t$ to $-\infty$ and $\infty$. To the leading order in our expansion we then get

$$I \approx \exp(\lambda w(z_0)) \exp(i\phi(z_0)) \int_{-\infty}^\infty \exp\left(-\frac{\lambda |w''(z_0)| t^2}{2}\right) dt \quad \text{(III.8.158)}$$

The integral in Eq. (III.8.158) is Gaussian, for which we have the exact expression. The latter is obtained through rescaling $y = \sqrt{\lambda |w''(z_0)|} x$ and using the result of Eq. (III.1.88), so that

$$I \approx \exp(\lambda w(z_0)) \exp(i\phi(z_0)) \sqrt{\frac{2\pi}{\lambda |w''(z_0)|}} \qquad \text{(III.8.159)}$$

**Exercise.** Use the specified integration path $(z(t) - z_0) = t \exp i\phi$ with $\phi = (\pi - \theta(z_0))/2$ to add evaluated correction terms, from Eq. (III.8.156) to Eq. (III.8.159). This expansion should have a similar structure to the one we considered for Laplace's method in Part II, Vol. 1.

**Example III.8.17 (Maths Practice ♪♪).** An integral representation of the Airy function for positive values of argument is given by

$$\text{Ai}(x) = \frac{1}{2\pi} \int_{-\infty}^{\infty} \cos\left(\frac{1}{3}t^3 + xt\right) dt \tag{III.8.160}$$

This is an integral on the real axis. Surprisingly (one of the beauties of mathematics!), it converges even if $x = 0$; careful analysis, which we will not go into, shows that $\text{Ai}(0) = \left\{3^{2/3}\Gamma\left(\frac{2}{3}\right)\right\}^{-1} \simeq 0.355$, where the second factor in the bracket is Euler Gamma function that we have already dealt with (Chapter 7, Part II, Vol. 1). The integral converges because with an increase of $t$ oscillations will take place faster and faster, and will tend to cancel one another out.

Let's try to find the asymptotic expansion to Eq. (III.8.160) for large positive values of $x$, and see if it matches with the one presented in Chapter 6 of this part of the book. There are no turning points on the real axis for positive values of $x$; so, in this case, we'll need to use the saddle point approximation. First, to have a clear expansion parameter (i.e. a form for $\lambda$), we rescale the integral by making the variable change $t = x^{1/2}s$ so that we can write it in the form of Eq. (III.8.152):

$$\text{Ai}(x) = \frac{x^{1/2}}{2\pi} \int_{-\infty}^{\infty} \exp(x^{3/2}\omega(s))ds = \frac{x^{1/2}}{2\pi} \int_{R} \exp(x^{3/2}\omega(z))dz \tag{III.8.161}$$

where now $\omega(z) = i\left(\frac{z^3}{3} + z\right)$. Let's determine the turning points:

$$\left.\frac{d\omega(z)}{dz}\right|_{z=z_0} = i(z_0^2 + 1) = 0 \quad \Rightarrow \quad z_0 = \pm i \tag{III.8.162}$$

The integrand is completely analytic in the complex plane, so we can simply deform the contour $X$ to path $\Gamma$ passing through the point $z_0 = i$. Next, we evaluate $\omega(z_0) = -2/3$ and the second derivative of this function at this point:

$$\omega''(z_0) = 2iz_0 = -2 \tag{III.8.163}$$

This corresponds to the value $\theta(z_0) = \pi$ as defined in Eq. (III.8.183). Now by expanding about the point $z_0 = i$ we can write

$$\text{Ai}(x) \approx \frac{x^{1/2}}{2\pi} \exp\left(-\frac{2}{3}x^{3/2}\right) \int_\Gamma \exp(-x^{3/2}(z-i)^2)dz \qquad \text{(III.8.164)}$$

We now need to specify the integration contour $\Gamma$, $z(t) - z_0 = t \exp\{i\phi\}$. The key, as always, is to choose $\Gamma$ such that $\omega''(z_0)(z-z_0)^2$ is real and negative, so that it lies along the direction of the steepest descent. Inspecting Eq. (III.8.164) we see that we should choose $\phi = 0$ for this to be so. Then, we obtain (for $x \gg 1$)

$$\int_\Gamma \exp(-x^{3/2}(z-i)^2)dz = \int_{-\infty}^{\infty} \exp(-x^{3/2}t^2)dt = \frac{\sqrt{\pi}}{x^{3/4}} \qquad \text{(III.8.165)}$$

Using the resulting formula of the saddle point approximation, Eq. (III.8.159), we are then able to write for large $x$ that

$$\text{Ai}(x) \approx \frac{1}{2x^{1/4}\sqrt{\pi}} \exp\left(-\frac{2}{3}x^{3/2}\right) \qquad \text{(III.8.166)}$$

This is what is known to be the leading order asymptotic behaviour of the Airy function should be.

**Exercise.** By considering the corrections to Eq. (III.8.159) work out the next to leading order correction to Eq. (III.8.166). Plot the obtained result together with the one of the leading order against the exact integral representation of the Airy function to see how large $x$ should be for these approximations to work.

Here, we must apologize for not giving any examples of physical applications in this section. This is because any such example, using saddle point approximation, would require a long introduction into the physics of the problem, and so cannot be presented succinctly. Nevertheless, the explained method should help you in the study of such physical cases.

## 8.5. Contour Integrals Using Hankel Functions

### *Introducing the Hankel functions*

In problems involving cylindrical geometry we may need to evaluate integrals involving Bessel functions $J_n(x)$ and $Y_n(x)$. In evaluating such integrals, it is sometimes useful to introduce the functions named after German mathematician and historian of mathematics Hermann Hankel, famous for his insights into the foundations of arithmetic and the theory of functions, who was

Hermann Hankel
(1839–1873)

one of the pioneers of solutions of the Bessel equation. These Hankel functions are defined as

$$H_n^{(1)}(x) = J_n(x) + iY_n(x) \quad H_n^{(2)}(x) = J_n(x) - iY_n(x) \quad \text{(III.8.167)}$$

The two Hankel functions are analytic in the complex plane, except for a cut that lies along the real axis from $x = -\infty$ to zero $x = 0$. For the discontinuity across the branch cut we have

$$H_n^{(1)}(-x + i\varepsilon) - H_n^{(1)}(-x - i\varepsilon) = -4(-1)^n J_n(x) \quad \text{(III.8.168)}$$

$$H_n^{(2)}(-x + i\varepsilon) - H_n^{(2)}(-x - i\varepsilon) = 4(-1)^n J_n(x) \quad \text{(III.8.169)}$$

In Fig. III.8.21, we show contour plots of $H_1^{(1)}(z)$ and $H_1^{(2)}(z)$, in some parts of the complex plane.

The Hankel functions have the following relations with the modified Bessel functions of the second kind.

$$K_n(z) = \frac{\pi(i)^{n+1}}{2} H_n^{(1)}(iz) \quad K_n(z) = \frac{\pi(-i)^{n+1}}{2} H_n^{(2)}(-iz)$$

$$\text{(III.8.170)}$$

In evaluating integrals, it is these properties of Hankel functions that we want to exploit. We'll utilize Hankel functions in two physically relevant examples.

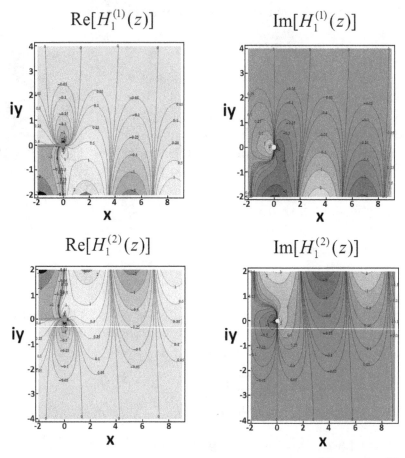

**Fig. III.8.21**   We show contour plots of the Hankel functions $H_1^{(1)}(z)$ and $H_1^{(2)}(z)$ for $z = x + iy$, top and bottom, respectively. Panels on the left show the real parts of each function, those on the right the imaginary part. We see some important features of these functions. First, both the real and imaginary parts of these functions oscillate as we vary $x$. For $H_1^{(1)}(z)$, when we increase $y$, the amplitude of the oscillations diminishes, whereas for $H_1^{(2)}(z)$ it increases. Also, we see discontinuities in both $\mathrm{Re}[H_1^{(1)}(z)]$ and $\mathrm{Re}[H_1^{(2)}(z)]$ due to branch cuts starting from $z = 0$, marked in orange. At the point $z = 0$, we also see a singularity in $\mathrm{Im}[H_1^{(1)}(z)]$ and $\mathrm{Im}[H_1^{(2)}(z)]$. All of these features are crucial when selecting integration contours with $H_1^{(1)}(z)$ and $H_1^{(2)}(z)$.

**Example III.8.18 (Physical Application ▸▸).** Let's obtain an expression for the electrostatic potential that solves the 3D linearized inhomogeneous Poisson–Boltzmann equation (written here in Gaussian units, used to the end of this example)

$$\nabla^2 \varphi(\mathbf{r}) - \kappa^2 \varphi(\mathbf{r}) = -\frac{4\pi}{\varepsilon} \rho(\mathbf{r}) \qquad \text{(III.8.171)}$$

in the medium with dielectric constant $\varepsilon$, non-confined in space, i.e. without boundary conditions. But we limit our analysis to a specific case where charge density, $\rho(\mathbf{r}) = \rho(R)$, has cylindrical symmetry and $R$ is the distance away from the cylinder axis. In this case, $\varphi(\mathbf{r})$ only depends on $R$. As discussed previously (Eq. III.4.206) to find a form for such a solution, we must solve the inhomogeneous Bessel equation for $n = 0$ (see Example III.4.10). In general, we should search for the solution in the form

$$\varphi(R) = \frac{4\pi}{\varepsilon} \int_0^\infty G(R, R')\rho(R')dR \qquad \text{(III.8.172)}$$

where Green's function to the inhomogeneous Bessel equation, $G_n(R, R')$ satisfies the equation

$$-\frac{\partial^2 G_n(R, R')}{\partial R^2} - \frac{1}{R}\frac{\partial G_n(R, R')}{\partial R} + \left(\frac{n^2}{R^2} + \kappa^2\right)G_n(R, R') = \delta(R - R')$$
$$\text{(III.8.173)}$$

Although for solving the problem in this example we will only need the solution for $n = 0$, it will be instructive to find the form of $G_n(R, R')$, for any value of $n$ (that could be useful, when the charge density depends also on azimuthal angle). So let us work on solving Eq. (III.8.173). To do this, we first rewrite it, using the inverse Hankel transform defined earlier in Chapter 4, Part III (Sec. 4.3), as

$$-\left[\frac{d^2}{dR^2} + \frac{1}{R}\frac{d}{dR} - \left(\frac{n^2}{R^2} + \kappa^2\right)\right]\int_0^\infty K\tilde{G}_n(K, R')J_n(KR)dK$$
$$= R'\int_0^\infty K J_n(RK)J_n(R'K)dK \qquad \text{(III.8.174)}$$

where we have used the orthogonality condition, Eq. (III.4.153). By using the form of the eigenvalue equation for Bessel functions, Eq. (III.4.125), we obtain

$$\int_0^\infty K(K^2 + \kappa^2)\tilde{G}_n(K, R')J_n(KR)dK$$

$$= R' \int_0^\infty K J_n(RK)J_n(R'K)dK \qquad \text{(III.8.175)}$$

Thus, we find

$$\tilde{G}_n(K, R') = \frac{R' J_n(KR')}{K^2 + \kappa^2} \qquad \text{(III.8.176)}$$

Therefore, we must have that

$$G_n(R, R') = R' \int_0^\infty K \frac{J_n(RK)J_n(R'K)}{K^2 + \kappa^2} dK \qquad \text{(III.8.177)}$$

Can we evaluate such integrals as Eq. (III.8.177) by complex contour integration? Yes, we can, and let us delve into it. $J_n(z)$ is fully analytic in the complex plane, and naively we may think to extend the function $f(K) = J_n(KR)J_n(KR')/(K^2 + \kappa^2)$ so that we have $f(z) = f(x + iy)$, where $x = K$. However, this is not a good idea, as

$$(-i)^n J_n(iy) = I_n(y) \quad \text{and} \quad (i)^n J_n(-iy) = I_n(y) \qquad \text{(III.8.178)}$$

and $I_n(y)$ exponentially increases with increasing real argument $y$. This means that we cannot find a suitable integration contour, through such a naïve choice, to evaluate Eq. (III.8.177). So, we will have to do something different, which is the trick often used in such situations, and which is definitely worth while to learn. To proceed, we rewrite Eq. (III.8.177) as

$$G_n(R, R') = R' \mathrm{Re} \int_0^\infty K \frac{J_n(RK)H_n^{(1)}(R'K)}{K^2 + \kappa^2} dK \quad \text{for } R' > R$$
$$\text{(III.8.179)}$$

$$G_n(R, R') = R' \mathrm{Re} \int_0^\infty K \frac{H_n^{(1)}(RK)J_n(R'K)}{K^2 + \kappa^2} dK \quad \text{for } R > R'$$
$$\text{(III.8.180)}$$

This in analogous to replacing $\cos x$ with the real part of $\exp(ix)$ or $\exp(-ix)$, which we utilized before to perform integrals by complex integration. We consider first the case $R' > R$.

To evaluate the integrals, we choose a closed contour $C$. This is made up of the following contributions: $C_\Lambda$ along the perimeter of the quarter segment of a circle, X along the real axis from $x = 0$ to $x = \Lambda$, $I_1$ along the imaginary axis from $z = iy = i\Lambda$ to $z = iy = i(\kappa+\varepsilon)$, $C_\varepsilon$ around the pole at $z = ik$, and $I_2$ along the imaginary axis from $z = iy = i(\kappa - \varepsilon)$ to $z = 0$, all shown in Fig. III.8.22.

For the integrand of the integral, with $R' > R$, we consider the function of a complex variable

$$f(z) = \frac{z J_n(Rz) H_n^{(1)}(R'z)}{z^2 + \kappa^2} \qquad (\text{III.8.181})$$

integrated over this contour. There are no poles contained within it. We need not worry about the singularity at $z = 0$, related to the logarithmic divergence of $H_n^{(1)}(z)$ at $z \to 0$, as it is negated by zeros

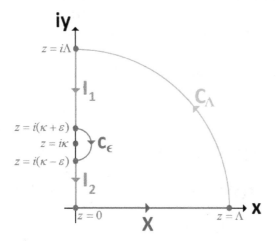

**Fig. III.8.22** The integration contour we use to evaluate the integral given by Eq. (III.8.166). This comprises integration along the real axis, contribution "X", (shown in purple), followed by, a path of integration about a quadrant of circle of radius $\Lambda$, contribution "$C_\Lambda$" (shown in green). Next we integrate along the imaginary axis with two paths, giving contributions $I_1$ and $I_2$ (shown in orange), as well as between them, to avoid the pole at $z = i\kappa$ (shown in blue) along a path $C_\varepsilon$ tracing out a semi-circle of radius $\varepsilon$ (shown in red).

$J_n(z)$, when $n \neq 0$, and is 'killed' anyway by the $z$-factor in the integrand. Thus we may write

$$\int_C f(z)dz = 0 \qquad \text{(III.8.182)}$$

or, in terms of the counterparts of this contour integral,

$$\int_X f(z)dz = -\int_{C_\Lambda} f(z)dz - \int_{I_1} f(z)dz - \int_{C_\varepsilon} f(z)dz - \int_{I_1} f(z)dz$$
$$\text{(III.8.183)}$$

The contribution to the contour along the real axis in the limit of $\Lambda \to \infty$ reads

$$\lim_{\Lambda \to \infty} \int_X f(z)dz = \int_0^\infty K \frac{J_n(RK)H_n^{(1)}(R'K)}{K^2 + \kappa^2}dK \qquad \text{(III.8.184)}$$

The integration contour $C_\Lambda$ is specified by $z(\theta) = \Lambda \exp(i\theta)$ where $0 \leq \theta \leq \pi/2$, but because of the choice of $H_n^{(1)}(R'K)$, for $R' > R$, this contribution vanishes when $\Lambda \to \infty$. By looking at the representative contour plots for $H_1^{(1)}(z)$ (but the same is true for other values of $n$) we can see why this is so: both its real and imaginary parts decay in the upper half of the complex plane as we increase the value of $y$ in $z = x + iy$. Also, along the real axis $(J_n(RK)H_n^{(1)}(R'K))/(K^2 + \kappa^2)$ reduces in size much faster than $1/K$ with $|x| \to \infty$, so again $C_\Lambda$ must vanish. Next we turn to the $I_1$-contribution for which the path is $z = iy$ with $\kappa + \varepsilon \leq y \leq \infty$, and the $I_2$ contribution for which the path is $z = iy$ with $0 \leq y \leq \kappa - \varepsilon$. For these we may write

$$\int_{I_1} f(z)dz = -\int_{k_z+\varepsilon}^\infty y \frac{J_n(iRy)H_n^{(1)}(iR'y)}{y^2 - \kappa^2}dy$$
$$= \frac{2i}{\pi} \int_{k_z+\varepsilon}^\infty y \frac{I_n(Ry)K_n(R'y)}{y^2 - \kappa^2}dy \qquad \text{(III.8.185)}$$

$$\int_{I_2} f(z)dz = -\int_0^{k_z-\varepsilon} y \frac{J_n(iRy)H_n^{(1)}(iR'y)}{y^2 - \kappa^2}dy$$
$$= \frac{2i}{\pi} \int_0^{k_z-\varepsilon} y \frac{I_n(Ry)K_n(R'y)}{y^2 - \kappa^2}dy \qquad \text{(III.8.186)}$$

These integrals end up to be purely imaginary, so they do not contribute to Eq. (III.8.179). Therefore, we are only left with $C_\varepsilon$, the semi-circle about the pole $z = i\kappa$. The simplest way of dealing with this is to write $f(z) = g(z)/(z - i\kappa)$ where $g(z) = zJ_n(Rz)H_n^{(1)}(R'z)/(z + i\kappa)$. Then, with the help of the residue theorem, we get

$$\lim_{\varepsilon \to 0} \int_{C_\varepsilon} f(z)dz = g(i\kappa) \lim_{\varepsilon \to 0} \int_{C_\varepsilon} \frac{1}{(z - i\kappa)} = -i\pi g(i\kappa)$$

Note that when we applied the residue theorem, we were integrating only over a half-circle and clock-wise, but not anti-clockwise, so that the factor multiplying $g(i\kappa)$ in the r.h.s was $-i\pi$ but not $+i2\pi$. Hence,

$$\Rightarrow \quad \lim_{\varepsilon \to 0} \int_{C_\varepsilon} f(z)dz = -\frac{i\pi}{2} J_n(i\kappa R)H_n^{(1)}(i\kappa R') = -I_n(\kappa R)K_n(\kappa R')$$
$$(III.8.187)$$

Combining Eqs. (III.8.183) and (III.8.187) and taking the real part, we find

$$\text{Re} \int_0^\infty K \frac{J_n(RK)H_n^{(1)}(R'K)}{K^2 + \kappa^2} dk_r = I_n(\kappa R)K_n(\kappa R') \quad (III.8.188)$$

We do not need to redo the derivation of the integral in (III.8.180), for $R > R'$ case—obviously, we can just swap $R'$ with $R$. Thus from Eqs. (III.8.179) and (III.8.180) we obtain

$$G_n(R, R') = R' \{ K_n(\kappa R')I_n(\kappa R)\theta(R' - R)$$
$$+ K_n(\kappa R)I_n(\kappa R')\theta(R - R') \} \quad (III.8.189)$$

where $\theta(R' - R)$ is the Heaviside theta function. Thus we may write an expression for the electrostatic potential

$$\phi(R) = \frac{4\pi}{\varepsilon} \left\{ K_0(\kappa R) \int_0^R R'I_0(\kappa R')\rho(R')dR' + I_0(\kappa R) \right.$$
$$\left. \times \int_R^\infty R'K_0(\kappa R')\rho(R')dR' \right\} \quad (III.8.190)$$

Of course, if $\rho(R') = 0$ when, for example, we look for the potential beyond the volume of localization of charges, only the first term in the bracket needs to be kept. Again, this form for $\phi(R)$ is valid provided that we have no boundary condition requirements of $\phi(R)$ through interfaces.

**Example III.8.19 (Physical Application ▸▸).** Let's consider the physical problem of two ions interacting in a slit nanopore. This is an important physical example (that applies to the nanotechnology of charge storage in electrochemical supercapacitors with nanostructured electrodes), which, in terms of maths, demonstrates a lot of what we've just learnt.

To start with, we consider a classical electrostatic problem: calculation of the electrostatic potential of a test charge positioned, here, for simplicity, in the middle of a gap between two semi-infinite 'ideal' metals, and evaluating the potential in a plane equidistant between the two surfaces, as sketched in Fig. III.8.23. By ideal we mean that there is no penetration by a static electric field into the metal, i.e. the static dielectric constant of the metal is equal to infinity. We suppose that inside the gap we have some uniform dielectric constant $\varepsilon$ (if it is a vacuum, $\varepsilon = 1$). A more general solution can be

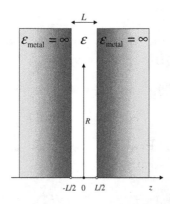

**Fig. III.8.23** A schematic picture of the system considered in Example III.8.20. We consider a slit of width $L$ that extends infinitely in the $x$–$y$ plane. On either side of the slit are metal walls modelled by an infinite dielectric constant (the simplest approximation, neglecting a finite screening length from the electrons in the metal). The slit geometry can be described through cylindrical polar coordinates $R = \sqrt{x^2 + y^2}$ and $z$. At the origin we place an elementary point charge describing an ion in the slit. The task in this example will be to find the value of the electrostatic potential $\varphi$ in the middle plane of the slit.

derived when the charge is not in the middle plane and so is the point we look at, but this is not essential to consider for demonstrating the power of the mathematical analysis that we focus on here.

In this example, we'll simply state the resulting solution of the electrostatic boundary value problem for the electrostatic potential $\varphi$, without deriving it, because that would distract attention from what want to demonstrate here. In Gaussian units the solution reads (as usual, to convert to SI units, multiply $\varepsilon$ by factor $4\pi\varepsilon_0$ in the denominator)

$$\varphi = \frac{e}{\varepsilon} \int_0^\infty dq\, J_0(qR) \tanh\left(\frac{Lq}{2}\right) \tag{III.8.191}$$

where $e$ is the elementary charge, and $J_0(qR)$ is the Bessel function of zero order. As we should have already appreciated in the analysis of homogeneous partial differential equations, this function regularly appears in problems with cylindrical symmetry (where the solutions doesn't depend on the polar angle in cylindrical polars). Let us again work neatly: rescale things by writing $t = qL/2$ and $p = 2R/L$, so that

$$e\varphi(R) = \frac{2e^2}{\varepsilon L} \int_0^\infty dt\, J_0(pt) \tanh(t) \equiv \frac{2e^2}{\varepsilon L} \boldsymbol{I}(p) \tag{III.8.192}$$

Now, we have reduced the problem to an integral, which is a function of just one parameter, $p(=2R/L)$:

$$\boldsymbol{I}(p) = \int_0^\infty dt\, J_0(pt) \tanh(t) \tag{III.8.193}$$

What are we going to do with this integral? Thought first, action second: as always, let us understand the behaviour of the integrand. The Bessel function $J_0(pt)$ is an oscillating function, shown in Fig. III.8.24 for three values of $p$. We will be particularly interested in the case where $p \gg 1$, i.e. when the charges are far away from each other, but we will also be interested in understanding $\boldsymbol{I}(p)$ in the domain of small and moderate $p$. Note that when $p$ is large, $J_0(pt)$ will oscillate rapidly. To deal with this problem, as we have demonstrated several times before (to re-reinforce a very important lesson), we need to extend the integration variable $t$ into the complex plane.

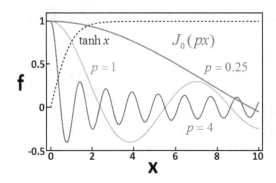

**Fig. III.8.24** We show plots of the Bessel function $J_0(pt)$ for the values $p = 0.25$ (red), $p = 1$ (green) and $p = 4$ (blue). We compare these plots with the function $\tanh t$, which is also in the integrand of Eq. (III.8.193). When $p$ is small, the dominant contribution to the integral will be at large values of $x$, and the integral can be well estimated by replacing $\tanh t$ with 1, which we see will work well for values of $p$ smaller than $p = 0.25$. But at large values of the integrand oscillates very rapidly and the estimation of the integral is less trivial: we need to extend the integration into the complex plane.

Before doing this, however, let us check if we recover an obvious limiting case, namely when $R \to 0$ (getting up close to the point charge) or $L \to \infty$ (moving the walls apart). In these cases, the effect of the walls should be negligible. Therefore, one must obtain the simple Coulomb law result, $\varphi(R) = e/(\varepsilon R)$, for a point charge. This law must, indeed, correspond to the limiting case $p \ll 1$ for $\boldsymbol{I}(p)$. The result described by Eq. (III.8.192) cannot be correct if it fails to recover this limiting case; as we have advocated many times such tests are useful in checking results. It is useful in this case to rescale the integral $\boldsymbol{I}(p)$ (Eq. (III.8.193)), by $t = px$ such that

$$\boldsymbol{I}(p) = \frac{1}{p} \int_0^\infty dy \, J_0(y) \tanh\left(\frac{y}{p}\right) \approx \frac{1}{p} \int_0^\infty dy \, J_0(y) \quad \text{when } p \ll 1$$

$$\text{(III.8.194)}$$

where we have used $\tanh(\frac{y}{p}) \approx 1$; this replacement will be valid for small $p$, as in this case this varies faster than $J_0(y)$. This can already be seen in Fig. III.8.24 for quite moderate values of $p$. Now, we have that

$$\int_0^\infty dy \, J_0(y) = 1 \quad \Rightarrow \quad I(p) = \frac{1}{p} \qquad \text{(III.8.195)}$$

Substitution of this result into Eq. (III.8.192) indeed yields the Coulomb law, $\varphi(R) = e/(\varepsilon R)$, as it should.

Having checked that the formula gives the correct limiting case for an isolated point charge, we will try to use complex integration to give a useful way of re-expressing $I(p)$. Again, the trick here is to use Hankel functions (defined by Eq. (III.8.167)) to write

$$I(p) = \text{Re}\left[\int_0^\infty dt\, H_0^{(1)}(pt)\tanh(t)\right] \qquad \text{(III.8.196)}$$

We'll then consider the function $f(z) = H_0^{(1)}(pz)\tanh(z)$ in the complex plane where $z = t + iy$. We note than on the imaginary axis $H_0^{(1)}(iy) = \frac{2}{\pi i}K_0(y)$ (from Eq. (III.8.170)), where $K_0(y)$ (as well as the other modified Bessel functions) has been discussed previously in Chapter 4 of this part of the book. The $K_0(y)$ function decreases exponentially for large values of $y$ (for its whole plot see Fig. III.4.14). Next, on the imaginary axis,

$$\tanh(it) = \frac{e^{it} - e^{-it}}{e^{it} + e^{-it}} = i\tan t \qquad \text{(III.8.197)}$$

The $\tanh(z)$ function in the integrand has first order poles at $z = i(k + 1/2)\pi$, where $k$ is any integer. It is crucial that $H_0^{(1)}(z)$ is fully analytic for $y > 0$, in the top right quadrant of the complex plane, and there are no other singularities in $f(z)$ than those of $\tanh(z)$. Therefore, if we take the contour of integration as shown in Fig. III.8.25, according to the Cauchy integral theorem, we can write

$$\int_{I_1} f(z)dz + \int_{I_2} f(z)dz = \int_{I_3} f(z)dz + \sum_{k=0}^\infty \int_{I_{r_k}} f(z)dz \qquad \text{(III.8.198)}$$

Here, we have the following contributions: $I_1$ along the real axis, $I_2$ about the quarter circle, $I_3$ along the imaginary axis, and $I_{r_k}$ semicircles skirting each of the poles of $\tan z$ (Fig. III.8.25). Now, since we closed the contour into the upper complex plane, when we send its radius to infinity $I_2$ vanishes due to the Hankel function. For each $I_{r_k}$ (in the limit where the radius of each semicircle goes to zero) we obtain

$$\int_{I_{r_k}} f(z)dz = i\pi \mathrm{res}\{f(z)\}_{z=i\pi(k+1/2)}$$

$$= i\pi \lim_{z \to i\pi(k+1/2)} \left\{ \frac{(z - i\pi(k+1/2))\sinh(z)}{\cosh(z)} H_0^{(1)}(z) \right\}$$

$$\Rightarrow \quad \int_{I_{r_k}} f(z)dz = -i\pi \sin(\pi(k+1/2)) H_0^{(1)}(i\pi(k+1/2))$$

$$\times \lim_{y \to \pi(k+1/2)} \left\{ \frac{y - \pi(k+1/2)}{\cos(y)} \right\}$$

$$\Rightarrow \quad \int_{I_{r_k}} f(z)dz = -\pi i H_0^{(1)}(i\pi(k+1/2)p) = 2K_0(\pi(k+1/2)p)$$

$$(\mathrm{III.8.199})$$

When we tend the radius of all semi-circles to zero we obtain from the contour $I_3$

$$\int_{I_3} f(z)dz = i \int_0^\infty dt H_0^{(1)}(ipt)\tanh(it) = \frac{2i}{\pi} \int_0^\infty dt K_0(pt)\tan t$$

$$(\mathrm{III.8.200})$$

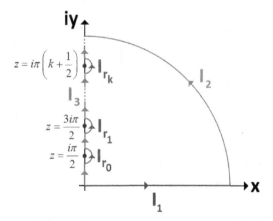

**Fig. III.8.25** By the Cauchy integral theorem the contour integral over path $I_1$ followed by $I_2$ is the same value as that along the path $I_3$ along the imaginary axis combined with the semicircular contours about each of the poles in $f(z) = H_0^{(1)}(pz)\tanh(z)$, $I_{r_k}$. This is because in the region enclosed by the integration contours is entirely analytic.

The integral Eq. (III.8.200) is purely imaginary and after taking the real part specified by Eq. (III.8.196) its contribution disappears and the only contributions from integration along the imaginary axis will be integration about the small semicircles, which after the application of the residue theorem will result in real values. So, on combining Eqs. (III.8.196), (III.8.199) and (III.8.200), we have that

$$\boldsymbol{I}(p) = \mathrm{Re} \int_{I_1} f(z)dz = 2 \sum_{k=0}^{\infty} K_0(\pi(k + 1/2)p) \qquad \text{(III.8.201)}$$

Now each of these terms in the sum given in Eq. (III.8.201) decays away with $p$, more rapidly at higher values of summation index $k$, so that each term in the series is progressively smaller.

Before exploring and applying this result, you might want to check that it is correct, by comparing it with numerical integration of $\boldsymbol{I}(p)$. We have done this in Fig. III.8.26, but keeping only the first term in the sum, and also showing there the result of simplification of the Bessel function $K_n(y)$ at large values of its argument, as discussed below.

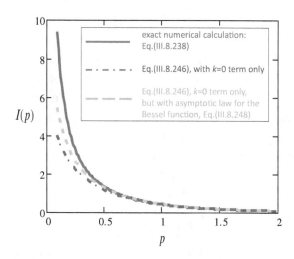

**Fig. III.8.26** Plotting the results of exact numerical calculation of the integral $\boldsymbol{I}(p)$ versus the approximate results, specified in the inset. This shows how amazingly well the latter reproduces the exact calculation already for for very moderate values of the argument, at $p > 0.5$.

We see how well both simplifications work for $p > 0.5$. Amazingly, the 'twice-simplified' formula that uses the asymptotic expansion of $K_n(y)$ works even better! Such things happen... Needless to say, if you keep more terms in the series, you will get prefect reproduction of the exact function, and the larger $p$, the smaller number of terms you will need to keep. We leave it to you to check, if you want to.

Substituting Eq. (III.8.201) back into Eq. (III.8.192), we obtain

$$e\varphi(R) = \frac{4e^2}{\varepsilon L} \sum_{k=0}^{\infty} K_0\left((2k+1)\pi\frac{R}{L}\right) \qquad \text{(III.8.202)}$$

For $R \gg L$, strictly speaking, but actually already for $R > L$, we may use only the leading order asymptotic approximation for the Bessel function

$$K_n(y) \approx \sqrt{\frac{\pi}{2y}} \exp(-y) \qquad \text{(III.8.203)}$$

as well as keeping only the first term in the sum (for the proof of this statement see Fig. III.8.26). Thus, we obtain an elegant and nontrivial formula:

$$e\varphi \approx \frac{e^2}{\varepsilon} \frac{4}{\sqrt{2RL}} \exp\left(-\frac{R}{L/\pi}\right) \qquad \text{(III.8.204)}$$

This means that the energy of electrostatic interaction between two point-charges in the slit between two metallic plates is not given by Coulomb law, but is exponentially screened by the metal walls of the slit. Most spectacular is that the screening length is not the width of the gap, but its value divided by 3.14. This means that large ions represented by point charges in the slit but whose physical diameter is just a bit smaller than the width of the gap will interact much more weakly than the Coulomb law would have suggested. In a dense array of such ions in such a pore, only the nearest neighbours will interact substantially with each other; next nearest neighbour interactions would be much less important. The result described in Eq. (III.8.203) was known, to our knowledge, already at least since 1920s, but unnoticed by many, it did not seem to receive the deserved engrossment in applications. It has recently drawn attention in rationalizing the electrical capacitance of nanoporous electrodes, charged with ionic liquids.

This last example provides an excellent illustration of the trick of converting oscillating integrands into exponentially decaying functions by 'moving' the integration onto the imaginary axis. It gave us an analytical formula in the form of a sum of known functions, which makes numerical evaluation easier. Even more importantly, from this formula it is straightforward to find an even simpler approximate law, valid for practical applications.

## 8.6. Key Points

- We may consider line integrals of functions $f(z = x + iy)$, in the complex plane, along paths specified by a curve (or contour) $C$ defined by $z(t) = x(t) + iy(t)$, where $t_1 \leq t \leq t_2$. These are called contour integrals and are written as $\int_C f(z)dz$ and they may be evaluated through Eq. (III.8.1).
- When we integrate $f(z)$ over a closed path in the complex plane, where $f(z)$ is entirely analytic within that enclosed region, the integral evaluates to zero, i.e. $\oint_C f(z)dz = 0$. This is known as Cauchy's integral theorem.
- Using Cauchy's theorem we may manipulate and distort integration contours, while maintaining the same value of the integral. The rules of contour manipulation are given at the end of Section 8.1; we suggest you learn them.
- When we integrate $f(z)$ over a closed path enclosing poles (points where $f(z)$ becomes singular) the residue theorem applies: $\oint_C f(z)dz = 2\pi i \sum_{k=1}^{n} \text{res}\{f(z)\}_{z=a_k}$ where we have poles at the points $a_k$ inside the integration contour oriented anti-clockwise and $\text{res}\{f(z)\}_{z=a_k}$ are the residues at those poles. For a pole of order $m_k$ (which diverges as $1/(z - a_k)^{m_k}$) near $a_k$, the residue is calculated as $\text{res}\{f(z)\}_{z=a_k} = \frac{1}{(m_k-1)!} \lim_{z \to a_k} \frac{d^{m_k-1}}{dz^{m_k-1}}\{(z - a_k)^m f(z)\}$.
- Similarly, clockwise integration over a semi-circle 'half-enclosing' a pole will give one half of the same result but with an opposite sign.

- Complex integration is very useful in dealing with integrals that have integrands that oscillate on the real axis. Oscillatory functions on the real axis may become monotonic on the imaginary axis. The trick of moving the contour of integration from the real axis to the imaginary one is sometimes referred to as rotation of contour of integration.
- Some integrals may be evaluated or modified by folding an integration contour around a branch cut, making use of the discontinuity. Examples of this technique are given in Section 8.3.
- A powerful method of approximation for definite integrals is the saddle point approximation. An integral that can be written as $I = \int_a^b \exp(\lambda w(x))dx$ where $\lambda \gg 1$ is amenable to this technique, if $w(z)$ has a saddle point(s), $z_0$, that satisfy $\frac{dw(z)}{dz} = 0$. Then, to be able to use the saddle point approximation, the integration contour should be deformed to pass through $z_0$, provided that the function $f(z) = \exp(\lambda w(z))$ remains analytic over the deformed contour. The (leading order) result of the saddle point approximation is

$$I \approx \exp(\lambda w(z_0)) \exp(i\phi(z_0)) \sqrt{\frac{2\pi}{\lambda |w''(z_0)|}}, \quad \text{with}$$

$$w''(z_0) = \left. \frac{d^2 w(z)}{dz} \right|_{z=z_0} = |w''(z_0)| \exp(i\phi(z_0))$$

- When considering integrals of the form $I = \int_0^\infty f(x) J_n(kx) dx$, which may arise in Hankel transforms (introduced in Chapter 4 of Part III, Vol. 2, and used there to solve the inhomogeneous Bessel equation) it is useful to consider the analytic properties of the Hankel functions, $H_n^{(1)}(z)$ and $H_n^{(2)}(z)$. These functions are defined as $H_n^{(1)}(z) = J_n(z) + iY_n(z)$ and $H_n^{(2)}(z) = J_n(z) - iY_n(z)$. For instance, we may deal with $I = \text{Re} \int_0^\infty f(x) H_n^{(1)}(kx) dx$ and attempt to use the analytic properties of $H_n^{(1)}(z)$ to devise an integration contour in the complex plane to either evaluate or convert the integral into a more manageable form.
- One important analytical property of these functions is that, as we change $x$ (which is the real part of $z$), the real and imaginary parts

of these functions oscillate. Another one is that as we increase $y$ (the imaginary part of $z$), $H_n^{(1)}(z)$ decays while $H_n^{(2)}(z)$ grows. For $z = iy$, where $y > 0$, $H_n^{(1)}(z)$ behaves as a modified Bessel function $K_n(z)$, and for $z = -iy$, where $y < 0$, $H_n^{(2)}(z)$ behaves the same way (the exact relationships are given in Eq. (III.8.170)). Lastly, both Hankel functions have a branch cut extending from 0 to $-\infty$. Representative plots for $H_n^{(1)}(z)$ and $H_n^{(2)}(z)$ are shown in Fig. III.8.20.

- We show two very instructive examples of how we may exploit Hankel functions to evaluate integrals in Section 8.5.

Chapter 9

# More Involved Methods for the Summation of Series

We will conclude this part of the book with a brief chapter introducing the reader to more sophisticated methods for the summation of series than those that were considered in Part I. For infinite sums, we extend our methods by first considering the series that we can sum through Fourier series. Also, we will look at a powerful method of summation of infinite series through the residue theorem, going into the complex plane. In Example III.8.20 we faced an opposite situation, when we managed to use the residue theorem to represent the integral under study by an infinite series. This gives us a hint that we can solve the reverse problem, and indeed in certain cases the residue theorem may give us simple formulae for summation of series. In the case of finite sums, we present a general method for approximating them which improves on the integral approximation, as a last resort before going to numerical summation.

## 9.1. Summation via Fourier Series

First, we consider summation of series through Fourier series by simply looking at a couple of examples. We have already introduced Fourier series in Chapter 2 of Part II (Vol. 1), and based on that the method itself is simple, and although it is a bit of an art finding a way to analytically perform such summations, it is a useful tool.

**Example III.9.1 (Maths Practice ♫♪).** We can sum the series

$$S = \sum_{n=1}^{\infty} \frac{1}{n^2} \tag{III.9.1}$$

which comes up in a lot of physical problems. The simplest way to do it is to consider the Fourier series

$$f(x) = \sum_{n=1}^{\infty} \frac{1}{n^2} \cos(nx) \tag{III.9.2}$$

where we have that $f(x) = f(x + 2\pi)$ and $f(x) = f(-x)$. In this case, one can show that Eq. (III.9.2) sums to the periodic function

$$f(x) = (x - \pi)^2/4 - \pi^2/12 \quad 0 \le x \le 2\pi \tag{III.9.3}$$

One can check this by showing that

$$\int_0^{2\pi} f(x)dx = 0 \quad \text{and} \quad \frac{1}{\pi}\int_0^{2\pi} f(x)\cos(nx)dx = \frac{1}{n^2} \quad \text{for } n \ge 1 \tag{III.9.4}$$

Hence, we can evaluate the series in Eq. (III.9.5), by simply setting $x = 0$ in Eqs. (III.9.2) and (III.9.3).

Thereby, we obtain the result

$$f(0) = \frac{\pi^2}{6} = \sum_{n=1}^{\infty} \frac{1}{n^2} \tag{III.9.6}$$

The art, here, is in finding such functions, as in Eq. (III.9.3), that enable us to sum a particular series. Let us give you a hint on how we deduced the form of Eq. (III.9.3). The term $(x - \pi)^2$ yields the integrand $x^2 \cos(nx)$, in the right-hand part of Eq. (III.9.4), on shifting the integration range to $-\pi$ and $\pi$, which would integrate to something proportional to $1/n^2$. To appreciate why this should be so, one should note that

$$\int_{-\pi}^{\pi} x \cos(nx)dx = \int_{-\pi}^{\pi} \cos(nx)dx = 0 \tag{III.9.7}$$

Then we are able to write our function $f(x) = b(x - \pi)^2 + a$ and determine $b$ and $a$ through substitution into Eq. (III.9.4). Here, note that constant $a$ can always be added to $b(x - \pi)^2$, as the contribution of $a$ into the right-hand part of Eq. (III.9.4) would be zero. We find the actual value $a = -\pi^2/12$ on substitution into Eq. (III.9.4).

**Example III.9.2 (Maths Practice♫♪).** Let us sum the series

$$S = \sum_{n=1}^{\infty} \frac{(-1)^n}{(2n + 1)^3} \tag{III.9.8}$$

Now again, we'll use some tricks. We might try to see if we could evaluate the sum by integrating the Fourier series of the previous example, i.e. Eq. (III.9.2). A reason why this might be a good thing to do is to try integrating each term in the series, so that the resulting terms would have $1/n^3$ as a factor in them. On integrating Eq. (III.9.2), we obtain

$$F(x) = \int_0^x [(x' - \pi)^2/4 - \pi^2/12]dx' = \sum_{n=1}^{\infty} \frac{1}{n^2} \int_0^x \cos(nx')dx'$$

$$\tag{III.9.9}$$

Equation (III.9.9) evaluates to

$$(x - \pi)^3/12 + \pi^3/12 - \pi^2 x/12 = \sum_{n=1}^{\infty} \frac{1}{n^3} \sin(nx) \tag{III.9.10}$$

Next, we set $x = \pi/2$, so that taking into account that $\sin m\pi = 0$ for any integer $m$, we obtain the result

$$\frac{\pi^3}{32} = \sum_{n=0}^{\infty} \frac{(-1)^n}{(2n + 1)^3} \tag{III.9.11}$$

You can check this result numerically.

## 9.2.  Summation of Infinite Series Through the Residue Theorem

**Example III.9.3 (Maths Practice♫♪).** We'll describe summation of series through the residue theorem by first considering an example. We will suppose that we need to sum the series

$$S = \frac{2\pi}{L} \sum_{n=-\infty}^{\infty} \frac{1}{\left(\frac{2\pi n}{L}\right)^2 + \alpha^2} \tag{III.9.12}$$

It is much easier to show how we solve this problem than describe in words the strategy for doing it, although we will try to summarize it at the end. So, now we go straight to the point.

Let us express Eq. (III.9.12) as the sum of residues from a complex contour integral about path $C$ of an integrand $\frac{1}{z^2+\alpha^2} \cot\left(\frac{zL}{2}\right)$ enclosing the poles of $\cot(zL/2)$. These poles lie at the values $zL/2 = n\pi$ on the real axis. This path of integration $C$ is shown in Fig. III.9.1. For the reason that you will see in a moment, we took this contour to be in the clockwise direction, but the residue theorem can be applied properly with a minus sign for this 'reverse' direction. The relationship between the sum and contour integral reads as (you should check this yourself be computing the residues)

$$S = \frac{i}{2} \int_C \frac{1}{z^2 + \alpha^2} \cot\left(\frac{zL}{2}\right) dz \tag{III.9.13}$$

Now the key here is to understand that contributions to $C$ at the ends, when $|z| = \infty$, are negligible. Indeed, Eq. (III.9.13) gives an integral representation of the sum through a contour on the complex plane, but it doesn't yet give us a simple answer. But we can now consider two new contours, shown in the right part of Fig. III.9.1, which are hemispheres in the bottom and top half of the complex plane enclosing either the $z = -i\alpha$ or $z = i\alpha$ pole. The integral over the sum of these two contours yields the same as that over $C$. From

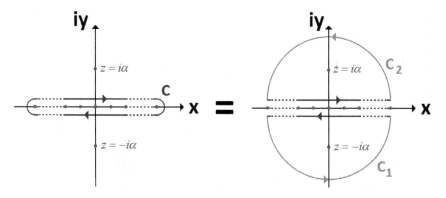

**Fig. III.9.1**   An illustration of how we can sum an infinite series by the residue theorem. To sum the series in Example III.9.3 we consider the function, $f(z) = \cot(zL/2)/(z^2+\alpha^2)$ in which $\cot(zL/2)$ has poles along the real axis at the values $z = 2n\pi/L$, shown as mid points. The series given by Eq. (III.9.10) may be generated by the integration over the contour $C$ from the residues of the poles that it encloses. Here the dashed lines are to suggest both the $x$ axis and integration contour extending out to $-\infty$ and $\infty$. Also, the function $(z^2 + \alpha^2)^{-1}$ has poles on the imaginary axis, at $i\alpha$ and $-i\alpha$. To sum the series we may break contour $C$ at both ends, as the end contributions are negligible. Then, we may form two new closed contours $C_1$ and $C_2$ to enclose $i\alpha$ and $-i\alpha$. The integral over the sum of these two contours is the same as over $C$, as the contributions of the paths about the two semicircles are zero. The contours $C_1$ and $C_2$ allow us to evaluate the sum by taking the residues of $f(z)$ at the two poles, at $i\alpha$ and $-i\alpha$.

the residues encircled by $C_1$ and $C_2$ we find

$$I_1 = \frac{i}{2} \int_{C_1} \frac{1}{z^2 + \alpha^2} \cot\left(\frac{zL}{2}\right) dz = \frac{i}{2} 2\pi i \operatorname{res}\left\{\frac{\cot\left(\dfrac{zL}{2}\right)}{z^2 + \alpha^2}\right\}_{z=i\alpha}$$

$$= \frac{\pi}{2\alpha} \coth\left(\frac{\alpha L}{2}\right) \tag{III.9.14}$$

and

$$I_2 = \frac{i}{2} \int_{C_2} \frac{1}{z^2 + \alpha^2} \cot\left(\frac{zL}{2}\right) dz = \frac{i}{2} 2\pi i \operatorname{res}\left\{\frac{\cot\left(\dfrac{zL}{2}\right)}{z^2 + \alpha^2}\right\}_{z=-i\alpha}$$

$$= \frac{\pi}{2\alpha} \coth\left(\frac{\alpha L}{2}\right) \tag{III.9.15}$$

Since we can take the radius of each hemisphere to go to infinity, we can neglect the contributions around the half-circles. Now as the remaining, contributing parts of these two contours coincide with the non-negligible parts of $C$, combining Eqs. (III.9.13), (III.9.14) and (III.9.15), we can write

$$S = I_1 + I_2 = \frac{\pi}{\alpha} \coth\left(\frac{\alpha L}{2}\right) \tag{III.9.16}$$

Let's check some limiting cases to see if Eq. (III.9.16) yields the correct behaviour, to make sure that no error had sneaked into our analysis.

(i) When $\alpha L \ll 1$, we expect the dominant contribution in the sum to be only from the $n = 0$ term. Thus, we should expect from Eq. (III.9.12) that here $S \approx 2\pi/(\alpha^2 L)$. But when $\alpha L$ is small, $\coth(\alpha L/2) \approx 2/(\alpha L)$, so Eq. (III.9.16) does indeed give us what we expected.

(ii) When $\alpha L \gg 1$ the sum is accurately approximated by an integral (the approach described in Chapter 8, Part I, Vol. 1), so that

$$S \approx \int_{-\infty}^{\infty} \frac{dk}{k^2 + \alpha^2} = \frac{\pi}{\alpha} \tag{III.9.17}$$

where we introduced the variable of integration $k = k_n = 2\pi n/L$, neglecting that $n$ is an integer, because the difference $\Delta_n = k_{n+1} - k_n = 2\pi/L << k_n$ at large $n$. But in the limit when $\alpha L \gg 1$, $\coth(\alpha L/2) \approx 1$, so we do get the same result from Eq. (III.9.16).

You can, of course, always check the result by performing the sum numerically.

### General formulae for summation through the residue theorem

The previous example illustrated the technique of summation by residues. Let's now formulate some general rules applicable to a broad class of series which can be summed this way. Let us consider a series of the form

$$S = \frac{2\pi}{L} \sum_{n=-\infty}^{\infty} f\left(\frac{2\pi n}{L}\right) \qquad \text{(III.9.18)}$$

with $f(z)$ analytical everywhere except at poles where $z = a_k$ (off the real axis), as well as fulfilling the requirement $Rf(iR\exp(i\theta)) \to 0$ when $R \to \infty$ (the latter has to be true for any convergent sum, anyhow). We can use exactly the same contours illustrated in Fig. III.9.1, with multiple poles for $f(z)$. The general result for the sum is then

$$S = -\pi \sum_k \text{res}\left\{f(z)\cot(zL/2)\right\}_{z=a_k} \qquad \text{(III.9.19)}$$

One can consider sums of many other forms which can be evaluated in this way. First, in similar fashion, we can sum a series of the form

$$S = \frac{2\pi}{L} \sum_{n=-\infty}^{\infty} (-1)^n f\left(\frac{2\pi n}{L}\right) \qquad \text{(III.9.20)}$$

This will be equivalent to $S = \frac{i}{2} \int_C \frac{f(z)}{\sin(zL/2)} dz$, as Eq. (III.9.20) will follow from this integral via the calculation of the residues of the poles of $1/\sin(zL/2)$ which lie at $z = 2\pi n/L$ and are equal $2(-1)^n/L$, provided that $f(z)$ is analytical everywhere except at the poles at $z = a_k$ (lying elsewhere, again off the real axis). Then, using the same logic as in the previous case, turning to contour integration over the contours $C_1$ and $C_2$ (provided that contributions on the semi-circle parts of $C_1$ and $C_2$ vanish), we obtain

$$S = -\pi \sum_k \text{res}\left\{\frac{f(z)}{\sin(zL/2)}\right\}_{z=a_k} \qquad \text{(III.9.21)}$$

**Exercise.** Show that one gets Eq. (III.9.21) by following the steps as in Eqs. (III.9.13)–(III.9.16).

Another example is

$$S = \frac{2\pi}{L} \sum_{n=-\infty}^{\infty} f\left(\frac{2\pi(n+1/2)}{L}\right) \qquad \text{(III.9.22)}$$

Again, $f(z)$ should be analytical everywhere except at poles at $z = a_k$, and we require that $Rf(iR\exp(i\theta)) \to 0$ when $R \to \infty$.

Performing similar operations as previously yields the summation formula

$$S = \pi \sum_k \mathrm{res}\left\{ f(z) \tan\left(\frac{zL}{2}\right) \right\}_{z=a_k} \qquad \text{(III.9.23)}$$

**Exercise.** Take $f(z) = \frac{1}{z^2+\alpha^2}$, and show the validity of the residue formula by calculating the sum numerically (it will be generally sufficient to run $-1000 < n < 1000$; of course the lager L, the slower the sum converges, and the more terms in it you will need to keep for an accurate result).

**Hint:** Make sure that for this form of $f(z)$ calculating the residue in (III.9.23), you will get $S = \frac{\pi}{\alpha} \tan h\left(\frac{\alpha L}{2}\right)$.

**Example III.9.4 (Maths Practice ♪♪).** To practise this technique, let us consider a slightly more complicated example, summing up the series

$$S = \sum_{n=-\infty}^{\infty} \frac{1}{\left(\frac{2\pi(n+1/2)}{L}\right)^4 + \beta^4} \qquad \text{(III.9.24)}$$

using directly the summation formula given by Eq. (III.9.23). For this, we need to find the poles of $f(z) = 1/(z^4 + \beta^4)$ by solving $z^4 = -\beta^4$. This equation for the poles can be solved by writing $z = \beta \exp(i\theta)$, as we did in Chapter 9 of Part I, in Vol. 1. We see that we have poles at the values

$$z = a_1 = \frac{\beta}{\sqrt{2}}(1+i) \quad z = a_2 = \frac{\beta}{\sqrt{2}}(1-i) \quad z = a_3 = -\frac{\beta}{\sqrt{2}}(1-i)$$

$$z = a_4 = -\frac{\beta}{\sqrt{2}}(1+i) \qquad \text{(III.9.25)}$$

Using the summation formula given by Eq. (III.9.23) we obtain

$$S = \frac{L}{2} \sum_{k=1}^{4} \lim_{z \to a_k} \left\{ \frac{(z-a_k)\tan(a_k L/2)}{(z-a_1)(z-a_2)(z-a_3)(z-a_4)} \right\} \equiv \frac{L}{2} \sum_{k=1}^{4} T_k$$

$$\text{(III.9.26)}$$

We find that

$$T_1 = \frac{1}{(a_1 - a_2)(a_1 - a_3)(a_1 - a_4)} \tan\left(\frac{La_1}{2}\right)$$

$$= \frac{1}{i\beta^3\sqrt{2}} \frac{1}{2+2i} \tan\left(\frac{L\beta(1+i)}{2\sqrt{2}}\right) \tag{III.9.27}$$

**Exercise.** Show that

$$T_1 = T_4 = T_2^* = T_3^* \tag{III.9.28}$$

and, by summing all four terms obtain the following expression for $S$, that

$$S = 2L\mathrm{Re}\left[ -\frac{1}{\beta^3\sqrt{2}} \frac{(1-i)}{4} \frac{e^{\frac{L\beta(i-1)}{2\sqrt{2}}} - e^{\frac{L\beta(1-i)}{2\sqrt{2}}}}{e^{\frac{L\beta(i-1)}{2\sqrt{2}}} + e^{\frac{L\beta(1-i)}{2\sqrt{2}}}} \right] \tag{III.9.29}$$

**Hint:** You'll need to express $\tan(x)$ in terms of $\exp(ix)$.

**Exercise.** To simplify things further, multiply top and bottom of what is contained in the square bracket in Eq. (III.9.29) by $e^{-\frac{L\beta(i+1)}{2\sqrt{2}}} + e^{\frac{L\beta(1+i)}{2\sqrt{2}}}$ to make the denominator real. Then, show that one obtains the final result:

$$S = \frac{L}{2\sqrt{2}\beta^3} \frac{\sinh\left(\frac{L\beta}{\sqrt{2}}\right) - \sin\left(\frac{L\beta}{\sqrt{2}}\right)}{\cosh\left(\frac{L\beta}{\sqrt{2}}\right) + \cos\left(\frac{L\beta}{\sqrt{2}}\right)} \tag{III.9.30}$$

Confession: what we did right after having obtained this exotic formula—we plotted numerically the sum as a function of $\beta$ for few values of $L$ and compared the results with those given by this formula. Do the same yourself to see that they coincide, even when running the sum in the interval of $-10 < n < 10$.

**Example III.9.4 (Continued).** Let's consider limiting cases. For $\beta L \ll 1$, the original sum in Eq. (III.9.24) reduces to

$$S \approx \frac{L^4}{(2\pi)^4} \sum_{n=-\infty}^{\infty} \frac{1}{(n+1/2)^4} = \frac{L^4}{48} \qquad \text{(III.9.31)}$$

You can check that $\sum_{n=-\infty}^{\infty} (n+1/2)^{-4} = \pi^4/3$ numerically, or [**Exercise**] you can cleverly think of the appropriate Fourier series to perform the sum (a challenge for you!). Now, let's check what our analytical result gives. In this limit we need to keep the first non-vanishing terms in the Maclaurin expansion of the numerator and denominator of Eq. (III.9.30). We have

$$\sinh\left(\frac{L\beta}{\sqrt{2}}\right) - \sin\left(\frac{L\beta}{\sqrt{2}}\right) \approx \frac{L^3\beta^3}{6\sqrt{2}}, \cos h\left(\frac{L\beta}{\sqrt{2}}\right) + \cos\frac{L\beta}{\sqrt{2}} \approx 2$$

$$\text{(III.9.32)}$$

where we used

$$\sin x \approx x - \frac{x^3}{6} \quad \sinh x \approx x + \frac{x^3}{6} \quad \cosh x \approx \cos x \approx 1 \quad \text{for} \quad x \ll 1$$

$$\text{(III.9.33)}$$

So that indeed we get $S \approx L^4/48$; complex analysis has got it right!

If you did not find this beautiful, you might not be the one who needs this book (but then it would be a wonder why you've stayed with us so far!).

**Exercise.** With that diversion in mind, we suggest you another test of certainty of your 'conversion'. Taking into account that in the limit of $L \gg 1$ the sum can be written as the integral

$$S = \frac{L}{2\pi} \int_{-\infty}^{\infty} \frac{dk}{k^4 + \beta^4} = \frac{L}{2\pi\beta^3} \int_{-\infty}^{\infty} \frac{dx}{x^4 + 1} \qquad \text{(III.9.34)}$$

evaluate the integral in Eq. (III.9.34) (you should find it equal to $\pi/\sqrt{2}$, but, please, show it yourself, referring if you need to the hint below) and show that Eq. (III.9.30) reproduces the limiting case of

Eq. (III.9.34) for $L \gg 1$, which is, of course ... $L/(2\sqrt{2}\beta^3)$, as the fraction there tends to 1 in this limit.

**Hint:** In order to evaluate the integral $\int_{-\infty}^{\infty} \frac{dx}{x^4+1}$ you may wish to use contour integration and utilize the residue theorem, evaluating first $\int_{-\infty}^{\infty} e^{ikx} \frac{dx}{x^4+1}$ and tending in the result $k \to 0$. Alternatively, you can consider this integral as $2 \int_0^{\infty} \frac{dx}{1+x^4}$ change variables from $x$ to $t$ through $x = e^t$, and thereby the domain of integration to $-\infty < t < \infty$, and then follow the method of Example III.8.13. Once you get this proved, we think you may experience a moment of pure mathematical pleasure as, perhaps, you might have had in checking the previous limit.

**Example III.9.5 (Physical Application ▶▶).** Let us consider the physical example of a positively charged semiflexible polymer of length $L$ of linear charge density $\rho$. The position of the molecular axis can be described by the vector $\mathbf{r}(s) \approx \mathbf{R}(s) + s\hat{\mathbf{k}}$, with $\mathbf{R}(s)$ standing for small displacements $\mathbf{R}(s) = x(s)\hat{\mathbf{i}} + y(s)\hat{\mathbf{j}}$ away from a vertical straight line. We will suppose that this macromolecule is trapped in $x$ and $y$ directions in a pore, in the potential having cylindrical symmetry with the minimum at $x = y = 0$. About such a minimum at $R = 0$, the potential can be written for small lateral displacements as $\varphi(\mathbf{r}) \approx \varphi_0 + \gamma(x^2 + y^2)/2$. It can be of electrostatic nature, if the polymer is charged. We'll consider the molecule thermally fluctuating, but we'll suppose that it is held between two end points where $x(L) = y(L) = x(0) = y(0) = 0$. Such a system is depicted in Fig. III.9.2. Our task will be to calculate the spatial mean of the thermal average of the squared displacement, $\langle R(s)^2 \rangle$, defined as

$$\overline{R^2} = \frac{1}{L} \int_0^L \langle R(s)^2 \rangle \, ds \quad \text{where} \quad \langle R(s)^2 \rangle = \langle x(s)^2 \rangle + \langle y(s)^2 \rangle$$

$$(\text{III.9.35})$$

First we can write (to satisfy the end constraints)

$$x(s) = \sum_{n=1}^{\infty} \tilde{x}_n \sin\left(\frac{\pi n s}{L}\right) \quad \text{and} \quad y(s) = \sum_{n=1}^{\infty} \tilde{y}_n \sin\left(\frac{\pi n s}{L}\right) \quad (\text{III.9.36})$$

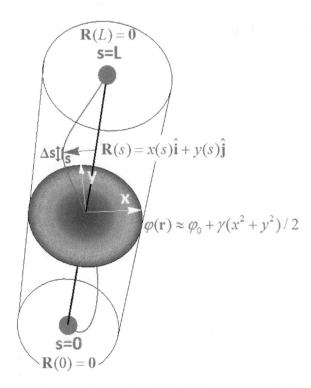

**Fig. III.9.2** A schematic picture of semiflexible linear macromolecule trapped in a harmonic potential. A possible configuration that the molecule may adopt is shown a red curve. The molecule is held at its two ends at the centre axis of a cylinder, chosen to be the z-axis. At any point $s$ along the molecule, the molecule has a displacement $\mathbf{R}(s) = x(s)\hat{\mathbf{i}} + y(s)\hat{\mathbf{j}}$ away from the z-axis. Note that at the fixed ends, $s = 0$ and $s = L$, we have $\mathbf{R}(0)0 = \mathbf{R}(L) = \mathbf{0}$. Shown in a cross-section is a density plot of the harmonic potential of electrostatic nature, $\varphi(\mathbf{r}) \approx \varphi_0 + \gamma(x^2 + y^2)/2$; blue represents low values, while red indicates high values. Note that the potential has cylindrical symmetry, so a cross-section at any other altitude would have the same density plot. Also shown, as a blue segment, a small segment of the molecule of length $\Delta s$. If the confining potential on the polymer is electrostatic and the polymer has linear charge density $\rho$ that section has charge $\rho \Delta s$. The electrostatic energy of that segment is roughly $\varphi(\mathbf{r}(s))\rho \Delta s$.

The Fourier coefficients $\tilde{x}_n$ and $\tilde{y}_n$ specify a particular configuration of the fluctuating polymer. We can then express $\overline{R^2}$ as

$$\overline{R^2} = \frac{1}{L} \sum_{n=1}^{\infty} \sum_{n'=1}^{\infty} \langle \tilde{x}_n \tilde{x}_{n'} + \tilde{y}_n \tilde{y}_{n'} \rangle \int_0^L ds \sin\left(\frac{\pi n s}{L}\right) \sin\left(\frac{\pi n' s}{L}\right)$$

$$(\text{III.9.37})$$

The integral in the r.h.s can be expressed through the Kroenecker delta, $\delta_{n,n'}$ (see Chapter 2, Part II, Vol. 1) as

$$\frac{1}{L} \int_0^L ds \sin\left(\frac{\pi n s}{L}\right) \sin\left(\frac{\pi n' s}{L}\right) = \frac{1}{2}\delta_{n,n'} \tag{III.9.38}$$

(you can check this inequality just by performing integration in its left-hand side). Then we can write

$$\overline{R^2} = \frac{1}{2}\sum_{n=1}^{\infty}\left(\langle \tilde{x}_n^2 \rangle + \langle \tilde{y}_n^2 \rangle\right) \tag{III.9.39}$$

The thermal averages or expectation values $\langle \tilde{x}_n^2 \rangle$ and $\langle \tilde{y}_n^2 \rangle$ are evaluated through

$$\langle \tilde{x}_n^2 \rangle = \int_{-\infty}^{\infty} \tilde{x}_n^2 P(\tilde{x}_n) d\tilde{x}_n \quad \langle \tilde{y}_n^2 \rangle = \frac{1}{Z_n}\int_{-\infty}^{\infty} \tilde{y}_n^2 P(\tilde{y}_n) d\tilde{y}_n \tag{III.9.40}$$

where $P_n(\tilde{x}_n)$ and $P_n(\tilde{y}_n)$ are Boltzmann probability distributions for each of the Fourier coefficients. Equation (III.9.40) simply comes from the definition of what an expectation value is (see Part I, Vol. 1). To obtain expressions for $P_n(\tilde{x}_n)$ and $P_n(\tilde{y}_n)$ we need to consider the energy of a polymer configuration.

Utilizing the worm like chain model (a classic theory of polymer physics and the biophysics of DNA, which we will revisit later in this book), the energy is given by the sum of elastic bending and electrostatic energies, $E = E_{els} + E_{\phi}$. The elastic bending energy, with $\mathbf{r}(s) \approx x(s)\hat{\mathbf{i}} + y(s)\hat{\mathbf{j}} + s\hat{\mathbf{k}}$, is given by

$$E_{els} = \frac{B}{2}\int_0^L ds \left(\frac{d^2\mathbf{r}(s)}{ds^2}\right)^2 \approx \frac{B}{2}\int_0^L ds \left[\left(\frac{d^2 x(s)}{ds^2}\right)^2 + \left(\frac{d^2 y(s)}{ds^2}\right)^2\right]$$

$$\tag{III.9.41}$$

where $B$ is the bending elastic modulus of the polymer. To get an expression for the electrostatic energy we think of a small, charged segment $\rho\Delta s$ positioned at $s$ with displacement $x(s)\hat{\mathbf{i}} + y(s)\hat{\mathbf{j}}$, illustrated in Fig. III.9.2. Its electrostatic energy $\varphi(\mathbf{r}(s))\rho\Delta s \approx \varphi_0\rho\Delta s + \frac{1}{2}\gamma\rho\Delta s[x(s)^2 + y(s)^2]$. If we sum up all the contributions from charged segments in the limit $\Delta s \to 0$ (treating the centre line of the polymer a smooth curve) we obtain

$$E_\phi = L\rho\varphi_0 + \frac{\rho\gamma}{2} \int_0^L ds(x(s)^2 + y(s)^2) \qquad \text{(III.9.42)}$$

Substituting Eq. (III.9.36) into Eqs. (III.9.41) and (III.9.42), we find

$$\Delta E = E - L\rho\varphi_0 = \sum_{n=1}^{\infty} \sum_{n'=1}^{\infty} \left[ \frac{B}{2} \left(\frac{\pi n}{L}\right)^2 \left(\frac{\pi n'}{L}\right)^2 + \frac{\rho\gamma}{2} \right]$$

$$\times (\tilde{x}_n \tilde{x}_{n'} + \tilde{y}_n \tilde{y}_{n'}) \int_0^L ds \sin\left(\frac{\pi n s}{L}\right) \sin\left(\frac{\pi n' s}{L}\right)$$

$$\Rightarrow \Delta E = \frac{L}{4} \sum_{n=1}^{\infty} \left[ B \left(\frac{\pi n}{L}\right)^4 + \rho\gamma \right] (\tilde{x}_n^2 + \tilde{y}_n^2)$$

$$\equiv \frac{k_B T}{2} \sum_{n=1}^{\infty} (\tilde{x}_n^2 + \tilde{y}_n^2)\Omega_n \qquad \text{(III.9.43)}$$

Note that we have defined $2k_B T \Omega_n = L[B(\pi n/L)^4 + \rho\gamma]$.

Now, in the worm like chain model, the Boltzmann probability (valid for thermal equilibrium) that a molecule will be in a particular configuration, described by a particular set of values of $\tilde{x}_n$ and $\tilde{y}_n$, is proportional to $\exp(-\Delta E/k_B T)$, with $\Delta E$ calculated for that particular configuration. This suggests that the fluctuations of each mode $\tilde{x}_n$ and $\tilde{y}_n$ should be weighted by the probability distributions

$$P_n(\tilde{x}_n) = \frac{\exp(-\Omega_n \tilde{x}_n^2/2)}{Z_n} \quad P_n(\tilde{y}_n) = \frac{\exp(-\Omega_n \tilde{y}_n^2/2)}{Z_n} \qquad \text{(III.9.44)}$$

where we have

$$Z_n = \int_{-\infty}^{\infty} \exp(-\Omega_n \tilde{x}_n^2/2)d\tilde{x}_n = \int_{-\infty}^{\infty} \exp(-\Omega_n \tilde{y}_n^2/2)d\tilde{y}_n \qquad \text{(III.9.45)}$$

Then, the expectation values defined by Eq. (III.9.40) are re-expressed as

$$\langle \tilde{x}_n^2 \rangle = \frac{1}{Z_n} \int_{-\infty}^{\infty} \tilde{x}_n^2 \exp\left(-\frac{\Omega_n}{2} \tilde{x}_n^2\right) d\tilde{x}_n$$

$$\langle \tilde{y}_n^2 \rangle = \frac{1}{Z_n} \int_{-\infty}^{\infty} \tilde{y}_n^2 \exp\left(-\frac{\Omega_n}{2} \tilde{y}_n^2\right) d\tilde{y}_n \qquad \text{(III.9.46)}$$

**Exercise.** By performing the integrals in Eqs. (III.9.45) and (III.9.46) show that

$$\langle \tilde{x}_n^2 \rangle = \langle \tilde{y}_n^2 \rangle = \frac{1}{\Omega_n} \qquad \text{(III.9.47)}$$

**Example III.9.5 (Continued).** Substitution of Eq. (III.9.47) in (III.9.49) leaves us with the sum

$$\overline{R^2} = \frac{2k_B T}{BL} \sum_{n=1}^{\infty} \frac{1}{\left(\frac{\pi n}{L}\right)^4 + \frac{\rho\gamma}{B}} \qquad \text{(III.9.48)}$$

First we rearrange Eq. (III.9.48) in a form for which we can use one of the summation formulae:

$$\overline{R^2} = \frac{k_B T}{BL} \sum_{n=-\infty}^{\infty} \frac{1}{\left(\frac{\pi n}{L}\right)^4 + \frac{\rho\gamma}{B}} - \frac{k_B T}{L\rho\gamma} \qquad \text{(III.9.49)}$$

The sum can be calculated in a similar fashion to the previous example. Here, we utilize the same poles given by Eq. (III.9.25) to compute the residues, but with $\beta = (\rho\gamma/B)^{1/4}$. This time the formula we use is Eq. (III.9.19) (with the replacement of $L$ by $2L$).

Hence, we come to:

$$\overline{R^2} = \frac{k_B T}{\sqrt{2}B\beta^3} \frac{\sinh\left(\sqrt{2}\beta L\right) + \sin\left(\sqrt{2}\beta L\right)}{\cosh\left(\sqrt{2}\beta L\right) - \cos\left(\sqrt{2}\beta L\right)} - \frac{k_B T}{L\rho\gamma} \qquad \text{(III.9.50)}$$

**Example III.9.5 (Continued).** Let's again consider limiting cases. Note that we may write $\beta L = L/L^*$ where $L^* = (B/\rho\gamma)^{1/4}$ is a characteristic confinement length. We start with $L/L^* \ll 1$; this represents the case of a very short or stiff molecule. In this case Eq. (III.9.48) reduces to

$$\overline{R^2} = \frac{2L^3 k_B T}{B\pi^4} \sum_{n=1}^{\infty} \frac{1}{n^4} \qquad (III.9.51)$$

It's worth commenting on the physics of this result. As the length of the molecule is short, we expect that the amplitudes of fluctuation $\overline{R^2}$ to be small, as sizable fluctuations in this case would cost too much bending energy. Thus, we expect that the electrostatic potential energy will have little effect in limiting the fluctuations for very short molecules, and the parameters $\gamma$ and $\rho$ shouldn't be in the answer—which is what we see here. As the length of the molecule is extended, initially the amplitude $\overline{R^2}$ grows rapidly $\sim L^3$, as on extending the length (from short molecules) the bending energy allows for a significantly larger value of $\overline{R^2}$. And so the result makes perfect sense.

**Exercise.** One can evaluate the sum in Eq. (III.9.51), by integrating both sides of Eq. (III.9.10) between $\pi/2$ and $x$ and then setting $x = 0$ in the resulting Fourier series. By doing this show that

$$\sum_{n=1}^{\infty} \frac{1}{n^4} = \frac{\pi^4}{90} \quad \text{and} \quad \overline{R^2} = \frac{k_B T L^3}{45B} \qquad (III.9.52)$$

Then, show that Eq. (III.9.50) agrees with this result by making Maclaurin series expansions of $\sinh(\beta L/\sqrt{2})$, $\sin(\beta L/\sqrt{2})$, $\cosh(\beta L/\sqrt{2})$ and $\cos(\beta L/\sqrt{2})$ for small $L$.

**Example III.9.5 (Continued).** When $L/L^* \gg 1$ one can approximate the sum in Eq. (III.9.49) with an integral such that

$$\overline{R^2} \approx \frac{k_B T}{\pi B} \int_{-\infty}^{\infty} \frac{dk}{k^4 + \beta^4} = \frac{k_B T}{\pi B^{1/4}} \left(\frac{1}{\gamma\rho}\right)^{3/4} \int_{-\infty}^{\infty} \frac{dx}{x^4 + 1}$$

$$= \frac{k_B T}{\sqrt{2} B^{1/4}} \left(\frac{1}{\gamma\rho}\right)^{3/4} \qquad (III.9.53)$$

You can check that the far r.h.s is true by evaluating the integral, using the residue theorem, but if you did an exercise on

Example III.9.4, there you should have dealt with this integral already (see there for a Hint on how to do it).

Let us briefly comment on the physics of the result in Eq. (III.9.52). When the molecule is very long the amplitude of fluctuations $\overline{R^2}$ no longer depends on $L$; it has saturated to a constant value as it is limited by the electrostatic interaction. The size of $\gamma$ and $\rho$ increases electrostatic forces, and so limits $\overline{R^2}$. Actually, we could have deduced the form of Eq. (III.9.53), to the accuracy of the numerical coefficient, using dimensional analysis (see Chapter 10, Part I, Vol. 1) from the variables in the problem, when $L \to \infty$. Clearly $\overline{R^2}$ has units of length squared, $k_B T$ has units of energy, while $B$ has units of energy times length, and $\gamma \rho$ has units of energy per unit length cubed [c.f. Eqs. (III.9.41) and (III.9.42), respectively]. Based on this analysis we could only write $\overline{R^2} \approx k_B T B^{-1/4} (\rho \gamma)^{-3/4}$, driven by the assumption that it is the electrostatics that limits the fluctuations.

**Exercise.** Show that the limiting case of $L/L^* \gg 1$ in Eq. (III.9.50) agrees with Eq. (III.9.53).

## 9.3.  Corrections to the Integral Estimate of Series

### Going beyond the integral approximation of finite series

Now, we can consider a more complicated summation formula for finite sums, which can also be used for infinite sums as a special case. Without its last, so-called 'correction' term, it was independently discovered by Euler and Maclaurin; the correction, which increases the accuracy of result, was later added by Poisson. The formula reads (we just state this result without deriving it)

$$S(N, M) = \sum_{n=N}^{M} f(n) \approx \int_{N}^{M} dx \, f(x) + \frac{1}{2} [f(M) + f(N)]$$

$$+ \sum_{m=2}^{k} \frac{b_m}{m!} \left[ f^{(m-1)}(M) - f^{(m-1)}(N) \right] - R_k \quad \text{(III.9.54)}$$

**Table III.9.1** The first non-vanishing Bernoulli numbers. Right of each values of $n$ is a Bernoulli number for that value.

| $n$ | $b_n$ | $n$ | $b_n$ |
|---|---|---|---|
| 0 | 1 | 6 | 1/42 |
| 1 | −1/2 | 8 | −1/30 |
| 2 | 1/6 | 10 | 5/66 |
| 4 | −1/30 | 12 | −691/2730 |

We can choose the number of terms $k$ to approximate the series in the Euler–Maclaurin approximation. The 'correction' term, $R_k$, calculated for the number of terms used, improves the accuracy. Here, $M > N$ are integers, $f(x)$ is a regular (non-diverging) function in the interval $\{M, N\}$, notation $f^{(m-1)}(x)$ stands here for the derivative of the function $f(x)$ of $(m-1)$-th order; and the $b_m$ are the Bernoulli numbers. The Bernoulli numbers are generated through

$$b_m = \frac{d^m}{dt^m}\left(\frac{t}{e^t - 1}\right)\bigg|_{t=0} \qquad \text{(III.9.55)}$$

A few of the first Bernoulli numbers are shown in Table. III.9.1 (you can easily calculate them yourself).

The first term in Eq. (III.9.54) is what we quite successfully used all the time back in Chapter 8 of Part I (Vol. 1). As for the Poisson residual term, $R_k$, it has a more complicated structure, and we will not focus on it, referring the reader to more specialized literature on the subject.

**Example III.9.6 (Maths Practice ♫♪).** We consider an example of the application of the Euler–Maclaurin formula without the Poisson correction. We will deliberately consider an example where we know the exact result, in order to check how many terms in the Euler–Maclaurin expansion we will need to involve, in order to get a decent result. This is

$$S(0, \infty) = s(z) = \sum_{n=0}^{\infty} e^{-nz}, \quad \text{where} \quad z > 0 \qquad \text{(III.9.56)}$$

This series reduces to a geometric progression and can be calculated exactly:

$$s(z) = 1 + \frac{1}{e^z} + \frac{1}{(e^z)^2} + \frac{1}{(e^z)^3} \cdots = \frac{1}{1 - e^{-z}} \qquad \text{(III.9.57)}$$

On the other hand the approximation, Eq. (III.9.54), gives us

$$s(z) \approx \int_0^\infty dx\, e^{-zx} + \frac{1}{2} + \sum_{m=2}^k \frac{b_m}{m!} \left[ \frac{d^{m-1}}{dn^{m-1}} e^{-zn} \right]_{n=\infty}$$

$$- \left. \frac{d^{m-1}}{dn^{m-1}} e^{-zn} \right|_{n=0} \right] = \frac{1}{z} + \frac{1}{2} - \sum_{m=2}^k \frac{b_m (-1)^m z^{m-1}}{m!}$$

$$\text{(III.9.58)}$$

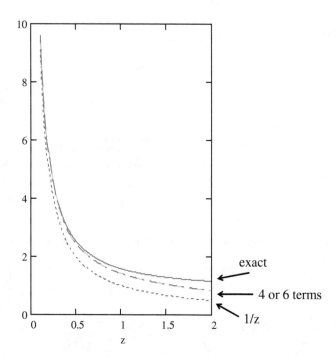

**Fig. III.9.3** We show a plot where we compare the exact value of the geometric series given by Eq. (III.9.54), with the integral estimate of the sum and the Euler-Maclaurin Method (without the Poisson term). The exact value of the geometric series is given by Eq. (III.9.55) and is shown as a red curve. The integral estimate is shown by the dotted blue curve. And the two merged dashed curves correspond to the Euler–Maclaurin method.

Figure III.9.3 shows a comparison between the exact result Eq. (III.9.57) and the approximate, Eq. (III.9.58). The comparison is good only for the domain of $z$-values where the integration method should work, namely at $z \ll 1$. Although the Euler–Maclaurin method does improve the result in the range of moderately small $z$, it is not accurate without the Poisson term. But together with the latter the result would not acquire a shape of a compact elegant formula, and in the particular case of interest it might be better to use a computer. The failure of the method for large $z$, for this example, is indicative of the ratio between successive terms in the series being small. Here, retaining the first few terms in the sum would be a better approximation.

## 9.4.  Key Points

- One way we may sum infinite series of the form $S = \Sigma_{n=1}^{\infty} f(n)$ is through Fourier series, for instance $y(x) = \Sigma_{n=1}^{\infty} f(n) \cos(n\pi x)$, where $y(0) = S$. This method certainly works for $1/n^m$ where $m$ is an integer for which $m \geq 2$.
- A way to sum a series of the form $S = \Sigma_{n=-\infty}^{\infty} f(\alpha(n))$ is through the residue theorem. This method relies on the fact $f(z)$ is analytic everywhere except at poles $z = a_k$ and we can multiply $f(z)$ by another function $g(z)$ that has first order poles at the values $z = \alpha(n)$ along the real axis.
- Some classic examples of summation formulae through residues are

$$\frac{2\pi}{L} \sum_{n=-\infty}^{\infty} f\left(\frac{2\pi n}{L}\right) = -\pi \sum_k \text{res}\left\{ f(z) \cot\left(\frac{zL}{2}\right)\right\}_{z=a_k}$$

$$\frac{2\pi}{L} \sum_{n=-\infty}^{\infty} (-1)^n f\left(\frac{2\pi n}{L}\right) = -\pi \sum_k \text{res}\left\{ \frac{f(z)}{\sin(zL/2)}\right\}_{z=a_k}$$

and a couple of others are given in the text. Here, $f(z)$ is analytic everywhere except at poles $z = a_k$.
- We may also approximate sums in a more accurate manner than by the simple integral estimates considered in Part I, Vol. 1. We

can do this with the Euler–Maclaurin method. The formula for it is given by Eq. (III.9.54) (including the additional Poisson residual term $R_k$). However, we show in an example that this extended approximation can be sometimes of limited use.

# Concluding Remarks

So, let us take a breath: we have reached
Camp 3. After having introduced you,
in Volume 1, to the basic skills needed
to be able to derive formulae, we started
Volume 2 on more advanced skills with
increasingly technical content. In the
spirit of the book, the presentation of
this material was intended to have minimal mathematical rigour and maximum common sense. As always, a great
deal of text was devoted to illustrative
examples—many being extended case
studies, interspersed with exercises—and delivering transparent and
intuitive explanations of presented methods and results. This part
completes the basis for our ascent to Camp 4. So, let's summarize
what we have learned on the way.

First we delved further into multivariable calculus and analytical
geometry. In Chapter 1, we reconsidered both surface and volume
integration (which was introduced in Part II of Vol. 1) over more
complicated shapes and how to change the coordinates we integrated
over in such integrals. We learnt, in Chapter 2, about notions of curvature of surfaces and how one could describe them in terms of vector
equations. We discussed vector-fields and their calculus in Chapter 3
of Part III, Vol. 2, finally introducing Stokes' and Gauss's theorems.

From this material, we understood how partial differential equations can arise in descriptions of physical systems.

The next four chapters focused on hard—and possibly less exciting, but absolutely necessary—manipulations of differential equations. In Chapter 4 we proceeded to differential eigenvalue equations, which led us quite naturally to the introduction of several special functions. We wanted to introduce you to only what we consider to be some of the most important ones, like Bessel functions and Legendre polynomials (see specialized textbooks and handbooks, for example Gradsteyn and Ryzhik, *Tables of integrals, series and products*, Academic Press (2015) or Bateman and Erdelyi, *Higher Transcendental Functions*, Vol. 2, McGraw Hill (1953), to get an idea of the range and application of special functions). We also introduced how, in general, one goes about solving linear second order inhomogeneous equations (first introduced in Part II, Vol. 1). Chapter 5 touches upon how to deal with systems of second order linear differential equations, putting in to play the matrix algebra we learnt (in Part II, Vol. 1), and a brief discussion of how to deal with differential equations of higher order. Some of you might have indeed found some dull moments here, sorry—we tried our best! Somewhat livelier, although not less challenging, were the techniques of approximate solutions of differential equations shown in Chapter 6: there we introduced perturbation theory, WKB approximation, and then gave you a flavour of the 'singular perturbation technique'. Lastly, in Chapter 7, we introduced partial differential equations, and their solution, ending up with some of the most important ones in the physical sciences: the wave equation, the diffusion equation and the 3D-Schrödinger equation.

Chapter 8 was a great relief for us: we guided you through beautiful landscapes, on the path of contour integration on the complex plane; presenting both the standard methods and less common tricks. This allowed us to introduce more involved asymptotic methods, stationary phase and saddle point approximations that would have not been possible to understand without complex analysis. All this smoothly led us to revisiting the summation of series, in Chapter 9, utilizing new methods based on complex integration.

In many places, it may have been much harder to digest the material than in most of Volume 1: sorry for repeating this. On the way to the summit, there were some deep crevasses to negotiate, followed by steep, treacherous snow slopes and exposed, jagged cliffs: we have come off the glacier and have started climbing up the face of the neighbouring peak, before Everest.

Now, having reached Camp 3; we'll need to push on to Camp 4, perched on the South Col before the summit (where, unfortunately, your weary guides will have to leave you, but with some idea of the route to take). On the way up, you'll note the air will get very thin (it was already getting thinner); we earnestly hope that we can provide you with enough extra oxygen and guidance...

Part IV

# From Camp 3: From Becoming at Ease with Inhomogeneous Partial Differential Equations to a Glimpse of Integral Equations, Functionals, Variation Calculus, Functional Integration, and a Snapshot of the 'Fractional World'

Now, we take on our last ascent in this book; some parts of it the reader will find the steepest. On this section of our long route we will first encounter monotonous, boring, but needed methods for the solution of inhomogeneous partial differential equations. Then we will take a quick look at the world of integral equations, and then proceed to the beautiful landscape of functionals, variational analysis, and calculation of functional integrals using different, yet relatively simple techniques, covering in the end few topics of less usual, fractional mathematics. Ready to go ahead?

# Chapter 1

# Dealing with Inhomogeneous Partial Differential Equations

Based on everything that we have learned before let us delve into few techniques for solving partial differential equations that have a nonzero 'r.h.s'. We will learn the basics starting with just a Poisson equation in free space, and then proceed to more complicated equations, and most importantly will become acquainted with how to handle boundary conditions.

## 1.1. Green's Function Technique for Solving Poisson's Equation in Free Space

### Poisson's equation

One of the simplest inhomogeneous second order partial differential equations is Poisson's equation in a medium with dielectric constant $\varepsilon$. We were introduced to this equation in Chapter 3 of Part III. As a reminder, the form it takes is

$$\vec{\nabla}^2 \varphi(\mathbf{r}) = \left[ \frac{\partial^2}{\partial x^2} + \frac{\partial^2}{\partial y^2} + \frac{\partial^2}{\partial z^2} \right] \varphi(\mathbf{r}) = -\frac{\rho(\mathbf{r})}{\varepsilon} \qquad \text{(IV.1.1)}$$

The solution can be written as a sum of the general solutions to the homogeneous equation and a particular solution of the inhomogeneous equation. Note that in this chapter we write the Poisson and similar electrostatic equations in the SI system of units (not the Gaussian units), but we include for brevity a dielectric constant of vacuum $\varepsilon_0$ into the value of $\varepsilon$, so that in vacuum $\varepsilon = \varepsilon_0$, and in water at room temperature $\varepsilon = 78\varepsilon_0$.

### *Solving Poisson's equation*

How do we find the particular solution of such an inhomogeneous equation? One method is to employ 3D Fourier transforms, introduced in Part II (Vol. 1). We may write

$$\varphi(\mathbf{r}) = \frac{1}{(2\pi)^3} \int \tilde{\varphi}(\mathbf{k}) \exp(i\mathbf{k}.\mathbf{r}) d\mathbf{k} = \frac{1}{(2\pi)^3} \int_{-\infty}^{\infty} dk_x$$
$$\times \int_{-\infty}^{\infty} dk_y \int_{-\infty}^{\infty} dk_z \tilde{\varphi}(k_x, k_y, k_z) \exp(ik_x x + ik_y y + ik_z z)$$

$$(\text{IV.1.2})$$

$$\rho(\mathbf{r}) = \frac{1}{(2\pi)^3} \int \tilde{\rho}(\mathbf{k}) \exp(i\mathbf{k}.\mathbf{r}) d\mathbf{k} = \frac{1}{(2\pi)^3} \int_{-\infty}^{\infty} dk_x$$
$$\times \int_{-\infty}^{\infty} dk_y \int_{-\infty}^{\infty} dk_z \tilde{\rho}(k_x, k_y, k_z) \exp(ik_x x + ik_y y + ik_z z)$$

$$(\text{IV.1.3})$$

Inserting Eqs. (IV.1.2) and (IV.1.3) into Eq. (IV.1.1) allows us to write

$$\int \mathbf{k}^2 \tilde{\varphi}(\mathbf{k}) \exp(i\mathbf{k}.\mathbf{r}) d\mathbf{k} = \frac{1}{\varepsilon} \int \tilde{\rho}(\mathbf{k}) \exp(i\mathbf{k}.\mathbf{r}) d\mathbf{k} \quad \text{where}$$
$$\mathbf{k}^2 = k_x^2 + k_y^2 + k_z^2 \qquad (\text{IV.1.4})$$

The only way to satisfy this is to require that

$$\tilde{\varphi}(\mathbf{k}) = \frac{\rho(\mathbf{k})}{\varepsilon \mathbf{k}^2} \equiv \tilde{G}(\mathbf{k})\rho(\mathbf{k}) \qquad (\text{IV.1.5})$$

Here, $\tilde{G}(\mathbf{k}) = \frac{1}{\varepsilon k^2}$ is a 3D Fourier transform of the Green's function for Poisson's equation; obviously $\mathbf{k}^2 = k^2$, where $k = |\mathbf{k}|$. We note (see Chapter 2 of Part III of this volume) that inverse Fourier transforming back we obtain

$$\varphi(\mathbf{r}) = \int G(\mathbf{r}\text{-}\mathbf{r'})\rho(\mathbf{r'})d\mathbf{r'} \qquad \text{(IV.1.6)}$$

This is simply the 3D convolution of the Green's function with the charge density, as it must be, because we had a product of their Fourier transforms.

### Evaluating the Green's function

Let's evaluate $G(\mathbf{r}\text{-}\mathbf{r'})$ for this elementary case and see if the result makes sense from a physics-based common sense perspective. We have

$$G(\mathbf{r}\text{-}\mathbf{r'}) = \frac{1}{(2\pi)^3\varepsilon} \int \frac{1}{k^2} \exp(-i\mathbf{k}.(\mathbf{r} - \mathbf{r'}))d\mathbf{k} \qquad \text{(IV.1.7)}$$

Now, instead of performing the integral blindly by integrating over $k_x$, $k_y$ and $k_z$, which is of course possible but messy, let's put into practice what we learnt from Chapter 1 of Part III, about changing variables of integration. Note that if we were to choose to integrate Eq. (IV.1.7) using spherical polar coordinates such that

$$k_z = k\cos\theta_k \quad k_y = k\sin\theta_k\sin\phi_k \quad k_x = k\sin\theta_k\cos\phi_k \qquad \text{(IV.1.8)}$$

then we have

$$\mathbf{k} = k\sin\theta_k\cos\phi_k\hat{\mathbf{i}} + k\sin\theta_k\sin\phi_k\hat{\mathbf{j}} + k\sin\theta_k\hat{\mathbf{k}} \quad \text{and} \quad \mathbf{k}^2 = k^2$$
$$\text{(IV.1.9)}$$

It is very important to note that we are also free to choose the orientation of these coordinates such that we can write $\mathbf{r} - \mathbf{r'} = |\mathbf{r} - \mathbf{r'}|\,\hat{\mathbf{k}}$. Thus, in polar coordinates (see Chapter 1 of Part III), we can write

$$G(\mathbf{r}\text{-}\mathbf{r'}) = \frac{1}{(2\pi)^3\varepsilon} \int_0^{2\pi} d\phi_k \int_0^\pi \sin\theta_k\,d\theta_k \int_0^\infty k^2 dk$$
$$\times \frac{\exp\left(-ik.|\mathbf{r} - \mathbf{r'}|\cos\theta_k\right)}{k^2} \qquad \text{(IV.1.10)}$$

We can change variables of the $\theta_k$ integration by writing $s = \cos\theta_k$ so that $ds = -\sin\theta_k d\theta_k$ to obtain

$$G(\mathbf{r}\text{-}\mathbf{r}') = \frac{1}{(2\pi)^3\varepsilon}\int_0^{2\pi} d\phi_k \int_0^\infty dk \int_{-1}^1 ds \exp\left(-ik\left|\mathbf{r}-\mathbf{r}'\right|s\right)$$

$$\text{(IV.1.11)}$$

Integration over both $\phi_k$ and $s$ yields

$$G(\mathbf{r}\text{-}\mathbf{r}') = \frac{1}{4\pi^2 i\varepsilon|\mathbf{r}-\mathbf{r}'|}\int_0^\infty dk\frac{1}{k}\left[\exp(ik|\mathbf{r}-\mathbf{r}'|)-\exp(-ik|\mathbf{r}-\mathbf{r}'|)\right]$$

$$= \frac{1}{4\pi^2 i\varepsilon\left|\mathbf{r}-\mathbf{r}'\right|}\int_{-\infty}^\infty dx\frac{1}{x}\exp\left(ix\right) \qquad \text{(IV.1.12)}$$

In writing the last line, we have used the substitutions $x = k\left|\mathbf{r}-\mathbf{r}'\right|$ and $x = -k\left|\mathbf{r}-\mathbf{r}'\right|$. In Chapter 8 of Part III, we learnt how to evaluate the remaining integral by contour integration, giving

$$\frac{1}{i}\int_{-\infty}^\infty dx\frac{1}{x}\exp\left(ix\right) = \pi \qquad \text{(IV.1.13)}$$

So our final result is:

$$G(\mathbf{r}\text{-}\mathbf{r}') = \frac{1}{4\pi\varepsilon\left|\mathbf{r}-\mathbf{r}'\right|} \qquad \text{(IV.1.14)}$$

It's useful to understand the physical significance of this result. One can show that $G(\mathbf{r}-\mathbf{r}')$ satisfies the equation

$$\vec{\nabla}^2 G(\mathbf{r}-\mathbf{r}') = -\frac{\delta(\mathbf{r}-\mathbf{r}')}{\varepsilon} \qquad \text{(IV.1.15)}$$

Here, $\delta(\mathbf{r}-\mathbf{r}')$ is the 3D delta function, which as we learnt previously may be defined by the product of 3D delta functions $\delta(\mathbf{r}-\mathbf{r}') = \delta(x-x')\delta(y-y')\delta(z-z')$.

Let's recall what Poisson's equation represents in electrostatics. It relates the electrostatic potential, $\varphi(\mathbf{r})$, with the charge density, $\rho(\mathbf{r})$, which creates that potential. It also accounts for the dielectric constant of the medium, $\varepsilon$, into which that charge density is embedded. A point charge, of charge $Q$, positioned at $\mathbf{r}'$ has the charge density $\rho(\mathbf{r}) = Q\delta(\mathbf{r}-\mathbf{r}')$, and so must have the electrostatic potential $QG(\mathbf{r}\text{-}\mathbf{r}') = \frac{Q}{4\pi\varepsilon|\mathbf{r}-\mathbf{r}'|}$.

## 1.2. Solving Other Inhomogeneous Partial Differential Equations in Free Space

Approaches to solving other linear inhomogeneous equations of this kind are similar to the discussed in the previous section. Let us consider a classical example.

### The inhomogeneous linear Poisson–Boltzmann equation

This equation reads as

$$-\vec{\nabla}^2\varphi(\mathbf{r}) + \kappa^2\varphi(\mathbf{r}) = \frac{\rho(\mathbf{r})}{\varepsilon} \qquad \text{(IV.1.16)}$$

Here, as always, $\kappa = 1/R_D$, where $R_D$ is the Debye screening length (recall Chapter 10, Part I, Vol. 1). Here, $\rho(\mathbf{r})$ are fixed charges or charges not accounted for by the Debye screening term $\kappa^2\varphi(\mathbf{r})$. The particular solution of Eq. (IV.1.16) can again be obtained through inverse Fourier transforms, Eqs. (IV.1.2) and (IV.1.3). This yields

$$\tilde{\varphi}(\mathbf{k}) = \frac{\rho(\mathbf{k})}{\varepsilon(\mathbf{k}^2 + \kappa^2)} \equiv \tilde{G}_{DH}(\mathbf{k})\rho(\mathbf{k}) \qquad \text{(IV.1.17)}$$

Again, we can write down Eq. (IV.1.6) for the solution of Eq. (IV.1.16), but with $G(\mathbf{r} - \mathbf{r}')$ now given by

$$G(\mathbf{r} - \mathbf{r}') = \frac{1}{(2\pi)^3\varepsilon} \int \frac{1}{k^2 + \kappa^2} \exp(-i\mathbf{k}.(\mathbf{r} - \mathbf{r}'))d\mathbf{k} \qquad \text{(IV.1.18)}$$

So, all that is left to do is to take the integral in Eq. (IV.1.18). Again it is most sensible, when dealing with no boundary conditions, to use spherical polar coordinates.

Following a similar analysis to that given in the previous subsection one obtains

$$G(\mathbf{r}\text{-}\mathbf{r}') = \frac{1}{(2\pi)^3\varepsilon} \int_0^\infty dk \frac{4\pi k^2}{k^2 + \kappa^2} \frac{\sin(k|\mathbf{r} - \mathbf{r}'|)}{k|\mathbf{r} - \mathbf{r}'|} \qquad \text{(IV.1.19)}$$

**Exercise.** Having rearranged the integral in this expression to

$$\frac{1}{|\mathbf{r} - \mathbf{r}'|} \frac{1}{2i} \int_0^\infty dk \frac{4\pi k}{k^2 + \kappa^2} \left\{ e^{ik|\mathbf{r}-\mathbf{r}'|} - e^{-ik|\mathbf{r}-\mathbf{r}'|} \right\}$$

$$= \frac{1}{|\mathbf{r} - \mathbf{r}'| \, 2i} \left\{ \int_0^\infty dk \frac{4\pi k}{k^2 + \kappa^2} e^{ik|\mathbf{r}-\mathbf{r}'|} - \int_0^\infty dk \frac{4\pi k}{k^2 + \kappa^2} e^{-ik|\mathbf{r}-\mathbf{r}'|} \right\}$$

$$= \frac{1}{|\mathbf{r} - \mathbf{r}'| \, 2i} \left\{ \int_0^\infty dk \frac{4\pi k}{k^2 + \kappa^2} e^{ik|\mathbf{r}-\mathbf{r}'|} \right.$$

$$\left. - \int_0^\infty d(-k) \frac{4\pi(-k)}{(-k)^2 + \kappa^2} e^{i(-k)|\mathbf{r}-\mathbf{r}'|} \right\}$$

$$= \frac{1}{|\mathbf{r} - \mathbf{r}'| \, 2i} \left\{ \int_0^\infty dk \frac{4\pi k}{k^2 + \kappa^2} e^{ik|\mathbf{r}-\mathbf{r}'|} - \int_0^{-\infty} dk \frac{4\pi k}{k^2 + \kappa^2} e^{ik|\mathbf{r}-\mathbf{r}'|} \right\}$$

$$= \frac{1}{|\mathbf{r} - \mathbf{r}'| \, 2i} \int_{-\infty}^\infty dk \frac{4\pi k}{k^2 + \kappa^2} e^{ik|\mathbf{r}-\mathbf{r}'|}$$

use what you have learned about complex contour integration in Chapter 8 of Part III and calculate the integral.

**Hint:** Since the denominator of the integrand has a simple pole in the upper complex plain at $k = i\kappa$, close the contour by a semi-circle in the upper complex plain and use the *residue theorem* to find the expression for the integral, thus obtaining

$$G(\mathbf{r}\text{-}\mathbf{r}') = \frac{1}{4\pi\varepsilon \, |\mathbf{r} - \mathbf{r}'|} \exp(-\kappa \, |\mathbf{r} - \mathbf{r}'|) \qquad \text{(IV.1.20)}$$

which is the Debye screened electrostatic potential due to a point charge in electrolytic solution.

## 1.3. Boundary Value Problems

### A simple approach to boundary value problems

Up to now, we have placed no restriction on $\varphi(\mathbf{r})$. Now, we consider the effect of a boundary, or interface, around where $\varphi(\mathbf{r})$ is defined. Over this boundary, we can specify the value of $\varphi(\mathbf{r})$ or the gradient normal to its surface $\hat{\mathbf{n}}.\vec{\nabla}\varphi(\mathbf{r})$. There are many approaches that we can take in solving this boundary value problem. One simple way to satisfy these conditions is to write

$$\varphi(\mathbf{r}) = \varphi_H(\mathbf{r}) + \int G(\mathbf{r} - \mathbf{r}')\rho(\mathbf{r}')d\mathbf{r}' \qquad \text{(IV.1.21)}$$

Here, $\varphi_H(\mathbf{r})$ satisfies the homogeneous equation, i.e. the partial differential equation when $\rho(\mathbf{r}) = 0$. We may try to solve for $\varphi_H(\mathbf{r})$ using the separation of variables method (if possible) and determine the coefficients from the general solution to satisfy the boundary condition.

### The induced charge technique

There are other ways we could satisfy the boundary conditions. A boundary in 3D space can be defined as a surface. We learnt in Chapter 2 of Part III that we could define the surface geometry through the vector equation $\mathbf{r}(\mathbf{R}) = \mathbf{r}(s, t)$. Now, we may want to find the solution to the inhomogeneous equation on one side of the surface. However, it is useful to extend the solution, $\varphi(\mathbf{r})$, over all space using a technique that mathematicians call analytic continuation. We then play a trick to ensure that $\varphi(\mathbf{r})$ satisfies the boundary conditions, which is introducing an effective/image charge density $\sigma_{img}(s, t)$ over the surface $\mathbf{r}(s, t)$. The trick then is to replace $\rho(\mathbf{r})$ with $\rho(\mathbf{r}) + \rho_{eff}(\mathbf{r})$, and treat $\varphi(\mathbf{r})$ as before as

$$\varphi(\mathbf{r}) = \int G(\mathbf{r} - \mathbf{r}')(\rho(\mathbf{r}') + \rho_{eff}(\mathbf{r}'))d\mathbf{r}' \qquad \text{(IV.1.22)}$$

Here, $\rho_{eff}(\mathbf{r}')$ is the effective charge density contribution from the surface, which can be written in terms of the surface charge density $\sigma_{img}(s, t)$ as a surface integral

$$\rho_{eff}(\mathbf{r}') = \int_S \sigma_{img}(\mathbf{R})\delta(\mathbf{r}' - \mathbf{r}(\mathbf{R}))d\mathbf{R}$$

$$= \int_{s_1}^{s_2} ds \int_{t_1}^{t_2} dt \sigma_{img}(s, t)\delta(\mathbf{r}' - \mathbf{r}(s, t)) |\mathbf{t}_s(s, t) \times \mathbf{t}_t(s, t)|$$

$$\text{(IV.1.23)}$$

where $\mathbf{t}_s(s, t) = \frac{\partial \mathbf{r}(s,t)}{\partial s}$, $\mathbf{t}_t(s, t) = \frac{\partial \mathbf{r}(s,t)}{\partial t}$.

The form of $\sigma_{img}(s,t)$ is to be chosen to satisfy the boundary conditions.

### Making the Green's function to satisfy the boundary conditions

Another way of doing things is to make the Green's function satisfy the boundary conditions. One way we can do this is to write

$$G(\mathbf{r}, \mathbf{r}') = G_H(\mathbf{r}, \mathbf{r}') + G_I(\mathbf{r} - \mathbf{r}') \qquad \text{(IV.1.24)}$$

$G_I(\mathbf{r} - \mathbf{r}')$ is the Green's function for free space without the boundary condition—in the case of Poisson's equation, it would just be $G_I(\mathbf{r} - \mathbf{r}') = 1/(4\pi\varepsilon |\mathbf{r} - \mathbf{r}'|)$. Here, $G_H(\mathbf{r}, \mathbf{r}')$ is then a solution to the homogeneous partial differential equation, which must be selected for $G(\mathbf{r}, \mathbf{r}')$ to satisfy the boundary condition. Once done, we may write our solution

$$\varphi(\mathbf{r}) = \int G(\mathbf{r}, \mathbf{r}')\rho(\mathbf{r}')d\mathbf{r}' \qquad \text{(IV.1.25)}$$

Another way to find $G(\mathbf{r}, \mathbf{r}')$ is again to introduce an effective charge density at the surface that describes the interface, which arises from $\rho(\mathbf{r}) = \delta(\mathbf{r} - \mathbf{r}')$, that contributes to the Green function, and we extend Green's function over all space. In the case of the linearized Poisson–Boltzmann equation $G(\mathbf{r}, \mathbf{r}')$ would now satisfy

$$-\vec{\nabla}^2 G(\mathbf{r}, \mathbf{r}') + \kappa^2 G(\mathbf{r}, \mathbf{r}') = \rho_{eff}(\mathbf{r}) + \delta(\mathbf{r} - \mathbf{r}') \qquad \text{(IV.1.26)}$$

Again $\rho_{eff}(\mathbf{r})$ is given by Eq. (IV.1.23) where both $\rho_{eff}(\mathbf{r})$ and $\sigma_{img}(s,t)$ depend also on $\mathbf{r}'$, and they must again be fixed to satisfy the boundary conditions.

The last way is to express the 3D Green's function in terms of its spectral representation. We'll again illustrate this for the linearized Poisson–Boltzmann equation. In this case we can write

$$G(\mathbf{r}, \mathbf{r}') = \sum_{j,k,l} \frac{\chi_{j,k,l}(\mathbf{r})\chi^*_{j,k,l}(\mathbf{r})}{E_{j,k,l}} \qquad \text{(IV.1.27)}$$

Here, $\chi_{j,k,l}(\mathbf{r})$ and $E_{j,k,l}$ are the eigenfunctions and values of the partial differential operator specific to the inhomogeneous equation that we want to solve. The eigenfunctions $\chi_{j,k,l}(\mathbf{r})$ satisfy the same boundary conditions as $G(\mathbf{r}, \mathbf{r}')$. In the case of the linearized Poisson–Boltzmann equation, these are obtained through the eigenvalue equation

$$-\vec{\nabla}^2\chi_{j,k,l}(\mathbf{r}) + \kappa^2\chi_{j,k,l}(\mathbf{r}) = E_{j,k,l}\chi_{j,k,l}(\mathbf{r}) \qquad \text{(IV.1.28)}$$

This can be simply written as the homogenous Helmholtz equation that we solved in Chapter 7 of Part III, where again $\chi_{j,k,l}(\mathbf{r})$ satisfy the same boundary conditions as both $G(\mathbf{r}, \mathbf{r}')$ and $\varphi(\mathbf{r})$.

### Computing a Green's function for particular interfaces

Here, we'll look at a couple of examples to make practical sense of what we have just stated.

**Example IV.1.1 (Maths Practice ♫♪).** Let's compute the linearized Poisson–Boltzmann Green's function for an 'infinitely' long cylinder of radius $a$ subject to the following boundary condition at the cylinder surface:

$$\hat{n}.\vec{\nabla}G(\mathbf{r}, \mathbf{r}')\Big|_{\mathbf{r}\,\in\,\text{surface of the cylinder}} = 0 \qquad \text{(IV.1.29)}$$

Such a condition might look artificial, but it is approximately satisfied in a system where the dielectric constant of the cylinder is much smaller that that of the surrounding medium (this will lead us to the next physical application). Obviously, it is best to choose cylindrical polar coordinates in which this boundary condition can be written simply as

$$\frac{dG(R, \varphi, z, R', \varphi', z')}{dR}\Big|_{R=a} = 0 \qquad \text{(IV.1.30)}$$

where $a$ is the radius of the cylinder.

Now, let's solve for the Green's function using the above described *induced charge technique*. The vector equation of the surface describing the cylinder can be written as (with $s = \varphi$ and $t = z$)

$$\mathbf{r}_s(\varphi, z) = a\cos\varphi\hat{\mathbf{i}} + a\sin\varphi\hat{\mathbf{j}} + z\hat{\mathbf{k}} \qquad \text{(IV.1.31)}$$

Then let us simplify Eq. (IV.1.23) that relates $\rho_{eff}(\mathbf{r}')$ to $\sigma_{img}(s, t)$. First, we obtain $\mathbf{t}_\varphi(\varphi, z) = -a\sin\varphi\hat{\mathbf{i}} + a\cos\varphi\hat{\mathbf{j}}$ and $\mathbf{t}_z(\varphi, z) = \hat{\mathbf{k}}$. Thus, we have

$$\left|\mathbf{t}_z(\varphi, z) \times \mathbf{t}_\varphi(\varphi, z)\right| = \left|-a\cos\varphi\hat{\mathbf{i}} - a\sin\varphi\hat{\mathbf{j}}\right| = a \qquad \text{(IV.1.32)}$$

and so Eq. (IV.1.23) reduces to

$$\rho_{eff}(\mathbf{r}; \mathbf{r}') = a\int_0^{2\pi} d\tilde{\varphi} \int_{-\infty}^{\infty} d\tilde{z}\,\sigma_{img}(\tilde{\varphi}, \tilde{z}; \mathbf{r}')\delta(\mathbf{r} - \mathbf{r}_s(\tilde{\varphi}, \tilde{z})) \qquad \text{(IV.1.33)}$$

Then, we can write the equation for the Green's function as Eq. (IV.1.26). Its solution can be written as

$$G(\mathbf{r}, \mathbf{r}') = G_0(\mathbf{r} - \mathbf{r}') + \int G_0(\mathbf{r} - \mathbf{r}'')\rho_{eff}(\mathbf{r}''; \mathbf{r}')d\mathbf{r}''$$

$$\Rightarrow G(\mathbf{r}, \mathbf{r}') = G_0(\mathbf{r} - \mathbf{r}') + a\int_0^{2\pi} d\tilde{\varphi} \int_{-\infty}^{\infty} d\tilde{z}\,\sigma_{img}(\tilde{\varphi}, \tilde{z}; \mathbf{r}')$$

$$\times\, G_0(\mathbf{r} - \mathbf{r}_s(\tilde{\varphi}, \tilde{z})) \qquad \text{(IV.1.34)}$$

Here, $G_0(\mathbf{r} - \mathbf{r}')$ satisfies Eq. (IV.1.26) with $\rho_{eff}(\mathbf{r}) = 0$.

To satisfy the boundary condition we will write the Green's functions in cylindrical coordinates: $G(\mathbf{r}, \mathbf{r}') = G(R, \phi, z, R', \phi', z')$ and $G_0(\mathbf{r} - \mathbf{r}') = G_0(R, R', \phi - \phi', z - z')$. First let us do it for $G_0(\mathbf{r} - \mathbf{r}')$, going back to Eq. (IV.1.18). In cylindrical polar coordinates we have

$$k_x = K\cos\phi_K, \quad k_y = K\sin\phi_K \quad \text{and} \quad k_z = k_z \qquad \text{(IV.1.35)}$$

so that $\mathbf{k}.\mathbf{r} = KR\cos(\phi - \phi_K) + k_z z$ and $\mathbf{k}.\mathbf{r}' = KR'\cos(\phi' - \phi_K) + k_z z'$. Then Eq. (IV.1.18) may be rewritten as

$$G_0(R, R', \phi - \phi', z - z')$$

$$= \frac{1}{(2\pi)^3 \varepsilon} \int_0^\infty K dK \int_0^{2\pi} d\phi_K \int_{-\infty}^\infty dk_z \frac{\exp(-ik_z(z - z'))}{K^2 + k_z^2 + \kappa^2}$$

$$\times \exp(-iKR\cos(\phi - \phi_K)) \exp(iKR'\cos(\phi' - \phi_K)) \quad \text{(IV.1.36)}$$

To go any further we should exploit an identity introduced in Chapter 4 of Part III in our discussion of Bessel functions, namely

$$\exp(iKR'\cos(\phi' - \phi_K)) = \sum_{n=-\infty}^\infty (i)^n J_n(KR') \exp(in(\phi' - \phi_K))$$

$$\text{(IV.1.37)}$$

Similarly we can write an identity for $\exp(-iKR\cos(\phi - \phi_K))$ by replacing in Eq. (IV.1.37) $i$, $R'$ and $\phi'$ with $-i$, $R$ and $\phi$, respectively. This then yields

$$G_0(R, R', \phi - \phi', z - z')$$

$$= \frac{1}{(2\pi)^3 \varepsilon} \sum_{n'=-\infty}^\infty \sum_{n=-\infty}^\infty \int_0^\infty K dK \int_0^{2\pi} d\phi_K \int_{-\infty}^\infty dk_z$$

$$\times \frac{\exp(-ik_z(z - z'))}{K^2 + k_z^2 + \kappa^2} (i)^{n-n'} J_n(KR) J_{n'}(KR')$$

$$\times \exp(-in'(\phi - \phi_K)) \exp(in(\phi' - \phi_K)) \quad \text{(IV.1.38)}$$

The integral over $\phi_K$ can be performed, as $\int_0^{2\pi} e^{i(n-n')\phi_K} d\phi_K = 2\pi \delta_{n,n'}$, where $\delta_{n,n'}$ is the Kronecker delta, yielding

$$G_0(R, R', \phi - \phi', z - z')$$

$$= \frac{1}{(2\pi)^2 \varepsilon} \sum_{n=-\infty}^\infty \int_0^\infty K dK \int_{-\infty}^\infty dk_z$$

$$\times \frac{\exp(-ik_z(z - z'))}{K^2 + k_z^2 + \kappa^2} J_n(KR) J_n(KR') \exp(-in(\phi - \phi'))$$

$$\text{(IV.1.39)}$$

Next, let us consider the integral over $K$, namely

$$S = \int_0^\infty K dK \frac{J_n(KR) J_n(KR')}{K^2 + k_z^2 + \kappa^2} \tag{IV.1.40}$$

In the last section of Chapter 8 of Part III we learnt how to evaluate such integral through contour integration. This yields

$$S = K_n \left( \sqrt{k_z^2 + \kappa^2} R' \right) I_n \left( \sqrt{k_z^2 + \kappa^2} R \right) \theta(R' - R)$$
$$+ K_n \left( \sqrt{k_z^2 + \kappa^2} R \right) I_n \left( \sqrt{k_z^2 + \kappa^2} R' \right) \theta(R - R') \tag{IV.1.41}$$

This then allows us to write

$$G_0(R, R', \phi - \phi', z - z') = \tilde{G}(R, R', \phi - \phi', z - z')\theta(R' - R)$$
$$+ \tilde{G}(R', R, \phi - \phi', z - z')\theta(R - R') \tag{IV.1.42}$$

where

$$\tilde{G}(R, R', \phi - \phi', z - z')$$
$$= \frac{1}{(2\pi)^2 \varepsilon} \sum_{n=-\infty}^{\infty} \int_{-\infty}^{\infty} dk_z I_n \left( \sqrt{k_z^2 + \kappa^2} R \right) K_n \left( \sqrt{k_z^2 + \kappa^2} R' \right)$$
$$\times \exp(-ik_z(z - z')) \exp(-in(\phi - \phi')) \tag{IV.1.43}$$

Now to satisfy the boundary conditions we suppose that the surface $\mathbf{r}_s(\tilde{\varphi}, \tilde{z})$ lies an infinitesimal distance inside the cylinder at which we want to apply the boundary conditions, while $R' > a$. Then the boundary condition given by Eq. (IV.1.34) translates into

$$0 = \left. \frac{\partial \tilde{G}(R, R', \phi - \phi', z - z')}{\partial R} \right|_{R=a} + a \int_0^{2\pi} d\tilde{\phi} \int_{-\infty}^{\infty} d\tilde{z} \sigma_{img}(\tilde{\varphi}, \tilde{z}; \mathbf{r}')$$
$$\times \left. \frac{\partial \tilde{G}(R, \phi - \tilde{\phi}, z - \tilde{z})}{\partial R} \right|_{R=a}$$

$\Rightarrow$

$$-\sum_{n=-\infty}^{\infty}\int_{-\infty}^{\infty}dk_z\sqrt{k_z^2+\kappa^2}I_n'\left(\sqrt{k_z^2+\kappa^2}a\right)K_n\left(\sqrt{k_z^2+\kappa^2}R'\right)$$

$$\times\exp(-ik_z(z-z'))\exp(-in(\phi-\phi'))$$

$$=a\int_0^{2\pi}d\tilde\phi\int_{-\infty}^{\infty}d\tilde z\,\sigma_{img}(\tilde\varphi,\tilde z;\mathbf{r}')\sum_{n=-\infty}^{\infty}\int_{-\infty}^{\infty}dk_z\sqrt{k_z^2+\kappa^2}$$

$$\times I_n\left(\sqrt{k_z^2+\kappa^2}R'\right)K_n'\left(\sqrt{k_z^2+\kappa^2}a\right)\exp(-ik_z(z-\tilde z))$$

$$\times\exp(-in(\phi-\tilde\phi))\tag{IV.1.44}$$

Hence,

$$\bar\sigma_{img}(n,k_z;\mathbf{r}')=\int_0^{2\pi}d\tilde\phi\int_{-\infty}^{\infty}d\tilde z\sigma_{img}(\tilde\phi,\tilde z;\mathbf{r}')\exp(ik_z\tilde z)\exp(in\tilde\phi')\tag{IV.1.45}$$

This allows us to obtain from Eq. (IV.1.44), by equating each of the coefficients that multiply $\exp(-ik_z z)\exp(-in\phi)$, the following result:

$$\bar\sigma_{img}(n,k_z;\mathbf{r}')=-\frac{1}{a}\frac{I_n'\left(\sqrt{k_z^2+\kappa^2}a\right)K_n\left(\sqrt{k_z^2+\kappa^2}R'\right)}{I_n\left(\sqrt{k_z^2+\kappa^2}a\right)K_n'\left(\sqrt{k_z^2+\kappa^2}a\right)}$$

$$\times\exp(ik_z z')\exp(in\phi')\tag{IV.1.46}$$

Now, we'll use this result. First, from Eq. (IV.1.34) we can write

$$G(R,\phi,z,R',\phi',z')=G_0(R,R',\phi-\phi',z-z')+a\int_0^{2\pi}d\tilde\phi$$

$$\times\int_{-\infty}^{\infty}d\tilde z\sigma_{img}(\tilde\phi,\tilde z;\mathbf{r}')G_0(R,a,\phi-\tilde\phi,z-\tilde z)\tag{IV.1.47}$$

Furthermore, substituting in our results for $G_0(R,R',\phi-\phi',z-z')$, Eqs. (IV.1.42) and (IV.1.43), into Eq. (IV.1.47) we obtain

$$G(R, \phi, z, R', \phi', z')$$

$$= \frac{1}{(2\pi)^2 \varepsilon} \sum_{n=-\infty}^{\infty} \int_{-\infty}^{\infty} dk_z \Bigg[ I_n \left( \sqrt{k_z^2 + \kappa^2} R \right) K_n \left( \sqrt{k_z^2 + \kappa^2} R' \right)$$

$$\times \theta(R' - R) + I_n \left( \sqrt{k_z^2 + \kappa^2} R' \right) K_n \left( \sqrt{k_z^2 + \kappa^2} R \right) \theta(R - R')$$

$$+ a \int_0^{2\pi} d\tilde{\phi} \int_{-\infty}^{\infty} d\tilde{z} \sigma_{img}(\tilde{\phi}, \tilde{z}; \mathbf{r}') I_n \left( \sqrt{k_z^2 + \kappa^2} a \right) K_n \left( \sqrt{k_z^2 + \kappa^2} R \right) \Bigg]$$

$$\times \exp(-ik_z(z - z')) \exp(-in(\phi - \phi'))$$

$$\text{(IV.1.48)}$$

Then, using Eqs. (IV.1.45) and (IV.1.46), we can write an expression for the Green's function

$$G(R, \phi, z, R', \phi', z')$$

$$= \frac{1}{(2\pi)^2 \varepsilon} \sum_{n=-\infty}^{\infty} \int_{-\infty}^{\infty} dk_z \Bigg[ I_n \left( \sqrt{k_z^2 + \kappa^2} R \right) K_n \left( \sqrt{k_z^2 + \kappa^2} R' \right)$$

$$\times \theta(R' - R) + I_n \left( \sqrt{k_z^2 + \kappa^2} R' \right) K_n \left( \sqrt{k_z^2 + \kappa^2} R \right) \theta(R - R')$$

$$- \frac{I_n' \left( \sqrt{k_z^2 + \kappa^2} a \right)}{K_n' \left( \sqrt{k_z^2 + \kappa^2} a \right)} K_n \left( \sqrt{k_z^2 + \kappa^2} R' \right) K_n \left( \sqrt{k_z^2 + \kappa^2} R \right) \Bigg]$$

$$\times \exp(-ik_z(z - z')) \exp(-in(\phi - \phi')) \qquad \text{(IV.1.49)}$$

If we want it, we can actually write this result in another form, recalling that $G_0(\mathbf{r} - \mathbf{r}')$ can be also presented in the form of Eq. (IV.1.20), taking into account that in cylindrical coordinates we can write $|\mathbf{r} - \mathbf{r}'|^2 = R^2 - 2R'R\cos(\phi - \phi') + R'^2 + z^2 + 2zz' + z'^2$. This yields

$$G(R, \phi, z, R', \phi', z')$$

$$= \frac{\exp\left( -\kappa \left( R^2 - 2R'R\cos(\phi - \phi') + R'^2 + z^2 + 2zz' + z'^2 \right)^{1/2} \right)}{4\pi\varepsilon \left( R^2 - 2R'R\cos(\phi - \phi') + R'^2 + z^2 + 2zz' + z'^2 \right)^{1/2}}$$

$$-\frac{1}{(2\pi)^2\varepsilon}\sum_{n=-\infty}^{\infty}\int_{-\infty}^{\infty}dk_z\frac{I_n'\left(\sqrt{k_z^2+\kappa^2}a\right)}{K_n'\left(\sqrt{k_z^2+\kappa^2}a\right)}K_n\left(\sqrt{k_z^2+\kappa^2}R'\right)$$

$$\times K_n\left(\sqrt{k_z^2+\kappa^2}R\right)\exp(-ik_z(z-z'))\exp(-in(\phi-\phi'))$$

Long, tedious? Yes, but you have got the result, and we show below how you can use it.

**Example IV.1.2 (Physical Application ▶▶).** Here, we'll apply the result of the previous example to the physical situation where we want to calculate the change in electrostatic/ solvation energy of a small spherical ion, due to other ions moving about it, when moving it close to a cylindrical region of low dielectric; this could represent a large macromolecule. The simplest way we can describe the effect of the other ions about the ion in question is by using the Debye–Hückel model; i.e. this supposes that the electrostatic potential $\varphi(\mathbf{r})$ in the presence of the small ion can be described by the linearized Poisson–Boltzmann equation (IV.1.16), where $\rho(\mathbf{r})$ is the charge density of an ion centred at $\mathbf{r}_0$.

The change in electrostatic energy of the ion is given by

$$E(R_0)-E(\infty)=\frac{1}{2}\int[\varphi(\mathbf{r};R_0)-\varphi(\mathbf{r};\infty)]\,\rho(\mathbf{r})d\mathbf{r}\quad\text{(IV.1.50)}$$

Here, $\varphi(\mathbf{r};R_0)$ is the value of the electrostatic potential of the small ion centred a distance $R=R_0$ away from the cylinder axis, and $\varphi(\mathbf{r};\infty)$ is what that value would have been if the ion was infinitely far away from the axis ($R=\infty$). Using the results of the previous example we can write

$$\varphi(\mathbf{r};R_0)=\int G(\mathbf{r},\mathbf{r}')\rho(\mathbf{r}')d\mathbf{r}'\quad\varphi(\mathbf{r};\infty)=\int G_0(\mathbf{r}-\mathbf{r}')\rho(\mathbf{r}')d\mathbf{r}'$$

$$\text{(IV.1.51)}$$

Where we have defined $G_0(\mathbf{r} - \mathbf{r}')$ as the form for $G(\mathbf{r}, \mathbf{r}')$ at $R_0 = \infty$. This allows us to write

$$E(R) - E(\infty) = \frac{1}{2} \iint \rho(\mathbf{r}) \left[ G(\mathbf{r}, \mathbf{r}') - G_0(\mathbf{r} - \mathbf{r}') \right] \rho(\mathbf{r}') d\mathbf{r} d\mathbf{r}'$$

$$(IV.1.52)$$

Or more explicitly, using the results of the previous section, this can be written as

$$E(R_0) - E(\infty)$$

$$= -\frac{1}{2(2\pi)^2 \varepsilon} \sum_{n=-\infty}^{\infty} \int_{-\infty}^{\infty} dk_z \int_0^{\infty} R \, dR \int_0^{\infty} R' \, dR'$$

$$\times \int_{-\infty}^{\infty} dz \int_{-\infty}^{\infty} dz' \int_0^{2\pi} d\phi \int_0^{2\pi} d\phi' \rho(R, \phi, z) \rho(R', \phi', z')$$

$$\times \frac{I'_n \left( \sqrt{k_z^2 + \kappa^2} a \right)}{K'_n \left( \sqrt{k_z^2 + \kappa^2} a \right)} K_n \left( \sqrt{k_z^2 + \kappa^2} R' \right) K_n \left( \sqrt{k_z^2 + \kappa^2} R \right)$$

$$\times \exp(-i k_z (z - z')) \exp(-i n (\phi - \phi')) \qquad (IV.1.53)$$

If the size of the ion is very small we can simplify things further through an approximation. In this case $\rho(R, \phi, z)$ changes very rapidly about the position that ion is centred, $(R_0, z_0, \phi_0)$, in the leading order approximation, by setting everywhere $R = R' = R_0$

$$E(R_0) - E(\infty)$$

$$= -\frac{1}{2(2\pi)^2 \varepsilon} \sum_{n=-\infty}^{\infty} \int_{-\infty}^{\infty} dk_z \int_0^{\infty} R \, dR \int_0^{\infty} R' \, dR' \int_{-\infty}^{\infty} dz \int_{-\infty}^{\infty} dz'$$

$$\times \int_0^{2\pi} d\phi \int_0^{2\pi} d\phi' \, \rho(R, \phi, z) \rho(R', \phi', z') \frac{I'_n \left( \sqrt{k_z^2 + \kappa^2} a \right)}{K'_n \left( \sqrt{k_z^2 + \kappa^2} a \right)}$$

$$\times \left[ K_n \left( \sqrt{k_z^2 + \kappa^2} R_0 \right) \right]^2$$

$$\Rightarrow E(R_0) - E(\infty) = -\frac{Q^2}{(2\pi)^2\varepsilon} \sum_{n=-\infty}^{\infty} \int_0^\infty dk_z \frac{I_n'\left(\sqrt{k_z^2 + \kappa^2}a\right)}{K_n'\left(\sqrt{k_z^2 + \kappa^2}a\right)}$$

$$\times \left[K_n\left(\sqrt{k_z^2 + \kappa^2}R_0\right)\right]^2 \qquad\qquad\text{(IV.1.54)}$$

where we have used that the integrand is an even function of $k_z$, and we have defined the charge of an ion as

$$Q = \int_0^\infty RdR \int_{-\infty}^\infty dz \int_0^{2\pi} d\phi\rho(R,\phi,z) \qquad\text{(IV.1.55)}$$

We must remember that the situation that we are considering here is when the distance of the ion from the axis is greater than the cylinder radius, i.e. $R_0 > a$. Secondly, we recall that the boundary condition, Eq. (IV.1.30), for which the formula for the Green's function in the previous example was obtained, corresponds to a boundary of a medium with high dielectric constant $\varepsilon$ and a cylinder of much lower dielectric constant. So looking at the integrand in Eq. (IV.1.54) we understand that it is always negative because all $I_n$ are growing functions, and all $K_n$ are decreasing functions; correspondingly all $I_n'$ are positive and all $K_n'$ are negative, so altogether the r.h.s of Eq. (IV.1.54) is positive. This makes perfect sense, because the charge in the medium of high dielectric constant will get repelled from the boundary with the medium of low dielectric constant, no matter what the sign of that charge is. Such force is called *image force*. The reader may be more familiar with attractive image forces acting on a charge near a metal surface, but there the situation is opposite: the ideal metal has an infinite dielectric constant, and the charge in the dielectric near the boundary with such metal will be attracted to the interface. But in the case considered above, the image force is repulsive!

We'll propose the metal case in an exercise below, but now let us get back to the result given by Eq. (IV.1.54). An important issue to notice is that when $R_0 = a$, the integral in this expression will diverge as the interface is sharp, and its energy approaching the interface will go to infinity. Of course, in reality, for a charge of a finite size and/or smeared interface the 'infinity' will get cut-off and the image potential energy will be large but finite. But we are not considering this case and should appreciate what we get. Related to the expected divergence comes the following obstacle: it will be difficult to use Eq. (IV.1.54) to plot at $R_0$ very close to $a$, because whereas the integral at $R_0 > a$ will converge, the series will be very slowly converging. [**Think:** Why this so? **Hint:** Go to textbooks/handbooks on special functions and look into the behaviour of the modified Bessel functions.] But this is clear from the general fact: as the image potential energy must diverge at $R_0 = a$, it must be very large when $R_0 \approx a$. Hence, if one uses a computer in an attempt to plot the result for $R_0 \approx a$, you will need to take increasingly large number of terms in the sum, the closer $R_0$ to $a$. In the region $(R_0 - a) \ll a$, the interface for a charge would locally look flat, like the surface of the ocean looks to us (although even young kids notice and ask for a meaning of the horizon).

Let us consider then the case of $\kappa = 0$ (which means no electrolyte concentration in the medium outside the cylinder). In the domain of $(R_0 - a) \ll a$, one would expect the classical image-repulsion of a charge from the flat interface. We will not derive this, but just draw that law referring the reader to textbooks on electricity; it reads

$$E(R_0) - E(\infty) = \frac{Q^2}{4\pi\varepsilon} \frac{1}{4(R_0 - a)} \qquad \text{(IV.1.56)}$$

Fig. IV.1.1 demonstrates what Eq. (IV.1.54) actually gives, and how well it reproduces Eq. (IV.1.56) at $(R_0 - a) \ll a$. We will present the results in the following way. Note that $\frac{Q^2}{4\pi\varepsilon} = k_B T \cdot \frac{Q^2}{4\pi\varepsilon k_B T} = k_B T \cdot L_B$ where $k_B T$ is the thermal energy, and $L_B$ is the *Bjerrum length* for a charge $Q$, $L_B = \frac{Q^2}{4\pi\varepsilon k_B T}$; recall Chapter 10 of Part I, Vol. 1. (We remind you that for unit elementary charge, $Q = e$, in a medium with a dielectric constant corresponding to that of water at room temperature, $L_B \approx 7\text{Å}$.) The results graphed in Figs. IV.1.1 and IV.1.2, present all distances in Angstroms for the case of $L_B = 7\text{Å}$, and in units of $k_B T$ for $E(R_0) - E(\infty)$. Figure IV.1.1 analyses the results at $\kappa = 0$, i.e. no electrolyte in the medium outside the cylinder (we will show the effect of the latter in another example, in Fig. IV.1.2).

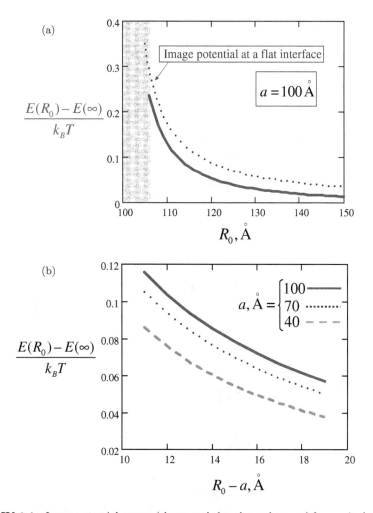

**Fig. IV.1.1** Image potential energy (shown scaled to thermal energy) for a unit elementary charge $e$ in the medium of high dielectric constant, corresponding to water, at the interface with the medium of much lower dielectric constant; the case of no electrolyte ($\kappa = 0$). (a) Solid curve shows the case of cylindrical interface following calculations via Eq. (IV.1.54); dotted shows the result for a flat interface, as described by the law $\frac{L_B}{4(R_0 - a)}$. Grey area indicates the domain of poor convergence of the series, i.e. leading to complications in the use of Eq. (IV.1.54). The results of calculations are plotted here for the number of terms in the sum, $n = \pm 65$. Taking more terms in the sum will bring the solid and dotted curves in this domain closer together. But image potential near the cylindrical surface is naturally smaller than near a flat surface. (b) Demonstration of the latter fact, showing the effect of the curvature of the surface, for the same distances from the interface: the smaller is the cylinder, the smaller is the image potential.

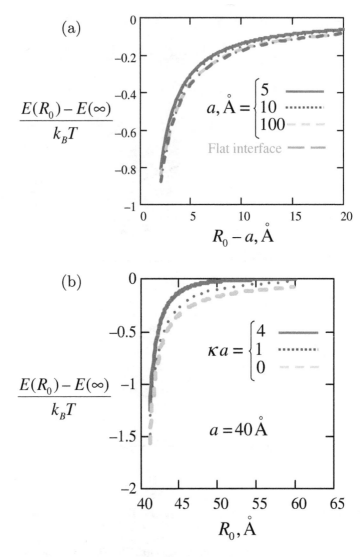

**Fig. IV.1.2** Image potential energy (shown scaled to thermal energy) for a unit elementary charge $e$ in the medium of dielectric constant corresponding to water at the interface with the ideal metal. The potential is attractive, as it should be for such interface (a) The case of no electrolyte ($\kappa = 0$). Plotted in the same domain of distances from the interface, the results show a weak effect of the curvature of the cylinder. At the 10 nm radius of the cylinder within the distance of 1 nm from the interface, the charge sees the interface as flat. (b) The graphs demonstrate how nonzero electrolyte concentrations, $c$, coming into play through the values of $\kappa \propto \sqrt{c}$, (see Chapter 10, Part I, Vol. 1) suppress/screen the image potential.

We will not be showing here the effect of $\kappa$: it will make the image potential exponentially decaying with the distance from the interface, with the decay length $\kappa^{-1}$. But, as we said, we will show this in the results of an exercise below.

**Exercise.** Applying the same method, find the image potential for another boundary condition,

$$G(a, \varphi, z, R', \varphi', z') = 0 \qquad (IV.1.57)$$

which corresponds to the case when the material of the cylinder is metallic. Show that in this case the image potential energy for a point charge near such interface takes the form

$$E(R_0) - E(\infty) = -\frac{Q^2}{(2\pi)^2 \varepsilon} \sum_{n=-\infty}^{\infty} \int_0^{\infty} dk_z \frac{I_n\left(\sqrt{k_z^2 + \kappa^2}\,a\right)}{K_n\left(\sqrt{k_z^2 + \kappa^2}\,a\right)}$$

$$\times \left[K_n\left(\sqrt{k_z^2 + \kappa^2}\,R_0\right)\right]^2 \qquad (IV.1.58)$$

Assuming that you have mastered the technique and have obtained this result, let us discuss it. Because the integrand is positive, this image potential is negative, and the closer the charge is to the interface (the closer $R_0$ to $a$) the larger its absolute value will be. Of course, this should be so, as the charge is attracted to the metal! Let us look at a couple of plots, generated via Eq. (IV.1.58), presenting the results for $L_B = 7\text{Å}$, and in the same way as in the previous example.

So, having learned this technique and applied it to particular problems, we have also learned something else. Not every formula that you obtain may be easy to handle. Some of them, like Eq. (IV.1.54), cannot be used blindfolded, as one may step in a dangerous area of poorly converging sums or integrals. Such situations will require you to think more seriously about the level of accuracy your computer should operate with such formulae, namely—how many terms in the sum will be enough, and at which upper-limit you truncate the integration. These decisions will depend on the domain of parameters you deal with in your case studies. All this is a part of your learning curve.

## 1.4.  Perturbation Theory Due to Small Terms in Partial Differential Equations: Introducing the 'Schwinger–Dyson' Equation for the Green's Function

Suppose now that we want to solve the inhomogeneous partial differential equation

$$-\vec{\nabla}^2\varphi(\mathbf{r}) + \lambda(\mathbf{r})\varphi(\mathbf{r}) = F(\mathbf{r}) \qquad (IV.1.59)$$

Again, if we can find the Green's function we can solve this equation such that

$$\varphi(\mathbf{r}) = \int G(\mathbf{r},\mathbf{r}')F(\mathbf{r}') \qquad (IV.1.60)$$

where now

$$-\vec{\nabla}^2 G(\mathbf{r},\mathbf{r}') + \lambda(\mathbf{r})G(\mathbf{r},\mathbf{r}') = \delta(\mathbf{r}-\mathbf{r}') \qquad (IV.1.61)$$

In what follows it will be useful to recast Eq. (IV.1.61) using the *inverse* Green's function $G^{-1}(\mathbf{r},\mathbf{r}'')$ such that

$$\int G^{-1}(\mathbf{r},\mathbf{r}'')G(\mathbf{r}'',\mathbf{r}')d\mathbf{r}'' = \delta(\mathbf{r}-\mathbf{r}') \qquad (IV.1.62)$$

In terms of $G^{-1}(\mathbf{r},\mathbf{r}'')$ it takes the form

$$G^{-1}(\mathbf{r},\mathbf{r}'') = -\vec{\nabla}^2\delta(\mathbf{r}-\mathbf{r}'') + \lambda(\mathbf{r})\delta(\mathbf{r}-\mathbf{r}'') \qquad (IV.1.63)$$

Indeed, by parts integration, it's not hard to show that writing both Eq. (IV.1.62) and Eq. (IV.1.63) is equivalent to writing Eq. (IV.1.61). Let's now suppose that we can write $\lambda(\mathbf{r}) = \lambda_0 + \Delta\lambda(\mathbf{r})$ where $\lambda_0$ does not depend on $\mathbf{r}$, and the absolute value of $\Delta\lambda(\mathbf{r})$ small: $|\Delta\lambda(\mathbf{r})| \ll \lambda_0$. We could then write

$$G^{-1}(\mathbf{r},\mathbf{r}'') = G_0^{-1}(\mathbf{r}-\mathbf{r}'') + \Delta\lambda(\mathbf{r})\delta(\mathbf{r}-\mathbf{r}'') \qquad (IV.1.64)$$

where $G_0^{-1}(\mathbf{r}-\mathbf{r}'') = -\vec{\nabla}^2\delta(\mathbf{r}-\mathbf{r}'')+\lambda_0\delta(\mathbf{r}-\mathbf{r}'')$. Using Eqs. (IV.1.62) and (IV.1.63), we can write

$$\int G_0^{-1}(\mathbf{r}-\mathbf{r}'')G(\mathbf{r}'',\mathbf{r}')d\mathbf{r}'' + \Delta\lambda(\mathbf{r})G(\mathbf{r},\mathbf{r}') = \delta(\mathbf{r}-\mathbf{r}') \qquad (IV.1.65)$$

Now, it is important to note that we can also write

$$\int G_0(\mathbf{r}''' - \mathbf{r})G_0^{-1}(\mathbf{r} - \mathbf{r}'')d\mathbf{r} = \delta(\mathbf{r}'' - \mathbf{r}''') \qquad \text{(IV.1.66)}$$

where

$$-\vec{\nabla}^2 G_0(\mathbf{r} - \mathbf{r}') + \lambda_0 G_0(\mathbf{r} - \mathbf{r}') = \delta(\mathbf{r} - \mathbf{r}') \qquad \text{(IV.1.67)}$$

But what is the use of these at a first glance meaningless 'back-and-forth' meandering manipulations? Be patient, please, we can benefit from them: we exploit Eq. (IV.1.66) to rewrite Eq. (IV.1.65) into a more useful form, by first multiplying both sides of Eq. (IV.1.65) by $G_0(\mathbf{r}''' - \mathbf{r})$ and then integrate over $\mathbf{r}$. This yields

$$\int G_0(\mathbf{r}''' - \mathbf{r})G_0^{-1}(\mathbf{r} - \mathbf{r}'')G(\mathbf{r}'', \mathbf{r}')d\mathbf{r}''d\mathbf{r}$$

$$+ \int G_0(\mathbf{r}''' - \mathbf{r})\Delta\lambda(\mathbf{r})G(\mathbf{r}, \mathbf{r}')d\mathbf{r}$$

$$= \int G_0(\mathbf{r}''' - \mathbf{r})\delta(\mathbf{r} - \mathbf{r}')d\mathbf{r}$$

$$\Rightarrow \int \delta(\mathbf{r}''' - \mathbf{r}'')G(\mathbf{r}'', \mathbf{r}')d\mathbf{r}'' + \int G_0(\mathbf{r}''' - \mathbf{r})\Delta\lambda(\mathbf{r})G(\mathbf{r}, \mathbf{r}')d\mathbf{r}$$

$$= G_0(\mathbf{r}''' - \mathbf{r}')$$

$$\Rightarrow G(\mathbf{r}''', \mathbf{r}') = G_0(\mathbf{r}''' - \mathbf{r}') - \int G_0(\mathbf{r}''' - \mathbf{r})\Delta\lambda(\mathbf{r})G(\mathbf{r}, \mathbf{r}')d\mathbf{r}$$

$$\text{(IV.1.68)}$$

Equation (IV.1.68) can be referred to as the *Schwinger–Dyson* equation, named after these two famous theoretical physicists, the first one of Nobel fame. We'll talk more about this later in its application to field theory and functional integration, and it is our first example of an integral equation, which we will consider a bit more systematically in a follow-up section.

Julian Schwinger
(1918–1994)

One can write an approximate solution to this equation by iteration; a perturbation expansion for small $\Delta\lambda(\mathbf{r})$. When $\Delta\lambda(\mathbf{r})$ is very small we can actually approximate $G(\mathbf{r}, \mathbf{r}') \simeq G_0(\mathbf{r} - \mathbf{r}')$. The first correction to this 'zero-order' solution due to $\Delta\lambda(\mathbf{r}) \neq 0$ we can find by substituting $G(\mathbf{r}, \mathbf{r}') = G_0(\mathbf{r} - \mathbf{r}')$ into the r.h.s of Eq. (IV.1.68), which yields

Freeman Dyson
(1923–2020)

$$G(\mathbf{r}''', \mathbf{r}') \simeq G_1(\mathbf{r}''', \mathbf{r}') = G_0(\mathbf{r}''' - \mathbf{r}') - \int G_0(\mathbf{r}''' - \mathbf{r})\Delta\lambda(\mathbf{r})G_0(\mathbf{r} - \mathbf{r}')d\mathbf{r}$$

(IV.1.69)

We can generate a next to leading order correction by substituting, Eq. (IV.1.69) again into the r.h.s of Eq. (IV.1.68) to get

$$G(\mathbf{r}''', \mathbf{r}') \simeq G_2(\mathbf{r}''', \mathbf{r}') = G_0(\mathbf{r}''' - \mathbf{r}')$$
$$- \int G_0(\mathbf{r}''' - \mathbf{r})\Delta\lambda(\mathbf{r})G_0(\mathbf{r} - \mathbf{r}')d\mathbf{r}$$
$$+ \int G_0(\mathbf{r}''' - \mathbf{r}_1)\Delta\lambda(\mathbf{r}_1)G_0(\mathbf{r}_1 - \mathbf{r}_2)$$
$$\times \Delta\lambda(\mathbf{r}_2)G_0(\mathbf{r}_2 - \mathbf{r}')d\mathbf{r}_1 d\mathbf{r}_2 \quad \text{(IV.1.70)}$$

Repeating this procedure, you, in principle, can calculate $G$ to any order in $\Delta\lambda$. Application of this is straightforward, and so, we will not entertain you here with any examples. You may invent one yourself, with some form of $\Delta\lambda(\mathbf{r})$, but of course $|\Delta\lambda(\mathbf{r})| << \lambda_0$.

## 1.5. Key Points

- A convenient way of solving linear inhomogeneous partial differential equations is to employ the method of Green's functions.
- Green's function is a solution of a differential equation whose r.h.s (the 'inhomogeneity') is a delta function. If the equation itself is linear, the solution of the initial equation can be expressed through the function standing in the r.h.s and the Green's function.

- This is elementary in free space, but far from that in a system with boundary conditions. Various methods exist for finding Green's functions that obey boundary conditions; some examples were considered in this chapter. One of them is the induced charge method.

- We have shown also how the Green's functions can be expressed through eigenfunctions of homogeneous equations obeying boundary conditions.

- You have been shown examples of the calculation of Green's functions for particular interfaces, including the calculation of image forces for cylindrical interfaces, showing how the limit of a flat interface can be recovered for large radii of the cylinders.

- Partial differential equations with variable coefficients can be calculated using perturbation theory if the variable part is small. We presented an example of such a theory, resulting in the solution for the Green's function described by the so-called Schwinger–Dyson equation.

# Chapter 2

# A Brief Look into the World of Integral Equations

In this chapter we introduce the reader to the basic concepts of integral equations, describe some classes of them, and consider few examples of the methods for their solution. Integral equations form an important, large field of mathematics that has been substantially developed during the last two centuries and continues to be so. Thus, all that we can to do here is to give a newcomer to this area a glimpse of what it is all about, and demonstrate some elementary methods of solutions of some of the equations most often met in applications.

## 2.1. What are Integral Equations? A Few Main Types

Integral equations are often defined as equations containing an unknown function (or functions) under the integral sign. It is not a rigorous or comprehensive definition, but will be sufficient for us, and instead of going deeper into its possible shortcomings, we will go on to list some most important classes of integral equations, limiting the discussion to linear ones.

### Fredholm equations

The most 'popular' and probably the most important class of integral equations bears the name
of Swedish mathematician Fredholm. In applications we most often meet with a *Fredholm equation of the second kind,* which in its simplest
version—for the case of unknown function of one
variable $u(x)$—reads:

Erik Ivar Fredholm
(1866–1927)

$$u(x) - \int_a^b K(x,s)u(s)ds = g(x) \qquad \text{(IV.2.1)}$$

The interval $[a,b]$ can be finite, or infinite (i.e. $a = -\infty, b = \infty$).
Function $K(x,s)$ is called the *kernel* of the equation, and it is
assumed to be well defined in the square $a \leq x \leq b$, $a \leq s \leq b$;
$g(x)$ is called the *free term* which must be defined in the interval
$a \leq x \leq b$. For further dealing with this class of equations the only
condition is assumed that

$$\int_a^b dx \int_a^b ds\, |K(x,s)|^2 < \infty, \quad \int_a^b |g(x)|^2\, dx < \infty \qquad \text{(IV.2.2)}$$

Generalization to higher dimensions is straightforward, e.g. for the
3D case,

$$u(\mathbf{r}) - \int_\Omega K(\mathbf{r},\mathbf{s})u(\mathbf{s})ds = g(\mathbf{r}) \qquad \text{(IV.2.3)}$$

and

$$\int_\Omega \int_\Omega d\mathbf{r}\, d\mathbf{s}\, |K(\mathbf{r},\mathbf{s})|^2 < \infty, \quad \int_\Omega d\mathbf{r}\, |g(\mathbf{r})|^2 < \infty \qquad \text{(IV.2.4)}$$

where $\Omega$ indicates the given volume integration domain.

A *Fredholm equation of the first kind* does not contain the
unknown function outside the integral sign; usually written as

$$\int_a^b K(x,s)u(s)ds = g(x) \qquad \text{(IV.2.5)}$$

and

$$\int_\Omega K(\mathbf{r}, \mathbf{s}) u(\mathbf{s}) ds = g(\mathbf{r}) \qquad \text{(IV.2.6)}$$

Uniqueness of solutions of the first-kind equations may be a problem, as for some forms of kernel and the free term one may satisfy this equality with different $u$ functions.

Actually, the second kind Fredholm equations can be considered as a particular case of the first kind. If in Eq. (IV.2.6)

$$K(\mathbf{r}, \mathbf{s}) = \delta(\mathbf{r} - \mathbf{s}) - F(\mathbf{r}, \mathbf{s}) \qquad \text{(IV.2.7)}$$

i.e. the kernel contains the $\delta$-function part and a purely nonlocal part, the equation becomes a second kind equation:

$$u(\mathbf{r}) - \int_\Omega F(\mathbf{r}, \mathbf{s}) u(\mathbf{s}) ds = g(\mathbf{r}) \qquad \text{(IV.2.8)}$$

Due to space limitations, we will concentrate on the equations of functions of a single variable. The principles of approach to solutions of functions of multiple variables are very similar, and the reader could learn them later from more specialized textbooks. However, delving into few general properties, we will sometimes present the results for functions depending generally on vectors, when it does not cost us any extra 'space' in this book.

### Volterra equations

These are named after another pioneer of the integral equations, Italian mathematician Vito Volterra, who was also a founder of functional analysis, which we will not dare to step in, although we will deal quite a lot with functionals in the remaining chapters of the book.

Vito Volterra
(1860–1940)

For simplicity of analysis we stick to functions of one variable, *Volterra equations of the second kind*, which read

$$u(x) - \int_a^x K(x, s) u(s) ds = g(x) \quad a \le x \le b \qquad \text{(IV.2.9)}$$

Equally, *Volterra equations of the first kind* are

$$\int_a^x K(x,s)u(s)ds = g(x) \qquad (IV.2.10)$$

An example of this class of equation is $K(x,s) = (x-s)^{-\alpha}$, which gives the *Abel equation* named after the brilliant Norwegian mathematician Niels Henrik Abel (who died aged 26 but made many pioneering, classical contributions in different fields of mathematics):

Niels Henrik Abel
(1802–1829)

$$\int_a^x \frac{u(s)ds}{(x-s)^\alpha} = g(x),\ 0 < \alpha < 1 \qquad (IV.2.11)$$

We will show its solution later in this chapter.

At a first glance Fredholm equations may look like particular cases of the Volterra equation, just the limit of integration being different. But this is not true, because the upper limit in the Volterra equation coincides with the first argument of the kernel.

### Convolution equations

Convolution equations can belong to the Fredholm or the Volterra class, of first or second kind, and their critical feature is that their kernel depends only on the difference between $x$ and $s$, i.e. $K(x,s) = K(x-s)$. The canonical form, when the integration in these equations extends to the full domain, reads

$$\underbrace{c \cdot u(x) - \int_{-\infty}^{\infty} K(x-s)u(s)ds = g(x)}_{\text{Fredholm}}, \quad \underbrace{c \cdot u(x) - \int_{-\infty}^{x} K(x-s)u(s)ds = g(x)}_{\text{Volterra}}$$

$$(IV.2.12)$$

where $c$ is a constant. Obviously these cover both the second order ($c \neq 0$), and first order ($c = 0$) cases. To the Volterra class of convolution equations, belongs the Abel equation (IV.2.11), and to Fredholm class, the famous *Wiener–Hopf* equation:

$$c \cdot u(x) - \int_0^\infty K(x-s)u(s)ds = g(x) \qquad \text{(IV.2.13)}$$

which these two great, but very different minds—Norbert Wiener, American mathematician, philosopher, and the father of cybernetics; and Eberhard Hopf, Austrian/American mathematician and astronomer—have studied together.

Norbert Wiener
(1894–1964)

### Coupled equations

In more complicated physical problems one may encounter systems of integral equation. For instance, we write below a pair of convolution equations:

Eberhard Frederich
Ferdinand Hopf
(1902–1983)

$$a \cdot u(x) - \int_{-\infty}^\infty K_1(x-s)u(s)ds = g(x), \ x > 0$$

$$\qquad \text{(IV.2.14)}$$

$$b \cdot u(x) - \int_{-\infty}^\infty K_2(x-s)u(s)ds = g(x), \ x < 0$$

where $a$ and $b$ are constants not equal to each other. One can imagine many other kinds of coupled equations.

There are, of course nonlinear equations, in which it is not $u(s)$ standing under the integral sign, but some function, e.g. $F(u(s),s)$. But stop! The reader must be tired already and impatient to learn how we can solve at least some of those equations. There are many excellent books on this subject, covering the field of integral equations in their full breadth, and some, surprisingly, written in a way understandable to non-mathematicians! (See. e.g. Zabreiko *et al, Integral equations*, Nordhoff Internat. Pbls, 1975). Below, we will just explain few general principles, and show some examples.

## 2.2.   Solving Integral Equations

### *Integral transformations as integral equations*

Needless to say, any integral transformations (cos-Fourier, sin-Fourier, complex-Fourier, Hankel, etc.) are the integral equations of Fredholm equations of the first kind, with respect to the function being transformed. You know already from previous chapters how to obtain the reverse transforms, which will be the solutions of such 'integral equations'. This fact, actually, gives us a hint on how we can solve some of the simplest integral equations.

### *The solution of a Fredholm convolution equation using the Fourier transform*

Let's start with the simplest case—solving linear Fredholm convolution Eq. (IV.2.12), $c \cdot u(x) - \int_{-\infty}^{\infty} K(x - s)u(s)ds = g(x)$, using complex Fourier transforms defined in Section 2.4, Part II, (Vol. 1). Substituting into in this equation

$$u(x) = \frac{1}{2\pi} \int_{-\infty}^{\infty} dk\, \tilde{u}(k)e^{-ikx}$$

$$g(x) = \frac{1}{2\pi} \int_{-\infty}^{\infty} dk\, \tilde{g}(k)e^{-ikx} \qquad \text{(IV.2.15)}$$

$$K(x - s) = \frac{1}{2\pi} \int_{-\infty}^{\infty} dk\, \tilde{K}(k)e^{-ik(x-s)}$$

we obtain

$$c \cdot \frac{1}{2\pi} \int_{-\infty}^{\infty} dk\, \tilde{u}(k)e^{-ikx} - \int_{-\infty}^{\infty} ds \frac{1}{2\pi} \int_{-\infty}^{\infty} dk\, \tilde{K}(k)e^{-ik(x-s)}$$

$$\times \frac{1}{2\pi} \int_{-\infty}^{\infty} dk'\, \tilde{u}(k')e^{-ik's} = \frac{1}{2\pi} \int_{-\infty}^{\infty} dk\, \tilde{g}(k)e^{-ikx}$$

Let us focus on the second term on the l.h.s. It has three integrations, and we do them in a smart order, rewriting this term as $-\frac{1}{2\pi} \int_{-\infty}^{\infty} dk\, \tilde{K}(k)e^{-ikx} \int_{-\infty}^{\infty} dk'\, \tilde{u}(k') \frac{1}{2\pi} \int_{-\infty}^{\infty} ds e^{i(k-k')s}$.

Since $\frac{1}{2\pi}\int_{-\infty}^{\infty} ds\, e^{i(k-k')s} = \delta(k-k')$, where $\delta(k-k')$ is the Dirac delta-function, ad proceeding then to integration over $k'$, we get $-\frac{1}{2\pi}\int_{-\infty}^{\infty} dk\, \tilde{K}(k)e^{-ikx}\int_{-\infty}^{\infty} dk'\, \tilde{u}(k')\delta(k-k') = -\frac{1}{2\pi}\int_{-\infty}^{\infty} dk\, \tilde{K}(k)\tilde{u}(k)e^{-ikx}$, and the whole equation can now be written as $c \cdot \frac{1}{2\pi}\int_{-\infty}^{\infty} dk\, e^{-ikx}\{\tilde{u}(k) - \tilde{K}(k)\tilde{u}(k) - \tilde{g}(k)\} = 0$. Obviously, this is satisfied when the expression in the curly brackets is zero, and hence

$$\tilde{u}(k) = \frac{\tilde{g}(k)}{c - \tilde{K}(k)} \tag{IV.2.16}$$

This means that we have found an expression of the Fourier transform of our unknown function, through the Fourier transforms, $\tilde{K}(k)$ and $\tilde{g}(k)$, of the given functions $K(x)$ and $g(x)$, which are to be calculated through taking the integrals,

$$\tilde{g}(k) = \int_{-\infty}^{\infty} dx\, g(x)\, e^{ikx}, \quad \tilde{K}(k) = \int_{-\infty}^{\infty} dx\, K(x) e^{ikx} \tag{IV.2.17}$$

Plugging in Eq. (IV.2.16) into the first of the Eqs. (IV.2.15), we obtain the solution of our integral equation as

$$u(x) = \frac{1}{2\pi}\int_{-\infty}^{\infty} dk\, \frac{\tilde{g}(k)}{c - \tilde{K}(k)}\, e^{-ikx} \tag{IV.2.18}$$

Of course, this result is meaningful, only when (a) the Fourier transforms $\tilde{g}(k)$ and $\tilde{K}(k)$ exist, i.e. integrals in the last two equation in (IV.2.15) do not diverge, and (b) the whole integral Eq. (IV.2.18) does not diverge.

It will, however, diverge, if at some real values of $k$ the denominator in the integrand vanishes, but the numerator does not, or the latter also tends to zero but slower than the denominator; the problem will equally arise if for large absolute values of $k$ the numerator grows faster than the denominator, or the ratio of the denominator to numerator grows not fast enough, namely as $k^{\alpha}$, $\alpha \leq 1$. In those cases, your integral equation may not have a solution! This happens, typically, when the physical problem for which such an integral equation has emerged has not been correctly formulated, or when it tends to describe a catastrophic situation.

**Example IV.2.1 (Maths Practice ♫♪).** Consider the equation

$$c \cdot u(x) + \lambda \int_{-\infty}^{\infty} e^{-p|x-s|} u(s) ds = g(x) \qquad \text{(IV.2.19)}$$

where the constant $\lambda$ is real, $p > 0$ is also real, and $g(x)$ is some specified real function (for which we will give a couple of examples). The solution for this equation can generally be given by Eq. (IV.2.18), in which

$$\tilde{K}(k) = -\lambda \int_{-\infty}^{\infty} dx \, e^{-p|x|} e^{ikx} \qquad \text{(IV.2.20)}$$

This integral is easy to calculate:

$$\int_{-\infty}^{\infty} dx e^{-p|x|} e^{ikx} = \int_{-\infty}^{0} dx e^{(p+ik)x} + \int_{0}^{\infty} dx e^{(-p+ik)x}$$

$$= \frac{1}{p+ik} e^{(p+ik)x} \Big|_{-\infty}^{0} + \frac{1}{-p+ik} e^{(-p+ik)x} \Big|_{0}^{\infty}$$

$$= \frac{1}{p+ik} - \frac{1}{-p+ik} = \frac{2p}{p^2 + k^2}$$

$$u(x) = \frac{1}{2\pi} \int_{-\infty}^{\infty} dk \, \frac{\tilde{g}(k)}{c + \dfrac{2p\lambda}{p^2 + k^2}} e^{-ikx} \qquad \text{(IV.2.21)}$$

This is already the answer: once we are given $g(x)$, we can calculate $\tilde{g}(k)$ subject to the first of the Eq. (IV.2.17), and then take the integral in Eq. (IV.2.21). Let us look at how it works for a couple of examples for $g(x)$:

(a) $g(x) = b \cdot \delta(x)$ where $b = \text{const} \quad \Rightarrow \quad \tilde{g}(k) = b$

(b) $g(x) = \frac{b}{2\pi} \cdot \cos(qx) \Rightarrow \tilde{g}(k) = \frac{b}{2} \left[ \delta(k-q) + \delta(k+q) \right]$

For case (a),

$$u(x) = \frac{b}{2\pi} \int_{-\infty}^{\infty} dk \, \frac{e^{-ikx}}{c + \dfrac{2p\lambda}{p^2 + k^2}} = \frac{b}{2c\pi} \int_{-\infty}^{\infty} dk \, \frac{\left(p^2 + k^2\right) e^{ikx}}{k^2 + \left(p^2 + 2\frac{p\lambda}{c}\right)}$$

$$= \frac{b}{2c\pi} \int_{-\infty}^{\infty} dk \, \underbrace{\frac{\left(p^2 + k^2\right) e^{ikx}}{\left[k + i\sqrt{p^2 + 2\frac{p\lambda}{c}}\right] \left[k - i\sqrt{p^2 + 2\frac{p\lambda}{c}}\right]}}_{I}$$

This integral is easy to calculate by contour integration. We will limit our analysis to the case when $p^2 + 2\frac{p\lambda}{c} > 0$. The denominator of the integrand has two poles on the imaginary axis at the points $k = \pm i\sqrt{p^2 + 2\frac{p\lambda}{c}}$. Using the method of contour integration, when, $x > 0$, we close the contour with the semi-circle in the upper complex plane and calculate the residue of the first order pole at $k = i\sqrt{p^2 + 2\frac{p\lambda}{c}}$:

$$I = 2\pi i \operatorname*{Res}_{k = i\sqrt{p^2 + 2\frac{p\lambda}{c}}} \left\{ \frac{\left(p^2 + k^2\right) e^{ikx}}{\left[k + i\sqrt{p^2 + 2\frac{p\lambda}{c}}\right] \left[k - i\sqrt{p^2 + 2\frac{p\lambda}{c}}\right]} \right\}$$

$$= 2\pi i \, \frac{p^2 - p^2 - 2\frac{p\lambda}{c}}{2i\sqrt{p^2 + 2\frac{p\lambda}{c}}} e^{-\sqrt{p^2 + 2\frac{p\lambda}{c}}\, x} = -2\pi \frac{p\lambda}{c} \frac{e^{-\sqrt{p^2 + 2\frac{p\lambda}{c}}\, x}}{\sqrt{p^2 + 2\frac{p\lambda}{c}}}$$

We can similarly deal with this integral for $x < 0$, but then we will need to close the contour with the semi-circle in the lower half of the complex plane, calculating the residue at $k = -i\sqrt{p^2 + 2\frac{p\lambda}{c}}$. Combining this analysis, we get $I = -2\pi \frac{p\lambda}{c} \frac{e^{-\sqrt{p^2 + 2\frac{p\lambda}{c}}\, |x|}}{\sqrt{p^2 + 2\frac{p\lambda}{c}}}$. It is easy to show that if you want to cover also the point of $x = 0$, you have to add to the right hand side of this expression the term $(b/c)\delta(x)$. Hence,

$$u(x) = -\frac{p\lambda b \, e^{-\sqrt{p^2 + 2\frac{p\lambda}{c}}\, |x|}}{c^2 \sqrt{p^2 + 2\frac{p\lambda}{c}}} + \frac{b}{c}\delta(x) \qquad \text{(IV.2.22)}$$

Case (b) is simpler:

$$u(x) = \frac{b}{4\pi} \int_{-\infty}^{\infty} dk \, \frac{[\delta(k-q) + \delta(k+q)]}{c + \dfrac{2p\lambda}{p^2 + k^2}} e^{-ikx} = \frac{b}{2\pi} \frac{\cos(qx)}{c + \dfrac{2p\lambda}{p^2 + q^2}}$$

**Exercise.** For each case, (a) and (b), take the obtained solution, and substitute it, together with the corresponding expression for $g(x)$ into the original Eq. (IV.2.19). Performing then the integration, show that the l.h.s becomes identical to the r.h.s (this is how we usually check that the obtained solution of the integral equation is error-free). **Hint:** Be prepared for a long calculation which will require your full concentration. Once you do this check error-free, both in case (a) and (b) you will encounter with a tricky cancellation of terms. Getting it right, you will be rewarded by seeing 'with higher resolution' how mathematics works.

### The concept of resolvent

Let us act integrate Eq. (IV.2.6) over $d\mathbf{r}$ with a function $R(\mathbf{s}', \mathbf{r})$

$$\int_{\Omega} d\mathbf{r} \, R(\mathbf{s}', \mathbf{r}) \int_{\Omega} d\mathbf{s} \, K(\mathbf{r}, \mathbf{s}) u(\mathbf{s}) = \int_{\Omega} d\mathbf{r} \, R(\mathbf{s}', \mathbf{r}) g(\mathbf{r}) \qquad \text{(IV.2.23)}$$

such that

$$\int_{\Omega} d\mathbf{r} \, R(\mathbf{s}', \mathbf{r}) K(\mathbf{r}, \mathbf{s}) = \delta(\mathbf{s} - \mathbf{s}') \qquad \text{(IV.2.24)}$$

Taking into account the latter condition, after integration of the l.h.s of Eq. (IV.2.23) over $\mathbf{r}$, we get

$$u(\mathbf{s}') = \int_{\Omega} d\mathbf{r} \, R(\mathbf{s}', \mathbf{r}) g(\mathbf{r}) \qquad \text{(IV.2.25)}$$

The function $R(\mathbf{r}, \mathbf{s}')$ is called *Resolvent*, because its knowledge immediately 'resolves' the problem, by giving us the solution for $u$ in the form of an integral, Eq. (IV.2.25).

But the equation defining the resolvent, Eq. (IV.2.24), is also an integral equation, which may be as difficult to solve as the initial Eq. (IV.2.6). So when this approach is useful? The answer sounds like a joke: when it is easy to solve Eq. (IV.2.24). For instance, if the space of integration $\Omega$ is unlimited, and the kernel is of convolution type $K(\mathbf{r}, \mathbf{s}) = K(\mathbf{r} - \mathbf{s})$, then the resolvent cannot be anything else than the convolution type, $R(\mathbf{s}', \mathbf{r}) = R(\mathbf{s}' - \mathbf{r})$. The rest is trivial! To be specific, let's consider a three-dimensional case; using Fourier transforms:

$$K(\mathbf{r} - \mathbf{s}) = \frac{1}{(2\pi)^3} \int d\mathbf{k}\, \tilde{K}(\mathbf{k}) e^{-i\mathbf{k}(\mathbf{r}-\mathbf{s})}$$

$$R(\mathbf{s}' - \mathbf{r}) = \frac{1}{(2\pi)^3} \int d\mathbf{k}'\, \tilde{R}(\mathbf{k}') e^{-i\mathbf{k}'(\mathbf{s}'-\mathbf{r})}$$

$$\delta(\mathbf{s} - \mathbf{s}') = \frac{1}{(2\pi)^3} \int d\mathbf{k}\, e^{-i\mathbf{k}(\mathbf{s}-\mathbf{s}')} \qquad \text{(IV.2.26)}$$

Plugging them into Eq. (IV.2.24), we get

$$\int d\mathbf{r}\, \frac{1}{(2\pi)^3} \int d\mathbf{k}'\, \tilde{R}(\mathbf{k}') e^{-i\mathbf{k}'(\mathbf{s}'-\mathbf{r})} \frac{1}{(2\pi)^3} \int d\mathbf{k}\, \tilde{K}(\mathbf{k}) e^{-i\mathbf{k}(\mathbf{r}-\mathbf{s})}$$

$$= \frac{1}{(2\pi)^3} \int d\mathbf{k}\, e^{-i\mathbf{k}(\mathbf{s}-\mathbf{s}')} = \frac{1}{(2\pi)^3} \int d\mathbf{k}\, e^{-i\mathbf{k}(\mathbf{s}'-\mathbf{s})}$$

Performing, in the l.h.s of this equation, integration over $\mathbf{r}$ first, we encounter the integral $\frac{1}{(2\pi)^3} \int d\mathbf{r}\, e^{-i\mathbf{r}(\mathbf{k}-\mathbf{k}')} = \delta(\mathbf{k} - \mathbf{k}')$; proceeding then to the integral over $\mathbf{k}'$ we get $\frac{1}{(2\pi)^3} \int d\mathbf{k}\, \tilde{R}(\mathbf{k})\tilde{K}(\mathbf{k}) e^{-i\mathbf{k}(\mathbf{s}'-\mathbf{s})} = \frac{1}{(2\pi)^3} \int d\mathbf{k}\, e^{-i\mathbf{k}(\mathbf{s}'-\mathbf{s})}$, which can be fulfilled only when $\tilde{R}(\mathbf{k})\tilde{K}(\mathbf{k}) = 1$. This finally give us,

$$\tilde{R}(\mathbf{k}) = \frac{1}{\tilde{K}(\mathbf{k})} \qquad \text{(IV.2.27)}$$

All in all, once we know

$$\tilde{K}(\mathbf{k}) = \int d\mathbf{r}\, K(\mathbf{r}) e^{i\mathbf{k}\mathbf{r}} \qquad \text{(IV.2.28)}$$

using the second of Eq. (IV.2.26) we can get the expression for the resolvent:

$$R(\mathbf{r}) = \frac{1}{(2\pi)^3} \int d\mathbf{k} \frac{e^{-i\mathbf{k}\mathbf{r}}}{\tilde{K}(\mathbf{k})} \qquad \text{(IV.2.29)}$$

We may enjoy the expression for $R(\mathbf{r})$ which we would thus obtain, plug it into our resolvent-based formula,

$$u(\mathbf{s}) = \int d\mathbf{r}\, R(\mathbf{s} - \mathbf{r}) g(\mathbf{r}) \tag{IV.2.30}$$

and take the integral to obtain $u(\mathbf{s})$, if we can do it. But it will be easier to proceed differently, as in the exercise below.

**Exercise.** By doing the Fourier transform of both functions in the integrand of Eq. (IV.2.30), performing then the integration over $\mathbf{r}$, and using the integral representation of the delta-function [third line of Eq. (IV.2.26)] show that

$$u(\mathbf{s}) = \frac{1}{(2\pi)^3} \int d\mathbf{k}\, \tilde{R}(\mathbf{k}) \tilde{g}(\mathbf{k}) e^{-i\mathbf{k}\mathbf{s}} = \frac{1}{(2\pi)^3} \int d\mathbf{k}\, \frac{\tilde{g}(\mathbf{k})}{\tilde{K}(\mathbf{k})} e^{-i\mathbf{k}\mathbf{s}}$$

$$\tag{IV.2.31}$$

**Example IV.2.2 (Physical Application ▶▶).** In this example, we will actually deal with an *integro-differential* equation, but you will see in a moment that it does not go much beyond what we have already considered in this section. Let us calculate electrostatic potential created by a spherical ion embedded into a medium (not disturbing it), characterized by a nonlocal dielectric tensor-function $\varepsilon_{\alpha\beta}(\mathbf{r} - \mathbf{r}')$, where the subscripts $\alpha, \beta$ label the Cartesian projections. The Poisson equation for the electrostatic potential created by the charge density $\rho(\mathbf{r})$ reads (for convenience we will use here *Gaussian units*, enthusiasts could convert the results into the SI units),

$$\nabla \mathbf{D} = 4\pi \cdot \rho(\mathbf{r}) \tag{IV.2.32}$$

where $\mathbf{D}$ is the electrostatic induction, and $\nabla$ stands for the gradient. In a medium with nonlocal dielectric polarizability, induction and electric field are related as

$$D_\alpha(\mathbf{r}) = \sum_\beta \int d\mathbf{r}'\, \varepsilon_{\alpha\beta}(\mathbf{r} - \mathbf{r}') E_\beta(\mathbf{r}') = -\sum_\beta \int d\mathbf{r}'\, \varepsilon_{\alpha\beta}(\mathbf{r} - \mathbf{r}') \frac{\partial}{\partial r_\beta} \phi(\mathbf{r}')$$

$$\tag{IV.2.33}$$

We have assumed here that the medium is homogeneous (translation-ally invariant) so that the kernel in the integrand depends on $(\mathbf{r} - \mathbf{r}')$, rather than $(\mathbf{r}, \mathbf{r}')$. Substitution of Eq. (IV.2.33) into Eq. (IV.2.32) gives us an integro-differential equation,

$$\sum_{\alpha,\beta} \frac{\partial}{\partial r_\alpha} \int d\mathbf{r}' \varepsilon_{\alpha\beta}(\mathbf{r} - \mathbf{r}') \frac{\partial}{\partial r'_\beta} \phi(\mathbf{r}') = -4\pi\rho(\mathbf{r}) \qquad \text{(IV.2.34)}$$

Since this equation is assumed to be valid in full space, we can use the method of Fourier transforms to solve it for $\phi(\mathbf{r}')$. Namely, by introducing

$$\varepsilon_{\alpha\beta}(\mathbf{r} - \mathbf{r}') = \frac{1}{(2\pi)^3} \int d\mathbf{k}\, e^{-i\mathbf{k}(\mathbf{r}-\mathbf{r}')} \tilde{\varepsilon}_{\alpha\beta}(\mathbf{k})$$

$$\tilde{\varepsilon}_{\alpha\beta}(\mathbf{k}) = \int d(\mathbf{r} - \mathbf{r}')\, e^{i\mathbf{k}(\mathbf{r}-\mathbf{r}')} \varepsilon_{\alpha\beta}(\mathbf{r} - \mathbf{r}') \qquad \text{(IV.2.35)}$$

$$\phi(\mathbf{r}') = \frac{1}{(2\pi)^3} \int d\mathbf{k}\, e^{-i\mathbf{k}\mathbf{r}'} \tilde{\phi}(\mathbf{k}), \quad \tilde{\phi}(\mathbf{k}) = \int d\mathbf{r}'\, e^{i\mathbf{k}\mathbf{r}'} \phi(\mathbf{r}')$$

$$\text{(IV.2.36)}$$

$$\rho(\mathbf{r}) = \frac{1}{(2\pi)^3} \int d\mathbf{k}\, e^{-i\mathbf{k}\mathbf{r}} \tilde{\rho}(\mathbf{k}), \quad \tilde{\rho}(\mathbf{k}) = \int d\mathbf{r}'\, e^{i\mathbf{k}\mathbf{r}} \tilde{\rho}(\mathbf{r}')$$

$$\text{(IV.2.37)}$$

and substituting these Fourier transform into Eq. (IV.2.34), we obtain

$$\sum_{\alpha,\beta} \frac{\partial}{\partial r_\alpha} \int d\mathbf{r}' \frac{1}{(2\pi)^3} \int d\mathbf{k}\, e^{-i\mathbf{k}(\mathbf{r}-\mathbf{r}')} \tilde{\varepsilon}_{\alpha\beta}(\mathbf{k}) \frac{\partial}{\partial r'_\beta} \frac{1}{(2\pi)^3} \int d\mathbf{k}'\, e^{-i\mathbf{k}\mathbf{r}'} \tilde{\phi}(\mathbf{k}')$$

$$= -4\pi \frac{1}{(2\pi)^3} \int d\mathbf{k}\, e^{-i\mathbf{k}\mathbf{r}} \tilde{\rho}(\mathbf{k})$$

$$\Rightarrow \sum_{\alpha,\beta} \int d\mathbf{r}' \frac{1}{(2\pi)^3} \int d\mathbf{k}\, (-ik_\alpha)\, e^{-i\mathbf{k}(\mathbf{r}-\mathbf{r}')} \tilde{\varepsilon}_{\alpha\beta}(\mathbf{k})$$

$$\times \int d\mathbf{k}'\, e^{-i\mathbf{k}'\mathbf{r}'} \left(-ik'_\beta\right) \tilde{\phi}(\mathbf{k}') = -4\pi \int d\mathbf{k}\, e^{-i\mathbf{k}\mathbf{r}} \tilde{\rho}(\mathbf{k})$$

On the r.h.s, let us take the integral over $\mathbf{r}'$ first:

$$\sum_{\alpha,\beta}(-ik_\alpha)\int d\mathbf{k}\, e^{-i\mathbf{k}\mathbf{r}}\tilde{\varepsilon}_{\alpha\beta}(\mathbf{k})\int \mathbf{k}'\,(-ik_\beta')\,\tilde{\phi}(\mathbf{k}')\underbrace{\frac{1}{(2\pi)^3}\int d\mathbf{r}'e^{i(\mathbf{k}-\mathbf{k}')\mathbf{r}'}}_{\delta(\mathbf{k}-\mathbf{k}')}$$

$$= -4\pi\int \mathbf{k}\, e^{-i\mathbf{k}\mathbf{r}}\tilde{\rho}(\mathbf{k})\;\Rightarrow$$

$$-\int \mathbf{k}\, e^{-i\mathbf{k}\mathbf{r}}\sum_{\alpha,\beta}k_\alpha k_\beta\tilde{\varepsilon}_{\alpha\beta}(\mathbf{k})\tilde{\phi}(\mathbf{k}) = -4\pi\int \mathbf{k}\, e^{-i\mathbf{k}\mathbf{r}}\tilde{\rho}(\mathbf{k})\;\Rightarrow$$

$$\tilde{\phi}(\mathbf{k})\sum_{\alpha,\beta}k_\alpha k_\beta\tilde{\varepsilon}_{\alpha\beta}(\mathbf{k}) = 4\pi\tilde{\rho}(\mathbf{k}) \qquad (IV.2.38)$$

Focus on $\sum_{\alpha,\beta}k_\alpha k_\beta\tilde{\varepsilon}_{\alpha\beta}(\mathbf{k})$. What is it? If the medium is not only homogeneous but also isotropic, as any liquid in the bulk would be (and we will consider here only such systems), there is no other vector that could determine the symmetric tensor-function, $\tilde{\varepsilon}_{\alpha\beta}(\mathbf{k})$, than vector $\mathbf{k}$. Hence,

$$\tilde{\varepsilon}_{\alpha\beta}(\mathbf{k}) = \tilde{\varepsilon}_{\parallel}(k)\frac{k_\alpha k_\beta}{k^2} + \tilde{\varepsilon}_{\perp}(k)\left(\delta_{\alpha\beta} - \frac{k_\alpha k_\beta}{k^2}\right) \qquad (IV.2.39)$$

and

$$\sum_{\alpha,\beta}k_\alpha k_\beta\tilde{\varepsilon}_{\alpha\beta}(\mathbf{k}) = k^2\varepsilon_{\parallel}(k) \qquad (IV.2.40)$$

Thus substituting Eq. (IV.2.40) into Eq. (IV.2.38) we get

$$\tilde{\phi}(\mathbf{k}) = 4\pi\frac{\tilde{\rho}(\mathbf{k})}{k^2\varepsilon_{\parallel}(k)} \qquad (IV.2.41)$$

and thereby,

$$\phi(\mathbf{r}) = \frac{1}{2\pi^2}\int d\mathbf{k}\, e^{-i\mathbf{k}\mathbf{r}}\frac{\tilde{\rho}(\mathbf{k})}{k^2\varepsilon_{\parallel}(k)} \qquad (IV.2.42)$$

Consider the charge distribution embedded into this medium to be spherically symmetric, so that $\rho(\mathbf{r}) = \rho(r)$. Then $\tilde{\rho}(\mathbf{k}) = \tilde{\rho}(k)$, and if we take into account that any integration of the kind

$$I(\mathbf{r}) = \int d\mathbf{k}\, e^{-i\mathbf{k}\mathbf{r}} f(k) \qquad (\text{IV.2.43})$$

can be recast, introducing integration over $\vartheta$, the angle between vector $\mathbf{k}$ and vector $\mathbf{r}$, and $\varphi$, the azimuthal angle around the $\hat{\mathbf{r}}$ axis, as

$$\begin{aligned}
I(\mathbf{r}) &= \int_0^{2\pi} d\varphi \int_0^{\pi} \sin\vartheta\, d\vartheta \int_o^{\infty} dk\, k^2 f(k) e^{-ikr\cos\vartheta} \\
&= 2\pi \int_0^{\infty} dk\, k^2 f(k) \int_0^{\pi} d\vartheta \sin\vartheta e^{-ikr\cos\vartheta} \\
&= 4\pi \int_0^{\infty} dk\, k^2 f(k) \frac{\sin(kr)}{kr} \qquad (\text{IV.2.44})
\end{aligned}$$

Hence, our result for the electrostatic potential reads

$$\phi(\mathbf{r}) = \frac{2}{\pi} \int_0^{\infty} dk\, \frac{\sin kr}{kr} \frac{\tilde{\rho}(k)}{\varepsilon_{\parallel}(k)} \qquad (\text{IV.2.45})$$

Equations (IV.2.42) or (IV.2.45) allow us to calculate the electrostatic potential for any given shape of charge density and any form of $\varepsilon_{\parallel}(k)$ (for those forms read specialized literature). Of course, this is valid only because we assumed that the embedded charge distribution does not disturb the spectrum of correlations in the medium, i.e. as long as our assumption that $\varepsilon_{\alpha\beta}(\mathbf{r},\mathbf{r}') = \varepsilon_{\alpha\beta}(\mathbf{r}-\mathbf{r}')$ is true.

Consider the case when the charge $q$ is homogeneously distributed over a sphere of radius $a$, centred at coordinate origin, i.e.

$$\rho(\mathbf{r}) = q\frac{\delta(r-a)}{4\pi a^2} \qquad (\text{IV.2.46})$$

Substitution of this equation into the second of Eqs. (IV.2.37) will give

$$\tilde{\rho}(\mathbf{k}) = q\frac{\sin ka}{ka} \qquad (\text{IV.2.47})$$

**Exercise.** Prove Eq. (IV.2.47), using the integration described in Eqs. (IV.2.43) and (IV.2.44).

Hence, for this case

$$\phi(\mathbf{r}) = q\frac{2}{\pi}\int_0^\infty dk \, \frac{\sin kr}{kr}\frac{\sin ka}{ka}\frac{1}{\varepsilon_\|(k)} \tag{IV.2.48}$$

For a point charge, $a \to 0$, $\tilde{\rho}(\mathbf{k}) = q\frac{\sin ka}{ka} \to q$.

**Exercise.** Show that in the 'dispersionless' limit (classical macroscopic electrostatics), i.e. in the case of $\varepsilon_{\alpha\beta}(\mathbf{r} - \mathbf{r}') = \varepsilon\,\delta_{\alpha\beta}$, and, thereby, $\varepsilon_\|(k) = \varepsilon$, one obtains the classical Coulomb potential, $\phi(\mathbf{r}) = \frac{q}{\varepsilon}\frac{2}{\pi}\int_0^\infty dk \, \frac{\sin kr}{kr}\frac{\sin ka}{ka} = \frac{q}{\varepsilon r}$, $r \ge a$, performing the last integral using contour integration.

**Exercise.** The classical model of Debye electrolyte corresponds to

$$\varepsilon_\|(k) = \varepsilon \cdot \left(1 + \frac{\kappa^2}{k^2}\right) \tag{IV.2.49}$$

where $\kappa$ is the inverse Debye length (see Chapter 10, Part I, Vol. 1). Equation (IV.2.48) will then read $\phi(\mathbf{r}) = \frac{q}{\varepsilon ar}\frac{2}{\pi}\int_0^\infty dk \, \frac{\sin kr \cdot \sin ka}{k^2+\kappa^2}$. Now, using contour integration, show that this latter gives

$$\phi(\mathbf{r}) = \frac{q}{\varepsilon r}e^{-\kappa r}\frac{\sinh(\kappa a)}{\kappa a}, \quad r \ge a \tag{IV.2.50}$$

which at $a \to 0$ results in the classical Debye potential $\phi(\mathbf{r}) = \frac{q}{\varepsilon r}e^{-\kappa r}$.

### Solving Volterra equations of the second kind

To get a glimpse of possibilities here, we consider an approach to the solution of the simplest case: the convolution type of Volterra equation with a 'difference' kernel:

$$u(x) - \int_0^x K(x - t)u(t)dt = g(x) \tag{IV.2.51}$$

Let us integrate both sides of this equation with an exponential function:

$$\int_0^\infty dx \, e^{-px}u(x) - \int_0^\infty dx \, e^{-px}\int_0^x K(x - t)u(t)dt = \int_0^\infty dx e^{-px}g(x) \tag{IV.2.52}$$

Note that

$$\int_0^\infty dx\, e^{-px} f(x) = \tilde{f}(p) \qquad \text{(IV.2.53)}$$

is another integral transform, called the Laplace transform. We did not touch it in this book, although it is very widely used in mechanical engineering and, importantly, in electrical engineering and the theory of communication systems. There are many excellent books on Laplace operational calculus, and since we cannot cover 'everything' we will just use Eq. (IV.2.53) as the definition of the Laplace transform $\tilde{f}(p)$, noting, without derivation, that the reverse transform, expressing $f(x)$ through $\tilde{f}(p)$ reads

$$f(x) = \frac{1}{2\pi i} \int_{c-i\infty}^{c+i\infty} dp\, e^{px} \tilde{f}(p) \qquad \text{(IV.2.54)}$$

Integration here is done along the vertical line $\operatorname{Re} p = c$ in the complex plane such that $c$ is greater than the real part of all singularities of $\tilde{f}(p)$, and $\tilde{f}(p)$ is bounded on this line. If all singularities or $\tilde{f}(p)$ are in the left half of the complex plane of $p$ then $c$ can be set to zero and the above inverse integral formula becomes identical to the inverse Fourier transform. In the end, computing this integral has to be done using the methods of contour integration, presented to you in Part III.

Going back then to Eq. (IV.2.52) we can rewrite it as

$$\tilde{u}(p) - \int_0^\infty dx \int_0^x dt\, e^{-px} K(x - t) u(t) = \tilde{g}(p) \qquad \text{(IV.2.55)}$$

Let us rearrange the double integral in this equation. In doing so, we will use the fact that

$$\int_0^\infty dx \int_0^x dt\, F(x, t) = \int_0^\infty dt \int_t^\infty dx\, F(x, t) \qquad \text{(IV.2.56)}$$

for any function $F(x, t)$, integrable over $x$ and $t$ in the considered domains of integration. **Think:** In order to understand why this is so, look at the $(x, t)$ plane of two-dimensional integration. We can then rewrite the integral as

$$
\int_0^\infty dx \int_0^x dt\, e^{-px} K(x-t) u(t)
$$

$$
= \int_0^\infty dt \int_t^\infty dx\, e^{-px} K(x-t) u(t)
$$

$$
= \int_0^\infty dt \int_t^\infty dx\, e^{-p(x-t)} e^{-pt} K(x-t) u(t)
$$

$$
= \int_0^\infty dt\, e^{-pt} u(t) \int_0^\infty dx\, e^{-pz} K(z) = \tilde{u}(p) \tilde{K}(p) \qquad \text{(IV.2.57)}
$$

(In Laplace analysis, this result is often called the *convolution* theorem, but we have just proved it in passing). Substituting it into Eq. (IV.2.55) transforms the latter to $\tilde{u}(p)\{1 - \tilde{K}(p)\} = \tilde{g}(p)$, so that

$$
\tilde{u}(p) = \frac{\tilde{g}(p)}{1 - \tilde{K}(p)} \qquad \text{(IV.2.58)}
$$

and

$$
u(x) = \frac{1}{2\pi i} \int_{c-i\infty}^{c+i\infty} dp\, e^{px} \frac{\tilde{g}(p)}{1 - \tilde{K}(p)} \qquad \text{(IV.2.59)}
$$

**Example IV.2.3 (Maths Practice ♫♪).** Let us solve Eq. (IV.2.51), for a somewhat 'devilish' example:

$$
K(x-t) = B\alpha e^{-\alpha(x-t)}, \quad \alpha > 0; \quad g(x) = A\sin\beta t \qquad \text{(IV.2.60)}
$$

Following the above described routine, we can find the Laplace transforms (Eq. (IV.2.53)) of these functions $\tilde{K}(p) = B\alpha/(p+\alpha)$; $\tilde{g}(p) = A\beta/(p^2 + \beta^2)$, so that following Eq. (IV.2.58) we get

$$
\tilde{u}(p) = \frac{A\beta}{p^2 + \beta^2} \cdot \frac{p+\alpha}{p+\gamma}, \quad \gamma \equiv \alpha(1 - B) \qquad \text{(IV.2.61)}
$$

We can now use the equation for the reverse Laplace transform to find Eq. (IV.2.54). That integration is rather a long story, and since this chapter is not about Laplace calculus, we just look into tables of reverse Laplace transforms [there are many handbooks about it, for example the famous H. Bateman and A. Erdelyi, *Table of Integral Transforms*, Vol. 1, McGraw-Hill, New York (1954)], and get

$$u(x) = \frac{A\beta}{\gamma^2 + \beta^2} \cdot \left\{ (\gamma - \alpha)\cos(\beta t) + \left(\beta + \frac{\gamma\alpha}{\beta}\right)\sin(\beta t) - (\alpha - \gamma)e^{-\gamma t}\right\}$$

$$(IV.2.62)$$

This is the solution!

**Exercise.** The best way to check the obtained solution of an integral equation is to substitute that solution back into the integral equation to check that the l.h.s does equal to the r.h.s. That simple! If it does not, you must have made an error in your calculations. Check the above obtained result, by substituting Eq. (IV.2.62) together with Eq. (IV.2.60) into Eq. (IV.2.51), and taking into account the definition of $\gamma$, as in Eq. (IV.2.61). You will see that you obtain the identity [that was the first thing we checked ourselves after presenting you Eq. (IV.2.62)].

### Solving Volterra equations of the first kind

Solving Volterra equations of the first kind is a big area, and we will show you just one example—solving the Abel equation, Eq. (IV.2.11). The way we show how to solve it *will not, unfortunately, teach you any standard method*! We apologize for that but, nevertheless, we think that you may enjoy following how it may be solved by a 'clever thought'. It is difficult to trace who was first to propose this solution, perhaps Abel himself; it is often presented in textbooks on integral equations, but frequently intermediate derivations are replaced there by words '... and after straightforward transformations you will find', etc. We will avoid this and navigate you all the way through.

Let us take Eq. (IV.2.11), replace $x$ in both right and l.h.ss by $w$, multiply the result by $(x - w)^{\alpha-1}$, and integrate over $w$ from $a$ to $x$. This results in a new equation

$$\int_a^x dw\,(x-w)^{\alpha-1} \int_a^w \frac{u(s)ds}{(w-s)^\alpha} = \int_a^x dw\,g(w)(x-w)^{\alpha-1},\ 0 < \alpha < 1$$

$$(IV.2.63)$$

Similarly to the 'geometrical principle' used in Eq. (IV.2.56), we will use the rule

$$\int_a^x dw \int_a^w ds\, F(w,s) = \int_a^x ds \int_s^x dw\, F(w,s) \qquad \text{(IV.2.64)}$$

to rewrite the l.h.s of Eq. (IV.2.63) as $\int_a^x ds \int_s^x dw\, \frac{u(s)}{(x-w)^{1-\alpha}(w-s)^{\alpha}}$. Using a new variable, $y = w - s$, this can be transformed to $\int_a^x ds\, u(s) \int_0^{x-s} dy\, \frac{1}{y^\alpha (x-s-y)^{1-\alpha}}$. Further introducing another variable $z \equiv \frac{y}{x-s}$, which varies here between 0 and 1, we can rewrite that double integral as

$$\int_a^x ds\, u(s) \int_0^{x-s} d\left(\frac{y}{x-s}\right) \frac{x-s}{(x-s)^\alpha \left(\frac{y}{x-s}\right)^\alpha (x-s)^{1-\alpha}\left(1-\frac{y}{x-s}\right)^{1-\alpha}}$$

$$= \int_a^x ds\, u(s) \cdot \int_0^1 dz\, \frac{1}{z^\alpha (1-z)^{1-\alpha}}$$

The key achievement of our efforts here is that the product of these two integrals is totally decoupled. The integral

$$I(\alpha) = \int_0^1 dz\, \frac{1}{z^\alpha (1-z)^{1-\alpha}}, \quad 0 < \alpha < 1 \qquad \text{(IV.2.65)}$$

is just a function of $\alpha$, which can be determined 'once and forever'. Thus, Eq. (IV.2.63) now looks as

$$\int_a^x ds\, u(s) = \frac{1}{I(\alpha)} \int_a^x dw \frac{g(w)}{(x-w)^{1-\alpha}} \qquad \text{(IV.2.66)}$$

Differentiating this over $x$, we get the solution for $u(x)$ to be:

$$u(x) = \frac{1}{I(\alpha)} \frac{d}{dx} \int_a^x dw \frac{g(w)}{(x-w)^{1-\alpha}} \qquad \text{(IV.2.67)}$$

But this is not the end of story. We can, actually, get a simple analytical expression for $I(\alpha)$. Indeed, it is the same integral we considered in Example III.8.14 in Chapter 8, Part III, of this Volume. To show this, however, we need to do the tricky substitution $z = \frac{1}{1+e^u}$, and hence $dz = \frac{-e^u du}{(1+e^u)^2}$, so that when $z = 0$, $u = \infty$; $z = 1$, $u = -\infty$.

It is then a straightforward exercise (for you) to substitute this into Eq. (IV.2.65) and show that $I(\alpha) = \int_{-\infty}^{\infty} du \frac{e^{\alpha u}}{1+e^u}$. This integral was analysed by contour integration in Example III.8.14 (our $\alpha$ here plays the role of $a$ in Example III.8.14), which gives us

$$I(\alpha) = \frac{\pi}{\sin(\alpha\pi)} \tag{IV.2.68}$$

Thus, all in all, we get the solution of the Abel equation:

$$u(x) = \frac{\sin(\alpha\pi)}{\pi} \frac{d}{dx} \int_a^x dw \frac{g(w)}{(x-w)^{1-\alpha}}, \quad 0 < \alpha < 1 \tag{IV.2.69}$$

which provides $u(x)$ for any given $g(w)$.

Of course, you will be lucky if $g(w)$ is such that you are able to get an analytical result for the integral in Eq. (IV.2.69); if you have to do it numerically, then the differentiation in the r.h.s of Eq. (IV.2.69) will also need to be done numerically. The latter must not necessarily be performed after the numerical integration, as Eq. (IV.2.69) may be rearranged, as in the exercise below.

**Exercise.** Show that Eq. (IV.2.69) can be rearranged via integration by parts to:

$$u(x) = \frac{\sin(\alpha\pi)}{\pi} \left\{ \frac{g(a)}{(x-a)^{1-\alpha}} + \int_a^x \frac{dw}{(x-w)^{1-\alpha}} \frac{dg(w)}{dw} \right\}, \quad 0 < \alpha < 1 \tag{IV.2.70}$$

Generally, this form of solution may be easier to handle.

**Exercise.** The usual test of substituting the solution for $u(x)$, obtained for a given input function $g(w)$, into the initial Abel equation, Eq. (IV.2.11), may prove difficult analytically. Indeed, to be able to do this, all integrals that you encounter must have a closed form. Still, for your confidence in the obtained solution, we suggest you check it for the simplest case, $g(w) = 1$. **Hint:** Show first that in this case $u(x) = \frac{\sin(\alpha\pi)}{\pi(x-a)^{1-\alpha}}$, then substitute it into Eq. (IV.2.11), perform integration, and check whether you get the identity (if you have not made any errors on the way, you will!).

## 2.3. Key Points

- We call *integral equations* the class of equations in which an unknown function stands under the integral sign. Integration can be over one variable, or many variables if the unknown is a function of many variables.
- There is a variety of classes of such equations, covered by excellent specialized mathematical textbooks, some of which—you would be surprised—are readable to non-mathematicians (!). They describe various sophisticated methods of analytical or semi-analytical solutions of such equations, in the first place when they are linear in the unknown function. For some of the equations complicated regular methods of solutions have been developed; some may look more 'heuristic'—and so may look even more beautiful to an aesthete.
- We have considered a few elementary classes of linear integral equations—Fredholm and Volterra equations—and have shown simple examples of how they can be solved when they belong to the so-called *convolution* class, using Fourier and Laplace transforms.
- To encourage you to risk going into unknown territory, we presented an example of a heuristic solution of the Abel equation, just to remind you that we may not always use motorways, but sometimes have to try winding narrow mountain roads, to reach a place of natural beauty.
- One simple lesson was how to check yourself—your result—after going into all that complexity. As for any equation, the best way to check the solution of your integral equation is to put back the solution—the found unknown function—back into the initial equation. If the solution is error-free, the r.h.s of the equation will be equal to the l.h.s. If it is not—search for an error!

# Chapter 3

# Principles of Variational Calculus

## 3.1. Getting Acquainted with Variational Problems: First Take

To introduce variational calculus, let us start from one of its first problems, publicly posed by the Swiss mathematician Johan Bernoulli in the seventeenth century. In response to his challenge, the problem was solved later independently by him (albeit not fully correctly) and by five other great minds, including German mathematician and philosopher Gottfried Leibniz, French aristocrat and mathematician Marquis de l'Hôpital, Johan's brother Jacob Bernoulli, German mathematician, physicist, physician, and philosopher Ehrenfried Walther Tschirnhaus and, faster than all of them, by Sir Isaak Newton. Notably, this was done long before regular methods of variational calculus were developed by Leonard Euler and matured later as a special mathematical discipline. (An apology: we violated here the convention to show a portrait of a particular scientist only once in our two volume project. Leibnitz already appeared in Vol. 1, but we could not resist showing another portrait, making it obvious to an unbiased reader how clever looking that man was!)

Johann Bernoulli
(1667–1748)

Gottfried Wilhelm
Leibniz
(1646–1716)

Guillaume François
Antoine de l'Hôpital
(1661–1704)

531

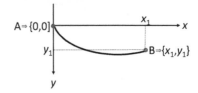

**Fig. IV.3.1**  The profile of a surface along which a particle of mass $m$ slides in gravitational field, friction free, between the heights $y(x = 0)$ and $y(x_1)$, with zero velocity at the initial point. What should be the shape of the solid line, for the particle to slide to point B in the shortest time?

The problem reads as follows. Let a particle of mass $m$ slide without friction and with zero initial velocity from height A to height B, as shown in Fig. IV.3.1 along the landscape the profile of which is depicted by the bold curve $y(x)$ (we assume the landscape to be translationally invariant in the direction normal to the figure, i.e. we consider a one-dimensional problem). What should be the shape of that curve to warrant the fastest trip from point A to point B? Note that in Fig. IV.3.1 we directed $y$-axis down, so $y$ is always positive, but we cannot presume the same for $y' = dy/dx$, as we yet do not know what that optimum curve would be. The sought curve received a name: *brachistochrone*.

We first write the first and crucial part of solving this problem, which will bring us to the notion of a functional—and we explain what a functional actually means, and then present a common-sense derivation of a Euler equation, and finally solve the problem stepping into Euler's shoes.

The energy conservation law (equalizing the acquired kinetic energy to the change in the potential energy) will determine us the velocity $v$ at any point of trajectory of motion: $\frac{mv^2}{2} = mgy$. Hence, like in Galileo experiments, the velocity does not depend on mass (as long as there is no friction or air resistance). Note that $v$ here is the absolute value of the velocity, and if we introduce the contour line of this trajectory, $s$, then this velocity will be defined as $v = \frac{ds}{dt}$. The horizontal component of this velocity reads is $\frac{dx}{dt} = \frac{dx}{ds}v$. But $\frac{dx}{ds} = \frac{dx}{\sqrt{(dx)^2+(dy)^2}} = \frac{1}{\sqrt{1+\left(\frac{dy}{dx}\right)^2}}$. Hence, $\frac{dx}{dt} = \sqrt{\frac{2gy}{1+(y')^2}}$ where

$y' = \frac{dy}{dx}$ is the varying slope along that curve. Correspondingly, $\frac{dt}{dx} = \sqrt{\frac{1+(y')^2}{2gy}}$, and thus for the time $T$ to reach point B moving from point A along the given trajectory $y(x)$, we should integrate the latter equation, to obtain:

$$T\sqrt{2g} = \int_0^{x_1} dx \sqrt{\frac{1 + (y'(x))^2}{y(x)}} \qquad \text{(IV.3.1)}$$

So far the precise shape of $y(x)$-curve has not been specified, but we have boundary conditions on it:

$$y(0) = 0, \ y(x_1) = y_1 \qquad \text{(IV.3.2)}$$

Now the Bernoulli problem is reduced to the question, when is the integral in Eq. (IV.3.1) minimal?

Here comes the notion of a *functional*. The integral in Eq. (IV.3.1) is a functional of the function $y(x)$. This is denoted as $U[y(x)]$. But what is a functional? A function of a single argument put in correspondence a number with another number. For instance $f(x) = x^2$ gives $f = 4$ for $x = 2$. A functional is put in correspondence a number to a function; for instance, in Eq. (IV.3.1), if you specify the shape of the function $y(x)$, after performing integration you will get a number for $T$. To complete this set of definitions: correspondence between the functions is provided by operators. Obviously, a function of several variables, $f(x, y, z)$ defines a correspondence of several numbers (values of $x, y, z$) to one number. Equally a functional may be a function of several functions. Generally, both the functions and functionals may not be single-valued; the situation when the function may jump from its one branch to another one with the variation of the argument, or the same happening with a functional, will correspond to abrupt transitions. We will avoid considering such cases, at least in the beginning. A functional may depend not only on function, but also on derivatives of any order; in Eq. (IV.3.1), it depends also on $y'$, so that $T = T[y(x), y'(x)]$.

A few more rigorous statements about the conditions on and terminology of functionals in parallel with similar statements on functions are summarized in the box. We will formulate them for functions of one variable of one function, but generalization to many variable functions and functionals of many functions is obvious and we will not waste space on it.

---

### On functions and functionals

| Functions $y(x)$ | Functionals $U[x, y(x), y'(x), y''(x), \ldots]$ |
|---|---|
| Difference of the argument, $\Delta x$ – a difference between two neighbouring points, $x - x_0$, where $x_0$ is a 'point of interest'. When this difference is infinitesimally small, $\Delta x = dx$. | Variation of the function in argument of a functional $\delta y(x)$ is a small difference $y(x) - y_0(x)$, where $y(x)$ is a varying function near the function of interest $y_0(x)$. |
| The function is called continuous if a small change of $x$ (small $\Delta x$) corresponds to small change of $y(x)$. | The functional is called continuous if a small variation of the function $y(x)$ leads to small variation of the functional $U$. |

One may consider functions $y(x)$ and $y_0(x)$ close if for any $x$, $|y(x) - y_0(x)| \ll |y(x)|, |y_0(x)|$ (0-order proximity) $+ |y'(x) - y_0'(x)| \ll |y'(x)|$, $|y_0'(x)|$ (1$^{\text{st}}$ order proximity), $\ldots + \left|y^{(n)}(x) - y_0^{(n)}(x)\right| \ll \left|y^{(n)}(x)\right|, \left|y_0^{(n)}(x)\right|$ ($n$th order proximity) A functional $U$ may be considered continuous at $y_0(x)$ in terms of $n$th order proximity if for any $\varepsilon$ ($>0$) we can find such $\delta(>0)$ that $|U[y(x)]| - |U[y_0(x)]| < \varepsilon$ at $|y(x) - y_0(x)| < \delta, |y'(x) - y_0'(x)| < \delta, \ldots \left|y^{(n)}(x) - y_0^{(n)}(x)\right| < \delta$.

---

But how to find a function that provides a minimum to the functional, for example, the one in Eq. (IV.3.1)? Such tasks are central to so-called *variational calculus*, and, in a moment, we will reveal the origin of this name.

Consider generally a functional

$$U = \int_a^b F(x, y, y') \qquad \text{(IV.3.3)}$$

where the $y(x)$, and $y' = dy/dx$, are some regular function, the values of which are fixed at the limits of integration, $y(a)$ and $y(b)$. We need to find an extremum of $U$ that can be delivered by $y(x)$, and thereby find which function will provide it. For this we will need to do a *variation* of this functional, which will be expressed through variation of $y$ and independently of $y'$:

$$\delta U = \int_a^b \left[ F_y'(x, y, y')\delta y + F_{y'}'(x, y, y')\delta y' \right] dx \qquad \text{(IV.3.4)}$$

where $F_y' = \frac{\partial F}{\partial y}$ and $F_{y'}' = \frac{\partial F}{\partial y'}$. If the function $y(x)$ delivers an extremum to $U$ as compared to any other close functions, then such variation must be equal to zero, i.e.

$$\delta U = 0 \qquad \text{(IV.3.5)}$$

If we apply this to Eq. (IV.3.4), we must conclude that

$$\int_a^b \left[ F_y'(x, y, y')\delta y + F_{y'}'(x, y, y')\delta y' \right] dx = 0 \qquad \text{(IV.3.6)}$$

and this must be satisfied for any variation of $\delta y$, $\delta y'$, but under the condition

$$\delta y(a) = \delta y(b) = 0 \qquad \text{(IV.3.7)}$$

as the values of the function $y$ have been fixed at those points. How can we satisfy Eq. (IV.3.6)?

It is time here to recall that although we varied the functional in terms of $y$ and $y'$ as if they are independent functions, of course they are not, and using a relationship between them we can find an equation on $y$ which will warrant satisfaction of Eq. (IV.3.6). Namely, using integration by parts, let us rearrange the integration over the second term in the brackets of Eq. (IV.3.6):

$$\int_a^b F_{y'}'(x, y, y')\frac{d\delta y}{dx}\, dx = F_{y'}'(x, y, y')\delta y \Big|_a^b - \int_a^b \frac{dF_{y'}'(x, y, y')}{dx}\delta y\, dx$$

Due to Eq. (IV.3.7), the first term in the r.h.s of this equation is zero, and thus we can rewrite Eq. (IV.3.6) as $\int_a^b \big[ F_y'(x, y, y') - \frac{dF_{y'}'(x,y,y')}{dx} \big] \delta y\, dx = 0$. But this must be satisfied for any variation of $\delta y$, so that we come to a requirement,

$$F_y'(x, y, y') - \frac{dF_{y'}'(x, y, y')}{dx} = 0 \qquad \text{(IV.3.8)}$$

This is the so-called *Euler equation*, a central equation of variational calculus. In differentiation over $x$, here, $y$ is considered as a function of $x$.

Often, however, it is more convenient to use another form of the Euler equation, which treats $x$, $y$ and $y'$ as independent variables:

$$\frac{\partial F(x,y,y')}{\partial y} - \frac{\partial^2 F(x,y,y')}{\partial x\, \partial y'} - \frac{\partial^2 F(x,y,y')}{\partial y\, \partial y'} y' - \frac{\partial^2 F(x,y,y')}{\partial y'^2} y'' = 0$$

$$(IV.3.9)$$

(in deriving this, we just differentiated over $x$ in the second term of Eq. (IV.3.8) over each argument of the function).

Let us use this equation to minimize the r.h.s of Eq. (IV.3.1). We deal here with the function

$$F(x,y,y') = \sqrt{\frac{1+(y')^2}{y}} \qquad (IV.3.10)$$

This does not explicitly depend on $x$, so that the second term in Eq. (IV.3.9) will be absent. One can of course calculate all the non-vanishing terms, by substituting Eq. (IV.3.10) into Eq. (IV.3.9), and summing them up to obtain a differential equation. The reader can do it, as an **Exercise**, but this straightforward route is rather cumbersome. We will do a trick, instead. Let us multiply what will be left in our case of Eq. (IV.3.9), the terms without explicit dependence on $x$, by $y'$. Then we get

$$\frac{\partial F(x,y,y')}{\partial y} y' - \frac{\partial^2 F(x,y,y')}{\partial y\, \partial y'} (y')^2 - \frac{\partial^2 F(x,y,y')}{\partial y'^2} y' y'' = 0$$

$$(IV.3.11)$$

It easy to show that the l.h.s of this equation is identical to $\frac{d}{dx}\{F - F'_{y'} y'\}$. For a proof, consider $\frac{d}{dx}\{F - F'_{y'} y'\} = \frac{\partial F}{\partial y} y' + \frac{\partial F}{\partial y'} y'' - y'\{\frac{\partial^2 F}{\partial y' \partial y} y' + \frac{\partial^2 F}{\partial y'^2} y''\} - \frac{\partial F}{\partial y'} y''$ (check this yourself as an **Exercise**.) The second and the last term in the r.h.s of this equation cancel each other, and thus we end up with the l.h.s of Eq. (IV.3.11). Hence, we get $\frac{d}{dx}\{F - F'_{y'} y'\} = 0$. The first integral of this equation gives:

$$F - F'_{y'} y' = C \qquad (IV.3.12)$$

where $C$ is a constant to be found from the boundary condition. Now, its' not a big deal to plug into Eq. (IV.3.12) the form of Eq. (IV.3.10), to get

$$\sqrt{\frac{1 + (y')^2}{y}} - (y')^2 \sqrt{\frac{1}{y\left[1 + (y')^2\right]}} = C$$

Summing up the two terms in the left hand side of this equation, we finally obtain a compact, but nontrivial equation

$$\frac{1}{\sqrt{y\left[1 + (y')^2\right]}} = C \qquad (IV.3.13)$$

It is tempting to square this equation, express $(y')^2$ through $y$, take a square root of it to express $y'$ through $y$, and solve the equation by separation of variables. But there is a trouble here, because we do not know which sign we should choose, when taking the square root. Indeed $y'$ need not always be positive, as $y$ may pass an extremum as, for example, in Fig. IV.3.1. Too much intellectual effort, too easy to make a conceptual error. So, we solve it differently, obtaining a *parametric solution*. Squaring Eq. (IV.3.13), will give

$$y = \frac{\tilde{C}}{1 + (y')^2} \qquad (IV.3.14)$$

where $\tilde{C} \equiv \frac{1}{C^2}$. Since $C$ is just an unknown constant, $\tilde{C}$ is a new free constant. Now, let us introduce a running parameter $t$, so that

$$y' = \cot t \qquad (IV.3.15)$$

Then, subject to Eq. (IV.3.14), $y = \frac{\tilde{C}}{1+(\cot t)^2} = \tilde{C}(\sin t)^2 = \frac{\tilde{C}}{2}(1 - \cos(2t))$, so that

$$y = \frac{\tilde{C}}{2}\left[1 - \cos(2t)\right] \qquad (IV.3.16)$$

What is now left is to find a parametric representation of $x$. Here, we can use that $y' = \frac{dy}{dx} \Rightarrow dx = \frac{dy}{y'}$ and hence $dx = \frac{2\tilde{C}\sin t \cos t}{\cot t}dt = 2\tilde{C}(\sin t)^2 dt = \tilde{C}(1 - \cos 2t)dt$.

Integrating this expression over $t$, we get, $x = \tilde{\tilde{C}} + \tilde{C}\left(t - \frac{1}{2}\sin 2t\right)$. But the new free constant of integration, $\tilde{\tilde{C}}$, must be equal to zero. Indeed, $y = 0$, when $t = 0$. Due to the boundary condition, $y|_{x=0} = 0$, $x$ must vanish when $t = 0$. This can be satisfied only when $\tilde{\tilde{C}} = 0$. Thus for $x(t)$ we obtain $x = \frac{\tilde{C}}{2}(2t - \sin 2t)$. Let us introduce a new running variable, denoting $2t \equiv p$, and redefine the free constant, $c \equiv \frac{\tilde{C}}{2}$. We can finally write down in a compact form the parametric definition of the curve of the brachistochrone,

$$\begin{cases} y = c[1 - \cos p] \\ x = c(p - \sin p) \end{cases} \qquad \text{(IV.3.17)}$$

Such a curve is called a *cycloid*, and $2c$ is called a cycloid radius. The constant (the radius) is found from the boundary condition: this curve must reach $y_1$ when $x = x_1$. This occurs at a point $p = p_1$:

$$\begin{cases} y_1 = c[1 - \cos p_1] \\ x_1 = c(p_1 - \sin p_1) \end{cases} \qquad \text{(IV.3.18)}$$

For given $y_1$ and $x_1$, this is a system of two nonlinear equations on $c$ and $p_1$, which is to be solved numerically. In Fig. IV.3.2 we show examples of such solutions and the corresponding plots of $y(x)$.

**Exercise.** Find $c$ and plot the $y(x)$ curves for $(x_1/y_1) > 1.5$. You will see that the curves will pass a maximum. In terms of sliding this means that you first slide down and then up, and it turns out this is faster than going, say, along the straight line to point $(x_1, y_1)$.

Let us check that moving along the straight line is slower than moving along the brachistochrone. For a straight line, plugging $y(x) = \frac{y_1}{x_1}x$ into Eq. (IV.3.1) we get

$$T\sqrt{2g} = \int_0^{x_1} dx \sqrt{\frac{1 + \left(\frac{y_1}{x_1}\right)^2}{\frac{y_1}{x_1}x}} = 2\sqrt{x_1} \cdot \sqrt{\frac{y_1}{x_1} + \frac{x_1}{y_1}} \qquad \text{(IV.3.19)}$$

For a brachistochrone, substituting Eqs. (IV.3.18) into Eq. (IV.3.1) and carrying out integration over $x$ and then integration over $p$, we get

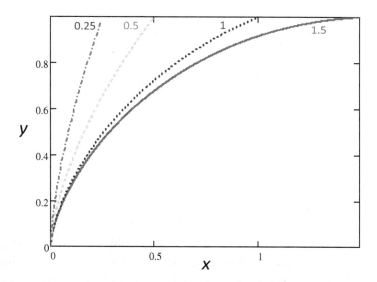

**Fig. IV.3.2** The plots of optimal $y(x)$ profiles warranting the fastest reach of drop depth $y_1$ positioned at point $x_1$. In this graph the drop $y_1$ is taken for the length unit, $y_1 = 1$. The curves correspond to the indicated values of $x_1/y_1$ ratio, for which we found the corresponding values of $c$. Namely, $\{(x_1/y_1) \Rightarrow c\} = \{0.25 \Rightarrow 3.859,\ 0.5 \Rightarrow 1.203,\ 1 \Rightarrow 0.573,\ 1.5 \Rightarrow 0.501\}$.

$$T\sqrt{2g} = \int_0^{p_1} dp \frac{dx}{dp} \sqrt{\frac{1 + \left(\frac{dy}{dp} \Big/ \frac{dx}{dp}\right)^2}{y}}$$

$$= \int_0^{p_1} dp\, c(1 - \cos p) \sqrt{\frac{1 + \left[\frac{c\sin p}{c(1-\cos p)}\right]^2}{c(1 - \cos p)}}$$

$$= \sqrt{2c} \int_0^{p_1} dp = p_1 \sqrt{2c} \tag{IV.3.20}$$

We compare the results of Eq. (IV.3.20) and Eq. (IV.3.19), for some three points in Table IV.3.1. Note that the output data in its columns has dimensionality of the square rout of length, and the length is measured in the units of $y_1$. To recover the sliding time in seconds, one should multiply the results by 0.226 (if the experiment takes place on Earth, $g = 9.81\,\mathrm{m/s^2}$.)

**Table IV.3.1**   A comparison of the results from Eq. (IV.3.20) and Eq. (IV.3.19).

| How far to go | $x_1 = 1.5$, $y_1 = 1$ $[c = 0.501,$ $p_1 = 3.052]$ | $x_1 = 0.25$, $y_1 = 1$ $[c = 3.859,$ $p_1 = 0.736]$ | $x_1 = 0.1$, $y_1 = 1$ $[c = 22.523,$ $p_1 = 0.299]$ |
|---|---|---|---|
| $T\sqrt{2g}$ (brachistochrone) | 3.055 | 2.045 | 2.007 |
| $T\sqrt{2g}$ (straight line) | 3.606 | 2.062 | 2.01 |

The way along the brachistochrone is always faster than on the straight line. As one may expect, the time difference gets smaller for shorter (steeper) ways—from left to right columns of Table IV.3.1. You can check, that the figures in the very right column are just a tiny bit larger than the one we would have had for a vertical drop, which makes perfect sense, as the lateral shift is just 0.1 of the height from which the object slides down. For such a steep slope, the difference would not be much larger, even if we had introduced friction into the problem, but generally, of course, all this analysis is accurate only for friction-free sliding.

Note that the shape of the brachistochrone does not depend on the gravitational constant $g$, but the absolute value of the time to reach the final point is $\propto 1/\sqrt{g}$, so that on the moon it will be $\sqrt{6}$-times longer than moving along the same track on Earth. We just doubt that someone would spend their precious time on the moon to test this conclusion experimentally, at least in near future.

**Exercise.** Consider $U[y(x), y'(x)] = \int_0^{\pi/2} dx\{(y(x)')^2 - y(x)^2\}$ with boundary conditions $y(0) = 0$, $y(\pi/2) = 1$. Using the Euler equation, Eq. (IV.3.9), show that the only function which provides an extremum to this functional, and which obeys the specified boundary conditions, is $y(x) = \sin x$. You will not need any parametric solutions, and this mathematical problem is designed to show that the analysis may not always be as complicated as that presented above!

**Example IV.3.1 (Physical Application ▶▶).**
We consider a system which in the bulk can
undergo second order phase transition. The system
is described by an order parameter: when it is zero
the system may be considered disordered, but when
it is nonzero, it is ordered. We have dealt with such

Vitaly Ginzburg
(1916–2009)

a system in Section 5.1, Example II.5.1 of Chapter 5, Vol. 1, based
on Landau's theory of phase transitions (we recommend the reader
to refresh their memory of the lessening analysis presented there).
But that example considered homogeneous systems. Now, we want
to study a system near an interface, and we will start its description
with the so-called Landau–Ginzburg functional. For simplicity, and
because we want to consider a system that can undergo second order
transition, unlike in Example II.5.1 in Vol. 1, we will not involve in
the consideration an external field. Such a *Landau-Ginzburg* energy
functional, written in terms of the order parameter $\varphi$ and its first
derivative, reads:

$$H = \int_0^\infty dx \, F[\varphi, \varphi'], \quad F[\varphi, \varphi'] = \left[\frac{1}{2}a\varphi^2 + \frac{1}{4}b\varphi^4 + \Lambda^2(\varphi')^2\right] \tag{IV.3.21}$$

with the boundary conditions:

$$\varphi(x = 0) = \varphi_0, \quad \varphi'(x = \infty) = 0 \tag{IV.3.22}$$

So, at the boundary the value of order parameter $\varphi$ is defined, and in
the bulk its value, whether zero or nonzero is assumed to be homogeneous, independent on $x$. The task of this problem is to find the
optimum profile of the order parameter in such a system, for brevity
called an *interfacial profile*.

Note that, generally, as in Example II.5.1 (Vol. 1), $a$ may change
sign, and when it is negative the order parameter in the bulk is
nonzero, whereas when it is positive it is zero. The point where $a$
changes sign is the point of the second order transition (see Example II.5.1). The last term in the brackets of Eq. (IV.3.21) accounts for

the cost of variation of the order parameter in space, with $\Lambda$ determining the characteristic scale of correlations of the order parameter (we will see what exactly determines the correlation length). This is all that characterizes the bulk. But what happens at the interface?

We will consider only the case of $a > 0$, as the most interesting one, but allow also for $a$ to be very close to zero. What should we expect? At the interface $\varphi = \varphi_0$ ($>0$, for definiteness) and in the bulk, as $a > 0$, $\varphi$ should vanish (see Example II.5.1, Vol. 1). So $\varphi$ should transit between $\varphi_0$ and 0. But how? This we should find out from minimization of the functional. Inserting this form of $F[\varphi, \varphi']$ into Eq. (IV.3.9) (here $\varphi$ and $\varphi'$ play the role of $y$ and $y'$, respectively), the Euler equation takes the form:

$$a\varphi + b\varphi^3 - \Lambda^2\varphi'' = 0 \qquad (IV.3.23)$$

Multiplying this equation by $\varphi'$, we can rewrite it as $\frac{d}{dx}\{\frac{a}{2}(\varphi^2) + \frac{b}{4}(\varphi^4) + \frac{\Lambda^2}{2}[(\varphi')^2]\} = 0$.

Taking the first integral of this equation, we get $(\varphi')^2 = \frac{a}{\Lambda^2}\varphi^2 + \frac{b}{2\Lambda^2}\varphi^4 + C$, where $C$ is the free constant of integration. But in the bulk $\varphi = 0$ and $\varphi' = 0$. Thus, it must be that $C = 0$. We therefore obtain $(\varphi')^2 = \frac{a}{\Lambda^2}\varphi^2 + \frac{b}{2\Lambda^2}\varphi^4$. Hence, $\varphi' = \pm\sqrt{\frac{a}{\Lambda^2}\varphi^2 + \frac{b}{2\Lambda^2}\varphi^4}$. Minus will be the correct sign to choose, because as we have set $\varphi_0 > 0$, the value of the order parameter should only go down, and thus $\varphi'$ must be negative. [This could have been more complicated, had we had higher order derivatives in the energy functional that could cause oscillations in the order parameter profile, but we are not considering such a case here]. Introducing, for brevity, new notations,

$$\kappa = \frac{\sqrt{a}}{\Lambda}, \quad \beta = \sqrt{\frac{b}{2}}\frac{1}{\Lambda}, \quad s = \frac{\kappa}{\beta} = \sqrt{\frac{2a}{b}} \qquad (IV.3.24)$$

we get

$$\frac{d\varphi}{dx} = -\sqrt{\kappa^2\varphi^2 + \beta^2\varphi^4} \qquad (IV.3.25)$$

From here. the next steps are straightforward. We solve this equation by separation of variables, first writing $-\frac{d\varphi}{\sqrt{\kappa^2\varphi^2 + \beta^2\varphi^4}} = dx$, and then integrating it, $\int_{\varphi_0}^{\varphi} \frac{d\varphi}{\sqrt{s^2\varphi^2+\varphi^4}} = -\beta x$. The integral in the l.h.s has an exact form. We will not be wasting time in deriving it; this chapter isn't about such trivialities—you may do it by the proper substitution of variables or get it from any of the famous tables of integrals, such as those by Ryzhik and Gradstein, or Dwight. Having done this, you will get:

$$\frac{1}{2s}\left\{\ln\left(\frac{\sqrt{s^2+\varphi_0^2}+s}{\sqrt{s^2+\varphi_0^2}-s}\right) - \ln\left(\frac{\sqrt{s^2+\varphi^2}+s}{\sqrt{s^2+\varphi^2}-s}\right)\right\} = -\beta x$$

(IV.3.26)

**Exercise.** Exponentializing this equation, and solving it for $\varphi$, derive

$$\varphi(x) = \frac{2s\sqrt{\frac{\sqrt{s^2+\varphi_0^2}+s}{\sqrt{s^2+\varphi_0^2}-s}}}{\frac{\sqrt{s^2+\varphi_0^2}+s}{\sqrt{s^2+\varphi_0^2}-s} - e^{-2\kappa x}}e^{-\kappa x}$$

(IV.3.27)

**Exercise.** Check that at $x = 0$, Eq. (IV.3.27) will return you $\varphi(x = 0) = \varphi_0$.

Let us consider the limiting forms of this result.

1. $\varphi_0 \to 0$

Here, $\frac{\sqrt{s^2+\varphi_0^2}+s}{\sqrt{s^2+\varphi_0^2}-s} = \frac{s\sqrt{1+\frac{\varphi_0^2}{s^2}}+s}{s\sqrt{1+\frac{\varphi_0^2}{s^2}}-s} \approx \frac{2}{\frac{\varphi_0^2}{2s^2}} = \frac{4s^2}{\varphi_0^2}$

so that

$$\varphi(x) \approx 2s\,e^{-\kappa x}\frac{\sqrt{\frac{4s^2}{\varphi_0^2}}}{\frac{4s^2}{\varphi_0^2} - e^{-2\kappa x}} = 2s\,e^{-\kappa x}\frac{\varphi_0}{2s}\frac{1}{1 - \frac{\varphi_0^2}{4s^2}e^{-2\kappa x}}$$

$$\approx \varphi_0 e^{-\kappa x}\left(1 + \frac{\varphi_0^2}{4s^2}e^{-2\kappa x}\right)$$

Thus, in this limiting case the profile of $\varphi(x)$ decays exponentially from the surface into the bulk:

$$\varphi(x) \approx \varphi_0 e^{-\kappa x}, \qquad \left(\frac{\varphi_0^2}{4s^2} \ll 1\right)$$

(IV.3.28)

## 2. $\varphi_0 \to \infty$

This case is a bit tricky. Here, $\frac{\sqrt{s^2+\varphi_0^2}+s}{\sqrt{s^2+\varphi_0^2}-s} \approx 1 + \frac{2s}{\varphi_0}$, so that

$$\varphi(x) \approx \frac{2s}{1 - e^{-2\kappa x} + \frac{2s}{\varphi_0}}\, e^{-\kappa x}, \qquad \left(\frac{\varphi_0^2}{s^2} \gg 1\right) \tag{IV.3.29}$$

Note that you, generally, cannot omit here the last term in the dominator, $2s/\varphi_0$, as a 'small one', because such simplification will not match the boundary condition at $x = 0$. Indeed, $\left.(1 - e^{-2\kappa x})\right|_{x=0} = 0$ and the denominator at $x = 0$ is equal to $2s/\varphi_0$, and so you get $\varphi(x = 0) = \varphi_0$, as you should. The omission is possible only for sufficiently large $x$, so that there

$$\varphi(x) = \frac{s}{\sinh(\kappa x)}, \qquad \left(\frac{\varphi_0^2}{s^2} \gg 1, \quad \frac{2s}{\varphi_0} \ll (1 - e^{-2\kappa x})\right) \tag{IV.3.30}$$

In Fig. IV.3.3 we plot the profiles given by our general result, Eq. (IV.3.27), and compare it, where appropriate, with the asymptotic laws, Eqs. (IV.3.28) and (IV.3.29). In panel (a) we see the general character of the profiles and how well the corresponding asymptotic laws reproduce the exact solution. Panel (b) demonstrates what happens if we get closer to the phase transition, although we have not crossed the transition line, and the bulk still prefers to be disordered: when $\kappa \propto \sqrt{a}$ approaches zero the profile of nonzero $\varphi$ will extend deeper and deeper into the bulk. In theoretical literature this is often called a 'wetting transition', as the surface imposes a preference on the order parameter that extends infinitely into the bulk only right at the transition point.

Actually, when $s \propto \kappa \propto \sqrt{a} \to 0$, the law of Eq. (IV.3.29) provides you the result. It is convenient to rewrite it as

$$\varphi(x) \approx \frac{\varphi_0 e^{-\kappa x}}{+1 + \frac{\varphi_0}{2s}(1 - e^{-2\kappa x})} \tag{IV.3.31}$$

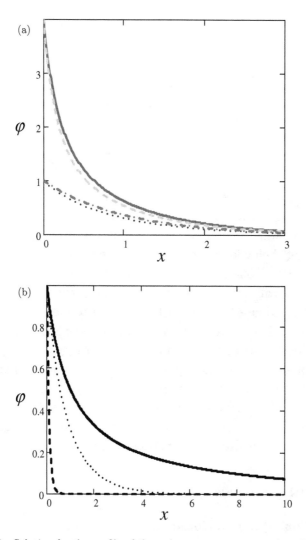

**Fig. IV.3.3** Solution for the profile of the order parameter near an interface in a system above the second order critical point. (a) Lines: solid—exact solution, Eq. (IV.3.27), dashed—asymptotic law for a large value of the order parameter at the boundary, Eq. (IV.3.29), both plotted for $s = 1, \kappa = 1, \varphi_0 = 4$. Dash-dotted—exact solution, Eq. (IV.3.27), dotted—asymptotic law for a small value of the order parameter at the boundary, Eq. (IV.3.28); both plotted for $s = 1$, $\kappa = 1$, $\varphi_0 = 1$. None of these values of $\varphi_0$ are too large or too small, but clearly here the corresponding asymptotic laws work well. (b) The effect of varying parameter $a$ [$s \propto \sqrt{a}$, $\kappa \propto \sqrt{a}$]. Lines: solid $\kappa = a = 0.1$, dotted $\kappa = a = 1$, dashed $\kappa = a = 10$; all curves plotted for $\varphi_0 = 1$.

For any finite $x$, when both $s \propto \kappa \propto \sqrt{a} \to 0$ (approaching the critical point) this formula reduces to

$$\varphi(x) \approx \begin{cases} \dfrac{\varphi_0 e^{-\kappa x}}{1 + \varphi_0 \dfrac{\kappa}{s} x} = \dfrac{\varphi_0 e^{-\frac{\sqrt{a}}{\Lambda} x}}{1 + \varphi_0 \sqrt{\dfrac{b}{2} \dfrac{x}{\Lambda}}}, & x < \dfrac{\Lambda}{2\sqrt{a}} \\[4mm] 2s e^{-\kappa x} = 2\sqrt{\dfrac{2a}{b}} e^{-\frac{\sqrt{a}}{\Lambda} x}, & x \gg \dfrac{\Lambda}{2\sqrt{a}} \end{cases}$$ 
(IV.3.32)

Had we put $b = 0$, the profile of nonzero $\varphi$ would have penetrated distances of the order of $\Lambda/\sqrt{a}$ into the bulk.

A note of warning: do not take too literally the obtained law for the 'wetting' behavior when $a$ tends to zero, i.e. close to the critical point. Here, one should go beyond the Euler equation (which theoreticians call going beyond 'mean-field') and consider fluctuations that dominate the system behavior near the critical point. To investigate their role would have required much more sophisticated theory, which exists and which we will only briefly touch upon in the next Chapter. However, as a part of a formal tutorial on variational analysis, this example serves it purpose.

This introduction has given you a glimpse of what variational calculus is about. We now feel safe to proceed to more rigorous and systematic representation of functionals and variational analysis.

## 3.2.   1D Functionals

### *The notion of a functional in more detail: 'Discretization'*

We will now give a more detailed definition of the functional. It is a mathematical object that has as its input a function $f(x)$; more precisely, its input is as set of infinite values $f_n = f(n\varepsilon)$, where the spacing, $\varepsilon$, between $x$-values, tends to zero. We can write a functional with an input function $f(x)$ defined in the domain $0 \le x \le b$ as

$$U[f(x)] \equiv \lim_{\substack{\varepsilon \to 0 \\ N \to \infty}} \{U(f_0, f_1, \ldots f_N)\} \quad \text{where } \varepsilon N = b \quad \text{(IV.3.33)}$$

One simple example of a functional would be

$$U[f(x)] = \lim_{\substack{\varepsilon \to 0 \\ N \to \infty}} \varepsilon \sum_{n=0}^{N} f_n^2 \qquad \text{(IV.3.34)}$$

In this case, the limit on the right is simply the definition of definite integration of $f(x)^2$ between $x = 0$ and $x = b$, which reads as

$$U[f(x)] = \int_0^b f(x)^2 dx \qquad \text{(IV.3.35)}$$

Functionals can also depend on other specified functions, not just by the set of input variables specified by $f(x)$. For instance, the Fourier transform operation can be written as

$$\tilde{f}(k) = U[f(x)] = \lim_{\substack{\varepsilon \to 0 \\ N \to \infty}} \varepsilon \sum_{n=-N}^{N} f_n \exp(i\varepsilon nk) \qquad \text{(IV.3.36)}$$

Functionals can also depend on derivatives of $f(x)$. For instance we could write

$$U[f(x)] = \lim_{\substack{\varepsilon \to 0 \\ N \to \infty}} \varepsilon \sum_{n=0}^{N} \frac{(f_{n+1} - f_n)^2}{\varepsilon^2} = \int_0^b \left(\frac{df(x)}{dx}\right)^2 dx \quad \text{(IV.3.37)}$$

### Functional differentiation

How do we minimize or maximize a functional? We require that in Eq. (IV.3.33), for the functional to be minimized

$$\frac{\partial U(f_0, f_1, \ldots f_N)}{\partial f_j} = 0 \quad \text{for all} \quad 0 \le j \le N \qquad \text{(IV.3.38)}$$

Now, let's define the functional derivative. It is taken to be

$$\frac{\delta U[f(x)]}{\delta f(x')} = \lim_{\substack{\varepsilon \to 0 \\ N \to \infty}} \left\{ \frac{1}{\varepsilon} \frac{\partial}{\partial f_j} U(f_0, f_1, \ldots f_N) \right\} \qquad \text{(IV.3.39)}$$

where $x' = \varepsilon j$. Our condition that the functional is maximized or minimized then corresponds to $\frac{\delta U[f(x)]}{\delta f(x')} = 0$.

Let's first examine what the functional derivative is in the case where

$$U[f(x)] = \int_0^b G(f(x))dx = \lim_{\substack{\varepsilon \to 0 \\ N \to \infty}} \varepsilon \sum_{n=0}^N G(f_n) \qquad \text{(IV.3.40)}$$

In this case we see that the functional derivative yields

$$\frac{\delta U[f(x)]}{\delta f(x')} = \lim_{\substack{\varepsilon \to 0 \\ N \to \infty}} \frac{dG(f_j)}{df_j} = \frac{dG(f(x'))}{df(x')} \qquad \text{(IV.3.41)}$$

How can we perform the functional differentiation in a more convenient way when we have to deal with other more complicated functionals that can be expressed as integrals? A way to do this is to consider the total derivative of $U$. This can be written as

$$\delta U[f(x)] = \int_0^b \frac{dG(f(x))}{df(x)} \delta f(x)dx = \lim_{\substack{\varepsilon \to 0 \\ N \to \infty}} \varepsilon \sum_{n=0}^N \frac{dG(f_n)}{df_n}df_n$$

$$\text{(IV.3.42)}$$

A functional minimum or maximum is then given by $\delta U[f(x)] = 0$, which in this case requires $\frac{dG(f(x))}{df(x)} = 0$ for all values of $f(x)$. To a more general situation we can extend Eq. (IV.3.42) so that

$$\delta U[f(x)] = \lim_{\substack{\varepsilon \to 0 \\ N \to \infty}} \varepsilon \sum_{n=0}^N \frac{\partial G(f_0, f_1 \dots f_N)}{\partial f_n}df_n \qquad \text{(IV.3.43)}$$

**Example IV.3.2 (Physical Application ▶▶).** Let's see why this second formulation of the functional derivative is useful. Consider the functional

$$S[x(t)] = \int_{-\infty}^\infty [T(\dot{x}(t)) - V(x(t))]dt$$

$$= \int_{-\infty}^\infty \left[ \frac{m}{2} \left( \frac{dx(t)}{dt} \right)^2 - \frac{k}{2} x(t)^2 \right] dt \qquad \text{(IV.3.44)}$$

This functional is of the same class as the one given by Eq. (IV.3.21), in Example IV.3.2; it is even simpler, as it does not have the fourth power term. But the sign of the coefficient of the quadratic term is opposite to the one considered in Example IV.3.2, and the physics that Eq. (IV.3.44) usually represents is different; it refers to classical mechanics. It is instructive to explore this case, as it demonstrates how minimization of a functional can yield an equation of motion.

As written, Eq. (IV.3.44) represents what is called in mechanics an *Action*. It describes 1D motion of a particle of mass $m$, the position of which is described by $x(t)$. The combination $T(\dot{x}(t)) - V(x(t))$ is referred to as the *Lagrangian*, named after a great French mathematician and astronomer Lagrange, born Italian as Giuseppe Luigi Lagrange [his portrait we have shown in Chapter 5 of Part II (Vol. 1)]. Here, $T(\dot{x}(t)) = m\dot{x}(t)^2/2$ is the particle kinetic energy, which depends only on its mass and the particle velocity $\dot{x} = dx/dt$ moving along the line $x(t)$; $V(x(t))$ is the potential energy; in Eq. (IV.3.44), that potential energy is chosen as the one for a mass on a spring: $V(x(t)) = kx(t)^2/2$, where $x = 0$ is the equilibrium spring position and $k$ is the spring constant.

Lagrange showed that one can get the classical equations of motion by minimizing the action from a given form of a Lagrangian. Not going any further into the details of the proof of this statement, we just proceed to minimization of this functional. As we have learned from the previous section, we will need to find the variation of the functional it terms of variations of $\delta x(t)$ and $\delta \dot{x}(t)$:

$$
\begin{aligned}
\delta S[x(t)] &= \int_{-\infty}^{\infty} \left[ \frac{dT(\dot{x}(t))}{d\dot{x}} \delta \dot{x}(t) - \frac{dV(x(t))}{dx} \delta x(t) \right] dt \\
&= \int_{-\infty}^{\infty} \left[ m \left( \frac{dx(t)}{dt} \right) \left( \frac{d\delta x(t)}{dt} \right) - kx(t)\delta x(t) \right] dt
\end{aligned}
$$

(IV.3.45)

Now, how do we convert $\delta \dot{x}(t)$ to $\delta x(t)$? The answer lies in integration by parts. This allows us to write

$$
\delta S[x(t)] = \left[ \frac{dx(t)}{dt} \delta x(t) \right]_{-\infty}^{\infty} + \int_{-\infty}^{\infty} \left[ -m \left( \frac{d^2 x(t)}{dt^2} \right) - kx(t) \right] \delta x(t) dt
$$

(IV.3.46)

To get rid of the first term we can set $\delta x(\infty) = \delta x(-\infty) = 0$. The functional derivative $\delta S[x(t)]/\delta x(t)$ is zero, when the square bracket in Eq. (IV.3.46) vanishes, i.e. when

$$m\left(\frac{d^2x(t)}{dt^2}\right) + kx(t) = 0 \qquad\qquad \text{(IV.3.47)}$$

We thus obtain an equation that describes simple harmonic motion.

According to Eq. (IV.3.41), the functional derivative of the functional Eq. (IV.3.44) is given by

$$\frac{\delta S[x(t)]}{\delta x(t)} = -m\left(\frac{d^2x(t)}{dt^2}\right) - kx(t) \qquad\qquad \text{(IV.3.48)}$$

**Exercise.** Show that, in general, for any potential energy function $V(x(t))$, minimizing the Action yields the following equation of motion:

$$m\left(\frac{d^2x(t)}{dt^2}\right) + \frac{\partial V(x(t))}{\partial x(t)} = 0 \qquad\qquad \text{(IV.3.49)}$$

Note here that the force on the particle is $F = -dV/dx$, and Eq. (IV.3.49) is simply a statement of Newton's second law!

You may ask, what is the point of all this if one could have simply written down Eq. (IV.3.49) from Newton's second law? There are two main arguments towards the usefulness of functionals in the physical sciences:

(a) If we wanted to find the equations of motion for a more complicated system, writing a Lagrangian is simpler than trying to identify resultant forces and apply Newton's second law. It is easier to start by writing down a kinetic energy and potential energy in the Lagrangian.

(b) Functionals become invaluable if we have to deal with systems that are not deterministic, but stochastic, where $x(t)$ can be influenced by fluctuations, be they thermal or quantum mechanical. In this case, we have to start thinking of physical behaviour determined by the function that is the saddle-point or minimum of a functional, averaged over functions that may or may not lie close to this, depending on the size of the fluctuations.

Fluctuations and functionals will be the key features in the last two chapters of this book.

### Stability and the functional Hessian matrix

Let us formally consider a functional $U[f(x)]$ for which some function $f^{(0)}(x)$ provides a zero functional derivative, $\delta U[f(x)]/\delta f(x)|_{f(x)=f^{(0)}(x)} = 0$. Now the question is whether $f^{(0)}(x)$ minimizes or maximizes $U[f(x)]$ or is, in fact a saddle point solution. Using our fundamental definition of a functional, Eq. (IV.3.33), we can write

$$U[f^{(0)}(x)] = \lim_{\substack{\varepsilon \to 0 \\ N \to \infty}} \left\{ U(f_0^{(0)}, f_1^{(0)}, \dots f_N^{(0)}) \right\} \qquad \text{(IV.3.50)}$$

where the function $f^{(0)}(x)$ is represented, as in Eq. (IV.3.33), by the $N$ independent variables $f_0^{(0)}, f_1^{(0)}, \dots f_N^{(0)}$. Now, we Taylor-expand $U$ in variables $f_0^{(0)}, f_1^{(0)}, \dots f_N^{(0)}$, representing the function $f(x)$ about the extremal points. Up to the second order in the expansion, this yields the following expression:

$$U[f(x)] = U[f^{(0)}(x)]$$

$$+ \lim_{\substack{\varepsilon \to 0 \\ N \to \infty}} \left\{ \sum_{j=0}^{N} \sum_{k=0}^{N} \frac{\partial^2}{\partial f_j^{(0)} \partial f_k^{(0)}} U(f_0^{(0)}, f_1^{(0)}, \dots f_N^{(0)}) \delta f_j \delta f_k \right\}$$

$$\text{(IV.3.51)}$$

The second order partial derivatives standing in Eq. (IV.3.51) are in fact entries in an $N \times N$ Hessian matrix (see Chapter 2 of Part III). As we learnt in that chapter, if all the eigenvalues of the matrix are negative, it is a maximum, while if they are all positive, we have a minimum. Otherwise it's a saddle point.

**Example IV.3.3 (Maths Practice ♫♪).** Let's consider the functional

$$U[x(s)] = \int_0^L \left[ \frac{K}{2} \left( \frac{dx(s)}{ds} \right)^2 + V(x(s)) \right] ds \qquad \text{(IV.3.52)}$$

for which the 'turning point', where $\delta F[x(s)]/\delta x(s) = 0$, is delivered by the function $x^{(0)}(s)$ that satisfies the Euler equation

$$-K\frac{d^2x(s)}{ds^2} + \frac{\partial V(x(s))}{\partial x(s)} = 0 \qquad \text{(IV.3.53)}$$

subject to the boundary value constraint $x(0) = 0$, $x(L) = \alpha$. Now our task is to expand $U[x(s)]$ about $U[x^{(0)}(s)]$ to find the Hessian matrix. To proceed it is easier to discretize Eq. (IV.3.52) such that

$$U[x(s)] = \lim_{\substack{\varepsilon \to 0 \\ N \to \infty}} \varepsilon \sum_{n=0}^{N} \left[ \frac{K}{2}\left(\frac{x_n - x_{n-1}}{\varepsilon}\right)^2 + V(x_n) \right]$$

Then by writing $x_n = x_n^{(0)} + \delta x_n$ we get

$$U[x(s)] = U[x^{(0)}(s)] + \varepsilon \sum_{n=0}^{N} \left[ \frac{K}{2}\left(\frac{\delta x_n - \delta x_{n-1}}{\varepsilon}\right)^2 + \frac{d^2V(x_n)}{dx_n^2}(\delta x_n)^2 \right]$$

$$= U[x^{(0)}(s)]$$

$$+ \varepsilon \sum_{m=0}^{N}\sum_{n=0}^{N} \left[ \frac{K}{2}\left(\frac{\delta_{n,m} - \delta_{n-1,m}}{\varepsilon}\right)^2 + \frac{d^2V(x_n)}{dx_n^2}\delta_{n,m} \right]\delta x_n \delta x_m$$

$$= U[x^{(0)}(s)] + \varepsilon \sum_{m=0}^{N}\sum_{n=0}^{N} S_{n,m}\delta x_n \delta x_m$$

So, $S_{n,m}$ are elements of the functional Hessian matrix. How do we find the eigenvalues and find out whether the solution $x^{(0)}(s)$ is a minimum, maximum or saddle point? The best way of doing this is to diagonalize the Hessian. Here, initially, we'll look at the easiest case in which all second derivatives of $V(x)$ are constant: $\frac{d^2V(x_n)}{dx_n^2} = \beta$.

For simplicity, we restrict the analysis to the case when $\delta x(s)$ is odd. Then the diagonalization can be achieved by a discrete (matrix) transform,

$$\delta x_n = \sum_{k=1}^{\infty} \delta\tilde{x}_k \sin\left(\frac{\pi k n}{N}\right) \qquad \text{(IV.3.54)}$$

Using this, we can write

$$U[x(s)] = U[x^{(0)}(s)] + \varepsilon \sum_{k=1}^{\infty} \delta\tilde{x}_k \sum_{k'=1}^{\infty} \delta\tilde{x}_{k'}$$

$$\times \sum_{n=0}^{N} \left[ \frac{K}{2\varepsilon^2} \left( \sin\left(\frac{\pi kn}{N}\right) - \sin\left(\frac{\pi k(n-1)}{N}\right) \right) \right.$$

$$\times \left( \sin\left(\frac{\pi k'n}{N}\right) - \sin\left(\frac{\pi k'(n-1)}{N}\right) \right)$$

$$\left. + \beta \sin\left(\frac{\pi kn}{N}\right) \sin\left(\frac{\pi k'n}{N}\right) \right]$$

As $N \to \infty$, we can expand

$$\sin\left(\frac{\pi kn}{N}\right) - \sin\left(\frac{\pi k(n-1)}{N}\right) \simeq \frac{\pi k}{N} \cos\left(\frac{\pi kn}{N}\right)$$

so that

$$U[x(s)] = U[x^{(0)}(s)] + \varepsilon \sum_{k=1}^{\infty} \tilde{x}_k \sum_{k'=1}^{\infty} \tilde{x}_{k'}$$

$$\times \sum_{n=0}^{N} \left[ \frac{K}{2\varepsilon^2} \left(\frac{\pi k}{N}\right) \left(\frac{\pi k'}{N}\right) \times \cos\left(\frac{\pi kn}{N}\right) \cos\left(\frac{\pi k'n}{N}\right) \right.$$

$$\left. + \beta \sin\left(\frac{\pi kn}{N}\right) \sin\left(\frac{\pi k'n}{N}\right) \right]$$

The next step is to take the limit of $\varepsilon \to 0$. This gives us

$$U[x(s)] = U[x^{(0)}(s)] + \sum_{k=1}^{\infty} \tilde{x}_k \sum_{k'=1}^{\infty} \tilde{x}_{k'}$$

$$\times \int_0^L \left[ \frac{K}{2} \left(\frac{\pi k}{L}\right) \left(\frac{\pi k'}{L}\right) \cos\left(\frac{\pi ks}{L}\right) \right.$$

$$\left. \times \cos\left(\frac{\pi k's}{L}\right) + \beta \sin\left(\frac{\pi ks}{L}\right) \sin\left(\frac{\pi k's}{L}\right) \right] ds$$

Now, we can do the $s$-integration, taking into account that

$$\int_0^L \cos\left(\frac{\pi ks}{L}\right) \cos\left(\frac{\pi k's}{L}\right) ds = \int_0^L \sin\left(\frac{\pi ks}{L}\right) \sin\left(\frac{\pi k's}{L}\right) ds = \frac{L}{2} \delta_{k,k'}$$

**Exercise.** Check this equality by direct integration, and then by showing that in the limit of $k \to k'$ the result delivers $L/2$, otherwise it is zero.

This then yields

$$U[x(s)] = U[x^{(0)}(s)] + \frac{L}{2} \sum_{k=1}^{\infty} \tilde{x}_k \sum_{k'=1}^{\infty} \tilde{x}_{k'} \left[ \frac{K}{2} \left( \frac{\pi k}{L} \right) \left( \frac{\pi k'}{L} \right) + \beta \right] \delta_{k,k'}$$

$$(\text{IV.3.55})$$

So, we have diagonalized the Hessian and obtained the spectrum of eigenvalues, which is given by

$$\frac{K}{2} \left( \frac{\pi k}{L} \right) \left( \frac{\pi k'}{L} \right) + \beta \qquad (\text{IV.3.56})$$

$\beta > 0$ will warrant that the function $x^{(0)}(s)$ will deliver a functional minimum.

**Exercise.** You may like to try to perform similar analysis for the case when $\delta x(s)$ is even. **Hint:** Instead of Eq. (IV.3.54) you should use a cos-Fourier transform.

### *Lagrange multipliers*

As already discussed in this chapter, we often seek to minimize (or maximize) a functional, subject to a constraint. For instance, a set of functions $f(x)$ might be constrained in such a way that

$$\alpha_0 = \alpha[f(x)] = \int_0^b f(x)dx \qquad (\text{IV.3.57})$$

How can we minimize a functional subject to such a constraint?

So far we have been accounting for constraints given by boundary conditions, doing it *ad hoc*, but there is a regular procedure for doing this for any kind of constraints—it was first formulated by Lagrange, albeit in a different form than the one we use nowadays. We will demonstrate this method for the constraint given by Eq. (IV.3.57).

First we add to our functional what we want to constrain, in this case the term $\alpha[f(x)]$ multiplied by the so-called *Lagrange multiplier,* $\lambda$. The next steps, and what comes out of it, we show on a particular functional,

$$U[f(x)] = \int_0^L \left[ \frac{a}{2} \left( \frac{df(x)}{dx} \right)^2 + \frac{b}{2} f(x)^2 \right] dx \quad \text{where } a, b > 0$$

(IV.3.58)

subject to the boundary value constraints $f(0) = 0$ and $f(L) = c$. The essence of the Lagrange multiplier method is that now instead of minimizing $U[f(x)]$ we minimize

$$\tilde{U}[f(x)] = U[f(x)] - \lambda \alpha[f(x)] \tag{IV.3.59}$$

Once we've done this, our result will depend on $\lambda$. The trick will be to find the value of $\lambda$ that satisfies the constraint.

**Exercise.** Show that the condition $\delta \tilde{U}[f(x)]/\delta f(x) = 0$ yields the equation

$$-a \frac{d^2 f(x)}{dx^2} + bf(x) = \lambda \tag{IV.3.60}$$

**Exercise.** Show, by any method you like, that the solution to Eq. (IV.3.60), subject to the boundary conditions $f(0) = 0$, and $f(L) = c$, is

$$f(x) = \frac{\lambda}{b} + \frac{\frac{\lambda}{b} \left( \exp\left( -\sqrt{\frac{b}{a}} L \right) - 1 \right) + c}{\left( \exp\left( \sqrt{\frac{b}{a}} L \right) - \exp\left( -\sqrt{\frac{b}{a}} L \right) \right)} \exp\left( \sqrt{\frac{b}{a}} x \right)$$

$$+ \frac{\frac{\lambda}{b} \left( 1 - \exp\left( \sqrt{\frac{b}{a}} L \right) \right) - c}{\left( \exp\left( \sqrt{\frac{b}{a}} L \right) - \exp\left( -\sqrt{\frac{b}{a}} L \right) \right)} \exp\left( -\sqrt{\frac{b}{a}} x \right) \tag{IV.3.61}$$

**Hint:** The easiest way to get it would be to divide the whole equation by a, then write the general solution of the homogeneous equation and add to it a particular solution of the inhomogeneous one, which in this case is just $\frac{\lambda}{b}$, and then find the two coefficients of the homogeneous solution through subjecting the whole solution to the two, above specified boundary conditions. This hint may be redundant for the reader who has reached this Chapter; please accept our apology then!

To keep things simple, we'll consider hereafter the case when $L\sqrt{b/a} \gg 1$. Omitting in both fractions of Eq. (IV.3.61) the terms smaller than the exponentially large terms $\propto \exp\left(\sqrt{\frac{b}{a}}L\right)$, we get

$$f(x) \approx \frac{\lambda}{b} + \left(c - \frac{\lambda}{b}\right) \exp\left(\sqrt{\frac{b}{a}}(x - L)\right) - \frac{\lambda}{b}\exp\left(-\sqrt{\frac{b}{a}}x\right)$$

$$\text{(IV.3.62)}$$

To find $\lambda$, we simply substitute our solution back into the constraint. Doing this for the limiting case—i.e. substituting Eq. (IV.3.62) into Eq. (IV.3.57)—we obtain an equation that relates $\lambda$ to $\alpha_0$:

$$\alpha_0 \approx \frac{\lambda}{b}L + \left(c - \frac{\lambda}{b}\right)\int_0^L \exp\left(\sqrt{\frac{b}{a}}(x - L)\right) dx$$

$$- \frac{\lambda}{b}\int_0^L \exp\left(-\sqrt{\frac{b}{a}}x\right) dx$$

$$\Rightarrow \alpha_0 \approx \frac{\lambda}{b}L + \left(c - \frac{\lambda}{b}\right)\sqrt{\frac{a}{b}}\left[\exp\left(-\sqrt{\frac{b}{a}}L\right) - 1\right]$$

$$- \frac{\lambda}{b}\sqrt{\frac{a}{b}}\left[1 - \exp\left(-\sqrt{\frac{b}{a}}L\right)\right]$$

$$\Rightarrow \alpha_0 \approx \frac{\lambda}{b}L - \left(2\frac{\lambda}{b} - c\right)\sqrt{\frac{a}{b}}$$

$$\Rightarrow \lambda \approx b\frac{\alpha_0 - c\sqrt{\frac{a}{b}}}{L - 2\sqrt{\frac{a}{b}}} \approx \frac{b}{L}\left(\alpha_0 - c\sqrt{\frac{a}{b}}\right) \qquad \text{(IV.3.63)}$$

**Exercise.** Find an expression for $\lambda$ in terms of $\alpha_0$ and $L$ for the exact solution, i.e. any value of $L\sqrt{b/a}$. It will be cumbersome, but doable! Then check that your result reduces to the limiting case, Eq. (IV.3.63), when $L\sqrt{b/a} \gg 1$.

**Example IV.3.4 (Physical Application ▶▶).** We will now investigate a more complicated case of applying the method of Lagrange multipliers. Consider an elastic rod of length $L$ bent in a 2D plane where the two ends are separated by a vector $\mathbf{R}_0 = R_0\hat{\mathbf{i}}$, i.e. both end points lie on the $x$-axis, and the rod is held so that so it is curved symmetrically about its midpoint (this condition corresponds to equal forces being applied at both ends balanced by the internal forces due to bending). What differential equation would describe its shape? What shape does solution of this equation, with the corresponding boundary conditions and the constraints, predict?

The shape of a curved rod lying in a 2D plane can first be described by position vector $\mathbf{R}(s)$ pointing to the axis of the rod at any point of its contour length $s$. We will use the coordinate system [slightly different from Fig. III.4.15 in the definition of the angle $\theta(s)$], as shown in Fig. IV.3.4.

Here, $\mathbf{R}(s) = x(s)\hat{\mathbf{i}} + y(s)\hat{\mathbf{j}}$ can be written in terms of its tangent vector as

$$\mathbf{R}(s) = \int_{-L/2}^{s} \hat{\mathbf{t}}(s)ds = \hat{\mathbf{i}}\int_{-L/2}^{s} \cos\theta(s)ds + \hat{\mathbf{j}}\int_{-L/2}^{s} \sin\theta(s)ds$$

$$(IV.3.64)$$

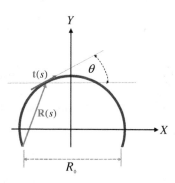

**Fig. IV.3.4** The geometry in which we describe a bent rod lying on an $XY$-plane, with end points kept at distance $R_0$.

where $\theta(s)$ is the angle at the contour point $s$ that the tangent vector to the axis of the rod makes with the $X$-axis. In Eq. (IV.3.64), the centre of the rod has been chosen to be $s = 0$ and at that point $\theta(0) = 0$. Here the constraint is

$$R_0 \hat{\mathbf{i}} = \int_0^L \hat{\mathbf{t}}(s) ds = \hat{\mathbf{i}} \int_{-L/2}^{L/2} \cos \theta(s) ds + \hat{\mathbf{j}} \int_{-L/2}^{L/2} \sin \theta(s) ds \quad \text{(IV.3.65)}$$

Because we're looking at a symmetric solution about $\theta(0) = 0$, $\theta(s)$ is odd and so is $\sin \theta(s)$. Thus, it follows that

$$\int_{-L/2}^{L/2} \sin \theta(s) ds = 0 \qquad \text{(IV.3.66)}$$

and our constraint becomes

$$R_0 = F[\theta(s)] = \int_{-L/2}^{L/2} \cos \theta(s) ds \qquad \text{(IV.3.67)}$$

We can also set the boundary conditions at the edges of our 'bow':

$$\theta(L/2) = -\theta(-L/2) = -\bar{\theta} \qquad \text{(IV.3.68)}$$

How do we now find an equation for $\theta(s)$? First we must minimize bending elastic energy subject to the constraint. The functional that describes this energy is

$$U[\theta(s)] = \frac{B}{2} \int_{-L/2}^{L/2} \frac{d\hat{\mathbf{t}}(s)}{ds} \cdot \frac{d\hat{\mathbf{t}}(s)}{ds} ds = \frac{B}{2} \int_{-L/2}^{L/2} \left( \frac{d\theta(s)}{ds} \right)^2 ds$$

$$\text{(IV.3.69)}$$

where $B$ is the bending elastic module—the larger $B$, the harder it would be to bend the rod.

Now, again using the method of Lagrange multipliers, to minimize $U[\theta(s)]$ subject to the constraint Eq. (IV.3.67) we minimize

$$\tilde{U}[\theta(s)] = U[\theta(s)] - \lambda F[\theta(s)] \qquad \text{(IV.3.70)}$$

where $\lambda$ is to be chosen in such a way that the solution that delivers the minimum to this functional must satisfy Eq. (IV.3.67).

**Exercise.** Show that minimizing $\tilde{U}[\theta(s)]$ yields the following equation:

$$\frac{d^2\theta(s)}{ds^2} - \frac{\lambda}{B}\sin\theta(s) = 0 \qquad \text{(IV.3.71)}$$

**Hint:** Use Euler equation (IV.3.8) for minimization of this functional.

You may recognize this equation from before; it's the sine Gordon equation that we met in Chapter 6 of Part II, Vol. 1; there, however, it described the motion of a pendulum, where $s$ would be replaced by time $t$ and $-\lambda/B$ with $g/l$. As we did it already many times, and in particular in the pendulum problem, we can reduce the order of this differential equation by multiplying it by $d\theta(s)/ds$ and integrating with respect to $s$, to obtain a first order differential equation that we can solve by the separation of variables. In general, however, the solution needs be written in terms of *elliptic integrals* (you may, if needed, read about them in other textbooks). But we won't go into them here. Instead, we will proceed, solving this equation approximately, for the case when $\theta(s)$ is small in the whole interval of variation of $s \in [-L/2, L/2]$. Obviously, in this case the bow will not look like the one depicted in Fig. IV.3.4: it will be much less curved, and we can replace Eq. (IV.3.71) with

$$\frac{d^2\theta(s)}{ds^2} - \frac{\lambda}{B}\theta(s) = 0 \qquad \text{(IV.3.72)}$$

We have already solved such equations in Vol. 1.

**Exercise.** Show that the solution of this equation satisfying the boundary condition prescribed by Eq. (IV.3.68) reads

$$\theta(s) = -\bar{\theta}\frac{\sinh(\kappa s)}{\sinh\left(\frac{\kappa L}{2}\right)} \qquad \text{(IV.3.73)}$$

where

$$\kappa = \sqrt{\frac{\lambda}{B}} \qquad \text{(IV.3.74)}$$

**Hint:** The solution of this equation is the liner combination of functions $e^{-\kappa s}$ and $e^{\kappa s}$ with coefficients from the boundary conditions and $\theta(s = 0) = 0$.

Equation (IV.3.73) makes perfect sense, but it is still not the end of story, as we need to find $\kappa$, i.e. $\lambda$. Substitution of Eq. (IV.3.73) into Eq. (IV.3.67) give an equation on $\kappa$:

$$R_0 = 2 \int_0^{L/2} \cos \left\{ \bar{\theta} \frac{\sinh(\kappa s)}{\sinh\left(\frac{\kappa L}{2}\right)} \right\} ds \qquad \text{(IV.3.75)}$$

There is no analytical form for this integral, but for given $L$ and $\bar{\theta}$ one can plot it numerically as a function of $\kappa$, and find the latter for a given $R_0$. But this is not necessary here. Indeed, we have already used the assumption that $\bar{\theta}$ is small, and since $|\sinh(\kappa s)| < |\sinh\left(\frac{\kappa L}{2}\right)|$ one may Maclaurin-expand the cosine, take the integral, and thereby obtain a transcendental equation on $\kappa$, which reads:

$$\frac{4}{\bar{\theta}^2} \left[ 1 - \frac{R_0}{L} \right] = \frac{\frac{\sinh(\kappa L)}{\kappa L} - 1}{\left[\sinh\left(\frac{\kappa L}{2}\right)\right]^2} \qquad \text{(IV.3.76)}$$

Had we known that $\kappa L \ll 1$, we could have expanded the r.h.s of Eq. (IV.3.76) and get an analytical solution for $\kappa$,

$$\kappa = \frac{1}{L} \sqrt{30 - \frac{180}{\bar{\theta}^2} \left( 1 - \frac{R_0}{L} \right)} \qquad \text{(IV.3.77)}$$

But we do not know this in advance, and so we generally have to solve Eq. (IV.3.76) for given values of $\bar{\theta}$, $R_0$, and $L$.

In fact, this mechanical 'experiment' is rather contrived, when you control independently both $\bar{\theta}$ and $R_0$. But if you do, and if you also keep $\bar{\theta}$ small but continue decreasing $R_0$, the shape of the rod will change: $\theta(s)$ may become a nonmonotonic function. Our approximation—linearization of Eq. (IV.3.71)—will no longer be valid. If you still tried to use this solution in such a regime, the argument of the square root would turn out negative, $\kappa$ would be imaginary, and Eq. (IV.3.73) would turn into $\theta(s) = -\bar{\theta}\dfrac{\sin(|\kappa|s)}{\sin\left(\frac{|\kappa|L}{2}\right)}$.

However, when we control only $R_0$, and the value of $\bar{\theta}$ adjusts in response to this action, the optimum at $(1 - R_0/L) \ll 1$ will be delivered by a segment of a circle, so that $\bar{\theta}$ can be found from the solution of

$$\frac{R_0}{L} = \frac{\sin\bar{\theta}}{\bar{\theta}} \tag{IV.3.78}$$

[you can prove this, e.g. by transforming Eq. (IV.3.67) into integration over $\theta$, taking into account, in this case, the constant derivative $\frac{ds}{d\theta} = \frac{L}{(2\bar{\theta})}$], and

$$\kappa = \frac{1}{L}\sqrt{30 - \frac{180}{\bar{\theta}^2}\left(1 - \frac{\sin\bar{\theta}}{\bar{\theta}}\right)} \tag{IV.3.79}$$

We will not obtain anything interesting, however, as in this case $\frac{\kappa L}{2} \ll 1$ and

$$\theta(s) = -\bar{\theta}\frac{2}{L}s \tag{IV.3.80}$$

**Exercise.** Solve Eq. (IV.3.76) numerically for $\kappa$ at small values of $\bar{\theta}$ (the condition that we used already) and 'smallish' values of $(1 - R_0/L)$ [otherwise we may be obtaining large values of $\theta(s)$ that are not allowed by the approximation Eq. (IV.3.72)]. For the found values of $\kappa$, plot graphs of $\theta(s)$, using Eq. (IV.3.73). Check under which conditions the rod takes a shape of a segment of a circle, so that you recover the linear law, Eq.(IV.3.80).

## 3.3. Multidimensional Functionals

### General remarks

The notion of a functional can easily be extended to dependence on multidimensional functions, as well as to multifunction functionals.

Let's start by considering the first of these, where our starting definition is easily extended to 3D-vector-dependent functions:

$$U[\psi(\mathbf{r})] = \lim_{\substack{\varepsilon \to 0 \\ N \to \infty \\ M \to \infty \\ L \to \infty}} \{U(\psi_{0,0,0}, \psi_{1,0,0}, \cdots \psi_{j,k,l} \cdots \psi_{N,M,L})\} \quad \text{(IV.3.81)}$$

Here, $\psi_{j,k,l}$ are values on a $N \times M \times L$ grid with spacing $\varepsilon$ that represents a function $\psi(\mathbf{r})$. One example of such a functional, similar to a 'one-dimensional' case, Example IV.3.1, but extended to functions depending on 3D arguments, and written for a general form of $V$ is

$$U[\psi(\mathbf{r})] = \int \left[ \frac{\alpha}{2} \left( \vec{\nabla} \psi(\mathbf{r}) \right)^2 + V(\psi(\mathbf{r})) \right] d\mathbf{r} \quad \text{(IV.3.82)}$$

The functional defined by Eq. (IV.3.82) crops up in the description of a lot of physical systems, and we'll consider it in detail later, in one of the examples.

It is also easy to extend the definition of the functional one that depends on more than one function. For instance,

$$U[f(x), g(x)] = \lim_{\substack{\varepsilon \to 0 \\ N \to \infty}} \{U(f_0, f_1, \ldots . f_N, g_0, g_1, \ldots . g_N)\} \text{ where } \varepsilon N = b$$

$$\text{(IV.3.83)}$$

for a functional of two functions. It is not hard to see how to extend the definition of a functional to depend on as many functions as you like.

### Functional differentiation

Let's see what functional differentiation yields. For Eq. (IV.3.82), the start of the procedure is to consider the small changes $\delta\psi(\mathbf{r})$, $\delta_x\psi(\mathbf{r}) = \partial\delta\psi(\mathbf{r})/\partial x$, $\delta_y\psi(\mathbf{r}) = \partial\delta\psi(\mathbf{r})/\partial y$ and $\delta_z\psi(\mathbf{r}) = \partial\delta\psi(\mathbf{r})/\partial z$ so that we can write

$$\delta U[\psi(\mathbf{r})] = \int \left[ \alpha \left( \frac{\partial \psi(\mathbf{r})}{\partial x} \delta_x \psi(\mathbf{r}) + \frac{\partial \psi(\mathbf{r})}{\partial y} \delta_y \psi(\mathbf{r}) + \frac{\partial \psi(\mathbf{r})}{\partial z} \delta_z \psi(\mathbf{r}) \right) \right.$$
$$\left. + \frac{\partial V(\psi(\mathbf{r}))}{\partial \psi(\mathbf{r})} \delta \psi(\mathbf{r}) \right] d\mathbf{r}$$

Again, to convert $\delta_x \psi(\mathbf{r})$, $\delta_y \psi(\mathbf{r})$ and $\delta_z \psi(\mathbf{r})$ into $\delta \psi(\mathbf{r})$ we need to perform integration by parts, but in a slightly more complicated way. We consider the volume integral over all space in Cartesian coordinates, i.e. $d\mathbf{r} = dx\,dy\,dz$, as well as the limiting case of a region of integration that is a cube of sides $L \to \infty$. Then, to convert $\delta_x \psi(\mathbf{r})$, $\delta_y \psi(\mathbf{r})$, and $\delta_z \psi(\mathbf{r})$ into $\delta \psi(\mathbf{r})$, we integrate by parts with respect to $x$, $y$ and $z$, respectively, to obtain $\delta U[\psi(\mathbf{r})] = \delta U_{surf}[\psi(\mathbf{r})] + \delta \tilde{U}[\psi(\mathbf{r})]$, where

$$\delta U_{surf}[\psi(\mathbf{r})] = \lim_{L \to \infty} \left\{ \int_{-L/2}^{L/2} dy \int_{-L/2}^{L/2} dz \left[ \frac{\partial \psi(\mathbf{r})}{\partial x} \delta \psi(\mathbf{r}) \right]_{x=-L/2}^{x=L/2} \right.$$
$$+ \int_{-L/2}^{L/2} dx \int_{-L/2}^{L/2} dz \left[ \frac{\partial \psi(\mathbf{r})}{\partial y} \delta \psi(\mathbf{r}) \right]_{y=-L/2}^{y=L/2}$$
$$\left. + \int_{-L/2}^{L/2} dx \int_{-L/2}^{L/2} dy \left[ \frac{\partial \psi(\mathbf{r})}{\partial z} \delta \psi(\mathbf{r}) \right]_{z=-L/2}^{z=L/2} \right\}$$

and $\delta \tilde{U}[\psi(\mathbf{r})] = \int \left[ -\alpha \vec{\nabla}^2 \psi(\mathbf{r}) + \frac{\partial V(\psi(\mathbf{r}))}{\partial \psi(\mathbf{r})} \right] \delta \psi(\mathbf{r}) d\mathbf{r}$. Now, we can ensure that $\delta U_{surf}[\psi(\mathbf{r})] = 0$ by fixing $\psi(\mathbf{r})$ at the boundaries, so that $\delta U[\psi(\mathbf{r})] = \delta \tilde{U}[\psi(\mathbf{r})]$. We then obtain the functional derivative of $U[\psi(\mathbf{r})]$ as

$$\frac{\delta U[\psi(\mathbf{r})]}{\delta \psi(\mathbf{r})} = -\alpha \vec{\nabla}^2 \psi(\mathbf{r}) + \frac{\partial V(\psi(\mathbf{r}))}{\partial \psi(\mathbf{r})} \qquad \text{(IV.3.84)}$$

The function that minimizes $U[\psi(\mathbf{r})]$ is then obtained from equalizing $\frac{\delta U[\psi(\mathbf{r})]}{\delta \psi(\mathbf{r})} = 0$, i.e. from

$$-\alpha \vec{\nabla}^2 \psi(\mathbf{r}) + \frac{\partial V(\psi(\mathbf{r}))}{\partial \psi(\mathbf{r})} = 0 \qquad \text{(IV.3.85)}$$

**Example IV.3.5 (Physical Application ▶▶).** To describe charged particles that distributed subject an applied electric field it may be useful to construct functionals containing an electrostatic energy functional as well as the other parts describing short range interaction between the particles and the host medium. Electrostatic energy functional alone, when minimized yields Poisson's equation. Such a functional can be written as Eq. (IV.3.82) with $\alpha = -\varepsilon$, the dielectric constant of the medium, and $V(\psi(\mathbf{r})) = \rho(\mathbf{r})\psi(\mathbf{r})$, where $\rho(\mathbf{r})$ is the local charge density. Substitution of these into Eq. (IV.3.85) indeed yields

$$\vec{\nabla}^2\psi(\mathbf{r}) = -\frac{\rho(\mathbf{r})}{\varepsilon} \qquad (IV.3.86)$$

This equation relates electrostatic potential with a 'given' distribution of charges. Of course charges do not exist themselves, they are related to particles that may be of different sorts, $i$, and the particle density $n_i(\mathbf{r})$. If we had a full energy functional with terms depending on $n_i(\mathbf{r})$ which will describes how energetically costly would be to distribute them this or that way, we would minimize the functionals with respect to $n_i(\mathbf{r})$ as well, and would have obtained more complicated equations than (IV.3.86); for instance with functionals accounting for limitations for accommodation of particles we could come to Poisson-Fermi equation. We will not go into this any further, but just note that such functions would be a typical example of functionals of many functions, each depending on 3D vector-coordinates.

**Example IV.3.6 (Physical Application ►►).** Consider now again the Landau-Ginzburg functional, like in Example IV.3.1, but functions depending on 3D coordinate. This deals with an order parameter $\psi(\mathbf{r})$, as discussed previously this can describe a net alignment of magnetic moments in a magnetic material. Also it can also describe superconductivity, where a $\psi(\mathbf{r}) \neq 0$ describes the appearance of an electron condensate, usually referred to as Cooper pairs, where the material is superconducting (there $\psi(\mathbf{r})$ needs to be described by a complex number, for reasons we won't go into here). Another application is the describing the hydration of macro-ions with charged surfaces, here a real-variable order parameter $\psi(\mathbf{r})$ will describe the orientation of the electrostatic dipole moments of water, when $\psi(\mathbf{r}) \neq 0$. If the dipoles in external field are aligned on average in a particular direction, in the direction of the local value of external field, $h_z(\mathbf{r})$ (subscript $z$ denotes that it is a projection along the direction of the field), we may use a scalar order parameter. The free energy functional is described by Eq. (IV.3.82), now with

$$V(\psi(\mathbf{r})) = -h_z(\mathbf{r})\psi(\mathbf{r}) + a\psi(\mathbf{r})^2 + b\psi(\mathbf{r})^4 \qquad \text{(IV.3.87)}$$

which leads to the Ginzburg-Landau equation,

$$-\alpha\vec{\nabla}^2\psi(\mathbf{r}) + a\psi(\mathbf{r}) + b\psi^3(\mathbf{r}) = h_z(\mathbf{r})$$

In the terminology of Example IV.3.1, $\alpha = \Lambda^2$.

As we already described, if in the absence of the external field, the system is disordered, like e.g. in water, $a > 0$. If $\psi(\mathbf{r})$ describes the mean magnetic moment inside a magnetic material and $h_z(\mathbf{r}) = h$ will play a role of an external magnetic field looking one direction. When $h_z(\mathbf{r})$ is small one can develop a perturbation expansion when $h_z(\mathbf{r})$, with the Green's function techniques we talked about in Chapter 1, Part III, coming into play.

**Example IV.3.7 (Physical Application ▶▶).** Consider the following functional

$$H[\Psi(\mathbf{r})] = \int \left[ \frac{\hbar^2}{4m}\vec{\nabla}\Psi^*(\mathbf{r}).\vec{\nabla}\Psi(\mathbf{r}) + \Psi^*(\mathbf{r})\left[V(\mathbf{r}) - E\right]\Psi(\mathbf{r}) \right] d\mathbf{r}$$

$$\text{(IV.3.88)}$$

One can show that when this functional is minimized one obtains the Schrödinger equation on $\Psi(\mathbf{r})$:

$$\frac{\hbar^2}{2m}\vec{\nabla}^2\Psi^*(\mathbf{r}) = (V(\mathbf{r}) - E)\Psi(\mathbf{r}) \qquad \text{(IV.3.89)}$$

**Exercise.** Prove the above. **Hint:** To do this you may like to write $\Psi(\mathbf{r}) = \psi_r(\mathbf{r}) + i\psi_i(\mathbf{r})$ and consider the minimization conditions

$$\frac{\delta H}{\delta \psi_r} = 0, \quad \frac{\delta H}{\delta \psi_i} = 0 \qquad \text{(IV.3.90)}$$

One can then obtain the Schrödinger equation for $\Psi(\mathbf{r})$ by appropriately combining the two resulting equations.

## 3.4. Finding the Minimum of Functionals via 'Variational Approximation': The Ritz Method

### *What is the variational approximation?*

For functionals that have a minimum (or a maximum) one can use a handy, simplifying approximation, called the *Ritz method*, after Swiss-German mathematician Walter Ritz. He died young, but he is also known for his contri-

Walter Ritz
(1878–1909)

bution to the theory of spectral lines of atoms. Be warned, however, if the functional has saddle points or a complicated set of more than one turning point, this method could fail: one should always test the results of this method.

The idea is to substitute into a functional $U[\psi(\mathbf{r})]$ a trial function $\psi_T(\mathbf{r}; a_1, \ldots a_N)$ of our choosing and invention, with $a_1, \ldots a_N$ being a set of parameters in this function. $U[\psi(\mathbf{r})]$ is, in fact, minimized by the function $\psi^{(0)}(\mathbf{r})$. Thus, if we say that $\psi^{(0)}(\mathbf{r})$ does minimize the functional, we have that

$$U_T(a_1, \ldots a_N) = U[\psi_T(\mathbf{r}; a_1, \ldots a_N)] \geq U[\psi^{(0)}(\mathbf{r})] \qquad \text{(IV.3.91)}$$

We are free to adjust the parameters $a_1, \ldots a_N$ to get the lowest value of $F[\psi_T(\mathbf{r}; a_1, \ldots a_N)]$, which would then give us the best approximation to $F[\psi^{(0)}(\mathbf{r})]$. Thus, we can derive a set of equations to find the best values of $a_1, \ldots a_N$ by differentiating $F_T(a_1, \ldots a_N)$ so that

$$\frac{\partial F_T(a_1, \ldots a_N)}{\partial a_j} = 0 \quad j = 1, \ldots, N \qquad \text{(IV.3.92)}$$

We will need then to solve the set of equations in Eq. (IV.3.92) to find the best choices of $a_1, \ldots a_N$.

Some important advice, before taking up this journey:

(a) Unless we are using variational approximation as a way to start to compute what $\psi^{(0)}(\mathbf{r})$ should be in a purely numerical/ empirical fashion, we should keep $N$, the number of parameters, small. Obviously the more parameters you involve, the closer the $U_T(a_1, \ldots a_N)$ will be to $U[\psi^{(0)}(\mathbf{r})]$. But don't be greedy and keep things simple!

(b) Also, in a physical problem, the variational approximation is most useful as description of what's going on if we can assign a physical meaning to each parameter; any parameter is not just in there to make the approximation more accurate.

(c) Lastly, the trial function used in the variational approximation should better be able to exactly reproduce some limiting case.

Let's look at some examples.

**Example IV.3.8 (Physical Application ▶▶).** Here, we consider a rare case of a lucky choice of a trial function: using variational approximation for finding the ground state energy and wave function of an electron in a hydrogen atom. In the ground state the wavefunction is spherically symmetric and thus depends only on the radial distance from the proton, $r$. The stationary Schrödinger equation is an eigenvalue equation, which reads

$$\hat{H}\psi(r) = E\psi(r) \qquad \text{(IV.3.93)}$$

Here, $E$ is the eigenvalue of energy, and $\hat{H}$ is the Hamiltonian operator

$$\hat{H} = -\frac{\hbar^2}{2mr^2}\frac{d}{dr}\left(r^2\frac{d}{dr}\right) - \frac{e^2}{r} \qquad \text{(IV.3.94)}$$

where $\hbar$ is the Planck constant, $m$ is the electron reduced mass of electron and proton approximately equal to the mass of the electron, and $e$ is the proton charge; the first term is the electron kinetic energy, the second term (written in Gaussian units) is the energy of Coulomb attraction of electron to the proton. The wave function is normalized to unity, i.e.

$$\int_0^\infty 4\pi r^2 \psi(r)\psi^*(r) = 1 \qquad \text{(IV.3.95)}$$

where $*$ stands generally for a complex conjugate. Multiplication of Eq. (IV.3.93) by $\psi^*(r)$ and subsequent spatial integration gives

$$\int_0^\infty dr\, 4\pi r^2 \psi^*(r) \hat{H} \psi(r) = \int_0^\infty dr\, 4\pi r^2 \psi^*(r) E \psi(r) \qquad \text{(IV.3.96)}$$

According to Eq. (IV.3.95), the r.h.s is simply equal to $E$ and so we get the energy as a functional of the wave function,

$$E = \int_0^\infty dr\, 4\pi r^2 \psi^*(r) \hat{H} \psi(r) \qquad \text{(IV.3.97)}$$

Let us apply the variational principle to find the ground state energy of an electron as a minimum of this functional, using the Ritz method. The exact solution of the Schrödinger equation, Eq. (IV.3.93), with the Hamiltonian operator given by Eq. (IV.3.94) exists and you can find it in textbooks on quantum mechanics. But for demonstration of the Ritz method, let us pretend that we know nothing about it, and propose a trial function which should make physical sense. We expect that a wave function must depend on distance, and so our first choice for the trial function would be an exponentially decaying one:

$$\psi(r) = A\, e^{-\alpha r} \qquad \text{(IV.3.98)}$$

(a real one, so complex conjugation hereafter will not be needed). To satisfy Eq. (IV.3.95) we must have

$$A = \sqrt{\frac{\alpha^3}{\pi}} \qquad \text{(IV.3.99)}$$

So, this trial function has only one adjustable parameter, $\alpha$, the inverse of the decay radius. Now, we plug Eq. (IV.3.98) together with Eq. (IV.3.99) into Eq. (IV.3.97) together with Eq. (IV.3.94) and preform all differentiations and integration.

**Exercise.** Show that as a result of these cumbersome but trivial manipulations you get

$$E = \frac{\hbar^2 \alpha^2}{2m} - e^2 \alpha \qquad \text{(IV.3.100)}$$

Now, we need to find the minimum of $E$ as a function of $\alpha$. From $\frac{dE}{d\alpha} = \frac{\hbar^2 \alpha}{m} - e^2 = 0$ we get

$$\alpha = \frac{me^2}{\hbar^2} \tag{IV.3.101}$$

and correspondingly, substituting this result back into Eq. (IV.3.100) we get

$$E = -\frac{me^4}{2\hbar^2} \tag{IV.3.102}$$

Astonishingly, Eqs. (IV.3.101) and (IV.3.102) are what one would obtain from the exact solutions of the Schrödinger equation for a hydrogen atom, where $\alpha^{-1} = \frac{\hbar^2}{e^2 m} = R_B$ is the Bohr radius (in Gaussian units)—see the result of our dimensionality analysis in Section 10.2 of Chapter 10, Part I, Vol. 1. How is it we were so lucky with the primitive variational approach? It is because our 'natural' guess for a trial function eventually coincided with the exact solution of the Schrödinger equation for the ground state wave-function and energy! Not always could we be that lucky.

**Exercise.** Take any other wave function decaying with $r \to \infty$, and do the same manipulations as above. Calculate the energy to see that it will always be higher than that given by Eq. (IV.3.102), to prove yourself by example the universal statement: *the correct, true function will always deliver the lowest minimum of the energy functional!*

**Example IV.3.9 (Physical Application ▶▶).** We now consider one more instructive example, related to linear combination of atomic orbitals (LCAO) theory, some elements of which we have already encountered in Chapters 3 and 4 of Part II, Vol. 1. Consider a molecule composed of two protons and one electron (i.e. $H_2^+$ ion). LCAO suggests we should write its wave-function as a linear combination of two hydrogen-like functions $\psi_1$ and $\psi_2$, each normalized to 1, i.e.

$$\int d\mathbf{r}\, \psi_1^2 = \int d\mathbf{r}\, \psi_2^2 = 1 \tag{IV.3.103}$$

$$\Phi = c_1 \psi_1 + c_2 \psi_2 \tag{IV.3.104}$$

Let us first not bother about normalization of $\Phi$, and just write the energy as an average:

$$E = \frac{\int d\mathbf{r}\, \Phi^* \hat{H}\, \Phi}{\int d\mathbf{r}\, \Phi^* \Phi} \qquad (IV.3.105)$$

We can neglect complex conjugation if we assume $\psi_1$ and $\psi_2$ to be real (in view of heading in the next steps towards the ground state), and write

$$E = \frac{c_1^2 H_{11} + c_1 c_2 H_{12} + c_2 c_1 H_{21} + c_2 H_{22}}{c_1^2 S_{11} + 2 c_1 c_2 S_{12} + c_2 S_{22}} \qquad (IV.3.106)$$

where

$$H_{11} = \int d\mathbf{r}\, \psi_1 \hat{H} \psi_1, \quad H_{22} = \int d\mathbf{r}\, \psi_2 \hat{H} \psi_2$$

$$H_{12} = \int d\mathbf{r}\, \psi_1 \hat{H} \psi_2, \quad H_{21} = \int d\mathbf{r}\, \psi_2 \hat{H} \psi_1$$

$$\qquad (IV.3.107)$$

$$S_{11} = \int d\mathbf{r}\, \psi_1 \psi_1, \quad S_{22} = \int d\mathbf{r}\, \psi_2 \psi_2$$

$$S_{12} = \int d\mathbf{r}\, \psi_1 \psi_2 = S_{21} = \int d\mathbf{r}\, \psi_1 \psi_2$$

Using notations

$$S_{12} = S_{21} \equiv S \qquad (IV.3.108)$$

and noting that following Eq. (IV.3.103)

$$S_{11} = S_{22} \equiv 1 \qquad (IV.3.109)$$

and considering that the atoms composing the molecule are the same (a *symmetric* system),

$$H_{11} = H_{22} \equiv \alpha, \quad H_{12} = H_{21} \equiv \beta \qquad (IV.3.110)$$

we get

$$E = \frac{c_1^2 \alpha + 2 c_1 c_2 \beta + c_2^2 \alpha}{c_1^2 + 2 c_1 c_2 S + c_2^2} \qquad (IV.3.111)$$

Now the simplest version of the variational approach is not to mess around with adjusting the shapes of $\psi_1$ and $\psi_2$, taking those as given, but consider only $c_1$ and $c_2$ as variational parameters. So, the next step will be to find equations on $c_1$ and $c_2$ from

$$\frac{\partial E}{\partial c_1} = \frac{\partial E}{\partial c_2} = 0 \qquad (IV.3.112)$$

One could straightforwardly differentiate the r.h.s of Eq. (IV.3.111), but this would be cumbersome. It is easier to rewrite that equation as $E\left(c_1^2 s + 2c_1 c_2 S + c_2^2 s\right) = c_1^2 \alpha + 2c_1 c_2 \beta + c_2^2 \alpha$, and then differentiate its both sides over $c_1$ and over $c_2$. In view of Eq. (IV.3.112), the derivatives of $E$ in the l.h.s both vanish, and the differentiation gives us $E\left(2c_1 + 2c_2 S\right) = 2c_1 \alpha + 2c_2 \beta$ and $E\left(2c_1 S + 2c_2\right) = 2c_1 \beta + 2c_2 \alpha$. Rewriting this system of equations on $c_1$ and $c_2$ in a canonical form, we obtain

$$\begin{cases} c_1(E - \alpha) + c_2(ES - \beta) = 0 \\ c_1(ES - \beta) + c_2(E - \alpha) = 0 \end{cases} \qquad (IV.3.113)$$

These are eigenvalue equations, which have nontrivial solutions only when

$$\begin{vmatrix} E - \alpha & ES - \beta \\ ES - \beta & E - \alpha \end{vmatrix} = 0 \qquad (IV.3.114)$$

(we have dealt with such equations in Chapters 3 and 4 of Part II, Vol. 1). Calculating the determinant and equalizing it to zero, we get

$$(E - \alpha)^2 - (ES - \beta)^2 = 0 \qquad (IV.3.115)$$

This equation has two solutions:

$$E = E_A = \frac{\alpha + \beta}{1 + S} \qquad (IV.3.116)$$

$$E = E_B = \frac{\alpha - \beta}{1 - S} \qquad (IV.3.117)$$

The substitution of Eq. (IV.3.116) into either of Eqs. (IV.3.113) gives the same result:

$$c_2 = -c_1 \equiv c \qquad (IV.3.118)$$

whereas the substitution of Eq. (IV.3.117) results in

$$c_2 = c_1 \equiv c \qquad \text{(IV.3.119)}$$

If we do the proper calculation of the integrals for $\alpha, \beta$ (a special task, beyond the scope of the analysis here), we will see that they are both negative, and as $S$ is substantially smaller than 1, $E_A > E_B$. The state with the energy $E_B$ and wave-function $\Phi = c(\psi_1 + \psi_2)$ is called a bonding state, and the one with $E_A$ and $\Phi = c(\psi_1 - \psi_2)$ is called antibonding. Constant $c$ is found by normalizing the wave function $\Phi$:

$$\int d\mathbf{r} \; \Phi^2 = 1 \qquad \text{(IV.3.120)}$$

which will then give

$$c = \begin{cases} \sqrt{\dfrac{1}{2(1+S)}}, & \text{bonding state} \\[3mm] \sqrt{\dfrac{1}{2(1-S)}}, & \text{antibonding state} \end{cases} \qquad \text{(IV.3.121)}$$

Thus, minimizing the functional we have obtained two solutions. One is a true ground state (bonding), while the other one is an excited (antibonding) state. If we forget about the physics, it is legitimate to ask a question, if the variational procedure could give us a minimum of the functional, how seriously should we take another solution, if it exists, as in this example? In fact we cannot be sure, and our advice would be to test in each case the result against the exact numerical calculations or experiments. But one should be ready for such situations, as they could reflect the inherent physics of the studied system.

## 3.5.  Key Points

- Functionals relate a function or a set of functions (of scalar or vectorial arguments) with a number.
- If a physical process may be realized by different functions (different ways to proceed), the number that the functional will deliver

for each function, will be a signature of what each of those functions could deliver.

- For instance, if a functional represents the energy (or free energy), you may wish to find out which functions will minimize it. Equally, if the functional stands for the time to travel along different routes, each route represented by a function, you may be interested in which route (which function) provides you the fastest travel time.
- In all those cases, an important role is played by boundary conditions, as the search for extrema of the functional is constrained by them.
- There could be additional constraints on the functions inherent to the functional under study. These are most conveniently treated by using the method of *Lagrange multipliers*.
- When finding the extrema of a functional, in many cases we are just interested in minima, and this leads to the so-called *Euler equations*. Depending on the complexity of the functional, these may have a different order and degree of nonlinearity. You may often be lucky to be able to solve your Euler equation analytically, but in many cases it may not be possible, unless the system parameters are such that you can do some simplifying approximations. But in any case, solving your Euler equation numerically, with the computer, will solve your variational problem.
- Minimizing the functional, you may obtain not one but several solutions of your Euler equation (or a system of equations, if the functional depends on several functions). In that case the first thing that you need to do is to plug the found function(s) back into the functional, calculate the value of the latter, and sort out the solutions in the *order* of delivering the lowest value of the functional. You may wish then to do the stability analysis of each of your solutions and, depending on the results, think which physics, if any, those different solutions represent. Sometimes they do!
- When you have several solutions, you may find that with a tiny change of parameters the solutions swap positions. Such situations take place, when you have phase, structural, or shape *transitions*, which are very interesting to investigate.

- A 'poor man' approach is a variational approximation, or Ritz method. In this approach you make a guess what the general form of your function should be, approximate it with a minimum possible number of parameters, then plug it into the initial functional, and perform corresponding operations of that functional (differentiation, integration, etc.—whatever it involves) and then get an expression which will no longer be a functional, but a function of parameters of your trial function. You can minimize then this function with respect to each of these parameters.

- The more parameters you have in your trial function, the lower will be the minimum of the functional, but the harder will be the 'algebra'. In any case the exact solution of the minimization procedure will always deliver the lowest minimum, lower than any trial function would do, unless its form coincides with the exact solution. You are lucky then, but if you know the exact solution, you do not need the variational approach.

- All such problems are handled by *variational calculus*. The basics of that discipline were set already in the 18th century by great mathematicians of that time. But what we discussed in this chapter was finding extrema of the functionals; in the next two chapters we will deal with the statistics of the fluctuations, considering examples of phenomena related to them. There we will be dealing with functional integrals, the science of which was developed in the 20th century. But even if you deal with the ground state solutions of functionals, it is often easier and more transparent to start the formulation of the physical problem setting the form of the functional, rather than immediately try to construct the corresponding Euler equations.

# Chapter 4

# Path Integrals—A First Look

## 4.1. The Meaning and Evaluation of Simple 1D Path Integrals

### What is a path (or functional) integral?

In Chapter 3 of this part of the book we investigated the idea of functionals and that their minimization could lead to deterministic physical laws encapsulated in Euler (Lagrange) equations. However, in statistical mechanics, dynamic systems, as well as quantum theory, we have to deal with processes that are not deterministic, but rather rely on probability in determining observable quantities.

One of the ways we can start to introduce the idea of a path integral is through the notion of probability distributions (for the 1D case). We can define the following functional

$$F[y(x)] = \lim_{\substack{\varepsilon \to 0 \\ N \to \infty}} \{F(y_0, y_1, \ldots .y_N)\} \qquad \text{(IV.4.1)}$$

Here again, as when we first defined functionals in Chapter 3, we have a set of variables $y_0, y_1, \ldots .y_N$, each corresponding to a particular value of the function $y(x)$ at $x = \varepsilon j$, where integer $0 \le j \le N$; we will just use slightly different notation. Now, we'll choose that $F(y_0, y_1, \ldots .y_N)$ defines an infinite set of probability distributions that determine the probability that each of the variables $y_j$ will take certain values. For illustration, if we supposed that the probability distributions were independent of each other, we could write

$$F(y_0, y_1, \ldots y_N) = F_0(y_0)F_1(y_1) \ldots F_N(y_N) \qquad \text{(IV.4.2)}$$

where, for instance, $F_0(y_0)$ is the probability distribution of the values of $y_0$. But, in most cases of physical interest, these probability distributions are conditional on each other, so we cannot make this factorization. It is this conditional probability of $y_j$ on the pervious value $y_{j-1}$ that necessitates a consideration of *correlations* and *correlation functions*.

Initially we'll be interested in confining the function $y(x)$ within the domain $0 \leq x \leq b$. Let's suppose for simplicity that the range of output values of function $y(x)$ can vary between $-\infty$ and $\infty$, for any input choice $x$. How do we find the probability that all of $y_0, y_1, \ldots y_N$ lie within a range $-a$ to $a$? In this case the probability is given by

$$P_{all} = \prod_{n=0}^{N} \int_{-a}^{a} dy_m F(y_0, y_1, \ldots y_N) \qquad \text{(IV.4.3)}$$

Now, what happens when we take the limits $\varepsilon \to 0$ and $N \to \infty$? We get what is called a *path or functional integral*, which is defined as

$$P_{all} = \int_{y(x)=-a}^{y(x)=a} F[y(x)] Dy(x)$$

$$= \lim_{\substack{\varepsilon \to 0 \\ N \to \infty}} \left\{ \prod_{m=0}^{N} \int_{-a}^{a} dy_m F(y_0, y_1, \ldots y_N) \right\} \qquad \text{(IV.4.4)}$$

Note that when $b = \infty$, we must have that $P = 1$: this normalizes our functional probability distribution $F[y(x)]$. In some physical applications, we may only be interested in the probability distribution of $y(b) = y_b$ given an initial value $y(0) = y_0$, and where all other values $y(x)$ are freely allowed to take any value. In this case, such a probability distribution would be given by

$$K(y_0, y_b) = \int_{y(0)=y_0}^{y(b)=y_b} F[y(x)]Dy(x)$$

$$= \lim_{\substack{\varepsilon \to 0 \\ N \to \infty}} \left\{ \prod_{m=1}^{N-1} \int_{-\infty}^{\infty} dy_m F(y_0, y_1, \dots, y_N) \right\} \quad \text{(IV.4.5)}$$

We talk about how one might evaluate Eq. (IV.4.5) in the next section within the context of statistical mechanics. Here, because of the limits on the functional integral, we require $y(b) = y_b = \lim_{N \to \infty} y_N$ and $y(0) = y_0$, but we may integrate over all other values of the function. We could also have boundary conditions where we deal with fixed $y(b)$ and $y(0)$, or the derivatives at those points; in the former case we may want to fix $K(y_0, y_b) = 1$.

If we know the form of $F[y(x)]$, we can calculate expectation values and correlation functions. The expectation value $\langle h(y(x')) \rangle$ evaluated at the point $x = x'$ is given by

$$\langle h(y(x')) \rangle = \int h(y(x'))F[y(x)]Dy(x)$$

$$= \lim_{\substack{\varepsilon \to 0 \\ N \to \infty}} \left\{ \prod_{n=0}^{N} \int_{-\infty}^{\infty} dy_m h(y_{m'})F(y_0, y_1, \dots, y_N) \right\} \quad \text{(IV.4.6)}$$

where $x' = \varepsilon n'$. For instance, in the simple case that $h(y(x')) = y(x')$ we have that

$$\langle y(x') \rangle = \int y(x')F[y(x)]Dy(x)$$

$$= \lim_{\substack{\varepsilon \to 0 \\ N \to \infty}} \left\{ \prod_{m=0}^{N} \int_{-\infty}^{\infty} dy_m y_{m'} F(y_0, y_1, \dots, y_N) \right\} \quad \text{(IV.4.7)}$$

Following on from the definition of correlation functions, we can write for the two-point correlation function and $n$-point correlation functions the following path integrals:

$$\langle y(x_1)y(x_2)\dots y(x_n)\rangle = \int y(x_1)y(x_2)\dots y(x_n)F[y(x)]Dy(x)$$

$$= \lim_{\substack{\varepsilon\to 0 \\ N\to\infty}} \left\{ \prod_{m=0}^{N} \int_{-\infty}^{\infty} dy_m y_{m_1} y_{m_2}\cdots y_{m_n} F(y_0, y_1, \dots y_N) \right\} \quad \text{(IV.4.8)}$$

### 1D path integrals in statistical mechanics

One of the most straightforward applications of path integrals is in equilibrium *statistical mechanics*, which we will be primarily referring to when illustrating the use of path integrals. We'll also discuss their application to *quantum field theory*, but not to any depth. We should also point out that functional integrals can also be used in (non-equilibrium) statistical dynamics, amongst other applications; we'll illustrate that in one of the examples. Here, we'll start with 1D path integrals, and extend this to 3D later in this chapter.

In statistical mechanics we can write down an explicit form for $F[y(x)]$ in terms of an energy functional $E[y(x)]$, which reads as

$$F[y(x)] = \frac{1}{Z}\exp\left(-\frac{E[y(x)]}{k_B T}\right) \quad\quad \text{(IV.4.9)}$$

where $k_B$ is Boltzmann constant and $T$ is absolute temperature. This statement means that the energy of each configuration $y(x)$ is Boltzmann-weighted, according to equilibrium statistical mechanics, where the partition function is given by

$$Z = \int \exp\left(-\frac{E[y(x)]}{k_B T}\right) Dy(x)$$

$$= \lim_{\substack{\varepsilon\to 0 \\ N\to\infty}} \left\{ \prod_{n=0}^{N} \int_{-\infty}^{\infty} dy_n \exp\left(-\frac{E(y_0, y_1, \dots, y_N)}{k_B T}\right) \right\} \quad \text{(IV.4.10)}$$

where we define

$$E[y(x)] = \lim_{\substack{\varepsilon\to 0 \\ N\to\infty}} \left\{ E(y_0, y_1, \dots y_N) \right\} \quad\quad \text{(IV.4.11)}$$

Note here that $E(y_0, y_1, \ldots .y_N)$ is the discretized form of the energy functional $\mathsf{E}[y(x)]$. One important thermodynamic quantity we might want to calculate is the Helmholtz free energy, a starting point for calculating thermodynamic observables, which we denote here by $A$ (in physical literature usually denoted by $F$, but we have just used this letter already):

$$A = -k_B T \ln Z \qquad (\text{IV.4.12})$$

The form of $E[y(x)]$ may be as complicated as the complexity of the system requires, but as one of the simplest examples, applicable to a number of physical problems, we will consider

$$\frac{\mathsf{E}[y(x)]}{k_B T} = \int_{-\infty}^{\infty} \left[ \frac{l}{2} \left( \frac{dy(x)}{dx} \right)^2 + V(y(x)) \right] dx \qquad (\text{IV.4.13})$$

Here, $y(x)$ might correspond to the azimuthal angle of a semi-flexible polymer, torsionally twisted (at a position $x$ along it), away from its relaxed configuration, and $V(y(x))$ is some external potential depending on that azimuthal angle of orientation. In this context, the authors have actually used Eq. (IV.4.13) in research into assemblies of DNA molecules.

**Think:** What is the form of $E(y_0, y_1, \ldots, y_N)$ corresponding to $\mathsf{E}[y(x)]$ given by Eq. (IV.4.13)?

In the next section, we'll also show a method that one might use to evaluate a 1D partition function up to a constant, with an energy functional of the form of Eq. (IV.4.13). Here, expectation values and correlation functions can be evaluated using Eqs. (IV.4.6)–(IV.4.9). But we can go further and introduce a trick for the evaluation of any of such correlation function. We define a generating functional

$$Z_G[j(x)] = \int \exp\left( -\frac{\mathsf{E}[y(x)]}{k_B T} \right) \exp\left( -\int_{-\infty}^{\infty} j(x) y(x) dx \right) Dy(x)$$
$$(\text{IV.4.14})$$

The correlation functions can then be written in terms of functional derivatives with respect to $j(x)$ in the following way:

$$\langle y(x_1)y(x_2)\ldots .y(x_n)\rangle = \frac{1}{Z}\left.\frac{\delta^n Z_G[j(x)]}{\delta j(x_1)\delta j(x_2)\ldots \delta j(x_n)}\right|_{j=0} \qquad \text{(IV.4.15)}$$

where the limit $j = 0$ on the $n$th functional derivative means that we set all $j(x)$ terms to zero after we have evaluated the functional derivatives. Note that also we have that $Z = Z_G[0]$; the partition function is the generating functional when $j(x) = 0$.

**Exercise.** Using the definition of functional differentiation in Chapter 3, show that

$$\frac{\delta}{\delta j(x_1)}\exp\left(\int_{-\infty}^{\infty} j(x)y(x)dx\right) = y(x_1)\exp\left(\int_{-\infty}^{\infty} j(x)y(x)dx\right)$$
$$\text{(IV.4.16)}$$

Hence, show that Eq. (IV.4.15) holds.

### On functional integration in quantum mechanics: 1D Feynman path integrals

Unfortunately, in quantum mechanics, the application of functional integration is not quite so straightforward, and it makes no sense to try to define $F[y(t)]$ as a probability distribution functional in the same way as in statistical mechanics. Instead of probability distributions and probability, path integrals are used to calculate *probability amplitudes* (we'll see what these are in a moment).

Let's consider a single particle with position $y(t)$ on a 'potential energy landscape' $V(y(t))$. Suppose we want to calculate the probability distribution of it being at position $y_f$ after a time $t = t_f$, initially starting at $t = 0$, namely $K(y_f; y_0, t_f)$. In quantum mechanics this is given by

$$K(y_f; y_0, t_f) = |\langle y_f|\psi(t_f)\rangle|^2 \qquad \text{(IV.4.17)}$$

Here, $\langle y_f |$ is the quantum state where the particle is at $y_f$, and $|\psi(t)\rangle$ is the quantum state that the state $|y_0\rangle$ at time $t = 0$ evolves into by time $t$, according to the time-dependent Schrödinger equation. We can write such an equation not just for wave functions, but also for quantum states:

$$i\hbar \frac{\partial}{\partial t} |\psi\rangle = \hat{H} |\psi\rangle \qquad \text{(IV.4.18)}$$

With the requirement that $|\psi(0)\rangle = |y_0\rangle$, we can write a formal solution to Eq. (IV.4.18). This reads as

$$|\psi\rangle = \exp\left(-\frac{i\hat{H}t}{\hbar}\right) |y_0\rangle \qquad \text{(IV.4.19)}$$

The value of $\langle y_f | \psi(t_f) \rangle$ is, in fact, the value of the time- and position-dependent wave-function $\Psi(y_f, t_f)$ at point $y_f$ and time $t_f$, i.e. $\langle y_f | \psi(t_f) \rangle = \Psi(y_f, t_f)$. At $t = 0$, this wave function is a delta-function-like spike centred at $y_0$, and at later times it spreads out and its maximum value moves according to what the form of $V(y)$ is. We can also consider $\langle y_f | \psi(t_f) \rangle$ as a *probability amplitude* between states $\langle y_f |$ and $|y_0\rangle$ at time $t = t_f$, given by $\langle y_f | \exp(-iHt_f/\hbar) | y_0 \rangle$.

As Feynman showed, this wave function, or probability amplitude, can be written in terms of a path integral:

$$\langle y_f | \exp(-iHt_f/\hbar) | y_0 \rangle = \int_{y(0)=y_0}^{y(t_1)=y_f} \exp\left(\frac{iS[y(t)]}{\hbar}\right) Dy(t) \qquad \text{(IV.4.20)}$$

Here, $S$ stands for what is called an *action* (the quantity discussed in Chapter 3 of Part III, Vol. 2):

$$S[y(t)] = \int_0^{t_f} dt \left[ \frac{m}{2} \left(\frac{dy(t)}{dt}\right)^2 - V(y(t)) \right] \qquad \text{(IV.4.21)}$$

We can also use path integrals to compute the normalized time-ordered expectation values $\langle 0 | \hat{T}[\hat{y}(t_1), \hat{y}(t_2) \ldots \hat{y}(t_n)] | 0 \rangle$. We'll briefly describe what these are (though knowing about this is not essential, when just learning the principles of mathematical methods; for a keen reader we'd recommend looking into textbooks on quantum mechanics to familiarize yourself with some of the basics, such as Heisenberg and Schrödinger pictures, operators, states and time evolution). First, in each expectation value we have Heisenberg picture operators $\hat{y}(t_j) = \exp\left(\frac{it_j\hat{H}}{\hbar}\right)\hat{y}_S \exp\left(-\frac{it_j\hat{H}}{\hbar}\right)$, where $\hat{y}_S$ is the Schrödinger picture position operator that satisfies $\hat{y}_S|y_j\rangle = y_j|y_j\rangle$, where $|y_j\rangle$ are stationary states that have definite positions $y_j$. Here, $\langle 0|$ and $|0\rangle$ represent the ground (vacuum) state, and $\hat{T}$ is a time-ordering operator, which orders the operators that act on the state $|0\rangle$, so that the one at the earliest time acts first, then the one at the next earliest time, and so on. Here, the physical details of such expectation values are not actually that important to us. In our discussion of the application of path integration, we focus on $\langle 0|\hat{T}[\hat{y}(t_1)\hat{y}(t_2)\ldots\hat{y}(t_n)]|0\rangle$. This may be considered as the quantum mechanical version of an $n$-point correlation function; it correlates the 'measured' positions of the particle at various times, which starts in its ground state and finally reverts back to its ground state. We can calculate these expectation values in a similar way as we did when calculating correlation functions in statistical mechanics, which is again to use a generating functional:

$$(i)^n \langle 0|\hat{T}[\hat{y}(t_1),\hat{y}(t_2)\ldots\hat{y}(t_n)]|0\rangle = \frac{\delta^n \ln \tilde{Z}[j(t)]}{\delta j(t_1)\delta j(t_2)\ldots\delta j(t_n)} \quad \text{(IV.4.22)}$$

Here, the generating functional is a little 'odder' than the one we considered before in statistical mechanics. It takes the form

$$\tilde{Z}[j(t)] = \int_C dt \left[\frac{m}{2}\left(\frac{dy(t)}{dt}\right)^2 - V(y(t))\right] + \int_{t_{\max}}^{t_{\min}} j(t)y(t)dt \quad \text{(IV.4.23)}$$

In this expression, $t$ is analytically continued into the complex plane and $t_{\min} < t_1, t_2 \ldots t_n < t_{\max}$. We can, however, introduce straightaway imaginary time $\tau = it$, and using the properties of analyticity in the complex plane, continue our calculated functions back to real time. In this case, we could have used similar forms to Eqs. (IV.4.13), (IV.4.14) and (IV.4.15), where $x$ would have been replaced by $\tau$, $l$ with $m$, and $k_BT$ by $\hbar$ to describe the quantum mechanics of a particle moving in 1D under the potential profile $V(y)$. One can then calculate the values of $\langle 0|\hat{T}[\hat{x}(\tau_1), \hat{x}(\tau_2) \dots \hat{x}(\tau_n)]|0\rangle$ through Eq. (IV.4.15).

### *Homogenous Gaussian path integrals in free space*

What we mean by *homogeneous path integrals* with respect to the functions $y(x)$ are those where all the $x$-dependence is contained only in those functions. By Gaussian we mean that the energy functional $\mathrm{E}[y(x)]$, which is the continuum limit of an $N$-variable function $E(y_0, y_1, \dots y_N)$, is a quadratic function of all its variables. One of the simplest, but non-trivial, examples of such integrals is

$$Z = \int \exp\left(-\frac{\mathrm{E}_0[y(x)]}{k_B T}\right) Dy(x) \qquad \text{(IV.4.24)}$$

where

$$\frac{\mathrm{E}_0[y(x)]}{k_B T} = \frac{1}{2} \int_{-\infty}^{\infty} \left[ l\left(\frac{dy(x)}{dx}\right)^2 + \alpha y(x)^2 \right] dx \qquad \text{(IV.4.25)}$$

We will now show you how you can evaluate such a functional integral, or related ones concerned with averaging the functions of $y(x)$ or with obtaining expressions for correlation functions. You will see that you will have to do a set of manipulations, none of which are 'rocket science', but performing them in the right order requires training and discipline.

In order to evaluate functional integral Eq. (IV.4.24) with $\mathrm{E}_0[y(x)]$ defined by Eq. (IV.4.25) we can use a Fourier transform:

$$y(x) = \frac{1}{2\pi} \int_{-\infty}^{\infty} \tilde{y}(k) \exp(-ikx) dk \qquad \text{(IV.4.26)}$$

We can then recast Eq. (IV.4.25) as

$$\frac{\mathrm{E}_0[y(x)]}{k_B T} = \frac{\tilde{\mathrm{E}}_0[\tilde{y}(k)]}{k_B T} = \frac{1}{4\pi} \int_{-\infty}^{\infty} \tilde{y}(k) \left[ lk^2 + \alpha \right] \tilde{y}(-k) dk$$

$$= \frac{1}{4\pi} \int_{-\infty}^{\infty} \tilde{y}(k) \left[ lk^2 + \alpha \right] \tilde{y}^*(k) dk \qquad \text{(IV.4.27)}$$

and we have that

$$\int \exp\left(-\frac{\mathrm{E}_0[y(x)]}{k_B T}\right) Dy(x) \propto \int \exp\left(-\frac{\tilde{\mathrm{E}}_0[\tilde{y}(k)]}{k_B T}\right) D\tilde{y}(k)$$

(IV.4.28)

In the path integral, we have supposed that $Dy(x) = D\tilde{y}(k)$, i.e. that the functional Jacobian can be set equal to an ill-defined constant that is not important to us. We'll justify Eq. (IV.4.28) later on, when we consider more generalized forms of the Gaussian integrals.

As we require $y(x)$ to be real we must also require that $\tilde{y}(-k) = \tilde{y}^*(k)$; this restricts the functional degrees of freedom that we integrate over. We can write $y(k) = a(k) + ib(k)$. Now $a(k)$ and $b(k)$, defined in the domain $0 < k < \infty$ for the path integration, can be treated independently. As $a(k) = a(-k)$ and $b(k) = -b(-k)$, the values for which $-\infty < k < 0$ cannot be treated as independent. This means that Eq. (IV.4.28) becomes

$$Z = \int_{k>0} \exp\left(-\frac{1}{4\pi}\int_0^\infty \tilde{a}(k)[lk^2 + \alpha]\tilde{a}(k)dk\right) D\tilde{a}(k)$$

$$\times \int_{k>0} \exp\left(-\frac{1}{4\pi}\int_0^\infty \tilde{b}(k)\left[lk^2 + \alpha\right]\tilde{b}(k)dk\right) D\tilde{b}(k)$$

Next, we can 'discretize' $k$ and write

$$\int_{k>0} \exp\left(-\frac{1}{4\pi}\int_0^\infty \tilde{a}(k)\left[lk^2 + \alpha\right]\tilde{a}(k)dk\right) D\tilde{a}(k)$$

$$= \lim_{\substack{L\to\infty \\ 2\pi N/L\to\infty}} \left\{\prod_{m=0}^N \int_{-\infty}^\infty da_m \exp\left(-\frac{1}{2L}a_m\left[l\left(\frac{2\pi m}{L}\right)^2 + \alpha\right]a_m\right)\right\}$$

(IV.4.29)

$$\int_{k>0} \exp\left(-\frac{1}{4\pi}\int_0^\infty \tilde{b}(k)\left[lk^2 + \alpha\right]\tilde{b}(k)dk\right) D\tilde{b}(k)$$

$$= \lim_{\substack{L\to\infty \\ 2\pi N/L\to\infty}} \left\{\prod_{m=0}^N \int_{-\infty}^\infty db_m \exp\left(-\frac{1}{2L}b_m\left[l\left(\frac{2\pi m}{L}\right)^2 + \alpha\right]b_m\right)\right\}$$

(IV.4.30)

Here, we've supposed that $-L/2 < x < L/2$, where $L$ is the 'length' of our system. So that when we take $L \to \infty$, we get back the original range of values that $x$ may take, as in Eq. (IV.4.25). We see that on the r.h.s we simply have a set of Gaussian integrals to do.

**Exercise.** Perform the integration

$$\int_{-\infty}^{\infty} da_m \exp\left(-\frac{1}{2L} a_m \left[l\left(\frac{2\pi m}{L}\right)^2 + \alpha\right] a_m\right) \qquad \text{(IV.4.31)}$$

to show that

$$Z = \lim_{\substack{\varepsilon \to 0 \\ \varepsilon N \to \infty}} \left\{ \prod_{m=0}^{N} \frac{(2\pi)L}{\left[l\left(\frac{2\pi m}{L}\right)^2 + \alpha\right]} \right\} \qquad \text{(IV.4.32)}$$

We can then compute the free energy

$$A = -k_B T \lim_{\substack{\varepsilon \to 0 \\ \varepsilon N \to \infty}} \left\{ \ln \left[ \prod_{m=0}^{N} \frac{(2\pi)L}{\left[l\left(\frac{2\pi m}{L}\right)^2 + \alpha\right]} \right] \right\}$$

$$= k_B T \lim_{\substack{L \to \infty \\ 2\pi N/L \to \infty}} \left\{ \sum_{m=0}^{N} \ln\left[l\left(\frac{2\pi m}{L}\right)^2 + \alpha\right] - (N+1)\ln(2\pi L) \right\}$$

$$\text{(IV.4.33)}$$

Obviously, if we were to take the required limits, the free energy per unit length $A/L$ would be divergent. This effect is called 'ultraviolet catastrophe'. But there are two ways of dealing with this.

One is to provide what is termed a cut-off, a maximum value of $k, k_{\max} = 2\pi N/L$ above which the sum in Eq. (IV.4.33) is truncated. This procedure is physically justified because the continuum variation of $y(x)$ is only an approximation to real statistical mechanical systems; in most applications, at the nanoscale level, we are dealing with molecular groups held together by bonds. Another way to deal with the divergence is to... 'disguise' it: we just measure the free energy with respect to the free energy $A_0$ of a reference state where $\alpha = 0$. In this second way of dealing with such divergences, we write the free energy difference per unit length as

$$\lim_{L\to\infty}\left\{\frac{A-A_0}{L}\right\}=k_BT\lim_{\substack{L\to\infty\\2\pi N/L\to\infty}}\left\{\frac{1}{L}\sum_{m=0}^{N}\ln\left[1+\frac{\alpha}{l}\left(\frac{L}{2\pi m}\right)^2\right]\right\}$$

$$=\frac{k_BT}{2\pi}\int_0^\infty\ln\left[1+\frac{\alpha}{lk^2}\right]dk \qquad (IV.4.34)$$

Though the integrand is singular at $k=0$, the singularity is harmless as it is logarithmic and can be integrated over.

**Exercise.** Using the Leibniz integral rule we introduced in Chapter 7 of Part II (Vol. 1), by differentiating with respect to $\alpha$ (and integrating it back) show that

$$\frac{1}{2\pi}\int_0^\infty\ln\left[1+\frac{\alpha}{lk^2}\right]dk=\frac{1}{2}\sqrt{\frac{\alpha}{l}} \qquad (IV.4.35)$$

### *Calculating expectation values with homogeneous Gaussian path integrals*

Let's now consider how we might calculate any expectation value $\langle h(y(x'))\rangle$ with Eqs. (IV.4.6), (IV.4.9) (with $E[y(x)]$ replaced by $E_0[y(x)]$ given by Eqs. (IV.4.25) and (IV.4.24), where the function $h(y(x'))$ is still arbitrary. First, combining these equations, the path integral for finding such expectation values reads as

$$\langle h\left(y\left(x\right)\right)\rangle=\frac{1}{Z}\int h(y(x))\exp\left(-\frac{E_0[y(x)]}{k_BT}\right)Dy(x) \qquad (IV.4.36)$$

Here, we can play a rather useful trick; we can rewrite this functional integral with the help of a delta-function, using later its integral representation. Namely,

$$\langle h(y(x))\rangle=\int_{-\infty}^\infty d\bar{y}\,h(\bar{y})\left\{\frac{1}{Z}\int\delta\left(\bar{y}-y(x)\right)\exp\left(-\frac{E_0[y(x)]}{k_BT}\right)Dy(x)\right\}$$

$$=\int_{-\infty}^\infty d\bar{y}\,h(\bar{y})\langle\delta(\bar{y}-y(x))\rangle \qquad (IV.4.37)$$

Let us now evaluate $\langle \delta(\bar{y} - y(x)) \rangle = \{ \frac{1}{Z} \int \delta(\bar{y} - y(x)) \exp(-\frac{E_0[y(x)]}{k_B T})$
$Dy(x) \}$, by using integral representation of the delta-function

$$\delta(\bar{y} - y(x)) = \frac{1}{2\pi} \int_{-\infty}^{\infty} \exp\{ip[\bar{y} - y(x)]\}dp \qquad \text{(IV.4.38)}$$

Thus, we can write

$$\langle \delta(\bar{y} - y(x)) \rangle = \frac{1}{2\pi Z} \int_{-\infty}^{\infty} dp \int \exp(ip(\bar{y} - y(x)))$$

$$\times \exp\left(-\frac{E_0[y(x)]}{k_B T}\right) Dy(x) \qquad \text{(IV.4.39)}$$

Using Eq. (IV.4.25), we see that we now need to evaluate the path integral that reads as

$$G(p) = \int \exp(-ipy(x)) \exp\left(-\frac{1}{2} \int_{-\infty}^{\infty} \left[ l\left(\frac{dy(x)}{dx}\right)^2 + \alpha y(x)^2 \right] \right) Dy(x)$$

$$\text{(IV.4.40)}$$

We can evaluate $G(p)$ by again expressing $y(x)$ in terms of its Fourier transform so that

$$G(p) = \int \exp\left\{ -\frac{ip}{4\pi} \left( \int_{-\infty}^{\infty} \left[ \tilde{y}(k)e^{ikx} + \tilde{y}(-k)e^{-ikx} \right] dk \right) \right\}$$

$$\times \exp\left\{ -\frac{1}{4\pi} \int_{-\infty}^{\infty} \tilde{y}(-k) \left[ lk^2 + \alpha \right] \tilde{y}(k) dk \right\} D\tilde{y}(k)$$

$$\text{(IV.4.41)}$$

We can then re-express $G(p)$ as

$$G(p) = \int D\tilde{y}(k) \exp\left( -\frac{1}{4\pi} \int_{-\infty}^{\infty} \left( \tilde{y}(-k) + e^{ikx}[lk^2 + \alpha]^{-1} \right) \right.$$

$$\times [lk^2 + \alpha](\tilde{y}(k) + e^{-ikx}[lk^2 + \alpha]^{-1})dk \bigg)$$

$$\times \exp\left( -\frac{p^2}{4\pi} \int_{-\infty}^{\infty} [lk^2 + \alpha]^{-1} dk \right) \qquad \text{(IV.4.42)}$$

Now, we can make the functional variable shift and write

$G(p)$ $=$ $\exp\left(-\frac{p^2}{4\pi}\int_{-\infty}^{\infty}[lk^2+\alpha]^{-1}dk\right)\int DY(k)\exp\left(-\frac{1}{4\pi}\int_{-\infty}^{\infty}\right.$ $(Y(-k))[lk^2+\alpha]Y(k)dk)$, from where, recalling the expression given in Eq. (IV.4.29), we get

$$G(p) = Z\exp\left(-\frac{p^2}{4\pi}\int_{-\infty}^{\infty}[lk^2+\alpha]^{-1}dk\right) \qquad \text{(IV.4.43)}$$

**Exercise.** By considering the definition of the path integral as the limit of $N$ multiple integrations, show that we can make a variable shift in each integral: $Y(k) = \tilde{y}(k) + e^{-ikx}\left[lk^2+\alpha\right]^{-1}$, and for $G(p)$ we get Eq. (IV.4.43).

**Exercise.** Show that

$$\frac{1}{2\pi}\int_{-\infty}^{\infty}[lk^2+\alpha]^{-1}dk = \frac{1}{2\sqrt{\alpha l}} \qquad \text{(IV.4.44)}$$

Then, using Eqs. (IV.4.44) and (IV.4.43) we can rewrite Eq. (IV.4.39) as

$$\langle\delta(\bar{y}-y(x))\rangle = \frac{1}{2\pi}\int_{-\infty}^{\infty}dp\exp(ip\bar{y})\exp\left(-\frac{p^2}{4\sqrt{\alpha l}}\right) \qquad \text{(IV.4.45)}$$

You already know how to do this integral, so you should be able to show (do this as an Exercise) that

$$\langle\delta(\bar{y}-y(x))\rangle = \sqrt{\left(\frac{1}{\pi}\right)}\alpha^{1/2}l^{1/2}\exp(-l^{1/2}\alpha^{1/2}\bar{y}^2) \qquad \text{(IV.4.46)}$$

Thus, we obtain

$$\langle h(y(x))\rangle = \sqrt{\left(\frac{1}{\pi}\right)}\alpha^{1/2}l^{1/2}\int_{-\infty}^{\infty}h(\bar{y})\exp\left(-l^{1/2}\alpha^{1/2}\bar{y}^2\right)d\bar{y} \qquad \text{(IV.4.47)}$$

The integral in Eq. (IV.4.47) is no longer a functional integral—it is just an ordinary integral; for any given form of $h(\bar{y})$ it will give you the result as a function of $\sqrt{l\alpha}$. Looking at it, you can immediately conclude that odd dependence on $\bar{y}$ will give you zero. This is the way it should be, because the $E_0[y(x)]$-functional did not involve any preferential sign of $y$; in particular, you, of course, get $\langle y(x) \rangle = 0$. For some of simple forms of $h(y)$ you will be able to get an analytical result. For others, the integral in Eq. (IV.4.47) will itself define a function.

**Exercise.** Get $\langle h(y(x)) \rangle$ for $h(y) = y^2$. **Hint:** you may start with the integral $\int_{-\infty}^{\infty} \exp(-p \cdot \bar{y}^2) d\bar{y}$ and differentiate it over $p$ once, to get the integral for which you need the answer.

### *Calculating correlation functions with homogeneous Gaussian path integrals*

Lastly, let's see how we can evaluate the two-point correlation function given by

$$\langle y(x')y(x) \rangle = \frac{1}{Z} \int y(x')y(x) \exp\left(-\frac{E_0[y(x)]}{k_B T}\right) Dy(x) \quad \text{(IV.4.48)}$$

Here, again, $E_0[y(x)]$ is given through Eq. (IV.4.25). There are quite a few ways of calculating this. We will present one in which we'll introduce sources $j(x)$, as was done using Eqs. (IV.4.14) and (IV.4.15). The generating functional is then given by

$$Z_G[j(x)] = \int \exp\left(-\int_{-\infty}^{\infty} j(x)y(x)dx\right)$$

$$\times \exp\left(-\frac{1}{2}\int_{-\infty}^{\infty}\left[l\left(\frac{dy(x)}{dx}\right)^2 + \alpha y(x)^2\right]\right) Dy(x)$$

$$\text{(IV.4.49)}$$

To evaluate the path integral given here, it is convenient again to use Fourier transforms, where now we have

$$j(x) = \frac{1}{2\pi}\int_{-\infty}^{\infty} dk \tilde{j}(k) \exp(-ikx) \quad \text{(IV.4.50)}$$

So that we may write

$$Z_G[j(x)] = \int \exp\left(-\frac{1}{4\pi}\left(\int_{-\infty}^{\infty}[\tilde{y}(k)\tilde{j}(-k) + \tilde{y}(-k)\tilde{j}(k)]\,dk\right)\right)$$

$$\times \exp\left(-\frac{1}{4\pi}\int_{-\infty}^{\infty}\tilde{y}(-k)[lk^2 + \alpha]\tilde{y}(k)\right)D\tilde{y}(k)$$

$$(\text{IV.4.51})$$

This can be reformatted as

$$Z_G[j(x)] = \int D\tilde{y}(k)\exp\left(-\frac{1}{4\pi}\int_{-\infty}^{\infty}(\tilde{y}(-k) + \tilde{j}(-k)[lk^2 + \alpha]^{-1})\right.$$

$$\times [lk^2 + \alpha](\tilde{y}(k) + \tilde{j}(k)[lk^2 + \alpha]^{-1})dk\Big)$$

$$\times \exp\left(\frac{1}{4\pi}\int_{-\infty}^{\infty}j(-k)[lk^2 + \alpha]^{-1}j(k)dk\right) \qquad (\text{IV.4.52})$$

By yet again changing functions in the functional integration, this time so that $\tilde{Y}(k) = \tilde{y}(k) + \tilde{j}(k)[lk^2 + \alpha]^{-1}$ we may perform the functional integration so that

$$Z_G[j(x)] = Z\exp\left(-\frac{1}{4\pi}\int_{-\infty}^{\infty}j(-k)[lk^2 + \alpha]^{-1}j(k)dk\right)$$

$$= Z\exp\left(\frac{1}{2}\int_{-\infty}^{\infty}\int_{-\infty}^{\infty}j(x)G(x-x')j(x')dx'dx\right) \quad (\text{IV.4.53})$$

where

$$G(x - x') = \frac{1}{2\pi}\int_{-\infty}^{\infty}dk\frac{\exp(ik(x-x'))}{lk^2 + \alpha} \qquad (\text{IV.4.54})$$

**Exercise.** Using contour integration in Eq. (IV.4.54), show that

$$G(x - x') = \frac{1}{2\sqrt{\alpha l}}\exp\left(-\sqrt{\frac{\alpha}{l}}\cdot|x - x'|\right) \qquad (\text{IV.4.55})$$

Now, through both Eqs. (IV.4.53) and (IV.4.15) we can write

$$\langle y(x')y(x)\rangle = \left.\frac{\delta^2 \ln(Z_G[j(x)])}{\delta j(x')\delta j(x)}\right|_{j=0} = \frac{\delta^2}{\delta j(x')\delta j(x)}$$

$$\times \exp\left(\frac{1}{2}\int_{-\infty}^{\infty}\int_{-\infty}^{\infty} j(\tilde{x})G(\tilde{x}-\tilde{x}')j(\tilde{x}')d\tilde{x}'d\tilde{x}\right)\Bigg|_{j=0}$$

$$(\text{IV.4.56})$$

**Exercise.** Using the definition of functional differentiation show that

$$\langle y(x')y(x)\rangle = G(x-x') \qquad (\text{IV.4.57})$$

What we see is that the correlation function for the Gaussian correlations is the Green's function that satisfies

$$-l\frac{d^2G(x-x')}{dx^2} + \alpha G(x-x') = \delta(x-x') \qquad (\text{IV.4.58})$$

Note that Eqs. (IV.4.36) and (IV.4.48) could describe quantum mechanical motion in a harmonic potential, in terms of imaginary time $x = \tau = it$, if we replace $k_BT$ with $\hbar$. As stated before, the same equation has application in the statistical mechanics of twisting semi-rigid polymers.

**Example IV.4.1 (Physical Application ▶▶).**
Let us consider overdamped one-dimensional Brownian motion along $x$-coordinate, in harmonic potential, $V(x)$, using 1D Gaussian path integration. Such motion can also be described through the Langevin equation (named after an outstanding French physicist Paul Langevin, known for his contributions to the theory of magentism and dynamics of random processes -to which this equation belongs, and also as active anti-Fascist).

Paul Langevin
(1872–1946)

$$kx + \gamma\frac{dx}{dt} = \hat{\xi}(t) \qquad (\text{IV.4.59})$$

where $\hat{\xi}(t)$ is an uncorrelated Gaussian random field, such that

$$\langle \hat{\xi}(t)\hat{\xi}(t')\rangle = 2\gamma k_BT\delta(t-t') \qquad (\text{IV.4.60})$$

We could find the correlation functions by finding the Green's function of Eq. (IV.4.59) and performing the appropriate averaging of $\hat{\xi}(t)$. But in the spirit of this chapter, we will show how to calculate the correlations functions through the following generating functional:

$$Z[j(t)] = \exp\left(-\frac{1}{4\gamma k_B T}\int_{-\infty}^{\infty}\left(kx + \gamma\frac{dx}{dt}\right)^2 dt - \int_{-\infty}^{\infty}j(t)x(t)\right)$$

$$(IV.4.61)$$

**Exercise.** Repeating the steps of Eqs. (IV.4.49)–(IV.4.53) and Eq. (IV.4.61) show that

$$Z[j(t)] = Z\exp\left(-\frac{1}{2}\int_{-\infty}^{\infty}\int_{-\infty}^{\infty}j(t)G(t-t')j(t')dxdx'\right) \quad (IV.4.62)$$

where

$$G(t - t') = \frac{\gamma k_B T}{\pi}\int_{-\infty}^{\infty}\frac{\exp(-i\omega(t-t'))}{k^2 + \gamma^2\omega^2} \quad (IV.4.63)$$

Then, using contour integration find the expression for $G(t - t')$.

It is easy to see from application of the procedure described by Eq. (IV.4.56) that $\langle x(t)x(t')\rangle = G(t-t')$. You may also have spotted that the cross-term $\int_{-\infty}^{\infty}2k\gamma x\frac{dx}{dt}dt$ vanishes; indeed, $\int_{-\infty}^{\infty}2k\gamma x\frac{dx}{dt}dt = 2k\gamma\int_{-\infty}^{\infty}\frac{dx^2}{dt}dt = 2k\gamma(x^2(\infty) - x^2(-\infty)) = 0$. So in practice we could have simply dealt with a path integral of the form of Eq. (IV.4.49).

We can generalize the 1D path integral approach to overdamped Brownian motion in any potential $V(x)$. In this case the generating functional reads as

$$Z[j(t)] = \exp\left(-\frac{1}{4\gamma k_B T}\int_{-\infty}^{\infty}\left(\frac{dV(x(t))}{dx} + \gamma\frac{dx(t)}{dt}\right)^2 dt\right.$$

$$\left. + \frac{1}{2\gamma}\int_{-\infty}^{\infty}\frac{d^2V(x(t))}{dx^2} - \int_{-\infty}^{\infty}j(t)x(t)\right) \quad (IV.4.64)$$

## 4.2. Another Approach to Evaluating 1D Path Integrals in Statistical Mechanics

In addition to what we have presented above, here we consider an alternative, more straightforward (but equivalent) method that can be used to evaluate path integrals. It will give us the same result, using less mathematical trickery.

We start by providing formulae that we can use to evaluate both Eq. (IV.4.12) and Eq. (IV.4.5). It will be useful to define the un-normalized distribution function

$$W(y_0, y_b) = Z \cdot K(y_0, y_b) \qquad \text{(IV.4.65)}$$

First we observe that we can write both $W(y_0, y_N)$ and the partition function $Z$ as

$$W(y_0, y_b) = \lim_{\substack{\varepsilon \to 0 \\ N \to \infty}} W_d(y_0, y_N) \qquad \text{(IV.4.66)}$$

$$Z = \lim_{\substack{\varepsilon \to 0 \\ N \to \infty}} \int_{-\infty}^{\infty} dy_0 \int_{-\infty}^{\infty} dy_N W_d(y_0, y_N) \qquad \text{(IV.4.67)}$$

where $W_d(y_{i'}, y_i)$ has the property

$$W_d(y_{i'}, y_i) = \int_{-\infty}^{\infty} W_d(y_{i'}, y_{i-1}) W_d(y_{i-1}, y_i) dy_{i-1} \quad \text{where } j' < j - 1 \qquad \text{(IV.4.68)}$$

Now what is $W_d(y_{i-1}, y_i)$? It is obtained from the form of $E(y_0, y_1, \ldots . y_N)$ and Eq. (IV.4.68), which you should have deduced; we left it as something for you to think about. We find that for the energy functional given by Eq. (IV.4.13) it is of the form

$$W_d(y_{i-1}, y_i) = \exp\left(-\frac{l(y_i - y_{i-1})^2}{2\varepsilon} - \varepsilon V(y_i)\right) \qquad \text{(IV.4.69)}$$

**Exercise.** Show that the path integral for $Z$ with $E[y(x)]$ given by Eq. (IV.4.13) can indeed be written from Eqs. (IV.4.67), (IV.4.68) and (IV.4.69).

Now first of all, we can re-express Eq. (IV.4.69) as

$$W_d(y_{i-1}, y_i) \propto \int_{-\infty}^{\infty} dp \exp\left(-\frac{\varepsilon p^2}{2l} + ip(y_i - y_{i-1}) - \varepsilon V(y_i)\right)$$

$$(IV.4.70)$$

The constant of proportionality is not important, so we have omitted it from Eq. (IV.4.70), though we suggest, as an additional exercise, that you find out what it is. We can expand out the exponential, for small $\varepsilon$, so that

$$W_d(y_{j-1}, y_j) \propto \int_{-\infty}^{\infty} dp \exp(ipy_{i-1}) \left[1 - \frac{\varepsilon p^2}{2l} - \varepsilon V(y_i)\right] \exp(-ipy_i)$$

$$= \int_{-\infty}^{\infty} dp \exp(ipy_{i-1}) \left[1 + \frac{\varepsilon}{2l} \frac{d^2}{dy_i^2} - \varepsilon V(y_i)\right] \exp(-ipy_i)$$

$$(IV.4.71)$$

Next we'll need eigenvalues and functions of the differential operator $\hat{E} = -\frac{\varepsilon}{2l} \frac{d^2}{dy_i^2} + \varepsilon V(y_i)$, which satisfy

$$\hat{E} \chi_j(y_i) = E_j \chi_j(y_i) \qquad (IV.4.72)$$

In terms of the eigenfunctions, we can express

$$\exp(-ipy_i) = \sum_{j=0}^{\infty} c_i(p) \chi_j(y_i) \qquad (IV.4.73)$$

where

$$c_i(p) = \int_{-\infty}^{\infty} dy_i \exp(-ipy_i) \chi_j^*(y_i) \qquad (IV.4.74)$$

Now, importantly, we have the following orthogonality condition between the eigenfunctions (as dictated by the Sturm–Liouville theory that we looked at in Chapter 4 of Part III):

$$\int_{-\infty}^{\infty} \chi_j^*(y_i) \chi_{j'}(y_i) dy_i = \delta_{j,j'} \qquad (IV.4.75)$$

This allows us to write

$$W_d(y_{i-1}, y_i) \propto \sum_{j=0}^{\infty} \sum_{j'=0}^{\infty} \int_{-\infty}^{\infty} dp\, c_j^*(p) \chi_j^*(y_{i-1}) \exp(-\varepsilon E_j) c_{j'}(p) \chi_{j'}(y_i)$$

$$\text{(IV.4.76)}$$

Now, using the definition of $c_{j'}(p)$, it is possible to show that

$$\int_{-\infty}^{\infty} dp\, c_j^*(p) c_{j'}(p) = 2\pi \delta_{j,j'} \qquad \text{(IV.4.77)}$$

This allows us to recast Eq. (IV.4.76) as

$$W_d(y_{i-1}, y_i) = \sum_{j=0}^{\infty} \chi_j^*(y_{i-1}) \exp(-\varepsilon E_j) \chi_j(y_i) \qquad \text{(IV.4.78)}$$

Next, we can compute $W(y_{i-2}, y_i)$, by using Eq. (IV.4.68). This gives us

$$W_d(y_{i-2}, y_i) = \sum_{j'=0}^{\infty} \sum_{j=0}^{\infty} \int_{-\infty}^{\infty} dy_{i-1} \chi_{j'}^*(y_{i-2})$$

$$\times \exp(-\varepsilon E_{j'}) \chi_{j'}(y_{i-1}) \chi_j^*(y_{i-1}) \exp(-\varepsilon E_j) \chi_j(y_i)$$

$$\text{(IV.4.79)}$$

Using Eq. (IV.4.75), we can then show that

$$W_d(y_{i-2}, y_i) = \sum_{j=0}^{\infty} \chi_{j'}^*(y_{i-2}) \exp(-2\varepsilon E_j) \chi_j(y_i) \qquad \text{(IV.4.80)}$$

Thus, from Eqs. (IV.4.66) and (IV.4.68) we get

$$W(y_0, y_b) \propto \lim_{\substack{\varepsilon \to 0 \\ N \to \infty}} \left\{ \sum_{j=0}^{\infty} \chi_j^*(y_0) \exp(-\varepsilon N E_j) \chi_j(y_N) \right\}$$

$$= \sum_{j=0}^{\infty} \chi_j^*(y_0) \exp(-b E_j) \chi_j(y_b) \qquad \text{(IV.4.81)}$$

where $b = N\varepsilon$. The partition function and $K(y_0, y_\infty)$ are then obtained by simply applying Eqs. (IV.4.65) and (IV.4.67). We see that due to the exponential factor, the larger eigenvalues, $E_j$, contribute less than the lower ones. When $b = \infty$ only the lowest eigenvalue contributes and we obtain

$$\lim_{b \to \infty} \{K(y_0, y_b)\} = \frac{\chi_0^*(y_0)\chi_0(y_b)}{\int_{-\infty}^{\infty} dy_0 \int_{-\infty}^{\infty} dy_b \chi_0^*(y_0)\chi_0(y_b)} \lim_{b \to \infty} \left\{ \frac{A}{b} \right\}$$

$$= k_B T E_0 + \Xi \qquad (IV.4.82)$$

The constant $\Xi$ is undetermined, but is unimportant as free energy is usually measured with respect to a reference one $A_0$, which defined as the one for $V(y(x)) = 0$.

**Exercise.** By working through similar steps as in Eqs. (IV.4.69)–(IV.4.81), and using the definitions of the path integrals, show that we can write the following expressions for the mean square expectation value and the two-point correlation function:

$$\langle y(x)^2 \rangle = \frac{1}{Z} \int_{-\infty}^{\infty} dy_0 \int_{-\infty}^{\infty} dy_x \int_{-\infty}^{\infty} dy_b W(y_0, y_x) y_x^2 W(y_x, y_b)$$

$$(IV.4.83)$$

$$\langle y(x)y(x') \rangle = \frac{1}{Z} \int_{-\infty}^{\infty} dy_0 \int_{-\infty}^{\infty} dy_x \int_{-\infty}^{\infty} dy_{x'}$$

$$\times \int_{-\infty}^{\infty} dy_b W(y_0, y_x) y_x W(y_x, y_{x'}) y_{x'} H(y_{x'}, y_b)$$

$$(IV.4.84)$$

where $W(y_x, y_{x'}) = \sum_{j=0}^{\infty} \chi_j^*(y_x) \exp(-(x' - x)E_j)\chi_j(y_{x'})$. Hence, show that in the limit $b \to \infty$, where $xE_j \gg 1$ and $x'E_j \gg 1$, we obtain

$$\langle y(x)^2 \rangle = \int_{-\infty}^{\infty} dy_x \chi_0(y_x) y_x^2 \chi_0^*(y_x) \qquad (IV.4.85)$$

$$\langle y(x)y(x') \rangle = \sum_{j=0}^{\infty} \int_{-\infty}^{\infty} dy_x \int_{-\infty}^{\infty} dy_{x'} \chi_0(y_x) y_x \chi_j^*(y_x) e^{-(x'-x)(E_j - E_0)}$$

$$\times \chi_j(y_{x'}) y_{x'} \chi_0^*(y_0) \qquad (IV.4.86)$$

**Example IV.4.2 (Physical Application ▶▶).** Let's look at an example where we can apply this method, as well as extend what we have learnt. Random Gaussian polymers (which we will encounter again in the last chapter of this book), and diffusing particles obey random walk statistics; for the former, the probability distribution function satisfies the equation which has the same form as diffusion equation

$$\frac{\partial P(n, \mathbf{r})}{\partial n} = \frac{b^2}{6} \vec{\nabla}^2 P(n, \mathbf{r}) \qquad \text{(IV.4.87)}$$

where $n$ marks the $n$th segment or 'monomer' in the polymer chain, and $b$ is the monomer length. Now, we suppose that we have initial condition

$$P(0, \mathbf{r}) = G(\mathbf{r}', \mathbf{r}, 0) = \delta(\mathbf{r} - \mathbf{r}') \qquad \text{(IV.4.88)}$$

This initial condition states that we already know the position of the zeroth segment of our polymer: it is at $\mathbf{r} = \mathbf{r}'$. But what will it likely be at $n \neq 0$? The reader should recall, from Chapter 4 of Part III, that $G(\mathbf{r}, \mathbf{r}', n)$ was what we defined as the *diffusion kernel* or *propagator*. Let's find it. Making a supposition that

$$G(\mathbf{r}', \mathbf{r}, n) = \sum_j \chi_j(\mathbf{r}) \exp(-nE_j) \chi_j^*(\mathbf{r}') \qquad \text{(IV.4.89)}$$

we substitute this form into Eqs. (IV.4.87) and (IV.4.88), which will yield a requirement

$$-\frac{b^2}{6} \vec{\nabla}^2 \chi_j(\mathbf{r}) = E_j \chi_j(\mathbf{r}) \quad \text{and} \quad \sum_j \chi_j(\mathbf{r}) \chi_j^*(\mathbf{r}') = \delta(\mathbf{r} - \mathbf{r}')$$
$$\text{(IV.4.90)}$$

for Eq. (IV.4.89) to be a solution to Eqs. (IV.4.87) and (IV.4.88).

This is all looking a bit familiar—indeed you should be asking: could we represent $G(\mathbf{r}, \mathbf{r}'n)$ with a path integral? The answer is indeed yes! Let's start by considering the following path integral

$$W(\mathbf{r}_0, \mathbf{r}_n) = \int \exp\left(-\frac{3}{2b^2}\int_0^n dn' \left(\frac{\partial \mathbf{r}}{\partial n'}\right)^2\right) D\mathbf{r}(n) \qquad \text{(IV.4.91)}$$

We can first write

$$W(\mathbf{r}_0, \mathbf{r}_n) = \lim_{\substack{\varepsilon \to 0 \\ N \to \infty}} \int_{-\infty}^{\infty} dy_0 \int_{-\infty}^{\infty} dy_n W_d(\mathbf{r}_0, \mathbf{r}_n) \qquad \text{(IV.4.92)}$$

where

$$W_d(\mathbf{r}_{i'}, \mathbf{r}_i) = \int_{-\infty}^{\infty} W_d(\mathbf{r}_{i'}, \mathbf{r}_{i-1}) W_d(\mathbf{r}_{i-1}, \mathbf{r}_i) d\mathbf{r}_{i-1} \quad \text{for} \quad i' < i - 1 \qquad \text{(IV.4.93)}$$

and

$$W_d(\mathbf{r}_{i-1}, \mathbf{r}_i) = \exp\left(-\frac{3}{2b^2\varepsilon}(\mathbf{r}_i - \mathbf{r}_{i-1})^2\right) \propto \int d\mathbf{p} \exp\left(-\frac{b^2\varepsilon}{6}\mathbf{p}^2\right)$$

$$\times \exp(i\mathbf{p}.(\mathbf{r}_i - \mathbf{r}_{i-1})) \qquad \text{(IV.4.94)}$$

Indeed, in this particular case of a non-interacting Gaussian chain without any confinement, we already have the eigenfunctions and values of Eq. (IV.4.90) in Eq. (IV.4.94):

$$\chi_{j_p}(\mathbf{r}_i) = \exp(i\mathbf{p}.\mathbf{r}_i) \quad \text{and} \quad E_{j_k} = \frac{\mathbf{p}^2 b^2}{6} \qquad \text{(IV.4.95)}$$

With confinement of the polymer things get less trivial, as Eq. (IV.4.95) no longer holds, and we'd be forced to expand out $\exp(i\mathbf{p}.\mathbf{r}_i)$ in terms of the functions that solve Eq. (IV.4.90) with the confining boundary condition.

**Exercise.** Find $W(\mathbf{r}_{i-2}, \mathbf{r}_i)$ from Eq. (IV.4.94) using that

$$W(\mathbf{r}_{i-2}, \mathbf{r}_i) = \int d\mathbf{r}_{i-1} W(\mathbf{r}_{i-2}, \mathbf{r}_{i-1}) W(\mathbf{r}_{i-1}, \mathbf{r}_i) \qquad \text{(IV.4.96)}$$

Hence, show that

$$W(\mathbf{r}', \mathbf{r}) \propto G(\mathbf{r}', \mathbf{r}, n) = \int d\mathbf{p} \exp(i\mathbf{p}.(\mathbf{r} - \mathbf{r}')) \exp\left(-\frac{n\, b^2 \mathbf{p}^2}{6}\right) \qquad \text{(IV.4.97)}$$

Then, show that this evaluates to

$$G(\mathbf{r}', \mathbf{r}, n) = \left(\frac{3}{2\pi b^2 n}\right)^{3/2} \exp\left(-\frac{3(\mathbf{r} - \mathbf{r}')^2}{2b^2 n}\right) \qquad \text{(IV.4.98)}$$

This is what one would expect for the propagator of the diffusion-type equation, Eq. (IV.4.87). Indeed, replace $n \Rightarrow t/\tau$, where $t$ is time and $\tau$ an elementary time-step diffusive random walk, and replace $b^2/6\tau \Rightarrow D$, where $b$ is the average length of the step, and $D$ is the diffusion coefficient, we get the classical diffusion propagator equation:

$$G(\mathbf{r}', \mathbf{r}, t) = \left(\frac{1}{4\pi Dt}\right)^{3/2} \exp\left(-\frac{(\mathbf{r} - \mathbf{r}')^2}{4Dt}\right) \qquad \text{(IV.4.99)}$$

You may check that Eq. (IV.4.98) directly satisfies both Eqs. (IV.4.87) and (IV.4.88). We have thus shown that we can represent a Gaussian chain with a path integral. Why should one bother? The answer is that the path integral approach allows us to easily add interactions between monomers in Eq. (IV.4.91) in terms of an additional Boltzmann weight, and thus study the effect using standard path integral techniques.

## 4.3. Path Integrals of Functions Dependent on More than One Variable

### *Definition*

It is straightforward to extend path/functional integration to any number of dimensions. In 3D integration of a functional $\Xi[\psi(\mathbf{r})]$ over a cubic region $-b \le x \le b, -b \le y \le b, -b \le z \le b$, is defined as

$$\int \Xi[\psi(\mathbf{r})]D\psi(\mathbf{r}) = \lim_{\substack{\varepsilon \to 0 \\ \varepsilon J = b \\ \varepsilon K = b \\ \varepsilon L = b}} \prod_{j=-J}^{J} \prod_{k=-K}^{K} \prod_{l=-L}^{L} \int_{-\infty}^{\infty} d\psi_{j,k,l}$$

$$\times \chi(\psi_{-J,-K,-L}, \cdots \psi_{0,0,-1}, \psi_{0,0,0}, \psi_{0,0,1} \cdots \cdots \psi_{J,K,L})$$

$$\text{(IV.4.100)}$$

where $\Xi[\psi(\mathbf{r})]$ is defined as

$$\Xi[\psi(\mathbf{r})] = \lim_{\substack{\varepsilon \to 0 \\ \varepsilon J = b \\ \varepsilon K = b \\ \varepsilon L = b}} \chi(\psi_{-J,-K,-L}, \cdots \psi_{0,0,-1}, \psi_{0,0,0}, \psi_{0,0,1} \cdots \psi_{J,K,L})$$

$$(IV.4.101)$$

The l.h.ss of both Eqs. (IV.4.100) and (IV.4.101) are defined on a 3D lattice, so that $\mathbf{r} = \varepsilon j \hat{\mathbf{i}} + \varepsilon k \hat{\mathbf{j}} + \varepsilon l \hat{\mathbf{k}}$, where the continuum limit corresponds to making the 'lattice spacing' $\varepsilon$ infinitesimally small and the number of lattice sites infinite.

## Two basic applications of path integration over multivariable functions

**Example IV.4.3 (Physical Application ▶▶).** This refers to an application of 3D path integrals in statistical mechanics. What is missing from the Ginzburg–Landau theory is the possibility that thermal fluctuations may create inhomogeneities in the order parameter. Far away from the phase transition these effects don't really matter, and the order parameter can simply be chosen to minimize the Ginzburg–Landau free energy functional, as in the 1D case study. However, very close to the phase transition such fluctuations play an important role, so the result that we obtained in Example IV.3.1 close to the transition point (in that example, at $a \to 0$) was actually misleading. These fluctuations modify the behaviour of thermodynamic quantities, such as the specific heat and susceptibility to an external (magnetic) field, calculated from the partition function. To start to take account of such fluctuations in the order parameter, we write down the following partition function

$$Z = \int \exp\left[-\frac{F[\psi(\mathbf{r})]}{k_b T}\right] D\psi(\mathbf{r}) \qquad (IV.4.102)$$

where

$$F[\psi(\mathbf{r})] = \int d\mathbf{r} \left[\frac{1}{2}(\vec{\nabla}\psi(\mathbf{r}))^2 + \frac{\alpha}{2}\psi(\mathbf{r})^2 + \frac{\beta}{4!}\psi(\mathbf{r})^4\right] \qquad (IV.4.103)$$

Eqs. (IV.4.102) and (IV.4.103) define what is referred to the Wilson–Ginzburg–Landau theory (using the names of three Nobel laureates).

Sometimes it is also useful to extend the theory to consider an $N$-component order parameter, $\boldsymbol{\Psi}(\mathbf{r}) = (\psi_1(r), \psi_2(r) \ldots \psi_N(r))$. The simplest extension, for such an order parameter, is to consider

Kenneth Wilson
(1936–2013)

$$Z = \int \exp\left[-\frac{F[\boldsymbol{\Psi}(\mathbf{r})]}{k_b T}\right] D\boldsymbol{\Psi}(\mathbf{r}) \qquad (IV.4.104)$$

$$F[\boldsymbol{\Psi}(\mathbf{r})] = \int d\mathbf{r} \left[\frac{1}{2}\vec{\nabla}\boldsymbol{\Psi}(\mathbf{r}).\vec{\nabla}\boldsymbol{\Psi}(\mathbf{r}) + \frac{\alpha}{2}\boldsymbol{\Psi}(\mathbf{r}).\boldsymbol{\Psi}(\mathbf{r}) + \frac{\beta}{4!}\left(\boldsymbol{\Psi}(\mathbf{r}).\boldsymbol{\Psi}(\mathbf{r})\right)^2\right]$$
$$(IV.4.105)$$

This theory is symmetric under rotations in the $N$-dimensional space spanned by the vector $\boldsymbol{\Psi}(\mathbf{r})$. More complicated $N$-component theories break this symmetry.

Functional integrals also have application in 4D. In quantum field theories (QFT) we have the additional *dimension of time*. We have already considered the idea of analytically extending time into the complex plane to do calculations; this is usually fine when just dealing with particle excitations above the ground vacuum state (absolute temperature $T = 0$) provided that the functions are analytic functions of a complex variable. The ground (zero chemical potential) state is the vacuum state with no (real) particles, and all excited states are those that contain particles. Again it's much better to work with purely imaginary time when doing calculations and then analytically continue back to real time when computing observables. When dealing with finite temperatures, in the so-called *Matsubara formalism*, this procedure can be a bit trickier; we won't have space to go into those problems.

Usually, in QFT, we are tasked with finding expectation values $\langle 0|T\{\hat{\varphi}(\mathbf{r}_1, t_1), \hat{\varphi}(\mathbf{r}_2, t_2) \ldots \hat{\varphi}(\mathbf{r}_n, t_n)\}|0\rangle$, where $\hat{T}$ stands for time ordering. These can be obtained by first calculating this average for real $\tau$, where $t_j = i\tau_j$ and then analytically continuing back to real times by replacing $\tau_j$ with $-it_j$.

The $\hat{\varphi}(\mathbf{r}, t)$ are referred to as field operators, which we will briefly describe. First we can write $\hat{\varphi}(\mathbf{r}, t) = \hat{\varphi}^+(\mathbf{r}, t) + \hat{\varphi}^-(\mathbf{r}, t)$. Here,

$\hat{\varphi}^-(\mathbf{r}_1, t_1)$ creates a single particle at position $\mathbf{r}_1$ and time $t_1$; when acting on the vacuum state $|0\rangle$ it promotes it into a one particle state $|\mathbf{r}_1, t_1\rangle$. If we act on $|\mathbf{r}_1, t_1\rangle$ with another operator $\hat{\varphi}^-(\mathbf{r}_2, t_2)$, we obtain the two-particle state $|\mathbf{r}_2, t_2; \mathbf{r}_1, t_1\rangle$, and so on. Now what about $\hat{\varphi}^+(\mathbf{r}_1, t_1)$? When acting to the right it acts as a destruction operator, destroying a particle at position $\mathbf{r}_1$ and time $t_1$. If it acts on the vacuum state, we get nothing, i.e. $\hat{\varphi}^+(\mathbf{r}_1, t_1)|0\rangle = 0$. On the other hand, if it acts on states to the left, it acts as a creation operator in the same way as $\hat{\varphi}^-(\mathbf{r}_2, t_2)$ acting to the right. Also note that $\hat{\varphi}^-(\mathbf{r}_2, t_2)$ acting to the left acts a destruction operator.

Why do we want to calculate such quantities as $\langle 0|\hat{T}\{\hat{\varphi}(\mathbf{r}_1, t_1), \hat{\varphi}(\mathbf{r}_2, t_2) \ldots \hat{\varphi}(\mathbf{r}_n, t_n)\}|0\rangle$? What do they tell us? In fact, such expectation values are probability amplitudes between various initial and final multiple particle states. We can use these to calculate the probability that a particular $n$-particle state will be in another $m$-particle state at later times. We can illustrate this by considering $\langle 0|T\{\hat{\varphi}(\mathbf{r}_1, t_1), \hat{\varphi}(\mathbf{r}_2, t_2)\}|0\rangle$, which is referred to as the propagator. We've used the same name for what emerged from the solution of the diffusion equation; this is by no means coincidental, and we invite you to think why that is so, by thinking about the similarities between the time-dependent Schrödinger equation and the diffusion equation. Using the fact that we can write $\hat{\varphi}(\mathbf{r}, t) = \hat{\varphi}^+(\mathbf{r}, t) + \hat{\varphi}^-(\mathbf{r}, t)$ and based on what we have discussed, it's not hard to see that we can write

$$\langle 0|\hat{T}\{\hat{\varphi}(\mathbf{r}_1, t_1), \hat{\varphi}(\mathbf{r}_2, t_2)\}|0\rangle = \langle 0|\hat{\varphi}^+(\mathbf{r}_1, t_1), \hat{\varphi}^-(\mathbf{r}_2, t_2)]|0\rangle\theta(t_1 - t_2)$$
$$+ \langle 0|\hat{\varphi}^+(\mathbf{r}_2, t_2), \hat{\varphi}^-(\mathbf{r}_1, t_1)]|0\rangle\theta(t_2 - t_1)$$
$$\text{(IV.4.106)}$$

If we suppose that $t_2 > t_1$, we can then write

$$\langle 0|\hat{T}\{\hat{\varphi}(\mathbf{r}_1, t_1), \hat{\varphi}(\mathbf{r}_2, t_2)\}|0\rangle = \langle 0|\hat{\varphi}^+(\mathbf{r}_2, t_2), \hat{\varphi}^-(\mathbf{r}_1, t_1)]|0\rangle$$
$$= \langle \mathbf{r}_2, t_2|\mathbf{r}_1, t_1\rangle \qquad \text{(IV.4.107)}$$

This is indeed the *probability amplitude* that starting with a particle at position $\mathbf{r}_1$ and time $t_1$, we'll end up with a particle at position

$\mathbf{r}_2$ and time $t_2$. The *probability distribution* of having a particle at positions $\mathbf{r}_2$ at time $t_2$, when we have a particle initially at position $\mathbf{r}_1$ at time $t_1$ is given by $P(\mathbf{r}_2, t_2; \mathbf{r}_1, t_1) = |\langle \mathbf{r}_2, t_2 | \mathbf{r}_1, t_1 \rangle|^2$.

We'll now discuss one of the simplest of QFTs, termed $\phi^4$-theory. $\phi^4$-theory is a field theory initially intended to describe relativistic particles, but unfortunately it is too simple to be a QFT describing the real world. However, it is useful in its relative simplicity to start with, to get to grips with field theoretical calculations. Also this field theory has a phase transition, which is interpreted as the vacuum-changing state (a generalized, more complicated version of such quantum field theory was applied to describe the legendary Higgs boson, in which below such a phase transition particles acquire mass, being 'massless' above the transition). Its analytic continuation into $\tau = it$ also has an important application in relation to the 3D Wilson–Ginzburg–Landau theory. This application is realized in terms of what is referred to as the $\varepsilon$-expansion, where $\varepsilon = 4-d$ (where $d$ stands for the dimension of space we work with). This expansion is used in so-called *renormalization group theory*, to obtain reasonable estimates of the singular behaviour of thermodynamic quantities close to the phase transition. However, we will not touch on this further, as we still have plenty of ground to cover. We are coming close to the end of the book, and testing your patience too much ... and our patience and strength in writing this lengthy tome.

In terms of the $\phi^4$ theory we can write

$$\langle 0 | \hat{T} \{ \hat{\varphi}(\mathbf{r}_1, \tau_1), \hat{\varphi}(\mathbf{r}_2, \tau_2) \dots \hat{\varphi}(\mathbf{r}_n, \tau_n) \} | 0 \rangle$$

$$= \int \varphi(\mathbf{r}_1, \tau_1) \varphi(\mathbf{r}_2, \tau_2) \dots \varphi(\mathbf{r}_n, \tau_n) \exp(-S[\varphi(\mathbf{r}, \tau)]) D\varphi(\mathbf{r}, \tau)$$

$$\text{(IV.4.108)}$$

where

$$S[\varphi(\mathbf{r}, \tau)] = \int_{-\infty}^{\infty} d\tau \int d\mathbf{r} \left[ \frac{1}{2} (\vec{\nabla}\varphi(\mathbf{r}, \tau))^2 + \frac{1}{2} \left( \frac{\partial \varphi(\mathbf{r}, \tau)}{\partial \tau} \right)^2 \right.$$

$$\left. + \frac{\alpha}{2} \varphi(\mathbf{r}, \tau)^2 + \frac{\beta}{4!} \varphi(\mathbf{r}, \tau)^4 \right] \qquad \text{(IV.4.109)}$$

Here the parameter $\alpha$ is the mass of particles when the mean field $\langle\varphi(\mathbf{r},t)\rangle$ is zero, and $\beta$ defines the strength of interactions between them. In writing Eqs. (IV.4.108) and (IV.4.109) we have chosen a system of units, often called natural units (although to a person less experienced with quantum electrodynamics they would not sound such!), where we set $\hbar = c = 1$; these are chosen for convenience, but such constants may be restored through dimensional analysis at the end of the calculation. In textbooks and papers it usually seen in a more compact notation:

$$\partial_\mu\varphi(x)\partial^\mu\varphi(x) = (\vec{\nabla}\varphi(\mathbf{r},\tau))^2 + \left(\frac{\partial\varphi(\mathbf{r},\tau)}{\partial\tau}\right)^2 \qquad \text{(IV.4.110)}$$

It is very important to note that here $x$ *is not a scalar*; it represents the 4-component vector $(\mathbf{r},\tau)$, the subscript $\mu$ is an index which numbers one of the four components of $(\mathbf{r},\tau)$ over which the summations in the l.h.s is implied, and $\partial_\mu\varphi(x)$ and $\partial^\mu\varphi(x)$ represent differentiation with a particular component of $(\mathbf{r},\tau)$. Traditionally, the $\mu = 0$ component is chosen to be $\tau$.

Indeed, we see that it is identical to Wilson–Ginzburg–Landau theory if we defined it in 4D (the only very small difference being that we have $\hbar = 1$ instead of $1/k_BT$). For the majority of the remaining examples, we'll be focusing on the 3D Wilson–Ginzburg–Landau theory. However, we simply need only add an extra dimension when considering this quantum field theory. For instance, as well as integrating in Fourier space over wave vector $\mathbf{k}$, we also have to integrate over frequency $\omega$. Though there is one caveat to this: going to 4D creates technical problems by causing more divergent integrals. Both in 3D and 4D, this has to be dealt with in a process called *renormalization*; in relation to 4D we'll invite you to study more advanced texts if you are eager to know more and really practise it. The ideas of the 4D renormalization process can also be very useful in 3D close to phase transitions, to deal with small-$k$ wave-number components, the so-called infra-red divergences, which appear in integrals at the phase transition. It is actually these divergent integrals that radically change the behaviour of thermodynamic quantities near a phase transition, causing what is called critical behaviour. Here, we're only interested in introducing some of the basics of path integration, so that again interested readers can consult more advanced texts.

## 4.4. The Saddle Point Approximation

### *The basic idea and saddle point free energy calculation*

Now, let's look at our Wilson–Ginsburg–Landau/$\phi^4$ theory. Generally, we expand both $F[\psi(\mathbf{r})]$ and $S[\varphi(\mathbf{r}, \tau)]$ about the saddle point solutions, $\psi_0(\mathbf{r})$ and $\varphi_0(\mathbf{r}, \tau)$. These correspond to

$$\frac{\delta F[\psi(\mathbf{r})]}{\delta \psi(\mathbf{r})}\bigg|_{\psi(\mathbf{r})=\psi_0(\mathbf{r})} = 0 \qquad \frac{\delta S[\varphi(\mathbf{r}, t)]}{\delta \varphi(\mathbf{r}, t)}\bigg|_{\phi(\mathbf{r},t)=\phi_0(\mathbf{r},t)} = 0 \quad \text{(IV.4.111)}$$

Actually, in spite of the name, these solutions don't really correspond to a saddle point, at least for both these functionals; rather they occur where the weights $\exp(-F[\psi(\mathbf{r})]/k_B T)$ and $\exp(-S[\varphi(\mathbf{r}, t)])$ are maximized. What we are going to do is, in fact, the functional analogue of Laplace's method that we became acquainted with and practised in Chapter 7 of Part II, Vol. 1: we will expand the functionals $F[\psi(\mathbf{r})]$ and $S[\varphi(\mathbf{r}, t)]$ up to the second order in $\delta\psi(\mathbf{r}) = \psi(\mathbf{r}) - \psi_0(\mathbf{r})$ and $\delta\varphi(\mathbf{r}, t) = \varphi(\mathbf{r}, t) - \varphi_0(\mathbf{r}, t)$, respectively. Below, we consider it for 3D, and look when this approximation is valid and when we may need corrections.

We'll demonstrate this approximation by working with the 3D Wilson–Ginsburg–Landau theory described by the functional given in Eqs. (IV.4.102) and (IV.4.103). Expanding out $F[\psi(\mathbf{r})]$ we find to second order in $\delta\psi(\mathbf{r})$

$$F[\psi(\mathbf{r})] = F[\psi_0(\mathbf{r})] + \Delta F[\psi_0(\mathbf{r}), \delta\psi(\mathbf{r})] \qquad \text{(IV.4.112)}$$

where

$$\Delta F[\psi_0(\mathbf{r}), \delta\psi(\mathbf{r})]$$
$$= \int d\mathbf{r} \left[\frac{1}{2}(\vec{\nabla}\delta\psi(\mathbf{r}))^2 + \frac{\alpha}{2}\delta\psi(\mathbf{r})^2 + \frac{\beta}{4}\psi_0(\mathbf{r})^2\delta\psi(\mathbf{r}_0)^2\right] \quad \text{(IV.4.113)}$$

For the moment, we'll suppose that there are no boundary conditions on $\psi_0(\mathbf{r})$ and it extends throughout all space. In this case, its Euler equation is trivially satisfied by the value

$$\psi_0(\mathbf{r}) = \overline{\psi} \qquad (IV.4.114)$$

which minimizes the Landau free energy function $F_0(\overline{\psi}) = \frac{\alpha\overline{\psi}^2}{2} + \frac{\beta\overline{\psi}^4}{4!}$. We have

$$\overline{\psi}^2 = 0 \quad \text{for} \quad \alpha \geq 0 \quad \text{and} \quad \overline{\psi}^2 = -6\alpha/\beta \quad \text{for} \quad \alpha < 0 \quad (IV.4.115)$$

We can then readily perform the functional integration. To do so we introduce the 3D Fourier transform

$$\delta\psi(\mathbf{r}) = \frac{1}{(2\pi)^3} \int_{-\infty}^{\infty} \tilde{\psi}(\mathbf{k}) \exp(-i\mathbf{k}.\mathbf{r}) d\mathbf{k} \qquad (IV.4.116)$$

(for brevity we omit symbol $\delta$ in front of $\tilde{\psi}(\mathbf{k})$). This allows us to write

$$\Delta F[\delta\psi(\mathbf{r})] = \Delta\tilde{F}[\tilde{\psi}(\mathbf{k})]$$

$$= \frac{1}{(2\pi)^3} \int d\mathbf{k} \left[\frac{1}{2}\mathbf{k}^2 + \frac{\alpha}{2} + \frac{\beta}{4}\overline{\psi}^2\right] \tilde{\psi}(-\mathbf{k})\tilde{\psi}(\mathbf{k})$$

$$(IV.4.117)$$

Note that as $\delta\psi(\mathbf{r})$ is real, we require that $\tilde{\psi}(-\mathbf{k}) = \tilde{\psi}^*(\mathbf{k})$ (where the star symbol stands, as usual, for a complex conjugate) and we can write $\tilde{\psi}(\mathbf{k}) = \psi_1(\mathbf{k}) + i\psi_2(\mathbf{k})$. Note that, because $\tilde{\psi}(-\mathbf{k}) = \tilde{\psi}^*(\mathbf{k})$, all the independent degrees of freedom of both $\psi_1(\mathbf{k})$ and $\psi_2(\mathbf{k})$ can either be chosen for $k_z > 0$ or for $k_z < 0$, i.e. $\mathbf{k}$ is defined in only the upper or lower half-hemisphere of all $k$-space. However, we can play a trick and introduce a new field $\Psi(\mathbf{k}) = \psi_1(\mathbf{k})\theta(k_z) + \psi_2(\mathbf{k})\theta(-k_z)$. Simply what this boils down to is replacing the complex fields $\tilde{\psi}(\mathbf{k})$ and $\tilde{\psi}(-\mathbf{k})$ with $\Psi(\mathbf{k})$.

This means we can re-express Eq.(IV.4.117) as

$$\Delta\tilde{F}[\Psi(\mathbf{k})] = \frac{1}{2(2\pi)^3} \int d\mathbf{k} \left[\frac{1}{2}\mathbf{k}^2 + \frac{\alpha}{2} + \frac{\beta}{4}\overline{\psi}^2\right] \Psi(\mathbf{k})^2 \qquad (IV.4.118)$$

where the path integral for the partition function is

$$Z = \int \exp\left(-\frac{\Delta F[\delta\psi(\mathbf{r})]}{k_B T}\right) D\psi(x) = \int \exp\left(-\frac{\Delta \tilde{F}[\Psi(\mathbf{k})]}{k_B T}\right) D\Psi(\mathbf{k})$$

(IV.4.119)

The path integral can then be discretized with the values

$$\mathbf{k} = \lim_{L\to\infty} \mathbf{k}_{l,m,n} \quad \text{where} \quad \mathbf{k}_{l,m,n} = \left(\frac{2\pi l}{L}\right)\hat{\mathbf{i}} + \left(\frac{2\pi m}{L}\right)\hat{\mathbf{j}} + \left(\frac{2\pi n}{L}\right)\hat{\mathbf{k}}$$

(IV.4.120)

Thus, we may write

$$Z = \lim_{L\to\infty} \prod_{l,m,n=0}^{\infty} \left[\int_{-\infty}^{\infty} \exp\left(-\frac{\delta F_{dis}(\Psi_{l,m,n})}{k_B T}\right) d\Psi_{l,m,n}\right] \quad \text{(IV.4.121)}$$

where

$$\delta F_{dis}(\Psi_{l,m,n}(\mathbf{k})) = \frac{1}{L^3}\left[\frac{1}{2}k_{l,m,n}^2 + \frac{\alpha}{2} + \frac{\beta}{4}\bar{\psi}^2\right]\Psi_{l,m,n}^2 \quad \text{(IV.4.122)}$$

and $\Psi_{l,m,n} = \Psi(\mathbf{k}_{l,m,n})$.

**Exercise.** By performing each of the Gaussian integrations in Eq. (IV.4.121), writing $\mathbf{k}$ in spherical polar coordinates $\{k, \theta_k, \phi_k\}$, and taking the limit $L \to \infty$, show that it is possible to write the free energy as

$$\frac{A - A_0}{L^3} = F_0(\bar{\psi}) + \frac{k_B T}{4\pi^2}\int_0^\infty k^2\left[\ln\left(k^2 + \alpha + \frac{\beta\bar{\psi}^2}{2}\right) - \ln k^2\right] dk$$

(IV.4.123)

This then gives two different forms above and below the transition:

$$\frac{A - A_0}{L^3} = \frac{k_B T}{4\pi^2}\int_0^\infty k^2 \ln\left(1 + \frac{\alpha}{k^2}\right) dk \quad \text{for} \quad \alpha \geq 0 \quad \text{(IV.4.124)}$$

$$\frac{A - A_0}{L^3} = -\frac{3\alpha^2}{\beta} + \frac{k_B T}{4\pi^2}\int_0^\infty k^2 \ln\left(1 - \frac{2\alpha}{k^2}\right) dk \quad \text{for} \quad \alpha < 0$$

(IV.4.125)

Here, $A_0$ is the value of $A$, in the saddle point approximation, at $\overline{\psi} = 0$, which is divergent but is a constant that does not depend on any of the parameters of the theory, $\alpha$ and $\beta$.

**Exercise.** Show that Eqs. (IV.4.124) and (IV.4.125) are still divergent.

So how do we deal with the remaining divergences? First, in real condensed matter systems we can cut off the $k$-integral, replacing infinity in the upper limit of integration with $Q \gg \sqrt{\alpha + \beta\overline{\psi}^2/2}$ in both Eqs. (IV.4.124) and (IV.4.125); this is because the continuous nature of $\psi(\mathbf{r})$ over $\mathbf{r}$ is only an approximation.

### *The effective potential and the Ginzburg criterion*

Actually, we can 'hide' dependence on the cut-off by doing a trick, which will give us something much more useful. Let's consider the derivative of Eq. (IV.4.123):

$$\chi(\overline{\psi}) = \frac{\partial A}{\partial \overline{\psi}} = \alpha\overline{\psi} + \frac{\beta}{3!}\overline{\psi}^3 + \frac{k_B T \beta \overline{\psi}}{4\pi^2} \int_0^\infty \frac{k^2}{k^2 + \alpha + \frac{\beta\overline{\psi}^2}{2}} dk$$

$$(IV.4.126)$$

The integral in Eq. (IV.4.126) still diverges, but there is a brave way to cope with this problem. We consider the last term as a leading order correction in $\beta$, arising due to the fluctuations. Then, we introduce a renormalized parameter $\alpha_R$, which to the leading order in $\beta$ is defined as

$$\alpha = \alpha_R - \frac{k_B T \beta}{4\pi^2} \int_0^Q dk \qquad (IV.4.127)$$

that is,

$$\alpha_R = \alpha + \frac{k_B T \beta}{4\pi^2} Q \qquad (IV.4.128)$$

But what is the meaning and the use of Eq. (IV.4.128), if we know that cut-off $Q$ is very large? A 'brave heart' would not worry too much about it, and just consider $\alpha_R$ as a new phenomenological parameter of the theory, identifying now $\alpha_R = 0$ as the point of phase transition through a redefinition of $T_c$ so that $\alpha_R \propto (T - T_c^*)$ close to the transition. This looks a bit like a chancer's approach: "Forget about your debts and creditors,

and continue investing into your business, as if there is no tomorrow." As yet, we never professed anything like that in our book, but let us try, and see what comes out of such adventurism.

Nominating $\alpha_R$ as the 'sign-changing body' about the critical point, we accept that due to thermal fluctuations the transition point is shifted from the point where $\alpha = 0$. This redefinition allows us, to leading order in $\beta$, to rewrite $\chi(\overline{\psi})$ in terms of an expression that *formally* does not diverge when $Q \to \infty$. Hiding your head in the sand? A bit like that. But then

$$\chi(\overline{\psi}) \simeq \alpha_R \overline{\psi} + \frac{\beta}{3!} \overline{\psi}^3 + \frac{k_B T \beta \overline{\psi}}{4\pi^2} \int_0^\infty \left[ \frac{k^2}{k^2 + \alpha_R + \frac{\beta \overline{\psi}^2}{2}} - 1 \right] dk$$

$$(\text{IV.4.129})$$

**Exercise.** Show that

$$\frac{1}{2\pi} \int_{-\infty}^\infty \left[ \frac{k^2}{k^2 + \alpha_R + \frac{\beta \overline{\psi}^2}{2}} - 1 \right] dk = -\frac{1}{2} \left( \alpha_R + \frac{\beta \overline{\psi}^2}{2} \right)^{1/2}$$

$$(\text{IV.4.130})$$

so that we may write

$$\chi(\overline{\psi}) \simeq \alpha_R \overline{\psi} + \frac{\beta}{3!} \overline{\psi}^3 - \frac{k_B T \beta \overline{\psi}}{8\pi} \left( \alpha_R + \frac{\beta \overline{\psi}^2}{2} \right)^{1/2}$$

$$(\text{IV.4.131})$$

We then integrate this expression over $\overline{\psi}$ back up to define a renormalized free-energy density as

$$\frac{A_R}{L^3} = \frac{\alpha_R \overline{\psi}^2}{2} + \frac{\beta}{4!} \overline{\psi}^4 - \frac{k_B T \beta}{12\pi} \left( \alpha_R + \frac{\beta \overline{\psi}^2}{2} \right)^{3/2} \tag{IV.4.132}$$

Now, Eq. (IV.4.132) should be interpreted as a *new Landau free energy function*, which is referred to as the *effective potential* (for $j(\mathbf{r}) = 0$). Thus, it makes sense to determine $\overline{\psi}$, the order parameter, through $\chi(\overline{\psi}) = 0$, and not through the saddle point equation. Note that there are higher order corrections to Eq. (IV.4.132) that must be evaluated through perturbation theory; Eq. (IV.4.132) is only the leading order correction due to thermal fluctuations.

Before continuing, it is more convenient to write Eq. (IV.4.132) in terms of a rescaled order parameter $\Phi$, such that $\overline{\psi}^2 = -\alpha_R \Phi^2 / \beta$ below the transition, and $\overline{\psi}^2 = \alpha_R \Phi^2 / \beta$ above the transition. We can then write for the rescaled free energy

$$A = \frac{\beta}{\alpha_R^2} \frac{A_R}{L^3} \tag{IV.4.133}$$

$$A = -\frac{\Phi^2}{2} + \frac{\Phi^4}{4!} - \lambda \left( \frac{\Phi^2}{2} - 1 \right)^{3/2} \quad \text{for } \alpha_R < 0 \tag{IV.4.134}$$

$$A = \frac{\Phi^2}{2} + \frac{\Phi^4}{4!} - \lambda \left( \frac{\Phi^2}{2} + 1 \right)^{3/2} \quad \text{for } \alpha_R > 0 \tag{IV.4.135}$$

where

$$\lambda = \frac{k_B T \beta}{12\pi |\alpha_R|^{1/2}} \tag{IV.4.136}$$

It is instructive to plot the functions described by Eqs. (IV.4.134) and (IV.4.135) for various values of $\lambda$, as shown in Fig. IV.4.1. The region $\Phi^2 < 2$ is not displayed, as the effective potential is no longer real here. For large enough values of $\lambda$, a $\Phi \neq 0$ solution minimizes the free energy both above the critical point (positive $\alpha_R$), and below

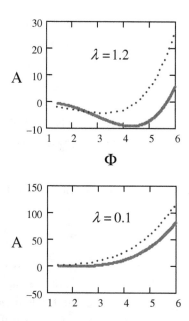

**Fig. IV.4.1** Renormalized free energy as a function of renormalized order parameter, plotted for indicated values of parameter $\lambda$. Solid curves: Eq. (IV.4.134); dotted curves: Eq. (IV.4.135).

it (negative $\alpha_R$). Whereas this is acceptable for the latter, for the former it cannot be true, and it emerges as an artefact of the approximation: for large $\lambda$ our approximation breaks down, as we have to consider higher order terms in the perturbation theory. The region of $\alpha_R \ll 1$ where $\lambda \gg \infty$, is defined as the critical region; there *perturbation theory breaks down* and we require other methods (like renormalization group scaling) to correctly obtain the behaviour of thermodynamic quantities; fluctuations here become very important! Conversely, we can only apply the Ginzburg–Landau theory when $\lambda \ll 1$; this condition is a reflection of the so called *Ginzburg criterion*, which will account for the Eq. (IV.4.136) demands that for the Gaussian approximation to be valid we need that

$$\frac{k_B T \beta}{12\pi |\alpha_R|^{1/2}} \ll 1 \qquad (\text{IV.4.137})$$

Below the transition temperature, $\Phi$ is determined through the following equation that minimizes Eq. (IV.4.134):

$$0 = -\Phi + \frac{\Phi^3}{3!} - \frac{3\lambda}{2} \left( \frac{\Phi^2}{2} - 1 \right)^{1/2} \qquad \text{(IV.4.138)}$$

Strictly speaking, we should only solve Eq. (IV.4.138) in a perturbative manner, when $\lambda \ll 1$, and we invite you to do this in the exercise below.

**Exercise.** Show that for small $\lambda$, the solution to Eq. (IV.4.138) reads

$$\Phi \approx \sqrt{6} + \frac{3\lambda}{2\sqrt{2}} \qquad \text{(IV.4.139)}$$

We did not learn much from this rather formal analysis, which was 'hiding divergences under the carpet', apart from the mere fact that fluctuations (a) shift the position of the critical point, (b) affect the average values of the order parameter, and (c) tell us that the mean-field theory will not work close to the critical point.

## *Two-point correlation function*

We can learn more about the behaviour of a system if we employ correlation functions to consider what happens near the critical point. Back to 3D: the two-point correlation function, $\langle \delta\psi(\mathbf{r}')\delta\psi(\mathbf{r})\rangle$, is given by

$$\langle \delta\psi(\mathbf{r}')\delta\psi(\mathbf{r})\rangle = \frac{1}{Z} \int \exp\left( -\frac{F[\psi(\mathbf{r})]}{k_B T} \right) \delta\psi(\mathbf{r}')\delta\psi(\mathbf{r}) D\delta\psi(\mathbf{r})$$

$$\text{(IV.4.140)}$$

In the saddle-point approximation, this reduces to

$$\langle \delta\psi(\mathbf{r}')\delta\psi(\mathbf{r})\rangle = \frac{1}{Z} \int \exp\left( -\frac{\Delta F[\psi_0(\mathbf{r}), \delta\psi(\mathbf{r})]}{k_B T} \right) \delta\psi(\mathbf{r}')\delta\psi(\mathbf{r}) D\delta\psi(\mathbf{r})$$

$$\text{(IV.4.141)}$$

Again, for the moment, we suppose that $\psi_0(\mathbf{r}) = \overline{\psi}$, which is true for an infinite unconstrained system. As in 1D, we can write correlation functions in terms of a generating functional. For instance, for the two-point function

$$\langle \delta\psi(\mathbf{r}')\delta\psi(\mathbf{r})\rangle = \frac{1}{Z}\frac{\delta^2 Z_G[j(\mathbf{r})]}{\delta j(\mathbf{r}')\delta j(\mathbf{r})} \tag{IV.4.142}$$

where the generating functional is given by

$$Z_G[j(\mathbf{r})] = \int \exp\left(-\frac{1}{k_B T}\int d\mathbf{r}\left[\frac{1}{2}(\vec{\nabla}\delta\psi(\mathbf{r}))^2\right.\right.$$
$$\left.\left.+\frac{\alpha}{2}\delta\psi(\mathbf{r})^2 + \frac{\beta}{4}\overline{\psi}^2\delta\psi(\mathbf{r})^2 - k_B T j(\mathbf{r})\delta\psi(\mathbf{r})\right]\right)D\delta\psi(\mathbf{r}) \tag{IV.4.143}$$

Utilizing Eq. (IV.4.116) and the Fourier transform

$$j(\mathbf{r}) = \frac{1}{(2\pi)^3}\int \tilde{j}(\mathbf{k})\exp(-i\mathbf{k}\cdot\mathbf{r})d\mathbf{k} \tag{IV.4.144}$$

We can rewrite the generating functional as

$$Z_G[j(\mathbf{r})] = \int \exp\left(-\frac{1}{4\pi k_B T}\int d\mathbf{k}[\delta\tilde{\psi}(-\mathbf{k}) + k_B T G(\mathbf{k})\tilde{j}(-\mathbf{k})]\right.$$
$$\times\left.\frac{1}{G(\mathbf{k})}[\delta\tilde{\psi}(\mathbf{k}) + k_B T G(\mathbf{k})\tilde{j}(\mathbf{k})]\right)D\delta\tilde{\psi}(\mathbf{k})$$
$$\times \exp\left(\frac{k_B T}{4\pi}\int d\mathbf{k}\delta\tilde{j}(-\mathbf{k})\tilde{G}_0(\mathbf{k})\delta\tilde{j}(\mathbf{k})\right) \tag{IV.4.145}$$

where

$$\tilde{G}_0(\mathbf{k}) = \frac{1}{\mathbf{k}^2 + \alpha + \frac{\beta\overline{\psi}^2}{2}} \tag{IV.4.146}$$

We can make the functional variable change $\delta\overline{\psi}(\mathbf{k}) = \delta\tilde{\psi}(\mathbf{k}) + k_B T\tilde{G}_0(\mathbf{k})\tilde{j}(\mathbf{k})$, so that the functional integral is then simply the partition function $Z$. Finally, for the generating functional, we find

$$Z_G[j(\mathbf{r})] = Z\exp\left(\frac{k_B T}{2}\int d\mathbf{r}\int d\mathbf{r}' j(\mathbf{r})G_0(\mathbf{r}-\mathbf{r}')j(\mathbf{r}')\right) \tag{IV.4.147}$$

where

$$G_0(\mathbf{r} - \mathbf{r}') = \frac{1}{(2\pi)^3} \int \frac{\exp(-i\mathbf{k}.(\mathbf{r} - \mathbf{r}'))}{k^2 + \alpha + \frac{\beta\overline{\psi}^2}{2}} d\mathbf{k} \qquad \text{(IV.4.148)}$$

on inversely Fourier transforming back $\tilde{j}(\mathbf{k})$ to $j(\mathbf{r})$. We already know from Chapter 1 of Part III (this volume) of how to do this 3D integral. We thus obtain

$$G_0(\mathbf{r} - \mathbf{r}') = \frac{1}{4\pi} \frac{\exp\left(-\left(\alpha + \frac{\beta}{2}\overline{\psi}^2\right)^{1/2} |\mathbf{r} - \mathbf{r}'|\right)}{|\mathbf{r} - \mathbf{r}'|} \qquad \text{(IV.4.149)}$$

To the lowest order in $\beta$ (pure saddle-point approximation), in the calculation of the correlation function, we may again replace $\alpha$ with $\alpha_R$, having $\overline{\psi} = 6\frac{\alpha_R}{\beta}\theta(-\alpha_R)$.

**Exercise.** Show, through Eqs. (IV.4.142) and (IV.4.147), that in the Gaussian approximation one obtains

$$\langle \delta\psi(\mathbf{r}')\delta\psi(\mathbf{r}) \rangle = k_B T \cdot G_0(\mathbf{r} - \mathbf{r}') \qquad \text{(IV.4.150)}$$

We can then define a characteristic length, called the correlation length $\xi$, which determines the length scale over which the correlation function decays. In the saddle-point approximation, we see that $\xi = \alpha_R^{-1/2}$ for $\alpha > 0$ and $\xi = (-2\alpha_R)^{-1/2}$ for $\alpha_R < 0$. [More accurate theories of critical behaviour in the critical region, computer simulations, and experimental measurements tells us that in fact, in 3D $\xi = f_+|\alpha_R|^{-\nu}$ above and $\xi = f_-|\alpha_R|^{-\nu}$ below the transition, where $\nu \simeq 0.64$].

Let us recall what $\langle \delta\psi(\mathbf{r}')\delta\psi(\mathbf{r}) \rangle$ characterizes. It quantifies how the fluctuations in $\psi(\mathbf{r}')$ are correlated with those of $\psi(\mathbf{r})$, as we move a point $\mathbf{r}'$ away from $\mathbf{r}$. One important thing we see is that the correlation length $\xi$ diverges when $\alpha_R \to 0$. What this means is that the effective size of a fluctuation in $\psi(\mathbf{r}')$ goes to infinity. We can also consider what $\langle \psi(\mathbf{r}')\psi(\mathbf{r}) \rangle$ is. It measures the linear response of the system to an external field $j(\mathbf{r})$, which could be a magnetic one, if is $\psi(\mathbf{r})$ the order parameter of a magnetic system.

## 4.5. Evaluating 3D Gaussian Functional Integrals with Boundary Conditions and in Inhomogeneous Systems

### *Inhomogeneous 1D Gaussian functional integrals*

For pedagogical reasons we will start again with 1D path integrals. Here, we'll begin with the functional integral of the following form:

$$Z = \int \exp\left(-\frac{1}{2}\int_{-\infty}^{\infty}\left[\left(\frac{dx(\tau)}{d\tau}\right)^2 + \alpha(\tau)x(\tau)^2\right]d\tau\right)Dx(\tau)$$

$$(\text{IV.4.151})$$

where now $\alpha(\tau)$ is any (regular) function of $\tau$. Equation (IV.4.151) could be useful in describing the quantum mechanics of a particle in a harmonic potential, the strength of which can vary in time $t = -i\tau$. We also want to consider this case as a warm-up for the 3D case.

To see how we should evaluate Eq. (IV.4.151), let's discretize it, so that we may write, with $x_k = x(\varepsilon k)$

$$Z = \lim_{\substack{\varepsilon \to 0 \\ bN \to \infty}} \tilde{Z}_\varepsilon \quad \text{where} \quad \tilde{Z}_\varepsilon = \prod_{j=-N}^{N}\int_{-\infty}^{\infty}dx_j$$

$$\times \exp\left(-\frac{1}{2}\sum_{k=-N}^{N}\sum_{l=-N}^{N}x_k M_{k,l}x_l\right)$$

$$(\text{IV.4.152})$$

Here, $M_{k,l}$ can be considered elements of a matrix $\mathbf{M}$ that take the form

$$M_{k,l} = (2 + \varepsilon\alpha_k)\delta_{k,l} - \delta_{k,l+1} - \delta_{k,l-1} \qquad (\text{IV.4.153})$$

We have already learned how discretization helps to evaluate functional integrals, but now we will go a bit further. We consider $x_l$ to be the components of some multidimensional vector $\mathbf{X}$. How can this help us to evaluate the integrals in Eq. (IV.4.152)? The answer is that we make the variable change

$$x_j = \sum_m O_{j,m} X_m \tag{IV.4.154}$$

which is equivalent to transforming the vector $\mathbf{x}$ to $\mathbf{X}$ through the matrix transformation $\mathbf{x} = \mathbf{O}.\mathbf{X}$. As the Jacobian of such a variable change is $|\det[\mathbf{O}]|$ we can first rewrite $\tilde{Z}_\varepsilon$:

$$\tilde{Z}_\varepsilon = \prod_{m=-N}^{N} \int_{-\infty}^{\infty} dX_m |\det[\mathbf{O}]| \exp\left( -\frac{1}{2} \sum_{n=-N}^{N} \sum_{p=-N}^{N} X_n D_{k,l} X_l \right) \tag{IV.4.155}$$

Here, $D_{k,l}$ are components of the transformed matrix $\mathbf{D} = \tilde{\mathbf{O}}.\mathbf{M}.\mathbf{O}$. So far we have said nothing about what matrix $\mathbf{O}$ is. Now, here comes the key point. We can choose the transformation matrix $\mathbf{O}$ so that $\mathbf{D}$ is diagonal with entries which are the eigenvalues of $\mathbf{M}$, as we showed in Chapter 4 of Part II, Vol. 1. Then, as we learnt, $\mathbf{O}$ is an orthogonal matrix when constructed from normalized eigenvectors of $\mathbf{M}$. In this case $\tilde{\mathbf{O}}.\mathbf{O} = \mathbf{I}$, so that $\det[\mathbf{O}] \cdot \det[\mathbf{O}] = 1$. Further using the fact that $\det[\tilde{\mathbf{O}}] = \det[\mathbf{O}]$, we find that $|\det[\mathbf{O}]| = 1$. All of these considerations allow us to write

$$\tilde{Z}_\varepsilon = \prod_{m=-N}^{N} \int_{-\infty}^{\infty} dX_m \exp\left( -\frac{1}{2} \sum_{l=-N}^{N} X_l E_l X_l \right) \tag{IV.4.156}$$

where $D_{k,l} = E_l \delta_{k,l}$. Next, it is simply a matter of performing each integration in Eq. (IV.4.156), for which the integrand is a Gaussian. We have that

$$\int_{-\infty}^{\infty} dX_l \exp\left( -\frac{X_l E_l X_l}{2} \right) = \sqrt{\frac{2\pi}{E_l}} \tag{IV.4.157}$$

Thus we obtain

$$\tilde{Z}_\varepsilon = \prod_{m=-N}^{N} \sqrt{\frac{2\pi}{E_m}} = \frac{(2\pi)^{2N+1}}{\det[\mathbf{D}]^{1/2}} \tag{IV.4.158}$$

Now, we need to understand how to take the continuum limit of Eq. (IV.4.158), so determining the continuum eigenvalues $E_m$. First, we can write

$$\exp\left(-\frac{1}{2}\int_{-\infty}^{\infty}\left[\left(\frac{dx(\tau)}{d\tau}\right)^2 + \alpha(\tau)x(\tau)^2\right]d\tau\right)$$

$$= \exp\left(-\frac{1}{2}\int_{-\infty}^{\infty}\int_{-\infty}^{\infty}x(\tau)M_c(\tau,\tau')x(\tau')d\tau d\tau'\right) \quad \text{(IV.4.159)}$$

where

$$M_c(\tau,\tau') = \left[-\frac{d^2}{d\tau^2} + \alpha(\tau)\right]\delta(\tau - \tau') \quad \text{(IV.4.160)}$$

Comparing Eq. (IV.4.159) with Eq. (IV.4.152), we see that $M_c(\tau,\tau')$ is the continuum version of $M_{j,k}$. Thus, the spectrum of eigenvalues $E_l$ is given through the eigenvalue equation

$$\int_{-\infty}^{\infty} M_c(\tau,\tau')\chi_l(\tau')d\tau' = E_l\chi_l(\tau) \quad \text{(IV.4.161)}$$

**Exercise.** Show that Eq. (IV.4.161) can be rewritten as

$$\left[-\frac{d^2}{d\tau^2} + \alpha(\tau)\right]\chi_l(\tau) = E_l\chi_l(\tau) \quad \text{(IV.4.162)}$$

We simply need to solve Eq. (IV.4.162) to find $E_l$, and substitute them back into Eq. (IV.4.158). Let's demonstrate that this recipe does indeed work for the continuum limit.

In Chapter 4 of Part III we learnt that eigenfunctions are equivalent to the eigenvectors of matrices such as those represented by $M_c(\tau,\tau')$. We already also know that the components of matrix $\mathbf{O}$ are the normalized eigenvectors. So the transformation $\mathbf{x} = \mathbf{O}.\mathbf{X}$ can be represented in the continuum limit as

$$x(\tau) = \sum_m \chi_m(\tau)X_m = \sum_m \chi_m^*(\tau)X_m^* \quad \text{(IV.4.163)}$$

where the normalized eigenfunctions satisfy the orthogonality condition

$$\int_{-\infty}^{\infty} \chi_m^*(s)\chi_{m'}(s)ds = \delta_{m,m'} \qquad (\text{IV.4.164})$$

Note that the Fourier transform (which we considered in Section 4.1 when evaluating path integrals) is a special case of Eq. (IV.4.154); however, the eigenfunctions $\chi_m(\tau) = \exp(i\tau\omega)$ are not normalized in infinite space (with the Jacobian of the transformation from $x(\tau)$ to its Fourier transform $X(\omega)$ being a constant that is unimportant). Now, using Eq. (IV.4.163) we can write

$$\exp\left(-\frac{1}{2}\int_{-\infty}^{\infty}\int_{-\infty}^{\infty} x(\tau)M_C(\tau,\tau')x(\tau')d\tau d\tau'\right)$$

$$= \exp\left(-\frac{1}{2}\sum_m\sum_{m'}X_m X_m^*\int_{-\infty}^{\infty}\int_{-\infty}^{\infty}\chi_m^*(\tau)M_C(\tau,\tau')\chi_{m'}(\tau')d\tau d\tau'\right)$$

Next we can utilize Eqs. (IV.4.161), (IV.4.163) and (IV.4.164) to find that

$$\exp\left(-\frac{1}{2}\int_{-\infty}^{\infty}\int_{-\infty}^{\infty} x(\tau)M_C(\tau,\tau')x(\tau')d\tau d\tau'\right)$$

$$= \exp\left(-\frac{1}{2}\sum_m X_m E_m X_m^*\right) \qquad (\text{IV.4.165})$$

If the spectrum of eigenvalues is discrete we can then write from Eq. (IV.4.151):

$$Z = \prod_m\left(\int_{-\infty}^{\infty}\exp\left(-\frac{1}{2}X_m E_m X_m^*\right)dX_m\right) = \prod_m\sqrt{\frac{2\pi}{E_m}} \qquad (\text{IV.4.166})$$

If the spectrum of eigenvalues is continuous, we need to write the integrals in Eq. (IV.4.166) as path integrals, using Fourier transforms, so that

$$Z = \int\exp\left(-\frac{1}{4\pi}\int X(k)E(k)(X(-k)dk\right)DX(k) \qquad (\text{IV.4.167})$$

**Exercise.** Following steps of Eqs. (IV.4.28)–(IV.4.34) show that from Eq. (IV.4.167), using the definition of the free energy in (Eq. (IV.4.12)), one obtains

$$\lim_{L \to \infty} \frac{A - A_0}{L} = \frac{k_B T}{2\pi} \int_0^\infty \ln \left[ \frac{E(k)}{k^2} \right] dk \qquad \text{(IV.4.168)}$$

where $A_0$ is the free energy evaluated when $\alpha(s) = 0$.

We can calculate expectation values of $h(x(\tau))$, an arbitrary function of $x(\tau)$ for the inhomogeneous case. In this case Eqs. (IV.4.39) and (IV.4.40) can be modified appropriately by Eq. (IV.4.151) to read

$$\langle \delta(\bar{y} - x(\tau)) \rangle = \frac{1}{2\pi} \int_{-\infty}^\infty dp \exp(ip\bar{y}) \tilde{\Gamma}(p, \tau) \qquad \text{(IV.4.169)}$$

where

$$\tilde{\Gamma}(p, s) = \frac{1}{Z} \int \exp(-ipx(\tau))$$
$$\times \exp\left( -\frac{1}{2} \int_{-\infty}^\infty \left[ \left( \frac{dx(\tau)}{d\tau} \right)^2 + \alpha(\tau) x(\tau)^2 \right] \right) Dx(\tau)$$

$$\text{(IV.4.170)}$$

To perform the path integration we need to utilize Eq. (IV.4.163) so that we may write

$$\tilde{\Gamma}(p, \tau) = \frac{1}{Z} \prod_m \left( \int_{-\infty}^\infty \exp\left( -\frac{1}{2} \left( X_m + \frac{ip\chi_m(\tau)}{E_m} \right) \right. \right.$$
$$\left. \left. \times E_m \left( X_m^* + \frac{ip\chi_m^*(\tau)}{E_m} \right) \right) dX_m \exp\left( -\frac{p^2 \chi_m(\tau)\chi_m^*(\tau)}{2E_m} \right) \right)$$

$$\text{(IV.4.171)}$$

By making the variable shift $\overline{X}_m = X_m + \frac{ip\chi_m(\tau)}{E_m}$, it's then easy to show that

$$\tilde{\Gamma}(p, \tau) = \exp\left( -\frac{p^2 G(\tau, \tau)}{2} \right) \qquad \text{(IV.4.172)}$$

where

$$G(\tau, \tau') = \sum_m \frac{\chi_m(\tau)\chi_m^*(\tau')}{E_m} \qquad \text{(IV.4.173)}$$

Perhaps a quick comment, at this stage: Eq. (IV.4.173) looks like the spectral representation of a Green's function, which indeed it is! You should be able to write down the differential equation that $G(\tau, \tau')$ satisfies.

Application of Eqs. (IV.4.169) and (IV.4.172) then yields

$$\langle h(x(\tau)) \rangle = \int_{-\infty}^{\infty} h(\bar{y}) \langle \delta(\bar{y} - x(\tau)) \rangle dy$$

$$= \sqrt{\frac{2\pi}{G(\tau, \tau)}} \int_{-\infty}^{\infty} h(\bar{y}) \exp\left(-\frac{\bar{y}^2}{2G(\tau, \tau)}\right) d\bar{y} \qquad \text{(IV.4.174)}$$

**Exercise.** Show from Eq. (IV.4.174) that $\langle x(\tau)^2 \rangle = G(\tau, \tau)$.

**Exercise.** Equation (IV.4.15) still applies for calculating the correlation functions, where now the generating functional is given by

$$Z_G[j(s)] = \int \exp\left(-\frac{1}{2} \int_{-\infty}^{\infty} \left[\left(\frac{dx(\tau)}{d\tau}\right)^2 + \alpha(\tau)x(\tau)^2\right] ds\right)$$

$$\times \exp\left(-\int_{-\infty}^{\infty} j(\tau)x(\tau)d\tau\right) Dx(\tau) \qquad \text{(IV.4.175)}$$

By using Eq. (IV.4.163) and writing

$$j(\tau) = \sum_m \chi_m(\tau)J_m \qquad \text{(IV.4.176)}$$

show that

$$Z_G[j(\tau)] = \exp\left(\frac{1}{2} \int_{-\infty}^{\infty} \int_{-\infty}^{\infty} j(\tau)G(\tau, \tau')j(\tau')d\tau d\tau'\right) \qquad \text{(IV.4.177)}$$

Hence, show (using Eq. (IV.4.15)) that

$$\langle x(\tau)x(\tau') \rangle = G(\tau, \tau') \qquad \text{(IV.4.178)}$$

**Hint:** Modify the analysis of Eqs. (IV.4.49)–(IV.4.54) to consider the inhomogeneous case. You will also need to use the orthogonality condition, Eq. (IV.4.164) to show

$$J_m = \int_{-\infty}^{\infty} \chi_m^*(\tau) j(\tau) d\tau \quad \text{and} \quad J_m^* = \int_{-\infty}^{\infty} \chi_m^*(\tau') j(\tau') d\tau'$$

$$\text{(IV.4.179)}$$

In the interests of space, we won't provide any examples here, as we normally would at this stage of a section, but will do so when we have considered the 3D case which is more relevant in real systems.

### Inhomogeneous 3D Gaussian functional integrals

As an illustration of inhomogeneous 3D Gaussian functional integrals, let's now consider the case where we have made the saddle point approximation for the Wilson–Ginzburg–Landau theory given by Eqs. (IV.4.112) and (IV.4.113). Here, we'll suppose that $\psi_0(\mathbf{r})$ is no longer constant—how might we attempt to do the functional integration? For our first task, we'll want to see how, potentially, we could evaluate

$$\overline{Z} = \int \exp\left(-\frac{\Delta F[\psi_0(\mathbf{r}), \delta\psi(\mathbf{r})]}{k_B T}\right) D\psi(\mathbf{r}) \qquad \text{(IV.4.180)}$$

Then, we'll consider how to calculate correlation functions and expectation values. First, we can express $\Delta F[\psi_0(\mathbf{r}), \delta\psi(\mathbf{r})]$ in the following way:

$$\frac{\Delta F[\psi_0(\mathbf{r}), \delta\psi(\mathbf{r})]}{k_B T} = \frac{1}{2} \int d\mathbf{r} \int d\mathbf{r}' \delta\psi(\mathbf{r}') M(\mathbf{r}', \mathbf{r}) \delta\psi(\mathbf{r}) \qquad \text{(IV.4.181)}$$

where

$$k_B T M(\mathbf{r}', \mathbf{r}) = \left[-\vec{\nabla}^2 + \alpha + \frac{\beta\psi_0(\mathbf{r})}{2}\right] \delta(\mathbf{r}' - \mathbf{r}) \qquad \text{(IV.4.182)}$$

Now as a 3D extension to Eq. (IV.4.161), we have the following eigenvalue equation:

$$k_B T \int M(\mathbf{r}, \mathbf{r}') \chi_{l,m,n}(\mathbf{r}') d\mathbf{r}' = E_{l,m,n} \chi_{l,m,n}(\mathbf{r}) \qquad \text{(IV.4.183)}$$

**Exercise.** Show that using the form of $M(\mathbf{r}', \mathbf{r})$ given by Eq. (IV.4.182), Eq. (IV.4.183) simplifies to

$$\left[ -\vec{\nabla}^2 + \alpha + \frac{\beta \psi_0(\mathbf{r})}{2} \right] \chi_{l,m,n}(\mathbf{r}) = E_{l,m,n} \chi_{l,m,n}(\mathbf{r}) \qquad \text{(IV.4.184)}$$

where $\chi_{l,m,n}(\mathbf{r})$ obey the orthogonality condition

$$\int \chi_{l',m',n'}^*(\mathbf{r}) \chi_{l,m.n}(\mathbf{r}) d\mathbf{r} = \delta_{l',l} \delta_{m',m} \delta_{n',n} \qquad \text{(IV.4.185)}$$

We can then write

$$\delta\psi(\mathbf{r}) = \sum_{l,m,n} \chi_{l,m,n}(\mathbf{r}) \Phi_{l,m,n} = \sum_{l,m,n} \chi_{l,m,n}^*(\mathbf{r}) \Phi_{l,m,n}^* \qquad \text{(IV.4.186)}$$

**Exercise.** Show that using Eqs. (IV.4.183), (IV.4.185) and (IV.4.186) we may write

$$Z = \prod_{l,m,n} \left( \int_{-\infty}^{\infty} \exp\left( -\frac{\Phi_{l,m,n}^* E_{l,m,n} \Phi_{l,m,n}}{2k_B T} \right) d\Phi_{l,m,n} \right)$$

$$= \prod_{l,m,n} \sqrt{\frac{2\pi k_B T}{E_{l,m,n}}} \qquad \text{(IV.4.187)}$$

As the determinant of a matrix is the product of its eigenvalues we may rewrite Eq. (IV.4.187) as

$$Z \propto \frac{1}{\sqrt{Det[\mathbf{M}]}} \qquad \text{(IV.4.188)}$$

Here, $\mathbf{M}$ is the matrix representation of the function $M(\mathbf{r}, \mathbf{r}')$.

For a continuous spectrum of eigenvalues we can replace Eq. (IV.4.187) with a functional integral that reads as

$$Z = \int \exp\left( -\frac{\int \Phi(\mathbf{k}) E(\mathbf{k}) (\Phi(-\mathbf{k}) d\mathbf{k}}{k_B T (2\pi)^3} \right) D\Phi(\mathbf{k}) \qquad \text{(IV.4.189)}$$

**Exercise.** Following the analysis of Eqs. (IV.4.116)–(IV.4.123) show from Eq. (IV.4.189) that $\overline{A} = -k_B T \ln \overline{Z}$ takes the form

$$\frac{\overline{A} - \overline{A}_0}{L^3} = \frac{k_B T}{(2\pi)^3} \int [\ln(E(\mathbf{k})) - \ln \mathbf{k}^2] d\mathbf{k} \qquad (IV.4.190)$$

where $\overline{A}_0$ is $\overline{A}$ evaluated at $\alpha = 0$, $\psi_0(\mathbf{r}) = 0$.

**Exercise.** Show that the correlation function for the theory described by Eqs. (IV.4.180), (IV.4.181) and (IV.4.182) takes the form

$$\langle \delta\psi(\mathbf{r})\delta\psi(\mathbf{r}') \rangle = k_B T \sum_{l,m,n} \frac{\chi_{l,m,n}(\mathbf{r})\chi_{l,m,n}^*(\mathbf{r})}{E_{l,m,n}} \qquad (IV.4.191)$$

### *Gaussian functional integration for systems with periodic boundary conditions*

Let us consider the simplest boundary condition constraint, where we can use all of the machinery for performing the path integration developed through Eqs. (IV.4.180)–(IV.4.188), the periodic one:

$$\delta\psi(\mathbf{r} + \mathbf{R}) = \delta\psi(\mathbf{r}) \qquad (IV.4.192)$$

where $\mathbf{R}$ is the periodicity vector given by

$$\mathbf{R} = n_x L_x \hat{\mathbf{i}} + n_y L_y \hat{\mathbf{j}} + n_z L_z \hat{\mathbf{k}} \qquad (IV.4.193)$$

Then, let's suppose that we have Eq. (IV.4.113) as our free energy functional, with $\psi_0(\mathbf{r}) = \overline{\psi}$, with $\delta\psi(\mathbf{r})$ defined in a cuboid region of dimensions $L_x \times L_y \times L_z$, but satisfying the periodic boundary condition for $\delta\psi(\mathbf{r})$. Then, $\chi_{l,m,n}(\mathbf{r})$ obeys the equation

$$\left[ -\vec{\nabla}^2 + \alpha + \frac{\beta\overline{\psi}^2}{2} \right] \chi_{m,n,l}(\mathbf{r}) = E_{m,n,l}\chi_{m,n,l}(\mathbf{r}) \qquad (IV.4.194)$$

and the periodic boundary condition

$$\chi_{l,m,n}(\mathbf{r} + \mathbf{R}) = \chi_{l,m,n}(\mathbf{r}) \qquad (IV.4.195)$$

so that Eq. (IV.4.192) is satisfied.

**Exercise.** Show that the normalized eigenfunctions that solve both Eqs. (IV.4.194) and (IV.4.195) are

$$\chi_{m,n,l}(\mathbf{r}) = \frac{1}{\sqrt{L_x L_y L_z}} \exp\left( i\left( \frac{2\pi m x}{L_x} + \frac{2\pi n x}{L_y} + \frac{2\pi l x}{L_z} \right) \right)$$

$$(IV.4.196)$$

which corresponds to the eigenvalues

$$E_{m,n,l} = \left( \frac{2\pi m}{L_x} \right)^2 + \left( \frac{2\pi n}{L_y} \right)^2 + \left( \frac{2\pi l}{L_z} \right)^2 + \alpha + \frac{\overline{\psi}^2}{2}\beta \quad (IV.4.197)$$

Thus, for the free energy we can write

$$\frac{A - A_0}{L_x L_y L_z} = \frac{\alpha}{2}\overline{\psi}^2 + \frac{\beta}{4!}\overline{\psi}^4$$

$$+ \frac{k_B T}{2 L_x L_y L_z} \sum_l \sum_m \sum_n \left[ \ln\left( \left( \frac{2\pi l}{L_x} \right)^2 + \left( \frac{2\pi m}{L_y} \right)^2 \right. \right.$$

$$\left. + \left( \frac{2\pi n}{L_z} \right)^2 + \alpha + \frac{\beta\overline{\psi}^2}{2} \right)$$

$$\left. - \ln\left( \left( \frac{2\pi l}{L_x} \right)^2 + \left( \frac{2\pi m}{L_y} \right)^2 + \left( \frac{2\pi n}{L_z} \right)^2 \right) \right] \quad (IV.4.198)$$

where again $A_0$ is the value of $A$ when $\alpha + \frac{\beta\overline{\psi}^2}{2} = 0$.

**Exercise.** Show that when $L_x \to \infty$, $L_y \to \infty$ and $L_z \to \infty$, from Eq. (IV.4.198) one can recover Eq. (IV.4.123) by utilizing the definition of definite integration as the limit of a sum.

Also, we can estimate the sums for terms for which $|l| > L$, $|m| > M$, and $|n| > N$ as integrals, provided that $L$, $M$ and $N$ are chosen to be sufficiently large. This observation indicates that Eq. (IV.4.198) contains the same divergence as Eq. (IV.4.123). To get rid of these divergences we need to perform similar steps as illustrated by Eqs. (IV.4.126)–(IV.4.132).

## Surface boundary conditions

We can consider a simple boundary condition on a 3D fluctuating field $\psi(\mathbf{r})$ on a closed surface $S$, which encloses the region $V$, specified by the vector $\mathbf{r}_S(s,t)$, where the fluctuating field is defined within the surface. In this case, we simply require that the eigenfunctions $\chi_{l,m,n}(\mathbf{r})$ satisfy the boundary conditions that we impose on $\psi(\mathbf{r})$. Then, to calculate functional integrals, we can do exactly the same procedure as before (Eqs. (IV.4.180)–(IV.4.188)), provided that the boundary condition is one that ensures the orthogonality condition.

**Example IV.4.4 (Maths Practice ♫♪).** Let's apply the Wilson–Ginzburg–Landau theory to a cylinder of radius $b$. We then work in cylindrical polars, so that $\delta\psi(\mathbf{r}) = \delta\psi(R, \phi, z)$. We will impose the boundary condition $\delta\psi(b, \phi, z) = 0$, but deal with a constant mean-field $\overline{\psi}$. In the example below this one, we'll extend the result we get here to the physically relevant case of fluctuations in the order parameter for a superconducting material above the super-conducting phase transition. In cylindrical coordinates, Eq. (IV.4.184), the eigenvalue equation becomes

$$-\frac{1}{R}\frac{\partial}{\partial R}\left(R\frac{\partial\chi_{l,m,n}(R,\phi,z)}{\partial R}\right) - \frac{1}{R^2}\frac{\partial^2\chi_{l,m,n}(R,\phi,z)}{\partial\phi^2}$$

$$-\frac{\partial^2\chi_{l,m,n}(R,\phi,z)}{\partial z^2} + (\alpha + \beta\overline{\psi}^2/2)\chi_{l,m,n}(R,\phi,z)$$

$$= E_{l,m,n}\chi_{l,m,n}(R,\phi,z) \qquad \text{(IV.4.199)}$$

where we now need to impose the boundary condition

$$\chi_{l,m,n}(b,\phi,z) = 0 \qquad \text{(IV.4.200)}$$

The reader should already know, from Chapter 4 of Part III, that we can write down the solution to Eqs. (IV.4.199) and (IV.4.200) as

$$\chi_{l,m,n}(R,\phi,z) = J_n\left(\frac{a_{n,m}R}{b}\right)\exp(in\phi)\exp(ik_z z) \qquad \text{(IV.4.201)}$$

$$E_{l,m,n} = E_{m,n}(k_z) = \left(\frac{a_{n,m}}{b}\right)^2 + k_z^2 + \alpha + \frac{\beta}{2}\overline{\psi}^2 \qquad \text{(IV.4.202)}$$

**Table IV.4.1** The values of $x$ that deliver zeros to Bessel functions (few first examples; you can find more in the Wolfram Language using the command BesselJZero.)

| m | $J_0(x)$ | $J_1(x)$ | $J_2(x)$ | $J_3(x)$ | $J_4(x)$ | $J_5(x)$ |
|---|----------|----------|----------|----------|----------|----------|
| 1 | 2.4048 | 3.8317 | 5.1356 | 6.3802 | 7.5883 | 8.7715 |
| 2 | 5.5201 | 7.0156 | 8.4172 | 9.7610 | 11.0647 | 12.3386 |
| 3 | 8.6537 | 10.1735 | 11.6198 | 13.0152 | 14.3725 | 15.7002 |
| 4 | 11.7915 | 13.3237 | 14.7960 | 16.2235 | 17.6160 | 18.9801 |
| 5 | 14.9309 | 16.4706 | 17.9598 | 19.4094 | 20.8269 | 22.2178 |

where

$$k_z = \lim_{L \to \infty} \left( \frac{2\pi l}{L} \right) \tag{IV.4.203}$$

Note here that $L$ is the length of the cylinder. You can verify both Eq. (IV.4.201) and (IV.4.202) by direct substitution into Eq. (IV.4.199). Also, for Eq. (IV.4.200) to be satisfied, we must choose $J_n(a_{n,m}) = 0$; here $a_{n,m}$ corresponds to the $m$th zero of the Bessel function $J_n(x)$, where we don't count any zeros at $x = 0$. Table IV.4.1 shows some of the values for $a_{n,m}$.

For $a_{n,m} \gg 1$, the zeros can be approximated by the formula

$$a_{n.m} = \left( m + \frac{1}{2}n - \frac{1}{4} \right) \pi - \frac{4n^2 - 1}{8 \left( m + \frac{1}{2}n - \frac{1}{4} \right) \pi}$$

$$- \frac{(4n^2 - 1)(28n^2 - 31)}{384\pi^3 \left( m + \frac{1}{2}n - \frac{1}{4} \right)^3} + \dots \tag{IV.4.204}$$

(eventually, this works fairly well for all positions in Table IV.4.1). Proceeding to the Helmholtz free energy we get

$$\lim_{L_z \to \infty} \frac{A}{L_z} = \pi b^2 \left( \frac{\alpha}{2} \bar{\psi}^2 + \frac{\beta}{4!} \bar{\psi}^4 \right) + \frac{k_B T}{2(2\pi)} \sum_{n=-\infty}^{\infty} \sum_{m=1}^{\infty}$$

$$\times \int_{-\infty}^{\infty} dk_z \ln \left( \left( \frac{a_{n,m}}{b} \right)^2 + k_z^2 + \alpha + \frac{\beta \bar{\psi}^2}{2} \right) \tag{IV.4.205}$$

Equation (IV.4.205) is again divergent without a cut-off. We can get rid of this divergence by again considering the function $\chi(\overline{\psi})$ defined through $\pi b^2 L_z \chi(\overline{\psi}) = \partial A / \partial \overline{\psi}$. This is now given by

$$\chi(\overline{\psi}) = \alpha \overline{\psi} + \frac{\beta}{3!} \overline{\psi}^3 + \frac{\beta k_B T \overline{\psi}}{4\pi b^2} \sum_{n=-\infty}^{\infty} \sum_{m=1}^{\infty}$$

$$\times \int_{-\infty}^{\infty} dk_z \frac{1}{\left(\frac{a_{n,m}}{b}\right)^2 + k_z^2 + \alpha + \frac{\beta}{2}\overline{\psi}^2} \qquad \text{(IV.4.206)}$$

Now to render the expression for $\chi(\overline{\psi})$ finite, as we did it previously, we define a renormalized parameter $\alpha_R$. In this case, however, one chooses

$$\alpha = \alpha_R - \frac{\beta k_B T}{4\pi^2 b^2} \sum_{n=-\infty}^{\infty} \sum_{m=1}^{\infty} \int_{-\infty}^{\infty} dk_z \frac{1}{\left(\frac{a_{n,m}}{b}\right)^2 + k_z^2} \qquad \text{(IV.4.207)}$$

Thus, we can, up to the leading order in $\beta$, write

$$\chi(\overline{\psi}) = \alpha_R \overline{\psi} + \frac{\beta}{3!} \overline{\psi}^3 + \frac{\beta \overline{\psi} k_B T}{4\pi^2 b^2} \sum_{n=-\infty}^{\infty} \sum_{m=1}^{\infty} \int_{-\infty}^{\infty} dk_z$$

$$\times \left( \frac{1}{\left(\frac{a_{n,m}}{b}\right)^2 + k_z^2 + \alpha_R + \frac{\beta}{2}\overline{\psi}^2} - \frac{1}{\left(\frac{a_{n,m}}{b}\right)^2 + k_z^2} \right)$$

$$\text{(IV.4.208)}$$

We can evaluate the integral over $k_z$ to obtain

$$\chi(\overline{\psi}) = \alpha_R \overline{\psi} + \frac{\beta}{3!} \overline{\psi}^3 + \frac{\alpha_R \overline{\psi} k_B T}{4\pi\xi} \Xi\left(\frac{b}{\xi}\right) \qquad \text{(IV.4.209)}$$

where the function $\Xi\left(\frac{b}{\xi}\right)$ is determined by the double sum

$$\Xi\left(\frac{b}{\xi}\right) \equiv \sum_{n=-\infty}^{\infty} \sum_{m=1}^{\infty} \left( \frac{1}{\sqrt{\left(\frac{b a_{n,m}}{\xi}\right)^2 + 1}} - \frac{\xi}{b} \frac{1}{a_{n,m}} \right) \qquad \text{(IV.4.210)}$$

Here the correlation length is given by

$$\xi = (\alpha_R + \beta\overline{\psi}^2/2)^{-1/2} \qquad (IV.4.211)$$

When the system size exceeds the correlation length, such that $b/\xi \ll 1$, from Eqs. (IV.4.209) and (IV.4.210) we recover Eq. (IV.4.132); i.e., as expected, the confinement does not affect the system's behaviour (and this is always true, not just to the level of the used approximation).

The effective potential is obtained by integrating Eq. (IV.4.209), with respect to $\overline{\psi}$.

What form does the two-point correlation function take? From Eqs. (IV.4.191), (IV.4.201) and (IV.4.202) it is not too difficult to show that it can be written as

$$\langle \psi(\mathbf{r})\psi(\mathbf{r}')\rangle = \frac{k_B T}{(2\pi)} \sum_{n=-\infty}^{\infty} \sum_{m=1}^{\infty} \int_{-\infty}^{\infty} dk_z$$

$$\times \frac{\exp(ik_z(z - z'))\exp\left(in(\phi - \phi')\right) J_n\left(\frac{a_{n,m}R}{b}\right) J_n\left(\frac{a_{n,m}R'}{b}\right)}{\left(\frac{a_{n,m}}{b}\right)^2 + k_z^2 + \alpha_R + \frac{\beta}{2}\overline{\psi}^2}$$

$$= \frac{k_B T}{2} \sum_{n=-\infty}^{\infty} \sum_{m=1}^{\infty}$$

$$\times \frac{\exp\left(-|z - z'|\left(\left(\frac{a_{n,m}}{b}\right)^2 + \alpha_R + \frac{\beta}{2}\overline{\psi}^2\right)^{1/2}\right)}{\sqrt{\left(\frac{a_{n,m}}{b}\right)^2 + \alpha_R + \frac{\beta}{2}\overline{\psi}^2}} \exp(in(\phi - \phi'))J_n\left(\frac{a_{n,m}R}{b}\right) J_n\left(\frac{a_{n,m}R'}{b}\right) \qquad (IV.4.212)$$

One thing we see clearly from Eq. (IV.4.212), is that for $R = b$ or $R' = b$ the correlation function is zero, because of our requirement $J_n(a_{n,m}) = 0$. This makes sense as we have imposed as boundary conditions that there can be no fluctuations at the cylinder walls, so a fluctuation inside the cylinder cannot cause, via correlations, a fluctuation at the surface. Let's look at the limiting case where $b/\xi = b(\alpha_R + \frac{\beta}{2}\overline{\psi})^{1/2} \ll 1$. Equation (IV.4.212) simplifies to

$$\langle \psi(\mathbf{r})\psi(\mathbf{r}')\rangle \approx \frac{k_B T}{2} \sum_{n=-\infty}^{\infty} \sum_{m=1}^{\infty}$$

$$\times \frac{\exp\left(-\frac{|z-z'|}{b}a_{n,m}\right)}{\left(\frac{a_{n,m}}{b}\right)} J_n\left(\frac{a_{n,m}R}{b}\right) J_n\left(\frac{a_{n,m}R'}{b}\right) e^{in(\phi-\phi')}$$

$$\text{(IV.4.213)}$$

What happens to this expression when $|z - z'| \gg b$? In this case, due to the factor $e^{-\frac{|z-z'|}{b}a_{n,m}}$ only the lowest zero $a_{0,1}$ contributes significantly to the sum. Hence, we can obtain a much simpler expression,

$$\langle \psi(\mathbf{r})\psi(\mathbf{r}')\rangle \approx \frac{k_B T}{2} \frac{\exp\left(-\frac{|z-z'|}{\lambda}\right) J_0\left(\frac{R}{\lambda}\right) J_0\left(\frac{R'}{\lambda}\right)}{\lambda} \qquad \text{(IV.4.214)}$$

where now we have a correlation length $\lambda = b/a_{0,1} = 0.42b$, which is finite and is totally determined by the size of the cylinder. It is essentially a quasi-1D theory limiting case! Generally, with systems like cylinders and slabs when the 3D correlations extend to much longer distances than the dimensions of the system, one can effectively reduce the 'dimensions' of the theory.

**Example IV.4.5 (Physical Application ▶▶).** The Wilson–Landau–Ginzburg theory of superconductors operates with the complex order parameter $\Psi(\mathbf{r}) = \psi_1(\mathbf{r}) + i\psi_2(\mathbf{r})$ and the partition function

$$Z = \int \exp\left(-\frac{F[\Psi(\mathbf{r})]}{k_B T}\right) D\Psi(\mathbf{r}) \qquad \text{(IV.4.215)}$$

in which the free energy functional, in the absence of an external electromagnetic field, is given by

$$F[\Psi(\mathbf{r})] = \int d\mathbf{r} \left(\frac{\hbar^2}{2m_c}\nabla\Psi(\mathbf{r})\nabla\Psi^*(\mathbf{r}) + \frac{\alpha}{2}|\Psi(\mathbf{r})|^2 + \frac{\beta}{4!}|\Psi(\mathbf{r})|^4\right)$$

$$\text{(IV.4.216)}$$

where $m_c$ is the effective mass of an electron pair, which may depend on the dispersion relation of electrons in the lattice and their interactions with phonons. Both $\alpha$ and $\beta$ can be calculated from the underlying microscopic theory, for instance the simplest one, the Bardeen–Cooper–Schrieffer (BSC) theory. This factor $\frac{\hbar^2}{2m_c}$ can be absorbed into $\alpha$ and $\beta$ by rescaling of the order parameter $\Psi(\mathbf{r})$. Note that $\Psi(\mathbf{r})$ can also be considered as some macroscopic wave function describing ground-state/condensate formed from the paired electrons. Here, we can apply the boundary condition $\Psi(b, \phi, z) = 0$ for a cylinder of radius $b$. The free energy functional given by Eq. (IV.4.216) is equivalent to Eq. (IV.4.105) for an $N$-component order parameter with $N = 2$. This boundary value problem would correspond to the case where we have cylinder of the superconducting material.

**Exercise.** Write down the saddle-point approximation for the free energy functional described by Eq. (IV.4.216) above the superconducting transition where the saddle-point/mean field is zero. Write this explicitly in terms of $\psi_1(\mathbf{r})$ and $\psi_1(\mathbf{r})$. From this approximation, repeat the analysis given in the previous example to derive an expression for $\langle \Psi(\mathbf{r})\Psi^*(\mathbf{r}') \rangle$.

## 4.6. The Delta-Functional and Hubbard–Stratonovich Transformation

Recall the methods of Lagrange multipliers, which we discussed in variational calculus (in Chapter 3), when some conditions or constraints are imposed on the variable functions with respect to which we will be looking for an extremum (usually—minimum) of the quantity that we study. Is there a technique that would help us in functional integration, under such additional constraints? Such does exist and it is named after Russian mathematician R.L. Stratonovich, who first introduced it and British physicist J. Hubbard (who by the way

Ruslan Stratonovich
(1930–1997)

John Hubbard
(1931–1980)

was a graduate of Imperial College) who demonstrated its use in physics. We present its most simple version below. This first requires an introduction to the *delta-functional*.

### The 1D delta functional

Consider two functions $u(x)$ and $v(x)$ in the domain $-b < x < b$. The 1D delta functional of their difference is defined as

$$\delta^{\infty}[u(x) - v(x)] = \lim_{\substack{\varepsilon \to 0 \\ N\varepsilon = b}} \prod_{j=-N}^{N} \delta(u_j - v_j) \qquad \text{(IV.4.217)}$$

where $u_j = u(\varepsilon j)$ and $v_j = v(\varepsilon j)$. As well as the ordinary delta-function, the delta functional is normalized,

$$\int \delta^{\infty}[u(x) - v(x)] Du(x) = 1 \qquad \text{(IV.4.218)}$$

and has a 'filter-property':

$$\int \delta^{\infty}[u(x) - v(x)] K[u(x)] Du(x) = K[v(x)] \qquad \text{(IV.4.219)}$$

where $K[u(x)]$ is any other 'smooth' functional. Such a property is useful if we want to impose a constraint. We'll illustrate this with one example.

**Example IV.4.6 (Physical Application ▶▶).** Let's consider a polymer described by a worm-like chain in free space (a more realistic model for semi-flexible polymers than the Gaussian chain model we considered in Example IV.4.2). The energy functional describing the worm like chain is

$$E[\mathbf{t}(s)] = \int_0^L \left[ \frac{l_p}{2} \left( \frac{d\mathbf{t}(s)}{ds} \right)^2 \right] ds \qquad \text{(IV.4.220)}$$

Here, we will want to impose the constraint that $|\mathbf{t}(s)| = 1$, so that the chain meanders but its overall length remains constant. This is, of course, an approximation, as the real chain can stretch around an equilibrium position, but we won't consider this here. We can use the delta-functional to account for this constraint:

$$Z = \int \delta^\infty (\mathbf{t}(s)^2 - 1) \exp\left(-\frac{E[\mathbf{t}(s)]}{k_B T}\right) D\mathbf{t}(s) \qquad \text{(IV.4.221)}$$

**Exercise.** Using similar analysis as used in Eqs. (IV.4.79)–(IV.4.81), show that one can write for the partition function of such a chain with free ends

$$Z \propto \int_0^{2\pi} d\phi_0 \int_0^{2\pi} d\phi_L \int_0^\pi \sin\theta_0 d\theta_0 \int_0^\pi \sin\theta_L d\theta_L$$

$$\times \sum_{l=0}^{\infty} \sum_{m=-l}^{l} Y_l^{(m)*}(\theta_0, \phi_0) \exp(-bE_l) Y_l^{(m)}(\theta_L, \phi_L)$$

$$\text{(IV.4.222)}$$

where the spherical harmonics, $Y_l^{(m)}(\theta_L, \phi_L)$, are needed here to satisfy the eigenvalue equation

$$\frac{1}{\sin^2\theta} \frac{\partial^2 Y_l^{(m)}(\theta, \phi)}{\partial\phi^2} + \frac{1}{\sin\theta} \frac{\partial}{\partial\theta}\left(\sin\theta \frac{\partial Y_l^{(m)}(\theta, \phi)}{\partial\theta}\right) = E_l Y_l^{(m)}(\theta, \phi)$$

$$\text{(IV.4.223)}$$

What are the values of $E_l$? For the free ends show that all the integrals in the sum of Eq. (IV.4.223) except that for $m = l = 0$ vanish, so that $Z \propto \exp(-bE_0)$. This would not be the case, if we fixed both ends at particular values of $\theta_0$, $\phi_0$, $\theta_L$ and $\phi_L$. **Think:** If we were to calculate the two-point correlation function for free ends $\langle \mathbf{t}(s).\mathbf{t}(s') \rangle$ would we have to consider values $m \neq l \neq 0$? **Hint:** To start with, you'll need to think about writing $\mathbf{t} = \hat{\mathbf{t}} = \sin\theta\cos\phi\hat{\mathbf{j}} + \sin\theta\sin\phi\hat{\mathbf{j}} + \cos\theta\hat{\mathbf{k}}$ and working with spherical polar coordinates to implement the constraint, $|\mathbf{t}(s)| = 1$.

### Extending the definition of the delta functional to 3D

We can easily extend delta-functional definition to 3D. The 3D delta functional for two functions $\psi(\mathbf{r})$ and $\varphi(\mathbf{r})$, defined in the cubic domain $-b \leq x \leq b$, $-b \leq y \leq b$ and $-b \leq z \leq b$, is given by

$$\delta^{\infty}[\psi(\mathbf{r}) - \varphi(\mathbf{r})] = \lim_{\substack{\varepsilon \to 0 \\ N\varepsilon = b}} \prod_{j=-N}^{N} \prod_{k=-N}^{N} \prod_{l=-N}^{N} \delta(\psi_{j,k,l} - \varphi_{j,k,l}) \quad \text{(IV.4.224)}$$

### The Hubbard–Stratonovich transform

Let's consider this transform first in 1D, where we might want to replace some function $v(x)$ with a new fluctuating function $u(x)$. Here, we may write

$$H[v(x)] = \int \delta^{\infty}[u(x) - v(x)] H[u(x)] Du(x)$$

$$\propto \int \exp\left(i \int_{-b}^{b} p(x)[u(x) - v(x)]\right) H[u(x)] Du(x) Dp(x)$$

$$\text{(IV.4.225)}$$

**Example IV.4.7 (Physical Application ▶▶).** In general, the statistical mechanics of a worm like chain in a confining potential can be described by an equation similar to Eq. (IV.4.221) but this can be shown more conveniently in terms of vector $\mathbf{r}$:

$$Z = \int \delta^{\infty}\left(\left(\frac{d\mathbf{r}}{ds}\right)^2 - 1\right) \exp\left(-\frac{E[\mathbf{r}(s)]}{k_B T}\right) D\mathbf{r}(s) \quad \text{(IV.4.226)}$$

with

$$E[\mathbf{r}(s)] = \int_0^L \left[\frac{l_p}{2}\left(\frac{d^2\mathbf{r}(s)}{ds^2}\right)^2 + V(\mathbf{r}(s))\right] ds \quad \text{(IV.4.227)}$$

However, let's consider such a polymer in 3D, but confined in a 2D constraining potential that depends on only two coordinates, for instance this could be a straight narrow tube. Here, we may write such a potential as $V(\mathbf{r}) = V_{2D}(x, y)$. If the tube is narrow enough we may suppose that the curvatures $\frac{d^2x}{ds^2}$, $\frac{d^2y}{ds^2}$ and the deflections $\frac{dx}{ds}$, $\frac{dy}{ds}$ are small, as well as

$$\frac{dz}{ds} \approx 1 \quad \text{(IV.4.228)}$$

With those facts in mind, let's exploit the constraint $\left(\frac{d\mathbf{r}}{ds}\right)^2 = 1$. If we differentiate this relationship with respect to $s$, we obtain $2\left(\frac{d\mathbf{r}}{ds}\right)\left(\frac{d^2\mathbf{r}}{ds^2}\right) = 0$, from where, going back to Cartesian coordinates $\mathbf{r} = \mathbf{i}x + \mathbf{j}y + \mathbf{k}z$, we obtain $\frac{dz}{ds}\frac{d^2z}{ds^2} + \left(\frac{dx}{ds}\frac{d^2x}{ds^2} + \frac{dy}{ds}\frac{d^2y}{ds^2}\right) = 0$. Now taking into account Eq. (IV.4.228), we get

$$\frac{d^2z}{ds^2} \approx -\left(\frac{dx}{ds}\frac{d^2x}{ds^2} + \frac{dy}{ds}\frac{d^2y}{ds^2}\right) \tag{IV.4.229}$$

Due to the smallness of the factors $\frac{dx}{ds}$ and $\frac{dy}{ds}$, it is obvious that $\frac{d^2z}{ds^2} \ll \frac{d^2x}{ds^2}, \frac{d^2y}{ds^2}$, i.e. although $\frac{d^2x}{ds^2}$ and $\frac{d^2y}{ds^2}$ are themselves small, $\frac{d^2z}{ds^2}$ is even smaller. Hence, introducing $\mathbf{R}(s) = x(s)\hat{\mathbf{i}} + y(s)\hat{\mathbf{j}}$, we may write

$$\left(\frac{d^2\mathbf{r}}{ds^2}\right)^2 \approx \left(\frac{d^2\mathbf{R}}{ds^2}\right)^2 = \left(\frac{d^2x}{ds^2}\right)^2 + \left(\frac{d^2y}{ds^2}\right)^2 \tag{IV.4.230}$$

and thus approximate the partition function as

$$Z = \int \exp\left(-\frac{1}{k_BT}\int_0^L \left[\frac{l_p}{2}\left(\frac{d^2\mathbf{R}(s)}{ds^2}\right)^2 + V_{2D}(\mathbf{R}(s))\right]ds\right)D\mathbf{R}(s) \tag{IV.4.231}$$

Note that $D\mathbf{R}(s)$ represents path integration over both $x(s)$ and $y(s)$.

All is ready now! Taking into account that

$$\int \delta^\infty\left(t_x(s) - \frac{dx(s)}{ds}\right)\delta^\infty\left(t_y(s) - \frac{dy(s)}{ds}\right)Dt_x(s)Dt_y(s)$$
$$\equiv \int \delta^\infty\left(\mathbf{t}(s) - \frac{d\mathbf{R}(s)}{ds}\right)D\mathbf{t}(s) = 1 \tag{IV.4.232}$$

we can do the Hubbard–Stratonovich transform, namely rewrite the partition function as

$$Z = \int\int \delta^\infty\left(\mathbf{t}(s) - \frac{d\mathbf{R}(s)}{ds}\right)$$
$$\times \exp\left(-\frac{1}{k_BT}\int_0^L \left[\frac{l_p}{2}\left(\frac{d\mathbf{t}(s)}{ds}\right)^2 + V(\mathbf{R}(s))\right]ds\right)D\mathbf{R}(s)D\mathbf{t}(s) \tag{IV.4.233}$$

where $l_p$ is the quantity proportional to *bending persistence length*: the larger $l_p$, the harder locally would it be for the polymer to bend. This can then be written through Eq. (IV.4.225) as

$$Z = \int \exp\left(i \int_0^L \left(\mathbf{t}(s) - \frac{d\mathbf{R}(s)}{ds}\right).\mathbf{p}(s)ds\right)$$

$$\times \exp\left(-\frac{1}{k_BT} \int_0^L \left[\frac{l_p}{2}\left(\frac{d\mathbf{t}(s)}{ds}\right)^2 + V_{2D}(\mathbf{R}(s))\right]ds\right)$$

$$\times D\mathbf{R}(s)D\mathbf{t}(s)D\mathbf{p}(s) \tag{IV.4.234}$$

To proceed we follow a method similar to that described by Eqs. (IV.4.65)–(IV.4.81). First we can write equations similar to Eqs. (IV.4.66)–(IV.4.68), which are

$$W(\mathbf{t}_0, \mathbf{R}_0, \mathbf{t}_L, \mathbf{R}_L) = \lim_{\substack{\varepsilon \to 0 \\ N \to \infty}} W_d(\mathbf{t}_0, \mathbf{R}_0, \mathbf{t}_N, \mathbf{R}_N) \tag{IV.4.235}$$

$$W_d(\mathbf{t}_{i-2}, \mathbf{R}_{i-2}, \mathbf{t}_i, \mathbf{R}_i)$$

$$= \int d\mathbf{t}_{i-1} \int d\mathbf{R}_{i-1} W_d(\mathbf{t}_{i-2}, \mathbf{R}_{i-2}, \mathbf{t}_{i-1}, \mathbf{R}_{i-1})$$

$$\times W_d(\mathbf{t}_{i-1}, \mathbf{R}_{i-1}, \mathbf{t}_i, \mathbf{R}_i) \tag{IV.4.236}$$

$$Z = \int d\mathbf{t}_0 \int d\mathbf{t}_L \int d\mathbf{R}_0 \int d\mathbf{R}_L W(\mathbf{t}_0, \mathbf{R}_0, \mathbf{t}_L, \mathbf{R}_L) \tag{IV.4.237}$$

where we have used the notations for the position $\mathbf{R}_i = \hat{\mathbf{i}}x_i + \hat{\mathbf{j}}y_i$ and tangent vectors $\mathbf{t}_i = t_{x,i}\hat{\mathbf{i}} + t_{y,i}\hat{\mathbf{j}}$. We then exploit the integral representation of $W_d$,

$$W_d(\mathbf{t}_{i-1}, \mathbf{R}_{i-1}, \mathbf{t}_i, \mathbf{R}_i) = \int d\mathbf{p} \exp(-i(\mathbf{R}_{i-1} - \mathbf{R}_i).\mathbf{p} + \varepsilon V_{2D}(\mathbf{R}_i))$$

$$\times \exp\left(-\frac{l_p(\mathbf{t}_{i-1} - \mathbf{t}_i)^2}{2\varepsilon} + i\mathbf{t}_i.\mathbf{p}\right) \tag{IV.4.238}$$

Further rewriting it as

$$W_d(\mathbf{t}_{i-1}, \mathbf{R}_{i-1}, \mathbf{t}_i, \mathbf{R}_i) \propto \int dk \int dp$$

$$\times \exp\left(-i(\mathbf{R}_{i-1} - \mathbf{R}_i).\mathbf{p} - \varepsilon V_{2D}(\mathbf{R}_i)\right)$$

$$\times \exp\left(i\varepsilon \mathbf{t}_i.\mathbf{p} - \frac{\varepsilon}{2l_p}\mathbf{k}^2\right) \exp(i(\mathbf{t}_i - \mathbf{t}_{i-1}).\mathbf{k}) \quad \text{(IV.4.239)}$$

Here the constant of proportionality does not matter, and we will not bother to calculate it. We can then expand out the exponential for small $\varepsilon$, which yields

$$W_d(\mathbf{t}_{i-1}, \mathbf{R}_{i-1}, \mathbf{t}_i, \mathbf{R}_i) \propto \int dk \int dp \exp(-i(\mathbf{t}_{i-1}.\mathbf{k} + \mathbf{R}_{i-1}.\mathbf{p}))$$

$$\times \left[1 + \frac{\varepsilon}{2l_p}\vec{\nabla}^2_{t_i} + \varepsilon \mathbf{t}_i.\vec{\nabla}_{R_i} - \varepsilon V_{2D}(\mathbf{r}_i)\right] \exp(i(\mathbf{t}_i.\mathbf{k} + \mathbf{R}_i.\mathbf{p}))$$

$$\text{(IV.4.240)}$$

where

$$\vec{\nabla}^2_{t_i} = \frac{\partial^2}{\partial t^2_{x,i}} + \frac{\partial^2}{\partial t^2_{y,i}} \quad \vec{\nabla}_{R_i} = \hat{\mathbf{i}}\frac{\partial}{\partial x_i} + \hat{\mathbf{j}}\frac{\partial}{\partial y_i} \quad \text{(IV.4.241)}$$

Next we can expand out $e^{i[\mathbf{t}_i.\mathbf{k} + \mathbf{R}_i.\mathbf{p}]}$ in terms of eigenfunctions $\chi_j(\mathbf{R}_i, \mathbf{t}_i)$ that satisfy

$$\hat{E}\chi_j(\mathbf{R}_i, \mathbf{t}_i) = \left[-\frac{1}{2l_p}\vec{\nabla}^2_{t_i} - \mathbf{t}_i.\vec{\nabla}_{R_i} + V_{2D}(\mathbf{R}_i)\right] \chi_j(\mathbf{R}_i, \mathbf{t}_i)$$

$$= E_j\chi_j(\mathbf{R}_i, \mathbf{t}_i) \quad \text{(IV.4.242)}$$

Such expansions can be written as

$$\exp(i(\mathbf{R}_i.\mathbf{k} + \mathbf{r}_i.\mathbf{p})) = \sum_j c_j(\mathbf{p}, \mathbf{k})\chi_j(\mathbf{r}_i, \mathbf{R}_i) \quad \text{(IV.4.243)}$$

These eigenfunctions, though they give complex eigenvalues, can obey an orthogonality condition [provided that $V(\mathbf{R}_i) = V(-\mathbf{R}_i)$]:

$$\int d\mathbf{R}_i \int dt_i \chi_{j'}(-\mathbf{R}_i, t_i) \chi_j(\mathbf{R}_i, t_i) = \delta_{j,j'} \qquad \text{(IV.4.244)}$$

Equation (IV.4.243) allows us to write

$$W(t_{i-1}, \mathbf{R}_{i-1}, t_i, \mathbf{R}_i) \propto \sum_{j,j'} \int d\mathbf{k}$$

$$\times \int d\mathbf{p} c_{j'}(\mathbf{p}, -\mathbf{k}) \chi_{j'}(-\mathbf{R}_{i-1}, t_{i-1})[1 - \varepsilon E_j]$$

$$\times c_j(\mathbf{p}, \mathbf{k}) \chi_j(\mathbf{R}_i, t_i) \qquad \text{(IV.4.245)}$$

Through Eqs. (IV.4.243) and (IV.4.244) we get

$$c_j(\mathbf{p}, \mathbf{k}) = \int dt_j \int d\mathbf{R}_i \exp(i(\hat{\mathbf{t}}_i.\mathbf{k} + \mathbf{R}_i.\mathbf{p})) \chi_j(-\mathbf{R}_i, t_i) \qquad \text{(IV.4.246)}$$

$$c_j(\mathbf{p}, -\mathbf{k}) = \int dt_j \int d\mathbf{R}_i \exp(-i(\hat{\mathbf{t}}_i.\mathbf{k} + \mathbf{R}_i.\mathbf{p})) \chi_j(\mathbf{R}_i, t_i) \qquad \text{(IV.4.247)}$$

and can show that

$$\int d\mathbf{k} \int d\mathbf{p} c_{j'}(\mathbf{p}, -\mathbf{k}) c_j(\mathbf{p}, \mathbf{k}) \propto \delta_{j,j'} \qquad \text{(IV.4.248)}$$

Then we can simplify Eq. (IV.4.245) to

$$W(t_{i-1}, \mathbf{R}_{i-1}, t_i, \mathbf{R}_i) \propto \sum_j \chi_j(-\mathbf{R}_{i-1}, t_{i-1}) \exp(-\varepsilon E_j) \chi_j(\mathbf{R}_i, t_i) \qquad \text{(IV.4.249)}$$

**Exercise.** Show that from Eqs. (IV.4.236), (IV.4.249) and the orthogonality condition, Eq. (IV.4.244), that

$$W(\mathbf{t}_0, \mathbf{R}_0, \mathbf{t}_L, \mathbf{R}_L) \propto \lim_{\substack{\varepsilon \to 0 \\ N \to \infty}} \sum_j \chi_j(-\mathbf{R}_0, \mathbf{t}_0) \exp(-\varepsilon N E_{j'}) \chi_j(\mathbf{R}_N, \mathbf{t}_N)$$

$$= \sum_j \chi_j(-\mathbf{R}_0, \mathbf{t}_0) \exp(-L E_{j'}) \chi_j(\mathbf{R}_L, \mathbf{t}_L)$$

$$(IV.4.250)$$

and

$$Z \propto \sum_j \int dt_0 \int dt_L \int d\mathbf{R}_0 \int d\mathbf{R}_L \chi_j(-\mathbf{R}_0, \mathbf{t}_0) \exp(-L E_{j'}) \chi_j(\mathbf{R}_L, \mathbf{t}_L)$$

$$(IV.4.251)$$

**Exercise.** From Eqs. (IV.4.250) and (IV.4.251) show similar forms to Eqs. (IV.4.82) in the limit when $L \to \infty$.

**Example IV.4.7 (Continued).** If we write $V_{2D}(\mathbf{R}) = V_x(x) + V_y(y)$, we can simplify Eq. (IV.4.242) through separation of variables by writing $\chi_j(\mathbf{R}_i, \mathbf{t}_i) = \sigma_n(x_i, t_{x,i}) \tau_m(y_i, t_{y,i})$ (where eigenvalues are now numbered) so that we obtain two separate equations

$$\left[ -\frac{1}{2l_p} \frac{\partial^2}{\partial t_x^2} - t_x \cdot \frac{\partial}{\partial x} + V_x(x) \right] \sigma_n(x, t_x) = E_n \sigma_n(x, t_x) \qquad (IV.4.252)$$

$$\left[ -\frac{1}{2l_p} \frac{\partial^2}{\partial t_y^2} - t_y \cdot \frac{\partial}{\partial y} + V_y(y) \right] \tau_m(y, t_y) = E_m \tau_m(y, t_y) \qquad (IV.4.253)$$

where the total eigenvalues are $E_j = E_n + E_m$.

We cannot use the separation of variables method any further in solving Eqs. (IV.4.252) and (IV.4.253). In general, finding analytical solutions to these partial differential equations can be hard, indeed for many forms of $V_x(x)$ and $V_y(y)$ simply impossible, unless some approximations are made. However, in the case of Harmonic potentials

$$V_x(x)/(k_B T) = k_x x^2/2, \quad V_y(y)/(k_B T) = k_y y^2/2 \qquad (IV.4.254)$$

these equations can straightforwardly be solved. What may be surprising is that it is much harder to solve this equation for infinite potential wells, which may go against a reader's intuitive understanding of simpler eigenvalue equations in quantum mechanics! Here, we'll not go into details of it, but just give you the result for the lowest eigenvalues $E_{n=0} = \sqrt{2}(k_x/l_P)^{1/4}$ and $E_{m=0} = \sqrt{2}(k_y/l_P)^{1/4}$.

Thus, the free energy of worm like chain in such a confining potential (up to an ill-defined constant, $A_0$) is given by

$$\lim_{L \to \infty} \left\{ \frac{A - A_0}{L} \right\} = 2\sqrt{2}k_B T[(k_x/l_P)^{1/4} + (k_y/l_P)^{1/4}] \quad \text{(IV.4.255)}$$

Note that if we wanted to model a polymer in a cylindrical tube or in the effective mean field of equally spaced neighbouring polymers we would set the 'spring constants' in Eq. (IV.4.254) $k_x = k_y = k$.

In the spirit of, both volumes of this book, Eq. (IV.4.255) is a classic example of a *formula*; all the rest in its derivation may be called *equations*. Its form is not complicated, but it looks nontrivial—all those $1/4$ exponents—and its derivation was rather 'painful'; it is worth taking the time to make sense of it. Of course, it is expedient to do this with any formula—it is just that we may not have enough space in the book to do it each time. But this case is particularly instructive.

So, let's understand and rationalize the obtained result. The free energy we have calculated here can be used to estimate configurational entropy loss due to confinement $\Delta S_{conf} = A/T$, where the infinite length unconfined polymer free energy can be set at zero. Note that as we increase both $k_x$ and $k_y$ the free energy in Eq. (IV.4.255) increases. Obviously it should be so, because the polymer chain is in a thermal bath and at nonzero temperature it 'wants' to fluctuate, and the steeper the confining potential, the harder it becomes for it to do so. The same comment refers to the dependence on $l_P$: larger bending persistence length, i.e. larger values of $l_P$, suppress fluctuations and so the effect of confinement is less pronounced. But how would you know, without calculation, that that free energy per unit length should end up scaling $\propto k^{1/4}$ and $\propto l_P^{-1/4}$?

Let us recall the lessons of dimensionality and scaling analysis, which we preached in Chapter 10 of Part I (Vol. 1). In a moment you will see how powerful this is! Indeed, if we assume that the free energy per unit length due to fluctuations under confinement must somehow be proportional to $k_B T$, then to get the free energy per unit length, we must build a combination of dimensionality of one over length from the only two other parameters of the problem, $k$ and $l_P$. To be able to do this, we need first to recognize the dimensionality of these parameters. The way that $l_P$ enters into the energy functional, Eq. (IV.4.227), tells us that its dimensionality is: $[l_p] = [\text{Energy} \cdot \text{Length}]$. Whereas from the same functional it is clear that the dimensionality of $V$ is: $[V] = \left[\frac{\text{Energy}}{\text{Length}}\right]$, and, as introduced in Eq. (IV.4.254), the dimensionality of $k$ comes out as $[k] = \frac{[\text{Energy}]}{[\text{Length}]^3}$. Hence, if want to make a combination of the dimensionality $\left[\frac{1}{\text{Length}}\right]$, we have to divide $k$ by $l_p$ and take the 1/4 root of that ratio, i.e. get $(k/l_P)^{1/4}$. Remarkably, unlike what we did in all the examples of Chapter 10 of Part I (Vol. 1), we did not even need here to guess first what should stand in numerator or denominator, $k$ or $l_p$. The dimensionality analysis did not give us a chance to misplace them. Isn't it beautiful?

Note that with a harmonic potential we could have calculated the free energy from Eq. (IV.4.231) through direct evaluation of the functional integral, by evaluating a Gaussian integral using spatial Fourier transforms. We have shown this technique already, so we leave this check to the reader, as an **Exercise**. However, for finite length chains, with free ends or fixed conditions at the ends, it can be useful to obtain both the eigenvalue spectrum and the eigenfunctions.

A less passionate reader (and you are certainly not among them, as you have reached this part of the book) may conclude, "Why should I go into the complexity of the functional integration, when— as you have demonstrated me—I can get the result at the tip of my pen using scaling analysis?" The answer is that you will not get the factor $2\sqrt{2}$ in Eq. (IV.4.255), and you may doubt (if you are particularly sceptical about everything) that the result is just proportional to $k_B T$.

## 4.7. A Bit More on Path Integrals

Definitely, just a bit. Indeed, the natural continuation of the story of the path integrals would be to introduce you to perturbation theory, say using for example Landau–Ginzburg–Wilson free energy and consider there $\frac{\beta}{4!}\psi(\mathbf{r})^4$ as perturbation, proceeding then to the Feynman diagrams representing the terms of the perturbation series, partial summation over the most important series of the diagrams, Hartree approximation, calculation of the correlation function and the like. But all this involves very cumbersome algebra and the paths to simple formulae are very long. Feeling that the patience of yours and of the publisher may be already running out, on this front we redirect you to more specialized books. Still, we will conclude this part with a few, perhaps at a first glance disconnected pieces of information and examples of application of functional integrals.

### The statistical mechanics of membranes encapsulated in a functional integral

**Example IV.4.8 (Physical Application ▶▶).** Let's look at a 2D case of functional integration. We may write a functional integral to describe the partition function of a fluctuating membrane in terms of its vertical displacement $f(x, y)$. Let's suppose for simplicity that the membrane's intrinsic curvature is zero and the membrane itself is restricted within the plane region, square $-a < x < a$, $-b < y < b$. As we described in Example. III.2.10, we can write the elastic energy of a membrane as

$$E_{el} = \int_S (2B_1\overline{k}^2 + B_2K)d\mathbf{R} = \int_{-b}^{b}\int_{-a}^{a} (2B_1\overline{k}^2 + B_2K)\sqrt{\det \mathbf{K}_1}\,dxdy$$

$$(IV.4.256)$$

For complete notations see Sec. 2.2 of Part III. Note that the surface is parametrized by $x, y$ coordinates (i.e. $t = y$ and $s = x$). We also derived forms for $\mathbf{K}_1$, $K$ and $\overline{k}$ for such a surface. We can describe the thermal fluctuations of a membrane by the following functional integral for the partition function:

$$Z = \int \exp\left(-\frac{E_{el}[f(x,y)]}{k_B T}\right) Df(x,y) \qquad (\text{IV.4.257})$$

**Exercise.** Use the results given by Eq. (III. 2.113), (III.2.120) and (III.2.121), for $\det \mathbf{K}_1$, $K$ and $\bar{k}$, to show that the energy functional for the case when gradients $\partial f/\partial x \ll 1$ and $\partial f/\partial y \ll 1$ can be written as

$$E_{el}[f(x,y)] \approx \int_{-a}^{a} dx \int_{-b}^{b} dy \left(\frac{B_1}{2}\left(\frac{\partial^2 f(x,y)}{\partial x^2} + \frac{\partial^2 f(x,y)}{\partial y^2}\right)^2\right.$$

$$\left. + B_2 \left(\frac{\partial^2 f(x,y)}{\partial x^2}\frac{\partial^2 f(x,y)}{\partial y^2} - \left(\frac{\partial^2 f(x,y)}{\partial x \partial y}\right)^2\right)\right)$$

$$(\text{IV.4.258})$$

and specify expressions for $B_1$ and $B_2$.

### A snapshot of path integrals describing Bardeen–Cooper–Schrieffer (BCS) superconductivity

**Example IV.4.9 (Physical Application ▶▶).** Functional integration and fluctuating field theories can be applied to the theory of metals. One classic application of this is the BCS (Bardeen–Cooper–Schrieffer) theory, a non-relativistic many-body/quantum field theory describing the pairing, through phonon interactions, of electrons of opposite spin to create a condensate of Cooper-pairs. In this case we may write down a path integral to describe the partition function to describe such a theory. In the simplest version of the theory the phonon interaction can be assumed to be short ranged (local), and its partition function can be written as (again in units where $\hbar = 1$)

$$Z = \int D\psi(\mathbf{r},\tau) \int D\psi^\dagger(\mathbf{r},\tau) \exp(-S_{BCS}[\psi(\mathbf{r},\tau), \psi^\dagger(\mathbf{r},\tau)])$$

$$(\text{IV.4.259})$$

where

$$\boldsymbol{\psi}(\mathbf{r}, t) = \begin{pmatrix} \psi_\uparrow(\mathbf{r}, \tau) \\ \psi_\downarrow^*(\mathbf{r}, \tau) \end{pmatrix} \quad \text{and} \quad \boldsymbol{\psi}^\dagger(\mathbf{r}, \tau) = (\psi_\uparrow^*(\mathbf{r}, \tau), \psi_\downarrow(\mathbf{r}, \tau))$$

(IV.4.260)

This is termed a Nambu spinor, where $\psi_\uparrow(\mathbf{r}, \tau)$ represents the field of a spin up electron and $\psi_\downarrow(\mathbf{r}, \tau)$ of a spin down one; and $S_{BCS}[\boldsymbol{\psi}(\mathbf{r}, \tau), \boldsymbol{\psi}^\dagger(\mathbf{r}, \tau)]$ is the BCS action. Again, $\tau$ is imaginary time. Importantly, the values of the fields $\psi_\uparrow(\mathbf{r}, \tau)$ and $\psi_\downarrow(\mathbf{r}, \tau)$, and their complex conjugates, are not described by normal numbers but by some weird numbers called Grassmann numbers (or variables) which, instead of commuting as normal numbers do, anti-commute; for instance $\psi_\uparrow(\mathbf{r}_1, \tau_1)\psi_\uparrow(\mathbf{r}_2, \tau_2) = -\psi_\uparrow(\mathbf{r}_2, \tau_2)\psi_\uparrow(\mathbf{r}_1, \tau_1)$. What is particularly weird is that from anti-commutation we must have $(\psi_\uparrow(\mathbf{r}_1, \tau_1))^2 = 0$. Why must we use such numbers? It is because electrons are fermions and their overall wave functions must be anti-symmetric, i.e. on swapping labels the wave function must change sign. Fields in the path integral that are Grassmann numbers are needed for us to satisfy this requirement.

The BCS action is then given in its simplest form by

$$S_{BCS}[\boldsymbol{\psi}(\mathbf{r}, \tau), \boldsymbol{\psi}^\dagger(\mathbf{r}, \tau)]$$

$$= \int_0^\beta d\tau \int d\mathbf{r} \left[ \sum_{\sigma=\uparrow,\downarrow} \psi_\sigma^*(\mathbf{r}, \tau) \left[ \frac{\partial}{\partial \tau} + \frac{1}{2m} \nabla^2 \right] \psi_\sigma(\mathbf{r}, t) \right.$$

$$\left. + g\psi_\uparrow^*(\mathbf{r}, \tau)\psi_\downarrow^*(\mathbf{r}, \tau)\psi_\downarrow(\mathbf{r}, \tau)\psi_\uparrow(\mathbf{r}, \tau) \right]$$

(IV.4.261)

where $g > 0$ is the strength of BCS pairing interaction between a spin up and spin down electron.

We should point out that we still would have a lot more to discuss if we were to consider the simplest of realistic relativistic field theories: quantum electrodynamics, quantum theory interaction between matter and the electromagnetic field, which can also be described by path integrals. If we wanted to consider only interactions between electrons and photons, we would still need to consider integration over a type of vector called a Dirac spinor containing four Grassmann fields to describe the spin states of electrons and their anti-matter partners (positrons), together with integration over a four-component vector field in space $A_\mu(\mathbf{r}, \tau) = (\varphi(\mathbf{r}, \tau), \mathbf{A}(\mathbf{r}, \tau))$; here $\varphi(\mathbf{r}, \tau)$ is the scalar (electrostatic) potential and $\mathbf{A}(\mathbf{r}, \tau)$ is the vector (magnetic) potential, describing quantum fluctuations—photons—of the electromagnetic field.

## 4.8.  Key Points

- Path (or functional) integrals are important in the description of the processes which may realize a continuum variety of pathways or properties that represent a continuum 'assortment' of options, of which some, of course, are more probable than others.
- Whereas variational calculus focuses on most probable paths, theories based on functional integrals can study the effects of fluctuations around those optional paths. In some situations, such as close to phase or structural transitions, fluctuations crucially determine the properties of the system.
- The variety of options and paths describe the correlation functions of the variables of interest in space and time. The method of *functional differentiation* with respect to fictitious functions inherent to specially constructed *generating functionals* offers a straightforward technique to calculate such correlation functions.
- For obvious reasons, path integrals find their most natural applications in statistical mechanics, classical stochastic dynamics, and, of course, in quantum mechanics where quantum particles follow the trajectories dispersed subject to the uncertainty principle. The formalism of path integrals in quantum mechanics operates initially with probability amplitudes rather than probability distributions. There is a link between these two classes of applications.
- The so-called homogeneous path integrals are those for which the dependence on the space-time coordinates comes into the integrand only through the 'argument' function(s).

- Homogenous *Gaussian* integrals (in which the dependence over the varying, 'argument' function(s) is quadratic), have a standard routine for calculation of *expectation values* for that (those) function(s), or any order of their *correlation functions*.

- Functional integrals can be applied to the Landau–Ginzburg theory, whose application shows the crucial importance of fluctuations near the critical point. The Ginzburg criterion specifies when they are essential.

- The concept of functional integration is most transparent in 1D (when the function(s) in question depends(depend) on one coordinate). But this concept extends also to 3D (and 4D, i.e. with time involved) both in free space and with boundary conditions. In the latter approach eigen-equation and eigenvalue analysis come into play, linked to the Sturm–Liouville theory introduced in Chapter 4 of Part III of this volume.

- The best pedagogical example that gives a glimpse of the power of functional integrals is diffusion, and indeed the path integrals were first introduced by Einstein and Smoluchowski in their theory of Brownian motion. Another example is the theory of worm-like chain polymers, particularly instructive when describing the latter in confinement (e.g. within a cylindrical tube): this demonstrates the role of boundary conditions.

- Delta-functionals and the so-called *Hubbard–Stratonovich transformation* are useful mathematical tricks, which help to calculate functional integrals if there are additional conditions for the functions that they must satisfy, which will affect the result of the functional integration.

- The complicated derivations using path-integral techniques may well result in compact analytical formulae, although having at a first glance exotic, 'unexpected form'. These are always good to check via dimensionality and scaling analysis, as demonstrated in Example IV.4.7 of this chapter. This example shows that the resulting law that may have seemed surprising should be exactly as obtained.

# Chapter 5

# In the Thin Air of the Fractional World

Now according to our plan, we may say that we have reached Camp 4. If you are a mountain lover you will know, however, that it is not recommended to stay there for long, but instead to move forward to the summit. But we still feel that you deserve some rest and entertainment at this point. Hence, we offer you in conclu-

sion this rather light chapter about fractional scaling and connections between the discrete and continuous worlds. Before going ahead with this we will need to introduce some very basic concepts of fractals and fractional dimensions, as we will be referring to them from time to time, when dealing with random walks and percolation.

## 5.1. Complexity, Euclidean and Non-Euclidean Worlds: Fractals and Fractal Dimensions

Fractals have become common place; even children know of them. Still, to settle few language issues we need to briefly speak about them, because the concepts related to fractals and fractal dimensions underpin what we would call here the world of fractional phenomena.

# The beauty of complexity

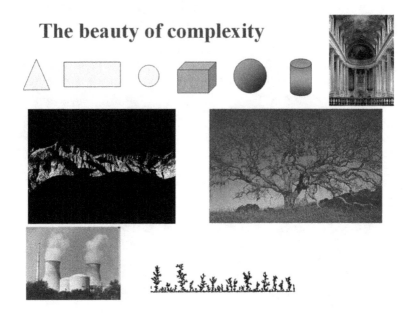

There is a saying, attributed to Albert Einstein, that a true scientist cannot not enjoy the beauty of Euclidean geometry which obeys strict and logical mathematical laws. But in the last quarter of the 20th century, scientists became massively excited by the search for common laws in much more complicated structures than regular Euclidean objects. The pictures above contrast the shape of some of such objects with what we see around us. But even in a cathedral, whereas the architecture of the building is quite Euclidean, the elements of the paintings on the ceiling are already quite complex. We see the same in trees or mountains, clouds above the power plant (although its buildings are quite Euclidean), and dendrites of metal deposition. Ancient Chinese wisdom says: "in a tiny rock-stone you can see the whole mountain." This already touches upon the concept of self-similarity, complex systems in which by increasing resolution you can see the same patterns at different scales. Anecdotal illustration of this concept is illustrated by photo-perspective, shown in the railway line box; should we take it seriously or not?

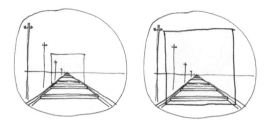

Self-similar as well as non-self-similar complex structures are a subject of what nowadays is called *complexity* science. That science deals with the following questions:

— How are such structures generated and how do they grow in time?
— How can they be described?
— How can they be experimentally characterized?
— What is special about the processes in/on these structures? (e.g. transport and reactions in complex media)

It also deals not only with structures but processes which leads to complex temporary patterns.

There are books and lecture courses on the science of complexity, and we have no capacity to go into it in detail here. But we will focus on what at a first glance seems a heretic question: can complexity lead in the end to some kind of simplicity? This chapter will show you that in some cases—yes; we will present a few examples and the corresponding mathematics which demonstrates this.

To proceed further, we need to touch upon few concepts related to *fractals*, the self-similar objects that look the same at different resolution. Many famous scientists and mathematicians have been involved in the science of fractals, but one name stands out—Benoit Mandelbrot, who is often called "the father of fractal geometry" (read his captivating book *The Fractal Geometry of Nature*).

Benoit Mandelbrot
((1924–2010))

According to Mandelbrot the story of fractals, at least for him, started from a British mathematician, physicists, and meteorologist Lewis Fry Richardson, although, as we will even see from the concise, distilled content of this chapter, the story goes back to the 1870s, to mathematicians who pioneered the theory of sets (a branch of mathematical logic that investigates the structures of collections of objects). There is no possibility and no need for us here to go into set theory, although we will refer to some outcomes of it from time-to-time.

Lewis Fry Richardson
(1881–1953)

So, back to Richardson. A pacifist, interested in the social theory of national conflicts, he came up with a hypothesis that one of the factors that influences the probability of conflicts between countries, if they are neighbours, is somehow related with the length of the border between them. So he started to gather information about the length of the borders between different countries, and then to his surprise discovered that the border between Spain and Portugal is reported smaller in Spanish books than in Portuguese ones. The same puzzle he found with Dutch and Belgian data, on their common border. He has explained this inconsistency: the 'yardsticks' that the two countries used to measure those borders were different—the one who reported it to be longer used a smaller yardstick. Mandelbrot has rationalized this inconsistency in his famous paper "How long is the Coast of Britain?"

Mandelbrot collected the data for different rugged coastlines to plot the log of the measured perimeter length vs the log of the length of the yardstick that has been used to measure it. They well fit to straight lines with negative slopes, steeper for the coast of Norway, less steep for the coast of Britain, and even less steep for the Australian coast. What did that mean, and what was that slope related to? Now the answer seems trivial, but it was Mandelbrot who made it obvious for everyone.

**Fig. IV.5.1** Measuring the perimeter of a rugged curve with a yardstick. The number of times the shorter (green) yardstick is applied is greater than for the longer (red) one.

As in Fig. IV.5.1, the number of times, $N$, that one needs to apply a yardstick of length $\varepsilon$ to measure the perimeter of that kind of curve is larger the smaller the stick. But by how much? If experimentally, in the log-log plot of $N(\varepsilon)$ we see a straight line, this would mean a power law, so that the dependence of the thus measured perimeter length on the yardstick length would be,

$$L(\varepsilon) = \varepsilon N(\varepsilon) \propto \varepsilon \left[\frac{L_0}{\varepsilon}\right]^D \propto \frac{1}{\varepsilon^{D-1}} \quad D > 1 \qquad \text{(IV.5.1)}$$

(where $L_0$ is the length of the base line, the exact value of which is not important here, since it is just a part of the proportionality factor in the dependence on $\varepsilon$). Hence,

$$\log L(\varepsilon) = \text{const} - (D-1)\log\varepsilon \qquad \text{(IV.5.2)}$$

If we made a similar measurement of the length of the straight line, see Fig. IV.5.2, the result would not depend on the length of the yardstick (possible incommensurability of the length of the yard-stick and the length of the line is unimportant for a line much longer than the yardstick). Thus had we pushed ourselves to write the result of this measurement in the form of Eq. (IV.5.2), we would get in this case $D = 1$. But this is the topological dimension of the straight line. So, what is $D(> 1)$ for a rugged curve? $D$ was called a *fractal dimension of the line*, or *perimeter dimension*, or the *Hausdorff–Besicovitch*

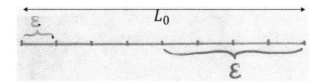

**Fig. IV.5.2** A smaller yardstick should be applied that many more times the smaller it (incommensurability of the length of the yard-stick and the length of the line is unimportant for a line much longer than the yardstick).

*dimension*, which is larger than the topological dimension and for a continuous curve would lie between 1 and 2. We will consider few examples in a moment.

The name of this dimension honours the fact that Cambridge mathematician Abram Besicovitch, born in the Russian Empire and trained in the top Russian mathematical school (of Andrey Markov), introduced the notion of fractal dimension in his studies of complex curves as early as 1929; even earlier, in 1918, it was proposed by German mathematician Felix Hausdorff in his studies of irregular sets, though in a less general form. But it was due to Mandelbrot that the idea of fractal dimension became known beyond the mathematical community.

Abram Besicovitch
(1881–1970)

Felix Hausdorff
(1868–1942)

### Deterministic fractals and calculation of fractal dimensions

We will now consider calculation of the perimeter dimension using a few examples of what Mandelbrot called *deterministic curves*, those built via a defined algorithm, starting with the so-called Koch coastline (see Fig. IV.5.3). This is named after Swedish mathematician Helge von Koch, who was the author of

Niels Fabian
Helge von Koch
(1870–1924)

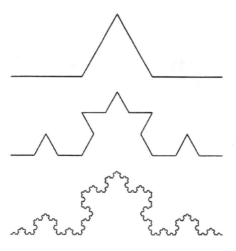

**Fig. IV.5.3** Generating the Koch coastline.

probably the first such curve, which he described in
1904. That 'coastline curve' is generated according to
the following algorithm. At every next generation a
straight-line segment is replaced by the form, here sketched on the
right.

Thus, each time we decrease the length of the yardstick by 3 times,
the measured perimeter will look 4/3 times longer, i.e.

$$L\left(\frac{\varepsilon}{3}\right) = \frac{4}{3}L(\varepsilon) \tag{IV.5.3}$$

The solution of this equation has the power law form,

$$L(\varepsilon) = A\varepsilon^{\mu} \tag{IV.5.4}$$

where $A$ is the proportionality constant (of no importance to us here)
and $\mu$ is the exponent which is to be found by the substitution
of Eq. (IV.5.4) into Eq. (IV.5.3), so that we get $A\frac{\varepsilon^{\mu}}{3^{\mu}} = \frac{4}{3}A\varepsilon^{\mu} \Rightarrow$
$\mu = 1 - (\ln 4/\ln 3) = 1 - D$, and thus the fractal dimension of the
curve generated according to this algorithm is given by

$$D = \frac{\ln 4}{\ln 3} = 1.26 \tag{IV.5.5}$$

**Koch snow-flake** $\quad D=1.26$

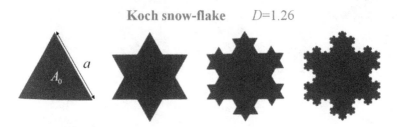

Its area is finite and Euclidean: $A = 1.6A_0$

Basic triangle area $\qquad$ Number and side length of new triangles in each $k$-generation

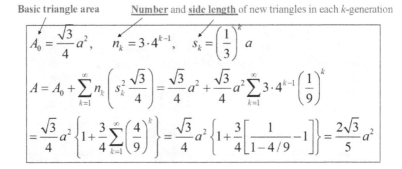

$$A_0 = \frac{\sqrt{3}}{4}a^2, \quad n_k = 3 \cdot 4^{k-1}, \quad s_k = \left(\frac{1}{3}\right)^k a$$

$$A = A_0 + \sum_{k=1}^{\infty} n_k\left(s_k^2 \frac{\sqrt{3}}{4}\right) = \frac{\sqrt{3}}{4}a^2 + \frac{\sqrt{3}}{4}a^2\sum_{k=1}^{\infty} 3 \cdot 4^{k-1}\left(\frac{1}{9}\right)^k$$

$$= \frac{\sqrt{3}}{4}a^2\left\{1 + \frac{3}{4}\sum_{k=1}^{\infty}\left(\frac{4}{9}\right)^k\right\} = \frac{\sqrt{3}}{4}a^2\left\{1 + \frac{3}{4}\left[\frac{1}{1-4/9}-1\right]\right\} = \frac{2\sqrt{3}}{5}a^2$$

**Fig. IV.5.4** Perimeter and the area of the Koch snowflake.

You can easily generalize this analysis by making the yardstick $b$ times smaller: you will have to apply $N$ times to measure the perimeter, and the equation similar to Eq. (IV.5.3) will read $L\left(\frac{1}{b}\right) = NL(1)$, the solution of which has the form of Eq. (IV.5.4) but with $\mu = 1 - D = 1 - \frac{\ln N}{\ln b}$, i.e.

$$D = \frac{\ln N}{\ln b} \tag{IV.5.6}$$

Obviously, the perimeter of Koch island (or the Koch snowflake), as in Fig. IV.5.4, has the same fractal dimension. Noteworthy, its area increases with each generation, but saturates to the finite maximal value for infinite number of generations; the area is 'Euclidean' as it scales proportionally to the square of the base length.

In Fig. IV.5.5 we show a few more instructive examples of fractal curves.

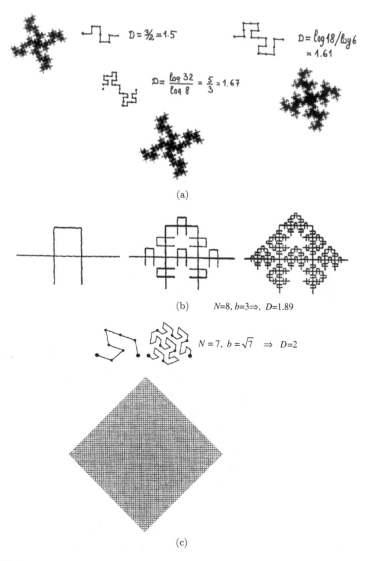

(a)

(b)     $N=8, b=3 \Rightarrow, D=1.89$

$N=7, b=\sqrt{7} \Rightarrow D=2$

(c)

**Fig. IV.5.5** Examples of the perimeter ramification algorithms and their corresponding fractal dimensions. (a) Area-conserving islands generated starting from a square, in which each of the four sides is replaced by the contour, as drawn near the corresponding cascade. In each successive generation every straight section of the resulting perimeter is further developed according to the corresponding algorithm. The fractal dimensions of the resulting perimeter are calculated following Eq. (IV.5.6). Three examples of the resulting constructions show that a dramatic ramification of the coastal line leads to a modest increase of the fractal dimension. (b) The "Mandelbrot-Given" curve. Its fractal dimension, following Eq. (IV.5.6), is $D = \ln 8/\ln 3 = 1.89$. (c) The "Peano" curve fills out the full plane and has perimeter dimension equal to topological dimension of the plane.

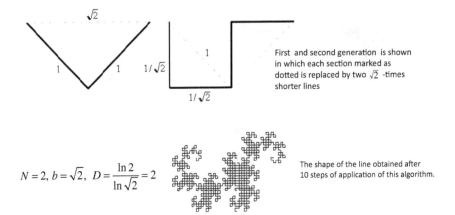

First and second generation is shown in which each section marked as dotted is replaced by two $\sqrt{2}$ -times shorter lines

$$N = 2,\ b = \sqrt{2},\ D = \frac{\ln 2}{\ln \sqrt{2}} = 2$$

The shape of the line obtained after 10 steps of application of this algorithm.

**Fig. IV.5.6**  Generating a "Dragon curve" of perimeter dimension equal to 2.

The example shown in Fig. IV.5.5(c) raises a question: can the perimeter dimension of an in-plane-confined curve ever be larger than 2? In principle, yes, for the so called "dragon"-curves. These are generated by lengthening seed segments and at the same time keeping the starting and ending points. Unlike other fractals, which

Karl Weirestrass
(1815–1897)

are limited to a confined space, fractals with 'calculated' dimension greater than 2 can continue to expand outwardly from the centre forever. Such curves are often called "dragon curves", but they are rather unphysical and we will not be dealing with them in this chapter. In Fig. IV.5.6 we show below a particular case of a dragon curve which still 'behaves well' and returns $D = 2$.

Some 'monstrous' curves can be generated by an analytical formula. One such is called the *Weierstrass function*. It bears the name of the outstanding German mathematician who invented it. Weierstrass's contribution to mathematics is on fundamentals of mathematical analysis, and he is called sometimes a 'father' of modern analysis. In the form adopted and modified by Mandelbrot, the *Weierstrass-Mandelbrot* function reads

$$C(t) = \sum_{n=-\infty}^{\infty} \frac{1 - \cos b^n t}{b^{n(2-D)}}, \quad 1 < D < 2 \qquad \text{(IV.5.7)}$$

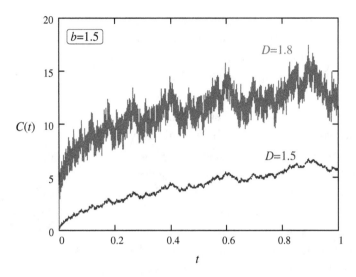

**Fig. IV.5.7** Plots of the curves determined by the Weierstrass function, as defined by Eq. (IV.5.7) for two different values of fractal dimension.

It is plotted in Fig. IV.5.7. As we will explain in a moment, parameter $D$ coincides with the fractal dimension of this curve.

This remarkable function is finite at any point, but its derivative

$$\frac{d}{dt}C(t) = \sum_{n=-\infty}^{\infty} \frac{\sin b^n t}{b^{n(1-D)}} \tag{IV.5.8}$$

is infinite at any point except for $t = 0$. Most importantly, this function obeys a scaling relationship,

$$C(bt) = \sum_{n=-\infty}^{\infty} \frac{1 - \cos b^{n+1}t}{b^{(n+1)(2-D)}} b^{2-D}$$

$$= b^{2-D} \sum_{n=-\infty}^{\infty} \frac{1 - \cos b^n t}{b^{(n)(2-D)}} = b^{2-D}C(t) \tag{IV.5.9}$$

For those unfamiliar with the story of Weierstrass analysis, this must sound amazing. But note that in the time of his discovery (1872) the fact that there may exist functions that can be continuous everywhere but not differentiable anywhere, caused a revolt; famous mathematicians described such functions as 'pathological' and they called Weierstrass' work "an outrage against common sense". But nowadays things look different. In Mandelbrot's world of fractals, this function looks...beautiful. It does describe a fractal curve. If you zoom in, the graph of this function will look the same (plot it for yourself, as an **Exercise**); the scaling law, Eq. (IV.5.9), warrants its self-similarity. Due to this scaling law, it is obvious that parameter $D$ in Eq. (IV.5.7) is the curve's Hausdorff–Besicovitch dimension.

Can we have lines with fractal dimension smaller than 1? Yes, we can. Here, we come to conceptually important fractal objects, which are generated by taking the material out of it at each generation, instead of adding it. The classical example is a *Cantor bar*, a particular case of a set, named after a famous German mathematician, the father of the "set theory",

Georg Ferdinand Ludwig Philipp Cantor (1845–1918)

Georg Cantor. A typical 'tutorial' type of a Cantor bar is shown in Fig. IV.5.8, a bar for which at each generation one third of it, in the middle, is removed. Its fractal dimension is less than 1. Indeed, at each increase of resolution (reduction of the yardstick) you see less of the continuous length, as the gaps become visible, and you see more of them the smaller is the yardstick. For such objects $b > N$, and $D = \ln N / \ln b < 1$.

For a peculiar object one obtains plotting the distribution of the mass along the Cantor bar, if its filled sections have constant density, see Fig. IV.5.9.

The Cantor bar inspires us to look at the Sierpiński gasket (or triangle), which is one of the constructed fractal objects created by and named after an outstanding and enormously productive Polish mathematician, W. Sierpiński. This construction starts from a big, initial triangle of which you take out its central triangular part which comprises $1/4$ of the area of the initial

Wacław Franciszek Sierpiński (1882–1969)

triangle; and you do the same at every next generation. With finer resolution, you will see more and more of smaller triangles. For the

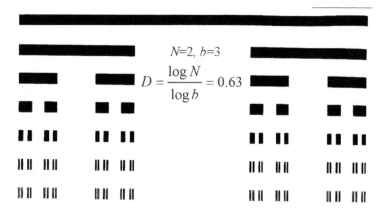

$$D = \frac{\log N}{\log b} = 0.63$$

$N=2, b=3$

**Fig. IV.5.8** An example of a Cantor bar, an object with fractal dimension smaller than the topological dimension.

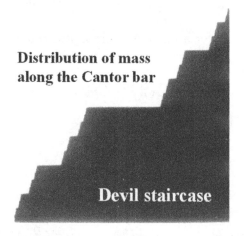

**Distribution of mass along the Cantor bar**

**Devil staircase**

**Fig. IV.5.9** The amount of mass does not change moving along empty sections of the Cantor bar; hence you see multiple plateaus on this horrifying, mystical staircase with never-ending steps at all scales...

resolution achieved by a yardstick $r$ shown in Fig. IV.5.10, you detect 9 triangles, whereas for a 4 times larger one (lower resolution), you would distinguish only one triangle.

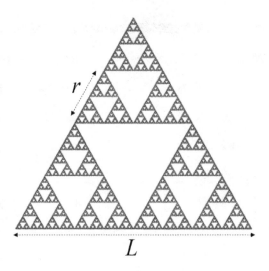

**Fig. IV.5.10**   The 'iconic' Sierpiński triangle.

The scaling law for a number of triangles seen with doubling the resolution is

$$N\left(\frac{r}{2}\right) = 3N(r) \tag{IV.5.10}$$

The solution of this equation is

$$N(r) = \left(\frac{L}{r}\right)^D, \quad D = \frac{\ln 3}{\ln 2} = 1.58 \tag{IV.5.11}$$

The area thus seen at such resolution is

$$S(r) \propto r^2 N(r) = r^2 \left(\frac{L}{r}\right)^D = r^{2-D} L^D \tag{IV.5.12}$$

$D < 2$, and with a smaller yardstick we see a smaller surface area. The density of that area,

$$\rho(r) \sim \frac{S(r)}{L^2} \propto r^2 N(r) = \left(\frac{r}{L}\right)^{2-D} \tag{IV.5.13}$$

This scaling law is called the *law of co-dimension*. Here, this law is manifested itself in the plane of 2D space. Independently of the specific value of $D$, $1 < D < 2$, it describes a fractal object embedded

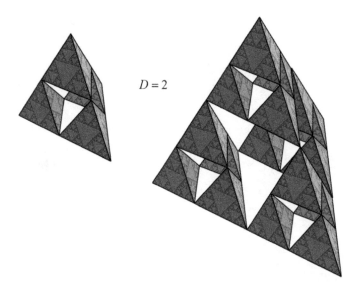

$D = 2$

**Fig. IV.5.11** Two successive generations of a Sierpiński tetrahedron. The second-generation construct is shown increased in size for better visibility.

in two-dimensional space, the density of which vanishes as resolution increases (decrease of $r$, or increase of $L$):

$$\rho(r) \underset{\substack{r \to 0 \\ L \to \infty}}{\propto} \left(\frac{r}{L}\right)^{2-D} \to 0 \qquad (IV.5.14)$$

A three-dimensional analogue of this construction is the Sierpiński pyramid, Fig. IV.5.11. For constructions of this kind, the scaling law for the number of smaller pyramids seen when resolution is increased $b$ times, (i.e. the yardstick length $r$, is $b$ times decreased), is given by

$$N(r/b) = \bar{N} \cdot N(r) \qquad (IV.5.15)$$

The number $\bar{N}$ tells us how many more small pyramids you will see after decreasing the yardstick $b$ times. As before, the solution of this equation is

$$N(r) = [L/r]^D \qquad (IV.5.16)$$

Substitution of Eq. (IV.5.16) into Eq. (IV.5.15) gives us $[Lb/r]^D = \bar{N} \cdot [L/r]^D \Rightarrow [b]^D = \bar{N}$, and thus

$$D = \ln \bar{N}/\ln b \qquad \text{(IV.5.17)}$$

For the Sierpiński pyramid, $\bar{N} = 4$, $b = 2$, and hence $D = \ln 4/\ln 2 = 2$.

Subject to Eq. (IV.5.16), the volume of these kind of objects scale as

$$V(r) \propto r^3[L/r]^D = L^D r^{3-D} \qquad \text{(IV.5.18)}$$

and the density as

$$\rho(r) \propto \frac{r^3[L/r]^D}{L^3} = \left(\frac{r}{L}\right)^{3-D} \qquad \text{(IV.5.19)}$$

which is, again, the law of co-dimension, but in a three-dimensional space.

The objects for which this law is valid are called *volume-* (or *mass-*) fractals. Sierpiński's gasket is a volume fractal in 2D space, a Sierpiński pyramid is a volume fractal in 3D space, whereas our starting object, the Cantor bar, is a volume fractal in 1D space. In each case the fractal dimension $D$ is smaller than the corresponding topological dimension $d$. This is in contrast with fractals formed by coastlines, that can be extended in 2D space to form a fractal surface. Their fractal dimensions are larger than the topological dimensions and these are called surface fractals. Fractal dimensions of perimeters ('coastlines') and surfaces are often called, the Hausdorff–Besicovitch coastline/perimeter dimensions, $D_c$ or $D_p$, and surface fractal dimensions, $D_s$, acquiring the corresponding subscripts. Except for some special cases, we will not be assigning such subscripts in this book where it is clear from the context of subject which fractal dimension we are dealing with—perimeter, surface, or volume.

**Fig. IV.5.12** Successive generations of a "Menger sponge".

**Exercise.** Find the value of the volume fractal dimension of another deterministic fractal, the "Menger Sponge" (Fig. IV.5.12), invented by Austrian-American mathematician K. Menger; the answer is shown in the picture).

Karl Menger
(1902–1985)

### *The scaling laws of fractal geometry*

We can thus summarize that for Euclidean objects, length $L$, perimeter $P$, surface area $S$, and volume $V$ are related as

$$L \propto P \propto S^{1/2} \propto V^{1/3} \qquad (IV.5.20)$$

but for 'surface fractals' it is

$$\begin{cases} L \propto P^{1/D} \propto S^{1/D}, & D > d - \text{for surface fractals} \\ L \propto P^{1/D} \propto S^{1/D} \propto V^{1/D}, & D < d - \text{for volume fractals} \end{cases}$$
$$(IV.5.21)$$

Without going any further, one may try to apply these laws to describe the objects in nature that look 'fractal' and treat the data in corresponding coordinates.

For instance, count distribution of matter in the aggregates of colloidal particles, or mass in galaxy clusters, within a varied radius $R$. If this agglomerate obeys the low of co-dimension (Eq. (IV.5.19), with $R$ now playing the role of $L$), then the calculated density $\rho$ and mass $m$ scale as $\rho(R) \propto \frac{1}{R^{3-D}} \Rightarrow m(R) \propto R^3 \rho(R) = R^D$, so that $\ln \rho = \text{const} - (3-D) \ln R$ and $\ln m = \text{const} + D \ln R$, which suggests log-log coordinates for treatment of the experimental data.

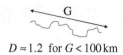

$$D \approx 1.2 \text{ for } G < 100 \,\text{km}$$

**Fig. IV.5.13**  Towards scaling of river lines against river basins.

For the perimeter $P$ of the clouds that look fractal but their area $S$ is not, the scaling law will read $P \propto (S)^{D/2}$, and the statistics of many clouds over the Pacific Ocean do show that the plots $\ln P = \text{cost} + \frac{D}{2} \ln S$ are straight with $D$ close to 1.45.

Another classical example is rivers. The perimeter of their central line is expected to be fractal, whereas area of the basin is not. If this is so then $P(G) \propto [S(G)^{1/2}]^D \Rightarrow \frac{S(G)^{1/2}}{P(G)^{1/D}} = \text{const}$, and the law $\ln S(G) = \text{const} + \frac{2}{D} \ln P(G)$ does appear to work, within certain domain of length $G$, as Mandelbrot presented it (Fig. IV.5.13).

Even for the relationship between the volume of brain (non-fractal) and surface area of brain cortex of mammalians, the scaling law $S \propto V^{D_s/3}$, $\ln S = \text{const} + \frac{D_s}{3} \ln V$, does seem to work (for clarity, we assign here a subscript $s$, to outline that this is a fractal dimension of the surface). The fractal dimension varies between mammalians, $2.73 < D_s < 2.79$, and it is believed to be higher for those which are more 'intelligent'.

Generally, fractal structures (volume or surface) have been found in all manner of places: curdling, fine structure of Saturn rings and galaxy clusters, granular media (sand-piles), porous media, rocks, percolation clusters, polymers, as well as in processes: cracking patterns, electrical discharge, pitting corrosion, electrodeposition, dynamics of colloidal cluster growth, liquid flow in porous media (viscous fingering), turbulent flows, deterministic chaos, epidemics, etc. In this chapter we pinpoint only few examples of such system.

### 'Random' fractals, patterns of noise, self-affine fractals

Our conclusions have so far been deduced from deterministic fractals, whereas in nature those are crude approximations of reality. We therefore discuss fractals built as a result of random processes.

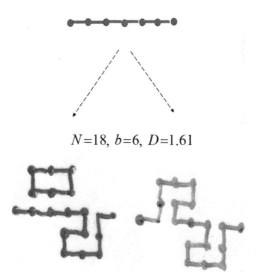

$N$=18, $b$=6, $D$=1.61

**Fig. IV.5.14** The semi-deterministic fractal: 'Koch archipelago' can be generated by random applications of the replacement of the straight lines (initial sides of a square containing six equal sections) by the shown generators. Can be applied with equal probability or otherwise. Since $N$ and $b$ are the same for both generators, the resulting coastline dimension will be the same.

A straightforward way to randomize fractal structure is to apply at random a few or even just two different generators, for example replacing each straight section (starting from the side of an initial square) with structures like those in Fig. IV.5.14.

But this is not very interesting. We concentrate instead on a coastlines, first, and then on surfaces, both generated by curves representing a random process or noise. Namely, we consider a random walk pattern, be it uncorrelated, correlated or anticorrelated. Let $Y(t)$ be the deviation from initial position after time $t$, and consider the mean square displacement,

$$\Delta(\tau) = \sqrt{\langle (Y(t+\tau) - Y(t))^2 \rangle} \qquad \text{(IV.5.22)}$$

where the symbol $\langle \rangle$ stands for time averaging over time $t$. This function thus depends only on $\tau$, and at long time it scales as

$$\Delta(\tau) \propto \tau^{\alpha} \qquad \text{(IV.5.23)}$$

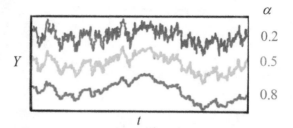

**Fig. IV.5.15** Noise trajectories for three different Holder exponents. For better visibility, the graphs are shifted with respect to each other in vertical direction.

Here, $\alpha$ is often called the Holder exponent, named after German mathematician Otto Holder (1859–1937). Unfortunately, he was quick to state his allegiance to the Nazi regime immediately in 1933, so in accordance with our deep anti-fascist beliefs we refrain from showing his portrait. A similar exponent was introduced by British hydrologist Harold Edwin Hurst, the explorer of river Nile. Mandelbrot, respecting parity, celebrated both scientists by using letter $H$ for $\alpha$, but with all due respect to Hurst we will stick to $\alpha$. In our context, the latter standardly lies in the interval $0 \leq \alpha \leq 1$ (although we'll see when this is not so). For a Gaussian process with $\alpha = 1/2$, which is also labelled as 'Brownian noise; for persistent, 'black' noise $\alpha = 1$, and for anticorrelated noise $\alpha = 0$. The snapshot graphs of $Y(t)$ for different-values $\alpha$ are shown in Fig. IV.5.15.

If we extend the process in the direction normal to $Y$, call it $X$, we can define a surface determined by $[X(t), Y(t)]$. It will look very much like a mountain landscape. In particular, the one generated by the algorithm with $\alpha = 1/2$ Mandelbrot called the "Brownian surface". Many beautiful coloured landscapes are shown in his *Fractal Geometry of Nature*, and you can play around with this idea yourself if you wish, with your computer, even varying $\alpha$ with time. But now we will do something different.

Let us try to estimate the fractal dimensions of curves generated in this manner, and then of the corresponding surfaces. For this we will use intuitive scaling analysis, which will bring us to the concept of *self-affine fractals*. If we rescale $t \Rightarrow pt$, $Y$ will be rescaled differently: $Y \Rightarrow qY$, and it is expected that $q = p^\alpha$. The term "self-affinity" means that when rescaling the argument of the function, the function itself scales differently; thus if $Y$ stands for the 'height' and $t$ for the 'width', rescaling the latter to retain the object isomorphic to itself is different in the two directions. Let us rewrite the scaling law for $\Delta$, Eq. (IV.5.23), in dimensional form:

$$\Delta(\tau) \approx \lambda \left(\frac{\tau}{\lambda}\right)^\alpha \tag{IV.5.24}$$

where we will call $\lambda$ the *correlation length*. In a moment we will see why such name suits this baby.

The perimeter of the curve as a function of the yardstick $\varepsilon$ with such a mean-square deviation from the straight line can be estimated as

$$P(\varepsilon) \approx \frac{L}{\varepsilon}[\varepsilon + \Delta(\varepsilon)] = \frac{L}{\varepsilon}\left[\varepsilon + \lambda \left(\frac{\varepsilon}{\lambda}\right)^\alpha\right] \tag{IV.5.25}$$

where $L$ is the overall interval of variation of the argument of the function $Y(t)$. Hence,

$$P(\varepsilon) \approx L \left[1 + \left(\frac{\lambda}{\varepsilon}\right)^{1-\alpha}\right] \tag{IV.5.26}$$

Because $\alpha \le 1$, this formula tells us that that at $\varepsilon \gg \lambda$, the perimeter will approach $L$, no longer depending on $\varepsilon$, i.e. with yardsticks larger than the correlation length, the curve will not be seen as a fractal, but with a smaller yardstick it will:

$$P(\varepsilon) \approx \begin{cases} L, & \varepsilon \gg \lambda \\ \propto \dfrac{1}{\varepsilon^{1-\alpha}}, & \varepsilon \ll \lambda \end{cases} \tag{IV.5.27}$$

Thus at the scales where it is a fractal, comparing this result with Eq. (IV.5.1), we get $D - 1 = 1 - \alpha$, i.e. the perimeter dimension of such curve is,

$$D = 2 - \alpha \qquad (IV.5.28)$$

For a surface generated, as explained, with the help of two such curves

$$D = 3 - \alpha \qquad (IV.5.29)$$

Generalizing it to any dimension, $d$, this result reads

$$D = d + 1 - \alpha \qquad (IV.5.30)$$

The perimeter of a cross-section of such a super-surface will have a fractal dimension $D = d - \alpha$.

### Spectral laws

Consider a correlation function,

$$S(\tau) = \langle Y(t)Y(t+\tau) \rangle - \langle Y(t) \rangle^2 \qquad (IV.5.31)$$

and its Fourier transform $\tilde{S}(f)$ defined as

$$S(\tau) = \int_0^\infty df\, \tilde{S}(f) \cos(2\pi f \tau) \qquad (IV.5.32)$$

$\tilde{S}(f)$ is called 'spectral function'. Assume for a moment that it has the form of a power law:

$$\tilde{S}(f) \propto \frac{1}{f^\beta} \qquad (IV.5.33)$$

where $\beta$ is called 'spectral exponent'. The substitution of Eq. (IV.5.33) into Eq. (IV.5.32), gives

$$S(\tau) \propto \int_0^\infty df \frac{1}{f^\beta} \cos(2\pi f \tau)$$

$$= \tau^{\beta-1} \int_0^\infty d(f\tau) \frac{1}{f^\beta \tau^\beta} \cos(2\pi f \tau)$$

$$= \text{const} \times \tau^{\beta-1} \qquad (IV.5.34)$$

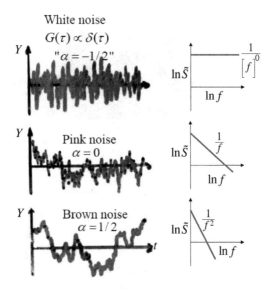

**Fig. IV.5.16** Noise patterns and the schematics of the corresponding spectra for the Fourier transforms of the correlation functions $S$.

But if our $S(\tau)$ obeys one of the noise patterns above, we have that

$$S(\tau) \propto \tau^{2\alpha} \tag{IV.5.35}$$

Comparing Eq. (IV.5.35) and Eq. (IV.5.34), we finally get the relationship between the spectral and noise exponents:

$$\beta = 1 + 2\alpha \tag{IV.5.36}$$

Taking into account Eq. (IV.5.30), we can thus establish the relationship between the spectral exponent and the perimeter or surface fractal dimension as

$$D = d + \frac{3 - \beta}{2} \tag{IV.5.37}$$

Since noises are often classified subject to the value of their of spectral exponents, let us introduce a couple of them, exemplified in Figs. IV.5.16 and IV.5.17. Following Eq. (IV.5.36) and Eq. (IV.5.37)

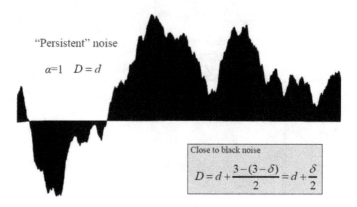

**Fig. IV.5.17** Black noise profiles, and fractal dimension of the noise that is 'almost' black.

we get $\beta = 1$, i.e. the famous $1/f$ noise, sometimes called "pink" noise, for an anti-correlated process, with $\alpha = 0$, the fractal dimension of which is $D = 2$ for a line, and $D = 3$ for a surface. For Brownian noise $\alpha = 1/2$, the spectral exponent $\beta = 2$, and $D = 1.5$ for a Brownian line, and $D = 2.5$ for a Brownian surface. Finally, for $\alpha = 1$ black noise, $\beta = 3$, $D = 1$ for a line and $D = 2$ for a surface—not fractal any more!

Interestingly, the time correlation of the white noise, which is proportional to the delta-function $\delta(\tau)$, and its Fourier transform will therefore be a constant independent of $f$, i.e. it gives $\beta = 0$. The only way to formally get $\beta = 0$, following Eq. (IV.5.36), is to require $\alpha = -1/2$.

An American theoretical physicist, Richard Voss, specializing in noises, random processes and growth patterns, has investigated spectra of classical and contemporary music, and has found that many of them follow $1/f$ patterns. Using computer simulations he created some musical tunes under the constraint that they should follow such patterns and they did sound 'pleasant' to the ear, whereas when he constricted them to obey Brownian $1/f^2$ law, the resulting compositions were perhaps reminiscent of minimalistic composers, but with lack of substance behind them they just sounded...boring. On the other hand, composing under constraint of $\beta = 0$ produced total chaos, 'compositions' that were unbearable to listen to—as it should be for the white noise. Based on this analysis, and noticing that the data for many compositions (of Bach, Beethoven, Richard Straus, Debussy, Beatles, American blues, folk-songs of Russia, Indian ragas, etc) lie more or less on the straight line of $\ln \tilde{S}$–$\ln f$ plot of the middle figure in the right column of Fig. IV.5.16, he philosophized on the mystery of music: "Music is imitating the characteristic way our world changes with time." Indeed, pink noise is seen in many natural phenomena, e.g. in electronic semiconductors devices, voltage fluctuations in biological membranes, fluctuations of traffic on busy motorways, etc. The origin of the pink noise has drawn a lot of attention, and the common opinion is that it emerges in systems with many "equal or almost equal opportunities". This may sound vague, so we illustrate this in the next example.

**Example IV.5.1 (Physical Application ▶▶).** Consider a Lorentzian spectrum that describe some relaxation process,

$$P(f, \tau) \propto \frac{4\tau P_0}{1 + (2\pi f \tau)^2} \qquad (IV.5.38)$$

where $f$ is the Fourier transform's frequency, and $\tau$ is the relaxation time. The latter often obeys the activation law, $\frac{1}{\tau(E)} = \frac{1}{\tau_0} e^{-E/k_B T}$, where $E$ is the activation energy of the process, and $k_B T$ is thermal energy. Hence, we have

$$\tau(E) = \tau_0 e^{E/k_B T} \qquad (IV.5.39)$$

If the relaxation times are described by the probability $p(\tau)$ then the average signal will be given by

$$P(f) = \int_0^\infty d\tau p(\tau) P(f, \tau) \qquad (IV.5.40)$$

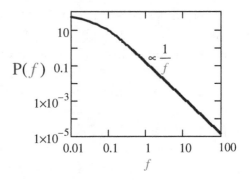

**Fig. IV.5.18** The frequency spectrum given by Eq. (IV.5.42), with continuous distribution of activation energies of relaxation times displays a broad band of $1/f$ behaviour.

We will now consider a special kind of a distribution function

$$p(\tau) = \begin{cases} \dfrac{k_B T}{E_2 - E_1} \dfrac{1}{\tau}, & \tau_1 \leq \tau \leq \tau_2 \\ 0, & \tau < \tau_1, \tau > \tau_2 \end{cases}$$

$$\tau_1 = \tau_0 e^{E_1/k_B T} \qquad \tau_2 = \tau_0 e^{E_2/k_B T} \qquad \text{(IV.5.41)}$$

How this expression comes about is shown in the derivation box below.

---

Probability density of having relaxation time $\tau$ of given activation energy $E$ is given by $p_E(\tau) = \delta(\tau - \tau(E))$. Hence, the probability of falling into the interval of $\tau_1 < \tau < \tau_2$ is $p(\tau) = \dfrac{\int_{E_1}^{E_2} p_E(\tau) dE}{E_2 - E_1} = \dfrac{\int_{E_1}^{E_2} p_E(\tau) \{\frac{d\tau}{dE}\}^{-1} d\tau}{E_2 - E_1}$. Because $\dfrac{d\tau}{dE} = \tau_0 \exp\{\frac{E}{k_B T}\} \frac{1}{k_B T} = \frac{\tau}{kT}$, we finally get $p(\tau) = \dfrac{k_B T}{(E_2 - E_1)\tau}$.

---

**Exercise.** Substitute Eq. (IV.5.41) into Eq. (IV.5.40), and perform the integration to obtain—

$$P(f) = \int_{\tau_1}^{\tau_2} d\tau \, P(f, \tau) p(\tau)$$

$$= \frac{2k_B T P_0 \{\arctan(2\pi f \tau_2) - \arctan(2\pi f \tau_1)\}}{\pi (E_2 - E_1) f} \qquad \text{(IV.5.42)}$$

The graph of this function is shown in Fig. IV.5.18. As we see, the $1/f$ law is reproduced by this formula, except for the range of small frequencies when the function levels off.

## For pedants: Definition of a fractal?

We have finished a basic, minimal introduction to fractals and fractal geometry, needed for the sections that follow, but you have probably noticed that we actually never gave a definition of the term "fractal". It may, perhaps, be arrogant to refer to the famous saying attributed to Louis Armstrong, who when asked "What is jazz?", answered: "If you are asking this question, you will never understand what it is". No, mathematics may be loved and intuitively understood (and this is what this book is about), but we still may wish to have some solid definitions of the items and concepts that we deal with. The father of fractals tried to give the answer several times.

1. First he defined it as "A fractal is a set for which the Housdorf–Besicovitch dimension strictly exceeds the topological dimension" (Mandelbrot, 1982). But this definition is not all-encompassing, valid only for surface fractals.
2. Later, he wrote "A fractal is a shape made of parts similar to the whole in *some way*." [Mandelbrot (1986)]. Isn't it a bit vague, even for pragmatic end users of mathematics? But you can also find his more precise statement on self-similarity: "when the sum of independent identically distributed random quantities has the same probability distribution as each random quantity distributed". He also rephrased it as "when the whole object looks like any of its parts".

Still, in the end, Mandelbrot proposed "to use *fractal* without a pedantic definition" and to focus on fractal dimension as a term applicable to all types of fractals. So people, particularly those who are less concerned about mathematical rigour, decided to settle on a compromise between the two Mandelbrot statements and agree that fractals are self-similar at different scales, iterated, mathematical constructs and physical objects having fractal dimensions. Many examples of these have been proposed and studied.

Whereas mathematically defined fractals can be self-similar at infinity, fractals emerging in nature are self-similar only within a certain range of scales, associated with different mechanisms of their growth or non growth beyond the self-similarity window. Massive collections of data performed for various systems and phenomena— colloidal aggregation, porous media, surfaces and fronts, fracture, critical phenomena, vibrations, turbulence, random walk, and even high energy physics–show that for many examples the scales of the observed self-similarity lie at best between 1.5–3 orders of magnitude. There are exceptions, of course, like a huge self-similarity range in the bronchial tree, which extends over 15 successive fractal generations.

For those who are not satisfied by the Louis Armstrong suggestion, let us get back, but following Mandelbrot's advice to forget about 'philosophy', focus on fractal dimensions—the cornerstones of fractal geometry. An illuminating approach to their definition has been presented in *The Geometry of Fractal Sets* (1985), a book written by British mathematician Kenneth John Falkoner of Saint Andrews University, an expert in fractal and combinatorial geometry and measure theory. He defined fractal dimension as a *critical dimension*. Here is what it is about.

Recall how the length, area, and volume are measured and related to each other in scaling terms in Euclidean geometry. $N(\varepsilon)$ will be again the number of yardsticks of size $\varepsilon$ applied to measure the object, $L$, $A$, and $V$ will stand for length, area, and volume, respectively. This notion is briefly explained in the box below.

---

**Euclid vs Fractal scaling and the definition of the critical dimension**

Recall how the length, area, and volume are measured and related to each other in scaling terms. $N(\varepsilon)$ will be again the number of yardstick of size $\varepsilon$ applied to measure the objects, length $L$, area $A$, volume $V$, and mass $M$.

---

Euclid geometry

*1D-object*:

$$L = N(\varepsilon)\varepsilon \underset{\varepsilon \to 0}{\Rightarrow} L_0\varepsilon^0, \quad A = N(\varepsilon)\varepsilon^2 \underset{\varepsilon \to 0}{\Rightarrow} L_0\varepsilon^1, \quad A = N(\varepsilon)\varepsilon^3 \underset{\varepsilon \to 0}{\Rightarrow} L_0\varepsilon^2$$

*2D-object*:

$$A = N(\varepsilon)\varepsilon^2 \underset{\varepsilon \to 0}{\Rightarrow} A_0\varepsilon^0, \quad V = N(\varepsilon)\varepsilon^3 \underset{\varepsilon \to 0}{\Rightarrow} A_0\varepsilon^1, \quad L = N(\varepsilon)\varepsilon \underset{\varepsilon \to 0}{\Rightarrow} A_0\varepsilon^{-1}$$

Fractal geometry

In full analogy with this above consideration, we can write

$$M(\tilde{d}) \equiv N(\varepsilon)\varepsilon^{\tilde{d}} \underset{\varepsilon \to 0}{\Rightarrow} \begin{cases} 0, & \tilde{d} > D \\ \infty, & \tilde{d} < D \end{cases}$$

where $D$ is the fractal dimension. So $D$ is a crossover value for $\tilde{d}$ when $\tilde{d} = D$,

$$N(\varepsilon)\varepsilon^{\tilde{d}} \underset{\varepsilon \to 0}{\Rightarrow} \text{const} \neq 0$$

Having set the 'language', we will now proceed to discuss several systems, for which the issues of 'fractionality', self-similarity, and scaling are central.

## 5.2. Ideal and Nonideal Gaussian Chains: From Random Walk to Polymers to Diffusion

We consider a random walk on a lattice the sites of which have a coordination number (the number of nearest neighbours) $z$. For example, as shown in Fig. IV.5.19, this could be a 2D square lattice, Each path will contain some $N$ step-vectors or 'bonds'. If $\mathbf{r}_n$ is the vector of the $n$th bond, then the end-to-end distance is $z = 4$. The trace of this walk is a simplistic model of a random polymer, in which $b$ is a bond length. As shown in Fig. IV.5.19, the walk is non-restricted.

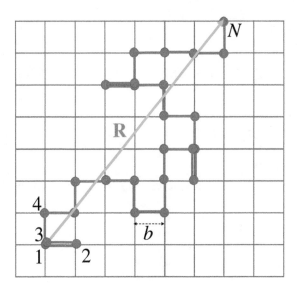

**Fig. IV.5.19** Random walk on square grid with lattice constant $b$. In a model of polymer $b$ plays the role of the bond length (thick red lines representing bonds between segments shown as circles). Return walks are not forbidden in this model, so if it is a polymer it can bend on itself. The latter is physically impossible, and the effect of such a constraint will be considered later in this section.

The return steps that we see in this figure are allowed. Each path will contain some $N$ step-vectors or 'bonds'. If $\mathbf{r}_n$ is the vector of the $n$th bond, then the end-to-end distance is

$$\mathbf{R} = \sum_{n-1}^{N} \mathbf{r}_n \qquad \text{(IV.5.43)}$$

As any $\mathbf{r}_n$ vector can look in any of the four directions on this grid, the average value

$$\langle \mathbf{r}_n \rangle = 0 \qquad \text{(IV.5.44)}$$

Hence, the average value of the end-to-end vector is also zero:

$$\langle \mathbf{R} \rangle = \sum_{n-1}^{N} \langle \mathbf{r}_n \rangle = 0 \qquad \text{(IV.5.45)}$$

This is not the case, however for average value of $\mathbf{R}^2$:

$$\langle \mathbf{R}^2 \rangle = \left\langle \left( \sum_{n=1}^{N} \mathbf{r}_n \right) \left( \sum_{m=1}^{N} \mathbf{r}_m \right) \right\rangle = \sum_{n=1}^{N} \sum_{m=1}^{N} \langle \mathbf{r}_n \mathbf{r}_m \rangle$$

$$= \sum_{n=1}^{N} \underbrace{\langle \mathbf{r}_n \mathbf{r}_n \rangle}_{b^2} + \sum_{\substack{n=1 \\ n \neq m}}^{N} \sum_{m=1}^{N} \underbrace{\langle \mathbf{r}_n \rangle}_{0} \underbrace{\langle \mathbf{r}_m \rangle}_{0} = N b^2 \qquad (\text{IV.5.46})$$

This means that the mean-square deviation from original point after $n$-steps

$$\sqrt{\langle \mathbf{R}^2 \rangle} = b \sqrt{N} \qquad (\text{IV.5.47})$$

scales with the number of steps: $\propto N^{1/2}$. In this way the end-to-end distance of a random (Gaussian) chain polymer depends on the number of bonds in the chain.

It easy to see that same result will be reached in the space of any dimension and for any type of lattice—it is not by chance that the coordination number $z$ did not enter the answer.

### Probability distribution function: The analogy with the diffusion probability

Let us consider the probability of reaching point $\mathbf{R}$ after $N$ steps, each of length $b$. To derive it, we will start with an obvious recurrence relation

$$P(\mathbf{R}, N) = \frac{1}{z} \sum_{i=1}^{z} P(\mathbf{R} - \mathbf{b}_i, N - 1), \quad i = 1, 2, \ldots z \qquad (\text{IV.5.48})$$

which relates the sought probability to reach the point $\mathbf{R}$ after $N$ steps from any of its neighbours reached after $N - 1$ steps, with the probability of reaching those neighbours. If the path is long, so that $R \gg b$, and $N \gg 1$, we Taylor-expand the r.h.s near the points $\mathbf{R}$ after $N$ to get

$$P(\mathbf{R} - \mathbf{b}_i, N - 1) = P(\mathbf{R}, N) - \frac{\partial P}{\partial N} - \frac{\partial P}{\partial R_\alpha} b_{i\alpha} + \frac{1}{2} \frac{\partial^2 P}{\partial R_\alpha \partial R_\beta} b_{i\alpha} b_{i\beta}$$

$$(\text{IV.5.49})$$

(Hereafter, we will imply summation over repeated Cartesian coordinate indices, $\alpha$ and $\beta$.) Plugging Eq. (IV.5.49) into the r.h.s of Eq. (IV.5.48) we get

$$P(\mathbf{R}, N) = \frac{1}{z} \sum_{i=1}^{z} \left\{ P(\mathbf{R}, N) - \frac{\partial P}{\partial N} - \frac{\partial P}{\partial R_\alpha} b_{i\alpha} + \frac{1}{2} \frac{\partial^2 P}{\partial R_\alpha \partial R_\beta} b_{i\alpha} b_{i\beta} \right\}$$

$$(\text{IV.5.50})$$

Now using the obvious relationships,

$$\frac{1}{z} \sum_{i=1}^{z} 1 = 1, \quad \frac{1}{z} \sum_{i=1}^{z} b_{i\alpha} = 0, \quad \frac{1}{z} \sum_{i=1}^{z} b_{i\alpha} b_{i\beta} = \frac{\delta_{\alpha\beta} b^2}{3} \quad (\text{IV.5.51})$$

we can arrive at

$$\frac{\partial P(\mathbf{R}, N)}{\partial N} = \frac{b^2}{6} \frac{\partial^2 P(\mathbf{R}, N)}{\partial \mathbf{R}^2} \qquad (\text{IV.5.52})$$

(we have already derived the same equation in the previous chapter by functional integration), for which we have an initial condition

$$\mathbf{R} = 0 \text{ for } N = 0 \qquad (\text{IV.5.53})$$

We now do replacements, denoting by $t$ the time needed to do $N$ steps, and by $\Delta t$ the time needed to make one step, and introducing $\mathcal{D}$ as diffusion coefficient—remember this is estimated as proportional to the square of length that is passed per the unit time step (see Section 10.6 of Part I, Vol. 1)—and obtain

$$N = \frac{t}{\Delta t}, \quad \frac{b^2}{6\Delta t} = \mathcal{D} \qquad (\text{IV.5.54})$$

[**Apology:** Everywhere else in this book letter $D$ has been solidly used for the diffusion coefficient, but in this chapter we had to use $D$ for the fractal dimension, a standard symbol for it in the literature. Thus, in this chapter we have chosen to use a special font for the diffusion coefficient, $\mathcal{D}$. It will appear only few times, whereas the fractal dimension will emerge almost on every page]. We then get a classical diffusion equation,

$$\frac{\partial P(\mathbf{R}, t)}{\partial t} = \mathcal{D} \frac{\partial^2 P(\mathbf{R}, t)}{\partial \mathbf{R}^2} \qquad \text{(IV.5.55)}$$

$$\mathbf{R} = 0 \quad \text{for } t = 0 \qquad \text{(IV.5.56)}$$

The solution of Eq. (IV.5.52) or Eq. (IV.5.55) with their corresponding initial conditions reads

$$P(\mathbf{R}, N) = \left(\frac{3}{2\pi N b^2}\right)^{3/2} \exp\left(-\frac{3R^2}{2Nb^2}\right) \qquad \text{(IV.5.57)}$$

$$P(\mathbf{R}, t) = \left(\frac{1}{4\pi Dt}\right)^{3/2} \exp\left(-\frac{R^2}{4Dt}\right) \qquad \text{(IV.5.58)}$$

**Exercise.** Prove that the solution given by Eq. (IV.5.57) does satisfy Eq. (IV.5.52), by the substitution of the former into the latter. **Hint:** You may wish to recall the Laplacian $\frac{\partial^2}{\partial \mathbf{R}^2} \equiv \frac{1}{R^2} \frac{\partial}{\partial R} R^2 \frac{\partial}{\partial R}$.

Again we remind you that we have obtained such equations by functional integration in the previous chapter.

**Example IV.5.2 (Maths Practice ♩♪).** Based on the result for the probability function, Eq. (IV.5.57), let us calculate the mean-square average distance to which the random walk will take the walker away from the origin, $R = 0$, after $N$ steps. Introducing for convenience notation $\bar{R} \equiv b\sqrt{N}$, we get

$$\langle \mathbf{R}^2 \rangle = \int d\mathbf{R}\, \mathbf{R}^2 P(N, \mathbf{R})$$

$$= \int_0^\infty dR\, 4\pi R^2 R^2 \left(\frac{3}{2\pi \bar{R}^2}\right)^{3/2} \exp\left(-\frac{3R^2}{2\bar{R}^2}\right)$$

$$= 4\pi \left(\frac{3}{2\pi}\right)^{3/2} \bar{R}^2 \int_0^\infty d\left(\frac{R}{\bar{R}}\right) \left(\frac{R}{\bar{R}}\right)^4 \exp\left(-\frac{3}{2}\left(\frac{R}{\bar{R}}\right)^2\right)$$

$$= \bar{R}^2 4\pi \left(\frac{3}{2\pi}\right)^{3/2} \left(\sqrt{\frac{2}{3}}\right)^{-5} \int_0^\infty d\left(\sqrt{\frac{3}{2}}\left(\frac{R}{\bar{R}}\right)\right) \left(\sqrt{\frac{3}{2}}\frac{R}{\bar{R}}\right)^4$$

$$\times \exp\left(-\left(\sqrt{\frac{3}{2}}\frac{R}{\bar{R}}\right)^2\right)$$

$$= \bar{R}^2 \left(\frac{1}{\pi}\right)^{1/2} \frac{8}{3} \underbrace{\int_0^\infty dx x^4 \exp\left(-x^2\right)}_{\frac{3}{8}\sqrt{\pi}} = \bar{R}^2$$

**Exercise.** Show that the integral in the last line, $\int_0^\infty dx\ x^4 \exp(-x^2) = \frac{3}{8}\sqrt{\pi}$. **Hint:** This can be obtained by bearing in mind the result for the Gaussian integral $I(p) = \int_0^\infty dx\ e^{-px^2} = \frac{1}{2}\sqrt{\frac{\pi}{p}}$, differentiating the latter over $p$ twice, and putting in the result $p = 1$.

Hence, $\langle \mathbf{R}^2 \rangle = \bar{R}^2$, i.e. again

$$\sqrt{\langle \mathbf{R}^2 \rangle} = b\sqrt{N} \tag{IV.5.59}$$

**Important.** In Example IV.5.2, we obtained exactly the same result as given by Eq. (IV.5.47) which obtained on a plain square grid. That means that this law does not depend on the dimensionality of the system or lattice, as long as $\sqrt{\langle \mathbf{R}^2 \rangle} \gg b$, i.e. $\sqrt{N} \gg 1$.

**Exercise.** Check that for the diffusion in $3d$ you get $\sqrt{\langle \mathbf{R}^2 \rangle} = 6Dt$.

**Example IV.5.3 (Physical Application ▶▶).** Based on what we have just learned we consider the elastic properties of an ideal Gaussian chain, using minimal knowledge of thermodynamics. Entropy of the chain is related to the probability $P(\mathbf{R}, N)$ as $S(\mathbf{R}, N) = k_B \ln P(\mathbf{R}, N)$, where $k_B$ is the Boltzmann constant. Hence, using Eq. (IV.5.57) we can write that the change of entropy if we evolve from point $\mathbf{R}_{f=0}$ reached without any applied force to point $\mathbf{R}$ which takes place due to applied force, should be given by $S(\mathbf{R}, N) = k_B \ln P(\mathbf{R}, N) = S(0) - \frac{3k_B}{2}\frac{\mathbf{R}^2}{Nb^2}$. Correspondingly the Helmholtz free energy is given by $F(\mathbf{R}, N) = E - TS = F(0) + \frac{3k_B T}{2}\frac{\mathbf{R}^2}{Nb^2}$. The force that would resist an attempt to stretch that chain will be given by $\mathbf{f} = -\frac{\partial F(\mathbf{R},N)}{\partial \mathbf{R}} = -\frac{3k_B T}{Nb^2}\mathbf{R}$. According to Eq. (IV.5.59), the mean-square displacement that will be there without any external force is given by $\langle \mathbf{R}^2 \rangle_{\mathbf{f}=0} = b^2 N$, and thus

$$\mathbf{f} = -\frac{3k_B T}{\langle \mathbf{R}^2 \rangle_{\mathbf{f}=0}}\mathbf{R} \qquad (IV.5.60)$$

For better visibility, we rewrite this equation, introducing a parameter of the dimensionality of the force, relative to which the force will be scaled. We call this *scaling entropic force*

$$f_0 \equiv \frac{3k_B T}{[\langle \mathbf{R}^2 \rangle_{\mathbf{f}=0}]^{1/2}} \qquad (IV.5.61)$$

Then Eq. (IV.5.60) can be rewritten as

$$\frac{\langle \mathbf{R} \rangle_{\mathbf{f}}}{[\langle \mathbf{R}^2 \rangle_{\mathbf{f}=0}]^{1/2}} = \frac{\mathbf{f}}{f_0} \qquad (IV.5.62)$$

Let us digest this formula. First, do not be surprised that with zero force, $\langle \mathbf{R} \rangle_{\mathbf{f}}$ vanishes. This must be so, because the average value of the-end-to-end vector (not its mean-square value!) in the absence of an external force cannot look in any direction. Secondly, this formula tells us how many times more we can, on average, stretch the chain relative to its natural extension, the external force giving a direction to an end-to-end vector. This will depend on how much larger the applied force is, compared to $f_0$. Note that according to Eq. (IV.5.61), the scaling force is larger with higher temperature. This also makes sense, because it is the entropy that resists stretching, and entropic effects become stronger the higher the temperature. Next, $\mathbf{f}$ as we calculated it is the resisting force. To stretch the chain we need to apply the force $\mathbf{f}_{appl} = -\mathbf{f}$; we can then present Eq. (IV.5.60) in a form of a Hooke's law,

$$\langle \mathbf{R} \rangle_{\mathbf{f}} = \frac{\langle \mathbf{R}^2 \rangle_{\mathbf{f}=0}}{3k_B T}\mathbf{f} \qquad (IV.5.63)$$

in which $\frac{\langle \mathbf{R}^2 \rangle_{\mathbf{f}=0}}{3k_B T}$ plays the role of the Hooke's force constant, which characterizes the elastic system to an external force.

It is legitimate, but naïve, to ask why there is any elasticity in a system of segments, about the interactions of which we did not say a word, because there was none in the chain except that each chain segment is for sure connected with its two neighbours. Well, the resistance is purely entropic: if you want to change something other than what the system spontaneously achieves, you will have to do work. Remember how much effort it takes to put back into a chest all toys that your toddlers dispersed over the floor, or how much they resist your requests/orders to clear up that mess! Growing up, they will learn what 'work' means and will be able to cope with it. But for some it takes to serve in the army to learn to make the bed in the morning.

Robert Hooke
(1635–1703)

Seventeenth-century polyphonic scientist Hooke, although known not to be loved by Newton, left remarkable traces in many fields—geometry, mechanics, optics, astronomy, geography, geology, palaeontology, and evolutionism. He would have been delighted to learn that his law applies to objects of which neither he nor his contemporaries could have had any idea.

### How will things change when stepping back is not allowed?

What we consider now is a kind of a short-range interaction. If in your random walk, after any step, your next step can be in any direction but not backwards (see Fig. IV.5.20), the algebra described above will have to be slightly changed. The analysis below shows by how much.

**Fig. IV.5.20** Restraining condition: reaching the point on the lattice you can step in any direction but not immediately back.

If you know that you have reached a certain point so that the vector for the $n$th-bond on the lattice is occupied and equals $\mathbf{r}_n$, then the average value of the next step vector is no longer zero: $\langle \mathbf{r}_{n+1} \rangle_{\mathbf{r}_n} \neq 0$. We can then write an equation $0 = \Sigma_{i=1}^{z} \mathbf{b}_i = (z-1)\langle \mathbf{r}_{n+1} \rangle_{\mathbf{r}_n} - \mathbf{r}_n$, from where we get that nonzero value:

$$\langle \mathbf{r}_{n+1} \rangle_{\mathbf{r}_n} = \frac{1}{z-1}\mathbf{r}_n \qquad (IV.5.64)$$

From here it follows that

$$\langle \mathbf{r}_{n+1}\mathbf{r}_n \rangle = (z-1)\langle \mathbf{r}_{n+1}\langle \mathbf{r}_{n+1} \rangle \rangle = (z-1)\langle \mathbf{r}_{n+1} \rangle \langle \mathbf{r}_{n+1} \rangle$$

$$= (z-1)\frac{\mathbf{r}_n^2}{(z-1)^2} = \frac{b^2}{(z-1)} \qquad (IV.5.65)$$

Similarly, with a non-obvious but correct assumption that $\langle \mathbf{r}_{n+2}\mathbf{r}_n \rangle = \langle \langle \mathbf{r}_{n+2} \rangle_{\mathbf{r}_{n+1}}\mathbf{r}_n \rangle$, we get

$$\langle \mathbf{r}_{n+2}\mathbf{r}_n \rangle = \frac{\langle \mathbf{r}_{n+1}\mathbf{r}_n \rangle}{(z-1)} = \frac{b^2}{(z-1)^2} \qquad (IV.5.66)$$

One can extend this further to

$$\langle \mathbf{r}_n\mathbf{r}_m \rangle = \frac{b^2}{(z-1)^{|n-m|}} \qquad (IV.5.67)$$

This means that different "steps" are now correlated, but correlations decay exponentially, except for the case of $z = 2$. **Think:** Why?

Equation (IV.5.67) allows us to obtain

$$\langle \mathbf{R}^2 \rangle = \sum_{n=1}^{N} \sum_{m=1}^{N} \langle \mathbf{r}_n \mathbf{r}_m \rangle = b^2 f(z, N) \qquad \text{(IV.5.68)}$$

where

$$f(z, N) = \sum_{n=1}^{N} \sum_{m=1}^{N} \frac{1}{(z-1)^{|n-m|}} \qquad \text{(IV.5.69)}$$

Thus, now we are left with the purely algebraic task of calculation of this double sum, as a function of $z$ and $N$, which will be convenient to rewrite in the form

$$f(z, N) = \sum_{n=1}^{N} \sum_{k=-n+1}^{N-n} \frac{1}{(z-1)^{|k|}} \qquad \text{(IV.5.70)}$$

(we have used replacements: $k = m - n$, $m = 1 \Rightarrow k = 1 - n$, $m = N \Rightarrow k = N - n$). We can, now, get an exact formula for $f(z, N)$, separately for $N \gg 1$, $z > 2$, and for any $N$, $z = 2$.

1.
$$f(2, N) = \sum_{n=1}^{N} \sum_{k=-n+1}^{N-n} 1 = N^2 \qquad \text{(IV.5.71)}$$

This means that

$$\sqrt{\langle \mathbf{R}^2 \rangle} = bN \qquad \text{(IV.5.72)}$$

This is an obvious result, because if you have only two neighbour points on a line, and you are not allowed to step back, you can only go one way and your deviation from the initial point will be proportional to the number of your steps.

2.
$$f(z > 2, N \gg 1) \approx \sum_{n=1}^{N} \sum_{k=-\infty}^{\infty} \frac{1}{(z-1)^{|k|}} = N \sum_{k=-\infty}^{\infty} \frac{1}{(z-1)^{|k|}}$$

To proceed from here we still need to do few more rearrangements. Namely,

$$\sum_{k=-\infty}^{\infty} \frac{1}{(z-1)^{|k|}} = 1 + 2 \sum_{k=1}^{\infty} \frac{1}{(z-1)^k},$$

$$\sum_{k=1}^{\infty} \frac{1}{(z-1)^k} = \sum_{k=1}^{\infty} \frac{1}{(z-1)(z-1)^{k-1}} = \frac{1}{(z-1)} \sum_{k=1}^{\infty} \frac{1}{(z-1)^{k-1}}$$

$$= \frac{1}{(z-1)} \sum_{m=0}^{\infty} \frac{1}{(z-1)^m}$$

As the last sum is the geometric progression, $\sum_{m=0}^{\infty} \frac{1}{(z-1)^m} = \frac{1}{1-\frac{1}{z-1}} = \frac{z-1}{z-2}$. So we have all expressions in place, and

$$f(z > 2, N \gg 1) \approx N \left\{ 1 + \frac{2}{z-2} \right\} = N \frac{z}{z-2} \qquad (IV.5.73)$$

Going back to Eq. (IV.5.68), we get

$$\sqrt{\langle \mathbf{R}^2 \rangle} \approx b \sqrt{\frac{z}{z-2}} \sqrt{N} \qquad (IV.5.74)$$

Thus the scaling law of dependence on $N$ has not changed, as compared to when stepping back is allowed. But the extension will be a tiny bit larger, by a factor of $\sqrt{z/(z-2)}$. One can say that the ban on stepping back just renormalizes the bond length $\Rightarrow b_{eff} = b\sqrt{z/(z-2)}$. Careful analysis, beyond the scope of this chapter, shows that this will be the case for any short-range interactions/correlations between the bonds. From Eq. (IV.5.67) we say that the case that we have considered here had exponential (i.e. short-range) correlations between the vectors of separated bonds. This class of chains we call quasi-ideal Gaussian chains.

### What is the fractal dimension of an uncorrelated/ short-correlated Gaussian random walk?

In a volume fractal, a number of points are embraced within the radius $R$, $N \propto R^D$, where $D$ is a volume fractal dimension. We know

now, after examining our random walk system, that in any dimension except for 1D, $\sqrt{\langle \mathbf{R}^2 \rangle} \propto \sqrt{N}$ in both an unrestricted and a short-range-restricted random walk. Thus the number of points visited, or the number of segments of the Gaussian chain, within the radius $R$ will be $N \propto R^2$. Note that the dimensionality of space, except for disallowing the one-dimensional case, did not appear in our analysis. This means that in any dimensional space, except for in 1D, ideal and quasi-ideal Gaussians chains have fractal dimension $D = 2$.

## A 'bead-spring' chain

One may ask, what is the probability $P(\{\mathbf{R}_n\})$ for all the segments of a chain to occupy a sequence of points $\{\mathbf{R}_n\} = (\mathbf{R}_0, \mathbf{R}_1, \dots \mathbf{R}_N)$? The probability to reach point $n$ from the point $n-1$, which, as we studied above, is: $P(\mathbf{R}_n; \mathbf{R}_{n-1}) = \left(\frac{3}{2\pi b^2}\right)^{3/2} e^{-\frac{3}{2b^2}(\mathbf{R}_n - \mathbf{R}_{n-1})^2}$, and so,

$$P(\{\mathbf{R}_n\}) = \left(\frac{3}{2\pi b^2}\right)^{3N/2} \prod_{n=1}^{N} e^{-\frac{3}{2b^2}(\mathbf{R}_n - \mathbf{R}_{n-1})^2}$$

$$= \left(\frac{3}{2\pi b^2}\right)^{3N/2} \exp\left(-\frac{3}{2b^2} \sum_{n=1}^{N} (\mathbf{R}_n - \mathbf{R}_{n-1})^2\right)$$

We can formally rewrite this result in the form of Boltzmann probability,

$$P \propto \exp\left\{-\frac{U(\{\mathbf{R}_n\})}{k_B T}\right\} \tag{IV.5.75}$$

with

$$U = \frac{\kappa}{2} \sum_{n=1}^{N} (\mathbf{R}_n - \mathbf{R}_{n-1})^2 \tag{IV.5.76}$$

where

$$\kappa = \frac{3k_B T}{b^2} \tag{IV.5.77}$$

Equations (IV.5.75) and (IV.5.76) look like a description of a system of harmonic springs, with identical *spring constant*, $\kappa$. But there are no interactions between segments, except for all of them belonging to one chain in which each segment is bonded with two of its nearest neighbours. Thus $\kappa$ is proportional to thermal energy because, as we have discussed, it is only entropy that determines 'elasticity' of these 'strings'.

### Statistical geometry, gyration radius and the Fourier transform of the correlation function

In a random chain of $N$ segments, an average density of segments at a position $\mathbf{r}$ from segment $n$ is given by the function $g_n(\mathbf{r}) = \sum_{m=1}^{N} \langle \delta(\mathbf{r} - (\mathbf{R}_m - \mathbf{R}_n)) \rangle$. Hence, the pair correlation function that characterize the density of a segment at distance $\mathbf{r}$ from another segment is given by,

$$g(\mathbf{r}) = \frac{1}{N} \sum_{n=1}^{N} g_n(\mathbf{r}) = \frac{1}{N} \sum_{n=1}^{N} \sum_{m=1}^{N} \langle \delta(\mathbf{r} - (\mathbf{R}_m - \mathbf{R}_n)) \rangle \qquad \text{(IV.5.78)}$$

Hence, the Fourier transform of this function, which is something that for real chains can be measured through light, X-ray, or neutron scattering, is given by:

$$g(\mathbf{q}) = \int e^{i\mathbf{q}\mathbf{r}} g(\mathbf{r}) = \frac{1}{N} \sum_{n=1}^{N} \sum_{m=1}^{N} \langle \exp[i\mathbf{q}(\mathbf{R}_m - \mathbf{R}_n)] \rangle \qquad \text{(IV.5.79)}$$

In scattering experiments, $\hbar\mathbf{q}$ plays the role of scattering particle momentum transfer upon scattering, so $\mathbf{q}$ is known and $g(\mathbf{q})$ gives the result for the scattering cross-section, a measurable quantity (we cannot go into further detail as we focus on the mathematics here).

Before calculating this function for the Gaussian chain in 3D, let us look at this function at small $q$ (the case of small angle scattering), expanding the exponential into a Maclaurin series and averaging each term in the expansion:

$$g(\mathbf{q}) = \frac{1}{N} \sum_{n=1}^{N} \sum_{m=1}^{N} [1 - iq_\alpha \langle (\mathbf{R}_n - \mathbf{R}_m)_\alpha \rangle$$

$$- \frac{1}{2} q_\alpha q_\beta \langle ((\mathbf{R}_n - \mathbf{R}_m)_\alpha (\mathbf{R}_n - \mathbf{R}_m)_\beta) \rangle + \cdots]$$

Here again, we implied summation over the repeated Cartesian indices, $\alpha$ and $\beta$. The second term in this expansion is obviously zero, whereas the third term is nonzero only for the product of the same Cartesian components, so that $\langle((\mathbf{R}_n - \mathbf{R}_m)_\alpha(\mathbf{R}_n - \mathbf{R}_m)_\beta)\rangle \propto \frac{1}{3}\delta_{\alpha\beta}$, and thus to the accuracy of the second term in expansion, we have $g(\mathbf{q}) \simeq \frac{1}{N}\Sigma_{n=1}^N\Sigma_{m=1}^N\left[1 - \frac{q^2}{6}\langle(\mathbf{R}_m - \mathbf{R}_n)^2\rangle\right]$. Taking into account that

$$g(0) = N \tag{IV.5.80}$$

and introducing notation,

$$R_g^2 \equiv \frac{1}{2N^2}\sum_{n=1}^N\sum_{m=1}^N \langle(\mathbf{R}_m - \mathbf{R}_n)^2\rangle \tag{IV.5.81}$$

we can write

$$g(\mathbf{q}) \simeq g(0)\left[1 - \frac{q^2 R_g^2}{3}\right] \tag{IV.5.82}$$

where the second term in the bracket is a small correction never greater than 1. $R_g$ has a name: it is called the *gyration radius*. We first find an expression for it for Gaussian chains and then explain why it was given such name.

For $N \gg 1$ the terms in the sum with $|n - m| \gg 1$ give the major contribution to it, and, thus, for the Gaussian chain we get $\langle(\mathbf{R}_m - \mathbf{R}_n)^2\rangle \approx b^2|n - m|$, so that $R_g^2 \approx \frac{b^2}{2N^2}\Sigma_{n=1}^N\Sigma_{m=1}^N|n - m|$. The sum in this expression can be calculated exactly. Since we have done this already in Chapter 8 of Part I, Vol. 1 (Example I.8.7), we will not be doing it again and just recall that the result valid for large $N$ is $\Sigma_{n=1}^N\Sigma_{m=1}^N|n - m| \approx \frac{N^3}{3}$. Hence, $R_g^2 = \frac{Nb^2}{6}$

$$R_g = b\frac{1}{\sqrt{6}}\sqrt{N} \tag{IV.5.83}$$

This states that the gyration radius is $\sqrt{6}$ times smaller than the average end-to-end distance is of the polymer of that number of segments.

Now, why the name? Let us introduce a vector pinpointing the position of a "centre of a distribution of sites" (see Fig. IV.5.21)

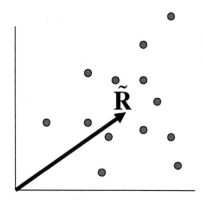

**Fig. IV.5.21**   Vector pointing to a centre of mass of a distribution of particles.

called for brevity a centre of mass of this distribution, defined as

$$\tilde{\mathbf{R}} = \frac{1}{N} \sum_{n=1}^{N} \mathbf{R}_n \qquad \text{(IV.5.84)}$$

The gyration radius should be defined with respect to $\tilde{\mathbf{R}}$ as

$$R_g^2 = \frac{1}{N} \sum_{n=1}^{N} \langle (\mathbf{R}_n - \tilde{\mathbf{R}})^2 \rangle \qquad \text{(IV.5.85)}$$

**Exercise.** Substituting Eq. (IV.5.84) into Eq. (IV.5.85) prove that you get Eq. (IV.5.81), our initial expression for gyration radius. **Hint:** You will need to use the following trivial identities: $(\mathbf{A} + \mathbf{B})^2 = \mathbf{A}^2 + \mathbf{B}^2 + 2\mathbf{A}\mathbf{B}$ and $\frac{1}{N}\Sigma_{n=1}^{N} 1 = 1$.

Let us now get the full expression for the Fourier transform of the correlation function. We first recall that we can get this Fourier transform directly from the real space expression for the correlation function, i.e.

$$\langle \exp[i\mathbf{q}(\mathbf{R}_m - \mathbf{R}_n)] \rangle = \left( \frac{3}{2\pi|n - m|b^2} \right)^{3/2} \int_0^\infty dr \, 4\pi r^2 \frac{\sin qr}{qr} e^{-\frac{3r^2}{2|n-m|b^2}}$$

$$= \exp\left( -\frac{q^2}{6}|n - m|b^2 \right) \qquad \text{(IV.5.86)}$$

Substitution of Eq. (IV.5.86) into Eq. (IV.5.79), and replacing summation by integration over $m$ and $n$, as taught in Chapter 8 of Part 1, Vol. 1, possible for large $N$, we get

$$g(\mathbf{q}) = \frac{1}{N} \int_0^N dn \int_0^N dm \exp\left(-\frac{q^2}{6}|n - m|b^2\right) = Nf(qR_g),$$

$$f(x) = \frac{2}{x^4}\left(e^{-x^2} - 1 + x^2\right) \qquad\qquad\qquad (IV.5.87)$$

**Exercise.** We did not derive for you the integrals in Eq. (IV.5.86) and Eq. (IV.5.87) but just presented the final result, leaving this for you to cope with them.

The limiting behaviour of $f(x)$ is trivial to obtain, so it is easy to find that

$$g(\mathbf{q} \to 0) = N, \quad g(\mathbf{q} \to \infty) = \frac{2N}{q^2 R_g^2} \qquad\qquad (IV.5.88)$$

Knowing this, one can propose an interpolation formula which obeys both limiting laws, of the so-called Ornstein–Zernicke type,

$$g(\mathbf{q}) = \frac{N}{1 + q^2 \left(\frac{R_g^2}{2}\right)} \qquad\qquad\qquad (IV.5.89)$$

In Fig. IV.5.22, we compare the two functions, representing the exact solutions and the Ornstein–Zernicke interpolation formula. So within any reasonable error bars in your experiments, you are not expected to see a difference.

Hence,

$$\frac{1}{g(\mathbf{q})} \approx \frac{1}{N} + q^2\frac{1}{N}\left(\frac{R_g^2}{2}\right) \qquad\qquad (IV.5.90)$$

and thus one may try to plot the experimental data for $\frac{1}{g(\mathbf{q})}$ versus $q^2$ to see a straight line. But for many systems that can be modelled as random chains, the slope is not straight and the exponent is not equal to 2 but has a fractional value. How can we explain this?

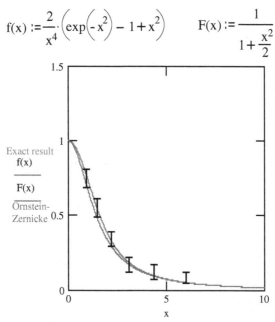

$$f(x) := \frac{2}{x^4} \cdot \left( \exp\left(-x^2\right) - 1 + x^2 \right) \qquad F(x) := \frac{1}{1 + \frac{x^2}{2}}$$

Exact result
f(x)

$\overline{F(x)}$

$\overline{\text{Ornstein-}\atop \text{Zernicke}}$

**Fig. IV.5.22** Comparing two functions.

## *Gaussian chains with excluded volume*

Explanation of this fact involves some physics, but sticking to the spirit of the book we may still present it as an aspect of statistical geometry and probability theory, showing thermodynamics-based analysis only at the very end of this section.

If the chain is a trace of random walk on a 2D or 3D lattice (or even higher dimensions), it will be a very contrived rule that one could not visit a site already visited, unless it is a search for lost property or the case of a tourist in a historical city who would try to see only new attractions on his path. It is of course different if the chain represents a polymer; that chain cannot map onto itself anywhere, at any segment. We therefore consider here the effect of chain self-avoidance.

The logic is as follows. Let us start first with an $N$-segment-long chain that is ideal, so that it can overlap on itself at any point.

The number of segments in an infinitesimally small volume element $4\pi R^2 dR$ would be given by

$$W_{\text{ideal}}(R)dR \propto P(\mathbf{R}, N)4\pi R^2 dR,$$

$$W_{\text{ideal}}(R) \propto 4\pi R^2 \left(\frac{3}{2\pi N b^2}\right)^{3/2} \exp\left(-\frac{3R^2}{2Nb^2}\right) \qquad \text{(IV.5.91)}$$

Next, the probability that no overlaps between any segments of the chain, each of volume $\Omega_s$, take place within the sphere of volume $\frac{4\pi}{3}R^3$ can be estimated as

$$p_{\#}(R) = (1 - \underbrace{\Omega/R^3}_{\substack{\text{Probability} \\ \text{of a segment} \\ \text{to overlap} \\ \text{with any} \\ \text{other one}}})\overbrace{N(N-1)/2}^{\substack{\text{Total number} \\ \text{of segment pair} \\ \text{combinations}}}$$

$$\equiv \exp\left[\frac{1}{2}N(N-1)\ln(1-\Omega/R^3)\right]\overbrace{\phantom{xxxxxxxx}}^{\substack{\text{Large } N \text{ and } R: \\ N(N-1)\approx N^2 \\ \ln(1-\Omega/R^3)\approx-\Omega/R^3 \\ \downarrow \\ \approx}}$$

$$\times \exp\left(-\frac{N^2\Omega}{2R^3}\right) \qquad \text{(IV.5.92)}$$

Here, we have introduced the reduced segment volume, defined as

$$\Omega \equiv \frac{3}{4\pi}\Omega_s \qquad \text{(IV.5.93)}$$

Now, the density of *excluded volume* chains in the interval $(R, RdR)$ is

$$W(R) = W_0(R)p(R) \propto R^2 \exp\left\{-\left[\frac{3R^2}{2Nb^2} + \frac{N^2\Omega}{2R^3}\right]\right\}$$

$$= \exp\left\{2\ln R - \frac{3R^2}{2Nb^2} - \frac{N^2\Omega}{2R^3}\right\} \qquad \text{(IV.5.94)}$$

We can find the maximum of that probability, which will give the most probable relation between $R$ and $N$. But first let's check where the maximum of $W_0(R) \propto \exp\left\{2\ln R - \frac{3R^2}{2Nb^2}\right\}$ lies. Differentiating this expression over $R$, and equalizing the result to zero, we get that for an ideal chain that probability is maximized by

$$R_0^* = b[2/3]^{1/2}\sqrt{N} \qquad (IV.5.95)$$

Doing the same for Eq. (IV.5.94), we get a more complicated equation on the value of $R^*$ which delivers the maximal value of $W(R)$. This equation reads $\frac{2}{R} - 3\frac{R}{Nb^2} + \frac{3N^2\Omega}{2R^4} = 0$. Using the notation of Eq. (IV.5.95), and understanding that $R$ can never be zero, this equation can be written in the form

$$\frac{R^5}{R_0^{*5}} - \frac{R^3}{R_0^{*3}} = \frac{9\sqrt{6}\sqrt{N}\Omega}{16b^3} \qquad (IV.5.96)$$

We call the solution of this equation $R^*$. Instead of trying to solve it, we can simplify it first. Indeed, for $\Omega = 0$ and $N \gg 1$ (strictly speaking, for $\frac{9\sqrt{6}\sqrt{N}\Omega}{16b^3} \gg 1$), we can omit the $(R^3/R_0^{*3})$-term, thus obtaining

$$\frac{R^5}{R_0^{*5}} \simeq \frac{9\sqrt{6}\sqrt{N}\Omega}{16b^3} \qquad (IV.5.97)$$

Hence, for $R^*$ we get

$$R^* \simeq R_0^* \left[\frac{\Omega}{b^3}\right]^{1/5} N^{0.1} \propto N^{3/5} \qquad (IV.5.98)$$

Even this simplified, non-rigorous analysis shows that the scaling of the characteristic extension of a chain with the number of its segments has changed. The exponent is no longer 0.5, but is now 0.6. Qualitatively, the effect of increase of the exponent is natural, because if the chain cannot cross the points of space where some of its segments already are, it has to be extended further. But the exact value of the increase of the exponent—this is a matter of theory. It is possible to do a more careful analysis for the gyration radius, and you will find the same result,

$$R_g \propto N^{3/5} \tag{IV.5.99}$$

Numerical simulations as well as experiments give $R_g \propto N^\mu$, $\mu = 0.588$. Pretty close, isn't it!

Generally, a long Gaussian chain is self-similar. When rescaling the bond length $\lambda^\mu$ times, a physical characteristic related to $f$ is expected to obey the scaling law. It would be tempting to assume that $f(\lambda^{-1}N, \lambda^\mu b) = f(N, b)$, but in fact the law was found to need a minor correction factor:

$$f(\lambda^{-1}N, \lambda^\mu b) = \lambda^s f(N, b) \tag{IV.5.100}$$

where the exponent $s$ is nonuniversal, and appears to depend on the physical quantity under consideration.

It is now easy to explain the problem with the deviation from the straight line in observed slopes of $\frac{1}{g(\mathbf{q})}$ versus $q^2$. Indeed, for $qR_g \gg 1$, $g(\mathbf{q})$ must not depend on $N$. For an ideal chain, the latter requirement is obviously satisfied, as $g(\mathbf{q} \to \infty) = \frac{2N}{q^2 R_g^2}$ and $R_g^2 \propto N$. Similarly, to satisfy independence on $N$ of $g(\mathbf{q} \to \infty)$ for an excluded volume chain, when $R_g \propto N^\mu$ one should require $g(\mathbf{q}) \propto N(qR_g)^{-1/\mu} \propto q^{-1/\mu}$, i.e. $\frac{1}{g(\mathbf{q})} \propto q^{1/\mu}$, and for $\mu \approx 3/5$, $\frac{1}{g(\mathbf{q})} \propto q^{\frac{5}{3}}$ instead of $\propto q^2$, and it is the fractional power law which is usually observed.

Note that by the character of derivation we have obtained it in 3D space. What is it going to be in other dimensions?

**Exercise.** Perform the same kind of analyses as in Eq. (IV.5.91)–(IV.5.98) for a polymer lying in the plane, to obtain that in two dimensions, $\mu = 3/4$. **Hint:** In the argument of the exponential function in Eq. (IV.5.94), the logarithmic term will arrive without a factor of 2 (which is not important for the resulting scaling law) and the term will be proportional to $R^{-2}$, but not to $R^{-3}$ (**Think:** Why?), and this will change the scaling exponent.

We thus find that using simplified statistical geometry-based calculations we get

$$\mu = \begin{cases} 3/5, & \text{in 3D} \\ 3/4, & \text{in 2D} \\ 1, & \text{in 1D} \end{cases} \qquad \text{(IV.5.101)}$$

It is also worth looking into another derivation of $\mu$ that uses physical arguments.

### Flory's mean-field theory

Very simple considerations were proposed by an American physical chemist, Paul Flory, who received the 1974 Nobel Prize in Chemistry "for his fundamental achievements, both theoretical and experimental, in the physical chemistry of macromolecules." In concentrated form we repeat them here.

Paul John Flory
(1910–1985)

He assumed that the free energy density of repulsion of monomers could be approximated by $f_{rep} \propto k_B T \Omega c^2$, where $\Omega$ is the volume of a monomer (in 3D), or the area that it would occupy on a surface if the chain is disposed in 2D. Here, $c$ is the net concentration of monomers (volume or surface concentration for 3D and 2D cases, respectively; $c \propto N/R^d$. The free energy of repulsion is then $F_{rep} = \int d^d R f_{rep} \propto k_B T \Omega \frac{N^2}{R^d}$. The free energy of 'elastic' interaction, without the excluded volume effect we have studied already in Example IV.5.3, is given by $F_{elastic} \propto k_B T \frac{R^2}{Nb^2}$. The total, free energy $F$ is the sum of these two terms, so that

$$\frac{F}{k_B T} =\propto \Omega \frac{N^2}{R^d} + \propto \frac{R^2}{Nb^2} \qquad \text{(IV.5.102)}$$

Differentiating this expression over $R$ and equalizing the result to zero, to find the value of $R$ that delivers the minimum to this function, we find that

$$R^* \propto N^{\frac{3}{d+2}} \qquad \text{(IV.5.103)}$$

This means that

$$\mu = \frac{3}{d+2} \tag{IV.5.104}$$

For $d = 1, 2, 3$ this formula reproduces Eq. (IV.5.101). This is an amazing success for this scaling argument, as more complicated theories and computer simulations gave very close values for $\mu$. An attentive reader will, of course, notice an internal similarity between the statistical geometry analysis we presented and Flory's arguments.

### Fractal dimension of an excluded volume polymer

There was serious analysis devoted to this issue in the literature, but if the average end-to-end distance (call it the characteristic size of a polymer) or its radius of gyration, scale as

$$R_g \propto N^\mu \tag{IV.5.105}$$

then $N \propto R_g^{1/\mu} = R_g^D$, so that

$$D = \begin{cases} 5/3, & \text{in 3D} \\ 4/3, & \text{in 2D} \\ 1, & \text{in 1D} \end{cases} \tag{IV.5.106}$$

The last line in this equation is, of course exact; the first two lines are results of simple calculations as discussed above, which are actually very well reproduced by Monte Carlo simulations.

### The role of environment and the collapse of the Gaussian chain

When we were solving Eq. (IV.5.96) we were able to neglect the second term in the l.h.s, but this was possible only because the r.h.s was large. We will now consider the situation when the latter can be small, due to a renormalization of excluded volume. Then keeping the neglected term allows us to obtain a second solution of that equation which corresponds to a smaller value of the extension of the chain, and the transition to that state is called collapse of the chain. We will show now how this occurs.

Rarely do chains exist in a vacuum or some 'abstract inert medium' with which the chains do not interact. Consider polymer chains dissolved in a liquid solvent. Let us try to take account of interactions of segments with the solvent. We will do it in the most primitive, mean-field way, but nevertheless we will have to involve free energy and Boltzmann probabilities, as in Flory's analysis. It will be clearer to do this on a lattice, although you already understand that for a long chain the lattice approximation is not important.

For some configuration "$i$", its energy is given by

$$E_i = -N_{pp}^{(i)} \varepsilon_{pp} - N_{ps}^{(i)} \varepsilon_{ps} - N_{ss}^{(i)} \varepsilon_{ss} \qquad \text{(IV.5.107)}$$

where $N_{pp}^{(i)}$, $N_{ps}^{(i)}$, and $N_{ss}^{(i)}$ are the numbers of polymer-polymer, polymer-solvent, and solvent-solvent pairs on the lattice, and $-\varepsilon_{pp}$, $-\varepsilon_{ps}$, and $-\varepsilon_{ss}$ are the corresponding energies that each of those pairs contributes to the system energy (see Fig. IV.5.23). Next, in a crude mean-field approximation, the probability of a polymer having end-to-end distance $R$ in the solvent can be evaluated as

$$W_s(R) \propto W(R) \exp\left\{-\frac{\langle E_i(R)\rangle}{k_B T}\right\} \qquad \text{(IV.5.108)}$$

Here, $W(R)$ is what we have studied before, the probability without taking account of polymer-solvent interaction, and in the mean-field approximation we are averaging the exponent, but not the exponential over all possible "$i$"-configurations. If we introduce the probability that a site is occupied by a polymer segment,

$$\phi = N\Omega/R^3 \qquad \text{(IV.5.109)}$$

we can write

$$\langle N_{pp}^{(i)}\rangle \simeq \frac{1}{2}zN\phi, \quad \langle N_{ps}^{(i)}\rangle \simeq zN(1-\phi),$$

$$\langle N_{ss}^{(i)}\rangle \simeq N_{ss}^{(0)} - \left[\frac{1}{2}zN\phi + zN(1-\phi)\right] \qquad \text{(IV.5.110)}$$

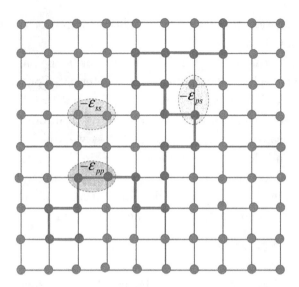

**Fig. IV.5.23** Lattice model of a polymer chain (red) and solvent molecules (blue): Each red bond between two monomers (segments) reduces energy by $\varepsilon_{pp}$, each interaction between two solvent molecules reduces the energy by $\varepsilon_{ss}$, and each interaction between polymer segment and solvent molecule reduces it by $\varepsilon_{ps}$.

Now, we introduce the key parameter characterizing the solubility of the polymer. This is

$$\Delta\varepsilon \equiv \frac{1}{2}(\varepsilon_{pp} + \varepsilon_{ss}) - \varepsilon_{ps} \tag{IV.5.111}$$

such that when $\Delta\varepsilon > 0$, the polymer is solvophobic—the polymer will tend to shrink, and when $\Delta\varepsilon < 0$ it solvophilic—the polymer will tend to swell, as compared with the neutral case when $\Delta\varepsilon = 0$. Then,

$$\langle E_i \rangle = zN\phi\Delta\varepsilon = -\frac{zN^2\Omega}{R^3}\Delta\varepsilon \tag{IV.5.112}$$

so that

$$W_s(R) \propto R^2 \exp\left\{-\left[\frac{3R^2}{2Nb^2} + \frac{N^2\Omega}{2R^3}(1 - 2\chi)\right]\right\} \tag{IV.5.113}$$

Here we introduced,

$$\chi = \frac{z\Delta\varepsilon}{k_B T} \tag{IV.5.114}$$

Equation (IV.5.113) looks like Eq. (IV.5.94) but includes the effective, renormalized, excluded volume,

$$\Omega_{eff} = \Omega(1 - 2\chi) \qquad \text{(IV.5.115)}$$

In a *good* solvent $\chi$ is small, or can be even negative, so that $\Omega_{eff} > 0$. In a *bad* solvent $\chi$ is positive and can be large, and $\Omega_{eff}$ may change sign. This takes place at $\chi > 1/2$. From equation $\chi = \frac{z\Delta\varepsilon}{k_B T_c} = \frac{1}{2}$, we find the value of the critical temperature to be

$$T_c = 2z\Delta\varepsilon/k_B \qquad \text{(IV.5.116)}$$

At this point $\Omega_{eff}$ turns zero.

And what happens then? To answer this question, we need to solve the same equation as Eq. (IV.5.96), but with $\Omega$ replaced by $\Omega_{eff}$,

$$\frac{R^5}{R_0^{*5}} - \frac{R^3}{R_0^{*3}} = \frac{9\sqrt{6}\sqrt{N}\Omega_{eff}}{16b^3} \qquad \text{(IV.5.117)}$$

in which the r.h.s can, in principle, become negative. When $\Omega_{eff}$ it is very small, we cannot neglect the second term in the l.h.s of Eq. (IV.5.117). You have to look for a solution of the whole equation.

This is illustrated in Fig. IV.5.24. When $\Omega_{eff}$ crosses zero and become negative, the second solution emerges, which corresponds to a *collapsed state* of a polymer. To observe this experimentally you do not necessarily need to change the solvent. You can instead reduce temperature. Indeed, the above arguments tell you that below the critical temperature, $\Omega_{eff}$ will be negative and the collapse will take place. These types of experiments are classical and demonstrate the so-called coil-to-globule transition of a polymer.

Similar thing would happen with a random walker, a tourist in an unknown town, if its environment got unfriendly to them or it just got very cold: if they still want fresh air they will just walk around their hotel.

Of course, the analysis of this effect as presented above is grossly simplified—totally 'mean-fieldish'. More involved descriptions of this effect based on field-theoretical methods have been developed that

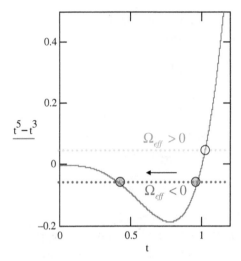

**Fig. IV.5.24** This illustrates the solution of Eq. (IV.5.117) where $t \equiv R/R_0^*$ and the dotted line represents the case of positive and negative values of the r.h.s of the equation. When it is negative we get a second solution, which corresponds to the collapsed state of the polymer, in which its segments like to interact with other segments more than with the solvent (bad solvent case). When the r.h.s is positive there is only one solution, which corresponds to the state in which the polymer is expanded (good solvent case) as compared to what it is in a neutral solvent.

allow for much more dramatic changes of the polymer gyration radius around the transition temperature, as normally experimentally seen. For these you will need to read specialized literature, but for grasping the qualitative picture of this effect, the above considerations are a good introduction.

## 5.3.    Percolation

Imagine a drunken worker who has to hammer nails into the sites of a marked square grid. Because they have consumed too much gin & tonic, each time they chose the site to nail in, they do it at random, immediately forgetting which site was previously nailed. But the number of nails is generally smaller than the number of lattice sites, so that their ratio ($<1$) serves a probability that a site is nailed. Now, we ask the worker's apprentice (who is sober as a judge), to draw a line/a path connecting the nearest nailed lattice

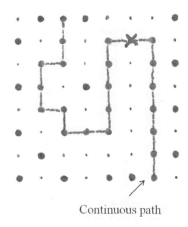

Continuous path

**Fig. IV.5.25** Continuous, percolation path connecting randomly occupied nearest neighbour sites (big dots) on a square grid from north to south. The continuity of the path is broken if we vacated one site on its way, as shown, e.g. by a crossed site.

sites on the grid (diagonally-proximal sites not qualified as 'neighbours') that comprises a continuous path from the north to the south, as in Fig. IV.5.25. What is the ratio, $p$, of the number of nails to the number of sites for such path to emerge? Which is the same as asking, what is the probability $p$ of a site to be nailed to warrant such a pathway? This critical value is denoted as $p_c$.

Fig. IV.5.25 shows a fragment of a lattice with a continuous path, but the same may take place on an infinite lattice. Moreover, the shortest percolation path may have satellite neighbours; connecting them to the path will display a percolation cluster. Such clusters are shown in Fig. IV.5.26.

The examples we have drawn refer to so-called *site percolation*, which should be distinguished from *bond percolation*. In the latter, one operates with the probability that the bond between sites is intact, and these two different percolation problems may have slightly different characteristics. For brevity, we will speak here only about site percolation.

The example drawn in Fig. IV.5.25 is on a plane square lattice. The same problem may result in the space of any dimensionality, any structure of the lattice, including disordered lattices, or in off-lattice descriptions. We will concentrate on 1D, 2D, and 3D

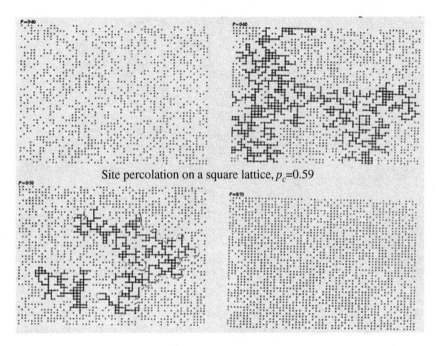

**Fig. IV.5.26** Occupation of sites on a squared grid and emergence of a percolation cluster. Dots show occupied sites. Black lines show the bonds between occupied sites in the largest clusters on a panel before the critical point and right after. [Reproduced from Fig. 2 of *Introduction to Percolation Theory* by D. Stauffer and A. Aharony, 2nd Edition, Taylor & Francis, London 1992, with permission of the Taylor & Francis Group].

systems, whereas mathematicians eagerly explore systems of higher dimensions (don't consider them egg-headed: whatever mathematicians discover appears to be useful, if not sooner then later!).

From various theoretical models and computer simulation, we know that $p_c$ depends the structure of the lattice (or more generally the graph representing the positions of the sites to be occupied) whereas the system properties for $p$ near $p_c$ appear to be universal, self-similar, and are characterized by the so called *critical exponents* (we will deal with them in this chapter). These properties are believed to depend only on the dimensionality of space in which the system is embedded; the various critical exponents and the fractal dimension of the clusters near $p = p_c$ do not depend on the lattice structure or even whether it is a site or bond percolation. This principle has been challenged by considering some exotic lattices, for which

**Table IV.5.1** Site-percolation thresholds for different lattices.

| Dimension | Lattice type | $p_c$ |
|---|---|---|
| 1 | 1D-line | 1 |
| 2 | Honeycomb | 0.696 |
| 2 | Square | 0.593 |
| 2 | Triangular | 1/2 |
| 3 | Diamond | 0.43 |
| 3 | Primitive cubic | 0.31 |
| 3 | BCC | 0.246 |
| 3 | FCC | 0.198 |
| 3 | Bethe lattice with coordination number $z$ | $\frac{1}{z-1}$ |

different characteristics have been obtained, although these lattices were embedded into the plane, and belonged to the same universality class as ordinary plane lattices. Well, in mathematics in order to disprove the universality of a principle, it is sufficient to present one example not obeying it. But let us leave this debate to the frontier line theorists and, for this introduction to percolation, all systems considered below will be assumed to obey this principle.

For your orientation, the values of site-percolation thresholds, some calculated exactly or approximately by theory, some evaluated via computer simulation, are presented in Table IV.5.1. Three of those listed (1D line, triangular, and Bethe) we will calculate using analytical theory. Percolation clusters close to $p = p_c$ are volume fractals, and where possible we will calculate their fractal dimensions.

The analysis presented below will be either exact, or approximate, or qualitative scaling—not rigorous. We will dare to present the results of the latter two, knowing how well they are reproduced by numerical simulation. But before diving into this analysis, it is worth thinking about why we should even bother about percolation. It is because percolation phenomena emerge, or better to say that the description of various systems and phenomena can be mapped on percolation models, in all manner of systems: electrical conductivity of composite materials (metal-insulator, metal-electrolyte), permeability and ionic conductivity of porous media, propagation of reaction fronts, forest fires, structure of water, etc. When such systems are close to percolation threshold, many laws in describing their properties/behaviour appear universal; it is just the physical variables in these laws are different.

But there should be no illusions, those laws are not universal away from the threshold, in the same way as the thermodynamic properties away from the critical points of continuous transitions. But often we are interested just in critical behaviour, and knowing that some laws are universal is balm for the souls of those who search for universal laws of nature.

Even in everyday life, percolation means a lot. Two miner teams digging a tunnel from the two sides of Montblanc between France and Italy must meet; getting your scientific paper accepted for publications must percolate through many hurdles; getting your kid accepted to Oxford, Harvard, or Tsinghua University, depending on where you are coming from, must overcome a threshold, and we had better not speak about receiving a Nobel Prize — how far your ideas should propagate, and how many hearts and minds your discoveries should conquer.

### *Random (uncorrelated) percolation on 1D lattice*

A percolation model in 1D (see for example Fig. IV.5.27) has an exact solution. It has, of course, $p_c = 1$, because it is a sufficient to vacate one side to block off percolation. But let us find it as a result of calculation, and simultaneously obtain the average size of the clusters as a function of $p$.

Let us introduce some notions first. Let $n_s$ be a *number of clusters of s-sites per lattice site*, which is obviously given by the formula

$$n_s = p^s(1-p)^2 \tag{IV.5.118}$$

in which the second factor accounts for the fact that two empty sides should be on the two sides of the cluster. The probability that a site

**Fig. IV.5.27** One dimensional grid with occupied (small circles) and empty (ticks) sites.

is a part of a cluster of $s$-sites is given by

$$W_s = n_s s \tag{IV.5.119}$$

What is then $\Sigma_s W_s$? By its meaning it is just the probability that a site is occupied, i.e. it should be equal to $p$. Let us check this:

$$\sum_s W_s = \sum_s n_s s = (1-p)^2 \sum_s p^s s = (1-p)^2 p \frac{d}{dp} \sum_s p^s$$

$$= (1-p)^2 p \frac{d}{dp} \frac{1}{1-p} = (1-p)^2 p \frac{1}{(1-p)^2} = p$$

where we used that the sum on the far right-hand side of the first line of this equation is that of an infinite geometric progression. So it is indeed true that $\Sigma_s W_s = p$.

Similarly we may calculate average cluster size:

$$S = \frac{\sum_s n_s s \cdot s}{\sum_s n_s s} \tag{IV.5.120}$$

$$\Rightarrow S = \frac{\sum_s n_s s \cdot s}{\sum_s n_s s} = \frac{(1-p)^2}{p} \sum_s p^s s^2 = (1-p)^2 \frac{d}{dp} \sum_s p^s s$$

$$= (1-p)^2 \frac{d}{dp} p \frac{d}{dp} \sum_s p^s = (1-p)^2 \frac{d}{dp} p \frac{d}{dp} \frac{1}{1-p}$$

$$= (1-p)^2 \frac{d}{dp} \frac{p}{(1-p)^2} = (1-p)^2 \frac{(1-p)^2 + p2(1-p)}{(1-p)^4}$$

$$= (1-p)^2 \frac{1-p^2}{(1-p)^4} = \frac{1+p}{1-p} \tag{IV.5.121}$$

Thus, $S$ diverges $\propto \frac{1}{1-p}$ as $p$ approaches 1, i.e.

$$S = \frac{1+p}{p_c - p} \tag{IV.5.122}$$

where

$$p_c = 1 \tag{IV.5.123}$$

Let us now look at the correlation function, $g(r)$, which is the probability that from an occupied given site, all $r$ sites are also occupied,

$$g(r) = p^r \qquad \text{(IV.5.124)}$$

We can formally rewrite Eq. (IV.5.124) in the form of an exponential function:

$$g(r) = \exp(-r/\xi), \quad \xi = -(\ln p)^{-1} \qquad \text{(IV.5.125)}$$

where $\xi$ is the correlation length. How does the correlation length depend at $p$ close to $p_c$? Here is the answer: $\xi = -\{\ln[1-(1-p)]\}^{-1} \approx -\{-(1-p)\}^{-1} = \frac{1}{1-p}$, i.e.

$$\xi = \frac{1}{(p_c - p)^\nu} \quad \nu = 1 \qquad \text{(IV.5.126)}$$

### General scaling hypothesis

In the spirit of what we see in 1D, it makes sense to mention the general hypothesis that close to percolation threshold, the correlation function should take a form

$$g(r) = \frac{1}{r^{d-D}} f(r/\xi), \quad f(x) = \begin{cases} \text{const}, & x \ll 1 \\ O(e^{-x}), & x \gg 1 \end{cases} \qquad \text{(IV.5.127)}$$

where $d$ is dimensionality of space, $D$ is the volume fractal dimension of the clusters, as the percolation cluster close to the percolation threshold is a volume fractal within the range of self-similarity determined by the correlation length $\xi$. The latter tends to infinity when the system approaches the percolation threshold. Right at the threshold, the whole infinite cluster is a fractal. For finite but large $\xi$ the clusters are fractal within the volume in 3D or area in 2D of radius $\xi$, and one may also write that the average size of the cluster scales as

$$S \propto \xi^{D'} \qquad \text{(IV.5.128)}$$

Numerical simulations and renormalization group theory (some elements of which you will get acquainted in the next sub-section) show that the exponent $D'$ is systematically smaller than $D$ (see Table IV.5.2).

**Table IV.5.2** Fractal dimensions of percolation clusters in space of different dimensionality.

| $d$ | 1 | 2 | 3 |
|-----|---|-----|-----|
| $D$ | 1 | 1.9 | 2.5 |
| $D'$ | 1 | 1.8 | 2.1 |

## Real space renormalization group (RG) approach

Using $RG$ one may evaluate both the critical exponents and for a given lattice also the percolation threshold. Since critical exponents (or at least for most known cases it is true) do not depend on the lattice but only on dimensionality of space, such results look general and could possibly be universally applied in different physical contexts. The example considered below is based on the 2D triangular lattice. Due to its general nature we will not classify the example as Maths practice... but neither it is a specific physical application. So we proceed straightaway to it in the main text.

The approach is based on the so-called Migdal–Kadanoff transformation. Driven by the fact of scale invariance when $p \to p_c$, $\xi \to \infty$, one can rescale the lattice, and then take into account that characteristics of the clusters on the rescaled lattice must not change. The idea behind the transformation is depicted in Fig. IV.5.28, with its algorithm explained in the caption. So we will not repeat it here but proceed straightaway to equations.

Alexander Migdal[1]
Born 1945

Following that algorithm, the probability that the site of the renormalized lattice is occupied is given by

Leo Kadanoff[2]
(1937–2015)

$$\tilde{p} = R(p) = p^3 + 3p^2(1-p) = 3p^2 - 2p^3 \quad \text{(IV.5.129)}$$

But at $p = p_c$ there must be no difference between $\tilde{p}$ and $p$, so that

$$R(p_c) = p_c \qquad \text{(IV.5.130)}$$

---

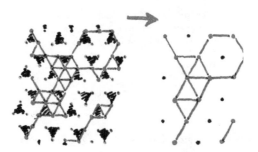

**Fig. IV.5.28** 'Blocking' transformation: every triangle on a triangular lattice which has two or three occupied vertices is replaced by a new site positioned in the middle of that triangle. The lattice constant of that renormalized lattice is $b = \sqrt{3}$ times smaller than of the original lattice. A site of a renormalized lattice is considered to be occupied if any two or three vertices of the prototype triangle of the original lattice are occupied.

Thus $3p_c^2 - 2p_c^3 = p_c$. Since for sure $p_c \neq 0$, from here we get a quadratic equation: $2p_c^2 - 3p_c + 1 = 0$. It has two solutions: $p_c = \frac{3 \pm 1}{4}$. Since obviously $p_c \neq 1$, we get for triangular lattice,

$$p_c = \frac{1}{2} \qquad (\text{IV.5.131})$$

which is exactly what we have in Table IV.5.2. Now, we use another self-similarity fact. Correlation length of an initial lattice close to $p = p_c$ is given by $\xi = \frac{\text{const}}{|p-p_c|^\nu}$, whereas for the rescale lattice $\tilde{\xi} = \frac{\text{const} \cdot b}{|\tilde{p}-p_c|^\nu}$. Since close to $p = p_c$ the correlation length must not change upon rescaling, we can write that $\frac{\text{const}}{|p-p_c|^\nu} = \frac{\text{const} \cdot b}{|R(p)-p_c|^\nu}$. Hence, $\frac{|R(p)-p_c|^\nu}{|p-p_c|^\nu} = b \implies \nu \ln \frac{|R(p)-p_c|}{|p-p_c|} = \ln b$. We can now obtain the correlation length exponent as

$$\frac{1}{\nu} = \lim_{p \to p_c} \frac{\ln \left| \frac{R(p)-p_c}{p-p_c} \right|}{\ln b} = \frac{\ln \left| \frac{R(p_c)+R'(p_c)(p-p_c)-p_c}{p-p_c} \right|}{\ln b}$$

and since $R(p_c) = p_c$, we finally get

$$\nu = \frac{\ln b}{\ln \left| \left( \frac{dR}{dp} \right)_{p=p_c} \right|} \qquad (\text{IV.5.132})$$

For our triangular lattice, $b = \sqrt{3}$, and $\frac{dR}{dp} = 6p - 6p^2 \Rightarrow$ $\left(\frac{dR}{dp}\right)_{p=p_c=1/2} = \frac{3}{2}$, and thus

$$\nu = \frac{\ln \sqrt{3}}{\ln(3/2)} = 1.354 \qquad (IV.5.133)$$

Some may say that this example is cheating, because one cannot carry the same calculations for any other lattice, e.g. a square one. Well... yes and no. This example is a beautiful demonstration of scaling ideas which practically lead to the exact value of a critical exponent as well of the percolation threshold. Something similar can be done for a square lattice, but for bond percolation (see, e.g., lecture notes on percolation by Imperial College Professor Kim Christensen, available online).

## Percolation on a Bethe lattice—Another exactly solvable case

Important structures in statistical mechanics are Cayley trees and Bethe lattices. They bear the names of famous British mathematician A. Cayley, one of the founders of modern algebra, and German-American physicist, Nobel Laureate H. Bethe, author of many formulae and discoveries, from astrophysics to nuclear physics to solid state physics and physical chemistry.

Arthur Cayley
(1821–1985)

Cayley trees and Bethe lattices are similar mathematical constructions, and to avoid the confusion as to 'who is who', let us define them properly first. Actually both are trees: connected undirected cycle-free graphs $G = (V, E)$, where $V$ stands for a number of vertices and $E$ for a number of edges (bonds).

Hans Bethe
(1906–2005)

A Cayley tree of order $z$ with $n$ shells is constructed from a root vertex (0), which is linked to $z$ new vertices by means of $z$ 'edges' (or 'bonds'). This first set of $z$ vertices forms the $n = 1$ shell of the tree. To build shell 2, each vertex of shell 1 is bonded to $z - 1$ new vertices, and off you go until this process is terminated, so that the

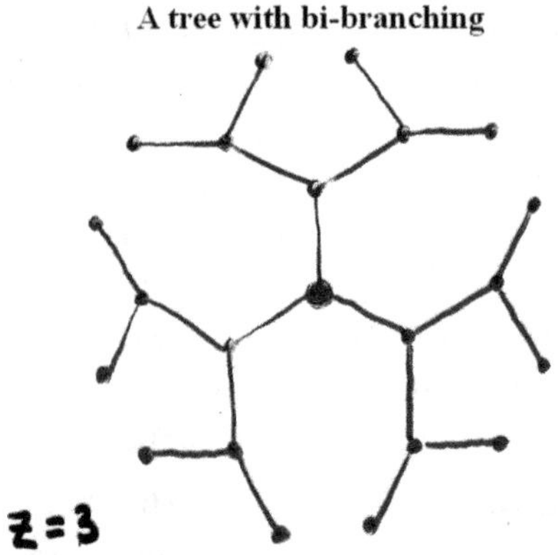

**A tree with bi-branching**

$z=3$

**Fig. IV.5.29** A fragment of a tree-like Bethe lattice with $z = 3$ bonds attached to each vertex, apart from the peripheral 12 vertices, for which the bonds branching from them outside the picture are not shown. Alternatively, you can consider it as a tiny 4-shell Cayley tree.

vertices of the last, frontier shell, $n = n_f$, each have only one bond attached to it; also, in a Cayley tree, the zero shell is represented by just the single 'root' vertex. The Bethe lattice is defined as a tree in which any vertex has $z$ bonds, i.e. there is no boundary to that lattice and there is no central, root vertex. In other words, the difference between the two structures is just that a Cayley tree is finite whereas a Bethe lattice is infinite. A fragment of the Bethe lattice or Cayley tree with just four shells ($n = 0, 1, 2, 3$) is shown in Fig. IV.5.29.

An interesting feature of these structures is their 'effective dimensionality'. Indeed, a number of sites in the sphere of radius $r$ (measured in a number of generations or shells) around any site of a Bethe lattice or the central site of a large Cayley tree, calculated for $z = 3$, is given by

$$N(r) = 1 + 3(1 + 2 + 4 + \ldots 2^{r-1}) = 1 + 3(2^r - 1) = 3 \cdot 2^r - 2$$

whereas the number of sites at the periphery of that radius is $N_S(r) = 3 \cdot 2^{r-1}$. Hence, $\underbrace{N(r)/N_S(r)}_{r \to \infty} \simeq 2$, which means that

$\underbrace{N_S(r) \propto N(r)}_{r \to \infty}$. This conclusion holds, of course, for $z \geq 3$. Now, recall the scaling between the surface area and the volume: Surface area $\propto$ (Volume$^{\frac{1}{d}}$)$^{d-1}$ = Volume$^{1-\frac{1}{d}}$. Such scaling, Surface area $\propto$ Volume, can only be true when $d \to \infty$. For this reason, and only in this context, Cayley and Bethe constructions are sometimes called 'infinitely dimensional'.

Let us not bother for now with edge effects, leave Cayley trees aside, and consider percolation on a Bethe lattice, as infinite as it is. Denote by $P$ the *probability that a site is occupied and belongs to an infinite cluster*, and by $Q$ that *a branch does not lead to infinity*. Consider first the case of $z = 3$. Then $p$, our main variable in the percolation theory—the probability that the site is occupied—and $Q$ are related by the following 'recurrent' relationship:

$$Q = 1 - p + pQ^2 \qquad (\text{IV.5.134})$$

This is a quadratic equation on $Q(p)$. It has two solutions $Q = \frac{1 \pm 2\left|p - \frac{1}{2}\right|}{2p}$, in which it is easy to see that the "+" solution is unphysical. Indeed, if we choose the "−" solution we get

$$Q = \frac{1 - 2\left|p - \frac{1}{2}\right|}{2p} = \begin{cases} 1, & p < \dfrac{1}{2} \\ \dfrac{1 - p}{p}, & p > \dfrac{1}{2} \end{cases} \qquad (\text{IV.5.135})$$

where $\frac{1}{2}$ here is the percolation threshold, $p_c = 1/2$. Thus, when $p > p_c$, the probability to not lead to infinity diminishes with the increase of $p$ and becomes zero at $p = 1$, whereas below the threshold that probability is 100%. Had you chosen the "+" solution, you would have obtained this result the other way around (**Exercise.** Check this yourself), which makes no sense.

Probability that a site is occupied, but it does not lead to $\infty$, is given by

$$p - P = pQ^3 \qquad (\text{IV.5.136})$$

Road sign: don't worry, the l.h.s here is always positive as $P$ is a product of a probability of being occupied and the probability to belong to...); perhaps it would be more comforting for you to write

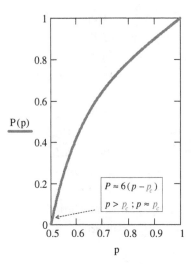

**Fig. IV.5.30** Probability that a site on a Bethe lattice, with $z = 3$, is occupied and belongs to an infinite cluster.

Eq. (IV.5.136) as $1 - P/p = Q^3$, because by definition $(P/p) < 1$. Solving this equation for $P$, and substituting Eq. (IV.5.135), we get

$$P = p(1 - Q^3) = \begin{cases} 0, (Q = 1), & p < p_c = \dfrac{1}{2} \\[2mm] p\left\{1 - \left[\dfrac{1-p}{p}\right]^3\right\}, & p > p_c = \dfrac{1}{2} \end{cases}$$

The graph of this function is shown in Fig. IV.5.30.

Thus, we have obtained the value of the percolation threshold, the full solution for $P(p)$, and of course its critical exponent, $\beta$, which describes the behaviour of $P(p)$ close to $p_c$:

$$P(p) \propto (p - p_c)^\beta; \quad \beta = 1 \tag{IV.5.137}$$

How will things change for $z > 3$?

**Example IV.5.4 (Maths Practice ♫♪).** We consider percolation on a Bethe lattice with $z = 4$. Instead of Eq. (IV.5.134) we have now

$$Q = 1 - p + pQ^3 \tag{IV.5.138}$$

Since formally one of the solutions of this equation is $Q = 1$, we can rewrite it as

$$(Q - 1)(pQ^2 + pQ - 1 + p) = 0 \qquad (\text{IV.5.139})$$

Because we are interested in the behaviour above the percolation threshold, $Q < 1$, and so we may rewrite this equation as

$$pQ^2 + pQ - 1 + p = 0 \qquad (\text{IV.5.140})$$

It is a quadratic equation for $Q$, and we can solve it, but first let us use it to get the relationship between $p$ and $Q$:

$$p = \frac{1}{1 + Q + Q^2} \qquad (\text{IV.5.141})$$

From here we immediately get the percolation threshold: we are at the threshold when $Q$ turns to 1. Substituting $Q = 1$ into Eq. (IV.5.141), we get

$$p_c = \frac{1}{3} \qquad (\text{IV.5.142})$$

Solving Eq. (IV.5.140) for $Q(p)$ we get

$$Q = \frac{1}{2}\left\{ \sqrt{1 + 4\frac{1-p}{p}} - 1 \right\} \qquad (\text{IV.5.143})$$

Taking into account that for $z = 4$,

$$P = p(1 - Q^4) \qquad (\text{IV.5.144})$$

we finally get

$$P = p\left\{ 1 - \frac{1}{16}\left[ \sqrt{1 + 4\frac{1-p}{p}} - 1 \right]^4 \right\} \qquad (\text{IV.5.145})$$

Taylor-expanding this expression near $p = p_c$, we obtain

$$P = 4(p - p_c), \quad p_c = 1/3 \qquad (\text{IV.5.146})$$

which means that Eq. (IV.5.137) again holds—only the value of $p_c$ is changed, but not the critical exponent $\beta$, just as it should be, because the latter is expected to not depend on the lattice.

Extension of this analysis for higher values of $z$ shows that there is a simple general expression

$$p_c = \frac{1}{z-1} \qquad \text{(IV.5.147)}$$

**Exercise.** Following the lines described in this example, prove it yourself. **Hint:** It will help you to again express $p(Q)$, and put there $Q = 1$, to get the value of $p_c$.

### *Scaling theory of percolation (scaling hypothesis)*

We have considered above some lattices for which exact or approximate solutions of the percolation problem exist and can describe the behaviour of percolation systems. These, and similar results utilising Monte Carlo simulations, suggest there may be some general percolation laws, depending only the dimensionality of space. Of course, such laws would hold only close to critical points—at the percolation threshold—but this is still valuable, if they are universal, i.e. applicable to any physical system where percolation-type behaviour is expected to take place. This scaling concept is defined in the following two tables. Table IV.5.3 presents the basic concept of three-exponent scaling, in which you just need to know three exponents to calculate the other three.

**Table IV.5.3**  Scaling construct near the percolation threshold.

| **Basic scaling assumption** | | |
|---|---|---|
| Number of clusters of $s$-sites per lattice site scales as $n_s = s^{-\tau} f([p - p_c]s^{\sigma})$. | | |
| Other exponents can be expressed through $\tau$ and $\sigma$, and fractal dimension $D$ | | |
| Average cluster size (i.e. number of sites in it) | $S \propto \dfrac{1}{\lvert p - p_c \rvert^{\gamma}}$ | $\gamma = \dfrac{3 - \tau}{\sigma}$ |
| Probability that a site is occupied and belongs to an infinite cluster | $P \propto (p - p_c)^{\beta}$ | $\beta = \dfrac{\tau - 2}{\sigma}$ |
| Correlation length | $\xi \propto \dfrac{1}{\lvert p - p_c \rvert^{\nu}}$ | $\nu = \dfrac{1}{\sigma D}$ |
| Cluster radius | $R_s \propto s^{1/D}$ | |

**Table IV.5.4** Critical exponents of percolation systems.

| $d$ | $\beta$ | $\gamma$ | $\nu$ | $\tau$ | $\sigma$ | $D$ |
|-----|---------|----------|-------|--------|----------|-----|
| 2 | 5/36 | 43/18 | 4/3 | 187/91 | 36/91 | 91/48 |
| 3 | 0.44 | 1.76 | 0.88 | 2.2 | 0.45 | 2.5 |

This concept has been moved even further by the so-called *hyper-scaling hypothesis* based on conjecture but tested by simulations for $d \leq 6$ (although as agreed, being practical, we are interested only in $d \leq 3$):

$$\nu d = \gamma + 2\beta \tag{IV.5.148}$$

With the equations of Table IV.5.3, this leads to

$$\frac{D}{d} = \frac{\gamma + \beta}{\gamma + 2\beta} = 1 - \frac{\beta}{\nu d} = \frac{1}{\tau - 1} \tag{IV.5.149}$$

This brings us to the idea of two-exponential scaling. The data in Table IV.5.4 all match this hypothesis.

We will now consider some qualitative ideas, which can also show-case the relationship between critical exponents, Eq. (IV.5.149), and estimate fractal dimensions.

## Percolation in finite systems

In a system of a finite size, even below the percolation threshold, but close to it, there will always be clusters large enough to extend from one boundary of the system to another. This would smear the percolation transition, as shown in Fig. IV.5.31. One can see the manifestations of such smearing in many real physical percolation structures. In addition to this fact, analysis of finite-sized systems can yield interesting results on critical exponents.

Let's examine a very simple qualitative scaling analysis. The percolation cluster is a self-similar volume fractal, if the correlation length is greater than the size of the system, $\xi \gg L$. The number of sites in such a cluster (call it 'mass', $M$) scales then as

$$M \propto L^D \tag{IV.5.150}$$

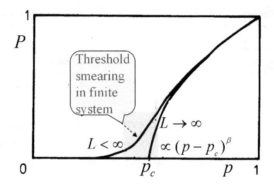

**Fig. IV.5.31**   The probability that a site is occupied and belongs to an infinite cluster. Percolation transition is no longer sharp when the system has boundaries, having characteristic finite size L. In such systems the cluster may not be infinite, but large enough to extend from one side to another across the system, thus providing an unperturbed path linking the two sides.

For $\xi \ll L$, the major part of the cluster is not self-similar, and one should expect

$$M \propto PL^d \qquad (\text{IV.5.151})$$

Let us try to blindly match the two asymptotic scaling laws. Having replaced $L$ by $\xi$ in Eqs. (IV.5.150) and (IV.5.151), when equalizing these two equations we get $\xi^D = P\xi^d$. Substituting in this equation $\xi \propto \frac{1}{|p-p_c|^\nu}$ and $P \propto (p-p_c)^\beta$, we get $\frac{1}{(p-p_c)^{\nu D}} \propto \frac{(p-p_c)^\beta}{(p-p_c)^{\nu d}}$, and hence we must have $-\nu D = \beta - \nu d$, and thus, as in Eq. (IV.5.149),

$$D = d - \beta/\nu \qquad (\text{IV.5.152})$$

If we use the values for $\beta$ and $\nu$ from Table IV.5.4, we get in the fractal dimensions of the two- and three-dimensional percolation clusters, respectively:

$$\begin{cases} 2d \;\Rightarrow\; D = 2 - \dfrac{5}{36} \Big/ \dfrac{4}{3} = 91/48 = 1.9 \\[2mm] 3d \;\Rightarrow\; D = 3 - \dfrac{0.44}{0.88} = 2.5 \end{cases} \qquad (\text{IV.5.153})$$

Monte Carlo simulations of the size of the largest cluster on a triangular lattice at $p = p_c = 1/2$ show that its logarithm versus the logarithm of size of the system is a straight line over 5 orders of magnitude interval of $L$, with a slope $D \approx 1.9$. Similar tests have been performed in three dimensions, verifying $D \approx 2.5$.

## The dimension of the shortest path

**Example IV.5.5 (Physical Application ▶▶).** Imagine a new age orienteering game, in which scouts, assisted by the navigator, have to run as fast as possible in the spiky bush, planted randomly (pink squares in Fig. IV.5.32) and close to the percolation threshold of the free sites between the bushes (white squares). The scouts' first task is to get quickly from point A to point B, unscratched. Their next task will be to get between two different points, A' to B', equally unscratched. The straight line separation $R$ between A' and B' is $\lambda$ times longer than between A and B. How much longer will the second task take them, assuming that (a) in both cases they can run with the same maximal speed and (b) their mobile phone navigator tells them which way is the shortest. With the scouts' running speed being the same for the two tasks, the time needed to perform each task will be proportional to the length of the path:

$$t \propto l_{\min} \qquad \text{(IV.5.154)}$$

Simulations tell us that

$$l_{\min}(\lambda R) = \lambda^{d_{\min}} l_{\min}(R) \qquad \text{(IV.5.155)}$$

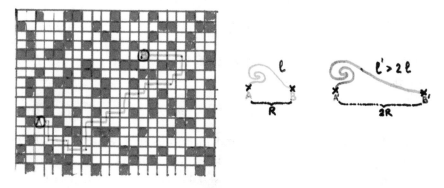

**Fig. IV.5.32** The shortest path between the two points along the percolation cluster is not proportional to the straight-line distance between the two points.

and that $d_{\min} > 1$. This new exponent is sometimes called the chemical dimension, because similar problems arise in propagation of chemical fronts through porous media. Let us try to estimate it.

First, Eq. (IV.5.155) means that $l_{\min} \propto R^{d_{\min}}$, and thus

$$R \propto l_{\min}^{1/d_{\min}} \qquad \text{(IV.5.156)}$$

Next, let us introduce propagation velocity

$$v = \frac{dR}{dt} \qquad \text{(IV.5.157)}$$

which is not the velocity of running along the 'minimal' path, but just the velocity of travelling away from the initial point on the map. Then combining Eqs. (IV.5.157) with Eq. (IV.5.154) and Eq. (IV.5.156), we get

$$v = \frac{dR}{dt} \propto \frac{dR}{dl_{\min}} = \left[\frac{dl_{\min}}{dR}\right]^{-1} \propto R^{1-d_{\min}} \qquad \text{(IV.5.158)}$$

If you want to deviate over the distance of the order of correlation length $R \sim \xi \propto (p - p_c)^{-\nu}$, then

$$v \propto (p - p_c)^{-\nu(1-d_{\min})} \qquad \text{(IV.5.159)}$$

Finally, if we assume that the velocity should be somehow proportional the probability of the site being occupied and belong to infinite cluster (otherwise, how it could reach the distance of the order of correlation length?), then it should also be that

$$v \propto P \propto (p - p_c)^{\beta} \qquad \text{(IV.5.160)}$$

Equalizing, Eqs. (IV.5.159) and (IV.5.160) we have that $v \propto P \propto (p - p_c)^{\beta}$ i.e.

$$d_{\min} = 1 + \beta/\nu \qquad \text{(IV.5.161)}$$

Is this true? In 2D, $\nu = 4/3$ and $d_{\min} = 1 + 5/48 \approx 1.1$, which is exactly what was obtained via simulations. In 3D, however, $\beta = -0.44$, $\nu = 0.88 \Rightarrow d_{\min} = 1 + 0.44/0.88 = 1.5$ which is larger than the 1.35 that has been reported from simulations. So we should be careful with bold scaling statements, and this is one of the reasons why people check their results by simulations, including those performed in fictitious higher dimensions, $d > 3$.

**Example IV.5.6 (Physical Application ▶▶).** Consider a dual metal insulator composite with identical size of metallic and dielectric grains, densely packed to form an FCC-lattice. You may wish to know what the conductivity is, close to the percolation threshold from electronic transport along the metallic pathways. It appears that generally the conductivity $\Sigma$ along a percolation cluster, and the diffusion coefficient $\mathcal{D}$ for the transport along the sites belonging to such clusters, scale as

$$\Sigma \propto \mathcal{D} \propto (p - p_c)^{\mu} \qquad \text{(IV.5.162)}$$

Different approximations were proposed for the exponent $\mu$. Most popular was that of Alexander and Orbach:

$$\mu = [(3d - 4)\nu - \beta]/2 \qquad \text{(IV.5.163)}$$

In 3D this gives $\mu = [5 \cdot 0.88 - 0.44]/2 = 1.98$.

Of course, the scaling law of Eq. (IV.5.162) should, strictly speaking, work only for the system close to the percolation threshold. But for some experimentally studied, specially designed dual composite samples, it was found that it may work far above the threshold, and thus dependence on the value of $p$, here standing for the volume portion of the metallic component, can be well approximated by this law in the whole interval of $p_c < p < 1$. Notably, $\mu$ is substantially larger than $\beta$, i.e. the conductivity with increasing $p$ increases slower than $P$. This is obviously because not all sites in the percolation cluster contribute equally to the passage of electrical current.

Note that a walk on sites of percolation clusters proceeds more slowly than a random walk on a regular lattice, as there are so many lacunas and loopholes that need to be bypassed. Gefen, Aharony and Alexander have shown that for percolation clusters, at $p = p_c$ deviation from origin of the path evolves with time (number of steps) as $R \propto t^k$, $p = p_c$, with $k = 1/3$ in 2D, and $k = 0.2$ in 3D (for more details see modern books on random processes and diffusion).

## 5.4.  Growth of Complex Objects and Formation of Complex Patterns

This subject has been studied in many papers and well described in a number of books. Going into that territory would take us too far, as the latter is first of all about computer simulations of models of growth (majorly on lattices) and the characterization of the geometry of the resulting structures, although quite sophisticated theories of these processes and the statistical geometry of resulting structures have also been developed. We therefore will just give a brief overview of such models, dwelling on a few of the most popular, and then speak about the phenomenological kinetics of cluster growth, which certainly belongs to this chapter.

There are patterns of growth that lead to non-fractal structures.

There is, for instance, the *Eden model* of non-cancer tumour growth. In this model, probability of site occupation is proportional to the number of occupied nearest neighbours, the process starting from a single seed site.

The *William-Bjerkens model* of skin cancer is more involved. Here, occupied points of the lattice represent cancerous cells, and free sites are healthy ones. When a cell divides, one of its neighbours is replaced by its daughter cell. If the dividing cell is cancerous and the neighbour is normal, the cancer grows; if the dividing cell is normal and the neighbour is cancerous, the cancer retreats. The process starts with one cancerous cell, and if cancerous cells divide with even a slightly higher probability than a healthy cell, cancer develops.

But many patterns of growth lead to fractal structures.

One such pattern is the *Sawada model*—essentially an Eden model but with preferential sticking to "tip-sites", the ones that have only one nearest neighbour occupied.

Another one is *ballistic aggregation*, in which particles one after another fly at random trajectories. The first particle that hits the initial immobile particle stops and forms a cluster with it. The next particle stops if it hits the cluster, etc. Volume fractal dimension of the resulting structures in in 2D is only slightly lower than the topological dimension; $D = 1.93$.

In *ballistic deposition* onto a surface (*Voss model*) particles are raining down from random starting positions and stop when they reach the surface or become neighbours of another stopped particle. The higher the sticking probability, the more 'fractal' the resulting structures are—they tend to form ramified dendrites. But with low sticking probability the particles tend to fill the volume near the surface more homogeneously.

In *cluster-cluster aggregation* (*Meakin and Kolb model*), particles are moving simultaneously on a grid, and any two of them may stick together if they find themselves on neighbouring sites. Further particles may stick to the clusters, and clusters may diffuse as a whole, and stick together forming larger clusters. Whether their diffusion coefficients depend on their size or not, the results appear to be the same! Fractal dimensions of these structures are slightly higher than for self-avoiding random polymer chains (see Eq. (IV.5.101)); in 2D it is $D = 1.4$ and in 3D it is $D = 1.8$.

The most famous model of fractal growth is *Witten and Sander diffusion limited aggregation* (DLA). Its algorithm is as follows. A seed particle is placed at the centre of the grid. The next one is introduced into the grid and allowed to random walk on the grid. It stops when it visits the neighbouring site of the seed. The next particle stops when it visits the neighbouring site of the cluster, etc. In 2D the resulting structures look like the one shown in Fig. IV.5.33. There has been a lot of work carried out on the theory of DLA; we will just mention here the basic idea proposed by Witten and Sander, based on the so-called *Laplace instability*. Consider its implementation in 2D. Let us call the field of concentration $c(\mathbf{r}, t)$ of particles at any point $\mathbf{r}$ on the plane at time $t$. The diffusion equation on this quantity reads $\frac{\partial c(\mathbf{r},t)}{\partial t} = \mathcal{D}\Delta c(\mathbf{r}, t)$, with the Laplacian in 2D being $\Delta = \frac{\partial^2}{\partial r^2} + \frac{1}{r}\frac{\partial}{\partial r} + \frac{1}{r^2}\frac{\partial^2}{\partial \theta^2}$. Let us take the absorbing boundary condition $c(\mathbf{r}|_s, t) = 0$ in the sense that a particle is out of the game when it reaches the surface, because it sticks to it and does not diffuse any more. The velocity of the particle before it gets absorbed is given by $-\mathcal{D}\hat{\mathbf{n}}\vec{\nabla}c(\mathbf{r}, t)|_s$, where $\hat{\mathbf{n}}$ is unit normal vector to the surface. Let us assume that the absorbing surface is a circle of radius $R$. Then the

**Fig. IV.5.33** The iconic two-dimensional Witten and Sander DLA pattern on a two dimensional square grid, with 100% sticking probability, returning the fractal dimension $D = 1.66$ (similar simulations based on off-lattice Brownian diffusion give slightly larger value of $D = 1.71$). Colours correspond to the time of arrival. With the first DLA pattern reported in T. A. Witten. Jr. & L. M. Sander, *Phys. Rev. Lett.* 47, 1400 (1981), this coloured version of the figure was generated for a larger number of particles (50,000). It highlights the 'hotter' regions of growth, showing that the 'interior' of the cluster is screened from growth even though the open valleys are very large. Reducing sticking probability will make the result of aggregation more homogeneous, 'valleys' more fully filled, thus bringing the fractal dimension of the pattern closer to 2.

*Source*: Courtesy of Prof. Leonard Sander.

speed with which this circle will grow with adsorption of particles on it will be equal to $V_s = \mathcal{D}\hat{n}\vec{\nabla}c(\mathbf{r}, t)|_s$. If the diffusion is slow, the Laplace equation will simply become $\Delta c(\mathbf{r}, t) = 0$. Its solution with the absorbing boundary condition on the surface of the circle $R$ is $c(\mathbf{r}, t) = A\{\ln r - \ln R\}$. From here one finds that the velocity of the circle growth, $V_s \equiv \frac{dR}{dt} = \frac{A\mathcal{D}}{R}$, which means that a smaller radius (smaller $R$) will grow faster.

Equally, one may show that the perturbation of the front, as shown in Fig. IV.5.34, will grow faster than the front itself. For cosinusoidal undulation

**Fig. IV.5.34** Perturbation of the front can be, for a test described by the cosinusoidal undulations $r = R + \delta \cos m\theta$.

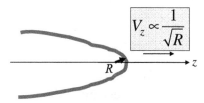

**Fig. IV.5.35** Langer and Mueller-Krumbhaar theory of propagation of a parabolic front: it propagates with the speed that is inversely proportional to the square root of the apex radius R of the front. Such fronts are usually unstable to development of fingers with smaller and smaller apex radii.

$$c(\mathbf{r}, t) = A \left\{ \ln r - \ln R - \delta R^{m-1} \frac{\cos m\theta}{r^m} \right\}$$

and $\frac{\partial \delta}{\partial t}\big|_{r=R} = (m-1)DA\frac{\delta}{R^2}$, which means that $\frac{\dot{\delta}/\delta}{\dot{R}/R} = (m-1)$, and for any $m > 1$, perturbation develops faster than the circular front, and the larger $m$, i.e. the shorter wave-length are the front undulations, the faster they develop.

Similarly, for a parabolic front moving along the symmetry axis (see Fig. IV.5.35), theory tells that the sharper the parabola, the faster it propagates. This is Laplace instability, often a source of ramified, fractal structures, which are observed in such phenomena as viscous fingering (penetration of one immiscible liquid into another under pressure in between flat-parallel plates separated by a small gap forming quasi-2D patterns), dielectric breakdown and electrical discharge, cracks, etc.

## 5.5. Kinetics of Cluster Growth

The dynamics of aggregation can be described by starting with discrete variables and ending up with a continuous picture. We demonstrate this, considering irreversible aggregation, based on Smoluchowski's system of nonlinear kinetic equations.

Marian von
Smoluchowski
(1972–1917)

The system operates with time-dependent variables, $c_k(t)$, which are the average concentrations of clusters of 'rank $k$' (those that contain $k$ individual particles) with an initial condition

$$c_k(0) = c\delta_{k1} \qquad (\text{IV.5.164})$$

where $\delta_{k1}$ is Kronecker-delta, which means that at the beginning we did not have any clusters but just individual particles with concentration $c$. Equations themselves have the simplest possible form:

$$\frac{dc_k}{dt} = \frac{1}{2} \sum_{i+j=k} K_{ij} c_i c_j - c_k \sum_j K_{kj} c_j \qquad (\text{IV.5.165})$$

Here, $K_{ij}$ are the *kinetic kernels* or *rate constants*. The first term in the r.h.s stands for aggregation of clusters of $i$ and $j$ into clusters of $k$ rank (factor $1/2$ is here to ensure we don't count things twice), i.e. producing $k$-large clusters. The second term describes the disappearance of $k$-clusters aggregating with any $j$-clusters to form clusters of larger size. The way that the kernels depend on their subscripts determines, as we will see, the dynamics of cluster growth. There are many models of such dependence studied in the literature, but we will consider only three classical ones (as an exercise we will suggest you study their combinations). Here they are with the names attributed to them:

$$K_{ij} = 2\alpha \ [\text{Smoluchowski kernel}] \qquad (\text{IV.5.166})$$

$$K_{ij} = 2\alpha(i + j) \ [\text{Branched polycondensation kernel}] \qquad (\text{IV.5.167})$$

$$K_{ij} = 2\alpha(i \cdot j) \ [\text{Flory kernel}] \qquad (\text{IV.5.168})$$

Here,

$$\alpha = \frac{1}{c\tau} \qquad (IV.5.169)$$

where $\tau$ is characteristic time. Equation (IV.5.166) assumes that the rate of aggregation does not depend on the size of the clusters; this assumption is easy to understand but not easy to justify. Equation (IV.5.168) assumes that the rate increases proportionally to the size of each aggregating cluster, which makes sense if the clusters are driven together by forces that depend on their masses, as for example in the case of Van der Waals forces. In Eq. (IV.5.167), the rate increases proportionally to the size of the cluster formed; this can be rationalized when aggregation requires just one element in the cluster to attach anywhere to another cluster.

### The method of moments

The way to quickly get useful information from Eq. (IV.5.165) is by the calculation of *moments*, defined as

$$M_n(t) = \sum_k k^n c_k(t) \qquad (IV.5.170)$$

What is their meaning, and what is the use of them? Zero moment ($n = 0$) tells you the concentration of all clusters that you have at a time moment $t$, or if the volume is fixed, how many clusters you have altogether:

$$M_0 = \sum_k c_k(t) \qquad (IV.5.171)$$

The first moment,

$$M_1 = \sum_k k c_k(t) \qquad (IV.5.172)$$

tells us about the total 'mass' of the clusters. We will see that this quantity is conserved, i.e. it does not depend on time. Knowing the values of these moments, we can calculate the average size of the cluster,

$$\langle S(t) \rangle = \frac{M_1}{M_0} = \frac{\sum\limits_{k} k c_k(t)}{\sum\limits_{k} c_k(t)} \tag{IV.5.173}$$

which is the main 'coarse-grain' characteristic of cluster growth.

Let us now 'convert' Eq. (IV.5.165) into an equation for moments so that we have

$$\sum_{k=1}^{\infty} k^n \left[ \frac{dc_k}{dt} = \frac{1}{2} \sum_{i+j=k} K_{ij} c_i c_j - c_k \sum_{j=1}^{\infty} K_{kj} c_j \right]$$

$$\Rightarrow \frac{dM_n}{dt} = \frac{1}{2} \left\{ \sum_{k=1}^{\infty} k^n \left[ \sum_{i+j=k} K_{ij} c_i c_j \right] - 2 \sum_{k=1}^{\infty} k^n c_k \sum_{j=1}^{\infty} K_{kj} c_j \right\}$$

$$= \frac{1}{2} \left\{ \sum_{i,j} (i+j)^n K_{ij} c_i c_j - \sum_{i}^{\infty} \sum_{j}^{\infty} i^n c_i K_{ij} c_j \right.$$

$$\left. - \sum_{i}^{\infty} \sum_{j}^{\infty} j^n c_i K_{ij} c_j \right\} \Rightarrow$$

so that

$$\frac{dM_n}{dt} = \frac{1}{2} \sum_{ij} K_{ij} c_i c_j \left[ (i+j)^n - i^n - j^n \right] \tag{IV.5.174}$$

From here we immediately see that, independently of the kernel, as expected

$$\frac{dM_1}{dt} = 0 \tag{IV.5.175}$$

and, as mass is conserved, that

$$M_1 = c \tag{IV.5.176}$$

We will now find the solutions for these two moments and these characteristics, solving Eq. (IV.5.174) for each of the three kernels of Eqs. (IV.5.166)–(IV.5.168), with the initial condition shown in Eq. (IV.5.164).

### Three modes of growth

(a) Smoluchowski's case.

From Eq. (IV.5.166), we get $\frac{dM_n}{dt} = \alpha \Sigma_{ij} c_i c_j [(i+j)^n - i^n - j^n]$. Hence,

$$\frac{dM_0}{dt} = -\alpha \sum_{ij} c_i c_j = -\alpha M_0^2 \qquad \text{(IV.5.177)}$$

The solution of this equation is obtained by separation of variables and with the given initial condition it reads

$$M_0 = \frac{c}{1 + t/\tau} \qquad \text{(IV.5.178)}$$

so that

$$\langle S \rangle = 1 + t/\tau \qquad \text{(IV.5.179)}$$

Thus, the clusters for this kernel grow linearly. As typically the clusters are fractals, we have that $\langle S \rangle \propto R^D$, where $R$ is the radius of the cluster, and therefore,

$$R \propto \langle S \rangle^{1/D} \qquad \text{(IV.5.180)}$$

Using Eq. (IV.5.179), at long times, $t \gg \tau$, we get

$$R \propto t^{1/D} \qquad \text{(IV.5.181)}$$

To check if this mechanism prevails, you may try to treat your data in $\log R \leftrightarrow \log t$ as a plot. If indeed $\ln R = \text{const} + \frac{1}{D} \ln t$, from the slope you could find the fractal dimension $D$.

(b) Branched polycondensation case.

Here, $\frac{dM_n}{dt} = \alpha \Sigma_{ij} c_i c_j (i+j)[(i+j)^n - i^n - j^n]$. Hence,

$$\frac{dM_0}{dt} = -\alpha \sum_{ij} c_i (i+j) c_j = -2\alpha M_0 M_1 = -2\alpha M_0 c \qquad \text{(IV.5.182)}$$

Solving Eq. (IV.5.182) for $M_0$ we get

$$M_0 = c e^{-2t/\tau} \qquad \text{(IV.5.183)}$$

and therefore

$$\langle S \rangle = e^{2t/\tau} \qquad \text{(IV.5.184)}$$

Correspondingly,

$$R \propto \exp\left\{\frac{2}{D}\frac{t}{\tau}\right\}$$

and the clusters grow exponentially. The data may be treated in $\log R \leftrightarrow t$ coordinates. If you get a straight line, you may hope to estimate the slope as $(2/D\tau)$.

(c) Flory case.

Can the growth be more dramatic than exponential? Actually, yes, as demonstrated by this case. Here, $\frac{dM_n}{dt} = \alpha\Sigma_{ij}c_ic_jij[(i+j)^n - i^n - j^n]$, and

$$\frac{dM_0}{dt} = -\alpha\sum_{ij}ic_ijc_j = -\alpha M_1^2 = -\alpha c^2 = -\frac{c}{\tau} \qquad (\text{IV.5.185})$$

Integrating this equation we get,

$$M_0 = c(1 - t/\tau) \qquad (\text{IV.5.186})$$

and thus,

$$\langle S \rangle = \frac{1}{1 - \frac{t}{\tau}} \qquad (\text{IV.5.187})$$

that is, cluster size diverges at $t = \tau$. This effect is called *gelation transition*; the Flory model predicts a critical phenomenon—a formation of infinite ("gel") clusters, whereas the finite size ("sol") clusters gradually disappear. The best way to treat your experimental data then would be in coordinates $\frac{1}{\langle S \rangle}$ vs $t$. The negative slope will give you $1/\tau$.

   If these elementary aggregation acts were to be diffusion controlled, then $\tau$ in the Smoluchowski case in 3D can be roughly estimated as $\frac{1}{8\pi\langle D \rangle\langle R \rangle c}$ where $\langle D \rangle$ is the average cluster diffusion coefficient and $\langle R \rangle$ is an average cluster radius during growth (we will not derive it here). Using Einstein–Stokes' formula, $\langle D \rangle \approx \frac{k_B T}{6\pi\eta\langle R \rangle}$ (see Example I.10.7, of Chapter I, Vol. 1), where $\eta$ stands for viscosity, we get $\tau \approx \frac{3}{4}\frac{\eta}{k_B Tc}$. For other kernels, even if the processes are diffusion limited, these estimates are more complicated.

The analysis of the 'fate' of clusters of specific rank, i.e. the calculation of $c_k(t)$, is a more complicated story, and we will refer the reader to more specialized literature. But the analysis of $\langle S(t) \rangle$ is very useful, as this quantity is typically measurable, and you may try to use the corresponding coordinates to treat the experimental data.

### Mixed modes of growth

We end this section with two exercises.

**Exercise.** Using the same technique, solve the equation of moments for a mixed kernel, $K_{ij} = 2\alpha + 2\beta(i \cdot j)$, and show that the average size of the cluster is given by the elegant, but at a first glance not obvious, formula

$$\langle S(t) \rangle = \frac{1 + \sqrt{\frac{\alpha}{\beta}} \tan\left(c\sqrt{\alpha\beta}t\right)}{1 - \sqrt{\frac{\beta}{\alpha}} \tan\left(c\sqrt{\alpha\beta}t\right)}.$$ Make sense of this formula by considering limiting cases of (a) $\frac{\beta}{\alpha} \to 0$, where you should recover the Smoluchowski law of growth, and (b) $\frac{\alpha}{\beta} \to 0$, when you should recover the Flory case. In the general case, show that the gelation point is given by the formula $t = t_g \equiv \tau\sqrt{\frac{\beta}{\alpha}} \arctan\left(\sqrt{\frac{\alpha}{\beta}}\right)$. Show that $t_g < \tau$ (the gelation is speeded up by the Smoluchowski contribution to the kernel).

**Exercise.** Try any other combinations of contributions to the kernel, from Eqs. (IV.5.166)–(IV.5.168) and calculate $\langle S(t) \rangle$ in a similar manner. Here, we do not give you the answers! If you correctly get the result of the previous exercise, you will cope with this one. Similarly to the previous exercise, you could check your results, analyzing them by switching off, one by one, the corresponding contributions to the kernels, and recovering the cases studied above. Think about the physical meaning of your results.

## 5.6.   Fractional Calculus

This discipline of mathematics considers derivatives and operators of *non-integer order*. Its history goes back to the work of Leibniz at the end of the seventeenth century. It was continued in the nineteenth century in the works of Abel, Liouville, Riemann, Heaviside (who had already found application of this formalism in the transmission of electrical signals), and others, and developed further by a number of outstanding mathematicians of the twentieth century in the context of functional analysis. As we have already stressed in this book, no matter how abstract the findings of great mathematical minds look, sooner or later they find—sometimes unexpected—practical applications. Today, elements of fractional calculus appear in various fields: treatment of electrochemical impedance and generally electrochemical kinetics, theory of anomalous diffusion, propagation of acoustic waves in complex media, multiple scattering of light in hot vapours, and even quantum mechanics based on fractional generalization of the Schrödinger equation.

Before introducing the basic ideas of fractional calculus, let us afford ourselves some 'philosophical' comments about contemporary motivation to use it. Already the ordinary Schrödinger equation has been 'engineered' to explain otherwise inexplicable experimental data. Later Feynman kind of derived it from his path integral formalism. In a similar way, people use fractional differential equations to describe the observed phenomena that display fractional exponents. Indeed, Brownian scaling $\sqrt{< (\Delta x)^2 >} \propto t^{1/2}$ is rarely observed; in most of the systems the exponent is not $1/2$. Since the Browning scaling comes out from the diffusion equation, it was natural to try to alter...the equation! If Einstein in his general relativity theory has derived his general equations through the concept of curved space, nowadays the disciples of the fractional calculus do the opposite: they change the equations (to contain fractional operators) to describe the processes in complex media. There is no independent, first-principles justification of such 'reverse engineering', so it is tested by comparison of the results of such theories with experiments. A pragmatist would say—"who cares, if it works?", but we leave this discussion aside and consider few elementary aspects of this 'mad' calculus.

Everybody knows that

$$\frac{d^n}{dx^n}x^m = \frac{m!}{(m-n)!}x^{m-n} \qquad \text{(IV.5.188)}$$

It may be less familiar to you that with $x > 0$,

$$\frac{d^n}{dx^n}x^\mu = \frac{\Gamma(\mu+1)}{\Gamma(\mu-n+1)}x^{\mu-n} \qquad \text{(IV.5.189)}$$

where $\Gamma(z) = \int_0^\infty dt\, e^{-t}t^{z-1}$ is the Euler Gamma function we discussed in Chapter 7 of Part II (Vol. 1), a spectacular feature of which is the recurrent relation $\Gamma(z+1) = z\Gamma(z)$, so that $\Gamma(n) = (n-1)!$ In the spirit of Eq. (IV.5.189), one may then suggest introducing a *fractional derivative* of the order $\alpha$ as,

$$\hat{D}_x^\alpha \equiv \frac{d^\alpha}{dx^\alpha} \qquad \text{(IV.5.190)}$$

such that

$$\hat{D}_x^\alpha x^\mu = \frac{\Gamma(\mu+1)}{\Gamma(\mu-\alpha+1)}x^{\mu-\alpha}, \qquad \mu > 0,\ x > 0 \qquad \text{(IV.5.191)}$$

and then develop the subsequent algebra of such differential operators, based on this principle. For instance, for a general function $f(x)$, and $0 < \alpha < 1$, a complete fractional derivative reads

$$\begin{aligned}
{}_a\hat{D}_x^\alpha f(x) &= \frac{1}{\Gamma(1-\alpha)}\frac{d}{dx}\int_a^x ds\frac{f(s)}{(x-s)^\alpha}, \qquad x > a \\[2mm]
{}_x\hat{D}_b^\alpha f(x) &= \frac{1}{\Gamma(1-\alpha)}\frac{d}{dx}\int_x^b ds\frac{f(s)}{(x-s)^\alpha}, \qquad x < b
\end{aligned} \qquad \text{(IV.5.192)}$$

with a simpler form for $x > 0$:

$${}_0\hat{D}_x^\alpha f(x) \equiv \hat{D}_x^\alpha f(x) = \frac{1}{\Gamma(1-\alpha)}\frac{d}{dx}\int_0^x ds\frac{f(s)}{(x-s)^\alpha}$$

Neither Eq. (IV.5.191), nor (IV.5.192) apply when the argument of the Gamma function appears negative, as the latter is not defined in that domain. To avoid this problem, a trick is used: you perform first all integer derivatives and then do the remaining fractional one:

$$\hat{D}_x^{7/3}f(x) = \hat{D}_x^{1/3}\hat{D}_x^2 f(x) = \hat{D}_x^{1/3}\left(\tfrac{d^2}{dx^2}f(x)\right).$$

Another philosophical diversion. Do not mix up what is proposed above with the idea of building a new algebra based on the postulate that "$2 + 2 = 5$", parodied by George Orwell in his 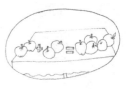 famous anti-utopia, dark satirical novel, *Nineteen Eighty-Four*: "In the end, the Party would announce that two and two made five, and you would have to believe it". Orwell was scoffing at and paraphrasing the anti-intellectualistic Nazi slogan "If the Führer wants it, two and two makes five!". Categorically, no! Equations (IV.5.191) and (IV.5.192) define a new operation, which does not dismiss the established result, Eq. (IV.5.189), which reproduces the known result when $\alpha = n$.

By analogy with the formula obtained via multiple integration by parts,

$$\int_a^x dy_1 \int_a^{y_1} dy_2 \ldots \int_a^{y_{n-1}} dy_n f(y) = \frac{1}{(n-1)!} \int_a^x dy(x-y)^{n-1} f(y)$$

$$(IV.5.193)$$

the Riemann–Liouville fractional integral, which is an operator inverse to fractional derivative that is determined by

$$_a\hat{D}_x^{-\alpha} f(x) = \frac{1}{\Gamma(\alpha)} \int_a^x dy(x-y)^{\alpha-1} f(y), \quad x > a$$

$$_x\hat{D}_b^{-\alpha} f(x) = \frac{1}{\Gamma(\alpha)} \int_x^b dy(y-x)^{\alpha-1} f(y), \quad x < b$$

$$(IV.5.194)$$

There is also a set of other definitions of fractional integrals that we will not dwell upon here.

Obvious relationships between fractional integrals and fractional derivatives are

$$_a\hat{D}_x^{\alpha} f(x) = \frac{d^n}{dx^n} {_a\hat{D}_x^{-(n-\alpha)}} f(x), \quad x > a$$

$$_x\hat{D}_b^{\alpha} f(x) = \frac{d^n}{dx^n} {_x\hat{D}_b^{-(n-\alpha)}} f(x), \quad x < b$$

$$(IV.5.195)$$

**Table IV.5.5** 'Zero-point' semi-integrals and semi-derivatives of the functions of positive argument.

| $f(x)$ | $_0\hat{D}_x^{1/2} f(x) = \dfrac{d^{1/2}}{dx^{1/2}}$ | $_0\hat{D}_x^{-1/2} f(x) = \dfrac{d^{-1/2}}{dx^{-1/2}}$ |
|---|---|---|
| Constant, $C$ | $\dfrac{C}{\sqrt{\pi x}}$ | $2C\sqrt{\dfrac{x}{\pi}}$ |
| $x$ | $2\sqrt{\dfrac{x}{\pi}}$ | $\dfrac{4}{3\sqrt{\pi}} x^{3/2}$ |
| $x^{1/2}$ | $\dfrac{\sqrt{\pi}}{2}$ | $\dfrac{\sqrt{\pi}}{2} x$ |
| $x^{-1/2}$ | $0$ | $\sqrt{\pi}$ |
| $x^\nu, \nu \geq -1/2$ | $\dfrac{\Gamma(\nu+1)}{\Gamma\left(\nu+\frac{1}{2}\right)} x^{\nu-\frac{1}{2}}$ | $\dfrac{\Gamma(\nu+1)}{\Gamma\left(\nu+\frac{3}{2}\right)} x^{\nu+\frac{1}{2}}$ |
| $e^x$ | $\dfrac{1}{\sqrt{\pi x}} + e^x \cdot \mathrm{erf}\left(\sqrt{x}\right)$ | $e^x \cdot \mathrm{erf}\left(\sqrt{x}\right)$ |
| $\ln x$ | $\dfrac{\ln(4x)}{\sqrt{\pi x}}$ | $2\sqrt{\pi}\dfrac{\ln(4x)-2}{\sqrt{x}}$ |

but perhaps the most important one is the rule for the Fourier transform operation, $\hat{F}$:

$$\hat{F}\{\hat{D}_t^\alpha f(t)\} = (i\omega)^\alpha \hat{F} f(t) \qquad \text{(IV.5.196)}$$

This relation was one of the inspirations for using fractional calculus in descriptions of fractional noises and impedance.

To show you how unusual the results of fractional integration and differentiation can be, as compared to those of ordinary 'integer' calculus, we present in Table IV.5.5, without any practical purpose (just for fun), the results for semi-integrals and semi-derivatives for several functions.

**Exercise.** Show that the fourth and fifth rows of this table are particular cases of the sixth row.

**Exercise.** You may wish to check, for any row of this table, that the rules $\dfrac{d^{1/2}}{dx^{1/2}} \dfrac{d^{1/2}}{dx^{1/2}} f(x) = \dfrac{d}{dx} f(x)$ and $\dfrac{d^{-1/2}}{dx^{-1/2}} \dfrac{d^{-1/2}}{dx^{-1/2}} f(x)) = \int dx f(x)$ are both satisfied.

As we have mentioned, there are many examples of applications of fractional calculus through designing fractional equations. The anomalous diffusion with the deviation from the origin $\mathbf{R}$ scales as a function of time as $\langle \mathbf{R}(t)^2 \rangle \propto t^\alpha$, with $\alpha \begin{cases} < 1 & \text{sub-diffusion} \\ > 1 & \text{super-diffusion} \end{cases}$. For ordinary diffusion the probability of diffusing to the point $\mathbf{R}$ is described by a standard diffusion equation, Eq. (IV.5.55), which corresponds to $\alpha = 1$. In order to describe anomalous diffusion and get the fractional scaling, it has been suggested that the equation in 3D should be modified to read

$$\frac{\partial^\alpha P(\mathbf{R}, t)}{\partial t^\alpha} = \mathcal{D}\Delta P(\mathbf{R}, t) \qquad (\text{IV}.5.197)$$

where $\Delta$ stands for the Laplace operator. Calculation shows that this does lead to the desired scaling law.

But this is relatively straightforward. You may have noticed that the second derivative (Laplace operator) was not 'fractionalized'—it was not needed to get the desired answer. This may look like pure engineering, but in fact can be justified in the random walk with fractional power law for distribution of hopping waiting times. But can you go further on this track? Brave hearts attack the r.h.s, modifying it in different ways. For instance, for diffusion and migration in an external field $\mathbf{f}$ they would modify the r.h.s of the classical Fokker–Planck equation, $\frac{\partial P(\mathbf{R}, t)}{\partial t} = \vec{\nabla}\{-\mu\mathbf{f} \cdot P(\mathbf{R}, t) + D\vec{\nabla}P(\mathbf{R}, t)\}$ introducing there fractional derivatives. Solving, for example, such engineered equations for the so-called Ornstein–Uhlenbeck process of diffusion in one-dimensional harmonic potential, with $f \propto x$, they could obtain the law of fractional diffusion

$$\langle x(t) \rangle = x(0)E_\alpha\left(-(t/\tau)^\alpha\right) \text{ with } E_\alpha\left(-(t/\tau)^\alpha\right) \approx \frac{e^{-(t/\tau)^\alpha}}{\Gamma(1 + \alpha)}$$

which shows a slow return from the initial position $x(0)$ to the bottom of the well.

We think it is enough to pinpoint for you some directions in this shaky area. Instead of engineering equations, it would be more convincing to build detailed, microscopic kinetic theories of this kind of processes, and you can find a number of attempts of this kind in the literature. Of course, they will be model-dependent; if you want to generalize the results you may have to go back to equation-engineering, but at least with some positive experience. Note, though, that building microscopic theories, whenever you deal with complicated distributions of 'options', you *will* enter the fractional wonderland. This is why fractionality emerges not only in physics, chemistry, biology, electrical and mechanical engineering, but also in economics, finance and social science. It was the genius of great mathematicians, starting with Leibniz, two centuries ago, to fearlessly pave the way into this territory.

## 5.7.  Key Points

- Complex objects, self-similar over a broad range of length scales, are called fractals, and characterized by fractal dimensions.
- Ramified fractal lines and rugged fractal surfaces have the fractal dimensions which are larger than corresponding topological dimensions. Such geometric structures are generally called 'surface fractals'.
- There are fractal lines, areas, and volumes, based on something being taken out of them (like Cantor Bars, Sierpiński gaskets, or porous volumes); their fractal dimension is smaller than the corresponding topological dimension. These are generally called 'volume fractals'. Their geometrical properties are characterized by the so-called co-dimension law.
- Whereas pedagogically it is good to introduce those concepts using deterministic fractal constructions (those built based on specific algorithms), in reality things are much more random, and it is important also to consider fractal structures formed by random processes.
- As such, we considered structures formed by different kind of noises, and related their spectral characteristics with their fractal dimensions.
- All those elementary concepts help us to set the language for the description and characterization of random structures.

- As the next step we considered various kinds of random work, and mapping on that mechanism the problems of long random chains (polymers) and the process of diffusion. We have learned that
  - Even Gaussian chains with no interactions between their segments exhibit elastic properties of an entropic nature.
  - Involving interactions changes the scaling law of gyration radius dependence on the number of segments in the chain and affects the fractal dimensions that these structures acquire.
  - Competition between the interaction of chain segments with the environment and with each other may lead to its abrupt transition, the collapse of the chain, triggered by a reduction of the temperature.
- We then presented a brief summary of percolation theory and described the properties of the percolation structures close to the percolation threshold.
  - We examined critical exponents characterizing different properties of these structures and analyzed the scaling theory of percolation.
  - We considered two exactly solvable models—1D model and quasi-infinitely dimensional model (Bethe lattice) and derived some critical exponents for them. We also presented a real space renormalization group approach to the calculation of percolation thresholds and critical exponents using the example of a 2D triangular lattice.
  - Consideration of the finite size of the percolation system also helps to calculate critical exponents and fractal dimensions of the percolation structures.
- Different mechanisms of aggregation can form non-fractal and fractal structures with different fractal dimensions. We have discussed some of the most 'popular'.
- Kinetics of cluster growth depends on how the aggregation rate depends on the size of the coagulating clusters. We have considered several laws of growth subject to examples of such dependence, including linear growth, exponential growth, the one resulting in gelation transition, and discussed how the combination of these mechanisms will affect the results.

- To describe the scaling laws with fractional exponents and their manifestation in real physical processes, scientists recalled the principles of *fractional calculus*—a mathematical discipline conceived towards the end of the seventeenth century and which has been developed ever since. It was proposed to replace the established equations of physical and chemical kinetics with their modifications that involve *fractional derivatives* and *fractional integrals*, hoping to delineate 'unconventional' laws describing processes in complex systems, which emerge in many areas of science and engineering.

- Attempts to directly derive such equations for description of physical processes, which would support such approaches at least in the case studies, are rare. So for most of the cases, the fractional calculus approach for description of such processes amounts to not much more than 'reverse engineering'. But it is interesting and sometimes bring results that accord with experimental observations.

- Fractional calculus concludes this book as an example how an 'odd' branch of mathematics, which evolved over more than three centuries, is being put into practise now in classical and even quantum mechanics. But, it is also something you may wish to dig into and use, when needed.

# Concluding Remarks

So, we have completed the last part of our planned journey.

We started with a detailed look at *inhomogeneous partial differential equations*, first in systems without boundaries and then accounting for boundary conditions. Having introduced methods for finding exact solutions to such equations, we then considered elements of *perturbation theory*.

Then we examined some simple *integral equations* and thought about ways of solving them.

Next, and much more demanding, we arrived at *variational calculous* and *functionals* (as before we dealt explicitly with functions) mainly and ways to find their minima. This opened for us a road to the next chapter, which presented complicated but necessary notions of *path* or *functional integrals* and various methods of their evaluation. We touched only lightly on these methods, just enough to prepare you to tackle various and much more advanced textbooks on this subject.

The last chapter in our journey was something of a bonus. Much lighter in algebra than the previous sections of Part IV, it overviewed various subjects that can be conventionally gathered under one roof: *fractionality*. We passed over a spectacular landscape of fractals, random walks and Gaussian chains, percolation, mathematical models of growth and their kinetics, and even a glimpse of fractional calculus. After a hard climb this might have seemed like a walk in the park, but for a reader unfamiliar with these topics, it demonstrated the requirement for a slightly different way of thinking, resonating with the approaches described in Chapter 10, Part I, Vol.1. Presenting this material in this book seemed to us to be a good idea for two reasons. First, to entertain you. And second, to familiarize you with less 'straightforward' routes of practical mathematics (although still miles away from the modern cutting edge!).

In Part IV—much more so than in previous parts—we have tried to prepare you for reading more specialized literature on each of the subjects presented in it and, as well, provoked you to do that. Have we succeeded or not—you are to judge.

# The Final Ascent to Everest

Now, we tell you the secret. Our metaphor that we have taken you to Camp 4, is conventional. In fact there is no Everest in these terms, as there are always more methods to learn. But for a student who feels like they have now reached Camp 4, it is worth mentioning what other fields of applied mathematics you may wish to study to reach your summit. Whatever your ambitions are, for getting analytical solutions of various physical problems, you will be even better equipped, if you studied additionally:

1. Laplace transforms and Laplace calculus.
2. Conformal transformations in complex variable calculus.
3. Various approximate methods for solution of nonlinear differential equations, including singular perturbation techniques (boundary layer method for matching asymptotic expansions).
4. Systems of nonlinear differential equations with 'feedback', leading to chaotic solutions, bifurcation points, attractors; theory of catastrophes.
5. Various methods of solutions of integral equations with specific sorts of kernels and boundary conditions.
6. Perturbation methods for calculating functional integrals, diagrammatic analysis, and partial summation over most important contributions.
7. Renormalization group theory.
8. Tensor analysis and elements of differential geometry.

And perhaps more. But with a nostalgic feeling we must leave now and say goodbye to you here. We hope you will not get lost on your way to the summit, whatever it is: you will derive your own beautiful formulae, as well as understand many others, using, as a start, some of the methods and tricks that we have done our best to share with you.

## Essential Textbooks in Physics

*(Continued from page ii)*

Printed in the USA
CPSIA information can be obtained
at www.ICGtesting.com
JSHW011210270723
45468JS00007B/13